Basic Discrete Mathematics
Logic, Set Theory, & Probability

Basic
Discrete
Mathematics
Logic, Set Theory, & Probability

Richard Kohar
Royal Military College of Canada

World Scientific

NEW JERSEY · LONDON · SINGAPORE · BEIJING · SHANGHAI · HONG KONG · TAIPEI · CHENNAI · TOKYO

Published by

World Scientific Publishing Co. Pte. Ltd.

5 Toh Tuck Link, Singapore 596224

USA office: 27 Warren Street, Suite 401-402, Hackensack, NJ 07601

UK office: 57 Shelton Street, Covent Garden, London WC2H 9HE

Library of Congress Cataloging-in-Publication Data

Names: Kohar, Richard.

Title: Basic discrete mathematics : logic, set theory, and probability / by
    Richard Kohar (Royal Military College of Canada, Canada).

Description: New Jersey : World Scientific, 2016. | Includes bibliographical references and index.

Identifiers: LCCN 2016014415 | ISBN 9789814730396 (hardcover : alk. paper) |
    ISBN 9789813147546 (softcover : alk. paper)

Subjects: LCSH: Logic, Symbolic and mathematical. | Proof theory. | Induction (Mathematics) |
    Set theory. | Probabilities.

Classification: LCC QA9.25 .K64 2016 | DDC 511/.1--dc23

LC record available at http://lccn.loc.gov/2016014415

British Library Cataloguing-in-Publication Data

A catalogue record for this book is available from the British Library.

Printed in Singapore

Dedicated to

James D. Stewart (1941–2014)

To the scholar who I only knew
through his high school textbooks,
but showed me the beauty of mathematics,
and inspired me to become a mathematician and a writer.

Contents

Preface

A textbook must be exceptionally bad if it is not more intelligible than the majority of notes made by students ... the proper function of lectures is not to give a student all the information he [or she] needs, but to rouse his [or her] enthusiasm so that he [or she] will gather knowledge himself [or herself], perhaps under difficulties.

—J. J. Thomson (1856–1940)

To the student

What kinds of problems can discrete mathematics solve? It can answer the following:

- What is the chance that you have HIV given that you have a positive test? (see p. 375)
- How do casinos and insurance companies make money in the long term? (see p. 404)
- How long do you expect to live if you are currently 20 years old? (see p. 351)
- How can you show whether a politician's argument is valid or invalid? (see p. 67)
- How can you show that you and another person in the city of London have the same number of strands of hair on your heads? (see p. 138)
- What is your chance of winning the lottery? (see p. 336)
- What is the maximum number of passwords you will have to attempt before you break into someone's computer account? (see p. 158)
- How many paths can you take from your home to school? (see p. 178)
- How can you quickly calculate $1 + 2 + 3 + \cdots + 100$? (see p. 222)

In this book, you will learn how to solve all of these types of problems and many more. But, what exactly is discrete mathematics?

Discrete mathematics is the study of discrete objects rather than continuous objects. An intuitive way of seeing the distinction between the discrete and continuous brings us into the realm of music. For example, a piano can play discrete notes: you can play a C or a C-sharp, but no notes in between. A violin, however, can play any sound between C and C-sharp: the finger can be placed anywhere between C and C-sharp on the continuous string.

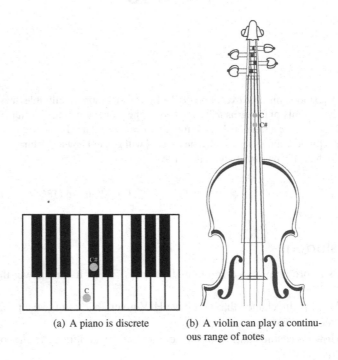

(a) A piano is discrete

(b) A violin can play a continuous range of notes

Fig. 0.1   Music can be discrete or continuous

In contrast with real numbers (which are continuous), discrete mathematics studies objects like integers. Integers are easy to count one by one—they are **countable**—while the real numbers are not. For example, there are five integer

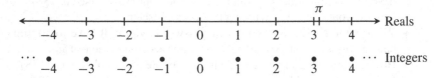

Fig. 0.2   The integers have gaps between them, while the reals do not. We can count all of the integers between 0 and 4, but we cannot count all the reals between 0 and 4.

numbers between 0 and 4 (inclusively), while we are unable to count the number of real numbers between 0 and 4. To see why, consider how close we can get to the number π without being equal to π.[1] You could say that a number close to π is 3.1, but we can still get closer with 3.14, or 3.1415, or 3.141 592 653 589 7 and so on and so forth, but we will never get to the true value. The real numbers have no gaps between them—they are continuous. Unlike the integers that have gaps between the numbers, we can keep repeating this process forever, and hence, we cannot count all of the number of real numbers between 0 and 4. (You could say, they're uncountable.) But what about numbers you can express as fractions—the rational numbers—can you count all of them too?[2]

> And it's time for our mathematics curriculum to change from analogue to digital, from the more classical, continuous mathematics, to the more modern, discrete mathematics—the mathematics of uncertainty, of randomness, of data—that being probability and statistics.
>
> —Arthur T. Benjamin (1961–)
>
> TED Talk given in February 2009

You may have struggled in the past with mathematics courses in high school, but consider this course a fresh start. This book is designed to stretch the understanding of mathematics that you achieved in high school, and to show you that mathematics indeed requires some creativity when solving problems. We will see that even some truly great mathematicians were either stumped or had to resort to ground-breaking new ways of thinking about a problem. John Waters (1946–) once said, "A poor person to me can have a big bank balance, but is stupid by choice—uncurious, judgmental, isolated, and unavailable to change."[3] In short, if by your own choice you are not open to these new ideas, I can't force them upon you; just put the book down now. If you're still reading, discuss the text with your classmates, get help from the instructor as soon as you need it, and promise yourself never to procrastinate and to complete your homework. With proper time management and some effort, this could very well be a class you enjoy.

Here are some general tips to help you use this book:

(a) *Read the book before the lecture.* Ask the instructor which section will be covered the following day, and read that section before the lecture. This will give you an overview of the next lecture's material, and you will be ready with questions if there is material that you didn't understand.

---

[1] The Greek letter π is pronounced as pie. If you think back to high school or elementary school, π is the mathematical constant that is equal to a circle's circumference divided by its diameter regardless of the circle's size (see Appendix B.2). Its decimal number representation never ends, and it never settles on any repeating pattern.

[2] Yes. See Theorem 4.2.9 on p. 126, but you might have to go back to the beginning of the Chap. 4 to understand why this is true. So if you thought we were just limited to integers, don't worry.

[3] From John Waters' 2015 Commencement Speech at Rhode Island School of Design on 30 May 2015.

(b) *Keep the book open during the lecture.* I found it helpful to keep the book open during a lecture. If I didn't feel comfortable asking my question, I could quickly skim through the text to find my answer. This could easily answer questions such as what a symbol means, or what a particular word means. Instructors sometimes gave hints in class about content of forthcoming examinations, so I would make a note in the relevant section of the textbook, and I could pay particular attention when studying and reviewing.

(c) *Read the book after lecture.* If you didn't read the textbook before the lecture, you should read the textbook after. I believe, just as did J. J. Thomson, that a good lecture should be motivating enough for the student to read the textbook and delve deeper into the details. During this second exposure to the material, I recommend that when you come up on an example, cover up the solution, and try to work it out yourself. Give it a zealous all-out effort. When you are finished, you can check to see if you got the right answer; if you did not, go back over your work and see where you made a mistake. If you get stuck, I mean really stuck, then take a peak and see if you can continue. By making mistakes, and correcting yourself, you will improve and gain an understanding of the material.

(d) *Read with a pencil/pen.* You solve mathematics problems with a pencil/pen, why not read mathematics with a pencil/pen as well? If there is something you don't understand, try figuring it out on a scrap of paper. If it's important, put your condensation of ideas in the margin. Don't be afraid to mark up the book if you own it—you can shed your high school mentality of not "defacing" your book. If you haven't noticed, the book is in black and white; feel free to add some color yourself.[4] By marking and adding your thoughts and drawings to the book, it helps you place mental index markers in the book, so you can find material quickly as you flip back and forth through the book. You should not be discouraged if you find your reading rate considerably slower than normal.

(e) *Skip ahead.* A book is linear: one page follows another. This does not mean that it has to be read in a linear way! If I have written about something and you don't understand the reason for me to present the idea, try skipping ahead and reading the next section. You may see why I had to go through some difficult material[5] to get to present an exciting topic or application. Read that topic or application, and then go back. You'll now understand the motivation for presenting the difficult material.

---

[4] Perhaps use a red pen to highlight the boxes of theorems or important equations. Try coloring the portrait of George Boole on p. 2 or Georg Cantor on p. 131.

[5] What one person finds boring, another may find exciting.

Mathematics is logical to be sure; each conclusion is drawn from previously derived statements. Yet the whole of it, the real piece of art, is not linear; worse than that its perception should be instantaneous.

—Artin (1953, p. 475)

(f) *Don't skip ahead.* There may be a hard example that you are trying to work through the solution on your own. Don't give up, and persevere! This is the hard part of learning. You will be spending a lot of time working on this subject, so do not worry if you feel you are slow.

Being a mathematician is a bit like being a manic depressive: you spend your life alternating between giddy elation and black despair. You will have difficulty being objective about your own work: before a problem is solved, it seems to be mightily important; after it is solved, the whole matter seems trivial and you wonder how you could have spent so much time on it.

—Krantz (1997, p. 78)

One way to overcome this is to celebrate the small successes. If you see a friend who is in your class walking down the hallway, tell your friend that you solved that problem you were struggling with. Your friend may be interested in how you solved the problem; your friend might have been stuck on the same problem too.

(g) *Study quietly by yourself.* We plug directly ourselves into social media feeds, music, videos, and the list goes on and on. However, to become good or even great at something, you need to learn to turn off technology (at least while studying). Take your phone, switch it off, and put it in a drawer. Even better, just *throw it out the window.* Why? Ericsson *et al.* (1993) found that to acquire expert performance as a chess grandmaster, a violinist, or a gymnast, required a substantial amount of deliberate practice. Best conducted in solitude, deliberate practice allows for intense focus, which allows you to identify gaps in your knowledge or skills, strive to fill in those gaps, monitor your performance, and correct accordingly (Cain, 2012, p. 81). Distractions hinder deliberate practice by causing us to lose focus, and when you are interrupted, it takes about 25 minutes to cycle back to the original task (Mark *et al.*, 2005). Two of the reasons I included the solutions at the back of the book are

(i) you can avoid the temptation of searching for solutions online if you become stuck and need a hint (and hence, you avoid being distracted by a text message or email alerts from others), and

(ii) you can extend your deliberate practice time longer as it does not depend on the availability of external resources, such as a professor or a tutor.

(h) *Read widely.* Mathematics is a wonderfully broad area of study. There is no single book that can contain all of its ideas, and this book is no different.

Another way to explore ideas that we will discuss is to read through the Bibliographic Remarks at the end of the chapters. You may find other books that you may want to read to further your study about a particular topic. A recent invention is the digital object identifier (DOI), which is a string of characters that uniquely identifies an electronic document. In the Bibliography, I have recorded the DOI for many of the articles and books listed. Unlike web addresses which change or break over time, DOIs remain fixed over the lifetime of the document. These DOIs will direct you to the article, and if you access them from the university, they are usually available free of charge to you.

Pólya's Approach to Problem Solving

This book has a lot of mathematical problems in it. Some of these problems you may never have seen before, and you'll have to solve them. The Hungarian mathematician George Pólya (1945) gave some strategies for tackling a mathematical problem, and we will use them extensively throughout the book. I summarize them below:

(a) *Understand the problem.* Read the problem, and determine what you are required to find. What are the unknowns? What is the given information? You might find that starting to *draw a diagram* will help (especially if you are a visual learner) to organize the information you have, and recognize the information that you need.

(b) *Devise a plan.* What is the relationship between the given to the unknown? If it is not immediately apparent, then try devising a plan based on the following:

  (i) *Draw a diagram.* This can help organize the data that is given in the problem. Our working memory is one of our limitations; we can hold about seven pieces of data.[6] By drawing a diagram, it allows us to extend and augment our memory. Sometimes, the solution will emerge or become obvious just from seeing a visual representation. See Example 5.3.1 on p. 147.

  (ii) *Divide into smaller problems.* If the problem is complex, break it up into small problems that are easier to solve. See Problem 7.1.1 on p. 277.

---

[6]Miller (1956) proposed that the limit for our working memory is 7 ± 2 "chunks" of information. A chunk could be collection of digits, letters, words, or objects. You can see that it is hard to remember the digits 2, 0, 2, 4, 5, 6, 1, 4, 1, 4. We can chunk them in the following way: 202 is the area code for Washington D.C., 456 is the successive numbers 4, 5, and 6, and finally two pairs of 14. Now you can easily remember this telephone number—which so happens to be the White House switchboard number: (202) 456-1414. We have the ability to keep more information in our working memory, but this requires us to find a way to chunk efficiently (to find a compact representation of the data in short-term memory by using knowledge from our long-term memory).

(iii) *Find something familiar.* You might recall a similar problem or example that you have already seen. You can see if you can apply a similar strategy to your current problem at hand. See Example 7.1.2 on p. 279, which is similar to Problem 7.1.1 on p. 277. You can also review the proofs of theorems for the key ideas that were used, and decide if these same ideas could be applied to your (specific) problem at hand. For example, Exercise 10.5.4 (b) on p. 464 requires an argument that is similar to the proof used to prove Theorem 10.5.4—we could try to adapt the proof of Theorem 10.5.4 to handle this specific situation.

(iv) *Find a pattern.* You might find a pattern that will continue, and you can generalize. For example, the Fibonacci sequence in Example 6.1.4 on p. 202.

(v) *Take cases.* We can divide the problem into several cases, and find a solution for each of those cases. See Example 3.4.6 on p. 101.

(vi) *Work backwards.* We assume the answer and work backwards until we arrive at the given data in the problem. For example, this is how we can prove Theorem 6.10.3 on p. 260. This can also be done with any of the exercises where you can look at the answer, and try to work backwards to get back to the original data in the exercise.

(vii) *Indirect reasoning.* Consider building a pyramid. There would be two basic ways you can do this. One way is to build the pyramid *directly*: you place one stone on top of another up you reach to the top. The other way is to build the pyramid *indirectly*: you begin with a stone cube, and you chisel and remove the stone pieces until the stone that remains is in the shape of a pyramid. This can be applied to a problem. If you are unable to solve the question in a direct fashion, consider if there is an indirect way; can you start with something and chisel away until you have the desired answer. See Example 5.4.4 on p. 153. You will see that you can usually solve a problem both ways, but one type of reasoning will often be easier to use than the other. Another example of indirect reasoning is proof by contradiction whereby we assume that the conclusion is false, and after encountering a contradiction, then we conclude that the conclusion must be true. For example, we can prove that there are infinitely many prime numbers by this method (see the proof of Theorem 4.1.2 on p. 120).

(c) *Carry out the plan.* Write out a detailed solution to prove each stage is correct.

(d) *Look back*, and ask yourself if your answer or conclusion makes sense. For example, if it is a question about probability, does it make sense that an event occurring has more than 100% chance? (No.) If your answer does not make sense, go back and double check your work, or try a different strategy. Trying a different strategy or method is helpful.

> In order to convince ourselves of the presence or of the quality of an object, we like to see and touch it. And as we prefer perception through two different senses, so we prefer conviction by two different [methods].

> —Pólya (1945, p. 15)

If we find two equivalent answers to the problem by two different methods, we increase our confidence that we have the correct answer.

Another suggestion that we often hear in writing essays is to put it away, and look at it later. The same advice can be applied to mathematics. The advantage of completing your homework assignment early is that it gives you a day or two to clear your mind, and look it over later with fresh eyes. You may be amazed at the mistakes that you can catch!

But most important of all, before giving up, try something!

For the more adventurous student, the exercises that are marked with † are meant to be challenging for you.

To the instructor

This book started out as a collection of handwritten lecture notes for a course that I gave in the Fall of 2013 and 2014 at the Royal Military College of Canada to humanities majors. The College required that these students take a course in introductory logic, sets, and discrete probability. The text in use at the time had placed logic as an afterthought in the form of an appendix, and other available logic textbooks for the humanities had tiptoed around the use of mathematical symbols for abstract reasoning and lacked diagrams. What a travesty! Without pictures and diagrams, visual learners (like me) felt isolated. I took it upon myself to create my own set of lecture notes.

A new textbook on basic discrete mathematics for the humanities cannot perpetuate the iniquity of earlier books. This new generation of students, infused by the spirit of the Internet, question authority, require transparency, and demand to know *why*. They expect reasons for any statements or formula that is stated. In this book, we give students what they want in the form of proofs. The proofs provide the rationale behind a concept or a formula, and all of them are all at the level of understanding for a humanities college freshmen.

For example, I could have stated the binomial coefficient formula

$$\binom{n}{r} = \frac{n!}{r!(n-r)!},$$

which is *how* to calculate the number of r-element subsets from a total of n elements, and walked away. But, this would have left the student scratching his or her head.

"But, *why* is this formula like this? I demand to know *why* we are dividing by $(n - r)!$ and $r!$. *Why* are we not subtracting?"

The student will become enraged (and rightly so) when met with the authoritative, "Because I say so" or even worse, "That's the way it's been done in mathematics for the past 400 years." In this case, the calming remedy for the student is a combinatorial proof—a proof which counts objects in one way, then counts it in another way, and then concludes by saying that the two counting methods are equivalent. (See Example 5.7.4 on p. 163 for the combinatorial proof of the binomial coefficient formula.)

I caution instructors against skipping proofs. This act signals to the students that the material is not important; this is detrimental. For without the reasoning, the mathematical formula is apt to be forgotten the moment the students burn their notes after their final examination. By going through the proofs, the notes may be gone, but the knowledge remains.

I must confess that when I was charged with the task of teaching to humanities majors at the College, I was nervous: I had grown up and spent many years learning and enjoying mathematics, and I wasn't oblivious to the fact that many students don't find it as enjoyable to say the least (especially the artsmen); these would be the kinds of students I would be teaching. To combat my nervousness and to ease the anxiety of many students, I made logic, sets, and probability more accessible by setting the topics in the context of history, and natural language, as well as in daily and professional life. Moreover, it became apparent to demonstrate the interplay between logic, sets, and probability as opposed to treating these isolated topics as separate islands of thought. By doing this, we can start to apply the concepts of associative learning from psychology to relate ideas of mathematics to one another and to concepts in life, thereby reinforcing this in the student's (long-term) memory.

The exercises are meant to help students gain skills in calculation, in problem solving, and in explaining concepts. Some of my favorite exercises require the student to give an explanation; *e.g.,* Exercise 8.3.10 (p. 318), or 8.5.1 (p. 326).

This book has many exercises for the students and solutions at the back of the book. I am sure there are many arguments against including solutions, but I believe it is to the student's advantage to include them. When I taught the course at the College, I ran an extra tutorial on Monday evenings for two or three hours. During this time, I did very little in terms of helping the students (I spent most of my time editing this book). I found that students were able to work quietly, and when they got stuck, they could ask me for help. Often, the questions were very minor, but could easily cause frustration and waste an hour of the student's time. (With this minimal effort, I received glowing reviews from the class for the extra help.) If we replace me with a set of solutions, the students can check the solutions when they get stuck. The only temptation that the student faces is not trying the problem and directly skipping to the solutions—they need to give it the ol' college try first.

The exercises that are marked with † are meant to be challenging for the student.

Another decision I made in this book was to avoid teaching technology to students; for example, teaching students to use a particular calculator, a particular software package, *etc.* There is such a plethora of technology, for me to pick a single tool would be a disadvantage. As soon as I would be done writing about a particular technology, this book would be out of date. So with this, you are free to select the tools you prefer, and not be hindered by nagging technology boxes found in many other books. I also hold the personal philosophy that everyone should be able to do a quick back-of-the-envelope calculation, so the book reflects this philosophy by emphasizing hand calculations; *sapiens omnia sua secum portat.*

Numbering scheme

The numbering scheme is designed to enable quick location of definitions, theorems, and examples, which are referred to in the text. Definitions, theorems, and examples are numbered sequentially within each Section (*e.g.,* Definition 3.1.2 is in Chapter 3, Section 1, and is the second highlighted item).

How the book was produced

This first edition was produced in LATEX using MikTEX 2.9 with TEXnicCenter 2. I used the Times 10/12 typeface with mathematics typeset in MathTime Professional 2 produced by Michael Spivak of Publish or Perish, Inc. The section headings are set in Avant Garde. Some of the book design came from the Legrand Orange book originally created by Mathias Legrand. The line illustrations were drawn by me with Tikz developed by Till Tantau using TikzEdt 0.2.3.0 by Thomas Willwacher.

Graphics credits

The following graphics are from `thenounproject.com`:

Ant by Jacob Eckert	*Fingers Crossed* by Richard Zeid
Ant by M. Turan Ercan	*Goat* by M. Turan Ercan
Bubble Question by Thomas Helbig	*House* by John Salzarulo
Building by Benoît Champy	*Light bulb* by Diego Naive
Bus by Pieter J. Smits	*Mouse* by Carla Gom Mejorada
Cheese by Nathan Thomson	*Rabbit* by Stewart Lamb Cromar
Coffee Bean by Jordan Díaz Andrés	*Telescope* by Dusan Popovic
Energy by riyazali	*Sea Turtle Head* by George Drums
Fountain by Mister Pixel	*Trojan Helmet* by Javier Triana

For the license plate graphics, I used the *License Plate* typeface by Dave Hansen. Credits for certain illustrations appear in the figure caption.

Mistakes

The mistake that you found—yes, the one you just circled—has been keeping me up at night. I know they exist, but it is difficult to catch them all.

A book is never finished; it is simply abandoned. With this in mind, I will continue to revise for the next printing (or the next edition), and I am always happy to receive corrections and comments. There are many people who helped point out the blunders, flaws, fallacies, miscalculations, inaccuracies, and other miscellaneous grammatical and mathematical howlers, but the rest remain my own fault; no mortal is wise at all times. If you spot an error, please send me an email with the correction and the page number to `richard@math.kohar.ca`. I will be maintaining a list of errata on my website, `www.kohar.ca`.

Website

You can visit my website, `www.kohar.ca`, to obtain supplementary information.

Math isn't so bad

Thank you for choosing to use this book. I hope that you will enjoy reading it. It will teach you some useful and exciting mathematics; it will be a good reference book in the future, and perhaps it might convince you that mathematics isn't so bad after all.

Richard Kohar
Royal Military College of Canada
Kingston, Ontario, Canada
May 2016

Acknowledgments

Although there is a single name on the front of this book, I was dependent on the extensive supporting network of friends, family, and colleagues.

I had been heavily influenced by my supervisors Dr François Rivest and Dr Alain Gosselin during our many discussions on education and exploring different teaching methods and styles.

Thank you to Mrs Diane McCombs, who in fact was my senior high school teacher in combinatorics and probability, for spending a Herculean amount of time with the manuscript—giving recommendations, clarifications, and insights into how students learn and the mindset of a student transitioning into college mathematics.

I would like to thank Ms E. H. Chionh, the editor of World Scientific Publishing, for giving me the chance to write my first textbook.

Thank you to the following who took time and give comments and suggestions on drafts of the manuscript (in alphabetical order):

- Dr Arthur T. Benjamin
  Harvey Mudd College
  Claremont, California, United States
- Mr Lucas Beyak
  Brock Solutions
  Kitchener, Ontario, Canada
- Dr Randy Elzinga
  Royal Military College of Canada
  Kingston, Ontario, Canada
- Mr Atul Kotecha
  Frontenac Secondary School
  Kingston, Ontario, Canada
- Mr Sachin Kotecha
  University of Waterloo
  Waterloo, Ontario, Canada

- Mr Terry Loveridge
  Royal Military College of Canada
  Kingston, Ontario, Canada
- Dr Miroslav Lovrić
  McMaster University
  Hamilton, Ontario, Canada
- Dr Gregory H. Moore
  McMaster University
  Hamilton, Ontario, Canada
- Dr Jamie Pyper
  Queen's University
  Kingston, Ontario, Canada
- Dr Peter Taylor
  Queen's University
  Kingston, Ontario, Canada

Of course, I cannot forget my parents Rick and Anneliese, and my brother Chris who always had words of encouragement: *Illegitimi non carborundum*. Besides them being supportive and reading through sections, they dealt tactfully with the never-ending growing piles of paper on the kitchen table where I had written the majority of this work.

No part of this work was supported financially by the Royal Military College of Canada.

1

Introduction to Logic

And I can confidently assure you that, as far as you may ever have occasion to exercise your reasoning powers upon any subject, a real acquaintance with the art of Logic will abundantly compensate the labor of acquiring it. Nor have I ever met a person unacquainted with it, who could state and maintain his arguments with facility, clearness, and precision.

—Walker (1847, p. 4)

The first notion in which a student can form a sense of logic is by viewing it as the examination of reasoning in arguments. Arguments consist of either true or false statements, and from these statements, we can decide if the reasoning that links these statements will yield a true conclusion. By using the knowledge that we know, and drawing conclusions, this process is called **logical inference**.

Nothing is better than eternal happiness.
A passing grade is better than nothing.

Therefore, a passing grade is better than eternal happiness.

Do you think just a passing grade will give you eternal happiness? Or am I trying to pass off bad logic to you? How would you show this is an incorrect argument?

Thus, the study of **logic** is the "study of the methods and principles used in distinguishing correct from incorrect arguments" (Copi, 1954). Of course, this definition does not say that you can only make distinctions of correct and incorrect arguments if you have undertaken the subject of logic, for it is not a necessary condition, but in the very least it will *help* you to distinguish between correct and incorrect arguments.

1.1 Historical Development of Logic

The historical origins of logic date back to antiquity. Many ancient civilizations such as the Greeks, the Persians, the Chinese, and the Indians had conceived the notion of logic in one form or another. Logic was first established as a formal discipline in the West by the Greek philosopher Aristotle (384 BC–322 BC). His method of analyzing and performing logic was dominant until the advent of modern predicate logic in the early 19th century. The *Organon*, the collective name used to describe the six texts of Aristotle's logical work, contains the heart of Aristotle's treatment of judgment and formal inference. Here, he introduces the basics and terminology of logic such as the **proposition** and the **syllogism**.

His system of logic remained highly influential and unchanged until the 19th century, when English mathematicians George Boole and John Venn (1834–1923) started to derive a system of manipulating symbols instead of manipulating words. The German mathematician Gottfried Wilhelm von Leibniz (1646–1716) had earlier attempted to create a distinctive logical calculus, but the majority of his work on logic remained unpublished until the turn of the 20th century.

George Boole (1815–1864) was born a shoemaker's son in Lincoln, England. He was not afforded an education beyond elementary school due to his family's small income, and hence was entirely self-taught in the areas of Greek and Latin. By age 16, with the necessity of supporting his poverty-stricken parents, Boole took up teaching in elementary schools, and this had led him to open his own school four years later.

The need to master the subject and prepare his students led Boole to delve into the works of the great mathematicians: Newton, Lagrange, and Laplace. At this time, he began to submit a steady stream of papers to the newly founded *Cambridge Mathematical Journal* and to

Fig. 1.1　George Boole (1815–1864)

the *Philosophical Transaction of the Royal Society*. Boole's manuscript *On a General Method in Analysis*, which won him an award for best paper from the Society, had almost been rejected, but one member steadfastly argued that the author's poverty and obscurity were no grounds for the paper's rejection.

In 1847, the award winning paper had been expanded into a book entitled *The Mathematical Analysis of Logic*. It is curious to note that one of his friends, and also a very famous logician Augustus De Morgan, published work similar to his in the same month, and some say even on the same day.

Boole was a pioneer in describing analysis because of the way he described it in his book: "The process of analysis does not depend on the interpretation of the symbols which are employed, but solely upon the laws of their combination."

De Morgan and his other friends had urged Boole to apply for a professorship in mathematics at the newly formed Queen's College in Ireland. Hesitant in applying for the position without a university degree, he secured the position on the basis of his research. He remained at his post until his premature death in 1864 from pneumonia, which he contracted after walking two miles in a drenching rain to lecture dutifully his class.

An interesting character in 19th century English mathematics is Augustus De Morgan (1806–1871). "He made no startling discoveries" (Burton, 2011, p. 649), yet he was able to exert considerable influence on public education through his gift of teaching and prolific output of writing about mathematics for the general public. He was born into a military family in Madra, India where his father worked for the East India Company (at the time the company was permitted to maintain an army). He was almost immediately brought back to England as an infant where he attended private schools and entered Cambridge at 16 years of age. De Morgan never hesitated to take a stance on principle, regardless of personal sacrifice. He refused to submit to the Church of England's religious test, which prevented him from proceeding to the masters degree or continuing as a fellow at Cambridge or Oxford (Burton, 2011, p. 649). At the time, the Church of England religious test required the person to follow the Anglican faith; this is a juxtaposition to the United States stance in which they enshrined a no religious test clause in their Constitution stating that no one be required to adhere to or accept any religion or belief. While contemplating to prepare for legal studies, De Morgan applied for a professorship at the newly created secular London University—later renamed University College, London in 1836—where he began lecturing at age 22 and remained there until his resignation in 1866, a lecturer for 16 years.[1]

De Morgan's chief contribution to the field of logic is his application of mathematical methods and the subsequent development of symbolic logic in the form of five papers he submitted to the *Transactions of the Cambridge Philosophical Society* between 1846–1862. The papers attempted to generalize traditional syllogistic reasoning; however, owing to his complicated notation and Boole's more readable *Mathematical Analysis of Logic*, his ideas were largely overlooked.

The German mathematicians Ernst Schröder (1841–1902) and Gottlob Frege (1848–1925) showed how logical systems were in fact far more comprehensive than those of which could be constructed from classical logic. With this, new sets of logical concepts and operations were quickly set up and investigated. Be-

---

[1] In 1976, the university changed to University College London, where it had abandoned the comma between College and London in its name after 140 years.

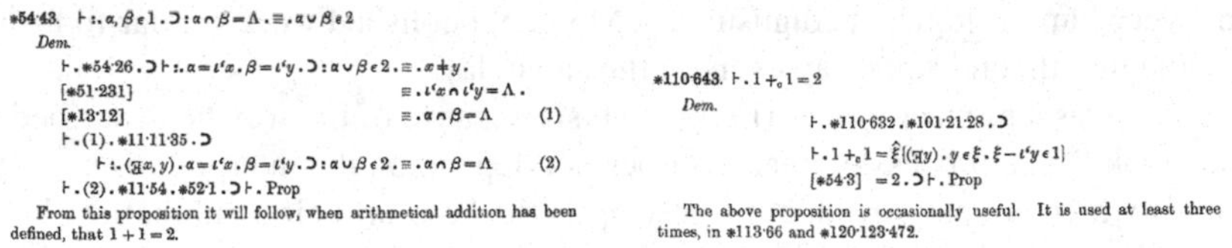

*54·43. $\vdash: \alpha, \beta \in 1 . \supset : \alpha \cap \beta = \Lambda . \equiv . \alpha \cup \beta \in 2$

Dem.

$\vdash . *54·26 . \supset \vdash : . \alpha = \iota'x . \beta = \iota'y . \supset : \alpha \cup \beta \in 2 . \equiv . x \neq y .$

[*51·231]　　　　　　　　$\equiv . \iota'x \cap \iota'y = \Lambda .$

[*13·12]　　　　　　　　$\equiv . \alpha \cap \beta = \Lambda$　　(1)

$\vdash . (1) . *11·11·35 . \supset$

$\vdash : . (\exists x, y) . \alpha = \iota'x . \beta = \iota'y . \supset : \alpha \cup \beta \in 2 . \equiv . \alpha \cap \beta = \Lambda$　　(2)

$\vdash . (2) . *11·54 . *52·1 . \supset \vdash . \text{Prop}$

From this proposition it will follow, when arithmetical addition has been defined, that $1 + 1 = 2$.

(a) An important step is given on p. 379 of Volume 1.

*110·643. $\vdash . 1 +_c 1 = 2$

Dem.

$\vdash . *110·632 . *101·21·28 . \supset$

$\vdash . 1 +_c 1 = \hat{\xi}\{(\exists y) . y \in \xi . \xi - \iota'y \in 1\}$

[*54·3]　$= 2 . \supset \vdash . \text{Prop}$

The above proposition is occasionally useful. It is used at least three times, in *113·66 and *120·123·472.

(b) On p. 86 of Volume 2, they finally prove $1 + 1 = 2$ with the small remark that it is "occasionally useful."

Fig. 1.2　The *Principia Mathematica* by Whitehead and Russell (1910, 1912, 1913) is a three-volume work that was published over a course of four years. The notation used in *Principia Mathematica* makes it difficult for a beginning logic student to read since it has been superseded by the notation of Hilbert and Ackermann (1959) (the notation we will use in this book).

tween the years 1910 and 1913, the English philosophers Alfred North Whitehead (1861–1947) and Bertrand Russell (1872–1970) published an encyclopedia of comprehensive logic, the **Principia Mathematica**. It was a monumental attempt—a book in three volumes of approximately 2000 pages—at trying to describe a set of axioms and inference rules in symbolic logic from which all mathematical truths could be proven. It takes a while for the exposition to get going before Whitehead and Russell (1912) were able to prove that $1 + 1 = 2$ after 752 pages; see Fig. 1.2. It was the first book to demonstrate the close ties between mathematics and formal logic.

Exercises

1.1.1. Define the term *logic*.

1.1.2. What aspect of Boole's work made him a pioneer in logic?

1.1.3. What was George Boole's cause of death?

1.1.4. Why was De Morgan's *Formal Logic* unsuccessful?

1.1.5. How was Augustus De Morgan a unique and influential 19th century figure?

1.1.6. Why did De Morgan not obtain a M.A. degree?

1.1.7. Why was the *Principia Mathematica* an important achievement in 20th century logic?

1.2 Propositions and Paradoxes

I know one thing: that I know nothing.

—Plato (c. 428 BC–c. 348 BC)

We can think of the rules of logic as simply the laws of English grammar (or the laws of any other language grammar). While there is considerable truth in this assertion, this is an over-simplification; however, semantic complexity (ambiguity in meaning of words and phrases) in human language (like English) can lead to errors in reasoning, and hence to errors in mathematics.

> "Then you should say what you mean," the March Hare went on.
>
> "I do," Alice hastily replied; "at least–at least I mean what I say–that's the same thing, you know."
>
> "Not the same thing a bit!" said the Hatter. "Why, you might just as well say that 'I see what I eat' is the same thing as 'I eat what I see'!"

—Carroll (1866, p. 98)

We will discuss further how these phenomena naturally occur in human language below (see Sec. 1.3.4, p. 11).

The **proposition** is the common object of logical analysis. Simply stated, the proposition is a sentence that declares or states something. Some examples are the following:

(a) John dislikes mathematics.

(b) Napoléon is dead.

(c) $4 + 5 > 3$.

(d) The world will end on 4 March 2045.

> **Definition 1.2.1 — Proposition.** A proposition is a meaningful, declarative statement that we can declare as either true or false, but never both.

When we evaluate the truth of a proposition, we assign it one of two labels, *true* or *false*—these labels are called the **truth values** of a proposition.

Interrogative statements such as, "Are you Canadian?" are not used in logic. Imperative statements such as "Report to the Commandant's office," are not used in logic either. Exclamatory statements such as, "Watch out!" are not used in logic. All of the previous examples cannot be propositions since one cannot decide if the statement is true or false.

Moreover, the truth value of a proposition can depend on the person making the judgment or on the time and circumstances in which the statement was made.

> For instance, on the planet Earth, man had always assumed that he was more intelligent than dolphins because he had achieved so much—the wheel, New York, wars and so on—whilst all the dolphins had ever done was muck about in the water having a good time. But conversely, the dolphins had always believed that they were far more intelligent than man—for precisely the same reasons.

—Adams (1979, p. 113)

Example 1.2.2

    (a) "Dolphins live in water" is a proposition.

    (b) "Humans are more intelligent than dolphins" is a proposition.

    (c) "$4 + 3 = 12$" is a proposition, albeit a false one.

    (d) "All hignals are progals" is a meaningless statement since we do not know what hignals or progals are. This can become a proposition if hignals and progals are defined.

    (e) "He enjoys mathematics" is not a proposition. The pronoun "he" prevents us from being able to state whether the sentence is true or false. The pronoun acts like a **variable** that may be replaced by the names of suitable elements from a replacement **set**. Such a set is the **domain** of the variable. The domain in this case, could be the set of all boys in your mathematics class. If we replaced "he" with one of those names, then the sentence would become a proposition.

    (f) "$x + 7 = 10$" is not a proposition because the domain of x is not defined. If we said "$x + 7 = 10$, where $x = 4$," then it would be a (false) proposition.

◀

Definition 1.2.3 — Prime Proposition. Prime propositions are simple statements that express a single complete thought, and whose truth value is true or false, but never both.

Example 1.2.4 The following are examples of prime propositions:

    (a) The sun is shining.

    (b) I walk to work.

    (c) Socrates is a man.

    (d) $17 + 25 = 42$.

    (e) Henry VIII had six wives.

    (f) The English Civil War took place in the 19th century.

◀

Definition 1.2.5 — Compound Proposition. Compound propositions are propositions that are made of two or more propositions, and whose truth value is true or false, but never both.

Example 1.2.6 The following are examples of compound propositions:

    (a) The sun is shining and I walk to work.

    (b) Socrates is a man or Socrates is a dog.

    (c) Henry VIII had six wives or the English Civil War took place in the 19th century.

◀

When we defined a proposition, we insisted that a proposition has a truth value that is either true or false but not both. One example of a statement where we cannot assign a truth value is a **paradox**—something that was known to the ancient Greeks.

Definition 1.2.7 — Paradox. A paradox is a statement that apparently contradicts itself and yet might be true.

Example 1.2.8 — Liar's Paradox. This is a famous example of a paradox that has been known since antiquity.

This proposition is false.

The proposition cannot be false and true at the same time. Trying to assign to this statement as true or false leads to a contradiction. If "this proposition is false" is in fact true, then the sentence is false, but then if "this proposition is false" is false, then the proposition is true. ◄

Fig. 1.3   The Pinocchio paradox: "My nose is growing."

Example 1.2.9 — Pinocchio paradox. This is another version of the Liar's paradox from Example 1.2.8.[2] From the *Adventures of Pinocchio*, Pinocchio is a wooden puppet who dreams of one day becoming a real boy. However, Pinocchio had such a nose that when he tells a lie, his nose grows. A paradox occurs when Pinocchio says, "My nose is growing."

Pinocchio's nose grows if (and only if) what he is saying is false, and Pinocchio says, "My nose is growing." So, Pinocchio's nose is growing if (and only if) it is not growing (Eldridge-Smith and Eldridge-Smith, 2010). Hence, it is a paradox. ◄

---

[2]This was devised by Veronique Eldridge-Smith when she was 11 years old.

Exercises

1.2.1. Which of the following are propositions?
  (a) Napoléon Bonaparte (1769–1821) lost the Battle of Waterloo in 18 June 1815.
  (b) Ludwig van Beethoven (1770–1827) called his third symphony, *Sinfonia Eroica*.
  (c) Who won the football game?
  (d) Stop!
  (e) $5 < 10$ (5 is less than 10)
  (f) $4 > 10$ (4 is greater than 10)
  (g) We'll always have Paris.
  (h) Mexico is located in South America.
  (i) Lock the door before you leave for the day.
  (j) The Toronto Maple Leafs won the next Stanley Cup.

1.2.2. Which of the following are propositions?
  (a) In 2015, Barack Obama was the president of the United States.
  (b) 5 is an even number or 17 is an even number.
  (c) What time is it?
  (d) It is raining right now, and I'm wet.
  (e) Is it raining?
  (f) Come to class!

1.2.3. Identify the prime propositions in Exercise 1.2.2.

1.2.4. Identify the compound propositions in Exercise 1.2.2.

1.2.5. Is Plato's proposition, "I know one thing: that I know nothing" a paradox?

1.2.6. Can you come up with your own version of the Liar's paradox?

1.3 Connectives

The words used to combine propositions are called **logical connectives**. In English, the words *and* and *or* are used to join two or more independent clauses together to form a compound sentence. Similarly in logic, there are symbols that we can use to join propositions together. We will show how these symbols work with the use of a mathematical table called a **truth table**, an invention developed in 1893 by the American logician Charles Sanders Peirce (1839–1914) and later adapted heavily in the works of Bertrand Russell (1872–1970) and Ludwig Wittgenstein (1889–1951).

1.3.1 Negation

We let

$$\neg p \qquad \text{denote} \qquad \text{"not } p\text{."}$$

Logically speaking, negation is a fundamental notion. Logical systems for truth tables can be constructed successfully even if we dispense with the logical concepts in the upcoming sections, but we cannot continue without negation. The negation of any proposition changes from one truth value to the other. For example, if the proposition is false, then the negation changes the truth value from F to T. For example, the negation of p is $\neg p$; the negation of $\neg p$ is $\neg(\neg p)$. The truth table for the negation is given in Table 1.1.

Table 1.1 $\neg p$

p	$\neg p$
T	F
F	T

Example 1.3.1 If we let the proposition p to be that Billy is exhausted, then $\neg p$ would mean that Billy is not exhausted. ◄

1.3.2 Conjunction

We let

$$p \wedge q \qquad \text{denote} \qquad \text{"}p \text{ and } q\text{."}$$

Since p and q are variables, we know that we may substitute simple propositions for them. In logic, $p \wedge q$ simply asserts that p and q are joined propositions, no matter how ridiculous it may sound in English. A perfectly acceptable conjunction would be: Pigs can fly and Bach composed the Brandenburg concertos.

Now, we turn our attention to the truth table in Table 1.2 for $p \wedge q$. The only time when $p \wedge q$ is true, is when p is true and q is true; all other instances, such as p is true and q is false, makes $p \wedge q$ false.

Table 1.2 $p \wedge q$

p	q	$p \wedge q$
T	T	T
T	F	F
F	T	F
F	F	F

Example 1.3.2 Consider the propositions

$$p = \text{Ángela passed logic}$$

and

$$q = \text{Ángela passed calculus.}$$

Suppose Ángela had passed logic, but failed calculus. What is the truth value of $p \wedge q$?

Solution The proposition $p \wedge q$ means in this case that Ángela passed both logic and calculus. Since Ángela failed calculus, the truth value of $p \wedge q$ is false. ◄

The standard English conjunction is *and*. While there is an abundance of other conjunction words that have different connotations, these conjunctions are the same logically: *but*, *yet*, *although*, *though*, *even though*, *moreover*, *furthermore*, *however*, and *whereas*.

Example 1.3.3 To illustrate, the following propositions are the same logically:
   (a) The costume party is tonight, *but* I cannot go.
   (b) *Although* the costume party is tonight, I cannot go.
   (c) The costume party is tonight, *however* I cannot go.
   (d) The costume party is tonight, *and* I cannot go.
Although these propositions have different connotations in English, they are all the same logically; they all can be represented by $p \wedge a$ (p for party tonight, and a for not attending). ◄

Conjunctions in English can also be disguised as relative pronouns like *who*, *which*, and *that*, or as compound adjectives.

Example 1.3.4 Rewrite the following using conjunctions:

Cadet Jones is an RMC exchange student who leads the marching band.

Solution We can rewrite this as the following:

Cadet Jones is an RMC student (r), *and* Cadet Jones is an exchange student (e), *and* Cadet Jones leads the marching band (b).

Symbolically, this is represented as $r \wedge e \wedge b$. ◄

1.3.3 Disjunction
We let

$p \vee q$ denote   "p or q or both."

In stating a disjunction, we simply assert that at least one of the propositions is true, and possibly both. You can see the corresponding truth table in Table 1.3.

The legal profession has invented an expression to circumvent this ambiguity by using the term *and/or*. Similarly, Latin uses two different words: *vel* expresses the inclusive sense of *or*, while the other, *aut*, expresses the exclusive sense.

Table 1.3 $p \vee q$

p	q	$p \vee q$
T	T	**T**
T	F	**T**
F	T	**T**
F	F	**F**

Example 1.3.5 Consider the propositions

$p =$ François passed logic

and

$q =$ François passed history.

Suppose François had failed logic, but passed history. What is the truth value of $p \lor q$?

Solution The proposition $p \lor q$ means in this case that François passed logic or history. The truth value would be true, since he had at least passed one of the subjects. ◄

1.3.4 Exclusive Disjunction

In ordinary conversation, the more common intention of "p or q" is "p or q and not both." We use this exclusive sense of "either ... or ..." when we say, "Either today is Monday or it's Tuesday" (but not both). We have observed, however, that this disjunction is not symbolized by $p \lor q$ but by one of these two forms:

$$(p \lor q) \land (\neg(p \land q)) \quad \text{or} \quad p \veebar q.$$

Both are read "either p or q, but not both p and q." Table 1.4 shows how to represent the exclusive disjunction in a truth table.

Table 1.4 $\;p \veebar q$

p	q	$p \veebar q$
T	T	F
T	F	T
F	T	T
F	F	F

Example 1.3.6 Consider the propositions

$$p = \text{The month is September}$$

and

$$q = \text{The month is October.}$$

Either the month is September or October, but not both. This can be represented symbolically as $p \veebar q$. ◄

1.3.5 Implication

We can use the words, "if ... then ..." to introduce an **implication**. For example,

If it is raining, *then* I'll bring an umbrella.

is an implication. The notation

$$p \to q \qquad \text{denotes} \qquad \text{"If } p, \text{ then } q\text{."}$$

The symbol $\to$ visually conjures the idea of direction or order; if p happens, then q will happen next.

For $p \to q$, this may be read as any of the following:

(a) If p, then q.

(b) p implies q.

(c) q, if p.

(d) p is **sufficient** for q.

(e) p, only if q.

(f) q is **necessary** for p.

In words, the implication is false only if the
antecedent is true and the consequent is false; otherwise it is true. Mathematically, the implication
$p \to q$ is only false when p is true and q is false;
otherwise, it is true. The truth table is given in
Table 1.5.

It is helpful to think of the implication in a
real world setting as

Table 1.5 $p \to q$

p	q	$p \to q$
T	T	**T**
T	F	**F**
F	T	**T**
F	F	**T**

$$\text{cause} \to \text{effect}.$$

Example 1.3.7 — Falling in Poison Ivy.
Consider the propositions,

$$p = \text{I fell into poison ivy,}$$

and

$$q = \text{I have a rash.}$$

The implication $p \to q$ would be if I fell into poison ivy then I will get a
rash. We know from personal knowledge that this is in fact true—after coming into
contact with poison ivy (p is true) one will always get a rash (q is true) [Row 1 of
Table 1.5]. However, if you have a rash (q is true), but you didn't step into poison
ivy (p is false) then this is still possible [Row 3]; you may have gotten a rash from
something else! If you didn't fall in poison ivy (p is false) and you didn't get a
rash (q is false), this is also a possibility [Row 4]. The only impossibility is when
you come into contact with poison ivy (p is true) and you didn't get a rash (q is
false) [Row 2]. ◀

Example 1.3.8 For each of the pair of propositions, express the implication in the
English language of the form *if... then....*
  (a) $p = $ You love me; $q = $ I shall be eternally happy.
  (b) $p = $ Arsenic is heated; $q = $ Arsenic sublimes.
  (c) $p = $ A wombat is tickled; $q = $ A wombat will laugh.

Solution
  (a) $p \to q$: If you love me, then I shall be eternally happy.
  (b) $p \to q$: If arsenic is heated, then it sublimes.
  (c) $p \to q$: If a wombat is tickled, then it will laugh.
                                                                    ◀

For the implication $p \to q$, we call p the **antecedent** and q the **consequent**.

Example 1.3.9 Identify the antecedent and the consequent for the following:

$$\text{If } x \text{ is human, then } x \text{ is mortal.}$$

Solution "x is human" is the antecedent, and the consequent is "x is mortal." ◀

The word *if* introduces the antecedent, and *then* introduces the consequent. Other synonyms of *if* that can introduce the antecedent are the following: *provided (that)*, *in case*, and *on the condition that*.

However, sometimes sentences have the form *p and q*, but what it really means in English is *p and then q*.

Example 1.3.10 For each of the sentences, represent them symbolically:

(a) I studied for the exam, and I passed the exam.

(b) I passed the exam, and studied for the exam.

Solution The two sentences do not express the same proposition, since *and* really means *and then*. In this case, these two propositions are not logically equivalent.

Let

$$p = \text{I studied for the exam}$$

and

$$q = \text{I passed the exam.}$$

We can represent the two sentences as

(a) $p \to q$: I studied for the exam, and then passed the exam.

(b) $q \to p$: I passed the exam, and then I studied for the exam.

Merely knowing that p is true and that q is true, one does not automatically know the order of the two events. You cannot study for the exam and pass the exam concurrently. You would study (p) and then write and pass the exam (q). ◄

1.3.6 Order of Precedence

Just as there is an order of precedence when we use the arithmetical operations $\times, \div, +$, and $-$, we must also observe an order of precedence when we use the logical connectives. The following list dictates the order of precedence for logical connectives.

$$(\ ), \neg, \wedge, \vee, \to \tag{1.1}$$

Thus, connectives inside of the parentheses ( ) should be applied first, followed by the connective $\neg$ next, and so on.

Example 1.3.11 Add parentheses to the following:

$$\neg p \vee q \to p \wedge r \vee s.$$

Solution First, we add parentheses around the negations

$$(\neg p) \vee q \to p \wedge r \vee s.$$

Next, we add parentheses around the conjunction

$$(\neg p) \vee q \to (p \wedge r) \vee s,$$

and finally

$$[(\neg p) \vee q] \to [(p \wedge r) \vee s].$$

◄

Exercises

1.3.1. State the negation of the following propositions:
- (a) It is not snowing.
- (b) Fish do not have gills.
- (c) The longest bridge in Great Britain is the Forth Bridge.
- (d) New York City is not the capital of New York.
- (e) I never gave or took any excuse.[3]
- (f) Every child is an artist.[4]
- (g) No baseball team wins two games in a row.
- (h) It is not the case that it rains and the wind blows.
- (i) Either the market is not growing, or the market is not being measured correctly.
- (j) Our national debt is not decreasing rapidly enough, and our government spending policies are not helping.
- (k) If sales taxes are lowered, then consumer spending will increase.

1.3.2. Let p be "House prices are high," and q be "Houses are in demand." Give a verbal translation for the following:

(a) $p \land q$	(c) $\lnot p \land \lnot q$	(e) $\lnot(p \land q)$
(b) $p \land \lnot q$	(d) $p \lor \lnot q$	(f) $\lnot(\lnot p \lor \lnot q)$

1.3.3. Let p be "I work hard," q be "I pass," and r be "I get promoted." Translate each of the following into symbolic form.
- (a) I get promoted, if I pass.
- (b) I work hard implies I pass.
- (c) Getting promoted is necessary for passing.
- (d) If I don't work hard, then I won't get promoted.

1.3.4. Assume that p is false, and q is false. Determine the truth value of the following:

(a) $p \lor q$	(c) $p \to q$	(e) $\lnot p \lor q$
(b) $p \land q$	(d) $\lnot q$	(f) $\lnot p \land q$

1.3.5. Assume that p is true, and q is false. Determine the truth value of the following:

(a) $p \lor q$	(c) $p \to q$	(e) $\lnot p \lor q$
(b) $p \land q$	(d) $\lnot q$	(f) $\lnot p \land q$

1.3.6. State the truth value of each of the following implications:
- (a) If $2 + 2 = 4$, then $7 > 2$.
- (b) If $2 + 3 = 5$, then $9 < -2$.
- (c) If $3 + 4 = 9$, then $-7 > -9$.

---

[3]Florence Nightingale (1820–1910)
[4]Pablo Picasso (1881–1973)

(d) if $3 + 4 = 9$, then $-7 < -9$.

1.3.7. Put the following into symbolic notation:
    (a) Personnel should be promoted if they are fully qualified.
    (b) Either Bacon or Marlowe wrote Shakespeare's plays, but not both.
    (c) Either the child is admitted to school immediately or he is sent home and notified of his admission.
    (d) If you don't talk to Marian, she won't talk to you.

1.3.8. Restore parentheses to the following propositions:
    (a) $r \vee \neg p \wedge q$ (b) $q \rightarrow \neg\neg\neg p \wedge r$

1.3.9. Rewrite each of the following propositions as an implication in the **if-then** form:
    (a) Practicing her serve daily is a sufficient condition for Darcy to have a good chance of winning the tennis tournament.
    (b) The computer is brand-new or I won't buy it.

1.3.10. Decide if the proposition is true or false: *if the moon is made of cheese, then Socrates is the Prime Minister of Canada.*

1.3.11. An important logical connective to computer scientists is **nand**, which we will denote with $\uparrow$. The *nand* connective produces a truth value of true if at least one of the propositions is false. Complete Table 1.6.

Table 1.6 $p \uparrow q$

p	q	$p \uparrow q$
T	T	
T	F	
F	T	
F	F	

1.4 Constructing Truth Tables for a Compound Proposition

As we saw in Sec. 1.3, we were able to create truth tables for each connective. Now, we can build more complicated propositions using these connectives, and with these more complicated propositions we would like to construct their corresponding truth tables.

Example 1.4.1 Construct a truth table for the formula $\neg(\neg p \wedge q)$.

Solution The first two columns of this table list the four possible combinations of truth values of p and q; see Table 1.7.

<table>
<tr><td colspan="2">Table 1.7 Step 1</td></tr>
<tr><td>p</td><td>q</td></tr>
<tr><td>T</td><td>T</td></tr>
<tr><td>T</td><td>F</td></tr>
<tr><td>F</td><td>T</td></tr>
<tr><td>F</td><td>F</td></tr>
</table>

<table>
<tr><td colspan="3">Table 1.8 Step 2</td></tr>
<tr><td>p</td><td>q</td><td>$\neg p$</td></tr>
<tr><td>T</td><td>T</td><td>**F**</td></tr>
<tr><td>T</td><td>F</td><td>**F**</td></tr>
<tr><td>F</td><td>T</td><td>**T**</td></tr>
<tr><td>F</td><td>F</td><td>**T**</td></tr>
</table>

Note that these columns have to be worked out in order, because each are used in computing the next. The third column (see Table 1.8) listing the truth values for the formula $\neg p$, is found by simply negating the truth values of p in the second column.

The fourth column (Table 1.9), for the formula $\neg p \wedge q$, is found by combining the truth values for $\neg p$ and q listed in the third and first columns, according to the truth value rule for $\wedge$. According to this rule, $\neg p \wedge q$ will be true only if both $\neg p$ and q are true. Looking in the first and third columns, we see that this happens only in row three of the table, so that the fourth column contains an F in the third row and T in all the other rows.

Table 1.9 Step 3			
p	q	$\neg p$	$\neg p \wedge q$
T	T	F	**F**
T	F	F	**F**
F	T	T	**T**
F	F	T	**F**

Table 1.10 Step 4				
p	q	$\neg p$	$\neg p \wedge q$	$\neg(\neg p \wedge q)$
T	T	F	F	**T**
T	F	F	F	**T**
F	T	T	T	**F**
F	F	T	F	**T**

Finally, the truth values for the formula $\neg(\neg p \wedge q)$ are listed in the fifth column (Table 1.10), which is found by negating the truth values in the fourth column. ◄

Exercises

1.4.1. Complete the truth table.

p	q	$\neg p$	$\neg q$	$p \wedge \neg q$	$\neg p \vee q$	$\neg p \rightarrow \dot{q}$
T	T					
T	F					
F	T					
F	F					

1.4.2. Construct a truth table for the following:

(a) $\neg(p \vee q)$

(b) $\neg(p \wedge q)$

(c) $(\neg p) \rightarrow q$

(d) $[(p \rightarrow q) \wedge (\neg p)] \rightarrow (\neg q)$

(e) $p \vee (q \wedge r)$

(f) $(p \vee q) \wedge (p \vee r)$

(g) $(p \wedge q) \vee (p \wedge \neg r)$

1.4.3. Construct a truth table for the following:

(a) $q \rightarrow p$

(b) $(r \wedge q) \rightarrow r$

(c) $(r \rightarrow s) \rightarrow r$

(d) $(p \vee q) \rightarrow p$

(e) $(p \rightarrow q) \rightarrow q$

(f) $(p \wedge q) \rightarrow (p \vee q)$

(g) $(p \vee q) \rightarrow [(\neg p) \vee q]$

(h) $(p \rightarrow r) \rightarrow [(\neg p) \vee r]$

(i) $(v \wedge w) \rightarrow [(\neg w) \rightarrow v]$

(j) $(p \vee q) \rightarrow (p \rightarrow q)$

(k) $(p \wedge q) \rightarrow [(\neg q) \rightarrow (\neg p)]$

(l) $(\ell \rightarrow k) \rightarrow [(\neg k) \rightarrow (\neg \ell)]$

(m) $(p \vee q) \rightarrow [p \wedge (\neg q)]$

(n) $(p \rightarrow q) \rightarrow r$

1.5 Switching Circuits

Claude Shannon (1916–2001), an American electrical engineer, developed switching circuit theory and published his ideas in 1938 while he was still a master's student at Massachusetts Institute of Technology.

These switching circuits are especially important, since their principles are the foundation of digital computers, telephone exchanges, or for any complex task that is performed automatically. In this section, we will look at Shannon's basic ideas about circuits, and just as he did, we will apply our knowledge of logic to this application.

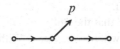

Fig. 1.4   A simple switch

First, we consider a **simple switch**, denoted by a lowercase letter, that is either open (or "off"), or closed (or "on"), but not both; see Fig. 1.4. We denote

- $p =$ T to be the switch is closed (or "on") so that the current is flowing through it.
- $p =$ F to be the switch is open (or "off") so that the current is not flowing through it.

A **switching circuit** is an arrangement of wires and switches such that it takes the energy from an input, like a battery, to an output, such as a light bulb.

There are two basic ways to arrange switches in a switching circuit.

(a) If the switches in the switching circuit is connected like Fig. 1.5, then it is a **series connection**.

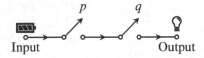

Fig. 1.5   A series connection: $p \wedge q$

We see that for current to flow through the circuit, both switches must be closed. If either switch or both are open, then the current has no path to flow from the input to the output.

Table 1.11   Truth table for a series connection

p	q	Output	$p \wedge q$
T	T	**T**	T
T	F	**F**	F
F	T	**F**	F
F	F	**F**	F

Notice that the output in the truth table, Table 1.11, is the same truth table as $p \wedge q$. *Thus, a series connection corresponds to the logical conjunction.*

(b) If the switches in the switching circuit is connected like Fig. 1.6, then it is a **parallel connection**.

In this circuit, we see that the only case the current does not flow to the output is when both switches are open. With the rest of the cases, current is able to travel through one or both of the closed switches to the output. Notice that the output truth table is the same truth table as $p \vee q$ in Table 1.12. *Thus, a parallel connection corresponds to the logical disjunction.*

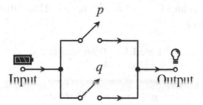

Fig. 1.6   A parallel connection: $p \vee q$

Table 1.12   Truth table for a parallel connection

p	q	Output	$p \vee q$
T	T	**T**	T
T	F	**T**	T
F	T	**T**	T
F	F	**F**	F

Example 1.5.1 For the switching circuit given in Fig. 1.7, construct a truth table for the output, and a proposition to determine the conditions which the current flows.

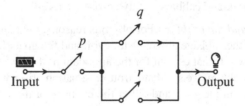

Fig. 1.7   Switching Diagram for Example 1.5.1

Solution In Table 1.13 on p. 20, we create a truth table for each possible case of switch p, q, and r being either on or off.

We can see a pattern in the table. In order for current to flow, switch p must be on, and either switch q or r or both to be on. Thus, the proposition for this switching circuit is $p \wedge (q \vee r)$. (We leave it in Exercise 1.5.1 for you to check that the truth table of $p \wedge (q \vee r)$ is the same as the output truth table.)

Another way to find the proposition is to break the switching circuit down into series and parallel components. The switches q and r are in parallel, which can be written as $q \vee r$. Then p and $(q \vee r)$ are in series. Thus, the switching circuit can be written as $p \wedge (q \vee r)$.

Table 1.13   Truth table for
Example 1.13

p	q	r	Output
T	T	T	T
T	T	F	T
T	F	T	T
T	F	F	F
F	T	T	F
F	T	F	F
F	F	T	F
F	F	F	F

◀

Example 1.5.2 — De lupo et capra et fasciculo cauli. A traveler has to take a wolf, a goat, and a bundle of cabbage across a river. The only boat he could find can only take two of them at a time. Unfortunately, if the traveler leaves the bundle of cabbage alone with the goat, the goat will eat the cabbage; and if the traveler leaves the goat alone with the wolf, the wolf will eat the goat. How can the traveler get the wolf, the goat, and cabbage safely across the river?

Solution *Understand the problem.* For obvious reasons, the man can never *either* leave the goat *and* the cabbage alone, *or* the wolf and the goat alone. The problem involves three keywords that cry out for the application of logic: *either-or* and *and*. By analyzing the problem, we see that a trouble situation arises when the man is on the opposite side from the goat, and either the cabbage or the wolf is on the same side as the goat.

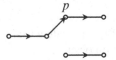

Fig. 1.8   An *either-or* switch

Devise a plan. The goal is to come up with a switching circuit such that a light that will signal trouble. If there is trouble, the circuit user can turn off the source and try again. The user has solved the puzzle if he or she can get all four items across the river without the trouble signal turning on. But, the puzzle uses an *either-or* (the exclusive disjunction). How can we represent this in a switching circuit diagram? One way is shown in Fig. 1.8.

Carry out the plan. The switching circuit is given in Fig. 1.9. This required a lot of trial-and-error to find the correct configuration of the circuit.

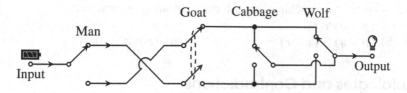

Fig. 1.9 Switching circuit to indicate trouble crossing the river

◀

Exercises

1.5.1. Create a truth table for $p \wedge (q \vee r)$, and confirm that your table matches the output truth table in Example 1.5.1.

1.5.2. Write the corresponding propositions for the following switching circuits:

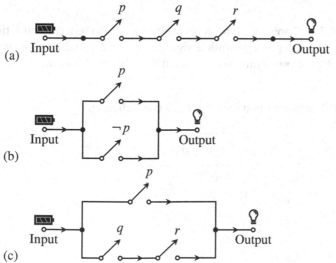

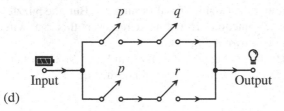

(d)

1.5.3. Construct a switching circuit for the following propositions.

(a) $(p \vee q) \vee r$ (c) $p \wedge (p \vee q)$

(b) $(p \vee q) \wedge (q \vee r)$ (d) $(p \wedge q) \vee (q \vee r)$

1.6 Tautologies and Contradictions

> **Definition 1.6.1 — Tautology.** A proposition is said to be a **tautology** if it is *true* under all possible truth assignments of the variables contained. We denote a tautology in this book by $\mathbb{T}$.

Example 1.6.2 The proposition $(p \vee \neg p)$ is a tautology.

p	$\neg p$	$p \vee \neg p$
T	F	T
F	T	T
		$\underbrace{}_{\mathbb{T}}$

◀

> **Definition 1.6.3 — Contradiction.** A proposition is said to be a **contradiction** if it is *false* under all possible truth assignments of the variables contained. In other words, the proposition is not satisfiable. In this book, we denote a contradiction by $\mathbb{F}$.

Example 1.6.4 The proposition $(p \wedge \neg p)$ is a contradiction.

p	$\neg p$	$p \wedge \neg p$
T	F	F
F	T	F
		$\underbrace{}_{\mathbb{F}}$

◀

> **Definition 1.6.5 — Contingency.** A proposition is said to be a **contingency** if it is neither a tautology nor a contradiction.

Example 1.6.6 The proposition $(p \to q)$ is a contingency.

p	q	$p \to q$
T	T	T
T	F	F
F	T	T
F	F	T

◄

Exercises

1.6.1. What are the differences between a tautology, a contraction, and a contingency?

1.6.2. Show that the following are tautologies by using truth tables:

    (a) $(p \wedge \neg p) \to q$ (b) $\neg(p \wedge \neg p)$ (c) $(\neg p \to p) \to p$

1.6.3. Which of the following propositions are tautologies?

    (a) $p \to (q \to r)$ (c) $[(p \vee q) \wedge (\neg q)] \to p$

    (b) $(q \vee r) \to (\neg r \to q)$ (d) $[(p \wedge \neg q) \vee (q \wedge \neg r)] \vee (r \wedge \neg p)$

1.6.4. Use truth tables to determine whether the following propositions are tautologies, contradictions, or contingencies:

    (a) $(p \wedge q) \vee (\neg p \wedge \neg q)$ (c) $\mathbb{T} \to (\neg p \wedge p)$

    (b) $(p \vee \neg q) \wedge (q \vee \neg p)$ (d) $\mathbb{F} \to (p \to r)$

1.7 Logical Equivalence and Derived Logical Implications

> **Definition 1.7.1 — Logical Equivalence.** Two propositions, p and q, are **logically equivalent** if and only if $p \leftrightarrow q$ is a tautology. We will denote logical equivalence by the meta-operator $\Leftrightarrow$.

In other words, $p \Leftrightarrow q$ if and only if p and q have the same truth table.

1.7.1 Biconditional

We let

$$p \leftrightarrow q \qquad \text{denote} \qquad \text{``}p \text{ if and only if } q.\text{''}$$

The biconditional $p \leftrightarrow q$ is true when both p and q are true, and when both p and q is false. In the truth table of the biconditional (Table 1.14), if only one of the propositions p or q is not true, then $p \leftrightarrow q$ is false.

Table 1.14 $p \leftrightarrow q$

p	q	$p \leftrightarrow q$
T	T	**T**
T	F	**F**
F	T	**F**
F	F	**T**

The biconditional proposition "$p \rightarrow q$ and $q \rightarrow p$" can be written as $p \leftrightarrow q$. Again, like the implication, this proposition may be written in various forms.

(a) If p, then q, and if q, then p.
(b) If p, then q, and conversely.
(c) If q, then p, and conversely.
(d) p, if and only if q.
(e) q, if and only if p.
(f) p is a necessary and sufficient condition for q.
(g) q is a necessary and sufficient condition for p.

Example 1.7.2 Consider the following statement: I am a member at the gym *if and only if* I have paid my gym fee.

If I am a member at the gym, then I paid my gym fee. Also, if I paid my gym fee, then I am a member. The statement also means that if I am not a member at the gym, then I did not pay my gym fee; conversely, if I did not pay my gym fee, then I am not a gym member.   ◄

Table 1.15   The biconditional $p \leftrightarrow q$ is equivalent to $(p \rightarrow q) \wedge (q \rightarrow p)$.

p	q	$p \rightarrow q$	$q \rightarrow p$	$(p \rightarrow q) \wedge (q \rightarrow p)$	$p \leftrightarrow q$
T	T	T	T	**T**	**T**
T	F	F	T	**F**	**F**
F	T	T	F	**F**	**F**
F	F	T	T	**T**	**T**

Since $p \rightarrow q$ is true except when p is true and q is false, and $q \rightarrow p$ is true except when q is true and p is false, $p \leftrightarrow q$ will be true except when only one of p and q is true and the other false. That is, $p \leftrightarrow q$ is a true proposition if both p

and q are true or if both p and q are false. Looking at the truth table in Table 1.15, we see the fifth and sixth columns are the same, and so $((p \rightarrow q) \wedge (q \rightarrow p))$ is equivalent $p \leftrightarrow q$.

1.7.2 Derived Logical Implications

There are three other easily derived implications for every given implication.

Definition 1.7.3 Given an implication

$$p \rightarrow q,$$

the three other derived implications are
  (a) $\neg q \rightarrow \neg p$ is the **contrapositive**;
  (b) $q \rightarrow p$ is the **converse**; and
  (c) $\neg p \rightarrow \neg q$ is the **inverse**.

The original implication is logically equivalent with the contrapositive;

$$p \rightarrow q \Leftrightarrow \neg q \rightarrow \neg p. \tag{1.2}$$

Also, the converse and the inverse are logically equivalent;

$$q \rightarrow p \Leftrightarrow \neg p \rightarrow \neg q. \tag{1.3}$$

To prove (1.2) and (1.3), by definition of logical equivalence it suffices to show that the implication on the left side has the same truth table as the other implication on the right side. We leave this as an exercise for the reader; see Exercises 1.7.7 and 1.7.8.

While many people make this mistake, it is important to note that the implication is *not* logically equivalent to the converse,

$$p \rightarrow q \not\Leftrightarrow q \rightarrow p, \tag{1.4}$$

and the implication is *not* logically equivalent to the inverse

$$p \rightarrow q \not\Leftrightarrow \neg p \rightarrow \neg q. \tag{1.5}$$

Example 1.7.4 Suppose we consider the following proposition:

> If a man can march, then he is a soldier.

State the contrapositive, the converse, and the inverse of the proposition.

Solution Let p be "A man can march," and q be "He is a soldier." Then the other three derived implications would be
  (a) the contrapositive ($\neg q \rightarrow \neg p$): If a man is not a soldier, then he cannot march;
  (b) the converse ($q \rightarrow p$): If a man is a soldier, then he can march; and
  (c) the inverse ($\neg p \rightarrow \neg q$): If a man cannot march, then he is not a soldier.

◀

Exercises

1.7.1. Given the following premises
  - $p =$ You love bananas
  - $q =$ You are a monkey's uncle
  - $r =$ Your nephew is a chimp

write the following in words:
  (a) $q \rightarrow p$ (c) $q \leftrightarrow r$
  (b) $\neg r \rightarrow \neg p$ (d) $q \wedge r$

1.7.2. Construct truth tables for the following:
  (a) $\neg p \leftrightarrow q$ (e) $(p \vee q) \leftrightarrow (\neg r \wedge \neg s)$
  (b) $\neg p \leftrightarrow (p \rightarrow \neg q)$ (f) $(p \vee q) \leftrightarrow (p \wedge q)$
  (c) $(p \vee q) \leftrightarrow (q \vee p)$ (g) $[p \rightarrow (q \vee r)] \wedge [p \leftrightarrow \neg r]$
  (d) $(p \rightarrow q) \leftrightarrow (\neg p \vee q)$

1.7.3. Use the symbols $\rightarrow$ or $\leftrightarrow$ to express each of the following.
  (a) p only if q (d) p is necessary and sufficient for q
  (b) p is necessary for q (e) p if and only if q
  (c) p is sufficient for q (f) q implies p

1.7.4. Prove by truth tables that the De Morgan Laws below are logically equivalent.
  (a) $\neg(p \vee q) \Leftrightarrow (\neg p \wedge \neg q)$ (c) $\neg(\neg p \vee q) \Leftrightarrow (p \wedge \neg q)$
  (b) $\neg(p \wedge q) \Leftrightarrow (\neg p \vee \neg q)$

1.7.5. What is the converse, inverse, and contrapositive for the following propositions?
  (a) If the traffic light is red, then cars must stop.
  (b) If the fruit is red, then it is an apple.
  (c) Unrest is sufficient for change.
  (d) A right implies a responsibility.[5]
  (e) The right to search for the truth implies also a duty.[6]
  (f) Speak only if it improves upon the silence.[7]

1.7.6. Show, using truth tables, that $\neg(p \wedge q)$ is not logically equivalent to $(\neg p \wedge \neg q)$.

1.7.7. Given the implication $p \rightarrow q$, show that the implication is equivalent to the contrapositive using a truth table; that is, show that

$$p \rightarrow q \Leftrightarrow \neg q \rightarrow \neg p.$$

---

[5] John D. Rockefeller (1839–1937)
[6] Albert Einstein (1879–1955)
[7] Mohandas Karamchand Gandhi (1869–1948)

1.7.8. Given the implication $p \rightarrow q$, show that the converse and the inverse are logically equivalent using a truth table; that is, show that

$$q \rightarrow p \Leftrightarrow \neg p \rightarrow \neg q.$$

1.7.9. State the contrapositive, the converse, and the inverse for the following implications:

(a) $q \rightarrow p$ (b) $\neg p \rightarrow \neg q$ (c) $\neg p \rightarrow q$.

1.7.10. Let proposition p be "I can think" and proposition q be "I am human."
(a) State the contrapositive, the converse and the inverse of the implication $p \rightarrow q$ in words.
(b) State the biconditional proposition $p \leftrightarrow q$ in words.
(c) In words, state under which conditions is $p \leftrightarrow q$ false.

1.7.11. Let proposition p be "an integer is divisible by 9," and proposition q be "an integer is divisible by 3." Are the following propositions true?

(a) $p \rightarrow q$ (b) $q \rightarrow p$ (c) $p \leftrightarrow q$

1.7.12. Naomi will graduate from high school if and only if she earns a high school diploma. Is this proposition true or false?

1.7.13. State whether the following is true or false.
(a) If it is nighttime, then it is 3:00 A.M.
(b) If it is 3:00 A.M., then it is nighttime.
(c) It is 3:00 A.M. if and only if it is nighttime.

1.7.14. Determine if the biconditional statement is true for each of the following conditional propositions:
(a) If the date is the 29th of the month, then the month is not February.
(b) If the computer is unplugged, then the computer will not run.
(c) If you download 12 songs for $11.88, then each song costs $0.99.
(d) If Jake lives in Las Vegas, then he lives in Nevada.
(e) If London is not in England, then $3 + 6 = 11$.

1.7.15. Prove by truth tables that the pair of propositions are logically equivalent.
(a) $p \wedge (q \vee r) \Leftrightarrow (p \wedge q) \vee (p \wedge r)$
(b) $p \vee (q \wedge r) \Leftrightarrow (p \vee q) \wedge (p \vee r)$

1.7.16. Show that $\neg(p \leftrightarrow q) \Leftrightarrow [(p \wedge \neg q) \vee (q \wedge \neg p)]$.

1.7.17. Express the proposition $p \rightarrow q$ using
(a) $\neg$ and $\vee$, but refraining from using $\rightarrow$.
(b) $\neg$ and $\wedge$, but refraining from using $\rightarrow$.
(*Hint*: You can test if you have found a correct logical equivalent using a truth table.)

1.8 Review

Summary

- **Logic** is the study of the methods and principles used in distinguishing correct from incorrect arguments.
- By using the knowledge that we know, and drawing conclusions, this process is called **logical inference**.
- A **truth value** is either true or false.
- To decide if a statement is a **proposition**, you must ask yourself if it is possible to assign a truth value. A proposition cannot be a question, an order, or an exclamation.
- A **prime proposition** is a simple complete thought, and when there are two or more prime proposition, this forms a **compound proposition**.
- A **paradox** is a statement that contradicts itself and yet might be true.
- Given an **implication** $p \to q$, then
  - the **contrapositive** is $\neg q \to \neg p$,
  - the **converse** is $q \to p$, and
  - the **inverse** is $\neg p \to \neg q$.
- Necessary and sufficient
  - (a) "p is sufficient for q" means $p \to q$.
  - (b) "p is necessary for q" means $q \to p$.
  - (c) If "p is necessary and sufficient for q," then $p \leftrightarrow q$.
- There is a lot of standard logic notation that has well accepted meaning in mathematics. For example, we summarize the logical connective meanings in Table 1.17. Unfortunately, many students may read the notation without remembering what it means; a helpful tip is to read aloud the meaning of the symbols to aid in remembering.

Table 1.16   Logical Connective Truth Tables

p	q	$\neg p$	$p \wedge q$	$p \vee q$	$p \veebar q$	$p \to q$	$p \leftrightarrow q$
T	T	F	T	T	F	T	T
T	F	F	F	T	T	F	F
F	T	T	F	T	T	T	F
F	F	T	F	F	F	T	T

Table 1.17   Summary of Logical Connectives Meanings

Name	Symbol	Meaning	Condition for a true truth value
Negation	$\neg p$	not p	p is false
Conjunction	$p \wedge q$	p and q	both p and q are true
Disjunction	$p \vee q$	p or q (or both)	at least one is true
Exclusive Disjunction	$p \veebar q$	p or q (not both)	either p is true or q is true, but not both
Conditional	$p \rightarrow q$	p implies q	q is true whenever p is true
Biconditional	$p \leftrightarrow q$	p if and only if q	p and q have same truth value

Exercises

1.8.1. Decide if the following are propositions?
  (a) She is Queen of England.
  (b) How old is he?
  (c) $7 + 5 = 12$
  (d) $6 > 2$
  (e) Some girls dislike sushi.
  (f) All boys dislike sushi.
  (g) She enjoys dancing and singing.
  (h) Why did you not do your homework?
  (i) Is $7 + 5 = 12$?
  (j) The Yukon is in Europe.

1.8.2. A crocodile stole a child, and promises the father the release of the child provided that the father can predict correctly whether or not the crocodile will release the child. The father says the crocodile will not give his child back. What should the crocodile do?

1.8.3. Write the negation of each of the following propositions.
  (a) $4 + 5 = 15$
  (b) -3 is a positive number.
  (c) Alexander the Great was a Roman leader and $2 + 1 = 3$.

1.8.4. Assume that p is false, and q is true. Determine the truth value of the following:
  (a) $p \vee q$
  (b) $p \wedge q$
  (c) $p \rightarrow q$
  (d) $\neg q$
  (e) $\neg p \vee q$
  (f) $\neg p \wedge q$

1.8.5. Rewrite the quote by Benjamin Franklin (1706–1790) as a conditional proposition: "Never put off till tomorrow what you can do today."

1.8.6. For each of the following propositions, state its converse, contrapositive, and negation:
   (a) If it's cold outside and it's not snowing, then I won't wear my coat.
   (b) If you live in Atlantis, then you'll need a snorkel.

1.8.7. Construct a truth table for each of the following propositions, and classify if each proposition as either a tautology, a contradiction, or a contingency:
   (a) $\neg p \wedge (p \to q)$ (d) $(p \vee q) \wedge (\neg p \wedge \neg q)$
   (b) $(p \to q) \vee (q \to p)$ (e) $\neg(p \wedge q) \leftrightarrow (q \wedge p)$
   (c) $(p \vee q) \wedge (p \vee \neg r)$

1.8.8. Show that $(p \leftrightarrow q) \Leftrightarrow \neg[(p \wedge \neg q) \vee (q \wedge \neg p)]$.

1.9 Bibliographic Remarks

The Pinocchio paradox was devised by Veronique Eldridge-Smith when she was 11 years old in February 2001. Her father had asked Veronique and her brother to come up with their own versions of the Liar paradox, and that is when Veronique came up with the Pinocchio paradox (Eldridge-Smith and Eldridge-Smith, 2010).

For years, it was thought that either Russell or Wittgenstein had developed the idea of truth tables. But recently, Anellis (2012) had found an unpublished 1893 manuscript by Charles Pierce with the implication truth table.

The notation for the logical connectives is consistent with the notation of Hilbert and Ackermann (1959). Other logic books may use Peano-Russell notation where negation is $\sim p$, the conjunction is $p \cdot q$, the implication is $p \supset q$, and the disjunction remains the same $p \vee q$ (Carruccio, 1964, p. 338).

The logic circuits section had been adapted from James *et al.* (2001), which in turn was based on the article by Shannon (1938). Example 1.5.2 of *De lupo et capra et fasciculo cauli*, or the wolf, the goat, and the bundle of cabbage is from a medieval Latin manuscript *Propositiones ad Acunedos Juvenes* by Alcuin of York (c. 735–804). The switching circuit solutions to this problem was devised by an avid puzzle-circuit-maker by the name of Harry Rudloe who at the time was a 16 year old boy from Brooklyn, New York (Ingalls, 1955, pp. 117–118).

2
Proofs and Arguments

"Take some more tea," the March Hare said to Alice, very earnestly.

"I've had nothing yet," Alice replied in an offended tone, "so I can't take more."

"You mean, you can't take *less*," said the Hatter: "it's very easy to take *more* than nothing."

—Carroll (1866, p. 106)

2.1 What is a Proof?

A proof is a proof. What kind of a proof? It's a proof. A proof is a proof. And when you have a good proof, it's because it's proven.

—Jean Chrétien (1934–)
20th Prime Minister of Canada

At the very heart of mathematics lies a fundamental concept: the proof. A proof is a convincing demonstration of the truth about a proposition, and it comes in two kinds—the formal proof, and the mathematical proof.

There are certain formulas that serve as the foundation of the formal system of mathematics—these fundamental formulas are called **axioms**. A **formal proof** is a sequence of formulas, each of which is either an axiom or follows from a formula that has been shown to be true by a rule of inference earlier. (Hilbert, 1931; Bundy *et al.*, 2005). The formal proof is largely a 20th century invention, predominately lobbied by the German mathematician David Hilbert (1862–1943) who wanted to place mathematics on a solid foundation by having a method that would avoid erroneous proofs.

Indeed, if you were to ask an undergraduate mathematics student what a proof is, his or her idea of a proof is a formal proof. In the mathematical community, this is not the case. There is another kind of proof which is preferred—the mathematical proof. For mathematicians, *they like to do it rigorously with a pencil.*

By *rigorously*, I mean that the proof is able to convince others that a certain statement or proposition is true. It really is a lot like a story; the story has to be heavily constrained, and the readers have to be familiar with the genre and even the language. Acquiring this familiarity with the genre and language takes time. Yet, while it takes time to learn, the reader becomes more comfortable and fluent in the nature of reading the material. This continues to evolve until at some point, to an outsider looking in, it looks like incomprehensible gibberish. In short, mathematics is about *why* things work, and not really about *what* things work, and mathematical proofs are the stories mathematicians tell their colleagues to explain why something is true using language understood amongst them.

By a *pencil*, I mean this in both the literal and the figurative sense. In the literal sense, we like to sit down with a piece of paper and try to work things out with a pencil. We try things. We scratch things out. We crumple the paper in anger and frustration and begin anew. In this process, we are trying to *understand the problem*[1] better by gaining more insight. The moment that a mathematician suspects that it can be easily done with a computer, they will walk away to look for another problem. Not that there is no challenge, but doing work for the sake of work does not yield any insight and hence, it provides no challenge to treat a computer like a calculator on steroids. Therefore, mathematicians are not concentrated solely on calculations, but on why these calculations work. In the figurative sense of a pencil, mathematical proofs are more about communication, and not so much about the truth. Communicating can employ a computer, but as a means of presentation— in a journal, blog, or forum—for comment by others or as a processor to construct a lecture presentation to share as if it was a big piece of paper.

In this book, you will encounter both kinds of proofs, formal and mathematical. We will begin by looking at formal proofs in this chapter.

2.2 Rules of Inference

In previous courses, you have seen the laws of arithmetic; see Table 2.2.[2] These same concepts from arithmetic can be applied to propositions yielding the rules of inference. The rules of inference in logic provides a systematic way of showing propositions are equivalent to each other.

2.2.1 Fundamental Logical Equivalences

The fundamental logical equivalences are listed below. They can all be formally justified using truth tables.

---

[1] This is one of the main steps in Pólya's approach (1945) to solving problems. Pólya's approach is summarized in the Preface starting on p. xviii.

[2] If you have not seen these before, you can see Appendix B.3.

Table 2.1 Correspondences between arithmetic and logic.

Arithmetic		Logic	
Addition	$+$	Disjunction (or)	$\vee$
Multiplication	$\times$	Conjunction (and)	$\wedge$

Table 2.2 Laws of Arithmetic

Name of Law	Operation	Law	Numerical Example
Associativity	Addition	$(a+b)+c = (a+b)+c$	$(3+4)+5 = 3+(4+5)$
	Multiplication	$(ab)c = a(bc)$	$(3 \times 4) \times 5 = 3 \times (4 \times 5)$
Commutativity	Addition	$a+b = b+a$	$3+4 = 4+3$
	Multiplication	$ab = ba$	$3 \times 4 = 4 \times 3$
Distributivity		$a(b+c) = ab+ac$	$3(4+5) = (3 \times 4)+(3 \times 5)$
		$(a+b)c = ac+bc$	$(3+4)5 = (3 \times 5)+(4 \times 5)$

Let p, q, and r be any three propositions (prime or compound), and $\mathbb{T}$ be a tautology and $\mathbb{F}$ be a contradiction:

$$p \wedge p \Leftrightarrow p \qquad \textbf{Idempotence} \qquad (2.1)$$

$$p \vee p \Leftrightarrow p \qquad (2.2)$$

$$(p \wedge q) \wedge r \Leftrightarrow p \wedge (q \wedge r) \qquad \textbf{Associativity} \qquad (2.3)$$

$$(p \vee q) \vee r \Leftrightarrow p \vee (q \vee r) \qquad (2.4)$$

$$p \wedge q \Leftrightarrow q \wedge p \qquad \textbf{Commutativity} \qquad (2.5)$$

$$p \vee q \Leftrightarrow q \vee p \qquad (2.6)$$

$$p \leftrightarrow q \Leftrightarrow q \leftrightarrow p \qquad (2.7)$$

$$p \wedge (q \vee r) \Leftrightarrow (p \wedge q) \vee (p \wedge r) \qquad \textbf{Distributivity} \qquad (2.8)$$

$$p \vee (q \wedge r) \Leftrightarrow (p \vee q) \wedge (p \vee r) \qquad (2.9)$$

$$\neg(p \vee q) \Leftrightarrow (\neg p) \wedge (\neg q) \qquad \textbf{De Morgan} \qquad (2.10)$$

$$\neg(p \wedge q) \Leftrightarrow (\neg p) \vee (\neg q) \qquad (2.11)$$

$$p \wedge \mathbb{T} \Leftrightarrow p \qquad \textbf{Identity} \qquad (2.12)$$

$$p \vee \mathbb{F} \Leftrightarrow p \qquad (2.13)$$

$$p \vee \mathbb{T} \Leftrightarrow \mathbb{T} \qquad \textbf{Domination} \qquad (2.14)$$

$$p \wedge \mathbb{F} \Leftrightarrow \mathbb{F} \qquad (2.15)$$

$$[(p \wedge q) \rightarrow p] \Leftrightarrow \mathbb{T} \qquad\qquad \textbf{Simplification} \qquad (2.16)$$
$$[(p \wedge q) \rightarrow q] \Leftrightarrow \mathbb{T} \qquad\qquad\qquad\qquad\qquad (2.17)$$
$$p \vee (\neg p) \Leftrightarrow \mathbb{T} \qquad\qquad \textbf{Excluded Middle} \qquad (2.18)$$
$$p \wedge (\neg p) \Leftrightarrow \mathbb{F} \qquad\qquad \textbf{Contradiction} \qquad (2.19)$$
$$[p \rightarrow (q \rightarrow r)] \Leftrightarrow [(p \wedge q) \rightarrow r] \qquad \textbf{Importation} \qquad (2.20)$$
$$[(p \wedge q) \rightarrow r] \Leftrightarrow [p \rightarrow (q \rightarrow r)] \qquad \textbf{Exportation} \qquad (2.21)$$
$$\neg(\neg p) \Leftrightarrow p \qquad\qquad \textbf{Double Negation} \qquad (2.22)$$
$$p \vee (p \wedge q) \Leftrightarrow p \qquad\qquad \textbf{Absorption} \qquad (2.23)$$

2.2.2 Logical Equivalences for Implication and Biconditional

Let p, q, and r be any three propositions (prime or compound):

$$(p \rightarrow q) \Leftrightarrow \neg p \vee q \qquad\qquad \textbf{Implication} \qquad (2.24)$$
$$[\neg(p \rightarrow q)] \Leftrightarrow [p \wedge (\neg q)] \qquad \textbf{Neg Implication} \qquad (2.25)$$
$$(p \leftrightarrow q) \Leftrightarrow [(p \rightarrow q) \wedge (q \rightarrow p)] \qquad \textbf{Biconditional} \qquad (2.26)$$
$$(p \leftrightarrow q) \Rightarrow (p \rightarrow q) \qquad\qquad\qquad\qquad (2.27)$$
$$(p \leftrightarrow q) \Rightarrow (q \rightarrow p) \qquad\qquad\qquad\qquad (2.28)$$
$$(p \rightarrow q) \Leftrightarrow [(\neg q) \rightarrow (\neg p)] \qquad \textbf{Contrapositive} \qquad (2.29)$$

Example 2.2.1 — Illustrating the Use of the Implication Rule. Apply the Implication Rule (2.24) to the statement

$$(p \wedge q) \rightarrow q.$$

Solution Notice at the beginning of the rules that we stated p, q, and r could be compound propositions. In fact, $p \wedge q$ is simply a compound proposition which we can denote by r.

$$\overbrace{(p \wedge q)}^{r} \rightarrow q \Leftrightarrow r \rightarrow q \qquad\qquad \text{Let } r \text{ be } (p \wedge q).$$
$$\Leftrightarrow \neg r \vee q \qquad\qquad\qquad \text{Apply (2.24).}$$
$$\Leftrightarrow \neg(p \wedge q) \vee q \qquad\qquad \text{Substitute } p \wedge q \text{ for } r.$$

◀

Example 2.2.2 Prove that $\neg(p \rightarrow q) \Leftrightarrow p \wedge \neg q$.

Solution By truth table:

p	q	$p \to q$	$\neg(p \to q)$	$\neg q$	$p \wedge \neg q$
T	T	T	F	F	F
T	F	F	T	T	T
F	T	T	F	F	F
F	F	T	F	T	F

Since the fourth and sixth columns are the same, by the definition of logical equivalence we can conclude $\neg(p \to q) \Leftrightarrow p \wedge \neg q$.

By the rules of inference:

$$\neg(p \to q) \Leftrightarrow \neg(\neg p \vee q) \qquad \text{By (2.24)}$$
$$\Leftrightarrow \neg(\neg p) \wedge \neg q \qquad \text{By (2.10)}$$
$$\Leftrightarrow p \wedge \neg q \qquad \text{By (2.22)}$$

◄

The application of the rules of inference in the order shown above may not necessarily be the only valid sequence. You will find that as you prove logical equivalences that there is more than one option at various points. The only way to find out if this option will lead you down the right path is to try it. This will likely involve a generous amount of scrap paper; mathematics is not known for being environmentally friendly. Here is an alternative proof for the above example.

Solution Instead of using De Morgan's first law (2.10), we used De Morgan's second law (2.11).

$$\neg(p \to q) \Leftrightarrow \neg(\neg p \vee q) \qquad \text{By (2.24)}$$
$$\Leftrightarrow \neg(\neg(p \wedge \neg q)) \qquad \text{By (2.11)}$$
$$\Leftrightarrow p \wedge \neg q \qquad \text{By (2.22)}$$

◄

Example 2.2.3 Prove that $\neg(p \leftrightarrow q) \Leftrightarrow (p \wedge \neg q) \vee (q \wedge \neg p)$.

Solution

$$\neg(p \leftrightarrow q) \Leftrightarrow \neg[(p \to q) \wedge (q \to p)] \qquad \text{By (2.26)}$$
$$\Leftrightarrow \neg(p \to q) \vee \neg(q \to p) \qquad \text{By (2.11)}$$
$$\Leftrightarrow \neg(\neg p \vee q) \vee \neg(\neg q \vee p) \qquad \text{By (2.24)}$$
$$\Leftrightarrow [\neg(\neg p) \wedge \neg q] \vee (q \wedge \neg p) \qquad \text{By (2.10)}$$
$$\Leftrightarrow (p \wedge \neg q) \vee (q \wedge \neg p) \qquad \text{By (2.22)}$$

◄

Theorem 2.2.4 — Simplification.

$$(p \wedge q) \to p \Leftrightarrow \mathbb{T} \tag{2.30}$$

Proof

$$
\begin{aligned}
(p \wedge q) \to p &\Leftrightarrow \neg(p \wedge q) \vee p & \text{By (2.24)} \\
&\Leftrightarrow (\neg p \vee \neg q) \vee p & \text{By (2.11)} \\
&\Leftrightarrow \neg p \vee (\neg q \vee p) & \text{By (2.4)} \\
&\Leftrightarrow \neg p \vee (p \vee \neg q) & \text{By (2.6)} \\
&\Leftrightarrow (\neg p \vee p) \vee (\neg q) & \text{By (2.4)} \\
&\Leftrightarrow (p \vee \neg p) \vee (\neg q) & \text{By (2.6)} \\
&\Leftrightarrow \mathbb{T} \vee \neg q & \text{By (2.18)} \\
&\Leftrightarrow \mathbb{T} & \text{By (2.14)}
\end{aligned}
$$

◄

The symbol ◄ denotes that the entire proof or demonstration is now complete.[3]

Corollary 2.2.5 — Simplification.

$$(p \wedge q) \to q \Leftrightarrow \mathbb{T} \tag{2.31}$$

The proof is similar to the proof of the previous theorem and is left as an exercise.

With Theorem 2.2.4, it would have been easier to prove this using a truth table. However, using a truth table becomes infeasible when proving more complicated propositions. The number of lines in a truth table grows exponentially[4] with the number of propositions: if n is the number of propositions to consider, then the number lines in the truth table is 2^n. More explicitly, with two propositions there are $2^2 = 4$ lines, and with three propositions there are $2^3 = 8$ lines, *etc.*

Example 2.2.6 Using the fundamental logical equivalences, prove the importation law:

$$[p \to (q \to r)] \Leftrightarrow [(p \wedge q) \to r].$$

Afterward, check the importation law using a truth table.

---

[3]Other mathematicians use the standard □, which is called a tombstone or Halmos mark named after the American mathematician Paul Richard Halmos (1916–2006) who popularized the use of the symbol and convention in mathematics. The □ replaced the Latin phase at the end of proofs *quod erat demonstrandum* (Q.E.D.) or in English "which had to be demonstrated." I use a different symbol because I am a rebel—you are free to be a rebel as well by selecting your own symbol to denote the end of your proofs.

[4]If you are unfamiliar with the exponential function, or what exponential growth looks like graphically, please refer to Fig. 10.5 on p. 436.

Solution

$$[p \to (q \to r)] \Leftrightarrow \neg p \vee (q \to r) \qquad \text{By (2.24)}$$
$$\Leftrightarrow \neg p \vee (\neg q \vee r) \qquad \text{By (2.24)}$$
$$\Leftrightarrow (\neg p \vee \neg q) \vee r \qquad \text{By (2.4)}$$
$$\Leftrightarrow \neg(p \wedge q) \vee r \qquad \text{By (2.11)}$$
$$\Leftrightarrow (p \wedge q) \to r \qquad \text{By (2.24)}$$

We can also check using a truth table, which is given in Table 2.3.

Table 2.3   Truth Table for the Importation Law

p	q	r	$q \to r$	$p \to (q \to r)$	$p \wedge q$	$(p \wedge q) \to r$
T	T	T	T	T	T	T
T	T	F	F	F	T	F
T	F	T	T	T	F	T
T	F	F	T	T	F	T
F	T	T	T	T	F	T
F	T	F	F	T	F	T
F	F	T	T	T	F	T
F	F	F	T	T	F	T

◀

Exercises

2.2.1. Suppose p, q, and r are propositions. State if each of the following illustrates an associative, a commutative, or a distributive property.

  (a) $(p \wedge q) \Leftrightarrow (q \wedge p)$
  (b) $[p \wedge (q \vee r)] \Leftrightarrow [p \wedge (r \vee q)]$
  (c) $[p \vee (q \wedge r)] \Leftrightarrow [(p \vee q) \wedge (p \vee r)]$
  (d) $[p \vee (q \vee r)] \Leftrightarrow [(p \vee q) \vee r]$

2.2.2. Let p, q, r, and s denote propositions. Determine which rule of inference each of the following propositions is illustrating.

  (a) $r \vee r$
  (b) $q \vee (r \wedge s) \Leftrightarrow q \vee (s \wedge r)$
  (c) $q \wedge (r \wedge s) \Leftrightarrow (q \wedge r) \wedge s$
  (d) $(q \wedge r) \vee (q \wedge s) \Leftrightarrow q \wedge (r \vee s)$
  (e) $((p \wedge q) \wedge q) \vee ((p \wedge q) \wedge r) \Leftrightarrow (p \wedge q) \wedge (q \vee r)$
  (f) $(q \wedge r) \vee s \Leftrightarrow s \vee (q \wedge r)$

2.2.3. Prove the following rules of inference: (*Hint*: Use a truth table.)
   (a) Distributive Law for Conjunction:

$$p \wedge (q \vee r) \Leftrightarrow (p \wedge q) \vee (p \wedge r)$$

   (b) De Morgan's Law:

$$\neg(p \vee q) \Leftrightarrow (\neg p) \wedge (\neg q)$$

   (c) Absorption Law:

$$p \vee (p \wedge q) \Leftrightarrow p$$

2.2.4. Apply the implication rule to
   (a) $q \rightarrow p$
   (b) $(p \wedge q) \rightarrow q$
   (c) $((p \wedge q) \wedge r) \rightarrow s$
   (d) $\neg p \vee (r \leftrightarrow s)$

2.2.5. Apply the implication rule twice to $(p \rightarrow q) \rightarrow r$.

2.2.6. Apply the associativity rules to
   (a) $((p \rightarrow q) \vee r) \vee s$
   (b) $u \wedge (v \wedge w)$
   (c) $(p \rightarrow q) \wedge (r \wedge s)$

2.2.7. Apply the commutative rules to
   (a) $j \wedge k$
   (b) $(p \rightarrow q) \wedge (q \rightarrow p)$
   (c) $(p \rightarrow q) \leftrightarrow (q \rightarrow p)$

2.2.8. Apply De Morgan's Laws to
   (a) $\neg((p \rightarrow q) \vee (q \rightarrow r))$
   (b) $\neg((p \wedge q) \wedge (q \wedge p))$
   (c) $\neg(p \vee q) \wedge \neg(q \vee p)$
   (d) $\neg(p \rightarrow q) \vee \neg(q \rightarrow r)$

2.2.9. Given the implication $p \rightarrow q$, show that the implication is equivalent to the contrapositive using the rules of inference.

2.2.10. Given the implication $p \rightarrow q$, show that the converse and the inverse are logically equivalent using the rules of inference.

2.2.11. Prove that $(\neg p) \wedge (\neg(p \vee q)) \Leftrightarrow \neg(p \vee q)$ using the rules of inference.

2.2.12. Prove that $(p \wedge \neg q) \vee q \Leftrightarrow p \vee q$ using the rules of inference.

2.2.13. Prove that $p \vee (q \vee r) \Leftrightarrow r \vee (q \vee p)$ using the rules of inference.

2.2.14. Prove that $p \rightarrow (p \vee q)$ is a tautology using the rules of inference.

2.2.15. Using the rules of inference, prove that $(p \wedge q) \rightarrow p$ is a tautology.

2.2.16. Prove that $\neg p \leftrightarrow q \Leftrightarrow p \leftrightarrow \neg q$.

2.2.17. Use the rules of inference to show that $\neg(p \vee \neg(p \wedge q))$ is a contradiction.

2.2.18. Using the rules of inference, prove that

$$\left[(p \land \neg(\neg p \lor q)) \lor (p \land q) \right] \to p$$

is a tautology.

2.2.19. Prove that $\neg(p \to q) \to p$ is a tautology. Do not use truth tables.

2.2.20. Find the negation of the implication: "If you are from the state of New York, then you will receive a scholarship."

2.3 What is an Argument?

If I asked you what an argument is, it may conjure images of squabbling people in the street, or politicians or celebrities howling at each other in the media. But in logic, an argument has a more precise meaning.

> **Definition 2.3.1 — Argument.** An **argument** consists of a set of propositions $p_1, p_2, \ldots, p_n$, called the **premises**, and a proposition q, called the **conclusion**. An argument is **valid** if and only if the conclusion is true whenever the premises are all true.

Example 2.3.2 An argument can be written in tabular form, such as the one given below.

All males are humans.	Major Premise
All boys are males.	Minor Premise
Therefore, all boys are humans.	Conclusion

The two premises in this argument can be referred as the major and minor premises. The **major premise** is a general proposition, such as "All males are humans," and it is usually an implication. The next line is the **minor premise** that is a proposition that is more specific, such as "All boys are males."

Determine whether the above argument is valid or not.

Solution By Definition 2.3.1, an argument is valid if and only if the conclusion is true whenever the premises are all true.

First, we will convert the argument into symbolic form. If we let m denote males, h denote humans, and b denote boys, then the argument can be represented as

$$m \to h \quad \text{Major Premise}$$
$$b \to m \quad \text{Minor Premise}$$

$$\therefore b \to h \quad \text{Conclusion}$$

The symbol $\therefore$ in mathematics means *therefore*. We can create the truth table in Table 2.4.

Table 2.4   Truth table for the argument in Example 2.3.2

m	h	b	$m \to h$	$b \to m$	$b \to h$
T	T	T	T	T	T
T	T	F	T	T	T
T	F	T	F	T	F
T	F	F	F	T	T
F	T	T	T	F	T
F	T	F	T	T	T
F	F	T	T	F	F
F	F	F	T	T	T

When we look at the rows where the premises $m \to h$ and $b \to m$ are both true, we see that the conclusion $b \to h$ is also true. Therefore, we can say that the argument is valid.   ◄

Definition 2.3.3 — Fallacy. An argument that is not valid is called a **fallacy**, or an invalid argument.

Theorem 2.3.4 Suppose that an argument consists of the premises $p_1, p_2, \ldots, p_n$ and conclusion q. Then, the argument is valid if and only if the proposition $[p_1 \wedge p_2 \wedge \cdots \wedge p_n] \to q$ is a tautology.

Notice in Theorem 2.3.4 the words "if and only if." Recall from Sec. 1.7.1, that the phrase "if and only if" indicates that this is a biconditional statement. Thus, in order to prove this proposition, we must show that
   (a) the implication is true: if an argument is valid, then the given proposition is a tautology; and
   (b) the converse is true: if the given proposition is a tautology, then the argument is valid.

Proof

(a) Suppose in order to derive a contradiction that the argument is valid, but $(p_1 \wedge p_2 \wedge \cdots \wedge p_n) \to q$ is not a tautology. For $(p_1 \wedge p_2 \wedge \cdots \wedge p_n) \to q$ not to be a tautology, this means that in the column of $(p_1 \wedge p_2 \wedge \cdots \wedge p_n) \to q$ of the truth table, there is at least one row with an F truth value.

p_1	p_2	$\cdots$	p_n	q	$p_1 \wedge \cdots \wedge p_n$	$(p_1 \wedge \cdots \wedge p_n) \to q$
$\vdots$	$\vdots$		$\vdots$	$\vdots$	$\vdots$	$\vdots$
?	?		?	?	?	F
$\vdots$	$\vdots$		$\vdots$	$\vdots$	$\vdots$	$\vdots$

For the implication $(p_1 \wedge \cdots \wedge p_n) \to q$ to be false, $(p_1 \wedge \cdots \wedge p_n)$ must have a T truth value, while q has an F truth value.

p_1	p_2	$\cdots$	p_n	q	$p_1 \wedge \cdots \wedge p_n$	$(p_1 \wedge \cdots \wedge p_n) \to q$
$\vdots$	$\vdots$		$\vdots$	$\vdots$	$\vdots$	$\vdots$
?	?		?	F	T	F
$\vdots$	$\vdots$		$\vdots$	$\vdots$	$\vdots$	$\vdots$

For $(p_1 \wedge p_2 \wedge \cdots \wedge p_n)$ to have a T truth value, each of the premises $p_1, p_2, \ldots, p_n$ must also be true; if any of the premises are false, this makes $(p_1 \wedge p_2 \wedge \cdots \wedge p_n)$ have an F truth value. Thus, for this particular assignment, each of the premises takes a T truth value while q takes an F truth value.

p_1	p_2	$\cdots$	p_n	q	$p_1 \wedge \cdots \wedge p_n$	$(p_1 \wedge \cdots \wedge p_n) \to q$
$\vdots$	$\vdots$		$\vdots$	$\vdots$	$\vdots$	$\vdots$
T	T		T	F	T	F
$\vdots$	$\vdots$		$\vdots$	$\vdots$	$\vdots$	$\vdots$

But this contradicts our initial assumption that the argument is valid (the conclusion is true whenever the premises are all true), and so $(p_1 \wedge p_2 \wedge \cdots \wedge p_n) \to q$ must be a tautology.

(b) Conversely, suppose in order to derive a contradiction that the proposition $(p_1 \wedge p_2 \wedge \cdots \wedge p_n) \to q$ is a tautology, and the argument is not valid. For an argument not to be valid, this means that the premises $p_1, p_2, \ldots,$ p_n are true while the conclusion q is false. So, there is a row in the truth table where $p_1, p_2, \ldots, p_n$ are all T truth values, but the conclusion q has an F truth value.

p_1	p_2	$\cdots$	p_n	q	$p_1 \wedge \cdots \wedge p_n$	$(p_1 \wedge \cdots \wedge p_n) \to q$
$\vdots$	$\vdots$		$\vdots$	$\vdots$	$\vdots$	$\vdots$
T	T		T	F	?	?
$\vdots$	$\vdots$		$\vdots$	$\vdots$	$\vdots$	$\vdots$

Since $p_1, p_2, \ldots, p_n$ have all T truth values, then $(p_1 \wedge \cdots \wedge p_n)$ must also have a T truth value.

p_1	p_2	$\cdots$	p_n	q	$p_1 \wedge \cdots \wedge p_n$	$(p_1 \wedge \cdots \wedge p_n) \to q$
$\vdots$	$\vdots$		$\vdots$	$\vdots$	$\vdots$	$\vdots$
T	T		T	F	T	?
$\vdots$	$\vdots$		$\vdots$	$\vdots$	$\vdots$	$\vdots$

Since $(p_1 \wedge \cdots \wedge p_n)$ is true, and q is false, the implication $(p_1 \wedge \cdots \wedge p_n) \to q$ is false.

p_1	p_2	$\cdots$	p_n	q	$p_1 \wedge \cdots \wedge p_n$	$(p_1 \wedge \cdots \wedge p_n) \to q$
$\vdots$	$\vdots$		$\vdots$	$\vdots$	$\vdots$	$\vdots$
T	T		T	F	T	F
$\vdots$	$\vdots$		$\vdots$	$\vdots$	$\vdots$	$\vdots$

But this contradicts our assumption that the proposition is a tautology, and so the argument must be valid.

◀

Having proved this proposition, we can now use it to determine the validity of an argument. If we are given an argument, we construct a truth table for

$(p_1 \wedge p_2 \wedge \cdots \wedge p_n) \rightarrow q$. If the truth table contains all the true values in its last column, then the argument is valid; otherwise it is invalid. This method of proof is illustrated in the next example.

Example 2.3.5 Determine whether the following argument below is valid:

$p \wedge q$	Major Premise
$\neg p$	Minor Premise
$\therefore q$	Conclusion

Solution Following Theorem 2.3.4, we can write the two premises together with the conjunction as

$$(p \wedge q) \wedge (\neg p) \tag{2.32}$$

with the implication such that

$$(p \wedge q) \wedge (\neg p) \rightarrow q \tag{2.33}$$

and we find the truth table for (2.33).

p	q	$p \wedge q$	$\neg p$	$(p \wedge q) \wedge (\neg p)$	$(p \wedge q) \wedge (\neg p) \rightarrow q$
T	T	T	F	F	T
T	F	F	F	F	T
F	T	F	T	F	T
F	F	F	T	F	T

From the truth table, we see that the last column is all true. Hence it is a tautology, and therefore by Theorem 2.3.4 the argument is valid. ◄

Exercises

2.3.1. For the following arguments, translate them into symbolic logic, and determine whether or not the argument is valid.

(a) If the Toronto Stock Exchange (TSX) index fails to increase this period, then the market will suffer a decline for several periods later. But, the TSX index is higher this period than the last. Therefore, there will be no decline for the next period.

(b) If a country is developing, it cannot devote much of its financial resources to technological development. However, if a country cannot devote much of its financial resources to technological development, then its economy will not grow. Therefore, a developing country will not have an economical growth.

2.4 Modus Ponens

Modus ponens comes from Latin for "the way that affirms by affirming." The argument is given in Table 2.5.

Table 2.5   *Modus Ponens*

$p \rightarrow q$	Major Premise
p	Minor Premise
$\therefore q$	Conclusion

It can be summarized as $p \rightarrow q$; p is asserted to be true, therefore q must be true. The origin of *modus ponens* goes back to antiquity. It is one of the most commonly used concepts in logic. The argument form has two premises. The first premise is the "if-then" statement, namely p implies q. The second premise is that p is true. From these two premises it can be logically concluded that q must be true as well.

Example 2.4.1 Given the major premise

"If I step in poison ivy, then I will have a rash,"

and the minor premise

"I step in poison ivy,"

what is the conclusion by *modus ponens*?

Solution We can rewrite this argument in tabular form.

If I step in poison ivy, then I will have a rash.	$p \rightarrow q$	Major Premise
I step in poison ivy	p	Minor Premise
Therefore, I will have a rash.	$\therefore q$	Conclusion

The conclusion by *modus ponens* is "I will have a rash."   ◄

Example 2.4.2 Given the major premise

"If I'm singing, then I'm happy,"

and the minor premise

"I'm singing,"

what is the conclusion by *modus ponens*?

Solution We can rewrite this argument in tabular form.

If I'm singing, then I'm happy.	$p \to q$	Major Premise
I'm singing.	p	Minor Premise

Therefore, I'm happy.	$\therefore q$	Conclusion

The conclusion by *modus ponens* is "I'm happy." ◄

Theorem 2.4.3 — Modus Ponens.

$$[(p \to q) \land p] \to q \Leftrightarrow \mathbb{T} \tag{2.34}$$

There are two ways that we can prove *modus ponens* is a tautology: either by using elementary rules of inference, or by using a truth table method.

Proof — 1. We will proceed using only elementary rules of inference.

$$[(p \to q) \land p] \to q$$

$\Leftrightarrow [(\neg p \lor q) \land p] \to q$	By (2.24)
$\Leftrightarrow [p \land (\neg p \lor q)] \to q$	By (2.5)
$\Leftrightarrow [(p \land \neg p) \lor (p \land q)] \to q$	By (2.9)
$\Leftrightarrow [\mathbb{F} \lor (p \land q)] \to q$	By (2.19)
$\Leftrightarrow [(p \land q) \lor \mathbb{F}] \to q$	By (2.6)
$\Leftrightarrow (p \land q) \to q$	By (2.13)
$\Leftrightarrow \mathbb{T}$	By (2.16)

◄

Proof — 2. We will prove *modus ponens* using a truth table and using Theorem 2.3.4.

p	q	$p \to q$	$(p \to q) \land p$	$[(p \to q) \land p] \to q$
T	T	T	T	**T**
T	F	F	F	**T**
F	T	T	F	**T**
F	F	T	F	**T**

◄

The words *imply* and *infer* are commonly confused in everyday language.

Example 2.4.4 — Implication versus Inference. Cory and Richard are talking about their mobile phones, when Richard makes the following (absurd) assertions:

If it ain't broke, don't fix it.

If a person updates their phone, they're stupid.

Richard's implications are being asserted for anyone, not just for a particular person. Unbeknownst to Richard, Cory had updated his phone recently. Cory thinks quickly:

If I update my phone, then I'm stupid.	$p \to q$	Major Premise
I updated my phone.	p	Minor Premise
Therefore, I'm stupid.	$\therefore q$	Conclusion

Cory exclaims, "I updated my phone! Are you *inferring* I'm stupid?"

Richard corrects him, "No, no, no. I *implied* it. You *inferred* it." ◀

From the previous example, we see that **implication** is to state a logical consequence, whereas **inference** is a conclusion that is reached based on implications.

Exercises

2.4.1. Analyze each of the following arguments and show that each is an instance of *modus ponens*.

(a) If Wei earns more than $1000, then she will go to France. Wei earns $1500. Hence, she goes to France.

(b) If Gabriel was not in school at 2 P.M. when the fire alarm was pulled, he couldn't possibly be the culprit. Gabriel was in the waiting room at his doctor's office at 2 P.M. Therefore, Gabriel is not guilty.

2.4.2. Using *modus ponens*, what can you conclude from the following sets of premises?

(a) If there is an examination tomorrow, then I will study. I have an examination at 2 P.M. tomorrow.

(b) If I don't wear my jacket while walking then I'll catch a cold. I couldn't find my jacket before I went for my evening walk.

(c) My improvement upon silence is necessary when I need to speak the truth. And after seeing these grave atrocities, I cannot restrain my tongue.

2.5 Modus Tollens

Modus tollens comes from the Latin for "the way that denies by denying." The first people known to say this explicitly were the Stoics, a school or following of Zeno (334 BC–264 BC) of Citium in the early third century BC. The argument is given in Table 2.6.

It can be summarized as the following: $p \rightarrow q$ and $\neg q$ are asserted to be true, therefore $\neg p$ must be true. The argument form has two premises. The major premise is the implication, namely p implies q. The minor premise is that $\neg q$ is true. From these two premises it can be logically concluded that $\neg p$ must be true as well.

Table 2.6 *Modus tollens*

$p \rightarrow q$	Major Premise
$\neg q$	Minor Premise
$\therefore \neg p$	Conclusion

Example 2.5.1 Given the major premise

"If I step in poison ivy, then I will have a rash"

and the minor premise

"I don't have a rash"

what is the conclusion by *modus tollens*?

Solution We can rewrite this argument in tabular form.

If I step in poison ivy, then I will have a rash.	$p \rightarrow q$	Major Premise
I don't have a rash.	$\neg q$	Minor Premise
Therefore, I didn't step in poison ivy.	$\therefore \neg p$	Conclusion

The conclusion by *modus tollens* is "I didn't step in poison ivy." ◀

Theorem 2.5.2 — Modus tollens.

$$[(p \rightarrow q) \wedge (\neg q)] \rightarrow (\neg p) \Leftrightarrow \mathbb{T} \tag{2.35}$$

Proof — 1. We will proceed using only elementary rules of inference.

$$[(p \rightarrow q) \wedge (\neg q)] \rightarrow (\neg p)$$

$\Leftrightarrow \neg[(p \rightarrow q) \wedge (\neg q)] \vee (\neg p)$	By (2.24)
$\Leftrightarrow \neg[((p \rightarrow q) \wedge (\neg q)) \wedge p]$	By (2.10)
$\Leftrightarrow \neg[(p \rightarrow q) \wedge ((\neg q) \wedge p)]$	By (2.3)
$\Leftrightarrow \neg[(p \rightarrow q) \wedge (p \wedge (\neg q))]$	By (2.5)
$\Leftrightarrow \neg(p \rightarrow q) \vee \neg(p \wedge (\neg q))$	By (2.11)
$\Leftrightarrow \neg(p \rightarrow q) \vee (\neg p \vee \neg(\neg q))$	By (2.11)
$\Leftrightarrow \neg(p \rightarrow q) \vee (\neg p \vee q)$	By (2.22)

$$\Leftrightarrow \neg(p \to q) \vee (p \to q) \qquad \text{By (2.24)}$$
$$\Leftrightarrow (p \to q) \vee \neg(p \to q) \qquad \text{By (2.6)}$$
$$\Leftrightarrow \mathbb{T} \qquad \text{By (2.18)}$$

◄

Proof — 2. We will prove *modus tollens* is a valid argument by using a truth table and Theorem 2.3.4.

Table 2.7   Truth table for *modus tollens*

p	q	$p \to q$	$\neg q$	$(p \to q) \wedge (\neg q)$	$\neg p$	$[(p \to q) \wedge \neg q] \to \neg p$
T	T	T	F	F	F	**T**
T	F	F	T	F	F	**T**
F	T	T	F	F	T	**T**
F	F	T	T	T	T	**T**

◄

Example 2.5.3 Given the major premise

"If I'm singing, then I'm happy,"

and the minor premise

"I'm not happy,"

what is the conclusion by *modus tollens*?

Solution We can rewrite this argument in tabular form.

If I'm singing, then I'm happy.	$p \to q$	Major Premise
I'm not happy.	$\neg q$	Minor Premise
Therefore, I'm not singing.	$\therefore \neg p$	Conclusion

The conclusion by *modus tollens* is "I'm not singing."   ◄

Example 2.5.4 Determine whether the following argument is valid:

$$\begin{array}{ll} p \vee (\neg q) & \text{Major Premise} \\ q & \text{Minor Premise} \\ \hline \therefore \ \neg p & \text{Conclusion} \end{array}$$

Solution We can simplify the first premise using the law of implication from our elementary rules of inference. In other words,

$$p \vee (\neg q) \Leftrightarrow \neg(\neg p) \vee (\neg q) \Leftrightarrow (\neg p) \to (\neg q).$$

So, we can rewrite the first premise as

$$\begin{array}{ll} (\neg p) \to (\neg q) & \text{Major Premise} \\ \neg(\neg q) & \text{Minor Premise} \\ \hline \therefore \ \neg p & \text{Conclusion} \end{array}$$

We have converted the argument into the same structure as *modus ponens*. Therefore, the argument is valid by *modus tollens*. ◀

2.6 Syllogism

A **syllogism** consists of three propositions, the last of which, called the conclusion, is a logical consequence of the two former called the major and minor premise.

2.6.1 Hypothetical Syllogism

The law of hypothetical syllogism takes two conditional statements and forms a conclusion by combining the hypothesis of one statement with the conclusion of another. In words, if p then q is true, and if q then r is also true, we conclude that if p then r is true as well. You can see that it forms a kind of logical chain: $(p \to q) \to r$. The argument is given in Table 2.8.

Table 2.8 Hypothetical Syllogism

$$\begin{array}{ll} p \to q & \text{Premise} \\ q \to r & \text{Premise} \\ \hline \therefore \ p \to r & \text{Conclusion} \end{array}$$

Example 2.6.1 Given the premises

"Socrates is a man,"

and

"All men are mortal,"

what is the conclusion by the hypothetical syllogism?

Solution We can rewrite this argument in tabular form.

Socrates is a man.	$p \to q$	Premise	
All men are mortal.	$q \to r$	Premise	
Therefore, Socrates is mortal.	$\therefore p \to r$	Conclusion	

The conclusion by hypothetical syllogism is "Socrates is mortal."   ◄

Theorem 2.6.2 — Hypothetical Syllogism.

$$[(p \to q) \land (q \to r)] \to (p \to r) \Leftrightarrow \mathbb{T} \qquad (2.36)$$

Proof — 1.

$[(p \to q) \land (q \to r)] \to (p \to r)$

$\Leftrightarrow [(p \to q) \land (\neg q \lor r)] \to (p \to r)$ By (2.24)

$\Leftrightarrow [((p \to q) \land \neg q) \lor ((p \to q) \land r)] \to (p \to r)$ By (2.9)

$\Leftrightarrow \Big[\big(((p \to q) \land \neg q) \lor (p \to q) \big)$

$\qquad \land \big(((p \to q) \land \neg q) \lor r \big) \Big] \to (p \to r)$ By (2.9)

$\Leftrightarrow \Big[(p \to q) \land \big(((p \to q) \land \neg q) \lor r \big) \Big] \to (p \to r)$ By (2.23)

$\Leftrightarrow \Big[(p \to q) \land \big(((\neg p \lor q) \land \neg q) \lor r \big) \Big] \to (p \to r)$ By (2.24)

$\Leftrightarrow \Big[(p \to q) \land \big(((\neg p \land \neg q) \lor (q \land \neg q)) \lor r \big) \Big]$

$\qquad \to (p \to r)$ By (2.8)

$\Leftrightarrow \Big[(p \to q) \land \big(((\neg p \land \neg q) \lor \mathbb{F}) \lor r \big) \Big] \to (p \to r)$ By (2.19)

$\Leftrightarrow [(p \to q) \land ((\neg p \land \neg q) \lor r)] \to (p \to r)$ By (2.13)

$\Leftrightarrow [(p \to q) \land ((\neg p \lor r) \land (\neg q \lor r))] \to (p \to r)$ By (2.9)

$\Leftrightarrow [(p \to q) \land ((\neg q \lor r) \land (\neg p \lor r))] \to (p \to r)$ By (2.8)

$$\Leftrightarrow \left[((p \to q) \wedge (\neg q \vee r)) \wedge (\neg p \vee r) \right] \to (p \to r) \qquad \text{By (2.3)}$$
$$\Leftrightarrow \left[((p \to q) \wedge (\neg q \vee r)) \wedge (p \to r) \right] \to (p \to r) \qquad \text{By (2.24)}$$
$$\Leftrightarrow \mathbb{T} \qquad \text{By (2.17)}$$

◀

Proof — 2. We will prove the hypothetical syllogism using a truth table and using Theorem 2.3.4. The truth table is given in Table 2.9.

Table 2.9   Hypothetical syllogism truth table

p	q	r	$[(p \to q)$	$\wedge$	$(q \to r)]$	$\to$	$(p \to r)$
T	T	T	T	T	T	T	T
T	T	F	T	F	F	T	F
T	F	T	F	F	T	T	T
T	F	F	F	F	T	T	F
F	T	T	T	T	T	T	T
F	T	F	T	F	F	T	T
F	F	T	T	T	T	T	T
F	F	F	T	T	T	T	T

◀

2.6.2 Disjunctive Syllogism

The law of disjunctive syllogism takes a single disjunction, and the negation of one of the propositions, and concludes that the alternate proposition in the disjunction must be true. In other words, if p or q is true, and it is not q, then it must be p that is true. The argument is given in Table 2.10.

Table 2.10   Disjunctive Syllogism

$p \vee q$	Major Premise
$\neg p$	Minor Premise
$\therefore q$	Conclusion

Example 2.6.3 Given the major premise

"Socrates is either living or dead,"

and the minor premise

"Socrates is not living,"

what is the conclusion by disjunctive syllogism?

Solution We can rewrite this argument in tabular form.

Socrates is either living or dead.	$p \lor q$	Major Premise
Socrates is not living.	$\neg p$	Minor Premise
Therefore, Socrates is dead.	$\therefore q$	Conclusion

The conclusion by disjunctive syllogism is "Socrates is dead."   ◄

Theorem 2.6.4 — Disjunctive Syllogism.

$$[(p \lor q) \land \neg p] \to q \Leftrightarrow \mathbb{T} \tag{2.37}$$

Proof — 1. We will proceed using only elementary rules of inference.

$$[(p \lor q) \land \neg p] \to q$$

$\Leftrightarrow [(\neg\neg p \lor \neg\neg q) \land \neg p] \to q$	By (2.22)
$\Leftrightarrow [\neg(\neg p \land \neg q) \land \neg p] \to q$	By (2.11)
$\Leftrightarrow \neg[(\neg p \land \neg q) \lor p] \to q$	By (2.10)
$\Leftrightarrow \neg\neg[(\neg p \land \neg q) \lor p] \lor q$	By (2.24)
$\Leftrightarrow [(\neg p \lor \neg q) \lor p] \lor q$	By (2.22)
$\Leftrightarrow (\neg p \land \neg q) \lor [p \lor q]$	By (2.4)
$\Leftrightarrow \neg(p \lor q) \lor [p \lor q]$	By (2.10)
$\Leftrightarrow [p \lor q] \lor \neg(p \lor q)$	By (2.6)
$\Leftrightarrow \mathbb{T}$	By (2.18)

◄

Proof — 2. We will prove the disjunctive syllogism using a truth table and using Theorem 2.3.4. The truth table is given in Table 2.11.

Table 2.11   Disjunctive syllogism truth table

p	q	$[(p \lor q)$	$\land$	$(\neg p)]$	$\to$	q
T	T	T	F	F	**T**	T
T	F	T	F	F	**T**	F
F	T	T	T	T	**T**	T
F	F	F	F	T	**T**	F

◄

2.7 Fallacies

You will be amused when you see that I have more than once deceived without the slightest qualm of conscience, both knaves and fools. As to the deceit perpetrated upon women, let it pass, for, when love is in the way, men and women as a general rule dupe each other. But on the score of fools it is a very different matter. I always feel the greatest bliss when I recollect those I have caught in my snares, for they generally are insolent, and so self-conceited that they challenge wit. We avenge intellect when we dupe a fool, and it is a victory not to be despised, for a fool is covered with steel, and it is often very hard to find his vulnerable part. In fact, to gull a fool seems to me an exploit worthy of a witty man. I have felt in my very blood, ever since I was born, a most unconquerable hatred towards the whole tribe of fools, and it arises from the fact that I feel myself a blockhead whenever I am in their company.

—Giacomo Girolama Casanova (1725–1798)

from *The Story of My Life*
translated by Arthur Machen

Nothing is better than eternal happiness.	Premise
A passing grade is better than nothing.	Premise
Therefore, a passing grade is better than eternal happiness.	Conclusion

The ability to spot an invalid argument is as important as constructing a valid one. As the 18th century playboy Casanova points out, if a fool puts before us an invalid argument and we are unable to spot it as such, then it makes fools of us as well.

Any argument which *deceives* us, by seeming to prove what it does not really prove, may be called a **fallacy** (derived from the Latin verb *fallo* "I deceive") (Carroll, 1896, p. 81). A fallacy is an argument that has an inherit flaw in the structure of the argument itself which renders the argument invalid.

Example 2.7.1 Let us return to poor Alice who has found herself at a mad tea party with the March Hare, the Mad Hatter, and the Dormouse.

"Then you should say what you mean," the March Hare went on.

"I do," Alice hastily replied; "at least—at least I mean what I say—that's the same thing, you know."

Alice let the emotion of the argument get to her, and *hastily* made the fallacy of stating that p (saying) $\to q$ (meaning) is equivalent to q (meaning) $\to p$ (saying). After hearing this, the Hatter, the March Hare, and the Dormouse—who all do not want to be made fools—all chime in to point out the absurdness of this fallacy substituting their own examples by switching p and q as Alice does.

> "Not the same thing a bit!" said the Hatter. "Why, you might just as well say that 'I see what I eat' is the same thing as 'I eat what I see'!"
>
> "You might just as well say," added the March Hare, "that 'I like what I get' is the same thing as 'I get what I like'!"
>
> "You might just as well say," added the Dormouse, who seemed to be talking in his sleep, "that 'I breathe when I sleep' is the same thing as 'I sleep when I breathe'!"

> —Carroll (1866, p. 98)

◄

The following three fallacies, affirming the disjunct, affirming the consequent, and denying the antecedent, are called **non sequiturs** which comes from Latin for "it does not follow."

2.7.1 Affirming the Disjunct

Affirming the disjunct occurs when an argument takes the following logical form in Table 2.12.

This fallacy is due to confusing the disjunction connective $\vee$ with the exclusive disjunction connective $\veebar$; the disjunction $\vee$ says that p or q or both can be true.

Table 2.12   Affirming the Disjunct

$p \vee q$	Major Premise
p	Minor Premise
$\therefore \neg q$	Conclusion

Proof Affirming the disjunct is not a valid argument. To see why, we can construct the truth table for the proposition $[(p \vee q) \wedge p] \to \neg q$; see Table 2.13.

Table 2.13   Truth table for affirming the disjunct

p	q	$\neg q$	$p \vee q$	$(p \vee q) \wedge p$	$[(p \vee q) \wedge p] \to \neg q$
T	T	F	T	T	**F**
T	F	T	T	T	T
F	T	F	T	F	T
F	F	T	F	F	T

When we look under the last column, we see that not all of the values are T; hence, $[(p \vee q) \wedge p] \rightarrow \neg q$ is not a tautology. Therefore, by Theorem 2.3.4, affirming the disjunct is not a valid argument. ◄

Example 2.7.2 Consider the following argument:

I am either going to listen to music or read a book.	$p \vee q$	Major Premise
I'll listen to music.	p	Minor Premise
Therefore, I cannot read my book.	$\neg q$	Conclusion

If we let p be "listen to music" and q be "read a book," then the symbolic argument has the same structure as Table 2.12. Hence, the argument is invalid. It is possible to listen to music and read a book at the same time. So, you can read a book and it does not preclude the possibility of listening to music. ◄

2.7.2 Affirming the Consequent

Recall from Sec. 1.7.2 that the implication is not logically equivalent to the converse; that is, $p \rightarrow q$ is not logically equivalent to $q \rightarrow p$. **Affirming the consequent**, sometimes known as the converse error or fallacy of the converse, is a fallacy of inferring the converse from the original implication. It is committed by reasoning of the form in Table 2.14.

Table 2.14   Affirming the consequent

$p \rightarrow q$	Major Premise
q	Minor Premise
$\therefore p$	Conclusion

Proof Affirming the consequent is not a valid argument. To see why, we can construct the truth table for the proposition $[(p \rightarrow q) \wedge q] \rightarrow p$; see Table 2.15. When we look under the last column, we see that not all of the values are T; hence,

Table 2.15   Truth table for affirming the consequent

p	q	$p \rightarrow q$	$(p \rightarrow q) \wedge q$	$[(p \rightarrow q) \wedge q] \rightarrow p$
T	T	T	T	T
T	F	F	F	T
F	T	T	T	F
F	F	T	F	T

$[(p \rightarrow q) \wedge q] \rightarrow p$ is not a tautology. Therefore, by Theorem 2.3.4, affirming the consequent is not a valid argument.   ◄

Example 2.7.3 Kingston is a small Canadian city in the province of Ontario located between Toronto and Ottawa. Consider the following argument:

If I'm in Kingston, then I'm in Ontario.	$p \rightarrow q$	Major Premise
I'm in Ontario.	q	Minor Premise
Therefore, I'm in Kingston.	$\therefore p$	Conclusion

Fig. 2.1   Map of Ontario. Modified from Natural Resources Canada (2002).

This is an invalid argument because it has same form as Table 2.14. We assume that the major and minor premise is true. Remember that if $p \rightarrow q$ is true, then it is possible that q is true, but p is false. Thus, in the major premise, it is possible that I'm in Ontario, but not in Kingston. The second premise establishes that I am in Ontario, but nothing more. Thus, it is not possible for us to conclude that I'm in Kingston. In fact, from the map of Ontario (Fig. 2.1), I might be in any city in Ontario—such as Ottawa or Toronto!   ◄

When an argument is invalid, even with true premises such as Example 2.7.3, we are unsure that the argument's conclusion it yields is true.

2.7.3 Denying the Antecedent

Denying the antecedent, sometimes also called the inverse error or fallacy of the inverse, is a formal fallacy of inferring the inverse from the major premise. It is committed by the form of reasoning in Table 2.16.

Table 2.16 Denying the antecedent

$p \to q$	Major Premise
$\neg p$	Minor Premise
$\therefore \neg q$	Conclusion

Proof Denying the antecedent is not a valid argument. To see why, we can construct the truth table for the proposition $[(p \to q) \land (\neg p)] \to (\neg q)$; see Table 2.17.

Table 2.17 Truth table for denying the antecedent

p	q	$p \to q$	$\neg p$	$(p \to q) \land (\neg p)$	$\neg q$	$[(p \to q) \land (\neg p)] \to (\neg q)$
T	T	T	F	F	F	T
T	F	F	F	F	T	T
F	T	T	T	T	F	F
F	F	T	T	T	T	T

When we look under the last column, we see that not all of the values are T; hence, $[(p \to q) \land (\neg p)] \to (\neg q)$ is not a tautology. Therefore, by Theorem 2.3.4, denying the antecedent is not a valid argument. ◄

Example 2.7.4 Consider the following argument below:

I mean what I say.	$p \to q$	Major Premise
I didn't mean it.	$\neg p$	Minor Premise
Therefore, I didn't say it.	$\neg q$	Conclusion

If we let p be "I mean it" and q be "I said it," then the symbolic argument has the same structure as Table 2.17. ◄

Example 2.7.5 — We Cannot be Machines. After the development of the digital computer, there were debates (and still continues to this day) on whether machines could "think" or not at the same level as humans. In the following example offered with apologies from one of the fathers of computer science, Alan Turing (1912–1954), for its lack of logical rigor tries to reproduce the proponents' argument that we cannot be machines:

If each man had a definite set of rules of conduct by which he regulated his life (r) he would be no better than a machine (m). But there are no such rules, so men cannot be machines

—Turing (1950, p. 452).

Symbolically, this is represented below:

$$r \to m \quad \text{Major Premise}$$
$$\neg r \quad \text{Minor Premise}$$

$$\therefore \neg m \quad \text{Conclusion}$$

However, men could still be machines that do not follow a definite set of rules. Thus, this argument (as Turing intends) is invalid; it is an example of denying the antecedent.

Exercises

2.7.1. Convert the following argument into symbolic form, and determine if the following argument is valid or a fallacy:

Coffee is energy.   Premise
Kerosene is energy.   Premise

Therefore, kerosene is coffee.   Conclusion

2.7.2. Consider the premise, "If I'm in Kingston, then I'm in Ontario" from Example 2.7.3. Rewrite the argument to follow *modus ponens*, and hence constructing a valid argument.

2.7.3. Identify the argument as one of the three non sequiturs.

If an animal meows, then it is a cat.   Premise
The animal does not meow.   Premise

Therefore, the animal is not a cat.   Conclusion

2.7.4. Identify the argument as one of the three non sequiturs.

If it is snowing, then it is winter.	Major Premise
It is winter.	Minor Premise
Therefore, it must be snowing.	Conclusion

2.7.5. The following argument is known as the **politician's syllogism**.[5] Identify the argument as one of the three non sequiturs.

If things are to improve, then things must change.	Major Premise
We're changing things.	Minor Premise
Therefore, we're improving things.	Conclusion

2.8 Deductive Reasoning and Inductive Reasoning

Deductive reasoning is the process of concluding that a specific case is true since a general principle is known to be true. It is logically *valid*, and it is the fundamental method in which mathematical facts are shown to be true. A formal proof and a rigorous proof are examples of deductive reasoning.

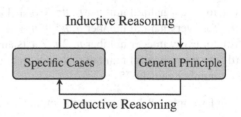

Fig. 2.2 Flow chart of inductive and deductive reasoning

Inductive reasoning is the process of concluding that a general principle must be true from the evidence of specific cases observed. It is not regarded to be final, but plausible only, whose purpose is to draw conclusions from the limited available information. But, it is logically *invalid*, and thus, the conclusions drawn may not be true. Inductive reasoning is not a substitute for a rigorous proof, but it is a useful first step in discovering a proposition that we can then try to prove by deductive reasoning.

---

[5]From the TV series *Yes, Prime Minister*, Season 2, Episode 5: Power to the People. The original syllogism is the following: Something must be done; this is something. Therefore, we must do it.

Example 2.8.1 Table 2.18 shows two types of arguments.

Table 2.18   Comparison of Deductive and Inductive Arguments

Deductive	Inductive	
All women are mortal.	Aspasia is a woman.	Premise
Aspasia is a woman.	Aspasia is mortal.	Premise
Therefore, Aspasia is mortal.	Therefore, all women are mortal.	Conclusion

The deductive argument begins with "All women are mortal," which is the general premise. Then, we have a specific premise of "Aspasia is a woman." Since she is a woman, we conclude by *hypothetical syllogism* that Aspasia is mortal.

The inductive argument begins with "Aspasia is a woman" and "Aspasia is mortal." This is a specific case and provides evidence that we use to generalize to all women, which we then conclude that all women are mortal.   ◀

Exercises

2.8.1. Use inductive reasoning (that is, taking specific cases), determine if the following propositions are true:

   (a) for every natural number n, $n^2 + n + 17$ is a prime number
   (b) for every natural number n, $n^2 + n + 41$ is a prime number
   (c) for every integer number n, $2n + 1$ is an odd number
   (d) for every natural number that ends in 5, that when it is squared, this new number ends in 25.

2.8.2. In Example 2.8.1, show that the inductive argument is invalid.

2.9 More Complex Arguments

Often an argument is presented in which many propositions are contained in the set of premises. The **dilemma** is an argument in which both the hypothetical syllogism and disjunctive syllogism are combined together.

Let us look at the argument in Table 2.19. Since there are four propositions (p, q, r, and s) involved in this argument, the truth table would have $2^4 = 16$ rows.

Table 2.19   A dilemma

$\neg(p \vee q)$	Premise
$\neg r \vee q$	Premise
$\neg s \rightarrow r$	Premise
$\therefore s$	Conclusion

Table 2.20 The dilemma in Table 2.19 is valid

p	q	r	s	$\neg(p \vee q)$	$\neg r \vee q$	$\neg s \to r$	$\neg(p \vee q) \wedge$ $(\neg r \vee q) \wedge$ $(\neg s \to r)$	$[\neg(p \vee q) \wedge$ $(\neg r \vee q) \wedge$ $(\neg s \to r)] \to s$
T	T	T	T	F	T	T	F	T
T	T	T	F	F	T	T	F	T
T	T	F	T	F	T	T	F	T
T	T	F	F	F	T	F	F	T
T	F	T	T	F	F	T	F	T
T	F	T	F	F	F	T	F	T
T	F	F	T	F	T	T	F	T
T	F	F	F	F	T	F	F	T
F	T	T	T	F	T	T	F	T
F	T	T	F	F	T	T	F	T
F	T	F	T	F	T	T	F	T
F	T	F	F	F	T	F	F	T
F	F	T	T	T	F	T	F	T
F	F	T	F	T	F	T	F	T
F	F	F	T	T	T	T	T	T
F	F	F	F	T	T	F	F	T

Of course for a diligent student this would be attainable, but once we start to have, say six propositions, then the truth table would have 64 rows and this is simply outside the attention span of most people.

Example 2.9.1 — Catch-22. A Catch-22, a phrase that was coined by Joseph Heller (1923–1999) in his 1961 novel of the same name, is a paradoxical situation. In the novel, John Yossarian is a United States Air Forces bombardier who does not wish to fly in an upcoming mission. The only way to remain grounded during the mission is for the flight surgeon to deem him unfit for service. A pilot is classified as unfit if one would have to be crazy enough to volunteer for such a dangerous mission. However, for John to be deemed unfit, he has to request first a mental health evaluation—an act that alone is sufficient for declaring him sane, for crazy people do not know they are crazy. Thus, the conditions make it impossible to be deemed unfit.

Table 2.21 Catch-22 argument

A person to be excused on the grounds of insanity (e), the person must be both insane (i) and request an evaluation (r).	$e \to (i \wedge r)$	Premise
An insane person (i) will not request an evaluation ($\neg r$), because they think they are sane.	$i \to \neg r$	Premise
Therefore, no person will be excused from flying ($\neg e$).	$\neg e$	Conclusion

Proof

$$[[e \to (i \wedge r)] \wedge (i \to \neg r)] \to \neg e$$

$$\Leftrightarrow \big[(\neg e \vee (i \wedge r)) \wedge (\neg i \vee \neg r)\big] \to \neg e \qquad \text{By (2.24)}$$

$$\Leftrightarrow \big[(\neg e \vee (i \wedge r)) \wedge \neg(i \wedge r)\big] \to \neg e \qquad \text{By (2.11)}$$

$$\Leftrightarrow \big[(\neg e \wedge \neg(i \wedge r)) \vee ((i \wedge r) \wedge \neg(i \wedge r))\big] \to \neg e \qquad \text{By (2.8)}$$

$$\Leftrightarrow \big[(\neg e \wedge \neg(i \wedge r)) \vee \mathbb{F}\big] \to \neg e \qquad \text{By (2.19)}$$

$$\Leftrightarrow \big[(\neg e \wedge \neg(i \wedge r))\big] \to \neg e \qquad \text{By (2.13)}$$

$$\Leftrightarrow \mathbb{T} \qquad \text{By (2.16)}$$

Since $[[e \to (i \wedge r)] \wedge (i \to \neg r)] \to \neg e$ is a tautology, the argument is valid. ◀

Example 2.9.2 — The Paradox of the Court. This is the most celebrated example of an argument from the Ancient Greeks. The famous teacher Protagoras (490 BC–420 BC) agreed to train a student named Euathlus in the art of logic, the condition being that only half the fee is required at the time of instruction and the remaining fee due when Euathlus won his first case in court. Should Euathlus fail, then the fee would be forfeited.

When Euathlus' training was completed, he delayed to undertake any case. Eventually, Protagoras could no longer wait for payment, and decided to expedite the process; Protagoras decided to sue Euathlus.

Protagoras argued (Joyce, 1908, p. 213):

If this case is decided in my favor, Euathlus must pay me by order of the court.	Premise
If it is decided in Euathlus' favor, Euathlus must pay me under the terms of the agreement.	Premise
But, it must be decided either in my favor, or in Euathlus' favor.	Premise
Therefore, Euathlus' is bound to pay me in any case.	Conclusion

Let p be "this case is decided in my favor," $\neg p$ be "this case is decided not in my favor," q be "you must pay me by order of the court," and r be "you must pay me under the terms of the agreement." The argument is represented symbolically in Table 2.22.

Euathlus rebutted with the following (Joyce, 1908, p. 214):

Table 2.22 Protagoras' argument

$p \rightarrow q$	Premise
$\neg p \rightarrow r$	Premise
$p \vee \neg p$	Premise
$\therefore q \vee r$	Conclusion

If the case is decided in favor of Protagoras, I am free by the terms of the agreement.	Premise
If it is decided in my favor, I am free by order of the court.	Premise
But, it must either be decided in Protagoras' favor or my favor.	Premise
Therefore, I am discharged of my debt in any case.	Conclusion

Let p be "the case decided in favor of Protagoras," q be "I am discharged of my debt." Euathlus' argument is represented symbolically in Table 2.22; the arguments are symbolically the same except for what the propositions p and q represent. Euathlus had used the same argument as his teacher—very clever. Since each of these arguments are valid (see Exercise 2.9.1), this leads to the tricky issue of selecting who should win the case. The judges left the case undecided. ◄

Exercises

2.9.1. Using a truth table, show that Protagoras' and Euathlus' argument in Table 2.22 is valid.

2.9.2. Determine whether the following argument is valid or invalid:

I study mathematics, or I study economics.	Premise
If I have to take English, then I study economics.	Premise
I study mathematics.	Premise
Therefore, I do not have to take English.	Conclusion

2.9.3. Using the rules of inference, prove that the following argument is valid:

$$
\begin{array}{ll}
p & \text{Premise} \\
q \to \neg s & \text{Premise} \\
p \to s & \text{Premise} \\
\hline
\therefore \neg q & \text{Conclusion}
\end{array}
$$

2.9.4. Show that the following dilemmas are valid.
  (a) Simple constructive dilemma

$$
\begin{array}{ll}
p \to r & \text{Premise} \\
q \to r & \text{Premise} \\
p \lor q & \text{Premise} \\
\hline
\therefore r & \text{Conclusion}
\end{array}
$$

  (b) Complex constructive dilemma

$$
\begin{array}{ll}
p \to r & \text{Premise} \\
q \to s & \text{Premise} \\
p \lor q & \text{Premise} \\
\hline
\therefore r \lor s & \text{Conclusion}
\end{array}
$$

  (c) Destructive dilemma

$$
\begin{array}{ll}
p \to r & \text{Premise} \\
q \to s & \text{Premise} \\
\neg r \lor \neg s & \text{Premise} \\
\hline
\therefore \neg p \lor \neg q & \text{Conclusion}
\end{array}
$$

2.9.5. Sir Richard Empson (c. 1450–1510), minister to King Henry VII of England, reportedly was able to demonstrate that any person was capable of paying a heavy tax.

If the accused lives at a small rate, his savings must make him rich.	Premise
If the accused maintains a large household, his expenditure proves he is rich.	Premise
But, either he lives at a small rate, or he maintains a large expenditure.	Premise
Therefore, he is rich, and consequently can pay heavily to the king.	Conclusion

This argument has been given the title of the **Emperor's fork**: whoever was found in the crosshairs of this argument was impaled on the prongs of this dilemma. Show that Sir Empson's argument is valid.

2.9.6. It is possible to rebut the Emperor's fork in Exercise 2.9.5. Consider the following argument:

If the accused lives at a small rate, his economy is evidence of his poverty.	Premise
If he has maintained a large expenditure, that must have impoverished him.	Premise
But, either he lives at a small rate, or has maintained a large expenditure.	Premise
Therefore, he is poor, and is incapable of paying heavily to the king.	Conclusion

Is the argument valid?

2.9.7. When the Roman Emperor Marcus Aurelius (121–180) persecuted early Christians, Tertullian (155–240) argued against the methods Aurelius employed.

Either Christians have committed crimes or not.	Premise
If they are guilty of crimes, your refusal to permit a public inquiry is irrational.	Premise
If they have committed no offense, it is unjust to punish them.	Premise
Therefore, your conduct is either unjust or irrational.	Conclusion

Show that Tertullian's argument is valid.

2.10 Truth and Validity

We have seen in previous sections ways to demonstrate if an argument is valid or invalid. However, there can be valid arguments that use false propositions, as in the following example from Augustine of Hippo (354–430).

Example 2.10.1 — A snail is not an animal. Suppose someone granted the first premise of the argument in Table 2.23.

Table 2.23   Snail is not an animal argument

If a snail is an animal (s), it has a voice (v). $s \rightarrow v$		Premise
A snail does not have a voice. $\neg v$		Premise
Therefore, a snail is not an animal. $\therefore \neg s$		Conclusion

With this granted, it is then shown that a snail does not have a voice, and the conclusion by *modus tollens* that a snail is not an animal (Augustine and Green, 2008, p. 60).   ◄

We know that a snail is an animal, and the derived conclusion is false; nonetheless, it had been derived by a valid argument! **Your inferences are only as good as your premises.**[6] We are beginning to see that there can be invalid deductions from true statements, just as there are valid arguments from false propositions. We now illustrate the difference between truth and validity with the following two definitions.

| **Definition 2.10.2 — Truth and Falsity.** Truth and falsity are attributes of *propositions*.

| **Definition 2.10.3 — Validity and Invalidity.** Validity and invalidity are attributes of *arguments*.

From Definitions 2.10.2 and 2.10.3, knowing the rules of valid deduction is not the same thing as knowing the truth of propositions. An argument with one or more false premises may still be valid. One example of this distinction comes from American History. In 1858, the 16th president of the United States, Abraham Lincoln (1809–1865), was in one of his debates with Stephen A. Douglas (1813–1861), the Senator from Illinois at the time.

Dred Scott (1799–1858) and his wife, Harriet, were African American slaves who attempted to sue for their freedom. The case of Dred Scott versus Sandford (1857) was a landmark decision by the Supreme Court of the United States; in

---

[6]In other words, the computer scientists would say, "*Purgamentum init, exit purgamentum*," which roughly translates to *Garbage in, garbage out*.

particular, as most legal scholars agree, it is the worst decision rendered by the Supreme Court. The decision would later be overturned by the 13th and 14th amendment to the United States Constitution.

The Supreme Court held that African Americans, whether slave or free, could not be American citizens, and therefore had no standing to sue in federal court. Moreover, the Court asserted that the federal government had no power to regulate slavery in the federal territories acquired after the creation of the United States. In a 7–2 decision written by Chief Justice Roger B. Taney (1777–1864), the Court denied Scott's request and in doing so, ruled an Act of Congress to be unconstitutional for the second time in its history. The decision would prove to be an indirect catalyst for the American Civil War (Finkelman, 2007, p. 3).

Lincoln attacked the Dread Scott decision, which obliged the return of slaves who had escaped into northern states to their owners in the South:

> I think it follows, [from the Dred Scott decision] and submit to the consideration of men capable of arguing, whether as I state it in syllogistic form the argument has any fault in it:

Nothing in the Constitution or laws of any State can destroy a right distinctly and expressly affirmed in the Constitution of the United States.	Premise
The right of property in a slave is distinctly and expressly affirmed in the Constitution of the United States.	Premise
Therefore, nothing in the Constitution or laws of any State can destroy the right of property in a slave.	Conclusion

> I believe that no fault can be pointed out in that argument; assuming the truth of the premises, the conclusion, so far as I have capacity at all to understand it, follows inevitably. There is a fault in it as I think, but the fault is not in the reasoning; but the falsehood in fact is a fault of the premises. I believe that the right of property in a slave is not distinctly and expressly affirmed in the Constitution, and Judge Douglas thinks it is. I believe that the Supreme Court and the advocates of that decision [the Dred Scott decision] may search in vain in the place in the Constitution where the right of property in a slave is distinctly and expressly affirmed. I say, therefore, that I think one of the premises is not true in fact.

Lincoln asserted the second premise—that the right of property in a slave is affirmed in the United States Constitution—is false. He argued that while the argument is valid, the truth of the conclusion is overshadowed with doubt when one

or more of its premises are false. For the validity of an argument, we emphasize
once again, depends only upon the relation of the premises to the conclusion.

There are many possible combinations of true and false premises and conclusions in both valid and invalid arguments. Consider the following illustrative
arguments:

(a) Some valid arguments contain only true propositions—true premises and
     a true conclusion:

All humans (h) can think (t).	$h \rightarrow t$	Premise
I am (i) human (h).	$i \rightarrow h$	Premise
Therefore, I can think. $\therefore i \rightarrow t$		Conclusion

(b) Some valid arguments contain only false propositions:

All cats (c) understand French (f).	$c \rightarrow f$	Premise
All dogs (d) are cats (c).	$d \rightarrow c$	Premise
Therefore, all dogs understand French. $\therefore d \rightarrow f$		Conclusion

(c) Some invalid arguments contain only true propositions—all their premises
     are true, and their conclusions are true as well:

If I win the lottery (ℓ), then I would be rich (r).	$\ell \rightarrow r$	Premise
I did not win the lottery.	$\neg \ell$	Premise
Therefore, I am not rich. $\therefore \neg r$		Conclusion

(d) Some invalid arguments contain only true premises and have a false
     conclusion. This can be illustrated with an argument exactly like the
     previous one in form, changed only enough to make the conclusion false:

If Donald Trump wins the lottery (ℓ), then Donald Trump would be rich (r).	$\ell \rightarrow r$	Premise
Donald Trump did not win the lottery.	$\neg \ell$	Premise
Therefore, Donald Trump is not rich. $\therefore \neg r$		Conclusion

The premises of this argument are true, but its conclusion is false. Such an argument cannot be valid because it is impossible for the premises of a valid argument to be true and its conclusion to be false.

(e) Some valid arguments have false premises and a true conclusion:

All fish (f) are mammals (m).	$f \to m$	Premise
All humans (h) are fish (f).	$h \to f$	Premise

Therefore, all humans are mammals.	$\therefore h \to m$	Premise

---

The conclusion of this argument is true, as we know; moreover, it may be validly inferred from these two premises, both of which are wildly false.

(f) Some invalid arguments also have false premises and a true conclusion.

All humans (h) have wings (w).	$h \to w$	Premise
All cats (c) have wings (w).	$c \to w$	Premise

Therefore, all cats are humans.	$\therefore c \to h$	Conclusion

---

From Examples (e) and (f) taken together, it is clear that we cannot tell from the fact that an argument has false premises and a true conclusion whether it is valid or invalid.

(g) Some invalid arguments, of course, contain all false propositions—false premises and a false conclusion:

All cats (c) have four legs (ℓ).	$c \to \ell$	Premise
All dogs (d) have four legs (ℓ).	$d \to \ell$	Premise

Therefore, all dogs (d) are cats (c).	$\therefore d \to c$	Conclusion

---

Table 2.24 and Table 2.25 below will make very clear the variety of possible combinations.

Table 2.24   Invalid Arguments

	True Conclusion	False Conclusion
True Premises	Example (c)	Example (d)
False Premises	Example (f)	Example (g)

Table 2.25   Valid Arguments

	True Conclusion	False Conclusion
True Premises	Example (a)	
False Premises	Example (e)	Example (b)

The one blank position in Table 2.25 exhibits a fundamental point: if an argument is valid and its premises are true, we may be certain that its conclusion is also true. To put it another way: if an argument is valid and its conclusion is false, not all of its premises can be true. Some perfectly valid arguments do have false conclusions—but any such argument must have at least one false premise.

> **Definition 2.10.4 — Sound argument.** When an argument is valid, and all of its premises are true, we call the argument sound.

The conclusion of a sound argument obviously must be true, and only a sound argument can establish the truth of its conclusion. If a deductive argument is not sound—that is, if the argument is not valid, or if not all of its premises are true—it fails to establish the truth of its conclusion even if in fact the conclusion is true. In the above examples, only Example (a) is a sound argument.

Exercises

2.10.1. Construct a series of arguments, on any subject of your choosing, each with only two premises, having the following characteristics:
- (a) A valid argument with one true premise, one false premise, and a false conclusion.
- (b) A valid argument with one true premise, one false premise, and a true conclusion.
- (c) An invalid argument with two true premises, and a false conclusion.
- (d) An invalid argument with two true premises, and a true conclusion.
- (e) An invalid argument with one true premise, one false premise, and a true conclusion.

2.10.2. Is the following argument valid or invalid? Is it sound or unsound?

Time is money.	Premise
All money is paper.	Premise
Therefore, all time is paper.	Conclusion

2.11 Review

Summary

- The rules of inference in logic provides a systematic way of showing statements are equivalent to each other.
- An **argument** consists of a set of propositions $p_1, p_2, \ldots, p_n$, called the **premises**, and a proposition q, called the **conclusion**. An argument is **valid** if and only if the conclusion is true whenever the premises are all true.
- A **fallacy** is an argument that has an inherent flaw in the structure of the argument itself which renders the argument invalid.
- A **syllogism** consists of three propositions, the last of which, called the conclusion, is a logical consequence of the two former called the major and minor premise. The two types of syllogism are the hypothetical syllogism and the disjunctive syllogism.
- The **dilemma** is an argument in which both the hypothetical syllogism and disjunctive syllogism are combined together.
- Truth and falsity are attributes of individual propositions.
- Validity and invalidity are attributes of arguments.
- When an argument is valid, and all of its premises are true, we call it a **sound argument**.

Table 2.26 Four main types of syllogisms

Modus Ponens	*Modus Tollens*	Hypothetical Syllogism	Disjunctive Syllogism
$p \to q$	$p \to q$	$p \to q$	$p \vee q$
p	$\neg q$	$q \to r$	$\neg p$
$\therefore q$	$\neg p$	$\therefore p \to r$	$\therefore q$

Table 2.27 Fallacies (non sequiturs)

Affirming a Disjunct	Affirming the Consequent	Denying the Antecedent
$p \vee q$	$p \to q$	$p \to q$
p	q	$\neg p$
$\therefore \neg q$	$\therefore p$	$\therefore \neg q$

Table 2.28   Three types of dilemmas

Simple Constructive	Complex Constructive	Destructive
$p \to r$	$p \to r$	$p \to r$
$q \to r$	$q \to s$	$q \to s$
$p \vee q$	$p \vee q$	$\neg r \vee \neg s$
$\therefore r$	$\therefore r \vee s$	$\therefore \neg p \vee \neg q$

Exercises

2.11.1. Let p, q, and r denote propositions. State the following arguments symbolically:

(a) *Modus ponens*   (c) Hypothetical syllogism

(b) *Modus tollens*   (d) Disjunctive syllogism

2.11.2. Consider the following:

(a) If I go to the store, then I will buy some milk. I am going to the store. What can you conclude I will do by *modus ponens*?

(b) If an animal eats a banana, then it is a monkey. The animal is not a monkey. What can you conclude by *modus tollens*?

(c) If Christine is Canadian, and all Canadians are nice, then what can you conclude about Christine by the hypothetical syllogism?

(d) Either Renata walks or bicycles to work. She did not walk to work. What can you conclude about Renata by the disjunctive syllogism?

2.11.3. Determine if the following arguments are valid or invalid and state why.

(a)

If $x = 2$, then $2x + 1 = 5$.	Premise
$2x + 1 \neq 5$.	Premise
Therefore, $x \neq 2$.	Conclusion

(b)

All men are mortal.	Premise
Victor is a man.	Premise
Therefore, Victor is mortal.	Conclusion

(c)

All women are mortal.	Premise
A penguin is mortal.	Premise

Therefore, a penguin is a woman. Conclusion

(d)

Homer is either at home or on a ship.	Premise
Homer is not on a ship.	Premise

Therefore, Homer is at home. Conclusion

(e)

If it is raining, then the ground is wet.	Premise
It is not raining.	Premise

Therefore, the ground is not wet. Conclusion

(f)

Simon is successful only if he studies.	Premise
Simon does not study.	Premise

Therefore, Simon is not successful. Conclusion

(g)

Lieselotte is successful if she studies.	Premise
Lieselotte does not study.	Premise

Therefore, Lieselotte is not successful. Conclusion

(h)

If Michelle studies, then she is successful.	Premise
If Michelle is successful, then she is happy.	Premise

Therefore, if Michelle is happy, then she studies. Conclusion

2.12 Bibliographic Remarks

The Emperor's fork in Exercise 2.9.5 was adapted from Joyce (1908, p. 210), and the rebuttal in Exercise 2.9.6 is from Joyce (1908, p. 213). Tertullian's argument in Exercise 2.9.7 is from Joyce (1908, p. 211).

For more recreational logic problems, you can read Smullyan (2015). Richard Smullyan (1919–) is an American logician who studied under Alonzo Church (Church was the same supervisor of Alan Turing). Over the course of his life, he has published many books on recreational mathematics: mathematical puzzles, and games. These types of puzzles often require little advanced mathematics, and it inspires those to continue to study the subject further. I would be remiss if I did not mention another great mind in recreational mathematics—Charles Lutwidge Dodgson (1832–1898) or known better by his pen name Lewis Carroll. If you have not read *Alice in Wonderland*, you should check it out of your library. The tale follows Alice as she falls into a fantasy world that plays with logic. In Example 2.7.1, we looked at a scene in Carroll (1866) where Alice is having tea with the March Hare, the Mad Hatter, and the Dormouse.

3

Sets and Set Operations

3.1 Introduction to Sets

Underlying the mathematics we study in algebra, geometry, and almost every other area of contemporary mathematics is the notion of a set. A flock of ducks, a troop of kangaroos, or a wardrobe of clothes are all examples of sets of things.

> **Definition 3.1.1 — Set.** A **set** is a well-defined collection of objects. The objects of a set are called **elements**. We denote sets with capital letters, such as A, B, C ..., and the elements are often denoted with lowercase letters, such as a, b, c, ...

We can describe sets in the following three ways:

(a) *Using a Venn diagram.* Graphically, we can represent sets with a **Venn diagram**, where we can represent a set as a circle, and an element can be represented by a dot.

> **Example 3.1.2** Consider the set A of the first three letters of the English alphabet: a, b, c. The Venn diagram is given in Fig. 3.1. ◄

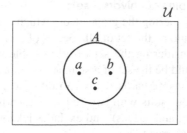

Fig. 3.1   A Venn diagram of the set A that contains the first three lowercase letters of the alphabet. The rectangle represents a set called $\mathcal{U}$, the universal set. In this case, the universal set could be the set of all the lowercase letters in the English alphabet.

(b) *Using Roster Notation.* **Roster notation** is writing the elements of a set by listing each element between curly brackets, { }, and separating each element with a comma. The order of the elements does not matter in the list.

Example 3.1.3 In roster notation, the set A of the first three lowercase letters of the English alphabet is $A = \{a, b, c\}$.   ◄

(c) *Using Set-builder Notation.* Another way to write a set is with **set-builder notation**. A rule is given that describes the definite property or properties an element x must satisfy to qualify to be in the set. For a set A,

$$x \in A \qquad \text{denotes "}x \text{ is an element of the set } A\text{,"}$$

and

$$y \notin A \qquad \text{denotes "}y \text{ is not an element of the set } A\text{."}$$

Example 3.1.4 Consider the set A written as

$$A = \{\, x \in \mathcal{U} \mid x \text{ is the first three letters} \,\}$$

where the (universal) set $\mathcal{U}$ is the set of all the lowercase letters in the English alphabet. The symbol $\in$ is read "is an element of." The vertical line $\mid$ within the set braces is read "such that." So the above would read "A is the set of all x which are elements of the English lowercase alphabet such that x is the first three letters."   ◄

Definition 3.1.5 — Universal Set. A **universal set** is a set that contains all of the required or necessary elements that can be chosen in dealing with a particular situation. We usually denote the universal set with $\mathcal{U}$.

For different problems, there are different associated universal sets, like the universal sets that we consider in the following example.

Example 3.1.6 — Examples of universal sets.
(a) If you were considering the grapes from which a wine is made, then the universal set could be the set of all species of grapes.
(b) If you were considering the set of all possible poker hands, then the universal set could be the set of a standard deck of 52 cards. A **standard deck of cards** has 52 cards, with 13 ranks (ace, 2, 3, 4, ..., 10, jack, queen, king), four suits with two red suits (hearts, diamonds) and two black suits (clubs and spades), and excludes jokers. The jack, queen, and king cards are referred to as face cards.
                    ◄

Definition 3.1.7 — Empty set. A set having no elements is said to be empty. We call this the **empty set**, or the **null set**, and denote this by the symbol $\varnothing$, or by empty curly brackets { }.

Sometimes, we are required to find the number of elements in a set. These types of problems are called **counting problems**. The number of elements in a **finite set** is a whole number[1] determined by simply counting the elements in the set; a set that is not finite is called **infinite**.

Definition 3.1.8 — The number of elements in a set. To denote the number of elements in A, or the **cardinality** of A, we will write $n(A)$.

Since we have already said that there are no elements in the empty set, $n(\varnothing) = 0$. Observe that the set $\{\varnothing\}$ is not the same as $\varnothing$. The set $\{\varnothing\}$ contains only one element (the empty set), and therefore $n(\{\varnothing\}) = 1$.

Example 3.1.9 Find the number of elements in A, for the following sets:
  (a) $A = \{2, 3, 5, 7, 11, 13, 17\}$, and
  (b) $A = \{\text{Macbeth, Macduff, Malcolm}\}$.

Solution
  (a) There are seven elements in A. Therefore, $n(A) = 7$.
  (b) There are three elements in A. Therefore, $n(A) = 3$.

◀

Example 3.1.10 — Finite or Infinite.
  (a) The set $E = \{\text{white, red, blue, orange, green, yellow}\}$ is the set of colors on a Rubik's Cube; $n(E) = 6$, so E is finite.
  (b) Let $A = \{1, 3, 5, 7, \ldots\}$ be the set of odd numbers. The symbol $\ldots$ means "and so on." Then A is infinite.
  (c) The set $E = \{2, 4, 6, \ldots\}$ of even numbers is also infinite.

◀

Definition 3.1.11 — Subset. If every element of a set A is also an element of a set B, we say that A is a **subset** of B, and write $A \subseteq B$. We could read this as "A is contained in or equal to B." If $A \subseteq B$, and if B contains at least one more element not in A, then A is a **proper subset** of B, and we write $A \subset B$. We could read this as "A is contained in B, but not equal to B."

---

[1] The set of whole numbers is $\mathbb{W} = \{0, 1, 2, \ldots\}$.

Example 3.1.12 Let $A = \{2, 4, 6\}$ and $B = \{1, 2, 3, 4, 5, 6\}$. Since A is a subset of B, $A \subseteq B$, and B has three elements not in A, then A is a proper subset of B: $A \subset B$. See Fig. 3.2.   ◄

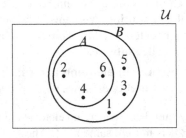

Fig. 3.2 $A \subset B$

Example 3.1.13

    (a) The set of natural numbers $\mathbb{N} = \{1, 2, 3, \ldots\}$ is contained in the set of whole numbers $\mathbb{W} = \{0, 1, 2, 3, \ldots\}$; thus, $\mathbb{N} \subset \mathbb{W}$: it is a proper subset since $0 \notin \mathbb{N}$.

    (b) The set $\mathbb{W}$ is contained in the set of integers $\mathbb{Z} = \{\ldots, -2, -1, 0, 1, 2, \ldots\}$; thus, $\mathbb{W} \subset \mathbb{Z}$: it is a proper subset since $-1, -2, \ldots$ are not elements of $\mathbb{W}$.

Therefore, $\mathbb{N} \subset \mathbb{W} \subset \mathbb{Z}$. See Fig. 3.3 for the Venn diagram.   ◄

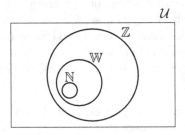

Fig. 3.3   A Venn diagram of the number systems that we commonly use: $\mathbb{N} \subset \mathbb{W} \subset \mathbb{Z}$.

Definition 3.1.14 — Equal sets. Two sets A and B are **equal** if and only if they have exactly the same elements. We write $A = B$.

It follows that if $A = B$, then A is a subset of B, and B is a subset of A. Conversely, if $A \subseteq B$, and $B \subseteq A$, then $A = B$. Moreover, every set A is a subset of itself, $A \subseteq A$, but A is not a proper subset of itself, $A \not\subset A$. (Take a moment to think about why $A \not\subset A$.) The empty set $\varnothing$ is a subset of every set, $\varnothing \subseteq A$.

Exercises

3.1.1. If two sets have the same cardinality, then are they equal? Explain with the aid of an example.

3.1.2. Can a set A and a proper subset $B \subset A$ have the same cardinality?

3.1.3. True or false.
    (a) $\{1, 3, 2\} = \{1, 2, 3\}$ (e) $\{3, 4, 8\} \subset \{1, 2, 3, \ldots, 8\}$
    (b) $2 \in \{22, 222\}$ (f) $\varnothing \in \{1, 2, 3, \ldots, 8\}$
    (c) $2 \in \{2^2, 2^3\}$ (g) $\varnothing \subset \{1, 2, 3, \ldots, 8\}$
    (d) $\{3, 4, 8\} \in \{1, 2, 3, \ldots, 8\}$

3.1.4. Which of the following sets are finite or infinite?
    (a) $\{1, 3, 7, 11\}$
    (b) $\{a, e, i, o, u\}$
    (c) Students absent from class.
    (d) The set of integers $\mathbb{Z} = \{\ldots, -2, -1, 0, 1, 2, \ldots\}$
    (e) All the living people on Earth.
    (f) All of the odd numbers $\{1, 3, 5, \ldots\}$

3.1.5. Write each of the following sets using roster notation.
    (a) $\{x \in \mathbb{Z} \mid 0 < x < 10\}$
    (b) $\{x \in \mathbb{Z} \mid 0 \le x \le 10\}$
    (c) $\{x \in \mathbb{Z} \mid x \text{ is odd and } 0 < x < 10\}$
    (d) $\{x \in \mathcal{U} \mid x \text{ is a vowel}\}$ where $\mathcal{U}$ is the set of uppercase letters in the English alphabet

3.1.6. List the elements of the following sets where $\mathbb{N} = \{1, 2, 3, \ldots\}$.
    (a) $A = \{x \in \mathbb{N} \mid 3 < x < 10\}$
    (b) $B = \{x \in \mathbb{N} \mid (x \text{ is even}) \wedge (x < 7)\}$
    (c) $C = \{x \in \mathbb{N} \mid 5 + x = 3\}$
    (d) $D = \{x \in \mathbb{N} \mid x \text{ is a multiple of } 3\}$

3.1.7. List all the subsets of $\{1, 2, 3\}$.

3.1.8. Why is $A = \{4, 5, 6, 7, 8\}$ not a subset of $B = \{x \in \mathbb{N} \mid x \text{ is even}\}$?

3.1.9. Show that $A = \{5, 6, 7, 8\}$ is a proper subset of $B = \{1, 2, 3, \ldots, 10\}$.

3.1.10. Which of the following are proper subsets of $\{1, 2, 3, 4\}$?
    (a) $\{1, 2, 3\}$ (c) $\{1, 2, 3, 4\}$ (e) $\{3\}$
    (b) $\{4, 2\}$ (d) $\{2, 3\}$ (f) $\varnothing$

3.1.11. Which of the following sets are equal?

$$\{a, c, b\}, \{b, c, a, b\}, \{c, b, c, a\}, \{b, a, b, c\}$$

3.2 Russell's Paradox

> Perhaps the greatest paradox of all is that there are paradoxes in mathematics.
>
> —Kasner and Newman (1949, p. 193)

In Sec. 3.1, we introduced the notion of a set. Sets, as they were conceived, have elements (see Definition 3.1.1). An element of a set can be a duck, a kangaroo, or a shirt. But it is also important to note that with this definition of a set, a set can also be an element of another set. For example, you can have a set of shirts (white, blue, red, *etc.*) which is part of a larger set of all your clothes. But, is there a problem with the way that we defined a set? Are there any issues that may arise? A paradox, perhaps?

Georg Cantor's (1845–1918) set theory was developed while he was trying to work out the cardinality of the real numbers. The young British philosopher Bertrand Russell (1872–1970) became interested in Cantor's set theory in 1900 when he attended the First International Congress of Philosophy and met Giuseppe Peano (1858–1932). Russell admired the Italian mathematician so much that when Russell returned home, he began immediately to adapt Peano's system of symbolic logic that had incorporated Cantor's set theory (Moore and Garciadiego, 1981, p. 325). After about a year, in June 1901, Russell stumbled upon the paradox that now bears his name:

> I examined [Cantor's] proof with some minuteness, and endeavored to apply it to the [sets] of all things there are. This led me to consider those [sets] which are not [elements] of themselves, and to ask whether the [set] of such [sets] is or is not [an element] of itself (Link, 2004, p. 352).

We state his paradox more explicitly.

Definition 3.2.1 — Russell's Paradox. Let S be the set of all sets that do not contain themselves. Does S contain itself? If so, then by the definition of S, it follows that S does not contain itself. Hence, by logic, S does not contain itself. But then this leads to a contradiction, since S, a set itself, does not contain itself, and hence by the definition of S, must contain itself.

Russell failed to realize the importance of the paradox for nearly a whole year, and remained silent[2] on the issue. The seriousness of the paradox was not realized until he entered into correspondence with Peano and Frege.

---

[2] In his voluminous correspondence with his wife and that with his friend Louis Couturat (1868–1914), there is no mention of this paradox (Moore and Garciadiego, 1981, p. 328), but this does not preclude the possibility that he discussed this with his collaborator Alfred North Whitehead (Link, 2004, p. 352).

Gottlob Frege (1848–1925) was a logician whose concern was to show that numbers, including the real numbers, were definable from purely logical concepts. This view is known as logicism, and was extended by Russell to all of pure mathematics in his later book, *Principia Mathematica*, coauthored with Alfred North Whitehead (1861–1947). When Russell sent a letter on 16 June 1902 to the deeply-admired Frege about this discovery, Russell received within 10 days Frege's horrified and devastated response; it was then Russell realized the seriousness of what he found.

"I have heard from Frege," Russell wrote to his wife, "a most candid letter; he says that my conundrum makes not only his Arithmetic, but all possible Arithmetics, totter" (Moore and Garciadiego, 1981, p. 328).

Over the course of the next couple of months, an exchange of nine long letters between Frege and Russell ensued in an attempt to patch up the mathematical framework and to resolve the paradox. While Russell's *The Principles of Mathematics* was being printed in 1902, he still believed that he could resolve the paradox in time. Reflecting in a letter to a friend, Russell wrote, "I believed I could avoid these contradictions, but now I see that I was mistaken, a fact which greatly diminishes the value of my book" (Moore and Garciadiego, 1981, p. 329).[3] This paradox later spurred Russell to collaborate with Whitehead to invent a new (and more complex) logical system that would correct this paradox; the result was the massive three-volume work, *The Principia Mathematica*.

Here is one way to illustrate what Russell's paradox means (Gonseth, 1936, pp. 255–257). Suppose that every public library in the United States is to prepare a database of all of its collections. Just before the librarian is about to send the file, he wonders if he should include the database as part of the library's collection. He decides against it since it being in the library's collection is self-evident.

From around the country, all the databases are submitted to the Library of Congress. Now as the national librarian starts to sort through the databases, she notices that some of the databases include themselves in the listings, while others do not. She compiles two master databases: one of all the databases that include themselves, and one of all of the databases that do not include themselves.

However, the national librarian is doomed with the database that does not list itself: she cannot include it in its own listing, because then it would include itself. But in that case, it should belong to the other database, that of databases that do include themselves. However, if she leaves it out, the database is incomplete. Either way, it can never be a true database of databases that do not list themselves.

---

[3] A quote from a letter to Russell's friend Louis Couturat dated 29 September 1902. The original French text reads: *Quand on a commencé à imprimer mon livre, j'ai cru pouvoir éviter ces contradictions, mais je vois à présent que je me trompais, ce qui diminue de beaucoup la valeur de mon livre* (Moore and Garciadiego, 1981, p. 345).

3.3 Set Operations

A visual diagram helps us to see how sets are related to one another. Figure 3.4 is an example of a **Venn diagram**, named after the mathematician John Venn (1834–1923); it shows the universal set $\mathcal{U}$ consisting of a rectangle, and the sets A and B as circles. It is possible to create new sets from previously existing sets within the universal set in different ways by combining or intersecting them. We will use a Venn diagram where the shaded areas indicate these new sets.

> **Definition 3.3.1 — Union.** The **union** of two sets A and B is denoted $A \cup B$ (read "A union B"), and represents the set of all elements that belong to A *or* B, *or* both. In mathematical notation,
>
> $$A \cup B = \{ x \in \mathcal{U} \mid x \in A \vee x \in B \}. \qquad (3.1)$$

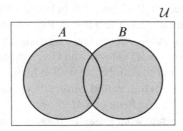

Fig. 3.4 $A \cup B$

Example 3.3.2 Let the universe $\mathcal{U} = \{1, 2, \ldots, 10\}$, $A = \{5, 6, 7, 8\}$, and $B = \{1, 2, 3, 4, 5, 6\}$. Find the union $A \cup B$.

Solution The union $A \cup B$ are all the elements that belong to A or B, or both in A and B. Therefore, the union $A \cup B = \{1, 2, 3, 4, 5, 6, 7, 8\}$. Notice that you have to write each element *only once*.   ◀

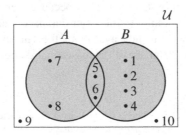

Fig. 3.5   The Venn diagram representing $A \cup B$ for Example 3.3.2

Example 3.3.3 Let the universe $\mathcal{U}$ be a single deck of 52 cards, A is the set of all diamond cards, and B is the set of all king cards. What is the union $A \cup B$?

Solution Using roster notation, the universe set is

$$\mathcal{U} = \{A\diamondsuit, 2\diamondsuit, \ldots, Q\diamondsuit, K\diamondsuit, A\heartsuit, 2\heartsuit, \ldots, Q\heartsuit, K\heartsuit,$$
$$A\spadesuit, 2\spadesuit, \ldots, Q\spadesuit, K\spadesuit, A\clubsuit, 2\clubsuit, \ldots, Q\clubsuit, K\clubsuit\}.$$

In roster notation, the set A of diamond cards is

$$A = \{A\diamondsuit, 2\diamondsuit, 3\diamondsuit, 4\diamondsuit, 5\diamondsuit, 6\diamondsuit, 7\diamondsuit, 8\diamondsuit, 9\diamondsuit, 10\diamondsuit, J\diamondsuit, Q\diamondsuit, K\diamondsuit\},$$

and the set B of all kings is

$$B = \{K\diamondsuit, K\heartsuit, K\spadesuit, K\clubsuit\}.$$

Then $A \cup B$ would be the set of cards that is either a diamond, *or* a king, *or both*. In roster notation,

$$A \cup B = \{A\diamondsuit, 2\diamondsuit, 3\diamondsuit, 4\diamondsuit, 5\diamondsuit, 6\diamondsuit, 7\diamondsuit, 8\diamondsuit, 9\diamondsuit,$$
$$10\diamondsuit, J\diamondsuit, Q\diamondsuit, K\diamondsuit, K\heartsuit, K\spadesuit, K\clubsuit\}.$$

Notice that for elements that are both in A and B we write them only once. For example, $K\diamondsuit$ is an element of A and B, but we write this element only once in the union $A \cup B$. ◀

> **Definition 3.3.4 — Intersection.** The **intersection** of two sets A and B is denoted by $A \cap B$ (read "A intersect B"), and represents the set of all elements common to both A *and* B. In mathematical notation,
> $$A \cap B = \{ x \in \mathcal{U} \mid x \in A \land x \in B \}. \tag{3.2}$$

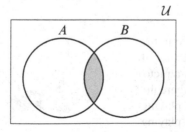

Fig. 3.6 $A \cap B$

Example 3.3.5 Let $\mathcal{U} = \{1, 2, 3, 4, 5\}$, $A = \{1, 3, 5\}$, and $B = \{2, 3\}$. Find the intersection $A \cap B$.

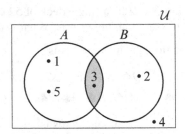

Fig. 3.7   A Venn diagram representing the set $A \cap B$ for Example 3.3.5; $A \cap B = \{3\}$.

Solution The intersection $A \cap B$ is the elements that are in both sets A and B. The only common element in A and B is 3. Therefore,

$$A \cap B = \{3\}.$$

The Venn diagram is given in Fig. 3.7.   ◀

Example 3.3.6 Let the universe $\mathcal{U}$ be a single set of 52 cards, A is the set of all heart cards, and B is the set of all face cards (jacks, queens and kings). What is the intersection $A \cap B$?

Solution In roster notation, the set A of heart cards is

$$A = \{A\heartsuit, 2\heartsuit, 3\heartsuit, 4\heartsuit, 5\heartsuit, 6\heartsuit, 7\heartsuit, 8\heartsuit, 9\heartsuit, 10\heartsuit, J\heartsuit, Q\heartsuit, K\heartsuit\},$$

and the set B of all queens and kings is

$$B = \{J\diamondsuit, J\heartsuit, J\spadesuit, J\clubsuit, Q\diamondsuit, Q\heartsuit, Q\spadesuit, Q\clubsuit, K\diamondsuit, K\heartsuit, K\spadesuit, K\clubsuit\}.$$

Then $A \cap B$ would be the set of cards that is a heart *and* a face card. In roster notation,

$$A \cap B = \{J\heartsuit, Q\heartsuit, K\heartsuit\}.$$

◀

     The elements that are in the intersection have properties of both sets. Figure 3.8 is a pictographic Venn diagram where the intersection of a guitar playing beaver and a keyboard playing duck is a keytar playing platypus. What do you think the universal set, $\mathcal{U}$, could be? The set of all instrument-playing animals?

Fig. 3.8   Graphics Credit: tenso GRAPHICS.

Definition 3.3.7 — Complement. If A is a subset of $\mathcal{U}$, the **complement** of A consists of the elements of $\mathcal{U}$ that are not elements of A. The complement of A is denoted by A'. In other words,

$$A' = \{\, x \in \mathcal{U} \mid x \notin A \,\}. \tag{3.3}$$

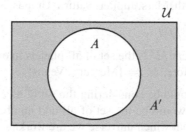

Fig. 3.9 A' is represented by the shaded area.

Example 3.3.8 Let the universe $\mathcal{U} = \{1, 2, 3, 4, 5\}$, and $A = \{1, 3, 5\}$. Find the complement A'.

Solution The complement A' are all the elements that are in $\mathcal{U}$, but not in A. The only elements that are not A are 2 and 4. Therefore, using roster notation,

$$A' = \{2, 4\}.$$

Graphically, the Venn diagram is given in Fig. 3.10.

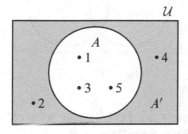

Fig. 3.10   The Venn diagram for $A' = \{2, 4\}$ for Example 3.3.8.

Example 3.3.9 Consider the universe $\mathcal{U}$ to be of all the planets[4] in our solar system;

$$\mathcal{U} = \{\text{Mercury, Venus, Earth, Mars, Jupiter, Saturn, Uranus, Neptune}\}.$$

Let M be the set of all planets with moons

$$M = \{\text{Earth, Mars, Jupiter, Saturn, Uranus, Neptune}\}.$$

What is the complement M'?

Solution The complement M' is the set of all planets in our solar system that do not have a moon. Therefore, $M' = \{\text{Mercury, Venus}\}$.   ◄

Another example would be considering the set of all even numbers, then the complement of this set would be the set of all odd numbers. We would have to make an assumption about which universe we are working in, and in this case the natural numbers $\mathbb{N}$ or the integers $\mathbb{Z}$ would suffice.

Notice that there are no elements shared in a set and the set's complement. In other words, in a specified universe $\mathcal{U}$, any set A and the intersection with its complement A' is empty: $A \cap A' = \varnothing$.

> **Definition 3.3.10 — Difference.** The **difference** of two sets A and B is denoted $B \smallsetminus A$ (read "B minus A") and represents the set of all elements belonging to B which are not in the set A. In mathematical notation,
>
> $$B \smallsetminus A = \{x \in B \mid x \notin A\}. \tag{3.4}$$

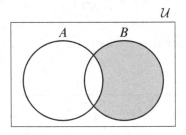

Fig. 3.11 $B \smallsetminus A$

Example 3.3.11 The set $A = \{1, 2, 4\}$ and $B = \{2, 3, 4, 5, 6\}$. Let $\mathcal{U} = \{1, 2, 3, 4, 5, 6, 7, 8\}$. What is the difference $B \smallsetminus A$?

---

[4] Pluto was considered to be a planet when it was first discovered in 1930 by Clyde W. Tombaugh. It remained classified as a planet until the International Astronomical Union (IAU) formally defined the term "planet" in 2006. This new definition excluded Pluto and reclassified it as a member of the newly defined "dwarf planet."

Solution Using roster notation,

$$B \smallsetminus A = \{2,3,4,5,6\} \smallsetminus \{1,2,4\}$$
$$= \{3,5,6\}.$$

Graphically, $B \smallsetminus A$ is represented in the Venn diagram in Fig. 3.12.

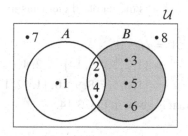

Fig. 3.12   The Venn diagram for $B \smallsetminus A$ for Example 3.3.11.

◄

Definition 3.3.12 — Mutually Exclusive (or Disjoint). If there are no elements that belong to a set A and a set B, then the intersection of A and B is the empty set $\varnothing$, and we say that the two sets are **mutually exclusive** (or disjoint). In other words, A and B are mutually exclusive if and only if $A \cap B = \varnothing$.

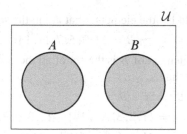

Fig. 3.13 $A \cap B = \varnothing$

For example, two sets that are mutually exclusive is the set A of even numbers $\{2,4,6,\dots\}$ and the set B of odd numbers $\{1,3,5,\dots\}$, since none of the elements of A are in B; in other words, $A \cap B = \varnothing$.

Example 3.3.13 In Example 3.3.9 with the planets in our solar system, the set of planets with moons and the set of planets without moons are mutually exclusive. In other words, $A \cap A' = \varnothing$.   ◄

Example 3.3.14 Let $\mathcal{U}$ be the set of all natural numbers less than 15, the set $A = \{2, 4, 6\}$ and the set B be the prime numbers[5] in $\mathcal{U}$. List the elements of the sets and draw a Venn diagram of the following:

(a) A' (c) $A \cup B$ (e) $A \cap B$

(b) B' (d) $(A \cup B)'$ (f) $B \smallsetminus A$.

Solution

(a) The complement of A is the set of all elements in $\mathcal{U}$ that are not elements of A; that is,

$$A' = \mathcal{U} \smallsetminus A$$
$$= \{1, 2, 3, \ldots, 14\} \smallsetminus \{2, 4, 6\}$$
$$= \{1, 3, 5, 7, 8, 9, 10, 11, 12, 13, 14\}.$$

The Venn diagram is given in Fig. 3.14.

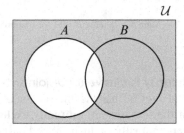

Fig. 3.14 A'

(b) Next, we should list the elements that are in the set B,

$$B = \{2, 3, 5, 7, 11, 13\}.$$

Then, the complement of B is the set of all the elements in $\mathcal{U}$ that are not elements of B; that is,

$$B' = \mathcal{U} \smallsetminus B$$
$$= \{1, 2, 3, \ldots, 14\} \smallsetminus \{2, 3, 5, 7, 11, 13\}$$
$$= \{1, 4, 6, 8, 9, 10, 12, 14\}.$$

The Venn diagram is given in Fig. 3.15.

(c) The union $A \cup B$ will be the set of all elements that are prime numbers less than 15, or are elements of A, or both. Thus,

$$A \cup B = \{2, 4, 6\} \cup \{2, 3, 5, 7, 11, 13\}$$
$$= \{2, 3, 4, 5, 6, 7, 11, 13\}.$$

The Venn diagram is the same as Fig. 3.4 on p. 82.

---

[5] A **prime number** is a natural number that is greater than 1 and has no divisors other than 1 and itself. For example, 3 is a prime number since 1 and 3 are its only factors.

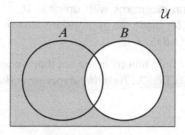

Fig. 3.15 B'

(d) The complement of $A \cup B$ is the set of all elements that are in $\mathcal{U}$ and are in neither A nor B (or both). Then,

$$
\begin{aligned}
(A \cup B)' &= \mathcal{U} \smallsetminus (A \cup B) \\
&= \{1, 2, 3, \ldots, 14\} \smallsetminus \{2, 3, 4, 5, 6, 7, 11, 13\} \\
&= \{1, 8, 9, 10, 12, 14\}.
\end{aligned}
$$

The Venn diagram is given in Fig. 3.16.

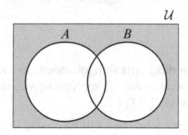

Fig. 3.16 $(A \cup B)'$

(e) The intersection $A \cap B$ is the set of all elements that are prime numbers below 15, and are elements of A. Thus,

$$
A \cap B = \{2, 4, 6\} \cap \{2, 3, 5, 7, 11, 13\} = \{2\}.
$$

The Venn diagram is the same as in Fig. 3.6 on p. 83.

(f) The difference $B \smallsetminus A$ is the set of all elements that are prime numbers below 15, and that are not elements of A. Thus,

$$
\begin{aligned}
B \smallsetminus A &= \{2, 3, 5, 7, 11, 13\} \smallsetminus \{2, 4, 6\} \\
&= \{3, 5, 7, 11, 13\}.
\end{aligned}
$$

The Venn diagram is the same one as in Fig. 3.11 on p. 86.

◄

Example 3.3.15 — Venn diagrams with unions. If A, B, and C are subsets of a given universal set $\mathcal{U}$, shade the appropriate area in the Venn diagram that represents the union $(A \cup B) \cup C'$.

Solution Since there are only unions in the set that we are interested in, we can simply shade $(A \cup B)$ (see Fig. 3.17(a)). We continue to shade in all of C', leaving us with Fig. 3.17(b).

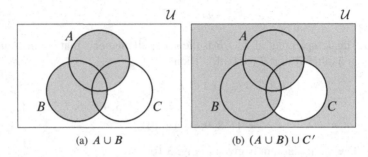

(a) $A \cup B$　　　　　(b) $(A \cup B) \cup C'$

Fig. 3.17　Venn diagrams for Example 3.3.15

◀

Example 3.3.16 — Venn diagrams with intersections. If A, B, and C are subsets of a given universal set $\mathcal{U}$, shade the appropriate area in the Venn diagram that represents the intersection $A \cap B \cap C'$.

Solution — 1. First, using a pencil (we will be erasing lines later) we draw northeast lines where the intersection $A \cap B$ is in the Venn diagram as seen in Fig. 3.18(a).

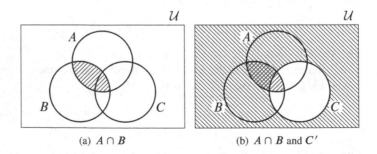

(a) $A \cap B$　　　　　(b) $A \cap B$ and C'

Fig. 3.18　Venn diagrams for Example 3.3.16

Next, using the same figure, we draw north-west lines for the complement of C: the north-west lines are drawn everywhere except on the inside of the set C; the result is Fig. 3.18(b).

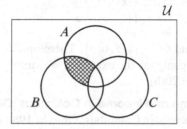

Fig. 3.19 Final Venn diagram for Example 3.3.16

The area where the north-west and north-east crisscross is $A \cap B \cap C'$. The rest of the lines surrounding the crisscross area can be erased, leaving you behind with the final Venn diagram, as seen in Fig. 3.19. ◄

If the previous solution made you dizzy by looking at all of those overlapping lines,[6] then consider the alternative solution below.

Solution — 2. We begin by drawing the three sets as circles in the Venn diagram, and labeling each of the sets including $\mathcal{U}$ for the universal set. After we have done this, we label each of the eight areas in the Venn diagram as shown in Fig. 3.20(a).

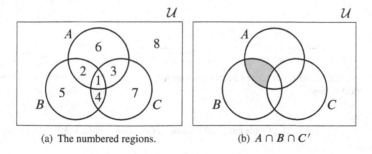

(a) The numbered regions. (b) $A \cap B \cap C'$

Fig. 3.20 An alternative solution to Example 3.3.16

---

[6]This dizzying effect by the "vibrating lines" is called a moiré pattern.

From Fig. 3.20(a), we can see that sets

$$A = \{1, 2, 3, 6\}$$
$$B = \{1, 2, 4, 5\}$$
$$C = \{1, 3, 4, 7\}$$
$$\mathcal{U} = \{1, 2, 3, 4, 5, 6, 7, 8\}.$$

Thus, $A \cap B = \{1, 2\}$ and $C' = \{2, 5, 6, 8\}$. Therefore, the desired set is given by $A \cap B \cap C' = \{2\}$. To complete the example, we simply shaded region 2 in the Venn diagram; see Fig. 3.20(b).   ◄

Example 3.3.17 — Crime and Income in Columbus, Ohio, 1980. Let us look at the crime rates and income in Columbus, Ohio in 1980 (Anselin, 1988, p. 189, Table 12.1).

By looking at the two maps in Fig. 3.21(a) and Fig. 3.21(b), we can see that there is a large area where high crime rates are occurring in areas with low income. Visually, what we are doing is trying to find the intersection of areas with high crime rate areas and low income. Geographers in a geographical information system (GIS) can pass in expressions such as (crime ≥ 34) ∩ (income < $15000) to get the map in Fig. 3.21(c).

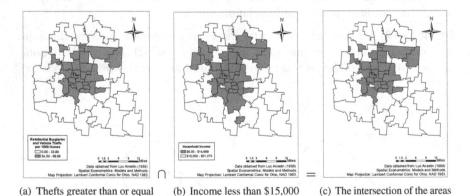

(a) Thefts greater than or equal to 34 per 1000 homes

(b) Income less than $15,000

(c) The intersection of the areas where there is high crime rates and low income

Fig. 3.21   Crime rates and income in Columbus, Ohio, 1980

This can help city planners and policy makers to designate where they should be concentrating more resources such as social and policing services. When geographers perform these type of set operations to create new maps, they call them **overlay** operations (Chang, 2010).   ◄

Definition 3.3.18 — Power set. The power set of a set A, written $\mathcal{P}(A)$, is the set of all subsets of A, including the empty set $\varnothing$ and A itself.

Theorem 3.3.19 The number of subsets of a set of n elements is 2^n.

Proof Let the set A have n elements; $A = \{a_1, a_2, a_3, \ldots, a_n\}$. Any subset of A may be formed by considering each element in turn and accepting or rejecting the element in the subset. All of the possible subsets can be formed by this process of accepting or rejecting. For element a_1, it may be accepted or rejected, and so there are 2 ways to dispose of the element a_1 when choosing a subset. Likewise, for element a_2, there are two ways to dispose a_2 when choosing a subset, and so on for the remaining elements. Therefore, there are

$$\underbrace{2 \times 2 \times \cdots \times 2}_{n \text{ times}} = 2^n$$

ways of forming subsets. ◄

Example 3.3.20 Suppose A is the set $\{1, 2, 3\}$. How many elements are in the power set of A, $\mathcal{P}(A)$? List the elements of $\mathcal{P}(A)$.

Solution Since the set A has three elements, or $n(A) = 3$, then by Theorem 3.3.19, the power set $\mathcal{P}(A)$ has $2^3 = 8$ elements. *Draw a diagram.* By using a tree diagram, see Fig. 3.22, we can find all of the possible subsets of A.

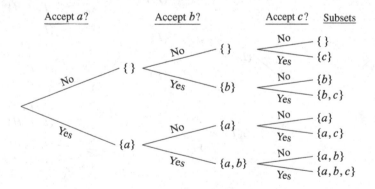

Fig. 3.22 Tree diagram for Example 3.3.20

The subsets of A are $\varnothing, \{1\}, \{2\}, \{3\}, \{1, 2\}, \{1, 3\}, \{2, 3\}$, and $\{1, 2, 3\}$. Therefore,

$$\mathcal{P}(A) = \{\varnothing, \{1\}, \{2\}, \{3\}, \{1, 2\}, \{1, 3\}, \{2, 3\}, \{1, 2, 3\}\}.$$

◄

Example 3.3.21 At Rick's Café Américain, you can order a cheeseburger with any of the following toppings: lettuce, tomato, pickles, and hot peppers. How many different ways can you order a cheeseburger at Rick's?

Solution Let T be the set of toppings:

$$T = \{L, T, P, H\}$$

where L, T, P, and H denote lettuce, tomato, pickles, and hot peppers, respectively. The number of elements in T is $n(T) = 4$. All of the subsets of T are all the possible ways of ordering a cheeseburger with toppings or without toppings. Therefore, the number of ways you can order a cheeseburger at Rick's is $2^4 = 16$.

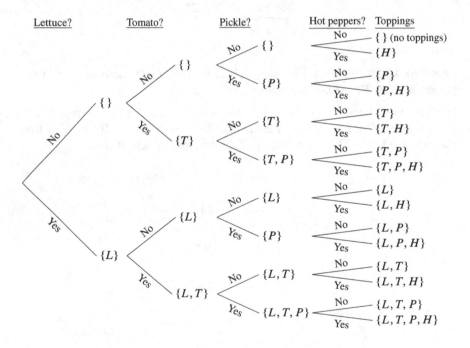

Fig. 3.23   Ordering a cheeseburger at Rick's Café Américain

Looking back, we can check this by listing all the possible ways to order a cheeseburger. We can use a tree diagram (see Fig. 3.23) to find all of the topping possibilities.   ◄

We will see in Chap. 5 that there are many more techniques to count without enumerating all of the possibilities.

Exercises

3.3.1. Let $A = \{1, 2, 3, 4, 5\}$ and $B = \{3, 4, 5, 6\}$, find
    (a) $A \cup B$ (b) $A \cap B$.

3.3.2. Let the set $A = \{\text{Rob, Rich, Rami}\}$, $B = \{\text{Rob, River, Rico}\}$, and $C = \{\text{River, Rico, Renée}\}$.
    (a) Find

(i) $A \cup B$	(iv) $B \cup C$	(vii) $A \cup C$
(ii) $A \cap B$	(v) $B \cap C$	(viii) $A \cap C$
(iii) $A \smallsetminus B$	(vi) $B \smallsetminus C$	(ix) $A \smallsetminus C$.

    (b) Let $\mathcal{U} = \{\text{Rob, Rich, Rami, River, Rico, Renée}\}$. Find

(i) A'	(v) $B' \cup C'$	(ix) $A' \cap C'$
(ii) B'	(vi) $A' \cup C'$	(x) $A' \smallsetminus B'$
(iii) C'	(vii) $A' \cap B'$	(xi) $B' \smallsetminus C'$
(iv) $A' \cup B'$	(viii) $B' \cap C'$	(xii) $A' \smallsetminus C'$.

3.3.3. Fill the missing entries in the following table.

$\cap$	$\varnothing$	$\{1\}$	$\{2\}$	$\{1, 2\}$
$\varnothing$		$\varnothing$		
$\{1\}$				$\{1\}$
$\{2\}$	$\varnothing$	$\varnothing$		$\{2\}$
$\{1, 2\}$	$\varnothing$			

3.3.4. Fill the missing entries in the following table.

$\cup$	$\varnothing$	$\{1\}$	$\{2\}$	$\{1, 2\}$
$\varnothing$		$\{1\}$		
$\{1\}$				
$\{2\}$	$\{2\}$	$\{1, 2\}$	$\{2\}$	$\{1, 2\}$
$\{1, 2\}$				

3.3.5. Define the universal set

$$\mathcal{U} = \{1, 2, 3, \ldots, 10\},$$

and the two sets $A = \{1, 2, 3, 4\}$ and $B - \{1, 2, 3, 4, 5\}$.
    Find the sets

(a) $A \cap B$	(c) $A' \cap B'$	(e) $A \cup B'$
(b) $A \cap B'$	(d) $A \cup B$	(f) $B \smallsetminus A$.

3.3.6. Draw the Venn diagram for $A \smallsetminus B$. Is it the same as in Fig. 3.11?

3.3.7. Draw the Venn diagram for the set $\{\, x \in \mathcal{U} \mid x \in A \veebar x \in B \,\}$. (*Hint*: Recall that $\veebar$ is the exclusive disjunction, and it is described in Sec. 1.3.4.)

3.3.8. Consider the Venn diagram given below.

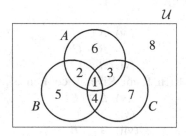

(a) Express the indicated regions in set notation.
  (i) Region 1
  (ii) Region 7
  (iii) Regions 2 and 3 together
  (iv) Regions 1, 2, and 4 together
  (v) Regions 2, 3, 4, 5, 6, and 7 together.
(b) State which regions would be shaded in the Venn diagram for the following sets
  (i) $A \cup B$ (xvi) $(A \cup B)'$
  (ii) $A \cup C$ (xvii) $(A \cup C)'$
  (iii) $B \cup C$ (xviii) $(B \cup C)'$
  (iv) $A \cap B$ (xix) $(A \cap B)'$
  (v) $A \cap C$ (xx) $(A \cap C)'$
  (vi) $B \cap C$ (xxi) $(B \cap C)'$
  (vii) $A \smallsetminus B$ (xxii) $(A \smallsetminus B)'$
  (viii) $A \smallsetminus C$ (xxiii) $(A \smallsetminus C)'$
  (ix) $B \smallsetminus C$ (xxiv) $(B \smallsetminus C)'$
  (x) $B \smallsetminus A$ (xxv) $(B \smallsetminus A)'$
  (xi) $C \smallsetminus A$ (xxvi) $(C \smallsetminus A)'$
  (xii) $C \smallsetminus B$ (xxvii) $(C \smallsetminus B)'$
  (xiii) A' (xxviii) $(A \cup B \cup C)'$
  (xiv) B' (xxix) $\big((A \cap B) \cup C\big)'$.
  (xv) C'

3.3.9. Consider the Venn diagram of four sets that was given in Venn's own book on *Symbolic Logic*. It is one of the simple and symmetrical diagrams that can be produced for four sets.

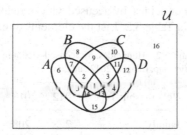

(a) Express the indicated regions in set notation.
  (i) Region 1, 2, 7, and 13 together
  (ii) Region 1, 2, 5, 6, 7, 13, 14, and 15 together
  (iii) Region 1, 2, 3, 4, 5, 6, 7, 8, 9, 13, 14, and 15 together
  (iv) Region 16
(b) State which regions would be shaded in the Venn diagram for the following sets.
  (i) $A \cup D$
  (ii) $B \cup C$
  (iii) $B \cap D$
  (iv) B'
  (v) $(A \cap D)'$
  (vi) $(B \cup C)'$
  (vii) $(A \cup B \cup C \cup D)'$
  (viii) $(A \cap B) \cap D'$
  (ix) $(A \cap B) \cap (C \cap D)$

3.3.10. The terminology of the British Isles is confusing to many people. This confusion is caused by many of the words carrying both geographical and political connotations. Using Table 3.1, draw a Venn diagram.

Table 3.1 The British Isles

	Great Britain	United Kingdom	Ireland	British Islands	British Isles
Scotland	✓	✓		✓	✓
England	✓	✓		✓	✓
Wales	✓	✓		✓	✓
Northern Ireland		✓	✓	✓	✓
Ireland			✓		✓
Isle of Man				✓	✓
Guernsey				✓	✓
Jersey				✓	✓

3.3.11. Verify the following set equalities with the use of Venn diagrams.
  (a) $A \cup B = (A' \cap B')'$
  (b) $A \cup (B \cap C) = (A \cup B) \cap (A \cup C)$

3.3.12. The number of homicides, by accused-victim relationship, in Canada for 2012 and 2013 is given in Table 3.2. List the elements of each set.

Table 3.2   Homicides, by accused-victim relationship, Canada, 2012 and 2013

Relationship Type	2012	2013
Family	228	208
Intimate	20	24
Acquaintance	158	149
Criminal	23	36
Stranger	65	49
Unknown	4	2

(a) The set of all relationship types in which there was more than 60 homicides in both 2012 and 2013.

(b) The set of all relationship types in which there was less than 50 homicides in 2012 or 2013.

(c) The set of all relationship types in which the number of homicides decreased from 2012 to 2013.

(d) The set of all relationship types in which the number of homicides increased from 2012 to 2013.

3.3.13. For the following searches of a website that helps you find a restaurant in a particular city, draw the corresponding Venn diagram. The restaurant

(a) serves vegetarian cuisine *or* halal food

(b) Italian *or* Greek cuisine but *not* sushi

(c) is a steakhouse *and* within five miles of a cinéma, but does *not* have average price of meals greater than $50

(d) either specializes in Thai *or* Cambodian but *not* both

(e) has an average user rating of 4 *or* 5 star rating *and* has more than 15 user ratings.

3.3.14. Suppose A is the set $\{1, 2, 3, 4\}$. How many elements are in the power set of A, $\mathcal{P}(A)$? List the elements of $\mathcal{P}(A)$.

3.3.15. How many subsets are possible with a set that has
  (a) 5 elements?   (b) 10 elements?   (c) 230 elements?

3.3.16. In how many ways can three coins land if they are tossed at the same time?

3.3.17. If a power set $\mathcal{P}(A)$ contains 128 elements, how many elements are in set A?

3.4 Principle of Inclusion and Exclusion

Consider the following motivation example.

Example 3.4.1 Given the following sets

$$A = \{3, 4, 5, 6\},$$

and

$$B = \{5, 6, 7, 8, 9, 10\}.$$

(a) What is the number of elements in each of the sets?
(b) What is the number of elements in $A \cap B$?
(c) What is the number of elements in $A \cup B$?

Solution

(a) $n(A) = 4; n(B) = 6$
(b) $A \cap B = \{5, 6\}; n(A \cap B) = 2$
(c) $A \cup B = \{3, 4, \ldots, 10\}; n(A \cup B) = 8.$

◀

Notice in the last example, that if we added the number of elements in the two sets together, and subtracted the number of elements in the intersection, then we have the number of elements in the union.

$$n(A) + n(B) - n(A \cap B) = 6 + 4 - 2 = 8$$

In fact, this is the case for any two finite sets. We can generalize this to what is called the Principle of Inclusion and Exclusion.[7]

Theorem 3.4.2 — Principle of Inclusion and Exclusion for Two Finite Sets. If sets A and B are finite, then

$$n(A \cup B) = n(A) + n(B) - n(A \cap B), \tag{3.5}$$

and $A \cup B$ and $A \cap B$ are also finite.

In words, $n(A \cup B)$ is equal to

the sum of the cardinalities of A and B,
minus   the cardinality of the intersection of A and B.

To find the number of elements in the complement of the set A, it is the number of elements that are in the universal set, but not in A; that is

$$n(A') = n(\mathcal{U}) - n(A). \tag{3.6}$$

What if the sets A and B are mutually exclusive? Then, that means $A \cap B = \emptyset$, so (3.5) reduces to

$$n(A \cup B) = n(A) + n(B). \tag{3.7}$$

Example 3.4.3 If $n(\mathcal{U}) = 100, n(A) = 70, n(B) = 60$, and $n(A \cap B) = 50$,

---
[7]You may see that some mathematicians refer to this principle as PIE. It is as easy as PIE!

   (a) what is $n(A \cup B)$?   (c) draw the Venn diagram.
   (b) what is $n\big((A \cup B)'\big)$?

Solution

   (a) We use the Principle of Inclusion and Exclusion for two finite sets (3.5).
$$n(A \cup B) = n(A) + n(B) - n(A \cap B)$$
$$= 70 + 60 - 50$$
$$= 80$$

   (b) $n\big((A \cup B)'\big) = n(\mathcal{U}) - n(A \cup B) = 100 - 80 = 20$
   (c) The Venn diagram is shown in Fig. 3.24.

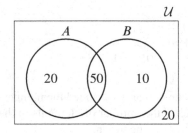

Fig. 3.24   Venn diagram for Example 3.4.3

◄

Example 3.4.4 How many from the set $\{1, 2, 3, \ldots, 9, 10\}$ are not divisible by 2 or 3 (or both)?

Solution We define the following sets:

   (a) $\mathcal{U} = \{1, 2, 3, \ldots, 9, 10\}$; $n(\mathcal{U}) = 10$.
   (b) A be the set of numbers in $\mathcal{U}$ divisible by 2: $A = \{2, 4, 6, 8, 10\}$, and so $n(A) = 5$.
   (c) B be the set of numbers in $\mathcal{U}$ divisible by 3: $B = \{3, 6, 9\}$, and so $n(B) = 3$.
   (d) $A \cap B$ be the set of numbers that are divisible by 2 and 3, or equivalently is divisible by 6: $A \cap B = \{6\}$, and so $n(A \cap B) = 1$.

   We are required to find $n\,((A \cup B)') = n(\mathcal{U}) - n(A \cup B)$, but we do not know $n(A \cup B)$. We can find $n(A \cup B)$ by the Principle of Inclusion and Exclusion,
$$n(A \cup B) = n(A) + n(B) - n(A \cap B) = 5 + 3 - 1 = 7,$$
and so,
$$n\,((A \cup B)') = n(\mathcal{U}) - n(A \cup B) = 10 - 7 = 3.$$
Therefore, there are three numbers that are not divisible by 2 or 3 in $\mathcal{U}$. To illustrate the solution visually, the Venn diagram is given in Fig. 3.25.

◄

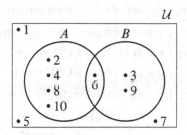

Fig. 3.25   Venn Diagram for Example 3.4.4

Theorem 3.4.5 — Principle of Inclusion and Exclusion for Three Finite Sets.
If sets A, B, and C are finite, then

$$n(A \cup B \cup C) = n(A) + n(B) + n(C) - n(A \cap B) - n(A \cap C)$$
$$- n(B \cap C) + n(A \cap B \cap C), \qquad (3.8)$$

and $A \cup B \cup C$, $A \cap B$, $A \cap C$, $B \cap C$, and $A \cap B \cap C$ are also finite.

In words, $n(A \cup B \cup C)$ is equal to

<blockquote>

the sum of the cardinalities of the individual sets,

minus   the cardinalities of the two-way intersections,

plus   the cardinality of the three-way intersection.

</blockquote>

Example 3.4.6 In a survey of 120 students, they were asked about their car's stereo system. The survey revealed the following data:

- 46 students said they have satellite radio
- 52 students said they have a CD player
- 50 students said they have a USB port
- 25 students said they have a satellite radio and a CD player
- 21 students said they have a satellite radio and a USB port
- 23 students said they have a CD player and a USB port
- 16 students said they have a satellite radio, CD player, and a USB port.

(a) How many of the students have at least a USB port, CD player, or a satellite radio?

(b) How many of the students have only one of the following: a satellite radio, CD player, or a USB port.

(c) How many of the students did not have a satellite radio, CD player, or a USB port.

(d) How many students only have a USB port?

Solution — 1. Let $\mathcal{U}$ denote the set of all students surveyed, and

$$S = \{\, x \in \mathcal{U} \mid x \text{ have a satellite radio} \,\}$$
$$C = \{\, x \in \mathcal{U} \mid x \text{ have a CD player} \,\}$$
$$P = \{\, x \in \mathcal{U} \mid x \text{ have a USB port} \,\}.$$

From the problem, we have the following data:

$$n(S) = 46$$
$$n(C) = 52$$
$$n(P) = 50$$
$$n(S \cap C) = 25$$
$$n(S \cap P) = 21$$
$$n(C \cap P) = 23$$
$$n(S \cap C \cap P) = 16.$$

(a) We require $n(P \cup C \cup S)$. Then, by the Principle of Inclusion and Exclusion,

$$\begin{aligned} n(P \cup C \cup S) &= n(P) + n(C) + n(S) - n(P \cap C) \\ &\quad - n(P \cap S) - n(C \cap S) + n(P \cap C \cap S) \\ &= 50 + 52 + 46 - 23 - 21 - 25 + 16 \\ &= 95. \end{aligned}$$

(b) *Take cases.*

(i) Car stereos that have only a satellite radio:

$$\begin{aligned} n(S) - n(S \cap C) - n(S \cap P) & \\ + n(S \cap C \cap P) &= 46 - 25 - 21 + 16 \\ &= 16. \end{aligned}$$

(ii) Car stereos that have only a CD player:

$$\begin{aligned} n(C) - n(C \cap S) - n(C \cap P) & \\ + n(S \cap C \cap P) &= 52 - 25 - 23 + 16 \\ &= 20. \end{aligned}$$

(iii) Car stereos that have only a USB port:

$$\begin{aligned} n(P) - n(P \cap S) - n(P \cap C) & \\ + n(S \cap C \cap P) &= 50 - 21 - 23 + 16 \\ &= 22. \end{aligned}$$

We add up all the cases. Therefore, the students that have exactly one of the components is $16 + 20 + 22 = 58$.

(c) The number of students that do not have any of the components would be the complement of the students having at least one of the components. Mathematically,

$$n\left((P \cup C \cup S)'\right) = n(\mathcal{U}) - n(P \cup C \cup S)$$
$$= 120 - 95$$
$$= 25.$$

(d) We can refer to part (b) to see that 22 students have only a USB port.

◀

Here is a second solution where we draw a Venn diagram. Remember, a diagram can help you organize the data that is given in the problem, and to help you spot what data is missing.

Solution — 2. Let $\mathcal{U}$ denote the set of all students surveyed, and

$$S = \{\, x \in \mathcal{U} \mid x \text{ have a satellite radio} \,\}$$
$$C = \{\, x \in \mathcal{U} \mid x \text{ have a CD player} \,\}$$
$$P = \{\, x \in \mathcal{U} \mid x \text{ have a USB port} \,\}.$$

Step 1. The datum that 16 students said they have a satellite radio, CD player, and a USB port means that $n(S \cap C \cap P) = 16$.

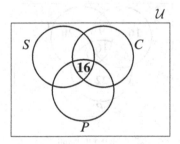

Fig. 3.26 Step 1 for Example 3.4.6

Step 2. The datum that 25 students said they have satellite radio and a CD player, which means $n(S \cap C) = 25$. This means that $25 - 16 = 9$ who have only a satellite radio and a CD player.

The datum that 21 students said they have satellite radio and a USB port, which means $n(S \cap P) = 21$. This means that $21 - 16 = 5$ who have only a satellite radio and a USB port.

The datum that 23 students said they have a CD player and a USB port, which means $n(C \cap P) = 23$. This means that $23 - 16 = 7$ who have only a CD player and a USB port.

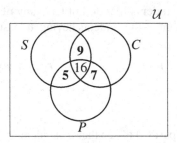

Fig. 3.27   Step 2 in Example 3.4.6

Step 3. The datum that 46 students said they have a satellite radio, which means that $n(S) = 46$. So, the number of students who only have a satellite radio is $46 - 16 - 9 - 5 = 16$.

The datum that 52 students said they have a CD player, which means that $n(C) = 52$. So the number of students who only have a CD player is $52 - 16 - 9 - 7 = 20$.

The datum that 50 students said they have a USB port, which means that $n(P) = 50$. So the number of students who have only a USB port is $50 - 16 - 7 - 5 = 22$.

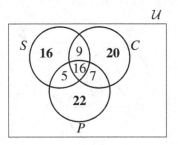

Fig. 3.28   Step 3 in Example 3.4.6

Now that we have completed the Venn diagram, we can answer the questions.

(a) Using Fig. 3.28, we sum the values that are in $S \cup C \cup P$. Therefore, the number of students who have at least a satellite radio, a CD player, or a USB port is

$$16 + 5 + 7 + 9 + 16 + 20 + 22 = 95.$$

(b) Using Fig. 3.28, we sum the values in the areas of the Venn diagram where there is no overlap. Therefore, the number of students who have only a satellite radio, or only a CD player, or only a USB port is

$$16 + 20 + 22 = 58.$$

(c) From part (a), we know the total number of students have a satellite radio, a CD player, or a USB port is 95. We also know that there were 120 students surveyed in total. Therefore, the number of students who do not have a satellite radio, CD player, or a USB port is

$$120 - 95 = 25.$$

We can add this new information to the Venn diagram as seen in Fig. 3.29.

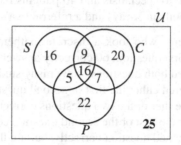

Fig. 3.29 Students who don't have a satellite radio, CD player, or USB port in Example 3.4.6

(d) From the Venn diagram in Fig. 3.29, 22 students have only a USB port.

◀

In general, we can extend the principle of inclusion and exclusion to any finite number of finite sets. If sets $A_1, A_2, A_3, \ldots, A_n$ are finite, then
$n(A_1 \cup A_2 \cup \cdots \cup A_n) =$

	the sum of the cardinalities of the individual sets,
minus	the cardinalities of the two-way intersections,
plus	the cardinalities of the three-way intersections,
minus	the cardinalities of the four-way intersections,
plus	the cardinalities of the five-way intersections,

$$\vdots \quad \vdots$$

and so on until you reach the cardinality of the n-way intersection.

Exercises

3.4.1. Give an example where $n(A \cup B) \neq n(A) + n(B)$.[8]

3.4.2. Suppose that 60 customers buy a desktop, 40 buy a laptop, and 30 buy both a desktop and a laptop. How many will buy a desktop or a laptop, but not both?

---

[8] An example that disproves a statement is called a **counterexample**. In this case, you are being asked to give a counterexample for $n(A \cup B) = n(A) + n(B)$.

3.4.3. To join a certain student club, members must be either a business major, or an accounting major or both. Of the 20 members in this club, 16 are business majors, and 10 are accounting majors. How many members in this club are both business and accounting majors.

3.4.4. In one day, 100 patients visited an emergency room at a hospital. The number of patients that visited the emergency room that resulted in neither an X-rays nor a referral to a specialist is 34. Of those going to the emergency room, 36 patients were referred to specialists and 40 patients required an X-rays. How many patients required both an X-rays and a referral to a specialist?

3.4.5. Of the 130 students who took a discrete mathematics examination, 90 correctly answered the first question, 60 correctly answered the second question, and 50 correctly answered both questions. How many students
  (a) correctly answered either the first or second question?
  (b) did not answer either of the two questions correctly?
  (c) answered either the first or the second question correctly, but not both?
  (d) answered the second question correctly, but not the first?
  (e) missed the second question?

3.4.6. On a track-and-field team,
  • 15 members run the 100 m
  • 17 members run the 200 m
  • 8 members run the 400 m
  • 7 members run both the 100 m and 200 m
  • 5 members run both the 200 m and 400 m
  • 6 members run both the 100 m and 400 m
  • 4 members run all three races.
  (a) Using the Principle of Inclusion and Exclusion, determine the number of runners.
  (b) Draw a Venn diagram to illustrate the given information. Clearly label and define all sets. Verify your answer using the Venn diagram.

3.4.7. To help plan a new radio station in the small city of Kingston, a telephone survey of local residences was conducted, and the following data was obtained:
  • 130 people listen to alternative rock
  • 180 people listen to classical
  • 275 people listen to jazz
  • 68 people listen to alternative rock and classical
  • 112 people listen to alternative rock and jazz
  • 90 people listen to classical and jazz
  • 58 people listen to all three genres.
  (a) Draw a Venn diagram to illustrate the given information. Clearly label and define all sets.

(b) How many people surveyed listen to
- (i) at least two of the genres?
- (ii) exactly one genre?
- (iii) only jazz?
- (iv) exactly two genres?

3.4.8. A survey of students who watch television produces the following data:
- 60% watch Nova
- 50% watch Horizon
- 50% watch The Nature of Things
- 30% watch Nova and Horizon
- 20% watch Horizon and The Nature of Things
- 30% watch Nova and The Nature of Things
- 10% watch all three shows.

What percentage of students view
- (a) at least one of these shows?
- (b) none of these programs?
- (c) Nova and Horizon, but not The Nature of Things?
- (d) exactly two of these shows?

3.4.9. A university newspaper survey found that of the students,
- 96% had laptops
- 65% had loans
- 51% had bicycles
- 63% had laptops & loans
- 49% had laptops & bicycles
- 31% had loans & bicycles
- 30% had all three.

What percentage of the students
- (a) had laptops only?
- (b) had student loans only?
- (c) had bicycles only?
- (d) do not have any of the three items (laptops, loans, or bicycles)?

3.4.10. For two mutually exclusive sets A and B, state the formulas for
- (a) $n(A \cup B)$
- (b) $n(A \cap B)$.

3.4.11. After a rough lacrosse game, it was reported in a newspaper that of the 10 players on the team, 8 hurt a hip, 6 hurt a hand, 5 hurt a knee, 3 hurt both a hip and a hand, 2 hurt both a hip and a knee, 2 hurt both a hand and a knee, and none hurt all three. The reporter was promptly fired. Why?

3.4.12. How many from the set $\{1, 2, 3, \ldots, 100\}$ are not divisible by 2, 3, or 5?

3.4.13. Express the principle of inclusion and exclusion for four finite sets: A, B, C and D.

3.5 Set Operations Revisited

In Table 3.3, notice how the set operations have a corresponding logical connective.

Table 3.3 Comparison of set operations and logical connectives

Set Operation	Symbol	Logical Connective	Symbol
Complement	A'	Negation	$\neg p$
Union	$A \cup B$	Disjunction	$p \vee q$
Intersection	$A \cap B$	Conjunction	$p \wedge q$
Subset	$A \subseteq B$	Implication	$p \rightarrow q$
Equality	$A = B$	Biconditional	$p \leftrightarrow q$

The set operations also follow the same structure as the rules of logical inference from Sec. 2.2. If A, B, and C are arbitrary subsets of the universal set $\mathcal{U}$, then

$$A \cap A = A \qquad \textbf{Idempotence} \qquad (3.9)$$

$$A \cup A = A \qquad\qquad (3.10)$$

$$(A \cap B) \cap C = A \cap (B \cap C) \qquad \textbf{Associativity} \qquad (3.11)$$

$$(A \cup B) \cup C = A \cup (B \cup C) \qquad\qquad (3.12)$$

$$A \cap B = B \cap A \qquad \textbf{Commutativity} \qquad (3.13)$$

$$A \cup B = B \cup A \qquad\qquad (3.14)$$

$$A \cap (B \cup C) = (A \cap B) \cup (A \cap C) \qquad \textbf{Distributivity} \qquad (3.15)$$

$$A \cup (B \cap C) = (A \cup B) \cap (A \cup C) \qquad\qquad (3.16)$$

$$(A \cup B)' = A' \cap B' \qquad \textbf{De Morgan's} \qquad (3.17)$$

$$(A \cap B)' = A' \cup B' \qquad\qquad (3.18)$$

$$A \cap \mathcal{U} = A \qquad \textbf{Identity} \qquad (3.19)$$

$$A \cup \varnothing = A \qquad\qquad (3.20)$$

$$A \cup \mathcal{U} = \mathcal{U} \qquad \textbf{Domination} \qquad (3.21)$$

$$A \cap \varnothing = \varnothing \qquad\qquad (3.22)$$

$$A \cup A' = \mathcal{U} \qquad \textbf{Complement} \qquad (3.23)$$

$$A \cap A' = \varnothing \qquad\qquad (3.24)$$

$$\mathcal{U}' = \varnothing \qquad\qquad (3.25)$$

$$\varnothing' = \mathcal{U} \qquad\qquad (3.26)$$

$$(A')' = A \qquad \textbf{Double Negation} \qquad (3.27)$$

Example 3.5.1 Show that the distributive law for sets

$$A \cap (B \cup C) = (A \cap B) \cup (A \cap C)$$

is true.

Proof Let A, B, and C be sets. Let x be an arbitrary element in $A \cap (B \cup C)$. Then,

$$
\begin{aligned}
x \in A \cap (B \cup C) &\Leftrightarrow (x \in A) \wedge (x \in B \cup C) && \text{By (3.2)} \\
&\Leftrightarrow (x \in A) \wedge (x \in B \vee x \in C) && \text{By (3.1)} \\
&\Leftrightarrow \big((x \in A) \wedge (x \in B)\big) \vee \big((x \in A) \wedge (x \in C)\big) && \text{By (2.8)} \\
&\Leftrightarrow (x \in A \cap B) \vee (x \in A \cap C) && \text{By (3.2)} \\
&\Leftrightarrow x \in \big((A \cap B) \cup (A \cap C)\big) && \text{By (3.1)}
\end{aligned}
$$

◄

Example 3.5.2 Show that De Morgan's Law for sets

$$
(A \cup B)' = A' \cap B'
$$

is true.

Proof First, we start with the observation that

$$
x \notin A \Leftrightarrow \neg(x \in A). \tag{3.28}
$$

Let x be an arbitrary element in $(A \cup B)'$. Then,

$$
\begin{aligned}
x \in (A \cup B)' &\Leftrightarrow x \notin (A \cup B) && \text{By (3.3)} \\
&\Leftrightarrow \neg\big(x \in (A \cup B)\big) && \text{By (3.28)} \\
&\Leftrightarrow \neg(x \in A \vee x \in B) && \text{By (3.1)} \\
&\Leftrightarrow \neg(x \in A) \wedge \neg(x \in B) && \text{By (2.10)} \\
&\Leftrightarrow (x \notin A) \wedge (x \notin B) && \text{By (3.28)} \\
&\Leftrightarrow (x \in A') \wedge (x \in B') && \text{By (3.3)} \\
&\Leftrightarrow x \in (A' \cap B') && \text{By (3.2)}
\end{aligned}
$$

Since every element of $(A \cup B)'$ is an element of $(A' \cap B')$, and vice versa, the two sets are equal. The other De Morgan's law is left as Exercise 3.5.1 for you. ◄

There are two ways you can prove a statement involving sets.
  (a) If the statement concerns one of the laws for sets, then you proceed by selecting an arbitrary element in the set on the left side, and show that it is also in the set on the right side. Since it is an arbitrary element, it holds for all elements in the sets.
  (b) If the statement does not concern one of the laws for sets, then you can use the laws to show that the set on the left side is equivalent to the set on the right side.

Example 3.5.3 Show that

$$
\big(A' \cap (B')'\big)' = A \cup B'.
$$

Solution

$$\left(A' \cap (B')'\right)' = \left(A' \cap B\right)' \qquad \text{By (3.27)}$$
$$= \left(A'\right)' \cup B' \qquad \text{By (3.18)}$$
$$= A \cup B' \qquad \text{By (3.27)}$$

◄

Example 3.5.4 Prove that

$$A \cup (B \cap C)' = (A' \cap B)' \cup C'.$$

Proof

$$A \cup (B \cap C)' = A \cup \left(B' \cup C'\right) \qquad \text{By (3.18)}$$
$$= \left(A \cup B'\right) \cup C' \qquad \text{By (3.12)}$$
$$= \left(A' \cap (B')'\right)' \cup C' \qquad \text{By (3.18)}$$
$$= (A' \cap B)' \cup C' \qquad \text{By (3.27)}$$

◄

Example 3.5.5 Derive the absorption law

$$A \cap (A \cup B) = A \qquad (3.29)$$

using the set operations and set complementation laws.

Solution

$$A \cap (A \cup B) = (A \cup \varnothing) \cap (A \cup B) \qquad \text{By (3.20)}$$
$$= A \cup (\varnothing \cap B) \qquad \text{By (3.16)}$$
$$= A \cup \varnothing \qquad \text{By (3.22)}$$
$$= A \qquad \text{By (3.20)}$$

◄

Exercises

3.5.1. Prove De Morgan's Law: $(A \cap B)' = A' \cup B'$.

3.5.2. Let A and B be sets. Show that $A \smallsetminus B = A \cap B'$.

3.5.3. Using the rules of set algebra, simplify the following:
   (a) $A \cap (A' \cup B)$ (c) $(A \cup B) \cap (A \cup B')$
   (b) $(A' \cup B') \cap (A \cap B)$ (d) $(A' \cap B') \cup (A \cup B)$.

3.5.4. Show that $(A \cap B)' \cup (A' \cap B' \cap C)' \cup A = \mathcal{U}$ is true.

3.6 Using Venn Diagrams to Check Categorical Syllogisms

When we use quantifiers such as *some* or *all* in syllogisms, these are called **categorical syllogisms**.

There are four main categorical logic forms. For each of these forms, we can represent them using a Venn diagram. (In fact, that is what John Venn had developed his diagrams to represent in the first place!)

(a) **All** *A* **is** *B*. This is equivalent to $A \subset B$; see Fig. 3.30.
For example, all *women* are *human*.

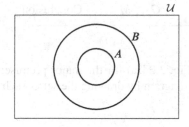

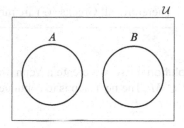

Fig. 3.30 $A \subset B$ Fig. 3.31 $A \cap B = \varnothing$

(b) **No** *A* **is** *B*. This is equivalent to $A \cap B = \varnothing$; see Fig. 3.31.
For example, no *reptiles* have *fur*.

(c) **Some** *A* **is** *B*. This is equivalent to $A \cap B \neq \varnothing$; see Fig. 3.32.
For example, some *websites* are *useful*.

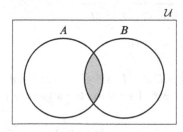

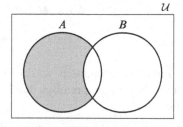

Fig. 3.32 $A \cap B \neq \varnothing$: Some *A* is *B*   Fig. 3.33 $A \smallsetminus B$: Some *A* is not *B*

(d) **Some** *A* **is not** *B*. This is equivalent to $A \smallsetminus B$; see Fig. 3.33.
For example, some *websites* are not *useful*.

Remember that a valid syllogism's conclusion asserts no more than what is implicit in the premises. If a syllogism is invalid, then the conclusion has been inferred incorrectly from the premises. In order to identify if a syllogism is valid, we draw a Venn diagram based on the major and minor premises. From the Venn diagram, we can visually see if the conclusion is true: if the conclusion is true in the diagram, then the syllogism is valid, otherwise it is invalid.

Example 3.6.1 Determine if the following argument, given in Table 3.4, is valid or invalid using Venn diagrams.

Table 3.4   Modus Barbara (aaa-1)

All people (P) are mortal (M).	$P \subset M$	Major Premise
All Greeks (G) are people (P).	$G \subset P$	Minor Premise
Therefore, all Greeks (G) are mortal (M).	$\therefore G \subset M$	Conclusion

Solution First, we create a Venn diagram in Fig. 3.34(a) for the major premise: $P \subset M$. The next step is to continue the Venn diagram by drawing the set G such that $G \subset P$.

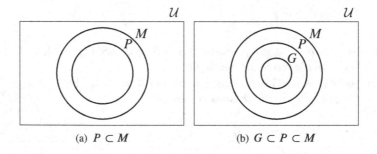

(a) $P \subset M$ (b) $G \subset P \subset M$

Fig. 3.34   Venn diagrams for Example 3.6.1

From the second step, we can visually see that $G \subset M$ in Fig. 3.34(b). So the inference from the premises that $G \subset M$ is correct. Therefore, the argument is valid.   ◄

Example 3.6.2 Determine if the following argument, given in Table 3.5, is valid or invalid using Venn diagrams.

Solution *Draw a diagram.* The first step is to draw the Venn diagram, shown in Fig. 3.35(a) for the first premise: $S \subset R$. The second step is to continue the Venn diagram, shown in Fig. 3.35(b), for the second premise: $R \cap F = \emptyset$. This means that R and F have no overlapping areas; sets R and F are mutually exclusive.

From the Venn diagram, we can visually see that $F \cap S = \emptyset$. So the inference from the premises that $F \cap S = \emptyset$ is correct. Thus, the argument is valid.   ◄

Table 3.5   Modus Calemes (aee-4)

All snakes (S) are reptiles (R).	$S \subset R$	Major Premise
No reptiles (R) have fur (F).	$R \cap F = \varnothing$	Minor Premise
Therefore, no furry animal (F) is a snake (S). $\therefore F \cap S = \varnothing$		Conclusion

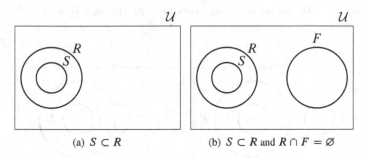

(a) $S \subset R$ (b) $S \subset R$ and $R \cap F = \varnothing$

Fig. 3.35   Venn diagrams for Example 3.6.2

Example 3.6.3 Determine if the following argument, given in Table 3.6, is valid or invalid using Venn diagrams.

Table 3.6   A categorical syllogism for Example 3.6.3

All murderers (M) are psychologically disturbed (P).	$M \subset P$	Major Premise
No saints (S) are murderers (M).	$S \cap M = \varnothing$	Minor Premise
Therefore, no saints are psychologically disturbed. $\therefore S \cap P = \varnothing$		Conclusion

Solution *Draw a diagram.* We begin with the first premise's Venn diagram in Fig. 3.36.

Now, we have an issue to consider when continuing the Venn diagram for the premise $S \subset M = \varnothing$. If we draw the set S, we are unsure if S will intersect a part of set P or not. In the first case, if S does not intersect P, then the conclusion is true: there are no saints that are psychologically disturbed; see the Venn diagram in Fig. 3.37(a). However if S does intersect P, then the conclusion is false and yet in Fig. 3.37(b), the Venn diagram correctly displays the major and minor premises. Since we found a (correct) Venn diagram that does not support the conclusion, the syllogism is invalid. ◄

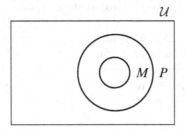

Fig. 3.36   The first step in creating a Venn diagram for Example 3.6.3: $M \subset P$.

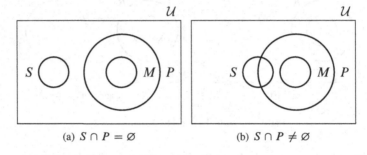

(a) $S \cap P = \varnothing$ (b) $S \cap P \neq \varnothing$

Fig. 3.37   Two possibilities for the final Venn diagram for Example 3.6.3.

Exercises

3.6.1. Is the following argument valid or invalid?

All unselfish people are generous.	Premise
No misers are generous.	Premise
Therefore, no misers are unselfish.	Conclusion

3.6.2. Is the following argument valid or invalid?

No reptiles have fur.	Premise
All snakes are reptiles.	Premise
Therefore, no snakes have fur.	Conclusion

3.6.3. Is the following argument valid or invalid?

No uncivilized nation is warlike. Premise
All unwarlike nations are uncivilized. Premise

Therefore, no nation is uncivilized Conclusion

3.6.4. Is the following argument valid or invalid?

All informative things are useful. Premise
Some websites are not useful. Premise

Therefore, some websites are not informative. Conclusion

3.6.5. Is the following argument valid or invalid?

All soldiers are brave. Premise
Some Englishmen are brave. Premise

Therefore, some englishmen are soldiers. Conclusion

3.6.6. Given the following premises, state a valid conclusion.
(a) No children are patient. No impatient person can sit still.
(b) All pigs are fat. No skeletons are fat.
(c) No monkeys are soldiers. All monkeys are mischievous.

3.7 Review

Summary

Table 3.7   Set Theory Symbols

Symbol	Read as	Definition
$\in$	is an element of	
$\notin$	is not an element of	
$\ldots$	and so on	
$\varnothing$	empty set or null set	$\varnothing = \{\ \}$ The set that contains no elements.
$\mathcal{U}$	universal set	The set from which all elements of interest are taken.
$n(A)$	cardinality of A	The number of elements in A

Table 3.8   Set Relations

Relation	Read as	Definition
$A \subseteq B$	A is contained or equal to B	Any element of A is also an element of B
$A \subset B$	A is strictly contained in B	All the elements of A are also elements of B, but there exists an element of B that is not in A.
$A = B$	A is equal to B	All the elements of A are contained in B, and all the elements in B are contained in A.

Table 3.9   Set Operations

Operation	Read as	Definition
$A \cup B$	A union B	The elements in A, *or* in B, or in both
$A \cap B$	A intersection B	The elements in both A *and* B
A'	A complement	All the elements in the universal set $\mathcal{U}$ which are not in A
$A \smallsetminus B$	A minus B	The elements in A which are not B
$\mathcal{P}(A)$	Power set of A	The collection of all subsets of A

Exercises

3.7.1. Determine if each of the following sets are finite or infinite.
   (a) The set A of all the provinces in Canada.
   (b) The set B of the months in a year.
   (c) The set C of the natural numbers less than one.
   (d) The set D of the odd integers.
   (e) The set E of the positive divisors of 14.
   (f) The set F of dogs living in Germany.

3.7.2. Is the following a paradox? Explain why.

   The Barber of Seville is a man in Seville who shaves all those, and only those, men in Seville who do not shave themselves. Who shaves the barber?

3.7.3. True or false: If two sets A and B are mutually exclusive, then $n(A) + n(B) = n(A \cup B)$.

3.7.4. For sets C and D, state the formula for $n(C \cup D)$ if
    (a) $C \cap D \neq \varnothing$ (b) $C \cap D = \varnothing$

3.7.5. Let $A = \{a, c, e, f\}$, $B = \{c, d, g\}$, $C = \{a, b, c\}$, and let the universal set $\mathcal{U} = \{a, b, c, d, e, f, g, h\}$.
    (a) What is $A \cap B$?   (d) What is $n(B)$?
    (b) What is $A \cup B$?   (e) What is A'?
    (c) What is $A \smallsetminus C$?   (f) What is $\mathcal{P}(B)$?

3.7.6. Table 3.10 gives the number of homicides, by city, in Canada for 2012 and 2013. List the elements of each set.

Table 3.10   Homicides, by city, Canada, 2012 and 2013

Cities	2012	2013
Toronto	81	79
Montréal	47	43
Winnipeg	33	26
Vancouver	37	42
Hamilton	6	15
Ottawa	7	9
Kingston	0	1

    (a) The set of cities where the number of homicides increased from 2012 to 2013.
    (b) The set of cities that had more than 30 homicides in 2012 or 2013.
    (c) The set of cities that had less than 9 homicides in 2012 and 2013.
    (d) The set of cities that had a decrease of 3 homicides or more from 2012 to 2013.

3.7.7. In a ward at a teaching hospital, there are 25 doctors, 35 females, and 19 supervisors. Amongst the females are 10 doctors and 11 supervisors; three of the supervising females are themselves doctors. Of the doctors, 15 are not supervisors. The hospital is considering an affirmative action program and needs to answer the following questions:
    (a) How many male supervisors are there?
    (b) What percentage of the supervisors are female?
    (c) How many people are employed as doctors or supervisors?

3.7.8. In a university, there are 250 freshmen who study calculus, 100 who study discrete mathematics, and 40 who study both subjects. What is the total number of students taking calculus or discrete mathematics?

3.7.9. Alice circles in red ink all of the even natural numbers from 1 to 100. Bob circles in blue ink all of the multiples of 3. Eve circles in green ink all of the multiples of 5. How many of the numbers are circled exactly
    (a) once?   (b) twice?

3.8 Bibliographic Remarks

Baron (1969) gives a historical overview of how Venn diagrams are used in logic. In particular, it appears that Aristotle had explored the possibility of representing logic by means of geometric figures and came up with representing the four standard categorical propositions as in Fig. 3.30, Fig. 3.31, and Fig. 3.32.

4

Infinity

4.1 Introduction

Douglas Adams (1952–2001) wrote a wholly remarkable book, *The Hitchhiker's Guide to the Galaxy*, which has been read and reread many times over many years by millions of people. If the *Hitchhiker's Guide to the Galaxy* had an entry for the term **infinity**, like it does for *space*, it would begin something like this:

> "Infinity," it would say, "is big. Really big. You just won't believe how vastly hugely mindbogglingly big it is. I mean you may think that the number of particles in the universe is a lot, but that's just peanuts to infinity."

After a while the style would settle down a bit, and would begin to tell you things that are interesting about the concept of infinity. Things you really need to know, and things that you should know about it. To be fair, when confronted by the sheer enormity of infinity, better minds than myself have faltered. The simple truth is that infinity will not fit into the human imagination—but we can try.

What is the largest possible number that you can think of? One million? A trillion? As of 15 June 2015, the United States debt[1] held by the public was $13 trillion, or to be exact $13 075 184 730 750.02. This is a huge number! The life expectancy for an American citizen[2] born in 2014 is 79.56 years or 2.51 billion seconds. To put the debt into perspective, a person born today would have to spend $5209.24 every second of their life; this person could buy a large screen monitor every second for their entire life. And yet, there are still numbers larger than all of the previous examples I have mentioned. Edward Kasner (1878–1955) asked his nine-year-old nephew, Milton Sirotta, to think of a name for a very big number, a

---

[1] Source: United States Department of the Treasury Bureau of the Fiscal Service http://www.treasurydirect.gov/NP/debt/current

[2] Source: https://www.cia.gov/library/publications/the-world-factbook/rankorder/2102rank.html

number with a 1 and a hundred zeros after it:

$$10\,000\,000\,000\,000\,000\,000\,000\,000\,000\,000\,000\,000\,000\,000\,000$$
$$000\,000\,000\,000\,000\,000\,000\,000\,000\,000\,000\,000\,000\,000$$
$$000\,000\,000\,000\,000$$

or 10^{100}. For comparison, the number of atoms in the (observable) universe is estimated to be 10^{80}. Milton thought for a moment, and that is when he suggested the name **googol**. Still, there could be a number that is bigger than a googol, and Milton suggested the name **googolplex**. A googolplex is the number 1 with a googol of zeros written after it,[3] or $10^{googol} = 10^{10^{100}}$ (Kasner and Newman, 1949, p. 23). However, even with these extremely large numbers, these numbers are still finite. "Above everything, we must realize that 'very big' and 'infinite' are entirely different" (Kasner and Newman, 1949, p. 34).

We will use the following definition for infinity.

> **Definition 4.1.1 — Infinity.** Infinity is an unbounded quantity, or a quantity that is not finite. The symbol (also called a lemniscate) to represent infinity in mathematics is ∞.

4.1.1 Euclid and Prime Numbers

Another famous author is the Greek mathematician Euclid of Alexandria. His book, *The Elements*, contains all of the geometric results that were known to the Ancient Greeks. Euclid presents proofs of these geometric results, and presents them in a logical, and systematic way—building upon each geometric result one after another. Although *The Elements* is known mainly for the geometric results, he indirectly included some results and properties about numbers. In particular, there is a very interesting property about prime numbers. A **prime number** is a natural number greater than one which has no divisors other than one and itself, and what Euclid demonstrates is that there are infinitely many of them. He derives this result using a technique called proof by contradiction. A **proof by contradiction** is a form of indirect reasoning: the truth of the proposition is shown by assuming that the proposition is false would imply a contradiction. This is the method we will use to show that there are infinitely many prime numbers.

> **Theorem 4.1.2 — Euclid's Theorem.** There are infinitely many prime numbers.

Proof Suppose in order to derive a contradiction, we will assume that there are not infinitely many prime numbers. This means that the set of prime numbers P is finite. Since we assumed that the set of prime numbers is finite, there is an element $p \in P$ that is the largest prime number; that is, $P = \{2, 3, 5, 7, 11, 13, 17, 19, 23, \ldots, p\}$. Let $q = 2 \times 3 \times \cdots \times p$. The number q is divisible by every prime

---

[3]There is simply not enough paper in the universe to write this number explicitly.

number from 2 to p, as it is the product of all of them. This means that $q + 1$ is not divisible by every integer from 2 to p, since it would give a remainder of 1 when divided by each number from 2 to p. Hence, $q + 1$ is either a prime number or there exists a prime number larger than p which divides $q + 1$. However, this contradicts our original assumption that the set of prime numbers P is finite and that p is the largest prime number. Therefore, the set of prime numbers P is not finite, and there is no element $p \in P$, which is the largest prime number; the set of prime numbers P is infinite. ◄

Exercises

4.1.1 (†). Give a proof by contradiction that $\sqrt{2}$ is irrational. (An **irrational number** is a real number that cannot be expressed as a fraction $\frac{a}{b}$, in which $a, b \in \mathbb{Z}$, and $b \neq 0$.) (*Hint*: Assume $\sqrt{2}$ can be written as $\frac{a}{b}$, and b does not divide a so that you can find a contradiction.)

4.1.2 (†). Prove that $\sqrt{3}$ is irrational using proof by contradiction.

4.1.3 (†). A **perfect square** is a natural number that is the square of a natural number. For example, 25 is a perfect square, since $25 = 5^2$. If $n \in \mathbb{N}$ is not a perfect square, prove that $\sqrt{n}$ is irrational using proof by contradiction. (*Hint*: Generalize from the previous exercise.)

4.2 Galileo and Cantor Counting Numbers Leads to Heresy (or The Integers and the Rationals are Countable)

> I would say here something that was heard from an ecclesiastic of the most eminent degree: "That the intention of the Holy Ghost is to teach us how one goes to heaven, not how heaven goes."
>
> From Galileo's 1615 *Letter to the Grand Duchess Christina*
> (Drake, 1957, p. 186)

In 1632, the Italian Renaissance scientist Galileo Galilei (1564–1642) published his famous work in Italian, *Dialogo sopra i due massimi sistemi del mondo* (English: Dialogue Concerning the Two Chief World Systems). Although contemporary books on science were published in Latin, Galileo wrote the book in Italian so that the layperson could read it, and in doing so allowed everyone to learn about the new heliocentric system that he was promoting: the solar system in which the earth revolves around the sun. Because of this wide dissemination of knowledge to the public, this raised the suspicion of the Roman Catholic Church, and led Galileo to be tried by the Inquisition. Galileo believed that he was not countering Scripture, since the Scripture was meant for how one goes to heaven, not how the heavens go. The Inquisition however found him to be "vehemently suspect of heresy," sentenced him to spend the remaining days of his life under house arrest, his works (and future works) were placed on the *Index Librorum Prohibitorum* (English: List of Prohibited Books), and forced him to remain silent on his heliocentric views.

> Galileo was no idiot. Only an idiot could believe that science requires martyrdom—that may be necessary in religion, but in time a scientific result will establish itself.

> —David Hilbert (1862–1943)

It was while he was living out his last days in his Tuscan villa, he wrote *Discorsi e Dimostrazioni Matematiche Intorno a Due Nuove Scienze* (English: Discourses and Mathematical Demonstrations Relating to Two New Sciences), which was later smuggled out of Italy to the Netherlands for printing. Although this work summarizes Galileo's work that he had done over the past 40 years or so, it contained a novel idea about infinity and infinite sets.

> ... there are as many square numbers, as there are numbers; for there are as many squares as there are roots, and as many roots as numbers ... (Galilei, 1638, p. 48).

> I see no other way, but by saying that all numbers are infinite; squares are infinite, their roots infinite, and that the number of squares is not less than the number of numbers... (Galilei, 1638, p. 49).

In more modern terms, he says the number of elements in the set of squares $S = \{1^2, 2^2, 3^2, \ldots\}$ is the same as the number of elements in the set of natural numbers $\mathbb{N} = \{1, 2, 3, \ldots\}$. To see why Galileo believed this, if we pair the numbers of the two sets so that each number in S is paired with another number in $\mathbb{N}$, and no numbers in either set are left alone or are paired to multiple numbers, then the two sets must have the same number of elements.

$$\mathbb{N} = \{\ 1,\ \ 2,\ \ 3,\ \ 4,\ \ 5,\ \ 6,\ \ 7,\ \ 8,\ \ 9,\ \ldots,\ n,\ \ldots\ \}$$
$$\updownarrow \ \ \updownarrow \ \ \updownarrow \ \ \updownarrow \ \ \updownarrow \ \ \updownarrow \ \ \updownarrow \ \ \updownarrow \ \ \updownarrow \qquad \updownarrow$$
$$S = \{\ 1^2, 2^2, 3^2, 4^2, 5^2, 6^2, 7^2, 8^2, 9^2, \ldots, n^2, \ldots\ \}$$

Galileo reasoned that since $\mathbb{N}$ is infinite, therefore S must be infinite too. This seemed rather odd to Galileo, and he did not want to accept this. Instead, he rejects the idea that $\mathbb{N}$ and S have the same cardinality—the idea of *less than*, *greater than*, or *equal to* does not apply to infinite sets—and quickly drops the subject.

In fact, his initial thinking was correct: the even numbers are in fact as many as the natural numbers, and to see this all you have to do is to pair one number from a set with a number with another set. This idea of pairing in mathematics is called establishing a 1-to-1 correspondence.

Definition 4.2.1 — 1-to-1 correspondence. There is a 1-to-1 correspondence between two sets A and B, if
  (a) every element in set A corresponds exactly to one element in set B, as shown in Fig. 4.1(a), and
  (b) every element in set B corresponds exactly to one element in set A, as shown in Fig. 4.1(b).

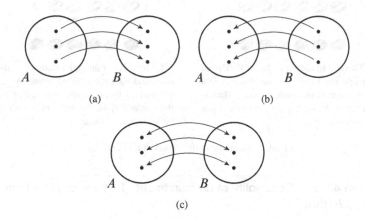

(a)   (b)

(c)

Fig. 4.1   Illustrating the concept of 1-to-1 correspondence

By pairing, or establishing a 1-to-1 correspondence, each element from one set with an element from another set, then they must have the same number of elements in each set; see Fig. 4.1(c).

Example 4.2.2 — Finite coffee beans. Suppose on a table there are dark and light roast coffee beans. Then I asked you, without counting them explicitly, is there an equal number of dark roast and light roast coffee beans? If you were able to match each dark roast bean with a light roast bean, like in Fig. 4.2(a), then you would conclude that there is an equal number of dark and light roast coffee beans; the cardinality of the dark and light roast coffee beans is the same. If you were unable to match each dark roast bean with a light roast bean, like in Fig. 4.2(b), then you would conclude that there is not an equal number of dark and light roast coffee beans; the cardinality of the dark and light roast beans is not the same. ◄

The interesting property about infinite sets that Galileo found is that the squared numbers, which is a proper subset of the natural numbers $\mathbb{N}$, can be paired with every natural number to form a 1-to-1 correspondence. Since there is a 1-to-1 correspondence, there must be the same number of squared numbers as there are natural numbers; hence, they must have the same cardinality. We will formally state this as a theorem.

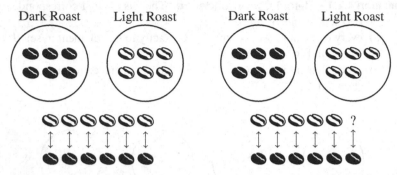

(a) There is an equal number of dark roast and light roast coffee beans. We can find a 1-to-1 correspondence; therefore, the cardinality of the dark and light roast coffee beans is the same.

(b) There is not an equal number of dark roast and light roast coffee beans. We cannot find a 1-to-1 correspondence; therefore, the cardinality of the dark and light roast beans is not the same.

Fig. 4.2   Counting coffee beans in Example 4.2.2

Theorem 4.2.3 — Cardinality of an infinite set. If A and B are infinite sets, and $A \subset B$, then

$$n(A) = n(B) \qquad (4.1)$$

(Dedekind, 1888).

Example 4.2.4 Show that there is as many even natural numbers as there are natural numbers.

Solution Since there exists a 1-to-1 correspondence between the even natural numbers E and the natural numbers $\mathbb{N}$, then the sets E and $\mathbb{N}$ must have the same cardinality.

$$\mathbb{N} = \{ 1, 2, 3, 4, 5, 6, 7, 8, 9, \ldots, n, \ldots \}$$
$$\updownarrow \updownarrow \updownarrow \updownarrow \updownarrow \updownarrow \updownarrow \updownarrow \updownarrow \qquad \updownarrow$$
$$E = \{ 2, 4, 6, 8, 10, 12, 14, 16, 18, \ldots, 2n, \ldots \}$$

Therefore, there are as many even natural numbers as there are natural numbers. ◄

But, this seems rather odd, since one would think, "Does not the set E have exactly half the number of elements of $\mathbb{N}$?" Intuitively, we can think of it in the following way as Galileo would have: if we take all of the even natural numbers, and we divide them all by 2, then we would get the set of natural numbers; hence, their cardinalities must be the same. This shows us that infinity is a strange concept that does not always follow our intuition that we have developed with finite sets. In the strange world of infinity, *a part may be equal to the whole!*

We can contrast Theorem 4.2.3 with a theorem about the cardinality of a finite set. A portion of a finite set is never equal to the whole of a finite set.

Theorem 4.2.5 — Cardinality of a finite set. If A and B are finite sets, and $A \subset B$, then

$$n(A) < n(B). \qquad (4.2)$$

Example 4.2.6 — Finite set. Consider the set $N = \{1, 2, 3, \ldots, 10\}$ and the set $E = \{2, 4, 6, 8, 10\}$. The set E is a proper subset of N; $E \subset N$. Therefore, $n(E) < n(N)$. To check, $n(E) = 5 < 10 = n(N)$.

After Richard Dedekind (1831–1916) gave Theorem 4.2.3, another German mathematician by the name of Georg Cantor (1845–1918) began to investigate the idea of infinite sets. He invented a new kind of number to serve his purposes, which he called a **transfinite number**. For example, he denoted the cardinality of the set of natural numbers $\mathbb{N}$ as

$$n(\mathbb{N}) = \aleph_0.$$

We read $\aleph_0$ as "aleph-null." The letter $\aleph$ is the first letter of the Hebrew alphabet.

> **Definition 4.2.7 — Countable.** If a set A has the same cardinality (the same number of elements) as the set of natural numbers $\mathbb{N}$, then we say A is **countable**; that is,
>
> $$n(A) = n(\mathbb{N}) = \aleph_0.$$

We use the term countable since the natural numbers are informally called the *counting numbers*—we can count them each one by one.

Recall that the set of integers $\mathbb{Z}$ is

$$\mathbb{Z} = \{\ldots, -4, -3, -2, -1, 0, 1, 2, 3, 4, \ldots\} \qquad (4.3)$$

Theorem 4.2.8 The set of integers $\mathbb{Z}$ is countable. In other words,

$$n(\mathbb{Z}) = \aleph_0.$$

Proof To show that the integers are countable, we need to find a function that will take a natural number $x \in \mathbb{N}$ and will map it in a one-to-one correspondence with an integer number $y \in \mathbb{Z}$.

$$\mathbb{N} = \{\ 1,\ \ 2,\ \ 3,\ \ 4,\ \ 5,\ \ 6,\ \ \ 7,\ \ \ 8,\ 9, \ldots\}$$
$$\updownarrow\ \ \updownarrow\ \ \updownarrow\ \ \updownarrow\ \ \updownarrow\ \ \updownarrow\ \ \ \updownarrow\ \ \ \updownarrow\ \ \updownarrow$$
$$\mathbb{Z} = \{\ 0,\ -1,\ 1,\ -2,\ 2,\ -3,\ -3,\ -4,\ 4, \ldots\}$$

So, a function that maps a natural number $x \in \mathbb{N}$ to an integer number $y \in \mathbb{Z}$ is

$$y = \begin{cases} \frac{x-1}{2} & \text{if } x \text{ is odd;} \\ -\frac{x}{2} & \text{if } x \text{ is even.} \end{cases} \qquad (4.4)$$

For a function that maps an integer number $y \in \mathbb{Z}$ to a natural number $x \in \mathbb{N}$, we have

$$y = \frac{x-1}{2}$$
$$2y = x - 1$$
$$x = 2y + 1 \qquad\qquad \text{if } y \text{ is positive}$$

and

$$y = -\frac{x}{2}$$
$$2y = -x$$
$$x = -2y \qquad\qquad \text{if } y \text{ is negative}$$

Therefore, a function that maps $y \in \mathbb{Z}$ to $x \in \mathbb{N}$ is

$$x = \begin{cases} -2y & \text{if } y \text{ is negative;} \\ 2y + 1 & \text{if } y \text{ is positive;} \\ 1 & \text{if } y = 0. \end{cases} \qquad (4.5)$$

◀

The set of rational numbers $\mathbb{Q}$ contains all numbers that can be expressed as a fraction of two integers.[4] In other words, the set of rational numbers $\mathbb{Q}$ is

$$\mathbb{Q} = \left\{ \frac{a}{b} \mid a, b \in \mathbb{Z}, b \neq 0 \right\}. \qquad (4.6)$$

Theorem 4.2.9 The set of rational numbers $\mathbb{Q}$ is countable. In other words,

$$n(\mathbb{Q}) = \aleph_0.$$

Proof To show that the set of rationals $\mathbb{Q}$ is countable, we have to show that there is a 1-to-1 correspondence between $\mathbb{Q}$ and $\mathbb{N}$.

Consider Fig. 4.3 that shows an array of the positive rationals denoted by $\mathbb{Q}^+$. To start, we will look at the positive rationals, but this can be extended to include the negative rationals too. The top row lists all rationals with numerator 1, the second row lists all the rationals with numerator 2, and so forth. This array is continued infinitely and will contain every positive rational number. Start counting at $\frac{1}{1}$ in Fig. 4.3 and continue to count by following the direction of the arrows. Every time we come to a number that we have counted before, such as $\frac{2}{2}, \frac{4}{2}, \ldots$, we do not count it again. Therefore, we can set up the following correspondence

---

[4]The notation $\mathbb{Q}$ was first used by the Italian mathematician Giuseppe Peano (1858–1932), since *quoziente* is Italian for quotient.

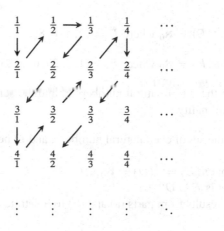

Fig. 4.3 Counting the positive rationals

$$\mathbb{N} = \{\, 1,\ 2,\ 3,\ 4,\ 5,\ 6,\ 7,\ 8,\ 9,\ \ldots \}$$
$$\updownarrow\ \updownarrow\ \updownarrow\ \updownarrow\ \updownarrow\ \updownarrow\ \updownarrow\ \updownarrow\ \updownarrow$$
$$\mathbb{Q}^{+} = \{\, \tfrac{1}{1},\ \tfrac{2}{1},\ \tfrac{1}{2},\ \tfrac{1}{3},\ \tfrac{3}{1},\ \tfrac{4}{1},\ \tfrac{3}{2},\ \tfrac{2}{3},\ \tfrac{1}{4},\ \ldots \}$$

Hence, every positive rational number corresponds to exactly one natural number, and every natural number corresponds to exactly one positive natural number; that is,

$$n(\mathbb{Q}^{+}) = n(\mathbb{N}) = \aleph_0. \tag{4.7}$$

Thus, the positive rationals are countable.

We can repeat the same process with the negative rationals. The negative rationals $\mathbb{Q}^{-}$ are countable, since there is a 1-to-1 correspondence with the negative rationals and the naturals;

$$\mathbb{N} = \{\, 1,\quad 2,\quad 3,\quad 4,\quad 5,\quad 6,\quad 7,\quad 8,\quad 9,\quad \ldots \}$$
$$\updownarrow\quad \updownarrow\quad \updownarrow\quad \updownarrow\quad \updownarrow\quad \updownarrow\quad \updownarrow\quad \updownarrow\quad \updownarrow$$
$$\mathbb{Q}^{-} = \{\, -\tfrac{1}{1},\ -\tfrac{2}{1},\ -\tfrac{1}{2},\ -\tfrac{1}{3},\ -\tfrac{3}{1},\ -\tfrac{4}{1},\ -\tfrac{3}{2},\ -\tfrac{2}{3},\ -\tfrac{1}{4},\ \ldots \}.$$

Thus, the negative rationals are countable;

$$n(\mathbb{Q}^{-}) = n(\mathbb{N}) = \aleph_0. \tag{4.8}$$

Then we can alternate between the negative rationals and the positive rationals like we did with the integers (in the proof of Theorem 4.2.8):

$$\mathbb{N} = \{\, 1,\quad 2,\quad 3,\quad 4,\quad 5,\quad 6,\quad 7,\quad 8,\quad 9,\quad \ldots \}$$
$$\updownarrow\quad \updownarrow\quad \updownarrow\quad \updownarrow\quad \updownarrow\quad \updownarrow\quad \updownarrow\quad \updownarrow\quad \updownarrow$$
$$\mathbb{Q} = \{\, 0,\ -\tfrac{1}{1},\ \tfrac{1}{1},\ -\tfrac{2}{1},\ \tfrac{2}{1},\ -\tfrac{1}{2},\ \tfrac{1}{2},\ -\tfrac{1}{3},\ \tfrac{1}{3},\ \ldots \}.$$

Therefore, the rationals are countable; that is,

$$n(\mathbb{Q}) = n(\mathbb{N}) = \aleph_0. \tag{4.9}$$

This completes the proof. ◄

Exercises

4.2.1. Show that $n(E) = \aleph_0$ where $E = \{2, 4, 6, 8, \ldots\}$.

4.2.2. Show that $n(F) = \aleph_0$ where $F = \{1, 8, 27, 64, 125, \ldots\}$.

4.2.3. Show that the set of natural numbers $\mathbb{N}$ and the set of whole numbers $\mathbb{W}$ have the same cardinality.

4.2.4. Let E be the set of even natural numbers, and O be the set of odd natural numbers.
   (a) Show that $n(E) = n(O) = \aleph_0$.
   (b) What set is $E \cup O$?
   (c) Use the results from parts (a) and (b) to illustrate that $\aleph_0 + \aleph_0 = \aleph_0$.

4.3 Hilbert's Hotel

One of the questions that David Hilbert (1862–1943) asked was "how does the infinite differ from the finite?" He wanted a way to illustrate to his students the distinguishing features between finite and infinite sets. To do this, he gave this famous example of which it is now called **Hilbert's Hotel** in his honor.

Example 4.3.1 — Finite Hotel. Suppose an innkeeper has a finite number of hotel rooms, say 100 rooms. All 100 rooms are occupied, and each room has a single guest. If a new traveler arrives at the hotel and would like a room, the innkeeper cannot accommodate him; there is simply no room available. ◄

Example 4.3.2 — Hilbert's Hotel. At Hilbert's Hotel, there is an infinite number of rooms numbered $1, 2, 3, 4, 5, \ldots$, in each of which a guest is staying. If a new traveler arrives at the hotel and would like a room, would the innkeeper still be able to accommodate him?

Solution Of course the innkeeper is able to accommodate this newly arrived guest, but the innkeeper will have to rearrange his guests that are already in their rooms. The innkeeper would get on the intercom[5] that broadcasts to all the rooms and say, "Ladies and gentlemen, I am sorry to disturb you, but we have a guest that has just arrived, and we need to accommodate him. Will you be so kind as to go to the room that is one number higher than your current room number?" Then, the guest in room 1 gets up and schlepps to room 2, the guest in room 2 lumbers to room 3, the guest in room 4 plods to room 5, and so on and so forth (see Fig. 4.4).

Now that room 1 is empty, the innkeeper can accommodate the new guest, and the innkeeper hands the key card to room 1 to the new guest with a smile. ◄

---

[5]I thought about this. If the innkeeper went himself to knock on every door, then he would take an infinite amount of time and would never come back!

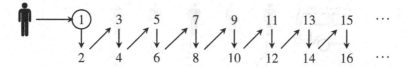

Fig. 4.4 Making room 1 available for the new guest in Hilbert's Hotel. The guest in room 1 moves to room 2, the guest in room 2 moves to room 3, and so on and so forth.

Of course, for any finite number of new guests, we can apply the same solution above to create room for the new guest. Therefore in a world with an infinite number of rooms and guests there would be no room shortage. This does mean though, if you were a guest at Hilbert's Hotel, it is unlikely you will get much sleep on account of you moving at any time to make more room for another guest.

Likewise, we could also have an infinite dance company where if each of the infinite number of men have a lady, and a new lady arrives, the dance roster can be rearranged so that no one remains without a partner. Yes, even for infinitely many new guests and ladies, it is possible to create space (Hilbert, 2013, p. 730).

Indeed, the mathematics with transfinite numbers behave strangely. In summary, the rules of addition with the infinite are the following:

$$\aleph_0 + 1 = \aleph_0$$
$$\aleph_0 + 2 = \aleph_0$$
$$\aleph_0 + 3 = \aleph_0$$
$$\vdots$$
$$\aleph_0 + \aleph_0 = \aleph_0.$$

Example 4.3.3 — An infinite bus arrives at Hilbert's Hotel. At Hilbert's Hotel with an infinite number of rooms, a bus arrives carrying an infinite number of people who each require a room. Would the innkeeper still be able to accommodate all of these infinitely many guests?

Solution Of course the innkeeper is able to accommodate this bus with an infinite number of passengers. The innkeeper would get again on the intercom that broadcasts to all the rooms and say, "Ladies and gentlemen, I am sorry to disturb you again, but we just had an infinite number of guests arrive, and we need to accommodate them. If your room number is n, can you be so kind as to move to room number $2n$?" Then, the guest in room 1 would move resentfully to room 2, the guest in room 2 would shift warily to room 4, and the guest in room 3 would stomp to room 6, and so on and so forth (see Fig. 4.5).

By doing this, all the guests that had a room will move into only the even number rooms. Each passenger on the bus will be numbered $1, 2, 3, \ldots$, and so the innkeeper tells the nth passenger that his room number is $2n - 1$. Then, passenger

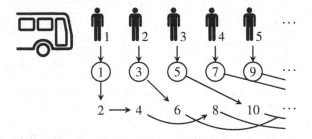

Fig. 4.5   Making the odd numbered rooms available for the infinite bus load of passengers that have arrived at Hilbert's Hotel. The guests who were already in their rooms were asked to move to the even numbered rooms: the guest in room n was asked to move to room $2n$. If the passengers are numbered $1, 2, 3, \ldots$, then passenger n will receive room $2n - 1$.

1 would be assigned to room 1, passenger 2 to room 3, passenger 3 to room 5, and so on and so forth. By doing this, all the passengers will be assigned to all of the odd number rooms.   ◄

In summary, the rules of multiplication with the infinite are the following:

$$\aleph_0 \times 1 = \aleph_0$$
$$\aleph_0 \times 2 = \aleph_0$$
$$\aleph_0 \times 3 = \aleph_0$$
$$\vdots$$
$$\aleph_0 \times \aleph_0 = \aleph_0.$$

In Cantor's assembly of the arithmetic for transfinite numbers analogous to the arithmetic of finite numbers, he realized the risk of being labeled a mathematical heretic, and even a religious one. In the 19th century, the concept of infinity was generally regarded to be in the domain of philosophers and theologians, and Cantor was starting to wander into their territory. Philosophers had long rejected the idea of the infinite since the time of Aristotle due to the apparent logical paradoxes they generate (like the one Galileo discussed on p. 122, or on the motion of objects examined by Zeno (c. 490 BC–c. 430 BC) at Elea in which we discuss later in Sec. 6.9). Mathematicians at the time also fell into this category with Cantor's former teacher, Leopold Kronecker (1823–1891), leading the opposition against transfinite theory.[6] By accepting infinity to be manipulated, it opened the door to paradoxes that could invalidate the entirety of mathematics. Theologians, on the other hand, did not discuss the infinite for an alternative reason: the concept

---

[6]Kronecker is famous for his remark in an 1886 lecture, "*Die ganzen Zahlen hat der liebe Gott gemacht, alles andere ist Menschenwerk*" as quoted by Weber (1893, p. 15) (English: God made the integers, everything else is the work of man). It is interesting to note that Kronecker selected the set of the integers—a set that is infinite.

of infinity was a direct challenge to the infinite nature of God (Dauben, 1977, pp. 85–89). These two groups had forbidden and barred all from crossing into this minefield for hundreds of years—Cantor was wandering right into it.

In order to avoid the repetition of Galileo's tragedy with the Roman Catholic Church, Cantor undertook an extensive correspondence with theologians, even writing a pamphlet directed to Pope Leo XIII. Cantor's hope was to convince the clerics and theologians that his ideas were not in variance with the Church's doctrine. He felt it was his duty "to make the Church accept the reality of a world God had made, and to face the fact that man had been given the capacity to understand it" (Dauben, 1977, pp. 95–96). Perhaps nothing encapsulated Cantor's views as the quote he adapted from the *Bible*: "The time will come when these things which are now hidden from you will be brought into the light" (Dauben, 1977, p. 107).[7]

Fig. 4.6   Georg Cantor (1845–1918)

Continuing our exploration of the infinite, you might start to wonder if there is any other answer besides $\aleph_0$. Are there any other transfinite numbers? The answer is yes. In order to show why this is true, we need the following theorem.

Theorem 4.3.4 For any set A, the cardinality of a set A is less than the cardinality of its power set; that is,

$$n(A) < n(\mathcal{P}(A)).$$

If the set A is finite, then we have two cases to consider.[8]

(a) If the set A is empty, $A = \varnothing$, then

$$n(A) = n(\varnothing) = 0 < 1 = n(\mathcal{P}(A)) = n(\{\varnothing\}).$$

(b) If the set A is nonempty, $A \neq \varnothing$, with m elements, then by Theorem 3.3.19 (on p. 93) and Exercise 6.10.14,

$$n(A) = m < 2^m = n(\mathcal{P}(A)).$$

---

[7] Based off of 1 Corinthians 4:5. "Therefore do not pronounce judgment before the time, before the Lord comes, who will bring to light the things now hidden in darkness and will disclose the purposes of the heart. Then each one will receive commendation from God" (NRSV).

[8] If A is infinite, then this requires more analysis that is beyond the scope of this book. For details, please refer to Fendel and Resek (1990, pp. 306–309).

If we apply Theorem 4.3.4 to the natural numbers, then

$$\aleph_0 = n(\mathbb{N}) < n(\mathcal{P}(\mathbb{N})).$$

Cantor thought about the power set of $\aleph_0$, the set of all possible subsets of $\aleph_0$. By Theorem 3.3.19, the number of elements in the power set is

$$n(\mathcal{P}(\mathbb{N})) = 2^{\aleph_0}, \tag{4.10}$$

and Cantor was able to prove that

$$\aleph_0 < 2^{\aleph_0}. \tag{4.11}$$

He defined a new transfinite number $\aleph_1$, by setting

$$\aleph_1 = 2^{\aleph_0}. \tag{4.12}$$

Thus, he continued in this manner, and thus built up a countable sequence of power sets of power sets:

$$T = \left\{ \underbrace{n(\mathbb{N})}_{\aleph_0}, \underbrace{n(\mathcal{P}(\mathbb{N}))}_{\aleph_1}, \underbrace{n(\mathcal{P}(\mathcal{P}(\mathbb{N})))}_{\aleph_2}, \ldots \right\} \tag{4.13}$$

Then, Cantor turned his attention to the real numbers. Is the set $\mathbb{R}$ countable too? That is, does $n(\mathbb{R}) = \aleph_0$? Or is the cardinality of the reals one of his newly found transfinite numbers $\aleph_1, \aleph_2, \ldots$.

The popular account of Cantor's final years is that he had a severe mental breakdown due to the attempts to disseminate the theory of transfinite numbers to the mathematical community and their subsequent rejection. They had harangued him to within an inch of his sanity, and he spent his remaining years, miserable, broken, and locked behind bars of an asylum (Bell, 1937). This is a tragic end to such a great mind—if it were true.

Cantor's life was "punctuated, rather than dominated, by periods of mental illness" (Grattan-Guinness, 1971, p. 369). After 44 years as a professor at Halle University, Cantor retired in April 1913, and lived quietly at home. During the outbreak of the First World War, it grew increasingly difficult to obtain food, and the hungry mathematician grew very thin and ill. He was taken to the psychiatric hospital in June 1917; he was forced to go, and frequently wrote to his wife to come and take him back. Although trapped in that hellhole, he continued his written correspondence with his colleagues. After six months, Cantor died suddenly on 6 January 1918 from a heart attack (Grattan-Guinness, 1971, pp. 373–374). He was 72.

4.4 The Real Numbers are Not Countable

> **Theorem 4.4.1** The set of real numbers, $\mathbb{R}$, is uncountable.

Proof — Cantor's Diagonal Argument. Suppose in order to derive a contradiction that it is possible to set up the real numbers and the natural numbers into a 1-to-1 correspondence. Then, we would be able to list all of the real numbers, so consider the following list of real numbers between 0 and 1:

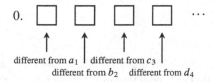

$$1 \quad \leftrightarrow \quad 0.a_1 \quad a_2 \quad a_3 \quad a_4 \quad a_5 \quad \cdots$$

$$2 \quad \leftrightarrow \quad 0.b_1 \quad b_2 \quad b_3 \quad b_4 \quad b_5 \quad \cdots$$

$$3 \quad \leftrightarrow \quad 0.c_1 \quad c_2 \quad c_3 \quad c_4 \quad c_5 \quad \cdots$$

$$4 \quad \leftrightarrow \quad 0.d_1 \quad d_2 \quad d_3 \quad d_4 \quad d_5 \quad \cdots$$

$$\vdots$$

If you draw a diagonal through the 1-to-1 correspondence as indicated, then we can construct a decimal number as follows:

$$0. \quad \square \quad \square \quad \square \quad \square \quad \cdots$$

different from a_1 | different from c_3 |
different from b_2 different from d_4

This new decimal number that we have created is different from any decimal number listed in the 1-to-1 correspondence since it differs in at least one decimal place. Thus, this new number we have created is not in the list. Since we found (or constructed) a number that is not in the correspondence between the reals and the natural numbers, we have arrived at our contradiction. Thus, the negation of our assumption is true: it is not possible to set up the reals and the natural numbers in a 1-to-1 correspondence. Therefore, the reals are not countable. ◄

This result is quite astonishing because this has other consequences. We can set up a 1-to-1 correspondence of all the points in the interior of a square, and the set of points in a line segment from 0 to 1 like we do in Fig. 4.7.

For example, the point $A = (0.34, 0.89)$ in the square can be matched with the point $B = 0.3849$ on the line segment (Fig. 4.7(a)) by alternating the digits in the coordinate (Fig. 4.7(b)). This means that $\mathbb{R} \times \mathbb{R} = \mathbb{R}^2$ has the same cardinality as

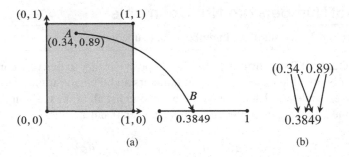

Fig. 4.7   Mapping a point in a square to a line segment.

$\mathbb{R}$; that is, $\aleph_1 \times \aleph_1 = \aleph_1$. Do you think this reasoning will hold if we try to map the points of a cube to the line segment?

Now that we know the difference between countable and uncountable sets, we should look back at the definition of discrete mathematics in the preface.

> **Definition 4.4.2 — Discrete mathematics.** Discrete mathematics is the study of finite and countable objects and sets.

This is in contrast with mathematics such as calculus that works with continuous objects and sets.

4.5 Review

Summary

- **Infinity** is an unbounded quantity. The symbol that denotes infinity is ∞.
- A **prime number** is a natural number greater than one that has no divisors other than one and itself.
- A **proof by contradiction** is a demonstration of a proposition's truth by assuming it to be false implies a contradiction.
- *Euclid's Theorem*: There are infinitely many prime numbers.
- A **1-to-1 correspondence** occurs when each element in two sets can be paired together.
- An infinite set is **countable** if it can be put into a 1-to-1 correspondence with the natural numbers. Sets that are countable are the set of natural numbers $\mathbb{N}$, the set of integers $\mathbb{Z}$, and the set of rational numbers $\mathbb{Q}$.
- An infinite set is **uncountable** if it cannot be put into a 1-to-1 correspondence with the natural numbers. A set that is uncountable is the set of real numbers $\mathbb{R}$.
- **Discrete mathematics** is the study of finite and countable objects and sets.

4.6 Bibliographic Remarks

I adapted the definition of countable from Rudin (1976): a set A is countable if there is a 1-to-1 correspondence to the natural numbers (Rudin, 1976, p. 25). Rudin also introduced the term **at most countable** to mean a set that is either finite or countable; I avoided this term in this book. There are some textbook authors, for example Lang (1993, p. 10), who have defined a countable set with the same meaning as Rudin's *at most countable*. Thus, in mathematics it is very important to state a definition explicitly and precisely; otherwise, some people may have a different notation about the term you are using, and both you and these people will not know there has been a miscommunication. Other synonyms used by mathematicians for a countable set are **denumerable** (Lang, 1993, p. 7), or **enumerable**. Other synonyms for uncountable are **non-denumerable**, and **non-enumerable**.

If you would like to see a proof of how to count the rationals without any duplicates in Fig. 4.3, look at the recent elegant and systematic proof by Calkin and Wilf (2000), or in Aigner and Ziegler (2014, pp. 124–127).

The first published account of Hilbert's Hotel was by Gamow (1961, pp. 17–18). Gamow jokingly cites the story from "the unpublished, and even never written, but widely circulating volume: 'The Complete Collection of Hilbert Stories' by R. Courant." Richard Courant (1888–1972) was a PhD student of David Hilbert who more than likely recited the story to Gamow. It was only recently that Hilbert's unpublished lectures were printed, and the original version of Hilbert's Hotel published (Hilbert, 2013, pp. 729–732). Hilbert's lecture notes were published in German, and hence the version in this book is based on my own translation of his works into English.

In an effort to avoid repeating Galileo's experience, Cantor spent a lot of time explaining his theory of the infinite to the Roman Catholic Church. The article by Dauben (1977) gives an in-depth look at this. Dauben (1977) describes Cantor's deeply religious roots, that Cantor believed transfinite numbers were revealed to him by God, and that this cemented his resolve not to abandon it.

While I was writing about Cantor's final years, I remembered that he suffered from mental illness. This recollection may be due to E. T. Bell's *Men of Mathematics* that romanticized his illness by attributing it to the mathematical community not accepting his theories. Bell's book is one of the most widely read books on the history of mathematics, with many notable 20th century mathematicians such as Julia Robinson,[9] and John Forbes Nash, Jr,[10] crediting Bell for sparking their interest in mathematics. But, "it is also one of the worst, it can be said to have done considerable disservice to the profession" (Grattan-Guinness, 1971, p. 350). I enjoyed Bell's *Men*, but I pass along my caveat that Bell was also a science-fiction novelist (under the name John Taine), and his writing style favors more of a lively written narrative than strictly adhering to the historical facts. On the other hand, the authoritative expert on the history of the mathematical development of probability

---

[9]Celebrated American logician, and the first female mathematician elected to the National Academy of Sciences in 1976. "The only idea of real mathematics that I had came from *Men of Mathematics*. In it I got my first glimpse of a mathematician per se" (Reid, 1996, p. 25).

[10]He won the Nobel prize in economics in 1994 for his contributions to non-cooperative game theory. "By the time I was a student in high school I was reading the classic *Men of Mathematics* by E. T. Bell and I remember succeeding in proving the classic Fermat theorem about an integer multiplied by itself p times where p is a prime" (Nash Jr, 2007, p. 6).

theory is Todhunter (1865) who follows the historical facts so closely that it is "just about as dull as any book on probability could be" (Kendall, 1963). Bell's *Men* covers important mathematicians from Zeno to Poincaré in approximately 600 pages. For a stricter historical adherence, narrative style, and comprehensive, consider reading Burton (2011), which covers mathematics up to this day in approximately 800 pages.

5

Elements of Combinatorics

Combinatorics is the study of counting various combinations or configurations. The ability to count, without directly listing or enumerating, all the combinations or configurations forms part of the basis of probability (see for example, Sec. 8.7 and Sec. 10.2).

5.1 The Pigeonhole Principle

Despite its simple statement, the Pigeonhole Principle is easy enough for anybody to understand, and yet, it is a powerful and invaluable tool in mathematics.

Fig. 5.1 A visual explanation of the Pigeonhole principle. There are 9 pigeons, but only 8 pigeonholes. The Pigeonhole Principle asserts that there must be a hole with 2 or more pigeons.

> **Theorem 5.1.1 — The Pigeonhole Principle.** Consider $n \in \mathbb{N}$. If there are $n + 1$ objects placed into n boxes, then there is a box that has two or more objects.

Proof If we have n objects and n boxes, then we can fill each box with one object. Then when we try to place one more object (for a total of $n + 1$ objects) we have no more empty boxes and must place it in a box that already has an object.   ◄

The German mathematician Peter Dirichlet (1805–1859) first came up with the Pigeonhole principle in 1834 which he called *Schubfachprinzip* or "drawer principle" in German. The term pigeonhole, in fact, refers to the thin vertical wooden partitions in which letters were filed in old-style writing desks.

Example 5.1.2 If the postman has 10 letters, and he is placing them in a communal mailbox that has 9 compartments, then there is a compartment that will receive at least two or more letters.   ◄

Example 5.1.3 Amongst a group of 366 students there are least two who have a birthday on the same day.   ◄

In the previous example, we had assumed that a year has 365 days. There could be a case where there is someone that is born on a leap year. How do you think you could modify the example in this case?

Example 5.1.4 In every 27 word sequence in *A Tale of Two Cities* by Charles Dickens, at least two words will start with the same letter.   ◄

Example 5.1.5 Show that if you pick six numbers from the set $\{1, 2, \ldots, 10\}$, then two of them must sum to 11.

Solution Every number can be paired with another to sum to 11. In all there are five pairs: $(1, 10)$, $(2, 9)$, $(3, 8)$, $(4, 7)$, and $(5, 6)$. Each of the six numbers you picked must belong to one of these pairs. Therefore, by the Pigeonhole Principle, two of the numbers that you picked must be from the same pair, which by construction sums to 11.   ◄

Exercises

5.1.1. How many students in class must there be to ensure at least two students are from the same American state?

5.1.2. How many numbers from the set $\{1, 2, \ldots, 20\}$ must you pick to ensure that two of the picked numbers sums to 21?

5.1.3. What is the contrapositive of the Pigeonhole Principle?

5.1.4. Prove that there are two non-bald males in the city of London, England who have the same number of hairs on their head. (*Hint*: Use the Internet to find out how many hairs a human head can contain, and the number of males that live in London.)

5.1.5. In your dresser drawer, there are three colors of socks: red, white, and blue. Due to a power failure, you have to get your clothes out in the dark. How many socks will you have to take to ensure that you have a matching pair?

5.1.6. Fifty pieces of sushi was ordered for the Department of Mathematics that has 20 people. Everyone received one piece. Then, leftover pieces were up for grabs—this is where it gets interesting. To keep it civilized, since everyone loves sushi in this department, once someone has picked up a piece then no one else may steal it out of his or her hand.[1] Lucky people may be able to pick up multiple pieces. Show that there must be some fortunate enough person who will end up with three or more pieces of sushi.

5.1.7. If there are two or more people in a cinéma, prove that there must be at least two people with same number of acquaintances at the cinéma.

5.1.8. Show that if there are six people at a party, then either three have met before the party, or three of them are utterly complete strangers before the party.

5.2 Fundamental Counting Principles

There are two fundamental counting principles: the **Multiplication Principle**, and the **Sum Principle**.

5.2.1 The Multiplication Principle

Example 5.2.1 I want to purchase a Macbook Air. There are two models: one has an 11" display and the other is 13". Each model can have a 128 GB (gigabyte[2]) SSD (solid-state drive[3]) or a 256 GB SSD. In how many ways can I order a Macbook Air?

Solution The first task is to select a screen size, and the second task is to pick an SSD storage size. These tasks are performed consecutively. From setting up Table 5.1 on the next page, there are two ways to select a screen size, and two ways to select an SSD, so there are $2 \times 2 = 4$ different laptop configurations. ◄

---

[1] A bad joke: A piece of sushi in the hand is worth two on the table.

[2] Seven minutes of high definition video (at 19.19 MBit/s) is approximately 1 gigabyte. A Blu-ray disc can hold about 25 GB.

[3] A solid-state drive (SSD) is a data storage device that uses integrated circuits instead of a hard drive which uses rotating disks.

Table 5.1   Ordering a Macbook in Example 5.2.1
Selection of Screen

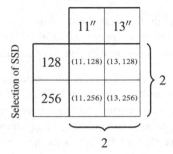

Theorem 5.2.2 — The Multiplication Principle. If the first task has n_1 outcomes, and no matter what the outcome of the first task, a second task has n_2 outcomes, then the two tasks performed consecutively have $n_1 \times n_2$ outcomes.

Proof Let $A, B, C, D, \ldots$ represent the different outcomes of the first task (taking as many letters as required to label all the different ways), and similarly let $a, b, c, d, \ldots$ represent the different outcomes of the second task.

Table 5.2   Proof of the Multiplication Principle
Outcomes of 1st Task

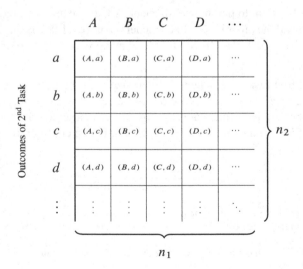

Then, we can create Table 5.2. We may regard each square in the table as representing the case in which the first task has the outcome marked at the head of the column, and the second task's outcome marked at the beginning of the row. For example, the first square denotes the case in which the first task has outcome A, and the second task has outcome a. Every square represents a different case, and that every case is represented by some square. Hence, the number of possible cases is the same as the number of squares. Therefore, the number of squares (or the number of all possible cases) is the product of the number of outcomes of the two tasks; that is, $n_1 \times n_2$. ◄

For three tasks, we can extend Table 5.2 to get Fig. 5.2. In Fig. 5.2, every cube represents a different case, and that every case is represented by some cube. For example, the cube marked with the asterisk will denote the case in which the first task has outcome A, the second task has outcome a, and the third task has outcome α. Therefore, the number of cubes is the product of the number of the outcomes of the three tasks.

In general, we can extend the principle to include as many tasks as we want.

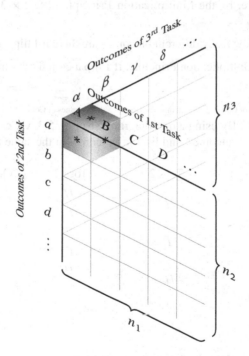

Fig. 5.2   The Multiplication Principle for three tasks

Theorem 5.2.3 — Generalized Multiplication Principle. If there is a total of r tasks to be performed consecutively, in which

the first task has n_1 outcomes,

the second task has n_2 outcomes,

$$\vdots$$

and the rth task has n_r outcomes,

then there are

$$n_1 \times n_2 \times \cdots \times n_r \qquad (5.1)$$

ways the r tasks performed consecutively.

Example 5.2.4 Continuing with Example 5.2.1 about the Macbook Air, I can also get a keyboard in English, French, or Arabic. How many configurations of a Macbook Air can I purchase?

Solution There are, by the Multiplication Principle, $2 \times 2 \times 3 = 12$ different configurations. ◀

Example 5.2.5 How many different outcomes are there if I flip a coin three times?

Solution By the Multiplication Principle, if I flip a coin three times, there are

$$\underbrace{2}_{\text{\# Outcomes of Coin 1}} \times \underbrace{2}_{\text{\# Outcomes of Coin 2}} \times \underbrace{2}_{\text{\# Outcomes of Coin 3}} = 2^3 = 8$$

different outcomes. By using a tree diagram, see Fig. 5.3, we can enumerate the number of different outcomes, and list the outcomes at the same time.

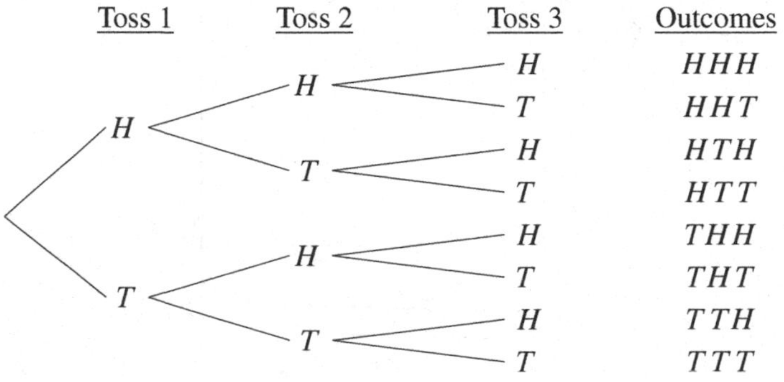

Toss 1	Toss 2	Toss 3	Outcomes
		H	HHH
	H	T	HHT
H		H	HTH
	T	T	HTT
		H	THH
	H	T	THT
T		H	TTH
	T	T	TTT

Fig. 5.3   Tree diagram for three flips of a coin.

◀

5.2.2 The Sum Principle

Theorem 5.2.6 — The Sum Principle. If one task can be performed in n_1 ways, a second task can be performed in n_2 ways, and these tasks cannot be performed simultaneously, then there are $n_1 + n_2$ ways in which the first *or* the second task, but not both, can be performed.

It is important to note that in the statement of the Sum Principle, the word *or* is used in the exclusive disjunctive sense (see Sec. 1.3.4).

Example 5.2.7 A student in New York state has decided to go on a holiday either down south or up north. If she goes up north her budget only allows for her to visit Toronto or Montréal. If she goes down south, she can only visit Myrtle Beach, Cape Cod, or New Orleans. How many ways can she go on holiday?

Solution *First task*: going up north for a holiday. There are 2 such ways (Toronto or Montréal).

Second task: going down south for a holiday. There are 3 such ways (Myrtle Beach, Cape Cod, or New Orleans).

Therefore, there are $2 + 3 = 5$ possible holiday destinations the student can visit. ◀

Theorem 5.2.8 — Generalized Sum Principle. If there is a total of r tasks to be performed, and these tasks cannot be performed simultaneously, in which

the first task has n_1 outcomes,

the second task has n_2 outcomes,

$$\vdots$$

and the rth task has n_r outcomes,

then there are

$$n_1 + n_2 + \cdots + n_r \qquad (5.2)$$

ways the first *or* the second *or* ... *or* the rth task can be performed, and *only one* of these tasks can be selected.

The interpretation of *or* as addition and *and* as multiplication in the fundamental counting principles will occur frequently throughout counting and probability, so it will be useful for you to remember this distinction. The problems that you will encounter may not explicitly use the words, but you may discover that the wording can imply *and* or *or*. Table 5.3 lists the correspondences between arithmetic, logic, and set theory.

Table 5.3   Correspondences between arithmetic, logic, and set theory.

Arithmetic		Logic		Set Theory	
Subtraction	−	Negation	¬	Complement	′
Addition	+	Disjunction (or)	∨	Union	∪
Multiplication	×	Conjunction (and)	∧	Intersection	∩

Exercises

5.2.1. The bookstore is giving away 10 different albums and 5 books. If Gabriel can select only one album or one book, how many choices does he have?

5.2.2. The student bar has 5 main dishes with meat and 6 without meat. How many possible single main dishes can be ordered?

5.2.3. If a dime and a quarter are tossed, in how many ways can they land?

5.2.4. If two dice are thrown together, in how many ways can they land?

5.2.5. How many ways can two distinct prizes be given to a class of 10 students, without giving both to the same student?

5.2.6. A classroom has four doors. In how many different ways can Chris enter and then leave the classroom?

5.2.7. A restaurant has 3 appetizers, 4 main courses, and 3 desserts. If you decide to order one appetizer, one main course, and one dessert, how many different meals can you order?

5.2.8. Sterling is planning to travel from Toronto to Halifax (see Fig. 5.4). If he can travel from his home in Toronto to Ottawa by car, bus or train, and then from Ottawa to Halifax by train or plane, then how many possible itineraries are there? Use a tree diagram to illustrate Sterling's possible itineraries.

5.2.9. How many double-digit natural numbers are not divisible by 2 or 5?

5.2.10. Suppose you are given the digits $\{0, 1, 2, 3, 4, 5, 6\}$. How many three digit numbers can be formed if
   (a) repetition of digits is allowed?
   (b) the three digit number must be an odd number and repetition is not allowed.

5.2.11. How many single-digit or double-digit natural numbers end in the digit 9?

5.2.12. A coin is flipped four times. How many different outcomes are possible?

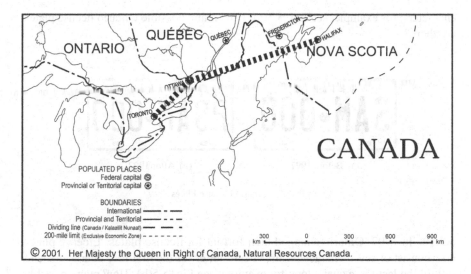

Fig. 5.4 Sterling's trip from Toronto to Halifax. Modified from Natural Resources Canada (2001).

5.2.13. A die is rolled six times. How many different outcomes are possible?

5.2.14. A Canadian postal code consists of six characters: the first, third, and fifth are letters, and the remaining characters are digits. For example, K7K 7B4 is a valid postal code. An American ZIP code consists of five digits. For example, 12345 is a valid ZIP code.
  (a) How many codes are possible for each country?
  (b) How many more possible codes does one country have over the other?
  (c) Check online to see how many postal codes are actually in use in Canada.

5.2.15. An extended ZIP+4 code was introduced in the 1980s, which included the five digits of the ZIP code, a hyphen, and four more digits that determined a more specific location within the ZIP code.
  (a) How many ZIP+4 codes are possible?
  (b) How many more ZIP+4 codes are there than ZIP codes?

5.2.16. A minivan holds seven people. How many ways can seven people be seated in the minivan if
  (a) there are no restrictions?   (b) only 3 people can drive?

5.2.17. If a sold-out concert hall has 1,000 seats, would there be two patrons that have the same first and last initials?

5.2.18. In the province of Ontario, license plates originally consisted of six characters: upper case letters for the first three characters, and 0-9 for the last three characters; for example, see Fig. 5.5(a). How many possible Ontario license plates are there?

(a) Before 1997   (b) After 1997

Fig. 5.5   Ontario License Plates

5.2.19. In 1997, Ontario changed their format for license plates. License plates now consist of seven characters: upper case letter for the first four characters, and 0-9 for the last three characters; for example, see Fig. 5.5(b). How many possible Ontario license plates are there? How many more license plates are possible than before 1997?

5.2.20. There are 12 ladies and 10 gentlemen, of whom 3 ladies and 2 gentlemen are sisters and brothers, the rest being unrelated. In how many ways can a marriage occur? (What assumptions do you make about an eligible marriage?)

5.2.21. The Chevrolet Camaro has the option of
- coupé, convertible, ZL1, or Z/28 model styles;
- standard V6, or supercharged V8 engine;
- nine paint colors: red rock metallic, ashen gray metallic, black, blue ray metallic, bright yellow, crystal red, hot red, silver ice, or summit white; and
- 18", 19", or 20" wheels.

How many different ways can a Camaro be ordered?

5.2.22. In Misa's wardrobe, she has five shirts, six skirts, seven pairs of pants, and eight dresses. Misa can select either a skirt or a pair of pants to go with a shirt, or she can wear only a dress. How many outfits can Misa create?

5.2.23. A user's password to a university's computer network consists of four letters from the alphabet followed by 3 or 4 digits. Find the total number of passwords
- (a) that can be created;
- (b) that can be created in which no digit repeats; and
- (c) that can be created in which no letter repeats.

5.2.24. In 1838, the American painter Samuel F. B. Morse (1791–1872) invented Morse code that was used to send messages by telegraph. A character in Morse code consists of a sequence of one to four dots or dashes. How many possible Morse code signals are there?

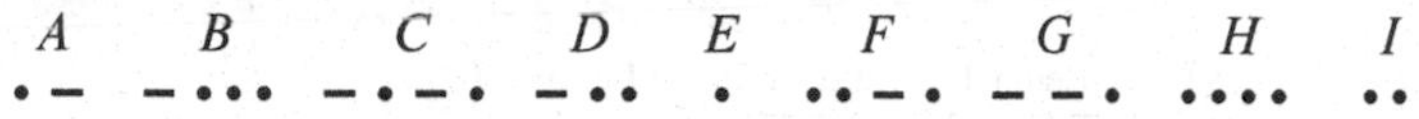

$$A \quad B \quad C \quad D \quad E \quad F \quad G \quad H \quad I$$
$$\bullet - \quad - \bullet \bullet \bullet \quad - \bullet - \bullet \quad - \bullet \bullet \quad \bullet \quad \bullet \bullet - \bullet \quad - - \bullet \quad \bullet \bullet \bullet \bullet \quad \bullet \bullet$$

Fig. 5.6 International Morse Code. Source: International Telecommunications Union (2009).

5.2.25. In a discrete mathematics class of n students, each student is given a choice of either one of x different logic problems, or one of y different combinatorics problems. How many different ways can the students select the problems?

5.3 Factorials and Permutations

There are two kinds of counting problems:
- problems where the order of the objects matter, and
- problems where the order of the objects does not matter.

A **permutation** of a set of objects is an arrangement of those objects into a particular order. A **combination** is a way of selecting objects out of a larger set, where the order does not matter. We will discuss permutations first; combinations are discussed later in Sec. 5.7.

Example 5.3.1 How many different permutations of the letters a, b, and c are possible?

Solution We can create a tree diagram. From Fig. 5.7, we see that there are 6 possible permutations of the letters a, b, and c: $abc, acb, bac, bca, cab,$ and cba.

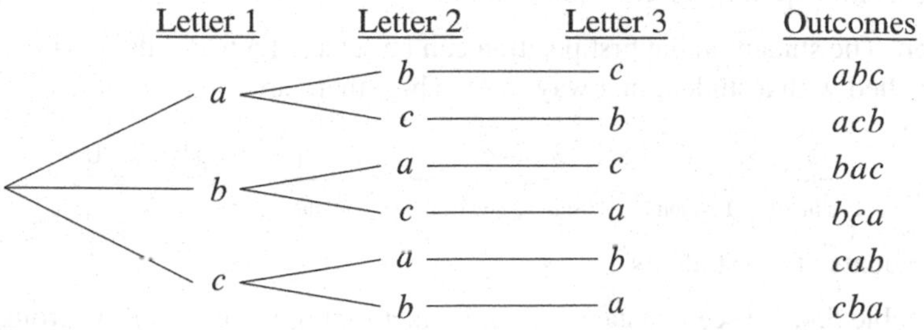

Letter 1	Letter 2	Letter 3	Outcomes
a	b	c	abc
	c	b	acb
b	a	c	bac
	c	a	bca
c	a	b	cab
	b	a	cba

Fig. 5.7 The permutations of a, b, and c using a tree diagram.

This result can also be obtained from the Multiplication Principle
- The 1st letter in the "word" can be any of the three letters.
- The 2nd letter in the "word" can be any of the remaining two letters (one of the letters was used as the first letter).
- The 3rd letter in the "word" can only be the one remaining letter.

$$\underbrace{\boxed{3}}_{\text{Letter 1}} \times \underbrace{\boxed{2}}_{\text{Letter 2}} \times \underbrace{\boxed{1}}_{\text{Letter 3}} = 6$$

Thus, there are $3 \times 2 \times 1 = 6$ different "words" that can be formed from the three letters. ◄

In general, n distinct objects can be arranged in

$$n \times (n-1) \times (n-2) \times \cdots \times 3 \times 2 \times 1. \tag{5.3}$$

In mathematics, because we encounter this type of product often, we have a special notation to denote this.

Definition 5.3.2 — Factorial. For any natural number n, the **factorial** of n is

$$n! = \begin{cases} n \times (n-1) \times (n-2) \times \cdots \times 3 \times 2 \times 1 & \text{if } n \neq 0 \\ 1 & \text{if } n = 0 \end{cases} \tag{5.4}$$

We read this as "n factorial."

Thus, in Example 5.3.1, we could express the result $3 \times 2 \times 1$ as $3!$ instead.

Example 5.3.3 Compute $5!$

Solution $5! = 5 \times 4 \times 3 \times 2 \times 1 = 120$. ◄

Example 5.3.4 — Five Students Lining Up for a Photo. In how many ways can 5 students line up in a row for a photograph?

Solution The student in the first position can be selected 5 ways, the next position can be filled with a student in 4 ways, *etc.* Thus, there are

$$\underbrace{5}_{\text{Position 1}} \times \underbrace{4}_{\text{Position 2}} \times \underbrace{3}_{\text{Position 3}} \times \underbrace{2}_{\text{Position 4}} \times \underbrace{1}_{\text{Position 5}} = 5! = 120$$

arrangements of the students. ◄

In Table 5.4, we see that calculating the factorial of the numbers 1 through 10, the result becomes rather large quickly.

We see that each value of $n!$ can be calculated by multiplying the numbers from 1 to n together, but this is not always necessary, because of the following result.

Table 5.4   Factorial values for the first 10 numbers

n	0	1	2	3	4	5	6	7	8	9	10
$n!$	1	1	2	6	24	120	720	5040	40 320	362 880	3 628 800

Example 5.3.5 Show that $n! = n \times (n-1)!$ where n is a natural number.

Solution For $n = 1$, it is true since $n! = 1 \times 0! = 1 \times 1 = 1$. If $n \geq 2$, then

$$
\begin{aligned}
n! &= n \times (n-1) \times (n-2) \times \cdots \times 2 \times 1 \\
&= n \times [(n-1) \times (n-2) \times \cdots \times 2 \times 1] \\
&= n \times (n-1)!.
\end{aligned}
$$

◀

Example 5.3.6 Given that $4! = 24$, compute $5!$.

Solution Using the result from Example 5.3.5,

$$
\begin{aligned}
5! &= 5 \times 4! \\
&= 5 \times 24 \\
&= 120.
\end{aligned}
$$

◀

Example 5.3.7 Compute $\dfrac{70!}{68!}$.

Solution

$$
\begin{aligned}
\frac{70!}{68!} &= \frac{70 \times 69 \times \overbrace{68 \times \cdots \times 2 \times 1}^{68!}}{68!} \\
&= \frac{70 \times 69 \times \cancel{68!}}{\cancel{68!}} \\
&= 70 \times 69 \\
&= 4830
\end{aligned}
$$

◀

In the previous example, it is helpful to know how to reduce $n!$ to a smaller product by cancellation with terms in the denominator. Most calculators are unable to compute $n!$ if n is greater than 69; they only have enough memory to hold a two-digit exponent otherwise it causes a memory overflow and crashes: $69! \approx 1.71 \times 10^{98}$ while $70! \approx 1.20 \times 10^{100}$.

Exercises

5.3.1. Compute the following expressions:

(a) $\dfrac{5!}{3!}$ (b) $\dfrac{20!}{17!}$ (c) $\dfrac{500!}{499!}$

5.3.2. Express the following using factorials:

(a) $6 \times 5 \times 4 \times 3 \times 2 \times 1$ (c) $\frac{20 \times 19 \times 18}{4 \times 3 \times 2 \times 1}$

(b) $10 \times 9 \times 8$ (d) $13 \times 12 \times 11 \times 10 \times 9$

5.3.3. Write each expression as a single factorial.

(a) $(n + 3)(n + 2)!$

(b) $(n^2 + 5n + 6)(n + 1)!$ (c) $\dfrac{(n + 2)!}{n + 2}$

5.3.4. Simplify the following where $n \in \mathbb{N}$:

(a) $\dfrac{n!}{(n - 1)!}$ (c) $\dfrac{n!}{(n - 2)!}$ (e) $\dfrac{(3n)!}{(3n - 1)!}$

(b) $n(n - 1)!$ (d) $(n + 1)n!$ (f) $n[n! + (n - 1)!]$

5.3.5. Write the set $A = \{\, x \in \mathbb{N} \mid x! \le 1000 \,\}$ using roster notation.

5.3.6. Can you explain why, by definition, $0! = 1$?

5.3.7. How many different ways can five people be arranged in a line?

5.3.8. How many different ways can 11 ladies and 11 gentlemen form themselves into couples for a dance?

5.3.9. A quarterback has a sequence of 6 plays. The coach instructs him to run through all 6 plays without repetition. In how many ways can the quarterback call the plays?

5.3.10. The *Complete Bob Major* holds legendary status amongst campanologists (bell-ringers). This is where eight bells are rung in every possible order using all eight bells exactly once. This feat has only been performed once: in 1963, Robert B. Smith and his fellow campanologists rang for 17 hours and 58 minutes.[4] How many ways can this be done?

5.3.11. In the 1920s, Arnold Schönberg (1874–1951) developed the 12-tone technique which was influential on composers of the mid-20th century. Through the use of *tone rows*, a sequence in which each of the 12 tones in an octave are played once, all 12 tones are given equal importance, and hence the music avoids being in any key. How many tone rows are possible for an octave?

---

[4]Source: Pound (2015)

5.3.12. In a **substitution cipher**, the letters of the plaintext are replaced by other letters. An example of a substitution cipher is given in Table 5.5. This graphically indicates that the letters of the plaintext are to be replaced by the cipher letters beneath them, and vice versa.[5] How many different substitution ciphers are possible?

Table 5.5 An example of a substitution cipher.

Plaintext	a	b	c	d	e	f	g	h	i	j	k	l	m
Ciphertext	K	O	H	A	R	M	T	I	Z	C	D	E	F

Plaintext	n	o	p	q	r	s	t	u	v	w	x	y	z
Ciphertext	U	J	B	P	Q	N	L	Y	X	G	W	V	S

5.4 Permutations without Repetition

Example 5.4.1 The number of permutations of four letters a, b, c, d is

$$4! = 4 \times 3 \times 2 \times 1 = 24.$$

Now, consider the number of permutations that are possible by taking two letters at a time from four. The different permutations are

$$ab, ac, ad, ba, bc, bd, ca, cb, cd, da, db, dc.$$

Using our Multiplication Principle, we have

$$n_1 \times n_2 = 4 \times 3 = 12.$$

◄

In general, n distinct objects taken r at a time can be arranged in

$$n \times (n-1) \times (n-2) \times \cdots \times (n-r+1) \text{ ways.} \tag{5.5}$$

Again, because mathematicians are lazy, we have a way of representing the above product.

Definition 5.4.2 — Permutations of distinct objects. The number of permutations of n distinct objects taken r at a time is

$$P(n,r) = n \times (n-1) \times (n-2) \times \cdots \times (n-r+1) = \frac{n!}{(n-r)!}. \tag{5.6}$$

---

[5]Kahn (1996, pp. xv-xvi).

Example 5.4.3 Evaluate the following expressions.

(a) $P(7,3)$ (b) $P(n,4)$

Solution (a) *Method 1*: $P(7,3) = \underbrace{\overbrace{7}^{\text{Begin at } 7} \times 6 \times 5}_{3 \text{ terms}} = 210$

Method 2: $P(7,3) = \dfrac{7!}{(7-3)!} = \dfrac{7!}{4!} = \dfrac{7 \times 6 \times 5 \times \cancel{4!}}{\cancel{4!}} = 210$

(b) *Method 1*: $P(n,4) = \underbrace{\overbrace{n}^{\text{Begin at } n} \times (n-1) \times (n-2) \times (n-3)}_{4 \text{ terms}}$

Method 2:

$$\begin{aligned} P(n,4) &= \frac{n!}{(n-4)!} \\ &= \frac{n(n-1)(n-2)(n-3)\cancel{(n-4)!}}{\cancel{(n-4)!}} \\ &= n(n-1)(n-2)(n-3) \end{aligned}$$

◀

We can read $P(n,r)$ as "n permute r." Note that the order of n and r in which we say "n permute r" is important since

$$\frac{n!}{(n-r)!} = P(n,r) \text{ is not equivalent to } P(r,n) = \frac{r!}{(r-n)!}$$

Table 5.6 $P(n,r)$, $0 \le n \le 10$, $0 \le r \le 10$.

	r										
n	0	1	2	3	4	5	6	7	8	9	10
0	1										
1	1	1									
2	1	2	2								
3	1	3	6	6							
4	1	4	12	24	24						
5	1	5	20	60	120	120					
6	1	6	30	120	360	720	720				
7	1	7	42	210	840	2 520	5 040	5 040			
8	1	8	56	336	1 680	6 720	20 160	40 320	40 320		
9	1	9	72	504	3 024	15 120	60 480	181 440	362 880	362 880	
10	1	10	90	720	5 040	30 240	151 200	604 800	1 814 400	3 628 800	3 628 800

There are two types of reasoning to think about how to solve a counting problem. **Direct reasoning** solves the problem by counting all the desired arrangements to give the final answer. **Indirect reasoning** starts with all the possible arrangements and removes the undesired arrangements leaving only the desired arrangements.

You will usually find that it is easier to solve a combinatorial problem with one type of reasoning over the other.

Example 5.4.4 — Steven's socks. Steven was doing his laundry, but he was running out of time before he had to go on a date. He has three pairs of socks that he did not match together before he threw them into his sock drawer. How many ways can he pull out two unmatched socks one after the other?

Solution — Direct reasoning. We can find the total number of ways that Steven can pick two unmatched socks by counting all the ways he can do this.

There are 6 socks in the drawer. Steven picks one sock leaving behind 5 socks. In those 5 remaining socks, there is the one sock that matches the sock that Steven picked—we do not want that sock. Then, there are only 4 choices of socks on the second pick that will ensure he does not have a match. Therefore, the number of ways Steven can pick two unmatched socks one after the other is

$$\underbrace{6}_{\text{Sock 1}} \times \underbrace{4}_{\text{Sock 2}} = 24.$$

◀

Solution — Indirect reasoning. We can find the total number of ways that Steven can pick two socks, remove the number of ways that Steven can pick where the two socks match, and this will leave us with the number of ways he can pick two unmatched socks one after another.

The total number of ways to pick two socks one after another is

$$\underbrace{6}_{\text{Sock 1}} \times \underbrace{5}_{\text{Sock 2}} = P(6,2) = 30.$$

The number of ways to pick two socks one after another that are matching is

$$\underbrace{6}_{\text{Sock 1}} \times \underbrace{1}_{\text{Sock 2}} = 6.$$

Once he selects sock 1, there is only one other sock that will match.

Therefore, Steven can pull out two unmatched socks in $30 - 6 = 24$ ways. ◀

Looking back at the two solutions, we see that we have found the same answer using the two different kinds of reasoning. We feel confident in our answer, and we can move onto the next example.

Example 5.4.5 — President and Treasurer. A president and a treasurer are to be chosen for the RMC Flying Club from the 50 officers in the club. Only one person can hold a position; that is, a person cannot be both the president and the treasurer. How many different choices of officers are possible if

(a) there are no restrictions?
(b) *A* will serve only if she is president?
(c) *B* and *C* will serve together or not at all?
(d) *D* and *E* will not serve together?

Solution

(a) *Direct reasoning.* The total number of choices of the officers, if there are no restrictions, is

$$P(50, 2) = \frac{50!}{(50 - 2)!}$$
$$= \frac{50!}{48!}$$
$$= \frac{50 \times 49 \times \cancel{48} \times \cancel{47} \times \cdots \times \cancel{1}}{\cancel{48} \times \cancel{47} \times \cdots \times \cancel{1}}$$
$$= 50 \times 49$$
$$= 2450.$$

(b) *Direct reasoning.* Since *A* will serve only if he is president, we have two cases here:

(i) *A* is selected as president, which yields 49 possible outcomes; and
(ii) Both officers are selected from the remaining 49 people which has the number of choices

$$P(49, 2) = \frac{49!}{(49 - 2)!}$$
$$= \frac{49!}{47!}$$
$$= \frac{49 \times 48 \times \cancel{47} \times \cancel{46} \times \cdots \times \cancel{1}}{\cancel{47} \times \cancel{46} \times \cdots \times \cancel{1}}$$
$$= 49 \times 48$$
$$= 2352.$$

Therefore, the total number of choices is

$$49 + 2352 = 2401.$$

(c) *Direct reasoning.* There are two cases:

(i) The number of selections when *B* and *C* serve together is 2; and

(ii) The number of selections when both B and C are not chosen is

$$P(48, 2) = \frac{48!}{(48-2)!}$$

$$= \frac{48!}{46!}$$

$$= \frac{48 \times 47 \times \cancel{46} \times \cancel{45} \times \cdots \times \cancel{1}}{\cancel{46} \times \cancel{45} \times \cdots \times \cancel{1}}$$

$$= 48 \times 47$$

$$= 2256.$$

Therefore, the total number of choices in this situation is

$$2 + 2256 = 2258.$$

(d) *Indirect reasoning.* Since D and E can only serve together in 2 ways, the answer is

$$P(50, 2) - 2 = 2450 - 2 = 2448.$$

◄

Exercises

5.4.1. Evaluate the expressions:

(a) $P(6, 3)$

(b) $P(8, 0)$

(c) $P(3, 2)$

(d) $P(6, 4)$

(e) $P(7, 6)$

(f) $P(12, 6)$

(g) $P(4, 3) \times P(3, 1)$

(h) $P(10, 2) \times P(5, 3)$

(i) $P(n, n-2)$

(j) $\dfrac{9!}{5!4!} \times \dfrac{5 \times 4!}{3!2!}$

5.4.2. Solve the equation for $n \in \mathbb{N}$.

(a) $\dfrac{n!}{(n-1)!} = 6$

(b) $\dfrac{n!}{(n-2)!} = 90$

(c) $P(n, 5) = 14 \times P(n, 4)$

(d) $P(n, 3) = 17 \times P(n, 2)$

(e) $\dfrac{n!}{(n-1)!} = 7$

(f) $\dfrac{n!}{(n-2)!} = 12$

(g) $8P(n, 3) = 7P(n+1, 3)$

(h) $3P(n, 4) = P(n-1, 5)$

5.4.3. How many ways can you rank 6 candidates who apply for a positions if you do not allow ties?

5.4.4. How many three letter permutations can be formed from the last 16 letters of the alphabet?

5.4.5. How many four-digit numbers have no repeated digits?

5.4.6. A record company evaluates 10 albums by sending them to a number of music critics, and asking them to name their top 5 picks in order. How many different lists are possible?

5.4.7. Excluding ties, how many different finishes in a horse race are possible for the first three places in an eight-horse race?

5.4.8. There are four stores in Hamilton, and the corporate office wishes to assign a manager to each store.
  (a) If there are 5 people available, then how many ways can they be assigned?
  (b) If there are 8 people available, then how many ways can they be assigned?

5.4.9. Suppose 10 cadets must line up in a row on the parade square. How many arrangements are there if Cadet James and Cadet Stewart cannot be placed side-by-side? Solve using
  (a) direct reasoning   (b) indirect reasoning.

5.4.10. A 7-volume encyclopedia is placed on a shelf. How many incorrect arrangements are there?

5.4.11. A bookshelf contains 5 distinct books of German, 6 of French, and 8 of English. How many ways can the books be arranged
  (a) without any restriction?
  (b) keeping all the German together, all the French together, and all the English together?

5.4.12. Eleven students line up for a team photo. How many different arrangements are possible if
  (a) there are no restrictions?
  (b) Trent must be in the middle?
  (c) Rhea is on the left end and Jerilyn is on the right end of the line?
  (d) best friends Brian and Stewart must stand beside each other?

5.4.13. In how many ways can four different music awards be distributed amongst 25 musicians if
  (a) no musician can receive more than one award?
  (b) there is no limit on how many awards a musician can win?

5.5 Circular Permutations

Example 5.5.1 Six people are invited to a dinner party and are seated around a circular table. How many permutations exist for the seating order around the table?

Solution We wish to count the number of circular permutations or the number of ways to seat A, B, C, D, E, and F around a table. Since the dinner guests are seated in a circle, there will be some duplication. For example, if each person moves one seat to their right, the arrangement is still considered the same because of the closed circle (see Fig. 5.8). Any circular permutation can be rotated so that A is in a fixed position; thus, we can fix person A. Then the only permutations

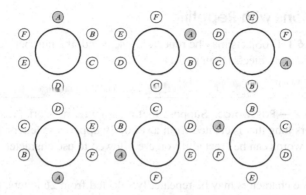

Fig. 5.8   The circular table permutation problem.

that we would have to consider is the permutations of B, C, D, E, and F: there are $5! = 120$ permutations. Thus, there are 120 different number of ways to sit six people around a circular table.   ◄

In summary, we have the following theorem.

Theorem 5.5.2 The number of circular permutations of n elements is $(n-1)!$.

Exercises

5.5.1. Suppose there are 5 players on a basketball team. In how many ways can they form a ring around their coach?

5.5.2. If you had 20 unique beads, how many different necklaces can you make?

5.5.3. What is the minimum number of beads required to ensure that there are 24 different possible necklaces?

5.5.4. Six people are invited to a dinner party, two of which do not want to sit beside each other, are seated around a circular table. How many circular seating arrangements are there?

5.5.5. Suppose there are 5 gentlemen and 5 ladies. In how many ways can they sit around a circular table if the gentlemen and the ladies are to alternate?

5.6 Permutations with Repetition

> **Theorem 5.6.1** If objects may be repeatedly selected, the number of permutations of n distinct objects taken r at a time is
>
> $$n^r \qquad (5.7)$$

Example 5.6.2 — Passwords. Suppose that a computer network password must be 8 characters long that is created from a set of 26 upper case letters. How many different passwords can be created if you are allowed to use characters more than once?

Solution Since characters may be repeatedly selected from 26 letters, then by the Multiplication Principle the number of passwords of 8 characters is

$$\underbrace{26}_{1^{st}\ letter} \times \underbrace{26}_{2^{nd}\ letter} \times \cdots \times \underbrace{26}_{8^{th}\ letter} = 26^8 = 208\,827\,064\,576.$$

◀

This may look like a large number of possibilities for a password, and it would take someone (or a computer) a long time to attempt and enter a potential password one at a time to find the correct one; this is called a **brute force attack**. Another password attack which uses the assumption that people may select a real English word for a password is called the dictionary attack; this technique tries words one at a time in an exhaustive list called a dictionary. At the moment, there is approximately one million words in the English language,[6] with about 14% being exactly eight-letters.[7] A computer can quickly try these 140 000 possibilities first before attempting to try the rest of the other 26^8 permutations.

Exercises

5.6.1. In Example 5.6.2, we calculated how many passwords were possible if repetition of the letters is allowed. How many passwords are possible if repetition is not allowed. Compare your answer to the answer given in Example 5.6.2.

5.6.2. How many possible outcomes are there with the throw of 12 dice?

5.6.3. If a light switch can be either on or off, and there are six switches on the wall, then what is the number of different switch configurations?

5.6.4. At Sandwich Express, you have the following choice of toppings: lettuce, tomatoes, red onions, green peppers, mushrooms, cheese, olives, cucumbers, ranch dressing and olive oil. How many ways can you select toppings for your sandwich?

---

[6]The number of words in the English language is 1 025 109. This is an estimate by the Global Language Monitor on 1 January 2014.

[7]http://norvig.com/mayzner.html

5.6.5. In genetics, DNA is made up of four bases: guanine (G), adenine (A), thymine (T), or cytosine (C). How many possible triplets of the four bases are possible if repetition is allowed?

5.6.6. How many ways can a manager assign five different tasks to six employees if it is possible to assign more than one task to an employee?

5.6.7. For a client to use an automatic teller machine (ATM) with their banking card, the bank or the client must first select a personal identification number (PIN).
   (a) How many different 4-digit PINs are possible using the digits $0, 1, 2, \ldots, 9$?
   (b) How many different 4-digit PINs are possible using the digits $0, 1, 2, \ldots, 9$, but the first digit of the PIN must not be 0?

5.6.8. The French educator Louis Braille (1809–1852) created the following system of reading and writing for use by the blind or visually impaired: each character is represented by a rectangular set of six raised or flat dots on paper. Examples are shown in Fig. 5.9. How many different configurations of bumps and flats can be made? Are there enough configurations that can encode all commonly used alpha-numeric symbols?

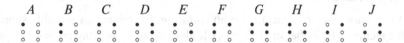

| A | B | C | D | E | F | G | H | I | J |

Fig. 5.9 First ten letters of Braille's alphabet

5.6.9. Consider this traditional English rhyme:

> As I was going to St Ives,
> I met a man with seven wives,
> Every wife had seven sacks,
> Every sack had seven cats,
> Every cat had seven kits:
> Kits, cats, sacks, and wives,
> How many were going to St Ives?

   (a) How many of the following did I meet
      (i) wives?   (ii) sacks?   (iii) cats?   (iv) kits?
   (b) How many did I meet all together?
   (c) How many were going to St Ives?

5.6.10. How many committees of 1, or 2, or 3, $\ldots$, or 6 members can be formed from six people?

5.6.11. Marilyn has an old briefcase with a 4-digit combination lock. She only remembers one of the digits is a 2, but she does not remember the position it occupied. How many, at most, will she have to try?

5.6.12. Assuming that every Canadian has three initials, show that at least two Torontonians have the same initials.

5.6.13. In the 18th century, *Musikalisches Würfelspiel* (German for musical dice game) was popular throughout western Europe. It made it possible for people who could not compose music to create minuets, marches, polonaises, waltzes and so forth by rolling a die to select clips of music and determine its order. For Mozart's[8] version of the game, there are 11 possibilities for each of the first 16 measures, and for the remaining 16 measures there are 6 possibilities. You are allowed to reuse music so that you can have repeating measures. In how many ways can 32 measures of music be "composed?"

5.6.14. A digital color (raster) image consists of $m \times n$ pixels, and each pixel has 3 layers (red, green, and blue) with 256 levels for each layer. How many digital images are possible?

5.7 Combinations

In this section, we are concerned with counting the number of possible subsets with r elements from the set $\{1, 2, \ldots, n\}$. Here is our motivating problem for this section.

> **Problem 5.7.1 — Lottery Ticket.** In how many ways can we pick 6 different numbers from $\{1, 2, \ldots, 49\}$ for a lottery ticket?

With a lottery ticket, none of the numbers repeat, and the order does not matter (since order does not matter, it is a convention to write the numbers in ascending order); see Fig. 5.10. This kind of selection of elements from a set, in which the order does not matter, is called a **combination**.

The irrelevance of order in a lottery ticket may seem obvious to those who play the lottery. However, for those of you who do not play the lottery, or who are learning about the lottery for the first

Fig. 5.10   Sample lottery ticket.

---

[8]There is some doubt over Mozart's authorship. It is quite possible the publisher of the game attributed it to Mozart in an effort to boost sales (Hedges, 1978, p. 183).

time, lotteries have a long history. Governments often use lotteries to raise revenue without imposing additional taxes[9] on the people for expensive projects. In the United Kingdom, Queen Elizabeth I sanctioned the first national lottery held in 1569 to raise funds for the crumbling Cinque Ports. In the last century, other countries such as Canada used lotteries to raise funds for the 1976 Olympics in Montréal, and Australia to finance partially the construction of the Sydney Opera House (Moore, 1997, p. 171).

Let us use the following notation to represent the number of ways we can perform this kind of selection in Problem 5.7.1.

Definition 5.7.2 — The binomial coefficient. The number of ways we can select r objects, in which order does not matter, from the collection of n distinct objects, is denoted by

$$\binom{n}{r}. \tag{5.8}$$

We read $\binom{n}{r}$ as "n choose r," and this is called the **binomial coefficient**.[10] But, what is the value of $\binom{n}{r}$? At the moment, we do not know, but it is important for us to figure this out so that we can solve Problem 5.7.1. For the time being, with our new notation in hand, we can denote the number of possible lottery tickets as $\binom{49}{6}$. The question still remains; what is the explicit value of $\binom{49}{6}$?

With Pólya's method of problem solving, what we have done is successfully introduced notation for the problem. The unknown has been identified. This unknown, $\binom{49}{6}$, by our intuition is quite a large number, otherwise it would not be possible for so many to play the lottery and not win. With so many possibilities for a lottery ticket, it is unlikely we can list or enumerate them all. From Pólya's list of suggestions, he tells us to *find something familiar*. We could try to recall a similar problem or example we have seen before. Example 5.3.4 (on p. 148) calculates the number of arrangements of five students who wish to line up for a photo, and it could be useful. Let us go back and revisit that example.

Example 5.7.3 — Five Students Lining Up for a Photo. In how many ways can 5 students line up in a row for a photograph?

---

[9]English economist William Petty (1623–1687) said, "Now in the way of Lottery men do also tax themselves in the general, though out of hope of Advantage in particular: A Lottery therefore is properly a Tax upon unfortunate self-conceited fools" in his *Treatise of Taxes and Contributions* (1662) (Hald, 1990, p. 34).

[10]The choice of the name will become apparent in Sec. 10.2. There are many other notations for $\binom{n}{r}$ such as $C(n,r)$, $_nC_r$, ^nC_r, C_n^r, or $C_{n,k}$ to name a few where C stands for combinations. In this textbook, we will restrict ourselves to using $\binom{n}{r}$.

Solution — 1. The student in the first position can be selected 5 ways, the next position can be filled with a student in 4 ways, *etc.* Thus, there are

$$\underbrace{5}_{\text{Position 1}} \times \underbrace{4}_{\text{Position 2}} \times \underbrace{3}_{\text{Position 3}} \times \underbrace{2}_{\text{Position 4}} \times \underbrace{1}_{\text{Position 5}} = 5! = 120$$

arrangements of the students.   ◄

Solution — 2. We could think about this problem in a different way.

(a) Suppose out of the five students, only three students showed up initially to the meeting area. The photographer starts to get them ready for the photo (since he is running on a tight schedule). The number of ways that 3 students can be selected from the original 5 students (where the order does *not* matter) is denoted by $\binom{5}{3}$. Next, how many ways can the photographer arrange those 3 students in a line? There are

$$\underbrace{3}_{\text{Position 1}} \times \underbrace{2}_{\text{Position 2}} \times \underbrace{1}_{\text{Position 3}} = 3!$$

ways. Thus, by the Multiplication Principle, there are $\binom{5}{3} \times 3!$ arrangements for the first three positions; see Step (a) in Fig. 5.11.

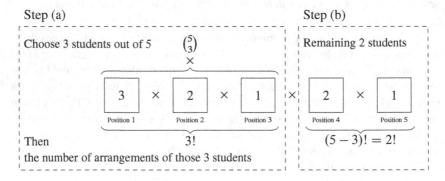

Fig. 5.11   The second solution for five students lining up for a photo.

(b) A couple of minutes later, the remaining two stragglers show up for the photo. They take their spots near the end of the line that has already formed. The number of arrangements with these remaining students is

$$\underbrace{2}_{\text{Position 4}} \times \underbrace{1}_{\text{Position 5}} = 2! = (5-3)!$$

ways; see Step (b) in Fig. 5.11.

Therefore, by combining the results from Steps (a) and (b) with the Multiplication Principle, the number of arrangements is $\binom{5}{3} \times 3! \times (5-3)!$.   ◄

Since solutions 1 and 2 are equivalent ways of thinking about the problem, then

$$5! = \binom{5}{3} \times 3! \times (5-3)!. \tag{5.9}$$

You might ask yourself now, "what is the value of $\binom{5}{3}$ (the number of ways that three students can be selected from five students, and the order does not matter)?" We can determine the value of $\binom{5}{3}$ by isolating it in Eq. (5.9).

$$\binom{5}{3} \times 3! \times (5-3)! = 5!$$

$$\binom{5}{3} = \frac{5!}{3! \times (5-3)!}$$

$$\binom{5}{3} = \frac{5!}{3! \times 2!}$$

$$\binom{5}{3} = \frac{120}{6 \times 2}$$

$$\binom{5}{3} = 10$$

Let us generalize Example 5.7.3 for an arbitrary number of students who are to be photographed, and for an arbitrary number of students who show up first.

Example 5.7.4 — n Students Lining Up for a Photo. In how many ways can n students line up in a row for a photograph?

Solution — 1. The student in the first position can be selected n ways, the next position can be filled with a student in $n - 1$ ways, *etc.* Thus, there are

$$\underbrace{n}_{\text{Position 1}} \times \underbrace{n-1}_{\text{Position 2}} \times \underbrace{n-2}_{\text{Position 3}} \times \cdots \times \underbrace{2}_{\text{Position } n-1} \times \underbrace{1}_{\text{Position } n} = n!$$

arrangements of the students.　◀

Solution — 2.
(a) Suppose out of n students, only r students show up initially to the meeting area. The number of ways that r students show up out of n students is $\binom{n}{r}$. Next, how many ways can the photographer line up r students? This can be done in $r!$ ways. Thus, by the Multiplication Principle, there are $\binom{n}{r} \times r!$ ways to arrange the students who showed up first; see Step (a) in Fig. 5.12.

(b) When the $(n - r)$ stragglers show up a couple of minutes later, they are appended to the line. The number of arrangements with these stragglers is $(n - r)!$; see Step (b) in Fig. 5.12.

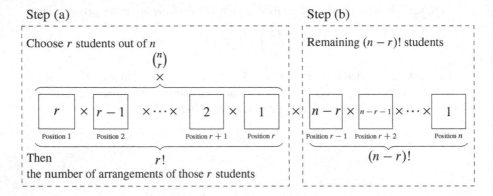

Fig. 5.12   The second solution for n students lining up for a photo.

Therefore, by combining the results from Steps (a) and (b) using the Multiplication Principle, the number of arrangements is

$$\binom{n}{r} \times r! \times (n-r)!.$$

◀

Since both of these solutions are equivalent ways of thinking about the problem, then

$$n! = \binom{n}{r} r!(n-r)!. \tag{5.10}$$

Together, the two solutions of Example 5.7.4 forms a combinatorial proof of Eq. (5.10). A **combinatorial proof** is a mathematical proof that uses a situation or a story about counting objects to convince you that the statement is true.[11] It demonstrates a combinatorial identity by counting the number of elements in two different ways, and concludes by setting the answers equal to obtain the desired identity. To write a combinatorial proof, we need to think of the correct set of objects to use, and think of how to answer the question, "In how many ways..." in two different ways. This simple technique is powerful enough to prove many identities and results in combinatorics. It has the advantage of being elegant, short, and intuitive to many people over an algebraic approach.[12]

---

[11] Remember that we talked about the differences between a formal proof and a mathematical proof in Sec. 2.1 on p. 31. Some people may have a hard time accepting a combinatorial proof because it is not what they are used to or expecting. Many people think that mathematical proofs are only algebraic in nature—this is not so.

[12] "You don't get to say you've proved something if you haven't explained it. A proof is a social construct. If the community doesn't understand it, you haven't done your job" —Cathy O'Neil (Chen, 2013).

Rearranging Eq. (5.10), we can solve for $\binom{n}{r}$:

$$n! = \binom{n}{r} r!(n-r)!$$

$$\binom{n}{r} r!(n-r)! = n!$$

$$\binom{n}{r} = \frac{n!}{r!(n-r)!}.$$

> **Definition 5.7.5 — Combinations of Objects.** The number of ways to select r objects from a collection of n distinct objects, where the order in which the objects are selected does not matter, is given by
>
> $$\binom{n}{r} = \frac{n!}{r!(n-r)!}. \tag{5.11}$$
>
> We say $\binom{n}{r}$ as "n choose r."

Note that the order of n and r in which we say "n choose r" is important since

$$\frac{n!}{r!(n-r)!} = \binom{n}{r} \text{ is not equivalent to } \binom{r}{n} = \frac{r!}{n!(r-n)!}.$$

Now with formula (5.11) for combinations in hand, we can calculate an explicit value for the number of possible lottery tickets in Problem 5.7.1.

Solution — Lottery Ticket Problem.

$$
\begin{aligned}
\binom{49}{6} &= \frac{49!}{6!(49-6)!} \\
&= \frac{49 \times 48 \times 47 \times 46 \times 45 \times 44 \times \cancel{43!}}{6!\cancel{43!}} \\
&= \frac{49 \times 48 \times 47 \times 46 \times 45 \times 44}{6!} \\
&= \frac{10\,068\,347\,520}{720} \\
&= 13\,983\,816.
\end{aligned}
$$

Therefore, a person can select from 13 983 816 different lottery tickets. ◀

There are approximately 14 million possible lottery tickets—that sure is a large number. If we instead decided that we wanted to enumerate all the possible tickets by writing them out, and say it took 3 seconds to write out each combination, then it would take us $3 \times 13\,983\,816 = 41\,951\,448$ seconds or about 485.5 days of continuous writing.[13] With the simple formula (5.11), we were able to calculate the number of possibilities in a couple of minutes instead of a few hundred days.

Example 5.7.6 How many five-card poker hands are possible from a standard 52 card deck?

Solution There are 52 distinct cards in a deck, and a person wants to pick 5 cards. Then, the number of possible poker hands is

$$
\begin{aligned}
\binom{52}{5} &= \frac{52!}{5!(52-5)!} \\
&= \frac{52!}{5! \times 47!} \\
&= \frac{52 \times 51 \times 50 \times 49 \times 48 \times \cancel{47!}}{5! \times \cancel{47!}} \\
&= \frac{52 \times 51 \times 50 \times 49 \times 48}{5 \times 4 \times 3 \times 2 \times 1} \\
&= \frac{311\,875\,200}{120} \\
&= 2\,598\,960.
\end{aligned}
$$

Therefore, there are $2\,598\,960$ different five-card poker hands.   ◄

Example 5.7.7 How many committees[14] of 2 men and 3 women can be formed in a company with 10 male and 13 female employees?

Solution Each committee is a subset of 2 men selected from a set of 10 men along with a subset of 3 women selected from a set of 13 women. The men can be selected in $\binom{10}{2}$ ways and the women can be selected in $\binom{13}{3}$ ways. Using the Multiplication Principle,

$$
\underbrace{\binom{10}{2}}_{\text{Men}} \times \underbrace{\binom{13}{3}}_{\text{Women}} = \frac{10!}{2!(10-2)!} \times \frac{13!}{3!(13-3)!}
$$

---

[13]There are 60 seconds in a minute, 60 minutes in an hour, and 24 hours in a day. That means there is $60 \times 60 \times 24 = 86\,400 \; \frac{\text{seconds}}{\text{day}}$. Then to find out how many days it would have taken us, $\frac{41\,951\,448 \; \cancel{\text{seconds}}}{86\,400 \; \frac{\cancel{\text{seconds}}}{\text{day}}} \approx 485.549$ days.

[14]"If you want to kill any idea in the world, get a committee working on it." —Charles F. Kettering (1876–1958), head of research at General Motors from 1920 to 1947.

$$= \frac{10!}{2!8!} \times \frac{13!}{3!10!}$$

$$= \frac{10 \times 9 \times \cancel{8!}}{2!\cancel{8!}} \times \frac{13 \times 12 \times 11 \times \cancel{10!}}{3!\cancel{10!}}$$

$$= \frac{10 \times 9}{2} \times \frac{13 \times 12 \times 11}{6}$$

$$= 12\,870.$$

Therefore, the number of possible committees is 12 870 consisting of 2 men and 3 women from the company. ◀

It is always a good idea to try out a formula on some special test cases. Here is an interesting property of combinations.

Theorem 5.7.8 — Symmetry of Combinations.

$$\binom{n}{r} = \binom{n}{n-r} \tag{5.12}$$

Proof — Algebraic approach.

$$\binom{n}{n-r} = \frac{n!}{(n-r)![n-(n-r)]!} \qquad \text{By (5.11)}$$

$$= \frac{n!}{(n-r)!(n-n+r)!} \qquad n-(n-r) = n-n+r$$

$$= \frac{n!}{(n-r)!r!} \qquad n-n = 0$$

$$= \frac{n!}{r!(n-r)!} \qquad \text{By commutativity.}$$

$$= \binom{n}{r} \qquad \text{By (5.11)}$$

◀

Proof — Combinatorial approach. Consider the situation of selecting objects out of a bag.

(a) The number of ways of picking r objects from n objects is $\binom{n}{r}$.

(b) The number of ways of not picking $n - r$ objects from n objects is $\binom{n}{n-r}$.
Since both of these methods are equivalent ways of counting the objects, therefore $\binom{n}{r} = \binom{n}{n-r}$. ◀

Recall that there are two types of reasoning to think about how to solve a counting problem. **Direct reasoning** solves the problem by counting all the desired arrangements to give the final answer. **Indirect reasoning** starts with all the possible arrangements and removes the undesired arrangements leaving only the desired arrangements.

Example 5.7.9 Suppose that I had five apples from which you will choose two of them. *Direct reasoning*: The number of ways for you to choose the two apples from the five that I am offering is

$$\binom{5}{2} = \frac{5!}{2!(5-2)!} = \frac{5!}{2! \times 3!} = \frac{5 \times 4 \times \cancel{3!}}{2! \times \cancel{3!}} = \frac{20}{2} = 10.$$

We can also give an equivalent way to think about this using *indirect reasoning*: the number of ways for you not to choose two apples from the five that I am offering is

$$\binom{5}{5-2} = \binom{5}{3} = \frac{5!}{3!(5-3)!} = \frac{5!}{3! \times 2!} = \frac{5 \times 4 \times \cancel{3!}}{\cancel{3!} \times 2!} = \frac{20}{2} = 10.$$

◀

The wonderful thing about proofs is that it allows you to see how something is true. But, it also affords you an opportunity to modify the proof to yield a different result.

I turn now our conversation upon painting for a moment. You might wonder how an artist did a painting—whether it be in oil, watercolor, or digital. To find out how he did it, you might find a video tutorial on his website. After finishing the video, you could try to paint the same picture that he painted exactly; the details are nicely done, but perhaps upon reflection you are dissatisfied with his painting as a whole. You may be more inclined to paint your own work, and to incorporate his techniques he demonstrated. In a conversation he was having with his personal secretary, the German writer Johann Wolfgang von Goethe (1749–1832) remarked, "If you see a great master, you will always find that he used what was good in his predecessors, and that it was this which made him great" (Oxenford, 1875, p. 187).

Proofs act in the same way as the video tutorial on painting. They give an explanation of why a theorem is true. A good proof should slowly construct an argument until it finishes with the final statement of the theorem. And in doing so, you may wonder if you can adapt or change parts of the argument to come up with a new theorem. Consider the following theorem.

Theorem 5.7.10

$$r\binom{n}{r} = n\binom{n-1}{r-1}$$

Proof We consider how many ways can we select r students from a class of n students for a committee with a president.

(a) We pick r students from a class of n students; the number of ways to do this is $\binom{n}{r}$. Then, there are r choices for the president. By the Multiplication Principle, there are $\binom{n}{r}r = r\binom{n}{r}$ ways to form the committee with a president.

(b) We first pick a president from the class; there are n choices. Then, we select the rest of the $r - 1$ committee members from the remaining $n - 1$ students; the number of ways to do this is $\binom{n-1}{r-1}$. By the Multiplication Principle, there are $n\binom{n-1}{r-1}$ ways to form the committee with a president.

Since both of these methods are equivalent ways of counting the ways of forming a committee with a president, therefore

$$r\binom{n}{r} = n\binom{n-1}{r-1}.$$

◀

Now looking back at the proof, you might ask yourself, what would happen if the committee needed a president *and* a treasurer? We essentially can replay the argument and adapt it so that we also pick a treasurer as well.

(a) We pick r students from a class of n students; the number of ways to this is $\binom{n}{r}$. Then, there are r choices for the president, and then $r - 1$ choices for the treasurer; by the Multiplication Principle, there are $r(r - 1)$ ways to do this. By the Multiplication Principle, there are $\binom{n}{r}r(r - 1) = r(r - 1)\binom{n}{r}$ ways to form the committee with a president and a treasurer.

(b) We first pick a president and a treasurer from the class; there are $n(n - 1)$ choices. Then, we select the rest of the $r - 2$ committee members from the remaining $n - 2$ students; the number of ways to do this is $\binom{n-2}{r-2}$. By the Multiplication Principle, there are $n(n - 1)\binom{n-2}{r-2}$ ways to form the committee with a president and a treasurer.

Since both of these methods are equivalent ways of counting the ways of forming a committee with a president and a treasurer, therefore

$$r(r - 1)\binom{n}{r} = n(n - 1)\binom{n-2}{r-2}.$$

We found another identity by modifying the previous proof, and simultaneously proved it. We can now state it also as a theorem.

Theorem 5.7.11

$$r(r - 1)\binom{n}{r} = n(n - 1)\binom{n-2}{r-2}$$

We conclude this section by generalizing the idea of a combinatorial proof into a template that you can use. All you have to do is to replace the text within the square brackets; this is often more easily said than done. The difficult part of a combinatorial proof is coming up with a situation and a set of objects where you count them in two ways such that it demonstrates the required combinatorial identity.

Template 5.7.12 — Combinatorial proof.
PROBLEM: Show that [combinatorial identity] is true.

Proof Consider [the following situation with a set of objects].
   (a) [Count the objects in a way that matches the left side of the combinatorial identity.]
   (b) [Count the objects in a way that matches the right side of the combinatorial identity.]
   Since both of these methods are equivalent ways of counting [the set of objects], therefore

$$[\text{restate combinatorial identity}]$$

◄

Be creative in choosing the situation or the set of objects for your proof. However, if you are having a difficult time coming up with your own, you can always try the following "standard" combinatorial objects and situations:
   (a) selecting toppings for a sandwich;
   (b) selecting men and women to form a committee;
   (c) selecting people who have a certain attribute and people who don't have that certain attribute to form a committee;
   (d) picking players (with and without certain attributes) to form a sports team; or
   (e) ordering different types of sushi from a menu.

Exercises

5.7.1. Evaluate the expression.

(a) $\binom{7}{3}$ (c) $\binom{7}{4}$ (e) $\binom{n}{2}$ (g) $\binom{7}{0}$

(b) $\binom{8}{8}$ (d) $\binom{10}{2}$ (f) $\binom{7}{r}$ (h) $\binom{n}{n-2}$

5.7.2. Which expression results in the smallest value?

$$\binom{10}{5} \qquad \binom{11}{7} \qquad \binom{15}{2}$$

5.7.3. Write an equivalent binomial coefficient for the following.

(a) $\binom{9}{3}$

(c) $\binom{8}{r}$ where $r > 0$.

(b) $\binom{43}{19}$

(d) $\binom{n}{3}$ where $n \geq 3$.

5.7.4. Using an algebraic proof, show that

(a) $\binom{10}{5} = \binom{9}{4} + \binom{9}{5}$

(c) $x\binom{n}{x} = n\binom{n-1}{x-1}$

(b) $5\binom{n}{5} = n\binom{n-1}{4}$

(d) $2\binom{2n-1}{n-1} = \binom{2n}{n}$

5.7.5. Determine if the following statement is true or false and state your justification:

$$P(n,r) = r!\binom{n}{r}$$

5.7.6. Solve for n, $\binom{n}{n-2} = 28$, where $n \in \mathbb{N}$.

5.7.7. If there are 8 points arranged on a circle, then how many different ways can 2 points be joined together to form a line?

5.7.8. For the 5-element set $\{a, b, c, d, e\}$, how many subsets are there that contain exactly three elements? List the subsets.

5.7.9. From a standard deck of 52 cards, how many poker hands (of 5 cards) consist of

(a) all red cards?

(c) all clubs?

(b) all face cards?

(d) all even rank cards?

5.7.10. From a standard deck of 52 cards, how many different five-card hands could be dealt if the hand included at least one card from each suit?

5.7.11. A bus shuttle takes 15 students at a time from the entrance of the university's library to the student parking lot. If there are 32 students waiting, how many ways can 15 students be chosen for the shuttle's

(a) first trip

(b) second trip

5.7.12. Alcide is packing his knapsack to take with him to school. He has 10 heavy textbooks, but he can carry at most 3 of them in his bag. What is the number of ways he could select the books that he brings to school?

5.7.13. A group of art experts are testing their ability of spotting a forgery. There are 20 paintings in which there are 19 legitimate paintings and one forgery. The judges of the competition allow each expert to select three paintings as a possible forgery.
  (a) How many ways can an expert make his or her selections?
  (b) How many ways can an expert make his or her selections if the forgery is one of the selected paintings?
  (c) How many ways can an expert select three paintings and all three paintings are legitimate?

5.7.14. There are 10 teams in a rugby tournament, and each of the teams must play each other team twice. How many games will be played in total?

5.7.15. Suppose in the House of Commons, there were 161 Conservatives, 97 New Democrats, 37 Liberals, and 7 Independent members of parliament. How many different committees could be formed with three Conservatives, three Liberals, three New Democrats, and one Independent member of parliament. (You can leave your answer as binomial coefficients.)

5.7.16. A final examination contains three sections, A, B, and C. In each section, there are 10 questions. If a student must answer all of the questions in section A, two questions from section B, and 3 questions from section C, then how many ways can a student select questions that he/she will answer on the examination?

5.7.17. In a package of 12 ballpoint pens, 4 are defective, but it is not known which pens are defective. In how many ways can you select
  (a) 4 pens from the box of 12?   (c) 4 pens that contain exactly two
  (b) 4 pens that are not defective?   defective pens?

5.7.18. At a business meeting, every person shakes each other's hand once. If there are 30 people, how many handshakes will take place?

5.7.19. At a business meeting, every person shakes each other's hand once. If there were 120 handshakes in total, then what is the number of people at this meeting?

5.7.20. A manager has 6 employees she can assign a task. In how many ways can she assign at least one employee to the task? Solve this using
  (a) direct reasoning.   (b) indirect reasoning.

5.7.21. How many different 5-card hands with at least 3 face cards can be dealt from a deck of 52 playing cards? Solve this using
  (a) direct reasoning.   (b) indirect reasoning.

5.7.22. Fong's Chop Suey House offers a combination plate where you can select any four different items from their menu for $11.75. They claim that there are 330 different combination plates. How many items are on the menu?

5.7.23. A young driver is writing a driver's license exam. The exam consists of 10 questions. In how many ways can this young driver choose the questions to answer if he decides to answer

(a) exactly eight questions? (c) at least eight questions?

(b) all of the questions?

5.7.24. How many ways can you select 12 members for a jury with an equal number of men and women from a group of 15 men and 12 women?

5.7.25. How many subsets with 7 elements of the set $\{1, 2, 3, \ldots, 50\}$ have

(a) 20 as the smallest element?

(b) 40 as the largest element?

(c) 20 as the smallest element and 40 as the largest?

(d) 25 as the middle element?

5.7.26. Using Theorem 5.7.8, show that

$$\binom{7}{0} + \binom{7}{2} + \binom{7}{4} + \binom{7}{6} = \binom{7}{1} + \binom{7}{3} + \binom{7}{5} + \binom{7}{7}$$

5.7.27. In a set A of 8 elements, let one element be marked as special.

(a) How many subsets of A with 5 elements that contain the special element are possible?

(b) How many subsets of A with 5 elements that do not contain the special element are possible?

(c) Using parts (a) and (b), show that $\binom{7}{4} + \binom{7}{5} = \binom{8}{5}$.

5.7.28. A contractor has built a neighborhood of n homes with all the same basic structure. To make them slightly different, the contractor painted them with different colors: r homes were red, b homes were blue, and the remaining y homes yellow. How many different ways could the contractor have painted these homes?

5.7.29. Create a scenario about selecting a committee where the calculation $\binom{8}{5}\binom{9}{2}$ would be appropriate.

5.7.30. Using Theorem 5.7.8,

(a) show that

$$\binom{7}{0}^2 + \binom{7}{1}^2 + \binom{7}{2}^2 + \cdots + \binom{7}{7}^2 = \binom{14}{7}$$

(*Hint*: You can use a combinatorial proof. Suppose you are picking a team of 7 students from 7 girls and 7 boys. How many ways can you do this? Can you solve this with two different approaches?)

(b) extend your reasoning to show that

$$\binom{n}{0}^2 + \binom{n}{1}^2 + \binom{n}{2}^2 + \cdots + \binom{n}{n}^2 = \binom{2n}{n}$$

(*Hint*: You can still use a combinatorial proof.)

5.8 Permutations with Repetition of Indistinguishable Objects

The permutations considered thus far have been those involving sets of **distinct** or **distinguishable** objects: given two objects we can discriminate between them based on some property of interest. In many situations, we are interested in finding the number of permutations of a set of objects in which not all of the objects are distinct.

Example 5.8.1 How many permutations can be made from the letters in the word DATA.

Solution There are $P(4, 4) = 4! = 24$ permutations of the four letters. We can do this by listing all of the possibilities.

DATA	DATA	DAAT	DAAT	DTAA	DTAA
TADA	TADA	TAAD	TAAD	TDAA	TDAA
ATAD	ATAD	AADT	AADT	AATD	AATD
ADAT	ADAT	ATDA	ATDA	ADTA	ADTA

But, why are there repeated permutations? It is because there are two indistinguishable letters; there are two As. To see this more clearly, we will replace one of the As with the symbol $\mathbb{A}$ and another with the symbol **A**.

DATA	DATA	DAAT	DAAT	DTAA	DTAA
TADA	TADA	TAAD	TAAD	TDAA	TDAA
ATAD	ATAD	AADT	AADT	AATD	AATD
ADAT	ADAT	ATDA	ATDA	ADTA	ADTA

We are over-counting using $P(4, 4)$. Instead, we need to refine what we want to find—what we want is to count the number of *distinguishable* permutations that we can form from the letters of DATA. To find the correct mathematical reasoning to count 12 distinguishable permutations, we apply direct reasoning and *take cases*. We break this problem into the following cases.

- The number of ways to select two As is $\binom{4}{2} = \frac{4!}{(4-2)!2!} = \frac{4!}{2!2!} = 6$.
- The number of ways to select one D from the remaining $4 - 2$ letters is $\binom{4-2}{1} = \binom{2}{1} = 2$.
- The number of ways to select one T from the remaining letter is $\binom{1}{1} = 1$.

Hence, by the Multiplication Principle there are

$$\binom{4}{2}\binom{4-2}{1}\binom{1}{1} = 6 \times 2 \times 1 = 12$$

distinguishable permutations from the letters of DATA. ◀

Example 5.8.2 The RMC Hockey team has scheduled six games during a season. How many ways can the season end in two wins, three losses, and one tie?

Solution — 1. *Take cases.* We can break this problem into the following cases:
  (a) Choose the games where the team would win. There are $\binom{6}{2}$ ways to do this.
  (b) Choose the games where the team would lose. To choose 3 loses amongst the remaining $6 - 2$ games is $\binom{6-2}{3} = \binom{4}{3}$ ways.
  (c) Choose the games where the team would tie. To choose 1 tie amongst the remaining $6 - 2 - 4$ games is $\binom{6-2-4}{1} = \binom{1}{1}$ ways.
By the Multiplication Principle,

$$\underbrace{\binom{6}{2}}_{\text{2 Wins}} \underbrace{\binom{6-2}{3}}_{\text{3 Loses}} \underbrace{\binom{6-2-4}{1}}_{\text{1 Tie}} = \binom{6}{2}\binom{4}{3}\binom{1}{1}$$

$$= 15 \times 4 \times 1$$
$$= 60.$$

Therefore, there are 60 ways that the season can end in two wins, three loses, and one tie. ◀

Theorem 5.8.3 summarizes what we have discovered in the previous examples. In fact, it gives us a handy formula that allows us to answer quickly these kinds of problems. After we have stated and proved the theorem, we will go back to Example 5.8.2 and apply the result.

Theorem 5.8.3 — Permutations of Objects, Not All Distinct. Given a set of n objects in which n_1 objects are alike and of one kind, n_2 objects are alike and of another kind, ..., and finally, n_r objects are alike and of yet another kind so that $n_1 + n_2 + \cdots + n_r = n$, then the number of permutations of these n objects taken n at a time is given by

$$\frac{n!}{n_1! n_2! \cdots n_r!} = \binom{n}{n_1, n_2, \ldots, n_r} \tag{5.13}$$

The symbol $\binom{n}{n_1, n_2, \ldots, n_r}$ is called a **multinomial coefficient**.

Proof Suppose we have n objects in which there are n_1 objects alike of one kind, n_2 objects alike of the second kind, $\ldots$, and finally n_r objects alike of the rth kind such that $n_1 + n_2 + \cdots + n_r = n$.

We can *take cases*.

- If we pick n_1 objects of one kind from n objects, then there are $\binom{n}{n_1}$ ways.
- If we pick n_2 objects of the second kind from the remaining $n - n_1$ objects, then there are $\binom{n-n_1}{n_2}$.
- If we pick n_3 objects of the third kind from the remaining $n - n_1 - n_2$ objects, there are $\binom{n-n_1-n_2}{n_3}$, and so on and so forth.
- If we pick n_r object of the rth kind from the remaining $n - n_1 - n_2 - \cdots - n_{r-1}$ objects, then there are $\binom{n-n_1-n_2-\cdots-n_{r-1}}{n_r}$.

By the Multiplication Principle, the number of ways is

$$
\binom{n}{n_1}\binom{n-n_1}{n_2}\binom{n-n_1-n_2}{n_3}\cdots\binom{n-n_1-n_2-\cdots-n_{r-1}}{n_r}
$$

$$
= \frac{n!}{(n-n_1)!n_1!}\frac{(n-n_1)!}{(n-n_1-n_2)!n_2!}\frac{(n-n_1-n_2)!}{(n-n_1-n_2-n_3)!n_3!}
$$

$$
\cdots\frac{(n-n_1-n_2-\cdots-n_{r-1})!}{(n-n_1-n_2-\cdots-n_r)!n_r!}
$$

$$
= \frac{n!}{n_1!n_2!n_3!\cdots\underbrace{\left(n-(n_1+n_2+\cdots+n_r)\right)}_{n}!n_r!}
$$

$$
= \frac{n!}{n_1!n_2!n_3!\cdots\underbrace{(n-n)!}_{0!=1}\,n_r!}
$$

$$
= \frac{n!}{n_1!n_2!n_3!\cdots n_r!}.
$$

◀

Solution — 2 of Example 5.8.2. Applying Theorem 5.8.3, the number of ways the team can end a season in two wins, three loses, and one tie is

$$
\binom{6}{2,3,1} = \frac{6!}{2!3!1!} = \frac{6\times5\times4\times3!}{2!3!1!} = \frac{6\times5\times4}{2} = 60.
$$

◀

Example 5.8.4 How many distinguishable permutations can be made from the letters in the word MISSISSAUGA?[15]

---

[15]Mississauga is a city in southern Ontario, Canada.

Solution In the word MISSISSAUGA, there are repeating letters; if we were not worried about the repetition, then by the Multiplication Principle, the number of ways to rearrange the word is 11!.

However, by Theorem 5.8.3, the number of distinguishable permutations that can be made from the letters MISSISSAUGA is

$$\binom{11}{1,2,4,2,1,1} = \frac{\overbrace{11!}^{\text{11 letters in total}}}{\underbrace{1!}_{\text{M}}\ \underbrace{2!}_{\text{I}}\ \underbrace{4!}_{\text{S}}\ \underbrace{2!}_{\text{A}}\ \underbrace{1!}_{\text{U}}\ \underbrace{1!}_{\text{G}}} = 415\,800.$$

Notice that the numbers in the denominator sum to 11:

$$1 + 2 + 4 + 2 + 1 + 1 = 11.$$

◀

Marin Mersenne (1588–1648) was a French music theorist (and mathematician) who is often called the father of acoustics. He was interested in counting the number of possible arrangement of musical notes when he stated Theorem 5.8.3 (Hald, 1990, p. 54). He was interested in the kind of problem like the one in the next example.

Example 5.8.5 How many distinct permutations of the notes (excluding the rest) in the opening bar of J. S. Bach's *Invention 2* are possible?

Fig. 5.13 The opening bar of *Invention 2* by Johann Sebastian Bach (1685–1750), BWV 773.

Solution As we can see in Fig. 5.13, there is a total of 14 notes in the opening bar, but some of the notes are repeated: ab4 is repeated twice, c5 three times, d5 twice, and eb5 twice. Using Theorem 5.8.3, there are

$$\binom{14}{1,1,2,1,1,3,2,2,1} = \frac{\overbrace{14!}^{\text{14 notes in total}}}{\underbrace{1!}_{\text{f4}}\ \underbrace{1!}_{\text{g4}}\ \underbrace{2!}_{\text{ab4}}\ \underbrace{1!}_{\text{bb4}}\ \underbrace{1!}_{\text{b4}}\ \underbrace{3!}_{\text{c5}}\ \underbrace{2!}_{\text{d5}}\ \underbrace{2!}_{\text{eb5}}\ \underbrace{1!}_{\text{f5}}} = 1\,816\,214\,400$$

distinguishable permutations of Bach's opening notes. ◀

Example 5.8.6 Victoria's apartment is 4 blocks south, and 9 blocks east from the university campus. How many paths can she take from her home to the university if she always travels either north or west?

Solution *Understand the problem.* To help us visualize and organize the data, we construct a diagram like Fig. 5.14(a). As we sketch, we begin to see acceptable paths from the apartment to campus, such as the path in Fig. 5.14(b). When we draw other acceptable paths, we notice that the length of all acceptable paths is 13, which consequentially is the sum of the grid dimensions $(9 + 4)$.

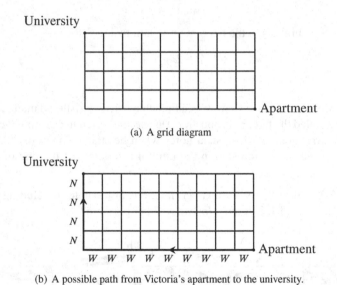

(a) A grid diagram

(b) A possible path from Victoria's apartment to the university.

Fig. 5.14   Paths from Victoria's apartment to the university.

Devise a plan. We can look for similarities to other problems that we have seen in this section. In fact, this problem looks equivalent to the number of arrangements of the possible decisions of either going north or west. Using direct reasoning, we can count all the possible paths of length 13 traveling north or west, and then remove the ways that will cause Victoria to move past the university too far west or too far north.

Carry out the plan. We introduce appropriate notation. Let N denote a north-going route past one block, and let W denote a west-going route past one block. Once we have done this, we can describe the path in Fig. 5.14(b) with $WWWWWWWWWNNNN$.

In general, for each possible path from her apartment to the university, Victoria must travel 4 blocks north and 9 blocks west.

Thus, the problem is equivalent to the number of arrangements of $9 + 4 = 13$ letters (like in Example 5.8.4), 4 of which are N and 9 of which are W. Therefore, Victoria can take any of the

$$\binom{13}{4,9} = \frac{13!}{4!9!} = 715$$

paths from her apartment to the university. ◀

Exercises

5.8.1. Does Example 5.8.1 use direct reasoning or indirect reasoning? Why?

5.8.2. For Example 5.8.1, construct a tree diagram. Is this direct reasoning or indirect reasoning? Why?

5.8.3. Compute the following multinomial coefficients:

(a) $\binom{3}{2,1}$ (b) $\binom{4}{2,0,2}$ (c) $\binom{15}{11,4}$ (d) $\binom{20}{15,2,3}$

5.8.4. Why is $\binom{20}{10,2,2,7}$ an invalid multinomial coefficient?

5.8.5. How many distinguishable "words" can be formed by rearranging the letters of the word

   (a) ACCEPTABLE   (i) PRESS
   (b) BAMBOO   (j) PRINCESSES
   (c) BOOKKEEPER   (k) QUINQUENNIUM
   (d) CANADA   (l) STATISTICS
   (e) CONTRAFIBULARITIES   (m) SUSPENDED
   (f) MATHEMATICS   (n) TORONTO
   (g) MATTRESSES   (o) WATERLOO
   (h) MISSISSIPPI   (p) SALADWARBLER

5.8.6. In how many ways can three employees divide 12 duties, each taking four?

5.8.7. The coach of the McMaster University rugby team needs to have 10 players standing in a row for a photo. Amongst these 10 players, there are 3 seniors, 4 juniors, 2 sophomores, and 1 freshman. How many different ways can they be arranged in a row if only their class level will be distinguished?

5.8.8. The rugby team played 11 games during the season. Their record was 6 wins and 5 losses. In how many ways could this occur?

5.8.9. In how many ways can eight dice land if there were three dice that landed on 1, two dice that landed on 4, and three dice that landed on 6?

5.8.10. In how many different ways are there to list the digits $\{1, 1, 1, 2, 2, 3, 4, 5\}$ so that the two 2s are in consecutive positions?

5.8.11. Ten coins are tossed. In how many ways can the coins land such that there are six tails and four heads?

5.8.12. A true or false quiz contains nine questions. How many different arrangements of the answers can the professor create if four answers are true, and five answers are false?

5.8.13. Given the word MAXIMUM,
    (a) how many distinguishable permutations are there?
    (b) how many of the permutations start with the letter X?
    (c) how many of the permutations begin with the letter M?

5.8.14. An art history class of 8 students is taking a trip to Paris. In how many ways can the students be assigned to
    (a) 4 double hotel rooms?   (b) 2 quad hotel rooms?

5.8.15. From Fig. 5.15, how many distinguishable permutations of the notes in J. S. Bach's *Invention 10* are possible?

Fig. 5.15   The opening bar of *Invention 10* by J. S. Bach (1685–1750), BWV 781.

5.8.16. From Fig. 5.16, how many distinguishable permutations of the notes (excluding the rest) in J. S. Bach's *Partita in A minor* are possible?

Fig. 5.16   The opening bar of *Partita in A minor* for solo flute by J. S. Bach (1685–1750), BWV 1013.

5.8.17. Lauren is visiting Manhattan and is staying at the hotel on 9th Avenue and West 57th Street; see Fig. 5.17. If Lauren can only walk either east or south, how many paths can she take

(a) from the hotel to the Museum of Modern Art (MoMA)?

(b) from the hotel to Rockefeller Center?

(c) from the hotel to Times Square?

(d) from Carnegie Hall to St. Patrick's Cathedral?

(e) from the Apple Store to St. Patrick's Cathedral?

Fig. 5.17   Map of Manhattan

5.8.18. For each of the following grids, how many paths are there if each path must only use north and west travel directions

(a) from point A to point B?

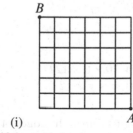

      (i)   (ii)

(b) if the grid size is

     (i) 10×10?   (ii) $x \times x$?   (iii) 7×14?   (iv) $x \times y$?

5.8.19. In Fig. 5.18, determine how many different paths there are from point *A* to point *B* if only north and west directions can be used.

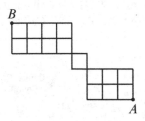

Fig. 5.18   Grid for Exercise 5.8.19

5.8.20. For each of the following grids made of cubes, how many paths are there if each path must follow the edges of the grid and must be as short as possible

    (a) from point *A* to point *B*?

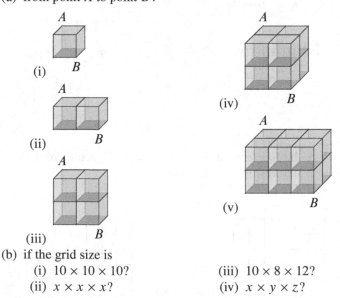

    (b) if the grid size is

        (i) $10 \times 10 \times 10$?   (iii) $10 \times 8 \times 12$?

        (ii) $x \times x \times x$?   (iv) $x \times y \times z$?

5.8.21. How many paths are there if each path must follow the edges of the grid (as shown in Fig. 5.19), and must be as short as possible from point *A* to point *B*? (*Hint*: Try exercises 5.8.19 and 5.8.20 first.)

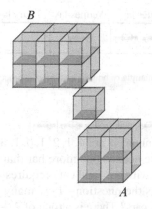

Fig. 5.19   Cube grid for Exercise 5.8.21

5.8.22 (†). Neoplasticism was a Dutch artistic movement in the 1920s and 1930s that was influenced by cubism and geometric forms. Suppose a neoplastic painter has separated his canvas into nine squares. He wants to paint the canvas as follows: three squares red, three blue, two yellow, and one black. How many ways can he complete his artwork if

(a) there are no other restrictions?   (b) there is a red square in every row?

(c) there is a red square in every row and in every column?

5.9 Combinations with Repetition of Indistinguishable Objects: An Application to Occupancy Problems

In some situations, it may be easier to think of distinguishable objects as indistinguishable. For example, you may be conducting a study on the number of accidents that occur in a week. Instead of looking at the individual accidents, you might be more interested in the number of accidents that occur in each of the weekdays. From this viewpoint, the accidents are indistinguishable. Such objects of study can be described by its **occupancy numbers** $r_1, r_2, \ldots, r_7$ where in this case r_i represents the number of accidents on the ith day. Every possible set of whole numbers that satisfies

$$r_1 + r_2 + r_3 + r_4 + r_5 + r_6 + r_7 = r \tag{5.14}$$

describes the possible assignment of accidents to each day, where r is the total number of accidents in a week.

Example 5.9.1 How many ways can 11 accidents occur over 7 days?

Solution One way that 11 accidents can occur over 7 days is given in Fig. 5.20.

We represent the accidents by cars and the days by bars. We can represent Fig. 5.20 using this notation, we would write

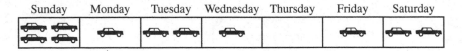

Sunday	Monday	Tuesday	Wednesday	Thursday	Friday	Saturday

Fig. 5.20   An example of how 11 accidents can occur over 7 days.

In this case, the occupancy numbers are $4, 1, 2, 1, 0, 1$, and 2. With this representation using cars and bars, we needed one more bar than the total number of days. With such a representation with bars and cars requires a bar at the beginning and a bar at the end. This begs the question—how many ways can we arrange the remaining 6 bars and the 11 cars? There is a total of $7 + 11 - 1$ symbols, in which we need to select 11 symbols; the number of ways to do this is

$$\binom{7 + 11 - 1}{11} = \binom{17}{11} = 12\,376.$$

Therefore, there are $12\,376$ ways for 11 accidents to occur over 7 days. ◄

Theorem 5.9.2 — Combinations with Repetition. The number of ways r indistinguishable objects can be distributed amongst n containers is given by

$$\binom{n + r - 1}{r} = \frac{(n + r - 1)!}{(n - 1)!\,r!} \tag{5.15}$$

We should point out that you do not need to memorize $\frac{(n+r-1)!}{(n-1)!r!}$. Instead, it is easier to remember $\binom{n+r-1}{r}$ and then use the expansion for $\binom{n}{r}$.

Proof We represent the objects as stars and the containers by bars. In Fig. 5.20, to represent using this notation, we would write

$$|\bigstar\bigstar\bigstar\bigstar|\bigstar|\bigstar\bigstar|\bigstar| \quad |\bigstar|\bigstar\bigstar|$$

In this case, the occupancy numbers are $4, 1, 2, 1, 0, 1$, and 2. With the representation using stars and bars, we needed one more bar than the total number of bins. In the general case, with r objects and n containers, we would need $n + 1$ bars and r stars. Such a representation with bars and stars requires a bar at the beginning and a bar at the end. This begs the question—how many ways can we arrange the remaining $n - 1$ bars and the r stars? There is a total of $n + r - 1$ symbols, in which we need to select r many symbols; the number of ways to do this is $\binom{n+r-1}{r}$. By the binomial coefficient definition 5.7.5,

$$\binom{n + r - 1}{r} = \frac{(n + r - 1)!}{((n + r - 1) - r)!\,r!} = \frac{(n + r - 1)!}{(n + \cancel{r} - 1 - \cancel{r})!\,r!} = \frac{(n + r - 1)!}{(n - 1)!\,r!}.$$

◄

Corollary 5.9.3 The number of combinations of n objects taken r at a time with repetition is given by

$$\binom{n+r-1}{r} = \frac{(n+r-1)!}{(n-1)!r!} \tag{5.15}$$

Example 5.9.4 A business supply store has several bins that contain different colors of markers: red, orange, yellow, green, blue and black. The store allows the customers to pick three markers for a fixed price. In how many ways can a selection of three markers be chosen?

Solution We assume that there is a sufficient number of markers of each color so that a customer will be permitted to select whatever markers he or she wants. For example, the customer may want all three markers to be blue; this is a combination with repetition. So,

$$\binom{6+3-1}{3} = \binom{8}{3} = \frac{8!}{(8-3)!3!} = \frac{8!}{5!3!} = \frac{8 \times 7 \times 6 \times 5!}{5!3!} = \frac{8 \times 7 \times 6}{3 \times 2 \times 1} = 56.$$

Therefore, there are 56 ways of selecting three markers from the bins, if repetition is allowed. ◄

Exercises

5.9.1. Why is $\binom{n+r-1}{r} = \binom{n+r-1}{n-1}$?

5.9.2. How many ways can a student select 2 cans of soda from a vending machine with 5 different flavors?

5.9.3. In how many ways can we assign 20 employees to five floors in a building?

5.9.4. At a buffet restaurant, there are 25 items available for selection. If you select the 5-item plate for $6.99,
    (a) how many selections can you make
        (i) with repetition of items?   (ii) without repetition of items?
    (b) Are there more choices if you allow repetition than without repetition? How much more or less?

5.9.5. In how many ways can you place 12 marbles of the same size in five different boxes if
    (a) all the marbles are black?
    (b) all the marbles are black and no container is empty?
    (c) each marble is of different color?

5.9.6. How many solutions to the equation $a + b = 4$ are there if a and b are whole numbers? List the solutions in the form (a, b).

5.9.7. How many solutions to the equation $a + b + c + d = 16$ are there if a, b, c, and d are whole numbers?

5.9.8. How many solutions to the equation

$$r_1 + r_2 + r_3 + r_4 = 20$$

such that $r_1 \in \{1, 2, 3, \ldots\}$, $r_2 \in \{3, 4, 5, \ldots\}$, and $r_3, r_4 \in \{0, 1, 2, \ldots\}$ are there? Why can you not apply Theorem 5.9.2 directly?

(*Hint: Find something familiar.* Try substituting $x_1 = r_1 - 1$. What are the elements of x_1? What does this substitution do? What will you replace r_2 with?)

5.9.9. The board game *Careers* was devised by the sociologist James Cook Brown (1921–2000) in 1955. Before the game begins, each player creates his or her own secret success formula by distributing 60 points in three categories: fame, fortune, and happiness. The first player to accumulate the minimum number of points in each category during play wins the game. In how many ways can a player create a secret success formula?

5.9.10. Curtis has \$100 to give to four of his favorite charities. Each donation must be in increments of \$10. One of the charities is very dear to him, so he gives the CNIB[16] \$20. How many ways can Curtis donate to the charities?

5.9.11. How many 4-digit numbers are there, the sum of whose digits is 8?

5.9.12. One of Dr Algebrix's least favorite duties is making up exams. For his discrete mathematics course, the exam has 8 questions and a total of 100 marks. Each question is worth at least 10 and at most 25 marks. In how many ways can Dr Algebrix assign marks to questions on the exam so that the marks for each question are a multiple of 5?

5.9.13. A bakery sells the following signature flavors of bagels: blueberry, French toast, cinnamon raisin, poppy seed, garlic Parmesan, and Swiss cheese. If you can make repeated selections of flavors, how many ways can you select
    (a) a dozen bagels?
    (b) a baker's dozen of bagels?
    (c) 24 bagels with at least two of each flavor?
    (d) 24 bagels with no more than two blueberry bagels?
    (e) 24 bagels with at least 5 blueberry and at least 3 French toast bagels?

---

[16]Formerly known as the Canadian National Institute for the Blind, it officially changed to its current name in 2006.

5.10 A Problem Solving Strategy for Counting Problems

As we near the end of the chapter, we have seen many tools and strategies to solve various counting-type problems. You have solved many problems in the previous sections, and I hope that you will agree that there is no single approach to solving a counting-type problem. The goal of this section is to review the tools we have accumulated thus far, and develop a problem solving strategy for solving counting-type problems.

We will develop our problem solving strategy based on Pólya's approach. His approach is summarized in the Preface starting on p. xviii, but we will review the four main steps and apply them to solving a counting problem: *understand the problem, devise a plan, carry out the plan, and look back.*

To develop our strategy, we will use the following problem.

Problem 5.10.1 A password with r letters is formed from the English alphabet. No letter may be used more than once in the password. How many of these passwords contain the letter a.

Understand the problem. Read the problem, and determine what you are required to find. With counting problems, we need to decide first if the problem requires us to find permutations (order matters) or combinations (order does not matter). Is the problem about finding the number of arrangements of objects (a permutation) or is it about trying to find the number of subsets of a set of objects (a combination)? With a complex counting-type problem, the problem may require dividing it into smaller problems, where

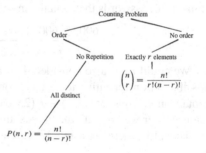

Fig. 5.21 Developing a strategy.

each of these smaller problems may require finding permutations, combinations, or both. With Problem 5.10.1, it is not obvious if it is the problem requires a combination or a permutation. It is foolish to give an answer to a question in which you do not understand the question in the first place. In order to help us understand the problem, we can look at a simpler related problem. If we are to form a password with $r = 3$ non-repeating letters, how many of those passwords would contain an a?

Devise a plan. Our plan is to solve the case when $r = 3$ and try to generalize afterwards to solve the original problem.

Carry out the plan. We first consider the case when $r = 3$. We need to select 3 non-repeating letters from the English alphabet. With this, it means that the order of the letters matter, so it must be a permutation. Our first guess would be

$$\underbrace{26}_{\text{Choices for 1}^{\text{st}}\text{ Letter}} \times \underbrace{25}_{\text{Choices for 2}^{\text{nd}}\text{ Letter}} \times \underbrace{24}_{\text{Choices for 3}^{\text{rd}}\text{ Letter}} = P(26,3) = 15\,600.$$

But, this does not ensure that at least one of the letters[17] is an a. Then, we fix one of the positions to be an a. We have the following three cases to consider:

- a is in the first position:

$$\underbrace{1}_{1^{\text{st}}\text{ Letter is }a} \times \underbrace{25}_{\text{Choices for 2}^{\text{nd}}\text{ Letter}} \times \underbrace{24}_{\text{Choices for 3}^{\text{rd}}\text{ Letter}} = 600$$

- a is in the second position:

$$\underbrace{25}_{\text{Choices for 1}^{\text{st}}\text{ Letter}} \times \underbrace{1}_{2^{\text{nd}}\text{ Letter is }a} \times \underbrace{24}_{\text{Choices for 3}^{\text{rd}}\text{ Letter}} = 600$$

- a is in the third position:

$$\underbrace{25}_{\text{Choices for 1}^{\text{st}}\text{ Letter}} \times \underbrace{25}_{\text{Choices for 2}^{\text{nd}}\text{ Letter}} \times \underbrace{1}_{3^{\text{rd}}\text{ Letter is }a} = 600.$$

By the Sum Principle, we add up the outcomes of all three cases to get the total number of passwords that contain an a:

$$\underbrace{600}_{\text{Case 1}} + \underbrace{600}_{\text{Case 2}} + \underbrace{600}_{\text{Case 3}} = 3(600) = 1800.$$

We could have also considered an indirect approach where we calculate all possible 3 non-repeating letter passwords and remove the passwords that do not contain an a. Thus, $P(26,3) - P(25,3) = 15\,600 - 13\,800 = 1800$, which is the same answer we got with direct reasoning.

Then, to generalize using direct reasoning, if there were r letters there would be

$$\underbrace{P(25,r-1) + P(25,r-1) + \cdots + P(25,r-1)}_{r\text{ times}} = r \times P(25,r-1).$$

Using indirect reasoning, there are $P(26,r) - P(25,r)$ passwords.

Look back. When we look at the two solutions, we need to ask ourselves if each solution is complete and correct in every detail of the reasoning. After doing so, we have good reason to believe that both solutions are correct. Nevertheless, errors are still possibly lurking especially if our reasoning is long and involved. "In order to convince ourselves of the presence or of the quality of an object, we like to see and touch it. And as we prefer perception through two different senses, so

---

[17]You may be thinking that this is a good opportunity to apply indirect reasoning; you would be correct. We will solve the problem by indirect reasoning after we first complete this constructive/direct reasoning solution.

we prefer conviction by two different [methods]" (Pólya, 1945, p. 15). This is the reason for solving this problem with two different methods. Although the answers appear to be different, we will now show that they are equivalent.

We will show that $P(26, r) - P(25, r) = r \times P(25, r - 1)$. Then,

$$
\begin{aligned}
P(26, r) - P(25, r) &= \frac{26!}{(26 - r)!} - \frac{25!}{(25 - r)!} \\
&= \frac{26 \times 25!}{(26 - r)!} - \frac{25!}{(25 - r)!} \frac{(26 - r)}{(26 - r)} \\
&= \frac{26 \times 25!}{(26 - r)!} - \frac{25! \times (26 - r)}{(26 - r)!} \\
&= \frac{26 \times 25! - 25!(26 - r)}{(26 - r)!} \\
&= (26 - 26 + r) \frac{25!}{(26 - r)!} \\
&= r \frac{25!}{(26 - r)!} \\
&= r \frac{25!}{(25 - (r - 1))!} \\
&= r \times P(25, r - 1)
\end{aligned}
$$

as required.

As stated before, it is not clear if this is a permutation or a combination problem. The previous two solutions used only permutations. We will give a third solution where we approach the problem from a different perspective.

Solution We will use Pólya's approach to solve Problem 5.10.1.

Understand the problem. We can think of the English alphabet as a set of distinct objects, and a password is a subset of the alphabet. From this perspective, we begin to think that we are counting subsets of the alphabet which leads us to think that this requires combinations.

Devise a plan. Using direct reasoning, we will look at the simple case when $r = 3$ letters, and then generalize.

Carry out the plan. For $r = 3$ letters, we need one of the letters to be a, and the rest can be any of the remaining 25 letters. Thus, there are $\binom{25}{2}$ possibilities for the remaining two letters. Now that we have selected the three letters from which we will form 3-letter passwords, how many ways can we form a password? This requires a permutation: for a set of 3 letters, there are 3! ways to permute them. Therefore, by the Multiplication Principle, there are $3! \times \binom{25}{2}$ passwords.

To generalize, if we pick r letters, there are $\binom{25}{r-1}$ possibilities for the remaining $r - 1$ letters (since one of the letters has to be an a). Once we have picked the r

letters, there are $r!$ ways to permute the letters. Therefore, by the Multiplication Principle, there are $r! \times \binom{25}{r-1}$ passwords.

Look back. We can compare our answer that we have found to one of the previous two answers to this problem. Again, our third answer looks different than the other two, but with some algebraic manipulation, we can see that they are equivalent.

$$r! \times \binom{25}{r-1} = r! \times \frac{25!}{[25 - (r-1)]!(r-1)!}$$

$$= r \times (r-1)! \times \frac{25!}{[25 - (r-1)]!(r-1)!}$$

$$= r \times \frac{25}{[25 - (r-1)]!}$$

$$= r \times P(25, r-1)$$

which is the same answer we found in the previous solution.   ◄

Notice that we found two different ways of expressing the answer using direct reasoning, and yet, we had different ways of thinking about the problem. This illustrates that we are able to count possibilities in many equivalent ways. When you are looking back at your work, and decide to compare with a friend, you should keep in mind that it may require some algebraic effort to show that the answers are equivalent, assuming neither you nor your friend made an error.

Exercises

5.10.1. How many subsets
   (a) are there for a set with seven elements?
   (b) from part (a) contain six elements?

5.10.2. How many subsets
   (a) are there for a set with n elements?
   (b) in part (a) contain $n - 1$ elements?

5.10.3. Improve Fig. 5.21 by adding more problem-types. For example, for counting problems with order and no repetition, a further subclass is that all elements are either distinct or some are alike.

5.10.4. Prove that $P(n, r) - P(n - 1, r) = r \times P(n - 1, r - 1)$ for $n, r \in \mathbb{N}$.

5.10.5. Six different signal flags can be selected to fly on a ship's flagpole. How many different signals can be made using at least two flags?

5.10.6. In a research laboratory at a university, there are 11 research technicians from which 6 need to be selected to attend a conference. Daniel and Mark are in the group of 11 research technicians, and their paper they submitted to the conference was accepted. They are required to give a presentation on this paper, and therefore they are required to be part of the group of 6 to attend. With this restriction, how many different ways can 6 research technicians from the laboratory attend the conference?

5.10.7. How many ways can 5 playing cards be selected from a deck of 52 cards that contain at least 4 hearts?

5.10.8. The prime factorization of 18 is $2 \times 3 \times 3$; see Fig. 5.22(a). Find the number of divisors of 18 other than 1 by finding all combinations of these numbers.

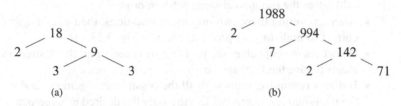

Fig. 5.22   Prime factorization

5.10.9. The prime factorization of 1988 is $2 \times 2 \times 7 \times 71$; see Fig. 5.22(b).
   (a) Find the number of divisors of 1988 other than 1.
   (b) Find the number of even divisors of 1988.

5.10.10. Every day for 10 days, a sales representative from a car dealership sells a vehicle. She sold 5 trucks, 3 crossovers, and 2 sedans. In how many ways could this have happened if after the first 4 days, she sold 3 trucks and a crossover?

5.10.11. In how many ways can you place n different books on r different book-shelves?

5.10.12. There are 25 married couples, and thus, there are 50 people in total. In how many ways can we form a committee of
   (a) 10 members that include at most 2 married couples?
   (b) 9 members and a Chairperson so that the Chair's partner is not a committee member?
   (c) 26 members that include no married couples?

5.10.13. At Fat King Burger, they claim in their advertisements that their "build your own burger" has over 1 billion (10^9) combinations. They have 5 choices of buns, 5 choices for meat/protein, and 8 choices of cheese. Assuming that you select a bun and a protein (you do not have to get it with cheese), how many toppings does

Fat King Burger need to have on their menu in order to support their advertisement claim?

5.11 Review

Summary

- When you are faced with a question that involves combinatorics, you need to ask yourself two very important questions:
    - (a) Is the problem concerned about the order of the objects or not?
    - (b) Does the problem allow repetition or not?
- After answering these two important questions, then you can apply the correct formula to your problem given in Fig. 5.23.
- **Direct reasoning** solves the problem by counting all the desired arrangements to give the final answer.
- **Indirect reasoning** starts with all the possible arrangements and removes the undesired arrangements leaving only the desired arrangements.
- You will usually find that it is easier to solve a combinatorial problem with one type of reasoning over the other.

Exercises

5.11.1. Pesto's restaurant offers the following menu in Table 5.7.

Table 5.7   Pesto's Menu

Appetizer	Main Course	Beverage
Wings	Canadian pizza	Soda
Chicken fingers	Meat Lover's pizza	Coffee
French fries	Hawaiian pizza	Milk
Onion fries	Meatball sub	Juice
Caesar salad	Calzone	
Poutine	Lasagna	
	Fettuccine Alfredo	

In how many ways can a single customer order a meal consisting of one choice from each category?

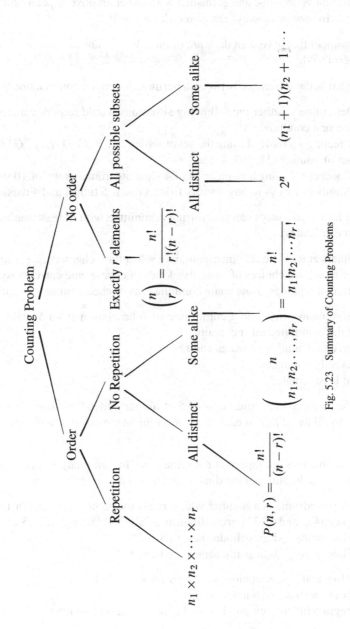

Fig. 5.23   Summary of Counting Problems

5.11.2. At the student bar, a martini[18] can be made from a selection of 4 different gins, 3 different vermouths, and garnished with either an olive, a pearl onion, or a lemon twist. In how many ways can you order a martini?

5.11.3. Padlocks that have digit dials are often called combination locks. Is this an accurate term? Why?

5.11.4. What is the difference between a permutation and a combination?

5.11.5. Determine whether the following situations would require calculating a permutation or a combination.
  (a) Creating a password using the set of letters $\{A, B, C, D, E, F, G\}$ and the set of numbers $\{1, 2, 3, 4, 5\}$.
  (b) Coaches selecting a swim team of 4 students from a group of 10 students.
  (c) Number of ways to buy flowers from 3 roses, 5 tulips, and 4 daisies.

5.11.6. In how many ways can you form a committee with at least one member from eight candidates?

5.11.7. Diners at Rick's Café Americain may select their cheeseburger from three varieties of meat, two choices of buns, five kinds of cheese, and can choose any of the ten different toppings. How many combinations of cheeseburgers are possible?

5.11.8. How many ways can a committee of 6 be chosen from 5 males and 7 females if the committee must contain
  (a) an equal number of males and females?
  (b) exactly 3 females?
  (c) at least 1 male?

5.11.9. The prime minister must choose 5 of his party members from a pool of 20 candidates to fill five different cabinet position. In how many different ways can he do this?

5.11.10. Seven dimes are flipped at the same time. In how many ways can 3 of the dimes land as heads, and 4 of the dimes land as tails?

5.11.11. A palindrome is a number which reads the same backward or forward. For example, 3443 and 12 321 are palindromes, while 3456 and 12 345 are not.
  (a) How many 4-digit palindromes are there?
  (b) How many 5-digit palindromes are there?

5.11.12. How many permutations of the numbers $\{1, 2, 4, 5\}$
  (a) begin with an even number?
  (b) begin with an even number, and end with an odd number?

---

[18]"A perfect martini should be made by filling a glass with gin, then waving it in the general direction of Italy" —Sir Noël Coward (1899–1973). Italy is a major producer of vermouth.

(c) begin with an even number, and end with an even number?

5.11.13. Determine how many 4-symbol passwords can be formed using the ten digits $0, 1, 2, \ldots, 9$ and the five vowels (a, e, i, o, u) of the alphabet if
   (a) repetitions are allowed but the password must begin with a vowel
   (b) repetitions are not allowed and the password must not begin with the digit 0.
   (c) repetitions are not allowed and the password must begin with a vowel.

5.11.14. What is the relationship between $P(n, r)$ and $\binom{n}{r}$?

5.11.15. Bingo is a game played with randomly drawn numbers which players match against numbers that have been preprinted on a 5×5 grid. Two typical Bingo cards are shown in Fig. 5.24.

B	I	N	G	O
2	16	31	46	61
6	18	35	60	65
15	20	Free	45	70
9	21	45	50	75
7	30	39	51	74

B	I	N	G	O
15	20	31	51	74
9	16	45	50	70
7	18	Free	60	75
2	21	35	45	61
6	30	39	46	65

Fig. 5.24   Examples of Two Bingo Cards

The numbers 1–15 are under column B, 16–30 are under column I, 31–45 are under column N, 46–60 are under column G, and 61–75 are under column O. Under the column N, there is a "free" space that the player automatically matches.
   (a) How many possible arrangements are possible under column
      (i) B?   (ii) I?   (iii) N?
   (b) How many possible Bingo cards are there?

5.11.16. An online video streaming service offers a selection of 300 movies. How many ways can you select 5 different movies for the weekend?

5.11.17. A manager needs 10 tasks to be completed, and wants two employees to do 5 tasks each. How many different ways can this division of duties be made?

5.11.18. How many three or four letter "words" can be formed from the letters in the word ALGORITHM?

5.11.19. Out of 50 Liberals and 42 Conservatives, how many ways are there of selecting a committee consisting of 4 Liberals and 4 Conservatives?

5.11.20. How many ways can 6 members of the hockey team line up at the blue line so that the 2 defensemen are not side by side?

5.11.21. How many ways can 5 playing cards be selected from a deck of 52 cards that contain at least 3 kings?

5.11.22. The Department of Mathematics receives a shipment of 24 mechanical pencils, including 4 that are defective. If Dr Pólya selects 5 mechanical pencils from the shipment,
  (a) how many selections can he make?
  (b) how many selections can be made where he selects none of the defective pencils?
  (c) how many selections can he make if he selects one or more defective pencils?

5.11.23. The members of the Emerson String Quartet composed of 2 violinists, a violist, and a cellist are to be selected from a group of 5 violinists, 3 violists, and 3 cellists, respectively.
  (a) How many ways can the string quartet be formed?
  (b) In how many ways can the string quartet be formed if one of the violinists is to be designated as the first violinist and the other is to be designated as the second violinist?

5.11.24. A football team played 12 games and has a record of 7 wins, 2 ties, and 3 losses. In how many ways could this happen if after the first 5 games, the team won 3 games and lost 2 games?

5.11.25. How many ways can ten students be assigned to 5 double hotel rooms?

5.11.26. If $P(n, r) = 6720$ and $\binom{n}{r} = 56$, then what is the value of r?

5.11.27. If $P(n, r) = 720$ and $\binom{n}{r} = 120$, then what is the value of r?

5.11.28. Show that

$$\binom{n}{r} = \frac{n - r + 1}{r} \binom{n}{r - 1}$$

where $n, r \in \mathbb{N}$.

5.11.29. Emma keeps many \$5, \$10, \$20, and \$50 bills in her bedside table. If she pulls out three bills from the drawer without looking, then how many assortments of bills could she have taken?

5.12 Bibliographic Remarks

There are many great exercises for the Pigeonhole Principle. The problem (Exercise 5.1.8) of showing that a gathering of any six people that either three of them are mutual acquaintances or complete strangers is by Bostwick (1958) with solution given by Bostwick *et al.* (1959). This is based on Ramsey Theory, which is the theory that is concerned with "how many elements of some structure must there be to guarantee some property will hold" problems. Ramsey Theory is named after the British mathematician Frank P. Ramsey (1903–1930) who published his ideas about these types of problems (Ramsey, 1930) shortly before he died from liver disease at age 26. Exercises 5.1.5 and 5.1.7 are from Walker (1977).

The proof of the Multiplication Principle was adapted from Whitworth (1870). In particular, Whitworth used the same table to visualize all possible outcomes of two tasks. His book was a classic for many years, but I have not seen any recent textbooks adapt his proof. I extended his table idea to three tasks.

Exercise 5.3.12 introduces the idea of cryptography. You can continue investigating the topic of cryptography with an elementary mathematical introduction by Sinkov and Feil (2009). For a comprehensive history of cryptography from Ancient Egypt up to about 1996, see Kahn (1996). In fact, it was so comprehensive that the National Security Agency tried to stop the first edition from being published while Kahn (1996) was writing it from 1964 to 1966. The Pentagon acquired the whole manuscript in 1966 and tried to pressure the publisher to remove sections and a whole chapter (Bamford, 1982, pp. 126–127).

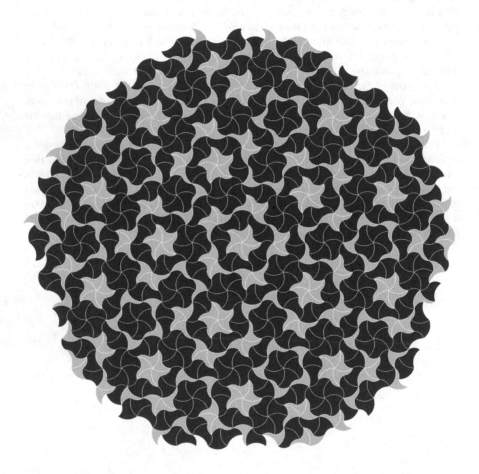

Fig. 6.1 A **Penrose tiling** is named after the English physicist and mathematician Sir Roger Penrose
(1931–). It is quite interesting that while it looks like it is a repeated and symmetrical pattern—a shifted
copy of a Penrose tiling will never match the original. Graphics Credit: Paul Gaborit.

6

Sequences and Series

6.1 Sequences

Humans are excellent at spotting patterns, and it is in our nature that we continue to seek out more of them; for example, see Fig. 6.1. It is this curiosity of spotting a pattern, making a guess what will happen next, and seeing if we are correct that is at the heart of Pólya's approach[1] to solving problems. In fact, humans are so good at pattern recognition that it is part of the measure for human intelligence. While this is something that we are able to perform easily, it has been surprisingly a difficult problem to teach computers, and thus a whole field of machine learning has developed.

Consider the following list of numbers

$$1, 1, 2, 3, 5, 8, 13, \ldots$$

what is the next number in the list? (Take a moment to think about this.)

Did you spot the pattern? It is an old pattern that was pointed out by the Italian mathematician Leonardo Bonacci (c. 1170–c. 1250), also known as Fibonacci of Pisa, while he was thinking about the patterns in rabbit breeding. When we write out a list of numbers, we call this a **sequence**.

> **Definition 6.1.1 — Sequence.** A **sequence** is an ordered list of numbers. The numbers in a sequence are called the **terms** of the sequence.

Example 6.1.2 We give some examples below of sequences.

    (a) $1, 3, 5, 7, \ldots$ (c) $\pi, 2\pi, 3\pi, 4\pi, \ldots$

    (b) $1, \frac{1}{2}, \frac{1}{3}, \frac{1}{4}, \ldots$ (d) $2, 3, 5, 7, 11, 13, 17, \ldots, 97$

The ellipsis ... (the three dots) can be read as "and so on," and denotes the idea[2] that the sequence will carry on for an infinite number of terms in parts (a)–(c); these are examples of infinite sequences. In part (d), the sequence is the prime numbers under 100, and in this case, the ellipses means "and so on until" we reach the prime number 97; this is an example of a finite sequence.   ◀

---

[1] His approach is summarized in the Preface starting on p. xviii.

[2] This is similar to the roster notation from set theory in Sec. 3.1.

To avoid confusion amongst other humans, it may be necessary to specify a rule in which to generate the next (infinite) terms.

Example 6.1.3 Consider the sequence

$$1, 3, 5, 7, 9, \ldots$$

Find a rule to find an arbitrary term in the sequence.

Solution We will use Pólya's approach to solve this problem.

Understand the problem. The problem is asking us to find a rule so that we can find an arbitrary term in the sequence; for example, how would we calculate the eighth term in the sequence? We notice that the sequence consists of the progression of odd natural numbers. This is an infinite sequence, and so this rule must hold for every natural number.

Devise a plan. We need to find a way to represent an arbitrary odd natural number. The idea is that the odd natural numbers and the even natural numbers are mutually exclusive sets; there are no common elements between the set of odd natural numbers, and the set of even natural numbers. If we find a rule for the even natural numbers, then it might be easier to find a rule for the odd natural numbers.

$$\underbrace{1}_{\text{odd}}, \overbrace{2}^{\text{even}}, \underbrace{3}_{\text{odd}}, \overbrace{4}^{\text{even}}, \underbrace{5}_{\text{odd}}, \overbrace{6}^{\text{even}}, \underbrace{7}_{\text{odd}}, \overbrace{8}^{\text{even}}, \underbrace{9}_{\text{odd}}, \overbrace{10}^{\text{even}}, \ldots$$

Since the natural numbers alternate between odd and even, if we find a rule for the even natural numbers, then we can just add or subtract 1 to yield the odd natural number rule!

Carry out the plan. We introduce appropriate notation. We will use t to represent a term, and t_n represent the nth term in the sequence. For even natural numbers, they are all multiples of 2; hence, the nth term in the even natural number sequence $2, 4, 6, \ldots$ is $t_n = 2n$ for every $n \in \mathbb{N}$. From our plan, we can form the odd natural number rule by adding 1:

$$t_n = 2n + 1 \qquad \text{for every } n \in \mathbb{N}.$$

Look back. It is important to check our rule is correct. We calculate a few terms in the sequence using the rule we created.

n	t_n
1	$2(1) + 1 = 3$
2	$2(2) + 1 = 5$
3	$2(3) + 1 = 7$
4	$2(4) + 1 = 9$

We see that *it is not* the correct sequence, since we did not include 1 in the sequence generated by the rule. Going back to our plan, we see that we could either add or subtract 1 from the even natural rule. If we use the following rule

$$t_n = 2n - 1 \qquad \text{for every } n \in \mathbb{N},$$

then we get the correct sequence.

n	t_n
1	$2(1) - 1 = 1$
2	$2(2) - 1 = 3$
3	$2(3) - 1 = 5$
4	$2(4) - 1 = 7$

◀

An interesting fact arose in Example 6.1.3: every integer n can be rewritten as either $n = 2k$ or $n = 2k + 1$ for some integer k. Those integers that have the form

$$2k \text{ are called } \textbf{even,}$$

and the integers that have the form

$$2k + 1 \text{ are called } \textbf{odd.}$$

Thus, an integer n is even if and only if n is not odd. In fact, the square of an even number is even

$$n^2 = (2k)^2 = 4k^2 = 2(2k^2),$$

and the square of an odd integer is odd

$$n^2 = (2k + 1)^2 = 4k^2 + 4k + 1 = 2(2k^2 + 2k) + 1.$$

The converse is true: if n^2 is even, then n is even; if n^2 is odd then n is odd. Thus, n is even *if and only if* n^2 is even; n is odd *if and only if* n^2 is odd.

If a sequence is alternating in sign such as

$$\underbrace{1}_{t_1}, \underbrace{-1}_{t_2}, \underbrace{1}_{t_3}, \underbrace{-1}_{t_4}, \ldots, \underbrace{(-1)^n}_{t_n}, \ldots$$

then you may want to incorporate $(-1)^n$ into your formula for the general term.

We will now return to our example of a sequence that was at the beginning of this chapter—the Fibonacci sequence. In Fibonacci's book, *Liber Abaci* (English: Book of Calculation) written in 1202, he considered an idealized rabbit population; we illustrate this in Fig. 6.2. In a field, there are a pair of newborn rabbits—one male and one female. When the rabbits are one month old, they have reached maturity, and they are able to reproduce. Thus, at the end of the second month of their life, they are able to produce another pair of rabbits. Fibonacci thought, if we assume that none of these rabbits died, and that each pair always produces another male-female pair of offspring, how many rabbits will there be in one year?

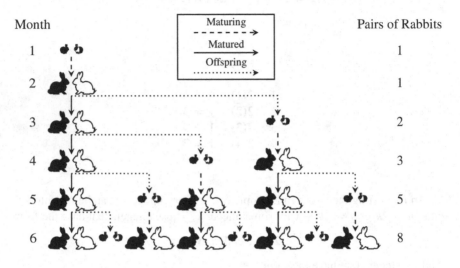

Fig. 6.2 (Idealized) Rabbits breeding generate the Fibonacci sequence.

To solve the problem, we will try to find a rule that generates the sequence in Fig. 6.2.

Example 6.1.4 — Fibonacci sequence. Find a rule to generate the Fibonacci sequence

$$1, 1, 2, 3, 5, 8, 13, \ldots$$

and determine the next two terms in the sequence.

Solution *Understand the problem.* We need to understand how the sequence is generated before we can write a rule. To start, you might *draw a diagram* like Fig. 6.2 of how many rabbits there are at each time step. After a couple of time steps, you may have noticed this pattern (Fig. 6.3): if we add the two previous terms, we get the next term.

Devise a plan. We have spotted the pattern, so we need to formalize the rule, and then use the rule to find the required eighth and ninth terms.

Carry out the plan. We introduce appropriate notation. We let t_n represent the nth term in the sequence. The previous term is t_{n-1}, and the term before t_{n-1} is t_{n-2}. Therefore, the rule for the Fibonacci sequence is

$$t_n = t_{n-2} + t_{n-1}. \tag{6.1}$$

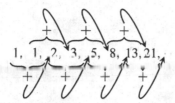

Fig. 6.3   If we add the two previous terms, we get the next term in the Fibonacci sequence.

The eighth term is t_8. By substitution in the given rule,

$$t_8 = t_6 + t_7 = 8 + 13 = 21$$

To find the ninth term,

$$t_9 = t_7 + t_8 = 13 + 21 = 34.$$

Look back. We should try to calculate some values using Eq. (6.1). There may be some issues if we handed this expression to another person. How would this person calculate t_1 or t_2? They cannot. In fact, we need to specify that $t_1 = 1$ and $t_2 = 1$. Then the Fibonacci sequence can be generated from the following rule

$$t_n = \begin{cases} 1 & \text{if } n = 1, 2 \\ t_{n-2} + t_{n-1} & \text{if } n = 3, 4, 5, \ldots \end{cases} \tag{6.2}$$

◄

Now, to answer Fibonacci's question: how many rabbits will there be after one year? Well, all we have to do is to find the 12th term in Fibonacci's sequence.

$$1, 1, 2, 3, 5, 8, 13, 21, 34, 55, 89, \mathbf{144}, 233, \ldots$$

Thus, after one year there would be 144 rabbits!

This type of rule in Example 6.1.4 to generate the Fibonacci sequence is called a recurrence relation. A **recurrence relation** is a rule that gives the value of the next term in the sequence using the previous terms in the sequence. The difficulty with a sequence given by recurrence relation is that you cannot calculate an arbitrary term without knowing the previous terms. In Example 6.1.4, we need to know t_{n-2} and t_{n-1} in order to calculate t_n. However, it is possible to find formulas for recurrence relations that do not depend on previous values. For example, the French mathematician Jacques Philippe Marie Binet (1786–1856) found a formula[3] to calculate any Fibonacci number:

$$t_n = \frac{1}{\sqrt{5}} \left[\left(\frac{1 + \sqrt{5}}{2} \right)^n - \left(\frac{1 - \sqrt{5}}{2} \right)^n \right]. \tag{6.3}$$

---

[3]For the proof of Binet's Formula, see Exercise 6.10.22 once we cover the Principle of Mathematical Induction later in this chapter.

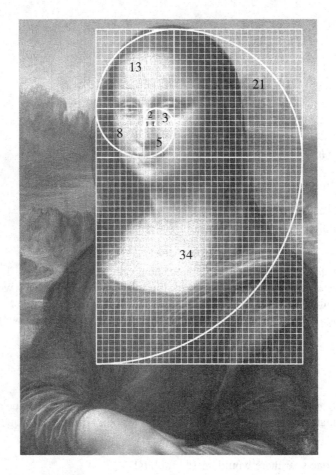

Fig. 6.4 *La Gioconda* or *Monna Lisa* is a painting by the Italian artist Leonardo da Vinci (1452–1519). The grid and Fibonacci sequence have been superimposed on the painting.

The number

$$\varphi = \frac{1 + \sqrt{5}}{2} = 1.618\,033\,988\ldots \tag{6.4}$$

is called the **golden ratio** (the Greek letter φ is phi).

The Fibonacci sequence is aesthetically pleasing to the eye, and it has been used deliberately by artists, designers, and architects in their designs. In Fig. 6.4 for example, I have superimposed a grid over *La Gioconda*, and you can see some interesting properties with relation to the curve the sequence has generated: the bottom of the eye touches it, the smile follows it, the temple grazes it, and the breasts are cupped by it. Although da Vinci is unlikely to have intentionally done this,[4] he may have subconsciously used these pleasing proportions in his work. Around 1498, he had illustrated his friend's book about the golden ratio called *De Divina Proportione* (English: On the Divine Proportion) by Luca Pacioli; da Vinci started painting *La Gioconda* around 1503.[5]

Figure 6.4 also illustrates another pattern that is found in the Fibonacci sequence:

$$1^2 + 1^2 = 2^2$$
$$1^2 + 2^2 = 3^2$$
$$2^2 + 3^2 = 5^2$$
$$3^2 + 5^2 = 8^2$$
$$\vdots$$
$$t_{n-2}^2 + t_{n-1}^2 = t_n.$$

Example 6.1.5 Find the last digit in the number 4^{1989}.

Solution *Understand the problem.* The first thing we notice is the sheer enormity of 4^{1989}. It really is a large number. You might have already pulled out your calculator (or smartphone) to try and calculate the answer, but you got a memory overflow error. It is too big for a calculator, so this requires you to demonstrate that you are smarter and more creative than it.

Devise a plan. If the problem is complex, break it up into smaller, but similar, problems that are easier to solve: *divide into smaller problems*. We could try investigating what the last digit is for smaller exponents.

Carry out the plan. Let us start with 4^1, 4^2, 4^3, ..., and see what happens.

Table 6.1

Number	Final Digit	Number	Final Digit
$4^1 = 4$	4	$4^5 = 1024$	4
$4^2 = 16$	6	$4^6 = 4096$	6
$4^3 = 64$	4	$4^7 = 16\,384$	4
$4^4 = 256$	6	$4^8 = 65\,536$	6

---

[4]In fact, many people would place the curve or the sequence in different locations of the painting.

[5]Markowsky (1992, pp. 10–12) argues that da Vinci did not use this aesthetic in his paintings, and that it is misleading to superimpose rectangles retroactively upon an artist's paintings when there is no evidence from the artist for doing so. We may be seeing patterns where none exist; for example, see the clustering illusion on p. 444.

By now, you should be able to *find a pattern* in Table 6.1. The final digits occur in a cycle of length two: 4, and 6. What is the 1989th term in the sequence

$$4, 6, 4, 6, 4, 6, \ldots?$$

If you divide 1989 by 2, the remainder is 1. So, the final digit is the first number in the cycle, namely, 4. Therefore, the final digit in the number 4^{1989} is 4.

Look back. After seeing this solution, we may be a little uneasy, for you might have spotted that we used *inductive reasoning*. Inductive reasoning tries to generalize from the observed cases, but this does not guarantee that this is true for all cases (especially the cases that we did not investigate).[6] We need a method that will be exhaustive in looking at all of the cases. Here is one way we could do that.

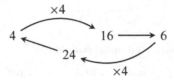

Fig. 6.5   The cycle of the last digit of repeatedly multiplying by 4.

For 4^1 equals 4. If we multiply our answer by 4, we get **16**. We keep the last digit, and we multiply it by 4, and we get **24**.   ◄

Now that we have investigated sequences, we will give the definitions of two major types of sequences: arithmetic sequences and geometric sequences. Informally, the next term in an arithmetic sequence is found by adding a constant to the previous term, whereas with a geometric sequence the next term is found by multiplying the previous term by a constant.[7]

Definition 6.1.6 — Arithmetic sequence. A sequence $t_1, t_2, t_3, \ldots$ is **arithmetic** if the difference between any two consecutive terms t_n and t_{n-1} is constant. An arithmetic sequence is determined by the first term a and the common (constant) difference d:

$$\underbrace{a}_{t_1}, \underbrace{a+d}_{t_2}, \underbrace{a+2d}_{t_3}, \underbrace{a+3d}_{t_4}, \ldots, \underbrace{a+(n-1)d}_{t_n}, \ldots \qquad (6.5)$$

The general term, or the nth term, is $t_n = a + (n-1)d$.

---

[6]For discussion about inductive reasoning and deductive reasoning, please refer to Sec. 2.8 on p. 59.
[7]You will prove this later in Exercise 6.10.18.

Example 6.1.7 Which of the following are arithmetic sequences?
  (a) $2, 5, 8, 11, 14, \ldots$
  (b) $2, 0, -2, -4, -6, -8, -10, \ldots$
  (c) $1, 1, 2, 3, 5, 8, 13, \ldots$
  (d) $2, 4, 8, 16, 32, \ldots$,

Solution
  (a) Yes. The common difference is $d = 3$.
  (b) Yes. The common difference is $d = -2$.
  (c) No. There is no common difference: $t_2 - t_1 = 0, t_3 - t_2 = 1, t_4 - t_3 = 1$, etc.
  (d) No. There is no common difference: $t_2 - t_1 = 2, t_3 - t_2 = 4, t_4 - t_3 = 8$, etc.

◀

Example 6.1.8 Find the tenth term in the arithmetic sequence

$$3, 11, 19, 27, \ldots$$

Solution The first term is $a = 3$, and the common difference is $d = 11 - 3 = 8$. Therefore, $t_{10} = 3 + (10 - 1)8 = 75$. ◀

Example 6.1.3 is also an arithmetic sequence. The first term is $a = 1$, and the common difference is $d = 2$. Therefore,

$$\begin{aligned} t_n &= 1 + (n - 1)2 \\ &= 1 + 2n - 2 \\ &= 2n - 1 \end{aligned}$$

We get the same t_n with both methods.

Definition 6.1.9 — Geometric sequence. A sequence $t_1, t_2, t_3, \ldots$ is **geometric** if the ratio, $r = \dfrac{t_n}{t_{n-1}}$, between any two consecutive terms is constant.
A geometric sequence is determined by the first term a, and the common (constant) ratio r:

$$a, ar, ar^2, ar^3, \ldots, ar^{n-1}, \ldots \qquad (6.6)$$

$$t_1 \quad t_2 \quad t_3 \quad t_4 \qquad t_n$$

The general term, or the nth term, is $t_n = ar^{n-1}$.

Example 6.1.10 Which of the following are geometric sequences?
  (a) $2, 4, 8, 16, 32, \ldots$
  (b) $-5, 10, -20, 40, -80, \ldots$
  (c) $4, \frac{8}{3}, \frac{16}{9}, \frac{32}{27}, \frac{64}{81}, \ldots$
  (d) $2, 4, 16, 256, 65\,636, \ldots$

Solution

    (a) Yes. The common ratio is $r = \frac{t_2}{t_1} = \frac{4}{2} = 2$.

    (b) Yes. The common ratio is $r = \frac{t_2}{t_1} = \frac{10}{-5} = -2$.

    (c) Yes. The common ratio is $r = \frac{t_2}{t_1} = \frac{\frac{8}{3}}{4} = \frac{2}{3}$.

    (d) No. The ratio is not common for all terms.

        For example, $\frac{t_2}{t_1} = \frac{4}{2} = 2$, $\frac{t_3}{t_2} = \frac{16}{4} = 4$, $\frac{t_4}{t_3} = \frac{256}{16} = 16$, *etc.*

                                                                                              ◄

Example 6.1.11 Find the tenth term in the geometric sequence

$$5, 10, 20, 40, \ldots$$

Solution The first term of the sequence is $a = 5$, and the common ratio is $r = \frac{10}{5} = 2$. Therefore, $t_{10} = 5(2)^{10} = 5120$.   ◄

Another interesting mathematical object is a fractal. **Fractals** are patterns that keep repeating themselves at every scale. You can keep zooming into a fractal, and it keeps appearing just as it did when you started. Fractals are interesting since they can produce very complex appearing patterns just with a simple rule. A Mandelbrot plot, named after the French-American mathematician Benoît Mandelbrot (1924–2010) who pioneered fractal research, is given in Fig. 6.6.

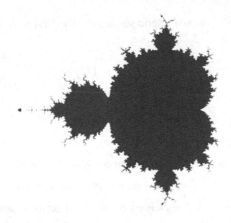

Fig. 6.6   A Mandelbrot plot

Example 6.1.12 — Koch's snowflake. One of the earliest examples of a fractal is from the Swedish mathematician Helge von Koch (1870–1924). The Koch snowflake can be constructed by starting with an equilateral triangle (Fig. 6.7(a)), then altering each line segment as follows:

    (a) Divide the line segment into three segments of equal length.

(b) Draw an equilateral triangle that has the middle segment from Step (a) as its base and points outward.

(c) Remove the line segment that is the base of the triangle from Step (b).

By altering each line segment on the equilateral triangle this way, this yields Fig. 6.7(b). If we repeat the same process on each line segment in Fig. 6.7(b), this yields Fig. 6.7(c), and so forth. This process can be repeated an infinite number of times. When this is repeated an infinite number of times, we call this Koch's snowflake, which is a fractal.

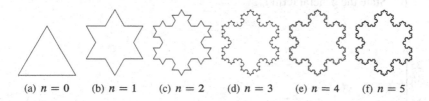

Fig. 6.7   Constructing Koch's snowflake. Koch's snowflake is when n goes to ∞.

Exercises

6.1.1. How many rabbits would there be in one year if Fibonacci started with
(a) two rabbits?   (b) three rabbits?

6.1.2. It is common today to express the Fibonacci sequence by the following expression

$$t_n = \begin{cases} 0 & \text{if } n = 1, \\ 1 & \text{if } n = 2, \\ t_{n-2} + t_{n-1} & \text{if } n = 3, 4, 5, \dots. \end{cases} \tag{6.7}$$

What is the difference between the Fibonacci sequence generated by expression (6.7), and expression (6.2) we found in Example 6.1.4?

6.1.3. Write the first six terms in each sequence.
(a) $t_n = 1 + (-1)^n$
(b) $t_n = (-3)^n$

(c) $t_n = \dfrac{n(n-1)}{2}$

(d) $t_n = (-1)^n 2n$

(e) $t_n = \dfrac{(-1)^n}{n}$

(f) $t_n = \dfrac{1}{n(n+1)}$

(g) $t_n = \dfrac{1}{n^2 + n}$

(h) $t_n = \begin{cases} n & \text{if } n \text{ is odd} \\ \dfrac{1}{n} & \text{if } n \text{ is even} \end{cases}$

6.1.4. Figure 6.8(a) shows a graph of an arithmetic sequence.
(a) List the first three terms of the sequence.
(b) Determine the 11th and 12th term of the sequence.
(c) State the general term t_n.

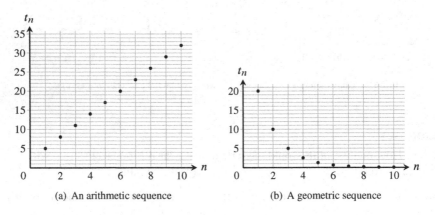

(a) An arithmetic sequence   (b) A geometric sequence

Fig. 6.8   Examples of an arithmetic and a geometric sequence.

6.1.5. Figure 6.8(b) shows a graph of a geometric sequence.
(a) List the first three terms of the sequence.
(b) State the general term t_n.

6.1.6. The first four terms of each sequence are given below. Give a rule for the nth term for each sequence.

(a) $5, 7, 9, 11, \ldots$

(b) $1, 4, 7, 10, \ldots$

(c) $\dfrac{1}{2}, \dfrac{2}{3}, \dfrac{3}{4}, \dfrac{4}{5}, \ldots$

(d) $1, 4, 9, 16, \ldots$

(e) $1, 8, 27, 64, \ldots$

(f) $\dfrac{1}{1}, -\dfrac{1}{2}, \dfrac{1}{3}, -\dfrac{1}{4}, \ldots$

6.1.7. Give a rule to find the next letter in the following sequence:

$$O, T, T, F, F, S, S, E, \ldots.$$

6.1.8. List the first six terms for each sequence given by the following recurrence relations.

(a) $t_1 = 1, t_{n+1} = 4t_n + 1$

(b) $t_1 = 2, t_{n+1} = \dfrac{t_n + 2}{2t_n}$

(c) $t_1 = 1, t_2 = 2, t_n = t_{n-2} + t_{n-1}$

(d) $t_1 = 3, t_2 = 5, t_{n+2} = \dfrac{t_{n+1} + t_n}{t_{n+1} - t_n}$

6.1.9. Explain the distinction between an arithmetic sequence and a geometric sequence.

6.1.10. Find the general term t_n for each of the arithmetic or geometric sequences below.

(a) $10, 1, 0.1, 0.01, \ldots$

(b) $-\dfrac{5}{6}, \dfrac{5}{9}, -\dfrac{10}{27}, \dfrac{20}{81}, \ldots$

(c) $2k, 2k + 2, 2k + 4, 2k + 6, \ldots$

(d) $\dfrac{a}{b}, -\dfrac{a}{b}, -\dfrac{3a}{b}, -\dfrac{5a}{b}, \ldots$

(e) $\dfrac{m^2}{n}, -m, n, -\dfrac{n^2}{m}, \ldots$

6.1.11. Suppose that the first two terms of a geometric sequence is $t_1 = x$ and $t_2 = y$. What is the tenth term in this sequence?

6.1.12. Suppose that an arithmetic sequence has the first term a, and a common difference d. Express this arithmetic sequence with a recursive relation.

6.1.13. Suppose that a geometric sequence has the first term a, and a common ratio r. Express this geometric sequence with a recursive relation.

6.1.14. In every generation, a bacteria divides by binary fission into two separate cells. Let t_n be the number of cells in the nth generation. If there are four cells in the first generation, what is the general term t_n that will describe this sequence?

6.1.15. After rinsing a glass, there is a little bit of soap residue that remains. Suppose that after each rinse, 95% of the residue is removed.

(a) How much soap residue remains after the first washing?

(b) What percentage of the original soap residue remains after 10 rinses?

(c) How many rinses of the glass will it take to ensure that 1% or less of the original soap residue remains?

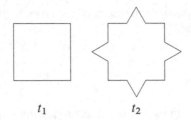

t_1 t_2

Fig. 6.9   What is the next diagram in the sequence?

6.1.16. Construct the next diagram in the sequence in Fig. 6.9.

6.1.17. Najma wrote repeatedly the sequence of symbols ♣◇◇♣♡♣♣♣♡ a total of 75 times. How many more ♣ symbols than ◇ symbols did she write?

6.1.18. Prove or disprove the following:
    (a) The sum of two even integers is an even number.
    (b) The sum of two odd integers is an odd number.

6.1.19 (†). Let $t_1, t_2, t_3, \ldots$ be an arithmetic sequence. Show that the terms $t_2, t_4, t_6, \ldots$ also form an arithmetic sequence.

6.2 Method of Finite Differences

The **method of finite differences** is used to find a general term of a sequence whose values are generated by a polynomial. A **polynomial** is a mathematical expression consisting of coefficients and variables with whole number exponents. An example of a polynomial is $3x^2 + 2x + 1$, where the **degree** (the highest exponent on a variable) is 2, and the coefficients are 3, 2, and 1. The general forms of polynomials are given in Table 6.2.

Table 6.2   Polynomial forms

Name	Degree	General Form
Linear	1	$ax + b$
Quadratic	2	$ax^2 + bx + c$
Cubic	3	$ax^3 + bx^2 + cx + d$
Quartic	4	$ax^4 + bx^3 + cx^2 + dx + e$
⋮	⋮	⋮

Example 6.2.1 Find the polynomial (or the general term) that generated the sequence in Table 6.3.

Table 6.3   Sequence values for Example 6.2.1

x	1	2	3	4	5	6
$f(x)$	1	4	9	16	25	36

Solution There are two steps: we need to find the degree of the polynomial, and then we need to determine the coefficients of the polynomial.

(a) *Find the degree of the polynomial by the method of finite differences.* In Fig. 6.10, we calculate all of the differences between the values; this is called the first difference, and we denote this with Δ^1. Next, we calculate the differences of the differences between the values; that is, we are finding the second differences, which is denoted by Δ^2. Notice that all of the second differences are the same; this indicates that the degree of the polynomial that generated this sequence is 2.

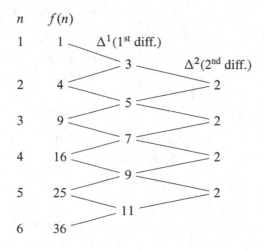

Fig. 6.10   Differences between sequence values in Example 6.2.1

Since we know that the polynomial is degree 2, it must have the form $ax^2 + bx + c$.

(b) *Solving three equations with three unknowns.* Using the data, we get

$$f(1) = 1 = a(1)^2 + b(1) + c = a + b + c$$
$$f(2) = 4 = a(2)^2 + b(2) + c = 4a + 2b + c$$
$$f(3) = 9 = a(3)^2 + b(3) + c = 9a + 3b + c$$

So, we have to solve 3 equations with 3 unknowns.[8]

$$\begin{cases} a + b + c = 1 \ ① \\ 4a + 2b + c = 4 \ ② \\ 9a + 3b + c = 9 \ ③ \end{cases}$$

Rewriting ① to isolate c on the left side, we have

$$\begin{cases} c = 1 - a - b \ ④ \\ 4a + 2b + c = 4 \qquad ② \\ 9a + 3b + c = 9 \qquad ③ \end{cases}$$

Substitute ④ into ② so that we have

$$4a + 2b + c = 4$$
$$4a + 2b + (1 - a - b) = 4$$
$$3a + b + 1 = 4$$
$$3a + b = 3$$
$$b = 3 - 3a \quad ⑤$$

and thus,

$$\begin{cases} c = 1 - a - b \ ④ \\ b = 3 - 3a \qquad ⑤ \\ 9a + 3b + c = 9 \qquad ③ \end{cases}$$

Substitute ⑤ into ④,

$$c = 1 - a - (3 - 3a)$$
$$c = 1 - a - 3 + 3a$$
$$c = -2 + 2a \quad ⑥$$

and thus,

$$\begin{cases} c = -2 + 2a \ ⑥ \\ b = 3 - 3a \qquad ⑤ \\ 9a + 3b + c = 9 \qquad ③ \end{cases}$$

---

[8]It is possible to do this quickly with a calculator. For instance, with the Casio fx991-MS, you can solve equations up to three degrees, and simultaneous linear equations with up to three unknowns. The TI-83 graphing calculator has a *quadratic regression* function to find polynomials of degree 2, and the *cubic regression* to find polynomials of degree 3. Check your calculator's manual for details.

Substitute ⑤ and ⑥ into ③, we have

$$9a + 3b + c = 9$$
$$9a + 3(3 - 3a) + (-2 + 2a) = 9$$
$$9a + 9 - 9a - 2 + 2a = 9$$
$$2a = 2$$
$$a = 1 \quad ⑦$$

We have

$$\begin{cases} c = -2 + 2a & ⑥ \\ b = 3 - 3a & ⑤ \\ a = 1 & ⑦ \end{cases}$$

and so to find b, we substitute ⑦ into ⑤,

$$b = 3 - 3(1)$$
$$b = 0$$

and to find c, substitute ⑦ into ⑥

$$c = -2 + 2(1)$$
$$c = 0$$

Hence, the coefficients are $a = 1$, $b = 0$, and $c = 0$. Therefore, the polynomial that generated the sequence of values in Table 6.3 is $f(n) = n^2$.

Looking back, we see that indeed the values follow a quadratic function.

◀

In summary,

(a) the method of finite differences allows us to determine the degree, D, of the polynomial if a polynomial did in fact generate the sequence;

(b) next, we set up $(D + 1)$ number of equations with $(D + 1)$ unknowns to determine the coefficients of the polynomial.

Exercises

6.2.1. Determine the degree of the following polynomials:

(a) $n^2 + n + 4$

(b) $3n^2 + 4n + 3$

(c) $10x^3 + 5x^2 + 9$

(d) $3x^4 + 2x^2 + 1$

(e) $3n^5 + 2n^{13} + 20n + 3$

(f) $x^{10} + 16x^{20} + 3x + x^3$

(g) $n^2 + n + 2^3$

(h) $n^3 + n^2 + 4^5$

6.2.2. Determine the next three terms in the following sequences. (*Hint:* You do not have to find the polynomial that generates the sequence. You can just use the method of finite differences.)

(a) $3, 8, 15, 24, 35, 48, \ldots$

(b) $9, 32, 91, 204, 389, 664, \ldots$

(c) $2, 25, 88, 209, 406, 697, \ldots$

(d) $45, 34, 11, -30, -95, -190, \ldots$

(e) $68, 82, 436, 2060, 6868, 18\,118, \ldots$

(f) $-5, -48, -81, 256, 1875, 6480, \ldots$

6.3 Series and Sigma Notation

A **series** is the sum of the terms in a sequence. For example,

$$1, 1, 2, 3, 5, 8, 13$$

is a sequence, and its corresponding series is

$$1 + 1 + 2 + 3 + 5 + 8 + 13.$$

If the sequence is finite, then its corresponding series is a finite series. If the sequence is infinite, then its corresponding series is an infinite series.

Suppose we have a finite sequence of n terms $t_1, t_2, t_3, \ldots, t_n$, then the nth **partial sum** of the corresponding series is

$$S_n = t_1 + t_2 + t_3 + \cdots + t_n, \tag{6.8}$$

which is the sum of the first n terms in the sequence. The partial sums can be visualized as the following:

$$t_1 \underbrace{+t_2 +t_3 +t_4 + \cdots + t_n}$$

$$S_1$$

$$S_2$$

$$S_3$$

$$S_4$$

$$S_n$$

Example 6.3.1 What is the 5th partial sum of the sequence with the general term $t_n = n\pi$.

Solution The sequence is $\pi, 2\pi, 3\pi, 4\pi, 5\pi, 6\pi, 7\pi, \ldots$. Then, the 5th partial sum of the sequence is $\pi + 2\pi + 3\pi + 4\pi + 5\pi = 15\pi$. ◀

We have been faced before with long sums that can take a long time to write. For example, the sum

$$1 + 2 + 3 + 4 + 5 + 6 + 7 + 8 + 9 + 10 + \cdots + 100 \tag{6.9}$$

or

$$\binom{10}{0} + \binom{10}{1} + \binom{10}{2} + \binom{10}{3} + \binom{10}{4} + \cdots + \binom{10}{9} + \binom{10}{10} \tag{6.10}$$

to write out is very time-consuming—not to mention a waste of paper and ink.

One way of writing long repetitive sums is to use the Greek letter $\sum$ (capital sigma, corresponding to the Roman letter S for sum). This is called **sigma notation**. Thus, we can rewrite (6.9) more compactly as

$$1 + 2 + 3 + \cdots + 99 + 100 = \sum_{i=1}^{100} i$$

and similarly with (6.10) as

$$\binom{10}{0} + \binom{10}{1} + \cdots + \binom{10}{10} = \sum_{i=0}^{10} \binom{10}{i}.$$

In general, to indicate the sum of the terms $t_1, t_2, t_3, \ldots, t_m, t_{m+1}, \ldots, t_n, \ldots,$ from the mth term up to and including the nth term (where $m \leq n$), we write

$$t_m + t_{m+1} + t_{m+2} + \cdots + t_{n-1} + t_n = \sum_{i=m}^{n} t_i. \tag{6.11}$$

$$\begin{array}{l} \text{This tells us to} \\ \text{end with } i = n. \end{array} \quad\longrightarrow\quad \overset{n}{\underset{i=m}{\sum}} t_i \quad\begin{array}{l} \text{This tells us that} \\ \text{it is a sum.} \end{array}$$

This tells us to start with $i = m$.

The letter i is called the **index of summation**,[9] the number m is called the **lower limit of summation**, and the number n is called the **upper limit of summation**.

Example 6.3.2 Write the following sums using sigma notation.

(a) $1 + 2 + 3 + 4 + \cdots + 10$

(b) $1 + \dfrac{1}{2} + \dfrac{1}{3} + \dfrac{1}{4} + \cdots + \dfrac{1}{10}$

(c) $\binom{10}{1} + \binom{10}{2} + \binom{10}{3} + \cdots + \binom{10}{10}$

---

[9] The i is also called a dummy variable—mathematics teachers love to call it this. Other letters such as $j, k, \ell, \ldots$ can be used as a dummy variable.

Solution

(a) This says to add up the numbers 1 through 10. So translating this into sigma notation, the lower limit of summation is 1 and the upper limit of summation is 10. Therefore,

$$1 + 2 + 3 + 4 + \cdots + 10 = \sum_{i=1}^{10} i.$$

(b) This says to add up the numbers $1 + \frac{1}{2} + \frac{1}{3} + \cdots + \frac{1}{10}$. We can translate this into sigma notation by the following: the ith term is $\frac{1}{i}$, the lower limit of summation is 1, and the upper limit of summation is 10; this is because we are adding the first term up to and including the tenth term. Therefore,

$$1 + \frac{1}{2} + \frac{1}{3} + \cdots + \frac{1}{10} = \sum_{i=1}^{10} \frac{1}{i}.$$

(c) $$\binom{10}{1} + \binom{10}{2} + \binom{10}{3} + \cdots + \binom{10}{10} = \sum_{i=1}^{10} \binom{10}{i}.$$

◄

Example 6.3.3 Write the sum of squared integers between 1 and 4 inclusively using sigma notation.

Solution $1^2 + 2^2 + 3^2 + 4^2 = \displaystyle\sum_{i=1}^{4} i^2 = 30.$ ◄

Exercises

6.3.1. Write out the first five terms of the following series:

(a) $\displaystyle\sum_{i=1}^{25} i^3$

(b) $\displaystyle\sum_{k=1}^{10} \frac{1}{k}$

(c) $\displaystyle\sum_{i=2}^{10} \frac{i-1}{i^2+3}$

(d) $\displaystyle\sum_{j=4}^{20} \binom{20}{j}$

6.3.2. State the sixth term of the following series:

(a) $\displaystyle\sum_{i=1}^{10} (i+2)$

(b) $\displaystyle\sum_{n=1}^{20} (n! + 1)$

(c) $\displaystyle\sum_{k=3}^{19} \frac{1}{k}$

6.3.3. Calculate the following:

(a) $\displaystyle\sum_{i=1}^{3} 2$ (b) $\displaystyle\sum_{k=2}^{4} \frac{1}{k}$ (c) $\displaystyle\sum_{n=1}^{3} n!$ (d) $\displaystyle\sum_{i=3}^{6} \binom{7}{i}$.

6.3.4. Calculate the following:

(a) $\displaystyle\sum_{i=1}^{4} i$ (d) $\displaystyle\sum_{k=3}^{9} k^3$ (g) $\displaystyle\sum_{i=1}^{5} i^3$

(b) $\displaystyle\sum_{i=1}^{5} 2^i$ (e) $\displaystyle\sum_{x=1}^{7} x$ (h) $\displaystyle\sum_{j=0}^{6} (j^2 + 1)$

(c) $\displaystyle\sum_{j=0}^{4} 3^j$ (f) $\displaystyle\sum_{i=1}^{5} 1$ (i) $\displaystyle\sum_{x=3}^{7} (2x - 1)$

6.3.5. Which of the following expresses $1 + x + x^2 + \cdots + x^{20}$?

(a) $\displaystyle\sum_{x=0}^{20} x^x$ (b) $\displaystyle\sum_{i=0}^{20} x^i$ (c) $\displaystyle\sum_{i=0}^{20} i^x$

6.3.6. Write the following in sigma notation.

(a) $1 + 2 + 3 + 4 + 5 + 6 + 7 + 8$

(b) $3 + 4 + 5 + 6 + 7$

(c) $4^2 + 5^2 + 6^2 + 7^2 + \cdots + 20^2$

(d) $4^3 + 5^3 + 6^3 + \cdots + n^3$

(e) $4 + 6 + 8 + 10 + \cdots + 2n$

(f) $3 + 3^2 + 3^3 + 3^4 + \cdots + 3^n$

(g) $1 + 3 + 5 + 7 + \cdots + (2n - 1)$

(h) $\dfrac{1}{k+1} + \dfrac{1}{k+2} + \cdots + \dfrac{1}{k+n}$

(i) $x_1 + 2x_2 + 3x_3 + \cdots + nx_n$

6.3.7. Write the following in sigma notation.

(a) $\dfrac{1}{2} + \dfrac{2}{3} + \dfrac{3}{4} + \dfrac{4}{5} + \dfrac{5}{6}$

(b) $0\dbinom{8}{0} + 1\dbinom{8}{1} + 2\dbinom{8}{2} + 3\dbinom{8}{3} + 4\dbinom{8}{4} + \cdots + 8\dbinom{8}{8}$

6.3.8. Express the following using sigma notation:

(a) $(1)(3)(7) + (3)(6)(11) + (5)(9)(15) + \cdots$ (to the nth term)

(b) $\dfrac{1}{(1)(2)} + \dfrac{1}{(2)(3)} + \dfrac{1}{(3)(4)} + \cdots + \dfrac{1}{n(n+1)}$

(c) $x + 3x^2 + 5x^3 + \cdots$ (to the nth term)

(d) $t_1 x_1^2 + t_2 x_2^3 + t_3 x_3^4 + \cdots$ (to the rth term)

(e) $f(x) + 2f(2x) + 3f(3x) + \cdots$ (to the $(k-1)^{\text{st}}$ term)

(f) $x + 3x^2 + 5x^3 + \cdots$ (to the $(n-1)^{\text{st}}$ term).

6.3.9. Express the following using summation notation.

(a) $x_1 y_1 + x_2 y_2 + x_3 y_3 + \cdots + x_n y_n$

(b) $x_1 + x_2^2 + x_3^3$

(c) $t_1 + 2t_2 + 3t_3 + \cdots + nt_n$

(d) $t_1 - 2t_2 + 3t_3 - \cdots + (-1)^{n+1}nt_n$

6.3.10. Show that $\displaystyle\sum_{i=1}^{n}(t_i - t_{i-1}) = t_n - t_0$.

6.4 Arithmetic Series

There is a legend that surrounds the famous Carl Friedrich Gauss (1777–1855), a great German mathematician, about his encounter with his teacher in elementary school.

After Gauss celebrated his seventh birthday, he first visited the Catharinen Elementary School in 1784 where elementary subjects were taught under the leadership of a certain man named Büttner. It was a musty, low schoolroom with an uneven worn floor, from which one could see the two slender Gothic towers of Catharinen Church, the stables, and poor backyard dwellings. Amongst the hundred students, Büttner went up and down the aisles with his leather whip[10] in hand, which was regarded as the final word in an argument with a student. And when the occasion arose, he used it. . . .

Here occurred an incident that we must not disregard completely, for later in life Gauss would often tell this story in his old age with great joy and vivacity. It was customary for the first student to finish his arithmetic solution and put his slateboard face down in the middle of a large table. The second student to finish would put his slate face down on top of the previous one, and so on. The young Gauss had barely entered the class when Büttner gave out a problem of adding[11] the numbers from 1 to 100. The task, however, was barely started when Gauss threw his slate on the table with words in lowly Braunschweig[12] dialect: *Ligget se*[13] (There it is). While the other students

Fig. 6.11   Carl Friedrich Gauss (1777–1855)

---

[10]Original text is *Karwatsche* which I have rendered as leather whip.

[11]The original text indicates that Büttner *die Summation einer arithmetischen Reihe aufgab* or gave the problem of the summation of an arithmetic progression. This arithmetic progression is not specified.

[12]Braunschweig (English: Brunswick) is a city in Lower Saxony, Germany. This is the city where Gauss was born.

[13]*Da liegt sie.*

calculated, busily multiplying and adding, Büttner, with conscious dignity, walked up and down only occasionally throwing an ironical, pitying glance toward the youngest of his pupils. The boy sat quietly knowing, already as much imbued with the firm unshakable awareness that filled him until the end of his days in any completed work, that his task was solved correctly, and that the result could be no other.

At the end of the hour, the slates were turned up, with one single solitary number laid on top by Gauss. As Büttner examined the answer, he found to the amazement of everyone present that Gauss was correct, whereas the remaining slates were wrong and were immediately rectified with the leather whip (von Waltershausen, 1856, pp. 12–13).

This legend has a natural appeal to young people; the hero is a child who outwits a brute. It is this appeal that the legend continues to be passed on from one generation to the next with each generation adding more embellishments to the wild tale (Hayes, 2006, p. 202). But you might ask yourself this: how did Gauss do it?[14] Here is one way we can solve it.

First, we denote $S_{100} = 1 + 2 + 3 + \cdots + 100 = \sum_{i=1}^{100} i$. Next, we can write out the series explicitly, and again in reverse, then add the two expressions together term by term:

$$
\begin{array}{rcccccccccc}
S_{100} & = & 1 & + & 2 & + & \cdots & + & 99 & + & 100 \\
+S_{100} & = & 100 & + & 99 & + & \cdots & + & 2 & + & 1 \\
\hline
2S_{100} & = & 101 & + & 101 & + & \cdots & + & 101 & + & 101
\end{array}
$$

$$\underbrace{\qquad\qquad\qquad\qquad}_{100 \text{ terms}}$$

So,

$$2S_{100} = 100(101)$$
$$S_{100} = \frac{100(101)}{2}$$
$$S_{100} = 5050.$$

Thus, the sum of the integers from 1 to 100 is 5050.

We can now use the same method that Gauss used to prove the following theorem.

---

[14]Many people have attributed this method that we show of adding the numbers from 1 to 100 to Gauss, but von Waltershausen (1856, pp. 12–13) does not mention or give any method. There are so many things attributed to Gauss that one wonders if he employed a public relations manager.

> **Theorem 6.4.1 — Gaussian summation formula.**
>
> $$\sum_{i=1}^{n} i = \frac{n(n+1)}{2} \tag{6.12}$$

Proof We follow the same method as Gauss. First, we denote

$$S_n = 1 + 2 + 3 + \cdots + n = \sum_{i=1}^{n} i.$$

Next, we can write out the series explicitly, and again in reverse, then add the two expressions together term by term:

$$
\begin{array}{ccccccccc}
S_n & = & 1 & + & 2 & + & \cdots & + & n-1 & + & n \\
+S_n & = & n & + & n-1 & + & \cdots & + & 2 & + & 1 \\
\hline
2S_n & = & n+1 & + & n+1 & + & \cdots & + & n+1 & + & n+1
\end{array}
$$

$$\underbrace{}_{n \text{ terms}}$$

So,

$$2S_n = n(n+1)$$
$$S_n = \frac{n(n+1)}{2}$$

as required. ◀

A visual explanation of Gauss' summation formula (Theorem 6.4.1) is given in Table 6.4. Table 6.4 helps to build formula (6.12) using the areas of triangles and squares.

By looking at the problem algebraically and visually, we have done the *look back* step in Pólya's problem solving approach.[15] Notice that the two expressions $\frac{n(n+1)}{2}$ and $\frac{n^2}{2} + \frac{n}{2}$ are not exactly the same, but they are equivalent:

$$\frac{n^2}{2} + \frac{n}{2} = \frac{n(n+1)}{2} = \frac{n^2 + n}{2} = \frac{n(n+1)}{2}.$$

It just required some algebraic effort to show that the answers from the algebraic and the visual method were equivalent.

Now that we feel confident that we have the correct formula, we can start to apply it to some examples.

Example 6.4.2 Find the sum of the integers from 1 to 5000.

---

[15]Pólya's approach is summarized in the Preface starting on p. xviii.

Table 6.4 Visual explanation of Gauss' summation formula. The shaded area represents the sum in the diagram.

n	$\displaystyle\sum_{i=1}^{n} i$	$\dfrac{n(n+1)}{?}$	$\dfrac{n^2}{2} + \dfrac{n}{2}$	Diagram
1	1	$\frac{1(2)}{2} = 1$	$\frac{1^2}{2} + \frac{1}{2} = 1$	
2	$1 + 2 = 3$	$\frac{2(3)}{2} = 3$	$\frac{2^2}{2} + \frac{2}{2} = 3$	
3	$1 + 2 + 3 = 6$	$\frac{3(4)}{2} = 6$	$\frac{3^2}{2} + \frac{3}{2} = 6$	
4	$1 + 2 + 3 + 4 = 10$	$\frac{4(5)}{2} = 10$	$\frac{4^2}{2} + \frac{4}{2} = 10$	

Solution This is a straightforward application of Gauss' summation formula.

$$\sum_{i=1}^{5000} i = \frac{5000 \times 5001}{2} = 12\,502\,500$$

◄

Example 6.4.3 Find the sum of the integers from 50 to 100.

Solution We will use Pólya's problem solving approach.

Understand the problem. The unknown sum that we are required to find is $\displaystyle\sum_{i=50}^{100} i$. We are not able to directly apply Gauss' summation formula to this because we are not starting at $i = 1$. The information we can find easily using Gauss' summation formula is $\displaystyle\sum_{i=1}^{49} i$ and $\displaystyle\sum_{i=1}^{100} i$.

$$\sum_{i=1}^{100} i$$

$$\underbrace{1 + 2 + 3 + \cdots + 49}_{\displaystyle\sum_{i=1}^{49} i} + \underbrace{50 + 51 + 52 + \cdots + 100}_{\displaystyle\sum_{i=50}^{100} i = ?}$$

Devise a plan. Our plan is to use indirect reasoning: the undesired portion is subtracted from the total sum to arrive at our desired sum. In this case, the total sum is $\sum_{i=1}^{100} i$ and we will subtract $\sum_{i=1}^{49} i$ to get our desired sum of $\sum_{i=50}^{100} i$.

Carry out the plan. Then,

$$\sum_{i=50}^{100} i = \sum_{i=1}^{100} i - \sum_{i=1}^{49} i$$

$$= \frac{100(101)}{2} - \frac{49(50)}{2}$$

$$= 5050 - 1225$$

$$= 3825.$$

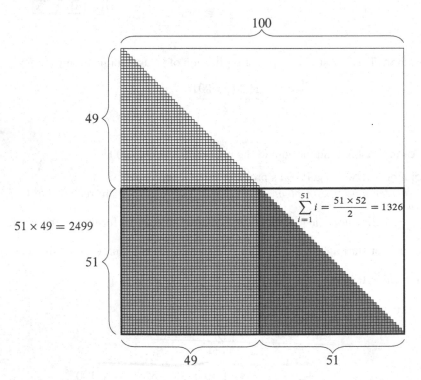

Fig. 6.12 A visual explanation to sum the integers from 50 to 100. We can break up the sum of integers from 1 to 100, and select the areas we want; this is the direct method. In this case, we can find the areas of the shaded regions. Adding these together, we have $2499 + 1326 = 3825$, which is the same answer we found by the indirect method.

Look back. We can confirm this using the direct method. We give a visual explanation in Fig. 6.12. ◄

We can use Gauss' method to derive a formula for the sum of the general arithmetic series.

Theorem 6.4.4 The nth partial sum of an arithmetic series

$$S_n = a + (a + d) + (a + 2d) + \cdots + [a + (n - 1)d]$$

can be determined by

$$S_n = \frac{n}{2}[2a + (n - 1)d] \qquad (6.13)$$

where

- n is the number of terms
- a is the first term
- d is the common difference.

Proof Using Gauss' method, we have

$$
\begin{array}{ccccccccc}
S_n & = & a & + & a + d & + & \cdots & + & a + (n-1)d \\
+ S_n & = & a + (n-1)d & + & a + (n-2)d & + & \cdots & + & a \\
\hline
2S_n & = & 2a + (n-1)d & + & 2a + (n-1)d & + & \cdots & + & 2a + (n-1)d
\end{array}
$$

$$\underbrace{\qquad\qquad\qquad\qquad\qquad\qquad}_{n \text{ terms}}$$

So,

$$2S_n = n[2a + (n - 1)d]$$
$$S_n = \frac{n}{2}[2a + (n - 1)d].$$

◄

Note that when $a = 1$ and $d = 1$, then Eq. (6.13) is the same as Eq. (6.12).

Exercises

6.4.1. For $2 + 4 + 6 + 8 + \cdots + 200$,
    (a) calculate the sum using Gauss' method, and
    (b) express the sum in sigma notation.

6.4.2. What is the sum of the first 20 terms of the arithmetic series

$$7 + 11 + 15 + \cdots ?$$

6.4.3. What is the sum of the first n even numbers: $2 + 4 + 6 + \cdots + 2n$?

6.4.4. Find the sum for each of the following arithmetic series.

(a) $6 + 9 + 12 + \cdots + 54$

(b) $11 + 22 + 33 + \cdots + 143$

(c) $9 + 4 + (-1) + \cdots + (-101)$

(d) $\frac{1}{7} + \frac{4}{7} + \frac{7}{7} + \cdots + \frac{37}{7}$

6.4.5. A contractor agrees to drill a well 30 meters deep at a cost of \$50 for the first meter, \$55 for the second meter and increasing by \$5 for each subsequent meter.

(a) What is the cost of drilling the last meter of the well?

(b) How much will it cost to drill the entire well?

6.4.6. You have a bunch of children who are selling chocolate bars where the first child has sold one chocolate bar, the second child has sold two chocolate bars, and the third child has sold three chocolate bars and so on and so forth. How many of these children are needed to have a total of 120 chocolate bars sold?

6.4.7. Solve for n.

(a) $\sum_{i=1}^{n} i = 78$

(b) $\sum_{i=1}^{n} (7i - 3) = 510$

(c) $\sum_{i=1}^{n} (9 - 4i) = -345$

(d) $\sum_{i=3}^{n} i = 88$

6.4.8. Show that $\sum_{i=1}^{n} i = \binom{n+1}{2}$.

6.4.9. Using the formula from Exercise 6.4.8, calculate

(a) $\sum_{i=0}^{100} i$

(b) $\sum_{j=45}^{100} j$

(c) $\sum_{k=30}^{45} k$.

6.4.10. In this exercise, we will be investigating the following question.

What is the sum of the first n odd numbers: $1 + 3 + 5 + \cdots + (2n - 1)$?

We will look at this using two approaches, but both require us to *find a pattern*.

(a) (i) Complete Table 6.5.

Table 6.5 The sum of odd numbers

1	$=$	1
$1 + 3$	$=$	4
$1 + 3 + 5$	$=$	9
$1 + 3 + 5 + 7$	$=$	
$1 + 3 + 5 + 7 + 9$	$=$	
$1 + 3 + 5 + 7 + 9 + 11$	$=$	

(ii) What pattern do you see in Table 6.5?

(iii) Test your guess to see if it correctly predicts the sum of $1 + 3 + 5 + 7 + 9 + 11 + 13$. Was it correct? If not, come up with another guess and try again.

(iv) In general, what is the sum of the first n odd numbers?

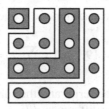

Fig. 6.13 How many circles are there in the square?

(b) (i) Using Fig. 6.13, what is the sum of
   i. $1 + 3$? (the sum of the first 2 odd numbers)
   ii. $1 + 3 + 5$? (the sum of the first 3 odd numbers)
   iii. $1 + 3 + 5 + 7$? (the sum of the first 4 odd numbers)

(ii) Can you *find a pattern* with Fig. 6.13? Guess the sum of $1 + 3 + 5 + 7 + 9$ (the sum of the first 5 odd numbers). Check by extending Fig. 6.13.

(iii) In general, what is the sum of the first n odd numbers?

6.4.11. How many rectangles can you find in

(a) Fig. 6.14.   (b) an $n \times n$ grid?

Fig. 6.14 Rectangles for Exercise 6.4.11

6.4.12. Evaluate

(a) $3 + 6 + 9 + \cdots + 3n$

(b) $4 + 8 + 12 + \cdots + 4n$

(c) $n + 2n + 3n + \cdots + n^2$

(d) $(1 + 2 + 3 + \cdots + 25)^2$

(e) $(1 + 2 + 3 + \cdots + n)^2$

(f) $(n + 2n + 3n + \cdots + n^2)^2$

(g) $(1 + 2 + 3 + \cdots + 25)^3$

(h) $(1 + 2 + 3 + \cdots + n)^3$

(i) $(n + 2n + 3n + \cdots + n^2)^3$

(j) $(1 + 2 + 3 + \cdots + 25)^k$

(k) $(1 + 2 + 3 + \cdots + n)^k$

(l) $(n + 2n + 3n + \cdots + n^2)^k$

6.4.13. Beginning with the sum of the first 500 natural numbers, $1 + 2 + 3 + \cdots + 500$, every third number is removed to create the new sum

$$1 + 2 + 4 + 5 + 7 + 8 + 10 + 11 + \cdots + 500.$$

Calculate this new sum.

6.4.14 (†). Suppose $t_1, t_2, \ldots$ is an arithmetic sequence, and

$$\sum_{i=1}^{2} t_i = 19 \quad \text{and} \quad \sum_{i=1}^{4} t_i = 50.$$

What are the first six terms of the series and the sum to 15 terms? (*Hint*: What is the first term a, and the common difference d?)

6.5 Properties of Series

Operations involving summation follow basic rules. Some of these rules are given below with the proofs.

Theorem 6.5.1

$$\sum_{i=1}^{n} c = cn \qquad (6.14)$$

where c is a constant.

Proof

$$\sum_{i=1}^{n} c = \underbrace{c + c + \cdots + c}_{n \text{ times}}$$

$$= cn$$

◄

Example 6.5.2 $\displaystyle\sum_{i=1}^{25} 4 = \underbrace{4 + 4 + \cdots + 4}_{25 \text{ times}} = 4 \times 25 = 100.$ ◄

Theorem 6.5.3

$$\sum_{i=1}^{n} ct_i = c \sum_{i=1}^{n} t_i \qquad (6.15)$$

where c is a constant.

Proof

$$\sum_{i=1}^{n} ct_i = ct_1 + ct_2 + \cdots + ct_n$$

$$= c(t_1 + t_2 + \cdots + t_n)$$

$$= c \sum_{i=1}^{n} t_i$$

◀

Example 6.5.4

$$\sum_{i=1}^{25} 4i = 4 \sum_{i=1}^{25} i \qquad \text{By Theorem 6.5.3}$$

$$= 4 \left(\frac{25 \times 26}{2} \right) \qquad \text{By Theorem 6.4.1}$$

$$= 4 \times 325$$

$$= 1300$$

◀

Theorem 6.5.5

$$\sum_{i=1}^{n} (a_i + b_i) = \sum_{i=1}^{n} a_i + \sum_{i=1}^{n} b_i \qquad (6.16)$$

Proof

$$\sum_{i=1}^{n} (a_i + b_i) = (a_1 + b_1) + (a_2 + b_2) + \cdots + (a_n + b_n)$$

$$= (a_1 + a_2 + \cdots + a_n) + (b_1 + b_2 + \cdots + b_n)$$

$$= \sum_{i=1}^{n} a_i + \sum_{i=1}^{n} b_i .$$

◀

Example 6.5.6 Calculate $\displaystyle\sum_{i=1}^{10} (i + 3)$.

Solution

$$\sum_{i=1}^{10} (i+3) = \sum_{i=1}^{10} i + \sum_{i=1}^{10} 3 \qquad \text{By Theorem 6.5.5}$$

$$= \underbrace{\frac{10(11)}{2}}_{\text{By Theorem 6.4.1}} + \underbrace{3(10)}_{\text{By Theorem 6.5.3}}$$

$$= 85.$$

◄

Example 6.5.7 — Arithmetic series. What is the nth partial sum of an arithmetic series?

$$S_n = a + (a+d) + (a+2d) + (a+3d) + \cdots + [a + (n-1)d]$$

Solution Recall that the general term of an arithmetic sequence is

$$t_n = a + (n-1)d \qquad \text{for } n = 1, 2, \ldots .$$

Then,

$$S_n = t_1 + t_2 + \cdots + t_n$$

$$= \sum_{i=1}^{n} [a + (i-1)d]$$

$$= \sum_{i=1}^{n} [a + id - d]$$

$$= \sum_{i=1}^{n} a + \sum_{i=1}^{n} id - \sum_{i=1}^{n} d \qquad \text{By Theorem 6.5.5}$$

$$= \sum_{i=1}^{n} a + \underbrace{d \sum_{i=1}^{n} i}_{\text{By Theorem 6.5.3}} - \sum_{i=1}^{n} d$$

$$= an + \underbrace{d \left(\frac{n(n+1)}{2} \right)}_{\text{By Theorem 6.4.1}} - \underbrace{dn}_{\text{By Theorem 6.5.1}}$$

$$= \frac{2an}{2} + \frac{d(n)(n+1)}{2} - \frac{2dn}{2} \qquad \text{Common denominator}$$

$$= \frac{n}{2} [2a + d(n+1) - 2d] \qquad \text{Factor } \frac{n}{2}$$

$$= \frac{n}{2} [2a + (n+1-2)d] \qquad \text{Factor } d$$

$$= \frac{n}{2}[2a + (n-1)d] \qquad (6.17)$$

$$= \frac{n}{2}[a + \underbrace{a + (n-1)d}_{t_n}]$$

$$= \frac{n}{2}(a + t_n). \qquad (6.18)$$

Therefore, the nth partial sum of an arithmetic series is

$$S_n = \frac{n}{2}[2a + (n-1)d] \quad \text{or} \quad S_n = \frac{n}{2}(a + t_n).$$

◀

Example 6.5.8 — Shifting the index. Are the following two expressions equivalent?

$$\sum_{i=0}^{10}(i + 1) \qquad \text{and} \qquad \sum_{i=1}^{11} i.$$

Solution — 1. By expanding each sum, we see that

$$\sum_{i=0}^{10}(i + 1) = 1 + 2 + 3 + 4 + \cdots + 11 = 66$$

and

$$\sum_{i=1}^{11} i = 1 + 2 + 3 + 4 + \cdots + 11 = 66.$$

Thus, the two expressions represent the same sum; that is, indeed

$$\sum_{i=0}^{10}(i + 1) = \sum_{i=1}^{11} i = 66.$$

◀

Solution — 2. By shifting *up* the term by one, and shifting *down* the index of summation by one, the net effect on the summation is zero.

$$\sum_{i=0}^{10} (i+1) = \overbrace{\sum_{i=0+1}^{10+1}}^{\substack{\text{Shift } up \\ \text{the index by 1:} \\ \text{add 1 to index.}}} \underbrace{((i-1)+1)}_{\substack{\text{Shift } down \\ \text{the term by 1:} \\ \text{replace } i \\ \text{with } i-1.}}$$

$$= \sum_{i=1}^{11} (i-1+1)$$

$$= \sum_{i=1}^{11} i.$$

◀

This is an algebraic manipulation that is useful to change the summation you are working with into an expression in which you can apply a formula. For example, we cannot apply directly the Gaussian summation formula to $\sum_{i=0}^{10} (i+1)$, but with some algebraic manipulation, like we did in the second solution of Example 6.5.8, we can convert it to $\sum_{i=1}^{11} i$ so we can apply the formula;

$$\sum_{i=0}^{10} (i+1) = \sum_{i=1}^{11} i = \frac{11(12)}{2} = 66.$$

Instead of shifting by 1, we could shift by k. Then, either we can shift *up* the index of summation by k,

$$\sum_{i=0}^{10} (i+1) = \overbrace{\sum_{i=0+k}^{10+k}}^{\substack{\text{Shift } up \\ \text{the index by } k: \\ \text{add } k \text{ to index.}}} \underbrace{((i-k)+1)}_{\substack{\text{Shift } down \\ \text{the term by } k: \\ \text{replace } i \\ \text{with } (i-k)}} = \sum_{i=k}^{10+k} (i-k+1)$$

or shift *down* the index by k,

$$\sum_{i=0}^{10}(i+1) = \underbrace{\sum_{i=0-k}^{10-k}((i+k)+1)}_{} = \sum_{i=-k}^{10-k}(i+k+1).$$

Shift *up*
the term by k:
replace i
with $(i+k)$.

Shift *down*
the index by k:
subtract k
from index.

Note that a common mistake made in using sigma notation is to assume that properties are true. Except in very special circumstances,

$$\sum_{i=1}^{n} t_i^2 \neq \left(\sum_{i=1}^{n} t_i\right)^2 \tag{6.19}$$

and

$$\sum_{i=1}^{n} a_i b_i \neq \left(\sum_{i=1}^{n} a_i\right)\left(\sum_{i=1}^{n} b_i\right) \tag{6.20}$$

(see Exercises 6.5.4 and 6.5.5).

Exercises

6.5.1. Evaluate each sum.

(a) $\displaystyle\sum_{i=1}^{9} 4$

(b) $\displaystyle\sum_{j=1}^{n} 5$

(c) $\displaystyle\sum_{x=1}^{n-1} 6$

(d) $\displaystyle\sum_{x=20}^{45} 6$

(e) $\displaystyle\sum_{i=1}^{17} 7i$

(f) $\displaystyle\sum_{j=6}^{17} 7j$

(g) $\displaystyle\sum_{j=6}^{n} 7j$

(h) $\displaystyle\sum_{i=1}^{15} (5i+3)$

(i) $\displaystyle\sum_{k=1}^{100} (10k+3)$

(j) $\displaystyle\sum_{j=1}^{30} (9j+6)$

(k) $\displaystyle\sum_{j=5}^{30} (9j+6)$

(l) $\displaystyle\sum_{j=5}^{n} (9j+6)$

6.5.2. Solve for n for each of the following:

(a) $\displaystyle\sum_{i=1}^{n} i = 91$

(b) $\displaystyle\sum_{i=1}^{n} (4i-1) = 1830$

6.5.3. The **average** of the numbers $x_1, x_2, \ldots, x_n$, denoted by $\bar{x}$, is given by
$$\bar{x} = \frac{1}{n} \sum_{i=1}^{n} x_i.$$

(a) Show that $\displaystyle\sum_{i=1}^{n} (x_i - \bar{x}) = 0$.

(b) Show that $\displaystyle\sum_{i=1}^{n} (x_i - \bar{x})^2 = \sum_{i=1}^{n} x_i^2 - n\bar{x}^2$.

6.5.4. Show by evaluation that $\displaystyle\sum_{i=1}^{4} i^2 \neq \left(\sum_{i=1}^{4} i \right)^2$.

6.5.5. Show by evaluation that $\displaystyle\sum_{i=1}^{4} i(i+1) \neq \left(\sum_{i=1}^{4} i \right)\left(\sum_{i=1}^{4} (i+1) \right)$.

6.5.6. Show that

$$\sum_{i=1}^{n} [af(i) + bg(i)] = a \sum_{i=1}^{n} f(i) + b \sum_{i=1}^{n} g(i)$$

where a and b are constants, and f and g are functions. (Assume that these functions $f(i)$ and $g(i)$ have values for $i = 1, 2, \ldots, n$.)

6.5.7. Show that $2^1 \times 2^2 \times 2^3 \times \cdots \times 2^n = (\sqrt{2})^{n(n+1)}$.

6.5.8. What is the sum of all the odd numbers in the set $\{1, 2, 3, \ldots, 99, 100\}$?

6.5.9. What is the sum of all the even numbers in the set $\{1, 2, 3, \ldots, 99, 100\}$?

6.5.10. What is the sum of all the numbers divisible by 5 in the set $\{1, 2, 3, \ldots, 100\}$?

6.5.11. Show that the following pairs of expressions represent the same sum:

(a) $\displaystyle\sum_{i=1}^{n} \frac{3i-1}{i+1}$ and $\displaystyle\sum_{i=2}^{n+1} \frac{3i-4}{i}$ (b) $\displaystyle\sum_{j=6}^{17} \frac{j^2}{2j+1}$ and $\displaystyle\sum_{k=3}^{14} \frac{(k+3)^2}{2k+7}$.

6.5.12 (†). What is the value of $\displaystyle\sum_{i=2}^{1000} \frac{1}{\log_i 1000!}$? (*Hint*: Use the change of base formula $\log_b n = \log n / \log b$; see Sec. B.6.)

6.6 Pi Notation

Similar to using sigma notation for expressing long sums, one way of writing long repetitive products is to use the Greek letter Π (capital π corresponding to the Roman letter P for product). This is called **pi notation**.

For example, we can write the following product as

$$1 \times 2 \times 3 \times 4 \times 5 \times \cdots \times 99 \times 100 = \prod_{i=1}^{n} i. \tag{6.21}$$

Moreover, for (6.21), we can express that product as 100!.

In general, to indicate the product of the terms $t_1, t_2, t_3, \ldots, t_m, t_{m+1}, \ldots, t_n$, $\ldots$ from the mth term up to and including the nth term (where $m \leq n$), we write

$$t_m \times t_{m+1} \times t_{m+2} \times \cdots \times t_{n-1} \times t_n = \prod_{i=m}^{n} t_i. \tag{6.22}$$

This tells us to end with $i = n$.

This tells us that it is a product.

$$\longrightarrow \prod_{i=m}^{n} t_i$$

This tells us to start with $i = m$.

In this case, the letter i is called the **index of multiplication**, the number m is called the **lower limit of multiplication** and the number n is called the **upper limit of multiplication**.

Example 6.6.1 Write the following products using pi notation.
  (a) $1 \times 2 \times 3 \times 4 \times \cdots \times 9 \times 10$
  (b) $1 \times \frac{1}{2} \times \frac{1}{3} \times \frac{1}{4} \times \cdots \times \frac{1}{10}$
  (c) $\binom{10}{1} \times \binom{10}{2} \times \binom{10}{3} \times \cdots \times \binom{10}{10}$

Solution
  (a) This says to multiply the numbers 1 through 10. So translating this into pi notation, the lower limit of multiplication is 1 and the upper limit of multiplication is 10:

$$1 \times 2 \times 3 \times \cdots \times 9 \times 10 = \prod_{i=1}^{10} i.$$

  (b) This says to multiply the numbers $1 \times \frac{1}{2} \times \frac{1}{3} \times \cdots \times \frac{1}{10}$. We can translate this into pi notation by the following: the ith term is $\frac{1}{i}$, the lower limit of multiplication is 1, and the upper limit of multiplication is 10. Hence,

$$1 \times \frac{1}{2} \times \frac{1}{3} \times \frac{1}{4} \times \cdots \times \frac{1}{10} = \prod_{i=1}^{10} \frac{1}{i}.$$

(c) $\dbinom{10}{1} \times \dbinom{10}{2} \times \dbinom{10}{3} \times \cdots \times \dbinom{10}{10} = \prod_{i=1}^{10} \dbinom{10}{i}.$

◀

Example 6.6.2 Write the product of squared natural numbers between 1 and 4 inclusively using pi notation.

Solution

$$1^2 \times 2^2 \times 3^2 \times 4^2 = \prod_{i=1}^{4} i^2 = 576.$$

◀

Exercises

6.6.1. For $n \geq 0$, express $n!$ using pi notation.

6.6.2. For $n \geq 0$, express $P(n, r)$ using pi notation.

6.6.3. Calculate the following:

(a) $\prod_{i=1}^{3} 2$ (b) $\prod_{i=2}^{4} \frac{1}{i}$ (c) $\prod_{k=3}^{6} 2i$ (d) $\prod_{k=2}^{4} \dbinom{4}{k}$

6.6.4. Prove the following properties of products:

(a) $\prod_{i=1}^{n} c = c^n$ where c is a constant.

(b) $\prod_{i=1}^{n} ct_i = c^n \prod_{i=1}^{n} t_i$ where c is a constant.

(c) $\prod_{i=1}^{n} (a_i \times b_i) = \prod_{i=1}^{n} a_i \times \prod_{i=1}^{n} b_i.$

6.7 The Geometric Series

The sum of the terms in a geometric sequence is called a **geometric series**. Consider the following geometric sequence

$$1, 4, 16, 64, 256$$

with the first term $a = 1$, the common ratio $r = 4$, and the number of terms is $n = 5$.

Now, suppose we wish to add up all of these terms,

$$S_5 = 1 + 4 + 16 + 64 + 256.$$

We can fall back on the method that was used to show the Gaussian summation formula. We will write out the series S_5 explicitly, multiply each term by the common ratio 4, and then subtract:

$$
\begin{array}{rcccccccccc}
S_5 & = & 1 & + & \cancel{4} & + & \cancel{16} & + & \cancel{64} & + & \cancel{256} \\
-4S_5 & = & & - & \cancel{4} & - & \cancel{16} & - & \cancel{64} & - & \cancel{256} & - & 1024 \\
\hline
-3S_5 & = & 1 & & & & & & & & & - & 1024
\end{array}
$$

Then, we solve for S_5

$$-3S_5 = 1 - 1024$$
$$S_5 = \frac{1 - 1024}{-3}$$
$$= 341.$$

Now, we will generalize this for arbitrary values of a, r and n with the following theorem and proof.

Theorem 6.7.1 — Geometric Series. For $n \in \mathbb{N}$,

$$S_n = a + ar + ar^2 + \cdots + ar^{n-1} = \sum_{i=1}^{n} ar^{i-1} = \begin{cases} an & \text{if } r = 1 \\ \dfrac{a(1 - r^n)}{1 - r} & \text{if } r \neq 1. \end{cases} \tag{6.23}$$

Proof *Take cases.* We have to consider two cases: when $r = 1$ and $r \neq 1$.

Case 1: If $r = 1$, then

$$a + a(1) + a(1)^2 + \cdots + a(1)^n = \underbrace{a + a + a + \cdots + a}_{n \text{ times}} = an$$

Case 2: If $r \neq 1$, then

$$
\begin{array}{rcccccccccc}
S_n & = & a & + & \cancel{ar} & + & \cancel{ar^2} & + & \cdots & + & \cancel{ar^{n-1}} \\
-rS_n & = & & - & \cancel{ar} & - & \cancel{ar^2} & - & \cdots & - & \cancel{ar^{n-1}} & - & ar^n \\
\hline
S_n - rS_n & = & a & & & & & & & & & - & ar^n
\end{array}
$$

Then, we solve for S_n

$$S_n - rS_n = a - ar^n$$
$$S_n(1 - r) = a(1 - r^n).$$

Since $r \neq 1$, we may divide by $1 - r$ ($\neq 0$), and thus

$$S_n = \frac{a(1 - r^n)}{1 - r}$$

$$a + ar + ar^2 + \cdots + ar^{n-1} = \frac{a(1 - r^n)}{1 - r}.$$

Now that we have considered all of the cases, the proof is complete. ◀

Example 6.7.2 Given the geometric series

$$3 + 12 + 48 + 192 + \cdots + 196\,608$$

find the sum using

(a) Gauss' method.   (b) applying the formula.

Solution Let $S = 3 + 12 + 48 + 192 + \cdots + 196\,608$; this is a geometric series with $a = 3$, and $r = 4$.

(a) Using Gauss' method,

$$
\begin{array}{rcccccccc}
S & = & 3 & + & 12 & + & 48 & + & \cdots & + & 196\,608 \\
-4S & = & & - & 12 & - & 48 & - & \cdots & - & 196\,608 & - & 786\,432 \\
\hline
-3S & = & 3 & & & & & & & & & - & 786\,432
\end{array}
$$

So,

$$-3S = 3 - 786\,432$$
$$S = \frac{3 - 786\,432}{-3}$$
$$= 262\,143$$

(b) In order to apply the formula, we need to know the number of terms, n. So, recall the general term of a geometric series is

$$t_n = ar^{n-1}$$

Substituting for $a = 3$ and $r = 4$, we have

$$t_n = 3(4)^{n-1}$$

but we know the last value of the sequence, $196\,608$,

$$196\,608 = 3(4)^{n-1}$$

and now we solve for n,

$$\frac{196\,608}{3} = \frac{3(4)^{n-1}}{3}$$
$$65\,536 = 4^{n-1}$$
$$4^8 = 4^{n-1}$$
$$8 = n - 1$$
$$n = 9.$$

The formula yields,

$$S_9 = \frac{3(1 - 4^9)}{1 - 4} = \frac{3(-262\,143)}{-3} = 262\,143,$$

which matches the answer we found in part (a).

◀

Example 6.7.3 Show that $\displaystyle\sum_{k=1}^{n} \left(\frac{1}{2^k} - \frac{1}{2^{k+1}} \right) = \frac{1}{2} - \frac{1}{2^{n+1}}.$

Solution — 1. Using geometric series formula. *Divide into smaller problems.*[16]
The series can be separated into two series, and then apply the geometric series formula to each of them.

$$\sum_{k=1}^{n} \left(\frac{1}{2^k} - \frac{1}{2^{k+1}} \right) = \sum_{k=1}^{n} \left(\frac{1}{2^k} \right) - \sum_{k=1}^{n} \left(\frac{1}{2^{k+1}} \right)$$

$$= \sum_{k=1}^{n} \left(\frac{1}{2} \right)^k - \sum_{k=1}^{n} \left(\frac{1}{2} \right)^{k+1}$$

$$= \sum_{k=1}^{n} \left(\frac{1}{2} \right)^k - \sum_{k=1}^{n} \left(\frac{1}{2} \right)\left(\frac{1}{2} \right)^k$$

$$= \sum_{k=1}^{n} \left(\frac{1}{2} \right)^k - \frac{1}{2}\sum_{k=1}^{n} \left(\frac{1}{2} \right)^k$$

$$= \frac{1}{2} \underbrace{\sum_{k=1}^{n} \left(\frac{1}{2} \right)^k}_{a=\frac{1}{2},\ r=\frac{1}{2}}$$

$$= \frac{1}{2} \left[\frac{\frac{1}{2}\left[1 - \left(\frac{1}{2} \right)^n \right]}{1 - \frac{1}{2}} \right]$$

$$= \frac{1}{2} \left[\frac{\cancel{\frac{1}{2}}\left[1 - \left(\frac{1}{2} \right)^n \right]}{\cancel{\frac{1}{2}}} \right]$$

$$= \frac{1}{2} \left[1 - \left(\frac{1}{2} \right)^n \right]$$

$$= \frac{1}{2} - \frac{1}{2}\left(\frac{1}{2} \right)^n$$

---

[16]For other problem solving strategies, see Pólya's approach, which is summarized in the Preface starting on p. xviii.

$$= \frac{1}{2} - \left(\frac{1}{2}\right)^{n+1}$$

$$= \frac{1}{2} - \frac{1}{2^{n+1}}$$

◀

Solution — 2. Using telescoping sums. Another problem solving approach we can try is to write a couple of terms, and see if we can *find a pattern*. In fact, after writing out the terms, if we shift the brackets to the right, or grouping the terms in such a way, many of the terms will cancel.

$$S_n = \sum_{k=1}^{n} \left(\frac{1}{2^k} - \frac{1}{2^{k+1}}\right)$$

$$= \left(\frac{1}{2} - \frac{1}{2^2}\right) + \left(\frac{1}{2^2} - \frac{1}{2^3}\right) + \cdots + \left(\frac{1}{2^n} - \frac{1}{2^{n+1}}\right)$$

$$= \frac{1}{2} + \left(-\frac{1}{2^2} + \frac{1}{2^2}\right) + \left(-\frac{1}{2^3} + \frac{1}{2^3}\right) + \cdots + \left(-\frac{1}{2^n} + \frac{1}{2^n}\right) - \frac{1}{2^{n+1}} \quad \text{Shift brackets.}$$

All these terms are zero. It folds together like a telescope.

$$= \frac{1}{2} - \frac{1}{2^{n-1}}$$

Since many of the terms collapse (like a telescope in Fig. 6.15) to zero, this is known as a **telescoping series**. Also note that the answer we got using the telescoping method is the same as the one in the previous solution. ◀

We will now generalize what we found in Example 6.7.3.

Fig. 6.15   A telescope

Theorem 6.7.4 — Telescoping Series.

$$\sum_{i=1}^{n} t_i = \underbrace{(T_1 - T_2)}_{t_1} + \underbrace{(T_2 - T_3)}_{t_2} + \cdots + \underbrace{(T_n - T_{n+1})}_{t_n} = T_1 - T_{n+1} \quad (6.24)$$

Using our example as a template for our proof, we can replace the term $\left(\frac{1}{2^k} - \frac{1}{2^{k+1}}\right)$ with $t_i = T_i - T_{i+1}$.

Proof

$$\sum_{i=1}^{n} t_i = t_1 + t_2 + t_3 + \cdots + t_n$$

$$= (T_1 - T_2) + (T_2 - T_3) + \cdots + (T_n - T_{n+1})$$

$$= T_1 + \underbrace{(-T_1 + T_1) + (-T_2 + T_2) + \cdots + (-T_n + T_n)}_{\text{This telescopes to 0}} - T_{n+1}$$

$$= T_1 - T_{n+1}.$$

◀

Exercises

6.7.1. Find the sum of n terms for the geometric series given.

(a) $a = 12, r = 3, n = 10$ (c) $a = \frac{1}{5}, r = -3, n = 12$

(b) $a = 7, r = \frac{1}{4}, n = 8$ (d) $a = 3, r = 1, n = 6$

6.7.2. Calculate.

(a) $27 + 9 + 3 + \cdots + \dfrac{1}{243}$ (b) $\dfrac{1}{3} + \dfrac{2}{9} + \dfrac{4}{27} + \cdots + \dfrac{128}{6561}$

6.7.3. Evaluate the following.

(a) $\displaystyle\sum_{i=1}^{12} 4^i$ (b) $\displaystyle\sum_{i=1}^{10} 3^{-i}$ (c) $\displaystyle\sum_{i=10}^{20} 3^i$

6.7.4. Each person has two natural parents, four grandparents, eight great-grandparents, and so on and so forth. How many direct ancestors do you have if you went back 12 generations?

6.7.5. Determine n for $3 + 3^2 + 3^3 + \cdots + 3^n = 3279$.

6.7.6. Suppose $t_1, t_2, \ldots$ is a geometric sequence and

$$\sum_{i=1}^{10} t_i = 244 \sum_{i=1}^{5} t_i.$$

Find the common ratio r.

6.7.7. Malware can spread quickly from device to device. Consider a piece of malware that spreads through smart phones. The hacker's phone passes the code to four other phones via bluetooth. Each of these phones passes it on to four more phones, which in turn each infects four more phones, *etc*.

(a) Write the corresponding series for the number of infected devices.

(b) How many devices are infected after 10 levels of this infection?

6.7.8. Evaluate each telescoping sum.

(a) $\displaystyle\sum_{i=1}^{500}(3^i - 3^{i-1})$ 　　　 (b) $\displaystyle\sum_{i=1}^{100}\left(\frac{1}{i} - \frac{1}{i+1}\right)$ 　　　 (c) $\displaystyle\sum_{i=1}^{n}[(i+1)^3 - i^3]$

6.7.9. Using the method of telescoping sums, show that

$$\sum_{i=1}^{n}\frac{1}{i(i+1)} = 1 - \frac{1}{n+1}.$$

(*Hint*: $\frac{1}{i(i+1)} = \frac{1}{i} - \frac{1}{i+1}$.)

6.7.10. Using the method of telescoping sums, find the sum of the series

$$\sum_{i=1}^{n}\log\left(\frac{i+1}{i}\right).$$

(*Hint*: You will need to use the properties of the logarithm. See Appendix B.6.)

6.7.11 (Arithmetico-Geometric Series). Using Gauss' method, show that

$$\sum_{i=1}^{n} i\,r^{i-1} = \frac{1-r^n}{(1-r)^2} - \frac{n r^n}{1-r}$$

provided that $r \neq 1$. (*Hint*: Let $S_n = \displaystyle\sum_{i=1}^{n} i\,r^{i-1}$.)

6.8 Investment: Applications of the Geometric Series

> Blessed are the young, for they shall inherit the national debt.
>
> —Herbert Hoover (1874–1964)

Money hidden away in some secure spot actually decreases in value due to inflation. **Inflation** is the sustained increase in the price of goods and services in an economy over time. For example, in Fig. 6.16 the average price for a pound of white bread in the United States was \$1.37 in January 2014, and the price continued to increase to \$1.48 in January 2015. This means that the inflation rate of bread from January 2014 to January 2015 was

$$\text{Inflation Rate} = \frac{1.48 - 1.37}{1.37} \times 100\% = 8.03\%.$$

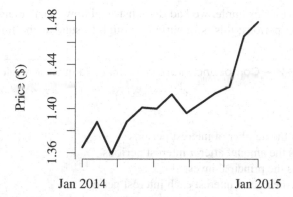

Fig. 6.16   Average price of one pound of white bread in the United States. Source: United States Department of Labor, Bureau of Labor Statistics. Series ID: APU0000702111

6.8.1 Compounded Interest

Example 6.8.1 For a principal amount P at an interest rate i, calculate the amount A after n interest payments.

Solution *Interest payment 1*:

$$A = P + Pi$$
$$= P(1 + i) \qquad\qquad \text{Factor.}$$

Interest payment 2: We now start with the amount $P(1 + i)$ and collect the second interest payment.

$$A = P(1 + i) + P(1 + i)i$$
$$= P(1 + i)(1 + i) \qquad\qquad \text{Factor.}$$
$$= P(1 + i)^2 \qquad\qquad \text{Simplify.}$$

Interest payment 3: We now start with the amount $P(1 + i)^2$ and collect the third interest payment.

$$A = P(1 + i)^2 + P(1 + i)^2 i$$
$$= P(1 + i)^2(1 + i) \qquad\qquad \text{Factor.}$$
$$= P(1 + i)^3 \qquad\qquad \text{Simplify.}$$

By continuing in this fashion, we can calculate interest payment n: we now start with the amount $P(1 + i)^{n-1}$ and collect the next interest payment.

$$A = P(1 + i)^{n-1} + P(1 + i)^{n-1}i$$
$$= P(1 + i)^{n-1}(1 + i) \qquad\qquad \text{Factor.}$$
$$= P(1 + i)^n \qquad\qquad \text{Simplify.}$$

◀

In the previous example, we had demonstrated how to calculate the amount after n interest periods; this is important enough to state as the following theorem.

Theorem 6.8.2 — Compound Interest. Compound interest is calculated by

$$A = P(1 + i)^n \qquad (6.25)$$

where
- n is the number of interest periods
- A is the amount after n interest periods
- P is the principal invested
- i is the rate of interest each interest period.

When you go to the bank, they will tell you what the annual interest rate or the rate *per annum* (Latin for "for each year") on the type of investment you wish to acquire. Furthermore, they will tell the number of times the interest is compounded. Compounding

- annually = once a year
- semi-annually = twice a year
- quarterly = 4 times a year
- monthly = 12 times a year
- biweekly = 26 times a year

- weekly = 52 times a year
- daily = 365 times a year (ignoring leap years that have 366 days).

Example 6.8.3 — Interest with a savings account. Samuel deposits $500 into a savings account at a bank that gives an interest rate of 2.25% per annum compounded monthly. He leaves the money in the account for $3\frac{1}{2}$ years. What is the amount in the savings account at the end of this time period?

Solution In this case, the principal amount $P = 500$, the interest rate is $i = \frac{0.0225}{12} = 0.001875$, and the number of interest periods is $n = 3\frac{1}{2} \times 12 = 42$. Then, using Eq. (6.25),

$$
\begin{aligned}
A &= P(1 + i)^n \\
&= 500(1 + 0.001\,875)^{42} \\
&= 500(1.001\,875)^{42} \\
&= 540.927.
\end{aligned}
$$

Therefore, the amount in the savings account after $3\frac{1}{2}$ years is $540.93 (to the nearest cent). In other words, Samuel made $40.93 in interest.

We can see visually how the amount in the saving account in Fig. 6.17. ◀

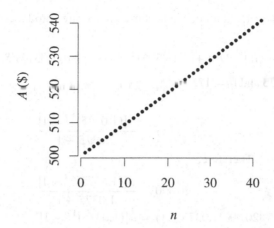

Fig. 6.17 The amount in the savings account in Example 6.8.3.

6.8.2 Annuities

> **Definition 6.8.4 — Annuity.** If a fixed amount of money is payable at regular time intervals, the sequence of payments is called an **annuity**. The time interval between two payments is the **period** of the annuity.

The word annuity itself suggests an annual payment.

Example 6.8.5 — Setting up a trust fund. Erika is planning for her son's future. She opens a trust fund which pays interest at 3.75% per year compounded annually. If the first deposit is made on her son's second birthday, what should be the amount of the annual deposit so that the fund reaches a total of $20 000 when the deposit is made on the boy's eighteenth birthday?

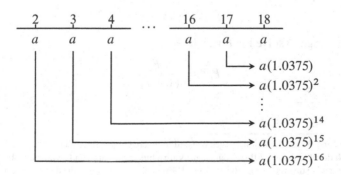

Fig. 6.18 The indicated sum of the accumulated values

Solution The indicated sum of the accumulated values produces the following geometric series

$$a + a(1.0375) + a(1.0375)^2 + \cdots + +a(1.0375)^{16},$$

where $r = 1.0375$ and $n = 17$.

Thus,

$$S_{17} = \frac{a[(1.0375)^{17} - 1]}{1.0375 - 1}.$$

But, $S_{17} = 20\,000$. Therefore,

$$20\,000 = \frac{a[(1.0375)^{17} - 1]}{1.0375 - 1}$$

$$20\,000(1.0375 - 1) = a[(1.0375)^{17} - 1]$$

$$a = \frac{20\,000(1.0375 - 1)}{(1.0375)^{17} - 1}$$

$$a \approx 862.256.$$

Therefore, the annual deposit is \$862.26 (to the nearest cent). ◄

6.8.3 Present Value

Frequently, it is necessary to invest money now in order to have enough money (with added interest) for future purchases. For example, suppose you are planning to make the mistake of getting married in the next couple of years.[17] You may want to know how much to invest now in order to have enough money for a down payment on a house or a car, or for alimony.

Example 6.8.6 What is the present value of F dollars due n periods from now if the interest rate is i per time period?

Solution Let PV be the present value of F dollars. Since,

$$F = PV(1 + i)^n \qquad (6.26)$$

we can rearrange to solve for PV, which is

$$PV = \frac{F}{(1 + i)^n}.$$

◄

---

[17]In England and Wales, 42% of marriages will end in divorce with half of them ending within 10 years (Office for National Statistics, 2014). Your chances are a little better in Canada with the probability that a marriage will end in divorce by the 50th year of 41.3% in 2004 (Statistics Canada, 2008). Unfortunately, there is no recent data from Statistics Canada since divorce data collection halted in 2008 due to budget cuts (Grant, 2011).

Theorem 6.8.7 — Present value. The present value, PV, is

$$PV = \frac{F}{(1+i)^n} \qquad (6.27)$$

where

- F is the future amount
- i is the interest rate per period
- n is the number of interest periods.

Example 6.8.8 — Present Value of an Annuity. Find the present value of an annuity of \$1000 which starts now and runs for 10 years, if money is worth 2.5% per year.

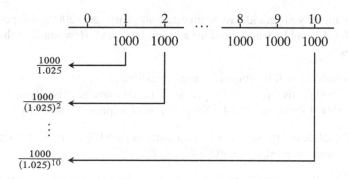

Fig. 6.19   Present value of an annuity of \$1000

Solution The present value of the annuity may be represented by the geometric series

$$\frac{1000}{1.025} + \frac{1000}{(1.025)^2} + \cdots + \frac{1000}{(1.025)^{10}}.$$

It will be convenient to rewrite this as

$$\frac{1000}{(1.025)^{10}} + \frac{1000}{(1.025)^9} + \cdots + \frac{1000}{1.025},$$

so that $a = \dfrac{1000}{(1.025)^{10}}$, $r = 1.025$, $n = 10$. Therefore,

$$\text{Present Value} = \frac{1000}{(1.025)^{10}} \times \frac{[(1.025)^{10} - 1]}{1.025 - 1} = 8752.0639.$$

The present value of the annuity is \$8752.06 (to the nearest cent). ◀

Exercises

6.8.1. For each of the following amounts, state the principal amount, the interest rate per period, and the number of interest periods:

(a) $100(1.014)^3$ (c) $500(1.18)^{10}$ (e) $15\,000(1.03)^9$

(b) $200(1.02)^6$ (d) $1000(1.0325)^{20}$ (f) $20\,000(1.0425)^{14}$

6.8.2. Calculate the amount for each of the following conditions:

(a) \$100 for three years at 1.4% per annum compounded annually

(b) \$200 for five years at 2% per annum compounded annually

(c) \$500 for five years at 2% per annum compounded semi-annually

(d) \$1000 for ten years at 3.25% per annum compounded quarterly

(e) \$1500 for seven years at 12% per annum compounded monthly

(f) \$20 000 for five years at 4.25% per annum compounded monthly

6.8.3. If one of your grandparents invested 40 years ago \$200 at 4% per annum compounded monthly, what would the amount be today? How much of this would be interest?

6.8.4. Which of the following is a better investment:

(a) \$500 for five years at 5% per annum compounded annually, or

(b) \$500 for five years at 4.5% per annum compounded monthly?

6.8.5. Calculate the present value of an annuity of \$1500 which starts now and runs for 5 years, if money is worth 3% per year.

6.8.6. Luca plans to get married in four years time from now. At that time, he wants to have saved \$50 000 for a down payment on a house. He decides to deposit the same amount of money in a savings account every month starting today and ending with a payment four years from today. If the bank pays interest at 3.25% per annum compounded quarterly, how much money must Luca deposit each month?

6.8.7. A mobile phone that costs \$500 will last three years, and then after that its scrap value is \$75. Find the regular semi-annual payment that must be made earning 2.5% per annum compounded semi-annually for 6 payments to replace the phone when it is worn out. (A series of deposits like this is called a **sinking fund**.)

6.8.8. The population in Amarillo, Texas is increasing at an annual rate of 1.01%. Assuming that this trend will continue, in how many months will the current population of 198 402 increase to 200 000?

6.8.9. The world's population in 2010 was estimated to be 6.884 billion people.[18]

(a) If the population increases at 1% per year, what will be the approximate world's population in 2020? 2030?

---

[18] Source: http://data.worldbank.org.

(b) If the population increases at 2% per year, what will be the approximate world's population in 2020? 2030?

(c) Complete the following table below for the projected world's population.

Rate of Growth	2010	2020	2030	2040	2050	2060	2070
1%	6.884						
2%	6.884						

(d) Make a graph from the completed table in part (c).

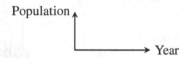

6.8.10. The world's population was estimated to be 6.490 billion in 2005 and 6.884 billion in 2010.[19] Find the approximate rate of yearly population increase for the planet.

6.8.11. Suppose that when a rumor is passed from person to person, only 90% of the truth is transmitted. If the person who first told the original story hears the rumor after it passed through five different people, how much truth remains of the original story?

6.8.12 (†). Wyatt borrows from the bank $15 000 and pays back $340 a month for 63 months. Wyatt's fiancée wants to know the rate of interest of the loan, but he won't tell her. If she found the above information scribbled on scrap paper while searching his desk, and she assumes that the last payment closes the account, what is the annual rate of interest Wyatt is paying?

6.9 The Infinite Geometric Series (A Trojan and a Tortoise)

Example 6.9.1 — Death and Taxes. Your great Uncle Vladimir from Pennsylvania wishes to give you an inheritance of $10 000. Unfortunately, state law requires you to pay an inheritance tax of 15%.[20] Your uncle wishes to give you $10 000, no less. How much should Uncle Vladimir give you such that your after-tax net (not your gross) is $10 000?

---

[19]Source: http://data.worldbank.org.
[20]Source: http://www.revenue.pa.gov/GeneralTaxInformation/Tax%20Types%20and%20Information/Pages/Inheritance-Tax.aspx#.Ve8bH5f09oM.

Solution Since Uncle Vladimir knows that you will pay the tax, he can give you more money to cover it; for example,

$$10\,000 + 10\,000(0.15).$$

However, then you will also have to pay tax on the amount $10\,000(0.15)$, so Uncle Vladimir can give you

$$10\,000 + 10\,000(0.15) + 10\,000(0.15)^2.$$

Again, you will have to pay tax on the amount $10\,000(0.15)^2$ on the added $10\,000(0.15)$. This process continues forever; that is, the gross amount Uncle Vladimir should give you is

$$S = 10\,000 + 10\,000(0.15) + 10\,000(0.15)^2 + 10\,000(0.15)^3 + \cdots \quad (6.28)$$

Then what is the value of S? Consider

$$\text{net} = \text{gross} - \text{tax}$$

then,

$$10\,000 = S - rS$$
$$10\,000 = S(1 - r)$$
$$S = \frac{10\,000}{1 - r}$$
$$S = \frac{10\,000}{1 - 0.15}$$
$$S = \frac{10\,000}{0.85}$$
$$S = 11\,764.71$$

to the nearest cent.

Looking back, we see that if Uncle Vladimir gives you $11\,764.71, then the amount of the tax to be collected is

Fig. 6.20   Uncle Vladimir's place

$$0.15 \times 11\,764.71 = 1764.71$$

leaving you with

$$11\,764.71 - 1764.71 = 10\,000.$$

◄

Equation (6.28) is an example of an **infinite geometric series**. We can generalize the previous example.

Theorem 6.9.2 — Infinite Geometric Series.

$$a + ar + ar^2 + \cdots = \sum_{i=1}^{\infty} ar^{i-1} = \frac{a}{1-r} \qquad (6.29)$$

whenever $-1 < r < 1.$

In the context of Example 6.9.1, a can be viewed as the net amount, and r as the tax rate.

Proof We let $S = a + ar + ar^2 + ar^3 + \cdots$. Then,

$$
\begin{array}{rclccccccccc}
S & = & a & + & ar & + & ar^2 & + & ar^3 & + & ar^4 & + & \cdots \\
-rS & = & & - & ar & - & ar^2 & - & ar^3 & - & ar^4 & - & \cdots \\
\hline
S - rS & = & a
\end{array}
$$

Then, we solve for S:

$$S - rS = a$$
$$S(1 - r) = a$$
$$S = \frac{a}{1-r}$$
$$a + ar + ar^2 + \cdots = \frac{a}{1-r}.$$

◄

In the span of three centuries, there were three great schools of thought that flourished in lower Italy—the school of Pythagoras (c. 570 BC–495 BC) at Croton, Archimedes (c. 287 BC–c. 212 BC) at Syracuse, and Zeno (c. 490 BC–c. 430 BC) at Elea. It is the philosopher Zeno who gave us the first important questions about the concept of infinity; he gave us the following paradox on motion.

Example 6.9.3 — Achilles Running a Stadion. Suppose the Greek hero Achilles is going to run a race course of one stadion.[21] Zeno argued that Achilles cannot traverse an infinite number of points in a finite amount of time. As seen in Fig. 6.21, Achilles must traverse

- $\frac{1}{2}$ of the stadion, then
- half of $\frac{1}{2}$, or $\frac{1}{4}$, of the stadion, then
- half of $\frac{1}{4}$, or $\frac{1}{8}$, of the stadion, then
- half of $\frac{1}{8}$, or $\frac{1}{16}$ of the stadion, then ...

---

[21] A stadion is an ancient Greek unit of measurement which is equivalent to 166.7 m (182.3 yd). There is much debate over the exact value of a stadion, but the 166.7 m value of a stadion was used by Eratosthenes (276 BC – 194 BC) of Cyrene in determining the circumference of the Earth (Gulbekian, 1987, p. 363).

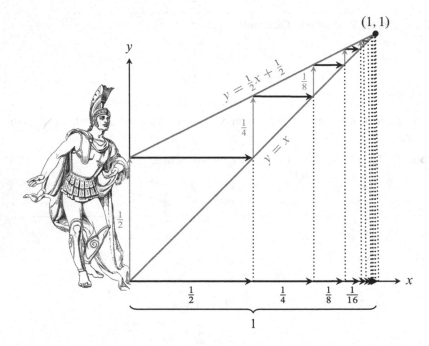

Fig. 6.21 A visual explanation of the infinite series for one stadion that Achilles must traverse. Bounded between $y = \frac{1}{2}x + \frac{1}{2}$ and $y = x$, the series $\frac{1}{2} + \frac{1}{4} + \frac{1}{8} + \frac{1}{16} + \frac{1}{32} + \cdots$ converges to 1. Graphics Credit: Achilles (1907) by Otho Cushing (1871–1942); Library of Congress.

This process continues on forever such that there are an infinite number of halfway points in any given space, and therefore as Zeno reasoned, Achilles cannot finish the race course in a finite amount of time (Burnet, 1908, p. 367).   ◄

I once used this reasoning with my mother when I was too lazy to fetch the remote control for the TV.

"Can't you do it yourself?" she asked.

"No, because for me to get to the remote, I will have to get halfway there, and then from that point to the remote is another half I must travel, and so on for an infinite number of halfway points. So you see Mom, I'll never get there. But you're already standing beside it, so can you just throw it to me?"

My mother being no fool responded, "But if I throw the remote, it will have to get halfway there, and then from that point to you is another half that it must travel, and so it will take an infinite number of halfway points. So you see Ricky, if I throw the remote to you, it will never get there."

We both were at an impasse; neither of us moved, because we both reasoned that we couldn't.

There is a well-known anecdote that Diogenes (c. 412 BC–c. 323 BC) the Cynic upon hearing Zeno's arguments refuted the claims by simply getting up and walking away (Cajori, 1920, p. 11).

Now by generalizing from Fig. 6.21, Fig. 6.22 shows a visual demonstration of the infinite geometric series. If $0 < r < 1$ and $a > 0$, then it can be visualized in two ways: for r and a it appears as an ascending staircase, and for $-r$ and $-a$ it gives the impression of a cobweb.

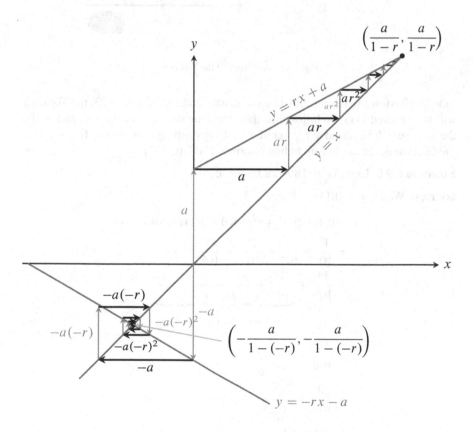

Fig. 6.22   In this visual explanation of the infinite geometric series $0 < r < 1$, and $a > 0$.

Example 6.9.4 — Achilles and the Tortoise. Achilles thought he could run faster than the Tortoise, so Achilles allowed the Tortoise a 100 foot (30.48 m) head start; in Fig. 6.23, Achilles starts at a_1, and the Tortoise at t_1. This only seems fair, right?

Alas, Achilles will never overtake the Tortoise for he must first reach the place from which the Tortoise started; that is, Achilles must reach a_2 in Fig. 6.23. By that

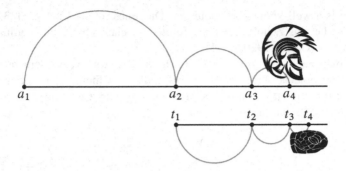

Fig. 6.23   Achilles and the Tortoise

time, the Tortoise will have moved some distance ahead; in Fig. 6.23, the Tortoise will have moved to t_2. Achilles must then make up that and reaches a_3, and again the tortoise will be ahead at t_3. Achilles is always getting close to the Tortoise, but the Grecian never catches up to him (Burnet, 1908, p. 367).   ◀

Example 6.9.5 Express $0.11\overline{89}$ as a fraction.

Solution We let $q = 0.11\overline{89}$. Then,

$$q = 0.1 + 0.01 + 0.008\,9 + 0.000\,089 + \cdots$$

$$= \frac{1}{10} + \frac{1}{10^2} + \frac{89}{10^4} + \frac{89}{10^6} + \cdots$$

$$= \frac{11}{100} + \underbrace{\frac{89}{10^4} + \frac{89}{10^6} + \frac{89}{10^8} + \cdots}_{a = \frac{89}{10^4},\, r = \frac{1}{10^2}}$$

$$= \frac{11}{100} + \frac{\frac{89}{10^4}}{1 - \frac{1}{10^2}}$$

$$= \frac{11}{100} + \frac{\frac{89}{10^4}}{\frac{10^2 - 1}{10^2}}$$

$$= \frac{11}{100} + \frac{\frac{89}{10^4}}{\frac{99}{10^2}}$$

$$= \frac{11}{100} + \frac{89}{10^4} \times \frac{10^2}{99}$$

$$= \frac{11}{100} + \frac{89}{10^2} \times \frac{1}{99}$$

$$= \frac{11}{100} + \frac{89}{9900}$$

$$= \frac{1089}{9900} + \frac{89}{9900}$$

$$= \frac{1178}{9900}$$

So $0.11\overline{89} = \frac{1178}{9900} = \frac{589}{4950}$. ◀

Exercises

6.9.1. Is $\displaystyle\sum_{i=1}^{\infty} ar^{i-1} = \sum_{i=0}^{\infty} ar^{i}$? Why?

6.9.2. Evaluate.

(a) $\displaystyle\sum_{i=1}^{\infty} 2\left(\frac{1}{3}\right)^{i-1}$

(b) $\displaystyle\sum_{i=1}^{\infty} 2(3)^{-i}$

(c) $\displaystyle\sum_{i=1}^{\infty} \left(-\frac{1}{2}\right)^{i-1}$

(d) $\displaystyle\sum_{i=1}^{\infty} 4(3)^{-i-1}$

(e) $7 + 3.5 + 1.75 + \cdots$

(f) $5 + 2.5 + 1.25 + 0.625 + \cdots$

(g) $\dfrac{1}{4} + \dfrac{1}{16} + \dfrac{1}{64} + \dfrac{1}{256} + \cdots$

(h) $\dfrac{1}{2} - \dfrac{1}{4} + \dfrac{1}{8} - \dfrac{1}{16} + \cdots$

(i) $\dfrac{1}{3} + \dfrac{1}{4} + \dfrac{3}{16} + \cdots$

(j) $\dfrac{1}{5} + \dfrac{1}{6} + \dfrac{5}{36} + \dfrac{25}{216} + \cdots$

6.9.3. Uncle Vladimir wants to give you $55 000 for a new sports car, but the state requires you to pay a 17% tax on gifts. How much should Uncle Vladimir give you on top of the $55 000 to cover the gift tax?

6.9.4. Evaluate

(a) $\left(\dfrac{1}{3} + \dfrac{1}{9} + \dfrac{1}{27} + \cdots\right) + \left(\dfrac{1}{4} + \dfrac{1}{16} + \dfrac{1}{64} + \cdots\right)$

(b) $\dfrac{1}{3} + \dfrac{1}{4} + \dfrac{1}{9} + \dfrac{1}{16} + \dfrac{1}{27} + \dfrac{1}{64} + \cdots$

6.9.5. Express each of the following repeating decimals as a fraction.

(a) $0.\overline{19}$

(b) $0.19\overline{53}$

6.9.6. Show that $0.\overline{9} = 1$.

6.9.7. Suppose that $x > 0$. Evaluate

$$(1 + x)^{-1} + (1 + x)^{-2} + (1 + x)^{-3} + \cdots$$

6.9.8. Show that $2^{\frac{1}{4}} \times 2^{\frac{1}{16}} \times 2^{\frac{1}{64}} \times \cdots = 2^{\frac{1}{3}}$.

6.9.9. A basketball is dropped from a height of 10 feet and bounces continuously. After each bounce, the basketball rises to 30% of the height on its previous bounce. What is the total distance, both up and down, the basketball traveled?

6.9.10. Suppose you are meeting your friend Fraser at a pub. You arrive early, and decide to order a pitcher of beer. A pitcher is 64 oz. While you are waiting for Fraser, you mindlessly drink half of the pitcher without noticing. When Fraser arrives finally, he drinks half of what remains in the pitcher, and passes it back to you. You drink half of what is left, and pass it back and forth until no more beer left. How much beer did each of you drink?

6.9.11. The infinite series $1 - 1 + 1 - 1 + \cdots$ is known as **Grandi's series** after the Italian monk Guido Grandi (1671–1742) who studied it in 1703.
  (a) Using Theorem 6.9.2, evaluate $1 - 1 + 1 - 1 + \cdots$.
  (b) Evaluate $(1 - 1) + (1 - 1) + (1 - 1) + \cdots$.
  (c) Why is there a discrepancy between the answers in parts (a) and (b)?

6.9.12. Observe the square in Fig. 6.24 with area of one. Demonstrate how Fig. 6.24 illustrates an infinite geometric series.

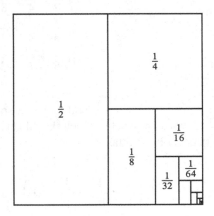

Fig. 6.24   An instance of the infinite geometric series.

6.9.13. Calculate the sum of the following series
$$1 + 2x + 3x^2 + 4x^3 + 5x^4 + \cdots$$
when $-1 < x < 1$.

6.10 Proofs by Mathematical Induction

The set of natural numbers $\mathbb{N}$ was originally created for counting—a process that leads naturally from one number n to its successor $n + 1$. This step by step procedure of passing from n to $n+1$ that generates the (countably) infinite sequence of the natural numbers is also the basis for one of the fundamental principles of mathematical reasoning—the Principle of Mathematical Induction.

.10.1 Inductive Reasoning versus Mathematical Induction

But before we begin to define mathematical induction, it is important to highlight the point—to emphasize—that this is completely different from induction reasoning. To recap what we had discussed previously (see Sec. 2.8), inductive reasoning takes a few cases and tries to abstract to the general case.

For example, while I was writing this section, I was standing in the hallway with my supervisor in the Mathematics Department when the power to the entire college was cut. My supervisor's mobile rang; the secretary was on the line who was driving.

She exclaimed, "The power is on at this traffic light, and oh! the power is on at the next traffic light, so the power must be on for every traffic light onwards!"

I cocked an eyebrow—she was using inductive reasoning.[22] Inductive reasoning is inherently uncertain for a variety of reasons, to be sure, which is why it seeks strong evidence rather than absolute proof. You cannot be absolutely sure that it will hold for everything in the future. The power grid during this time is unstable, and so there would be sections of the city that would have power, and others that would be without it.

Sure enough, within a couple of seconds, she hit a traffic light that was not functioning, and so away went her claim that all of the traffic lights were working now from a certain distance from the college.

Inductive reasoning is different from deductive reasoning. With inductive reasoning you cannot *prove* that it holds in general from the few observed cases; however, deductive reasoning allows you to be certain of your conclusion as it has been reasoned logically from a set of premises. The Principle of Mathematical Induction is just that—a deductive tool used in demonstrating a proposition is true for all natural numbers. It is just rather unfortunate that both have similar names, since they mean different things logically.

10.2 The Principle of Mathematical Induction

Axiom — The Principle of Mathematical Induction. A proposition involving the natural numbers n is true provided that
    (a) the proposition is true in the special case $n = 1$ (this is the **base step**);
    (b) the truth of the proposition for $n = k$, $k \in \mathbb{N}$ implies the truth of the proposition for $n = k + 1$ (this is the **inductive step**).

---

[22]Others may know this as *anecdotal evidence* or the **fallacy of incomplete evidence**.

The Principle of Mathematical Induction can also be stated symbolically. A proposition p_n is true for every natural number n, provided that

(a) the proposition p_1 is true (base step), and

(b) the implication $p_k \rightarrow p_{k+1}$ is true for every $k \in \mathbb{N}$ (inductive step).

The assumption that the proposition p_k is true is called the **inductive hypothesis**.

Since we have stated that this is an axiom,[23] we shall not hesitate to accept the Principle of Mathematical Induction—we simply accept it as one of the basic principles of mathematical reasoning.

The Principle of Mathematical Induction is analogous to climbing a ladder with (countably) infinite rungs in which each rung is labeled $1, 2, 3, \ldots$; see Fig. 6.25.

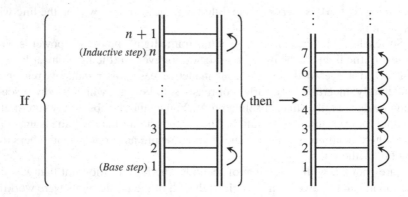

Fig. 6.25   The ladder analogy for the Principle of Mathematical Induction

If it is possible

(a) to step on the first rung (*base step*), and

(b) you know how to climb from rung k to rung $k + 1$, for every $k \in \mathbb{N}$, (*inductive step*),

then it is possible for you to climb the entire (countably) infinite ladder.

6.10.3 Proofs Using the Principle of Mathematical Induction

In using the Principle of Mathematical Induction to prove a proposition, the use of this principle falls into two steps: the base step and the inductive step. Template 6.10.1 breaks down the requirements in each step.

---

[23]In mathematics, we use the term *axiom* to mean a statement or proposition that is regarded as being an irreducible fact that is self-evidently true. The Principle of Mathematical Induction is an axiom in Peano's theory for the natural numbers. However, in other systems such as Zermelo-Fraenkel set theory, it can be derived as a theorem.

Template 6.10.1 — Proof by Principle of Mathematical Induction.

PROBLEM: Prove that for every natural number n, the proposition p_n is true.

Proof

    (a) *Base step.* Verify that the proposition is true for $n = 1$.

    (b) *Inductive step.*

        (i) Assume that the proposition is true for $n = k$, $k \in \mathbb{N}$.

        (ii) Prove that the proposition is true for $n = k + 1$.

Therefore, by the Principle of Mathematical Induction, the proposition is true for every natural number n. ◄

In Sec. 6.4, we saw the ingenious young Gauss add up the numbers between 1 and 100 for his teacher. We revisit the problem, and use the Principle of Mathematical Induction to prove his answer.

Theorem 6.10.2 — Gaussian summation formula.

$$\sum_{i=1}^{n} i = \frac{n(n + 1)}{2} \tag{6.30}$$

Proof We will prove this using the Principle of Mathematical Induction.

    (a) *Base step.* For $n = 1$, we have on the left side is $\sum_{i=1}^{1} i = 1$, and on the right side is $\dfrac{1(1 + 1)}{2} = \dfrac{2}{2} = 1$. Thus, it is true for $n = 1$.

    (b) *Inductive step.*

        (i) We *assume* that the proposition is true for $n = k$, $k \in \mathbb{N}$. That is, we assume that

$$\sum_{i=1}^{k} i = \frac{k(k + 1)}{2}.$$

        This is our inductive hypothesis.

        (ii) We need to *prove* that the proposition is true for $n = k + 1$. That is, we need to prove that

$$\sum_{i=1}^{k+1} i = \frac{(k + 1)(k + 2)}{2}. \tag{6.31}$$

        So, starting on the left side of Eq. (6.31),

$$\sum_{i=1}^{k+1} i = 1 + 2 + 3 + \cdots + k + (k + 1)$$

$$= \left(1 + 2 + 3 + \cdots + k\right) + (k + 1)$$

$$= \sum_{i=1}^{k} i + (k + 1)$$

$$\underbrace{\qquad}_{\frac{k(k+1)}{2}}$$

$$= \frac{k(k + 1)}{2} + (k + 1) \qquad\qquad \text{By the induction hypothesis}$$

$$= \frac{k(k + 1)}{2} + \frac{2(k + 1)}{2} \qquad\qquad \text{Common denominator}$$

$$= \frac{k(k + 1) + 2(k + 1)}{2} \qquad\qquad \text{Factor}$$

$$= \frac{(k + 1)(k + 2)}{2}$$

which equals the right side of Eq. (6.31), and this concludes the inductive step.

Therefore, $\displaystyle\sum_{i=1}^{n} i = \frac{n(n + 1)}{2}$ is true for every natural number n by the Principle of Mathematical Induction. ◀

Theorem 6.10.3

$$\sum_{i=1}^{n} i^2 = \frac{n(n + 1)(2n + 1)}{6} \qquad\qquad (6.32)$$

Here is another example of proving a summation formula.

Proof We will prove this using the Principle of Mathematical Induction.

(a) *Base step.* For $n = 1$, Eq. (6.32) is true, since on the left side of Eq. (6.32),

$$\sum_{i=1}^{1} i^2 = 1^2 = 1$$

and on the right side of Eq. (6.32),

$$\frac{1(1 + 1)(2 + 1)}{6} = \frac{1(2)(3)}{6} = \frac{6}{6} = 1.$$

(b) *Inductive step.*

    (i) We *assume* that the proposition is true for $n = k$, $k \in \mathbb{N}$. That is, we assume that

$$\sum_{i=1}^{k} i^2 = \frac{k(k + 1)(2k + 1)}{6} \qquad\qquad (6.33)$$

is true.

(ii) We need to *prove* that the proposition is true for $n = k + 1$. That is, we need to show that

$$\sum_{i=1}^{k+1} i^2 = \frac{(k+1)(k+2)\big(2(k+1)+1\big)}{6} \tag{6.34}$$

is also true. Starting on the left side of Eq. (6.34),

$$\sum_{i=1}^{k+1} i^2 = \underbrace{\sum_{i=1}^{k} i^2}_{\frac{k(k+1)(2k+1)}{6}} + (k+1)^2$$

$$= \frac{k(k+1)(2k+1)}{6} + (k+1)^2 \qquad \text{By the induction hypothesis (6.33)}$$

$$= \frac{k(k+1)(2k+1)}{6} + \frac{6(k+1)^2}{6}$$

$$= \frac{2k^3 + 3k^2 + k}{6} + \frac{6(k^2 + 2k + 1)}{6}$$

$$= \frac{2k^3 + 3k^2 + k}{6} + \frac{6k^2 + 12k + 6}{6}$$

$$= \frac{2k^3 + 9k^2 + 13k + 6}{6} \tag{6.35}$$

$$= \frac{(k+1)(k+2)(2k+3)}{6}$$

$$= \frac{(k+1)(k+2)\big(2(k+1)+1\big)}{6} \tag{6.36}$$

which is the same as the right side of Eq. (6.34). Hence, we have shown that Eq. (6.34) is also true, and this concludes the inductive step.

Therefore, $\sum_{i=1}^{n} i^2 = \dfrac{n(n+1)(2n+1)}{6}$ is true for every natural number n by the Principle of Mathematical Induction. ◀

Often, the difficult step in a proof using the Principle of Mathematical Induction is proving the inductive step. Perhaps, you got stuck at (6.35). Maybe you do not know how to factor a cubic polynomial. In this instance, we are working forwards from the initial situation on the left side of the equation until we reach the desired result on the right side—from the data to the unknown. But, suppose that we got stuck at the second equality and we made several failed attempts. At this point, we must change our strategy and try a different strategy. If we are unable to find the path in a *forward* manner, what is to stop us working *backwards*?—from the unknown to the known?

Let us invoke a helpful problem solving strategy—*work backwards*.[24] We know that we need to get from (6.36) to (6.35); we can start with (6.36), and expand:

$$\frac{(k+1)(k+2)\big(2(k+1)+1\big)}{6} = \frac{(k+1)(k+2)(2k+3)}{6}$$

$$= \frac{2k^3 + 9k^2 + 13k + 6}{6}.$$

Now, we successfully bridged the gap, and have completed the proof.

6.10.4 Abusing Mathematical Induction: More Fallacies

It is important to not overlook proving either the base step or the inductive step, even if they may seem to be obvious or clear. I have tried to avoid the words obviously, clearly, or trivial. They are often the kiss of death for further thought because they encourage the easy acceptance of fallacious proofs. Salesmen know this. They are quick to point out that an obviously clever dude like you can see *clearly* that their product is superior, despite the trivial issues pointed out by previous consumers. There is a legend that has been passed down from mathematics professors to their students through the ages, and it goes something like this:

> A professor is giving a lecture to his students when he proclaims, "This is clearly trivial." (Proof by authority?) One of the humble students raises her hand and says, "Professor, I don't understand why this is true." The professor pauses and begins to stroke his beard pensively. He walks over to a corner of the chalkboard and begins to scribble—his back obscuring the view from the rest of the class. He erases it with his hand a couple of times; his face is smeared with chalk dust as he continues to stroke his beard. After a couple of attempts, he erases completely what he has written, and bolts for the door. The students sit bewildered at the sight of their professor running out of the lecture hall. After twenty minutes, the professor re-emerges, announces, "Yes! Yes! This is indeed trivial!" and then proceeds with the rest of the lecture.

The great communicator of modern physics, Richard Feynman (1918–1988), once wrote down his observations on how mathematicians spoke:

> We [physicists] decided that "trivial" means "proved." So we joked with the mathematicians: "We have a new theorem—that mathematicians can prove only trivial theorems, because every theorem that's proved is trivial" (Feynman and Leighton, 1985).

---

[24]For other problem solving strategies, see Pólya's approach, which is summarized in the Preface starting on p. xviii.

Of course, mathematicians were not too happy with this—for then, mathematicians could only prove things that are obvious. The next two examples will show that skipping either the base step or the inductive step is detrimental.

Example 6.10.4 Is the following true?

$$2 + 4 + 6 + \cdots + 2n = \left(n + \frac{1}{2}\right)^2 \qquad (6.37)$$

Solution Let us begin with the inductive step. We assume that Eq. (6.37) is true for $n = k$; that is,

$$2 + 4 + 6 + \cdots + 2k = \left(k + \frac{1}{2}\right)^2.$$

Now, assuming that Eq. (6.37) is true for $n = k$, we prove that this implies that Eq. (6.37) is true for $n = k + 1$; that is

$$2 + 4 + 6 + \cdots + 2k + 2(k + 1) = \left((k + 1) + \frac{1}{2}\right)^2. \qquad (6.38)$$

We begin with the left side of (6.38), and manipulate it so that we get the right side.

$$\underbrace{2 + 4 + 6 + \cdots + 2k}_{\left(k + \frac{1}{2}\right)^2} + 2(k + 1) = \left(k + \frac{1}{2}\right)^2 + 2(k + 1)$$

$$= \left(k + \frac{1}{2}\right)^2 + 2(k + 1)$$

$$= k^2 + k + \frac{1}{4} + 2k + 2$$

$$= k^2 + 3k + \frac{9}{4}$$

$$= \left(k + \frac{3}{2}\right)^2$$

$$= \left(k + 1 + \frac{1}{2}\right)^2$$

This completes the inductive step.

If we just complete the inductive step, and then walked away from the proof believing that Eq. (6.37) is true, we would be mistaken. In fact, Eq. (6.37) is not true since if we check the base case for $n = 1$, on the left side of Eq. (6.37) is 2 and the right side is $\left(1 + \frac{1}{2}\right)^2 = \left(\frac{3}{2}\right)^2 = \frac{9}{4}$. Since $2 \neq \frac{9}{4}$, the base step is not proven.

Therefore by mathematical principle, for all $n \in \mathbb{N}$

$$2 + 4 + 6 + \cdots + 2n \neq \left(n + \frac{1}{2}\right)^2$$

◄

Example 6.10.5 Is $n^2 + n + 17$ a prime number for any $n \in \mathbb{N}$?

Solution Let us begin with the base step. For $n = 1$,

$$n^2 + n + 17 = 1^2 + 1 + 17$$
$$= 19$$

which is prime. In fact, if we check the statement for $n = 2, 3, 4, \ldots, 15$, we find that this statement is true for all of these numbers; see Table 6.6.

Table 6.6   Checking if $n^2 + n + 17$ is prime for $n = 2, 3, \ldots, 16$

n	1	2	3	4	...	9	10	11	12	13	14	15	16
$n^2 + n + 17$	19	23	29	37	...	107	127	149	173	199	227	257	289
Prime?	✓	✓	✓	✓	...	✓	✓	✓	✓	✓	✓	✓	✗

However, the statement is not true for $n = 16$. If $n = 16$, then

$$n^2 + n + 17 = 16^2 + 16 + 17$$
$$= 289$$
$$= 17^2$$

which is not a prime number. Therefore, for any $n \in \mathbb{N}$, the number $n^2 + n + 17$ is not necessarily prime.   ◄

Example 6.10.5 demonstrates the importance of proofs and illustrates the difference between inductive and deductive reasoning.[25] We may look at many cases and see that the proposition is true, but just around the corner—like that next non-working traffic light—lurking—could be the case that shows the proposition is false. Examples 6.10.4 and 6.10.5 show that checking only the inductive step or the base step is not sufficient to prove the proposition. Therefore, if the base step or inductive step is missing in the proof, it is absolutely incorrect. Full stop.

When mathematical induction is applied incorrectly, it can lead to fallacies.

Example 6.10.6 Show that any group of n people, every person has the same name.

Solution Let p_n denote the proposition that any group of n people have the same name.

(a) *Base step.* This is true for p_1. There is only one person in this group and hence, he or she has the same name.

---

[25]See Sec. 2.8 beginning on p. 59 for a discussion on the difference between inductive and deductive reasoning.

(b) *Inductive step.* We assume that p_k is true; that is, any group of k people have the same name. (This is the inductive hypothesis.) We need to prove that any group of $(k + 1)$ people have the same name. Let $(k + 1)$ people be given, and we line them up in a row; see Fig. 6.26.

k people

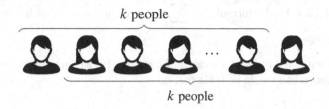

k people

Fig. 6.26 $(k + 1)$ people line up in a row

The first k people all have the same name, and the last k people have the same name, by the inductive hypothesis. So the first person and the last person have the same name.

Since the base step and the inductive step have been verified, therefore every person in a group of n people have the same name by mathematical induction!

We have to reconcile this seemingly paradoxical result; we know from our own personal experiences that not every group we have encountered has people with the same name. The gap in the reasoning is due to the abuse of the definition of a "group." A social group is defined as two more people who interact with each other. Thus, the base step is not correct when it looks at a group of just one person. If the base case looks at a group of two people, then it is not always true that these two people will have the same name. ◄

10.5 Guessing the Formula

... a good guess is quite as important as a good proof.

—Paul R. Halmos (1916–2006)

From Halmos (1955) on his review of Pólya (1954).

Once the correct proposition has been found, often a proof using the Principle of Mathematical Induction is sufficient. But there is something I always find unsatisfying about this kind of proof. Yes, I can see that the proof is correct, but *how did you discover it?* I want to know your technique of how you discovered it so I can use it to discover other similar things.[26] "Inasmuch as such a proof does not give a clue to the act of discovery, it might more fittingly be called a *verification*" (Courant and Robbins, 1941, p. 15).

---

[26]Do you remember what Goethe said? See p. 168.

The question of discovery returns us to the realm of Pólya's problem solving approach and inductive reasoning. Using inductive reasoning, we have to investigate, and guess, and then when we think we have our answer, we switch to deductive reasoning to prove our answer.

Example 6.10.7 Find a formula for $\sum_{i=1}^{n} i(i+1)$, and prove that it is true for all natural numbers n.

Solution First, we will see if we can *find a pattern* by writing out the first couple of partial sums, like we do in Table 6.7.

Table 6.7   Partial sums of $f(n) = \sum_{i=1}^{n} i(i+1)$

n	1	2	3	4	5	6	7	8	9	10	11	12
$f(n)$	2	8	20	40	70	112	168	240	330	440	572	728

Although you might have spotted it, I was not able to spot a pattern immediately. This was my approach to coming up with a formula, but you may have found a different method.[27] As part of my *understanding the problem* process, I looked at the differences between the values to see if I could find a common difference or a common ratio (from Sec. 6.1); this leads me to use the method of finite differences (from Sec. 6.2).

By this method, I could see from Fig. 6.27 that a third degree polynomial could have generated this sequence; the column of the third differences is all the same. If the polynomial is degree 3, then it must have the form $ax^2 + bx^2 + cx + d$ (according to Table 6.2). Using the data, I needed to come up with 4 equations with 4 unknowns:

$$f(1) = 2 = a(1)^3 + b(1)^2 + c(1) + d = a + b + c + d$$
$$f(2) = 8 = a(2)^3 + b(2)^2 + c(2) + d = 8a + 4b + 2c + d$$
$$f(3) = 20 = a(3)^3 + b(3)^2 + c(3) + d = 27a + 9b + 3c + d$$
$$f(4) = 40 = a(4)^4 + b(4)^2 + c(4) + d = 64a + 16b + 4c + d$$

---

[27]For a little while, I will switch to saying *I* instead of *we* since this was my thinking and approach to the problem which may be different from how you spotted the pattern.

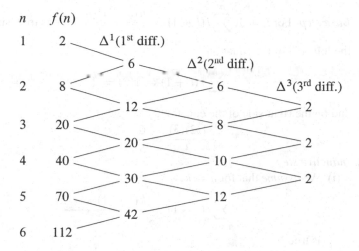

Fig. 6.27   Differences between sequence values in Example 6.10.7

and I need to solve

$$\begin{cases} a + b + c + d = 2 & ① \\ 8a + 4b + 2c + d = 8 & ② \\ 27a + 9b + 3c + d = 20 & ③ \\ 64a + 16b + 4c + d = 40 & ④ \end{cases}$$

After solving this, I found that the coefficients of the polynomial are $a = \frac{1}{3}$, $b = 1$, $c = \frac{2}{3}$, and $d = 0$. Therefore, the polynomial that could have generated the sequence of values is

$$\begin{aligned} f(n) &= \frac{1}{3}n^3 + n^2 + \frac{2}{3}n \\ &= \frac{n^3 + 3n^2 + 2n}{3} \\ &= \frac{n(n^2 + 3n + 2)}{3} \\ &= \frac{n(n+1)(n+2)}{3} \end{aligned}$$

By my inductive reasoning, I used a couple of values of the sequence to find a polynomial that fits. Thus, my guess for the formula is

$$\sum_{i=1}^{n} i(i+1) = \frac{n(n+1)(n+2)}{3}.$$

Now, *we* need to prove this using deductive reasoning to show this holds for *all* natural numbers.

We will prove this using the Principle of Mathematical Induction.

(a) *Base step.* For $n = 1$, $\sum_{i=1}^{n} i(i + 1) = \dfrac{n(n + 1)(n + 2)}{3}$ is true, since on the left side of the equation,

$$\sum_{i=1}^{1} i(i + 1) = 1(2) = 2$$

and on the right side of the equation,

$$\frac{1(2)(3)}{3} = \frac{6}{3} = 2.$$

(b) *Inductive step.*

(i) We *assume* that for $n = k$,

$$\sum_{i=1}^{k} i(i + 1) = \frac{k(k + 1)(k + 2)}{3} \tag{6.39}$$

is true.

(ii) We need to *prove* that if we assume this, then this implies that for $n = k + 1$,

$$\sum_{i=1}^{k+1} i(i + 1) = \frac{(k + 1)(k + 2)(k + 3)}{3} \tag{6.40}$$

is also true. Starting on the left side of Eq. (6.40),

$$\sum_{i=1}^{k+1} i(i + 1) = \underbrace{\sum_{i=1}^{k} i(i + 1)}_{\frac{k(k+1)(k+2)}{3}} + (k + 1)(k + 2)$$

$$= \frac{k(k + 1)(k + 2)}{3} + (k + 1)(k + 2) \qquad \text{By the induction hypothesis (6.39)}$$

$$= \frac{k(k + 1)(k + 2)}{3} + \frac{3(k + 1)(k + 2)}{3}$$

$$= \frac{k(k + 1)(k + 2) + 3(k + 1)(k + 2)}{3}$$

$$= \frac{(k + 1)(k + 2)(k + 3)}{3},$$

which is the same as the right side of Eq. (6.40). Hence, we have shown that Eq. (6.40) is also true, and this concludes the inductive step.

Therefore, by the Principle of Mathematical Induction,

$$\sum_{i=1}^{n} i(i + 1) = \frac{n(n + 1)(n + 2)}{3}$$

is true for every natural number n. ◀

Exercises

6.10.1. State the $(k + 1)^{\text{st}}$ term in each of the following series where the kth term has been indicated.

(a) $3 + 6 + 9 + \cdots + 3k + \cdots$

(b) $4 + 7 + 10 + \cdots + (3k + 1) + \cdots$

(c) $\dfrac{1}{2} + \dfrac{1}{3} + \dfrac{1}{4} + \cdots + \dfrac{1}{k} + \cdots$

(d) $1 + 3 + 3^2 + 3^3 + \cdots + 3^{k-1} + \cdots$

6.10.2. Given the summation formulas from $i = 1$ to n, state the formula for $i = 1$ to $n + 1$.

(a) $\displaystyle\sum_{i=1}^{n} 2^{n-1} = 2^n - 1$ $\qquad$ $\displaystyle\sum_{i=1}^{n+1} 2^{n-1} = \,?$

(b) $\displaystyle\sum_{i=1}^{n}(4n + 1) = n(2n + 3)$ $\qquad$ $\displaystyle\sum_{i=1}^{n+1}(4n + 1) = \,?$

(c) $\displaystyle\sum_{i=1}^{n}(2n - 1)^2 = \frac{1}{3}n(4n^2 - 1)$ $\qquad$ $\displaystyle\sum_{i=1}^{n+1}(2n - 1)^2 = \,?$

Using the Principle of Mathematical Induction, prove each of the following propositions or formulas.

6.10.3. $1 + 3 + 5 + \cdots + (2n - 1) = n^2$

6.10.4. $2 + 6 + 10 + \cdots + (4n - 2) = 2n^2$

6.10.5. $1 + 4 + 7 + \cdots + (3n - 2) = \dfrac{n(3n - 1)}{2}$

6.10.6. $1 \cdot 2 + 2 \cdot 2^2 + 3 \cdot 2^3 + \cdots + n2^n = 2 + (n - 1)2^{n+1}$

6.10.7. $1 \cdot 3 + 2 \cdot 5 + 3 \cdot 7 + \cdots + n(2n + 1) = \dfrac{n(n + 1)(4n + 5)}{6}$

6.10.8. $\left(1 + \dfrac{3}{1}\right)\left(1 + \dfrac{5}{4}\right)\left(1 + \dfrac{7}{9}\right) \cdots \left(1 + \dfrac{2n + 1}{n^2}\right) = (n + 1)^2$

6.10.9. $\displaystyle\sum_{i=1}^{n} i^3 = \left[\dfrac{n(n + 1)}{2}\right]^2$

6.10.10. $\displaystyle\sum_{i=1}^{n}(2i - 1) = n^2$

6.10.11. $\displaystyle\sum_{i=1}^{n}(2i-1)^2 = \frac{n(2n-1)(2n+1)}{3}$

6.10.12. $\displaystyle\sum_{i=1}^{n}\frac{1}{i(i+1)(i+2)} = \frac{1}{2}\left[\frac{1}{2} - \frac{1}{(n+1)(n+2)}\right]$

6.10.13. $\displaystyle\sum_{i=1}^{n}(2i)^3 = 2n^2(n+1)^2$

6.10.14. For every natural number n, $n < 2^n$.

6.10.15. For every natural number n, $n! \le n^n$.

6.10.16. For every natural number n, $n^2 + n$ is even.

6.10.17. For every natural number n, $n(n+5)$ is divisible by 2. (*Hint:* A number n is **divisible** by m if there is no remainder when n is divided by m. More formally, an integer n is divisible by a nonzero integer m if $n = m\ell$ for some integer ℓ.)

6.10.18. Given the arithmetic sequence $a, a+d, a+2d, \ldots$, show that the nth term of the sequence is $a + (n-1)d$.

6.10.19. Consider the following formula

$$S_n = \sum_{i=1}^{n}\frac{1}{i(i+1)}.$$

(a) Calculate S_1, S_2, S_3, and S_4.
(b) Guess a formula for S_n
(c) Prove that the formula is correct for every natural number n.

6.10.20. Consider the following formula

$$S_n = \sum_{i=1}^{n}\frac{i}{(i+1)!}.$$

(a) Calculate S_1, S_2, S_3, and S_4.
(b) Guess a formula for S_n.
(c) Prove that the formula is correct for every natural number n.

6.10.21. Consider the following

$$1 = 1$$
$$3 + 5 = 8$$
$$7 + 9 + 11 = 27$$
$$13 + 15 + 17 + 19 = 64$$
$$21 + 23 + 25 + 27 + 29 = 125$$

$$\vdots$$

Guess a formula, express it in suitable mathematical notation, and then prove that your guess is correct.

6.10.22 (Binet's Formula for Fibonacci Numbers). In Example 6.1.4, we investigated the Fibonacci sequence. Prove that the nth Fibonacci number is given by Binet's formula,

$$t_n = \frac{1}{\sqrt{5}}\left[\left(\frac{1 + \sqrt{5}}{2}\right)^n - \left(\frac{1 - \sqrt{5}}{2}\right)^n\right].$$

(*Hint*: Show for the base case that $t_1 = 1$ and $t_2 = 1$ is true. Let $a = \frac{1+\sqrt{5}}{2}$ and $b = \frac{1-\sqrt{5}}{2}$, and show that $a + 1 = a^2$ and $b + 1 = b^2$. Then, for the inductive step, show that $t_{n+1} = t_{n-1} + t_n$ is true.)

6.10.23. Consider the formula

$$S_n = \sum_{i=1}^{n} i! = 3^{n-1}$$

(a) Is the formula true for S_1, S_2, and S_3?
(b) Is this formula true for all $n \in \mathbb{N}$?

6.10.24. What is the flaw in the following argument?

> There is no such thing as a heap of sand.
>
> (a) *Base step.* For $n = 1$, the proposition is true since no one would say that a grain of sand is a heap,
> (b) *Inductive step.* Assume that the proposition is true for $n = k$; that is, k grains of sand is not a heap. We consider now the proposition for $n = k + 1$. If we do not have a heap with k grains of sand, then adding one more grain of sand still will not make it a heap. So the proposition is true for $n = k + 1$, and this concludes the inductive step.

6.11 Review

Summary

- A **sequence** is an ordered list of numbers. The numbers in a sequence are called the **terms** of the sequence.
- A sequence $t_1, t_2, t_3, \ldots$ is **arithmetic** if the difference between any two consecutive terms t_n and t_{n-1} is constant.
- A sequence $t_1, t_2, t_3, \ldots$ is **geometric** if the ratio, $r = \dfrac{t_n}{t_{n-1}}$, between any two consecutive terms is constant.
- A **series** is the sum of the terms in a sequence.
- The nth **partial sum** of the corresponding series is

$$S_n = t_1 + t_2 + t_3 + \cdots + t_n,$$

which is the sum of the first n terms in the sequence.

- Sigma notation: $t_m + t_{m+1} + t_{m+2} + \cdots + t_{n-1} + t_n = \displaystyle\sum_{i=m}^{n} t_i$

- Pi notation: $t_m \times t_{m+1} \times t_{m+2} \times \cdots \times t_{n-1} \times t_n = \displaystyle\prod_{i=m}^{n} t_i$

This tells us to end with $i = n$.

This tells us that it is a **sum**. $\longrightarrow \displaystyle\sum_{i=m}^{n} t_i$

This tells us to start with $i = m$.

This tells us to end with $i = n$.

This tells us that it is a **product**. $\longrightarrow \displaystyle\prod_{i=m}^{n} t_i$

This tells us to start with $i = m$.

- Gaussian summation formula: $\displaystyle\sum_{i=1}^{n} i = \dfrac{n(n+1)}{2}$

- $\displaystyle\sum_{i=1}^{n} i^2 = \dfrac{n(n+1)(2n+1)}{6}$

- Properties of Series

  (a) $\displaystyle\sum_{i=1}^{n} c = nc$, where c is a constant

  (b) $\displaystyle\sum_{i=1}^{n} ct_i = c \sum_{i=1}^{n} t_i$, where c is a constant

  (c) $\displaystyle\sum_{i=1}^{n} (a_i + b_i) = \sum_{i=1}^{n} a_i + \sum_{i=1}^{n} b_i$

- Arithmetic series:

$$S_n = \frac{n}{2}[2a + (n-1)d] \quad \text{or} \quad S_n = \frac{n}{2}(a + t_n).$$

- Geometric series: For $n \in \mathbb{N}$,

$$S_n = a + ar + ar^2 + \cdots + ar^{n-1} = \sum_{i=1}^{n} ar^{i-1} = \begin{cases} an & \text{if } r = 1 \\ \dfrac{a(1-r^n)}{1-r} & \text{if } r \neq 1 \end{cases}$$

- Investment applications
  (a) Compound interest: $A = P(1+i)^n$
  (b) Present value: $PV = \dfrac{F}{(1+i)^n}$

- Infinite geometric series: $a + ar + ar^2 + \cdots = \displaystyle\sum_{i=1}^{\infty} ar^{i-1} = \dfrac{a}{1-r}$

  whenever $-1 < r < 1$
- **Principle of Mathematical Induction**: A proposition involving the natural numbers n is true provided that
  (a) the proposition is true in the special case $n = 1$ (this is the **base step**);
  (b) the truth of the proposition for $n = k$, $k \in \mathbb{N}$ implies the truth of the proposition for $n = k + 1$ (this is the **inductive step**).

Exercises

6.11.1. Write each series using sigma notation.
  (a) $7 + 10 + 13 + \cdots + 28$
  (b) $2 + 1 + \frac{1}{2} + \frac{1}{4} + \cdots + \frac{1}{32}$
  (c) $1 + 2 \cdot 2! + 3 \cdot 3! + 4 \cdot 4! + \cdots + 10 \cdot 10!$
  (d) $1 \cdot 2 + 2 \cdot 3 + 3 \cdot 4 + \cdots + 15 \cdot 16$
  (e) $\dfrac{1}{4(7)} + \dfrac{1}{7(10)} + \dfrac{1}{10(13)} + \cdots + \dfrac{1}{31(34)}$
  (f) $1 + x + x^2 + \cdots + x^{20}$

6.11.2. Write the sum $\displaystyle\sum_{i=4}^{10} \frac{i}{i+1}$ in expanded form.

6.11.3. What is the sum of the even (natural) numbers from 100 to 200 inclusively?

6.11.4. Evaluate each of the following sums.
  (a) $1 + 2 + 3 + \cdots + 149 + 150$
  (b) $1^2 + 2^2 + 3^2 + \cdots + 149^2 + 150^2$
  (c) $\dfrac{1}{2} + \dfrac{1}{4} + \dfrac{1}{8} + \cdots + \left(\dfrac{1}{2}\right)^9 + \left(\dfrac{1}{2}\right)^{10}$

(d) $100 + 105 + 110 + \cdots + 150$

(e) $1 \cdot 2 + 2 \cdot 3 + 3 \cdot 4 + \cdots + 149 \cdot 150 + 150 \cdot 151$

(f) $\left(1 - \dfrac{1}{2}\right) + \left(\dfrac{1}{2} - \dfrac{1}{3}\right) + \left(\dfrac{1}{3} - \dfrac{1}{4}\right) + \cdots + \left(\dfrac{1}{149} - \dfrac{1}{150}\right)$

6.11.5. Let $\displaystyle\sum_{i=1}^{60} a_i = -4$ and $\displaystyle\sum_{i=1}^{60} b_i = 7$. Evaluate $\displaystyle\sum_{i=1}^{60}(17a_i + 8b_i)$.

6.11.6. Write the product $\displaystyle\prod_{i=5}^{12} \dfrac{1}{i+1}$ in expanded form.

6.11.7. Helen's starting wage as a local museum curator is \$45 000. She will receive a \$500 raise every three months.

(a) How much will she be earning (per annum) at the beginning of her $(n+1)^{\text{st}}$ year of employment?

(b) How much did she earn in her first n years?

6.11.8. Evaluate $\displaystyle\sum_{i=1}^{10} i2^{i-1}$.

6.11.9. What is the sum of the arithmetic series $t_1 + t_2 + \cdots + t_n$ with $t_1 = 4$, $t_n = 25$, and $n = 8$.

6.11.10. Find the sum of all the multiples of 7 between 49 and 119.

6.11.11. Evaluate $\displaystyle\sum_{i=1}^{\infty} 3^{1-i}$.

6.11.12. Evaluate $\displaystyle\sum_{i=1}^{n} \left(\dfrac{1}{i} - \dfrac{1}{i+2}\right)$ by using telescoping sums.

6.11.13. Prove using the Principle of Mathematical Induction that

$$2^1 \times 2^2 \times 2^3 \times \cdots \times 2^n = (\sqrt{2})^{n(n+1)}$$

for all natural numbers n.

6.11.14. Using the Principle of Mathematical Induction, prove the geometric series

$$a + ar + ar^2 + \cdots + ar^{n-1} = \dfrac{a(1 - r^n)}{1 - r}$$

is true for every natural number n.

6.11.15. Guess and prove a formula for $\displaystyle\sum_{i=1}^{n} \dfrac{2i + 1}{[i(i+1)]^2}$.

6.12 Bibliographic Remarks

In Sec. 6.1, I mentioned the field of machine learning. It is a very active field of research in having machines recognize patterns and to learn. Many applications of machine learning that we use in our daily life include spam filtering, search engines, and speech recognition (speech to-text). One book recommendation to learn more about machine learning is Kurzweil (2013). Ray Kurzweil (1948–) is a director of engineering at Google, and hence is more than qualified to give intuitive explanations to how some of the technology works, which he in some cases invented.

This story of the young Gauss surprising his schoolmaster with the correct answer to the difficult task of adding up numbers comes from von Waltershausen (1856, pp. 12–13) in German. I translated the text with the help of the English translation by Gauss' great granddaughter Helen Worthington Gauss. The visual explanation of $1 + 2 + 3 + \cdots + n = \frac{n^2}{2} + \frac{n}{2} = \frac{n(n+1)}{2}$ was adapted from Richards (1984).

To my knowledge, the illustration of the convergent infinite geometric series as seen in Fig. 6.22 is due to Milankovitch (1909), and later published in the textbook of Carslaw (1914, p. 243). Others who have independently found the illustration are Gibbins (1944), Deadman (1970), and Cohen *et al.* (2001).

The opening example in Sec. 6.9 with Uncle Vladimir that yields an intuitive and informal way of explaining the infinite geometric series formula was adapted from Ecker (1998). This has a certain appeal to students who are more business oriented, or perhaps students that may be inheriting money from a deceased relative in the near future.

Zeno's paradox can be applied to other areas of mathematics, not just infinite geometric series. For example, I recommend reading the short note, *What the Tortoise said to Achilles*, by Carroll (1895). Not to spoil the surprise, Achilles and the Tortoise are having a conversation when the Tortoise leads the poor Grecian warrior up the garden path through an (countably) infinite set of hypothetical syllogisms.

The origin of the Principle of Mathematical Induction has been traced back to the mathematicians Francesco Maurolico (1494–1575), Jacob (also known as James) Bernoulli (1654–1705), Blaise Pascal (1623–1662), and Pierre Fermat (1601–1665) (Cajori, 1918, p. 197). However, the origin of the term *mathematical induction* appeared in print for the first time in a *Penny Cyclopædia* (1838) entry written by Augustus De Morgan—the same man who played a vital role in promoting logic to the public (Cajori, 1918, p. 200).

Coleman *et al.* (1973), Ernest (1984), and Gunderson (2011) have heavily shaped my presentation on the Principle of Mathematical Induction. Ernest (1984, pp. 185–186) provided criteria to evaluate textbooks on teaching this topic such as if there was a discussion on the ambiguity of induction definition, different analogies, or showing faulty proofs. For further study on the Principle of Mathematical Induction, I highly recommend Gunderson (2011); it contains several hundred worked-out exercises and applications of this proof technique to other areas of mathematics.

A very well-written and brief introduction to the Principle of Mathematical Induction can be found in many books, but I quite like Courant and Robbins (1941, pp. 9–19) who stresses the use of intuition and discovery of mathematics. In particular, I liked their statement that proofs by mathematical induction were *verifications*, and did not yield an idea of how the proposition was discovered. The book itself is still considered a classic by many mathematicians. Other books that stressed the importance of guessing and then proving the guess is correct are Pólya (1945) and Pólya (1954). In fact, the fallacy in Example 6.10.6 was adapted from Pólya (1954, p. 120).

7

The Binomial Theorem

7.1 Pascal's Triangle

Let us consider the following motivating problem.

> **Problem 7.1.1 — The Path Problem.** Suppose we have a given grid, like the one in Fig. 7.1(a). From start to finish, how many paths are there if at each step you are going down one row diagonally to the left or right?

One such path is shown in Fig. 7.1(b). How many paths do you think there are? More than 50? Less than 500? But, how would you determine what is the correct number of paths?

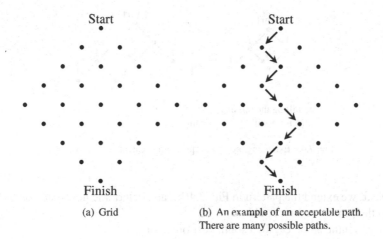

(a) Grid

(b) An example of an acceptable path. There are many possible paths.

Fig. 7.1   The path problem

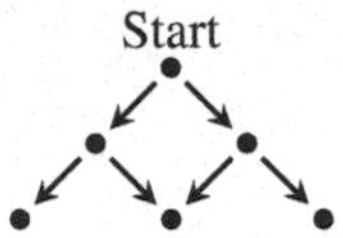

Start

Fig. 7.2   A smaller related problem

The Hungarian mathematician George Pólya (1887–1985) stressed, if you cannot solve the proposed problem, *find something familiar*—first, solve a suitably related problem. One way to do this is to break the problem into a simpler problem; see Fig. 7.2. We now consider how many paths are there traveling only southeast and southwest.

There is only one path to each of the corners on the southwest edge, namely, the path consisting of traveling always southwest and never southeast. Similarly, there is only one path to each of the corners on the northeast edge. We label these on the diagram in Fig. 7.3.

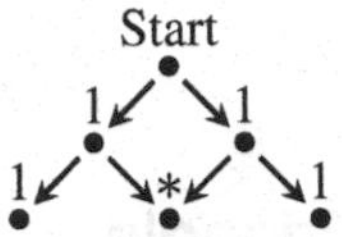

Start

Fig. 7.3   Edges labeled with 1.

What about the point labeled by ∗ in Fig. 7.3? How many paths are there to the point ∗? You can get to that point by going

- one block southwest then one block southeast; or
- one block southeast then one block southwest.

So there are 2 paths to the point ∗, and we label these on the diagram in Fig. 7.4(a).

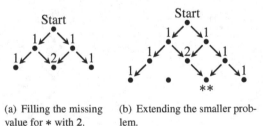

(a) Filling the missing value for ∗ with 2.

(b) Extending the smaller problem.

Fig. 7.4   Extending the smaller problem to calculate more values.

Next, we extend the pattern in Fig. 7.4(b), and determine how many paths there are to the point ∗∗. You could go

- southeast twice, then southwest once; or
- southwest, then southeast twice; or
- southeast, then southwest, then southwest.

So, there are three paths. By symmetry, the corresponding point on the western side, we can keep filling in some more values in Fig. 7.5.

Start

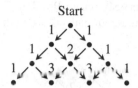

Fig. 7.5   Filling in more values.

Continuing in this fashion is becoming tedious. We make the following observation that the two numbers added together, gives the number in the row below; see Fig. 7.6.

Fig. 7.6

This observation helps us to fill in the rest of the points in Fig. 7.7. Therefore, we get our final answer of 70 paths to the finish point.

Start

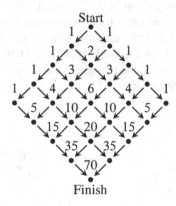

Finish

Fig. 7.7   The completed problem

Example 7.1.2 How many different paths will spell EULER, starting from the top and proceeding to the next row below by moving diagonally to the left or to the right?

```
                E
             U     U
          L     L     L
        E     E     E     E
      R     R     R     R     R
```

Solution This problem is very similar to Problem 7.1.1 at the beginning of the chapter; we can count the number of paths down to the bottom rule.

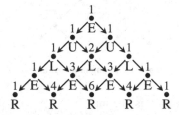

The difference between this problem and Problem 7.1.1 is that there are many finish points. We can remedy this by adding up the numbers on the bottom row. Therefore, there are $1 + 4 + 6 + 4 + 1 = 16$ paths to form the word EULER. ◄

The numbers at each point (Fig. 7.8(a)) we were calculating were the binomial coefficients (Fig. 7.8(b)), and this pattern forms what is known as **Pascal's triangle**, named after the French mathematician Blaise Pascal (1623–1662).[1] Pascal actually referred to it not as his triangle, but as the "arithmetical triangle." The arithmetical triangle has been rediscovered many times throughout history. In Europe, Niccolò Tartaglia (c. 1499–1557) in 1556 gave the binomial coefficients up to the 12th row and had arranged them in a triangle (Hald, 1990, p. 40), but Pascal's name may have stuck since he studied the triangle extensively for its properties and gave proofs of them (Hald, 1990, p. 46).

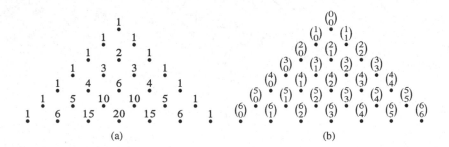

Fig. 7.8   Pascal's triangle

We see the boundaries of the triangle are equal to 1; that is

$$\binom{n}{0} = \binom{n}{n} = 1. \tag{7.1}$$

Notice the pattern again,

---

[1] "Please pardon us for not showing all of Pascal's triangle; infinite tables use up too much paper" Pólya *et al.* (1983, p. 20).

$$\binom{4}{1} + \binom{4}{2} = \binom{5}{2}.$$

In general,

$$\binom{n}{r} + \binom{n}{r+1} = \binom{n+1}{r+1}.$$

The general rule to find the values is better known as **Pascal's Theorem**.

Theorem 7.1.3 — Pascal's Theorem.

$$\binom{n}{r} + \binom{n}{r+1} = \binom{n+1}{r+1}$$

where $0 < r < n + 1$.

Proof — Algebraic approach.

$$\binom{n}{r} + \binom{n}{r+1}$$

$$= \frac{n!}{(n-r)!r!} + \frac{n!}{[n-(r+1)]!(r+1)!}$$

$$= \frac{n!}{(n-r)!r!} + \frac{n!}{(n-r-1)!(r+1)!}$$

$$= \frac{n!}{(n-r)(n-r-1)!r!} + \frac{n!}{(n-r-1)!(r+1)r!}$$

$$= \frac{n!(r+1)}{(n-r)(n-r-1)!r!(r+1)} + \frac{n!(n-r)}{(n-r-1)!(r+1)r!(n-r)}$$

$$= \frac{n!(r+1) + n!(n-r)}{(n-r)(n-r-1)!r!(r+1)}$$

$$= \frac{n!(\cancel{r} + 1 + n - \cancel{r})}{\underbrace{(n-r)(n-r-1)!}_{(n-r)!} \; \underbrace{r!(r+1)}_{(r+1)!}}$$

$$= \frac{n!(n+1)}{(n-r)!(r+1)!}$$

$$= \frac{(n+1)!}{(n-r)!(r+1)!}$$

$$= \frac{(n+1)!}{[(n+1)-(r+1)]!(r+1)!}$$

$$= \binom{n+1}{r+1}$$

◄

Proof — Combinatorial approach. Suppose we have to select a committee with $(r+1)$ members from a class of $(n+1)$ students. In how many ways can we form a committee?

(a) (Left side of the identity.) Suppose Diane is in the class. *Take cases.* There are two cases: either Diane is on the committee *or* she is not.

    (i) If Diane is on the committee, then she takes one of the positions, and there are r remaining positions to be filled by the remaining n students. Hence, the number of ways to form a committee with Diane is $\binom{n}{r}$.

    (ii) If Diane is not on the committee, then there are $(r+1)$ members to be selected from amongst the n students. Hence, the number of ways to form a committee without Diane is $\binom{n}{r+1}$.

By the Sum Principle, we can add the two cases together to get

$$\binom{n}{r} + \binom{n}{r+1}.$$

(b) (Right side of the identity.) By definition of the binomial coefficient, the number of ways to select $(r+1)$ members from $(n+1)$ students is $\binom{n+1}{r+1}$.

Since parts (a) and (b) are equivalent ways of thinking about the problem, we have shown that

$$\binom{n}{r} + \binom{n}{r+1} = \binom{n+1}{r+1}$$

as required. ◄

Example 7.1.4 Express the sum of two binomial coefficients as a single binomial coefficient.

(a) $\binom{8}{3} + \binom{8}{4}$ (b) $\binom{64}{13} + \binom{64}{14}$ (c) $\binom{125}{6} + \binom{125}{5}$

Solution Using Theorem 7.1.3, we get the following

(a) $\binom{8}{3} + \binom{8}{4} = \binom{9}{4}$

(b) $\dbinom{64}{13} + \dbinom{64}{14} = \dbinom{65}{14}$

(c) $\dbinom{125}{6} + \dbinom{125}{5} = \dbinom{125}{5} + \dbinom{125}{6} = \dbinom{126}{6}$

◀

Pascal's triangle is quite remarkable since it contains so many interesting patterns and properties. Recall the symmetry of combinations property (Theorem 5.7.8), $\binom{n}{r} = \binom{n}{n-r}$. This property is present in Pascal's triangle, and visually, we can see in Fig. 7.9 that there is a line of symmetry.

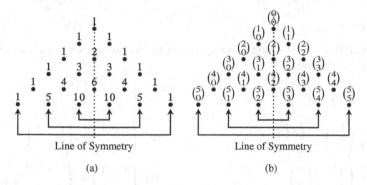

(a)   (b)

Fig. 7.9   The symmetry property in Pascal's triangle

Another example of a property present is if we look at Fig. 7.10, we can find the terms of the Fibonacci sequence, which is a sequence that we had investigated earlier in Example 6.1.4 (on p. 202). In the exercises, you will investigate other properties in Pascal's triangle. You will have to use Pólya's method: make a guess, see if you are correct, and if so then prove it is true.

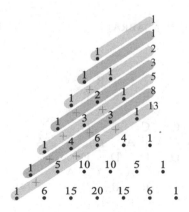

Fig. 7.10   The Fibonacci sequence in Pascal's triangle

Exercises

7.1.1. Write an equivalent expression for each of the following binomial coefficients.

(a) $\dbinom{10}{2}$　　　　(b) $\dbinom{20}{18}$　　　　(c) $\dbinom{25}{0}$　　　　(d) $\dbinom{26}{26}$

7.1.2. Extend Pascal's triangle to 10 rows.

7.1.3. Fill in the missing numbers for parts of Pascal's triangle.

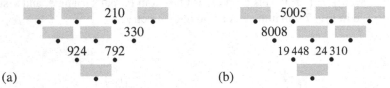

(a)　　　　　　　　　　　　　　　　　(b)

7.1.4. Use Pascal's Theorem to write each of the following expressions as a single binomial coefficient.

(a) $\dbinom{8}{3} + \dbinom{8}{4}$　　　　　　　　(d) $\dbinom{15}{11} + \dbinom{15}{5}$

(b) $\dbinom{17}{6} + \dbinom{17}{7}$　　　　　　　　(e) $\dbinom{n-1}{r-1} + \dbinom{n-1}{r}$

(c) $\dbinom{9}{5} + \dbinom{9}{4}$　　　　　　　　(f) $\dbinom{n-1}{r-1} + \dbinom{n-1}{n-r-1}$

7.1.5. How many different paths will spell each of the following, starting from the top and proceeding to the next row below by moving diagonally to the left or the right?

(a) ABRACADABRA　　　　　　　　(b) RENAISSANCE

```
                    A
                 B     B
              R     R     R
           A     A     A     A
        C     C     C     C     C
     A     A     A     A     A     A
        D     D     D     D     D
           A     A     A     A
              B     B     B
                 R     R
                    A
```

```
                    R
                 E     E
              N     N     N
           A     A     A     A
              I     I     I
                 S     S
              S     S     S
           A     A     A     A
              N     N     N
                 C     C
                    E
```

7.1.6. The diagram in Fig. 7.11 represents the routes offered by a new bus company between various cities and towns in England.

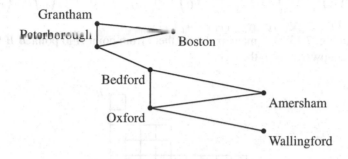

Fig. 7.11   Bus routes for Exercise 7.1.6

(a) How many routes are there from Wallingford to Grantham?
(b) How many routes are there from Amersham to Boston?
(c) If the directions were reversed, would the number of bus routes be the same for parts (a) and (b)? Explain.

7.1.7. A checkerboard is given in Fig. 7.12, with a checker placed on the fifth square along the bottom row. The checker is able to move one square at a time diagonally left or diagonally right to the row above. How many paths can the checker take to the top row?

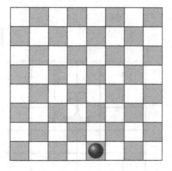

Fig. 7.12   Checkerboard with a checker

7.1.8. Explain why the following equation is true, without explicitly performing a calculation.

$$\binom{9}{0} + \binom{9}{2} + \binom{9}{4} = \binom{9}{5} + \binom{9}{7} + \binom{9}{9}$$

7.1.9. Show algebraically that

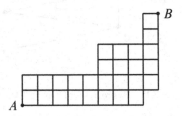

(a) $\binom{n}{0} = \binom{n}{n}$ (b) $\binom{r}{0} = \binom{r+1}{0}$ (c) $\binom{r}{r} = \binom{r+1}{r+1}$

7.1.10. In Fig. 7.13, how many paths are there from point A to point B if you can only travel upwards or to the right?

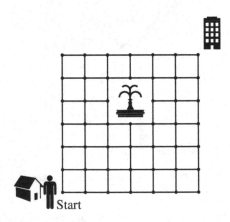

Fig. 7.13   A grid for Exercise 7.1.10

7.1.11. Suppose that streets in a particular city are constructed in a rectangular grid as shown below, running north-south and east-west (see Fig. 7.14). Dr Algebrix wants to walk from his home to the university, which is six blocks north, and six blocks east. How many ways can he make this journey if he only travels north and east?

Fig. 7.14   Dr Algebrix wants to walk from his home to the university

7.1.12. A pattern in Pascal's triangle: *the sum of the entries in a row.*
(a) Complete the chart below.

n		Sum
0	1	1
1	$1+1$	2
2	$1+2+1$	4
3	$1+3+3+1$	
4	$1+4+6+4+1$	
5	$1+5+10+10+5+1$	
6	$1+6+15+20+15+6+1$	
7	$1+7+21+35+35+21+7+1$	

(b) *Find a pattern.* What would you predict the sum of row $n = 8$? Check your prediction.
(c) Evaluate the following using what you learned from parts (a) and (b).

(i) $\binom{9}{0} + \binom{9}{1} + \binom{9}{2} + \binom{9}{3} + \cdots + \binom{9}{9}$

(ii) $\binom{10}{0} + \binom{10}{1} + \binom{10}{2} + \binom{10}{3} + \cdots + \binom{10}{10}$

(iii) $\sum_{i=0}^{11} \binom{11}{i}$

(d) Generalize your findings to find an expression for the sum of row n. In other words, what is the value of $\sum_{i=0}^{n} \binom{n}{i}$?

(e) Prove your formula in part (d) using a combinatorial proof. (*Hint*: Suppose you are at a sandwich shop that offers n toppings. Can you think of two approaches to counting the number of ways you can order a sandwich?)

(f) Prove your formula in part (d) using the Principle of Mathematical Induction.

7.1.13. If $\binom{n}{0} + \binom{n}{1} + \binom{n}{2} + \cdots + \binom{n}{n} = 32$, then what is the value of n?

7.1.14. If $\sum_{i=0}^{n} \binom{n}{i} = 256$, then what is the value of n?

7.1.15. Another pattern in Pascal's triangle: *the sum of the squared entries in a row*.

(a) Complete the chart below.

n		Sum
0	1^2	1
1	$1^2 + 1^2$	2
2	$1^2 + 2^2 + 1^2$	6
3	$1^2 + 3^2 + 3^2 + 1^2$	
4	$1^2 + 4^2 + 6^2 + 4^2 + 1^2$	
5	$1^2 + 5^2 + 10^2 + 10^2 + 5^2 + 1^2$	
6	$1^2 + 6^2 + 15^2 + 20^2 + 15^2 + 6^2 + 1^2$	
7	$1^2 + 7^2 + 21^2 + 35^2 + 35^2 + 21^2 + 7^2 + 1^2$	

(b) Find these numbers in Pascal's triangle.

(c) *Find a pattern.* What would you predict the sum of the squared entries in row $n = 8$? Check your prediction.

(d) Evaluate the following using what you learned from parts (a)-(c).

(i) $\dbinom{9}{0}^2 + \dbinom{9}{1}^2 + \dbinom{9}{2}^2 + \cdots + \dbinom{9}{9}^2$

(ii) $\dbinom{10}{0}^2 + \dbinom{10}{1}^2 + \dbinom{10}{2}^2 + \cdots + \dbinom{10}{10}^2$

(e) Generalize your findings: what is another expression in the form $\dbinom{n}{r}$ for

$$\dbinom{n}{0}^2 + \dbinom{n}{1}^2 + \dbinom{n}{2}^2 + \cdots + \dbinom{n}{n}^2 = \sum_{i=0}^{n} \dbinom{n}{i}^2 ?$$

(f) Prove your formula in part (e) using a combinatorial proof. (*Hint*: Suppose that you are picking a team of n people from n males and n females.)

(g) Prove your formula in part (e) using the Principle of Mathematical Induction.

7.1.16. If $\displaystyle\sum_{i=0}^{n} \dbinom{n}{i}^2 = 252$, then what is the value of n?

7.1.17. Another pattern in Pascal's triangle: *the sum of the alternating sign entries in a row*.

(a) Complete the chart below.

n		Sum
0	1	1
1	$1 - 1$	0
2	$1 - 2 + 1$	0
3	$1 - 3 + 3 - 1$	
4	$1 - 4 + 6 - 4 + 1$	
5	$1 - 5 + 10 - 10 + 5 - 1$	
6	$1 - 6 + 15 - 20 + 15 - 6 + 1$	
7	$1 - 7 + 21 - 35 + 35 - 21 + 7 - 1$	

(b) *Find a pattern.* What would you predict the sum of the alternating sign entries in row $n = 8$? Check your prediction.

(c) Evaluate the following using what you learned from parts (a)-(b).

  (i) $\binom{9}{0} - \binom{9}{1} + \binom{9}{2} - \cdots - \binom{9}{9}$

  (ii) $\binom{10}{0} - \binom{10}{1} + \binom{10}{2} - \cdots - \binom{10}{9} + \binom{10}{10}$

  (iii) $\displaystyle\sum_{i=0}^{11}(-1)^i \binom{11}{i}$

(d) Generalize your findings to find an expression for

$$\binom{n}{0} - \binom{n}{1} + \binom{n}{2} - \binom{n}{3} + \cdots + (-1)^n \binom{n}{n} = \sum_{i=0}^{n}(-1)^n \binom{n}{i}$$

  Note: There are two cases to consider.

7.1.18. Using the properties of Pascal's triangle, evaluate the following.

(a) $\binom{16}{9} + \binom{16}{10}$

(b) $\binom{21}{18} + \binom{21}{19}$

(c) $\binom{14}{4} + \binom{14}{3}$

(d) $\displaystyle\sum_{r=11}^{12} \binom{17}{r}$

(e) $\binom{11}{0} + \binom{11}{1} + \binom{11}{2} + \cdots + \binom{11}{11}$

(f) $\binom{12}{0} - \binom{12}{1} + \binom{12}{2} - \cdots - \binom{12}{12}$

(g) $\binom{15}{0}^2 + \binom{15}{1}^2 + \binom{15}{2}^2 + \cdots + \binom{15}{15}^2$

7.1.19. Pennies can be arranged in an equilateral triangle[2] as shown in Fig. 7.15.

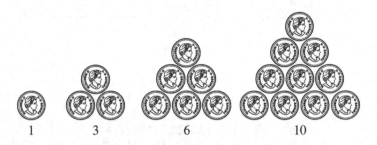

Fig. 7.15   Illustrating triangular numbers with pennies.

(a) How many pennies are there in the next step?
(b) These numbers are known as **triangular numbers**. Locate these numbers on Pascal's triangle. Where are these numbers located?
(c) Generalize your results to find an expression for the nth triangular number in the form $\binom{n}{r}$. Check your formula by drawing a diagram of pennies for the sixth triangular number and seeing if you get the same answer from your drawing and from your formula.
(d) Use Pascal's triangle to determine a formula for the sum of the first n triangular numbers. Express your formula using sigma notation. Check your formula by testing it with the sum of the first four triangular numbers.
(e) Prove your formula you found in part (d) using the Principle of Mathematical Induction.

7.1.20. Show that

$$\binom{n+2}{r+2} = \binom{n}{r} + 2\binom{n}{r+1} + \binom{n}{r+2}$$

for $r = 2, 3, \ldots, n$
(a) by an algebraic proof. (*Hint*: Use Pascal's Theorem on $\binom{n+2}{r+2} = \binom{n+1+1}{r+2}$.)
(b) by a combinatorial proof. (*Hint*: Look at the combinatorial proof of Pascal's Theorem. Instead of just Diane, consider Diane and her friend Shelly.)

7.1.21 (†). Prove that $\displaystyle\sum_{i=1}^{n} i \binom{n}{i} = n(2^{n-1})$ using
(a) a combinatorial approach.
(b) the Principle of Mathematical Induction. (Hint: $i \binom{n+1}{i} = (n+1)\binom{n}{i-1}$.)

---

[2] An **equilateral triangle** is a triangle that has three equal sides.

7.2 The Binomial Theorem

Suppose we have an urn[3] that contains 6 white balls and 6 black balls. If one ball is drawn, then there are two possible events: either a white ball is drawn denoted by W, *or* a black ball is drawn denoted by B; thus, by the Sum Principle (Theorem 3.2.6), the number of ways of selecting ball is the number of white balls plus the number of black balls. If we put the ball back, and draw again, then by the Multiplication Principle (Theorem 5.2.2) we have

$$(W + B) \times (W + B) = (W + B)^2. \tag{7.2}$$

If we expand expression (7.2),

$$\begin{aligned}
(W + B)^2 &= (W + B)(W + B) \\
&= W^2 + WB + BW + B^2 \\
&= W^2 + 2WB + B^2
\end{aligned}$$

we get a compact representation of all the ways to draw two balls from an urn containing an equal number of white and black balls with replacement. To see why this is a compact representation, we turn to our trusted visual diagram, the tree diagram, in Fig. 7.16, and we note the following:

- W^2 says there is one way 2 white balls can be drawn,
- $2WB$ says there are two ways in which 1 white ball and 1 black ball can be drawn, and
- B^2 says there is one way 2 black balls can be drawn

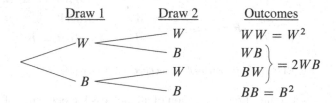

Fig. 7.16   Tree diagram for drawing two balls with replacement

The coefficient tells us how many ways an outcome can happen. For example $2WB$, the coefficient is 2, and this means that there are 2 ways in which 1 white ball and 1 black ball is drawn.

---

[3]An **urn** is a large vase that usually has an ornamental pedestal. Students (and even instructors) ask me, "Why do we use the term *urn* instead of a vase or a box to hold the balls? Are not urns for holding ashes of the cremated dead?" Of course, urns can hold ashes; see Fig. 8.23 on p. 352. But, urns can hold other things as well. In mathematics, we have been using the term to describe these type of problems for a long time. James Bernoulli (1654–1705) is likely to be the first mathematician to mention problems using the term urn in his treatise *Ars Conjectandi* (1713). It was quite fashionable to have an urn as part of your décor in Bernoulli's time, and so Bernoulli might have used it since everyone knew what it was. Urn models help us to imagine objects of real interest (*e.g.*, atoms, *living* people, cars) by representing them as colored balls.

The exponent says how many balls of the same color in the same outcome. For example W^2, the exponent is ^{2}, and this means that there are 2 white balls in the outcome. (The coefficient is 1, and there is only one way for this outcome to occur.)

Now that we have done this for drawing 2 balls, what if we made 3 drawings from the urn? By expanding the expression $(W + B)^3$, we have

$$
\begin{aligned}
(W + B)^3 &= (W + B)(W + B)(W + B) \\
&= (W + B)(W^2 + WB + BW + B^2) \\
&= (W + B)(W^2 + 2WB + B^2) \\
&= W^3 + 2W^2B + WB^2 + W^2B + 2WB^2 + B^3 \\
&= W^3 + 3W^2B + 3WB^2 + B^3.
\end{aligned}
$$

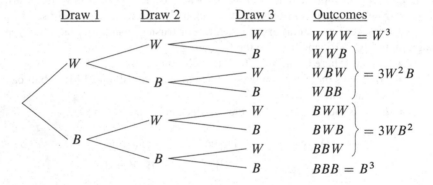

Fig. 7.17   Tree diagram for drawing three balls with replacement

Again, this is a compact representation of all the possible outcomes of drawing 3 balls from the urn, which we can see in Fig. 7.17. For this compact representation of drawing three balls,

- W^3 says there is one way 3 white balls can be drawn,
- $3W^2B$ says there are three ways 2 white balls and 1 black ball can be drawn,
- $3WB^2$ says there are three ways 1 white ball and 2 black balls can be drawn, and
- B^3 says there is one way 3 black balls can be drawn.

What if we continue this process, and we made 4 drawings from the urn? Five drawings from the urn? Then, from Fig. 7.18, we have

$$
(W + B)^4 = W^4 + 4W^3B + 6W^2B^2 + 4WB^3 + B^4
$$
$$
(W + B)^5 = W^5 + 5W^4B + 10W^3B^2 + 10W^2B^3 + 5WB^4 + B^5.
$$

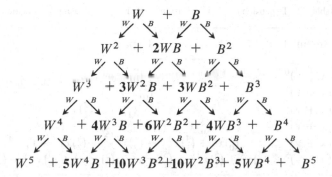

Fig. 7.18 Extending the tree diagram of drawing black and white balls from an urn with replacement

Do you recognize a pattern with the coefficients[4] of the terms when we expand? If they seem familiar, that is because they are from Pascal's Triangle (compare with Fig. 7.8 on p. 280)! We can see this in Table 7.1 when we expand $(x + y)^n$.

Table 7.1 Expansion of $(x + y)^n$ for $n = 0, 1, 2, \ldots, 5$. Notice that the coefficients of the binomial expansion are the same as the entries in Pascal's triangle (see Fig. 7.8).

n	Binomial	Expansion	Coefficients
0	$(x + y)^0$	1	1
1	$(x + y)^1$	$x + y$	1 1
2	$(x + y)^2$	$x^2 + 2xy + y^2$	1 2 1
3	$(x + y)^3$	$x^3 + 3x^2y + 3xy^2 + y^3$	1 3 3 1
4	$(x + y)^4$	$x^4 + 4x^3y + 6x^2y^2 + 4xy^3 + y^4$	1 4 6 4 1
5	$(x + y)^5$	$x^5 + 5x^4y + 10x^3y^2 + 10x^2y^3 + 5xy^4 + y^5$	1 5 10 10 5 1

With this observation that the coefficients are the entries from Pascal's triangle, we can rewrite the coefficients using $\binom{n}{r}$ notation as seen in Table 7.2. Now it should be evident why we call $\binom{n}{r}$ the **binomial coefficient**. Another observation about the pattern of a binomial expansion is that from left to right, the exponents of x are decreasing by 1 and the exponents of y are increasing by 1.

Now, we are prepared to summarize all of the observations we have made, and state what mathematicians call the **Binomial Theorem**.

---

[4]The coefficients are in bold.

Table 7.2   Expansion of $(x + y)^n$ for $n = 0, 1, 2, \ldots, 5$ using $\binom{n}{r}$ notation.

n	Binomial	Expansion
0	$(x + y)^0$	$\binom{0}{0}x^0 y^0$
1	$(x + y)^1$	$\binom{1}{0}x + \binom{1}{1}y$
2	$(x + y)^2$	$\binom{2}{0}x^2 + \binom{2}{1}xy + \binom{2}{2}y^2$
3	$(x + y)^3$	$\binom{3}{0}x^3 + \binom{3}{1}x^2 y + \binom{3}{2}xy^2 + \binom{3}{3}y^3$
4	$(x + y)^4$	$\binom{4}{0}x^4 + \binom{4}{1}x^3 y + \binom{4}{2}x^2 y^2 + \binom{4}{3}xy^3 + \binom{4}{4}y^4$
5	$(x + y)^5$	$\binom{5}{0}x^5 + \binom{5}{1}x^4 y + \binom{5}{2}x^3 y^2 + \binom{5}{3}x^2 y^3 + \binom{5}{4}xy^4 + \binom{5}{5}y^5$

Theorem 7.2.1 — The Binomial Theorem.

$$(x + y)^n = \binom{n}{0}x^n + \binom{n}{1}x^{n-1}y + \binom{n}{2}x^{n-2}y^2 + \cdots$$

$$+ \binom{n}{r}x^{n-r}y^r + \cdots + \binom{n}{n}y^n \tag{7.3}$$

$$= \sum_{r=0}^{n}\binom{n}{r}x^{n-r}y^r \tag{7.4}$$

R   Using $\binom{n}{r} = \binom{n}{n-r}$ (Theorem 5.7.8), we can rewrite (7.4) as

$$\sum_{r=0}^{n}\binom{n}{r}x^{n-r}y^r = \sum_{r=0}^{n}\binom{n}{r}x^r y^{n-r}. \tag{7.5}$$

See Exercise 7.2.14.

Another way to look at the Binomial Theorem is to think about the expansion

$$(x + y)^n = \underbrace{(x + y)(x + y) \cdots (x + y)}_{n \text{ times}}$$

and to ask yourself, "How many ways can one create an $x^{n-r}y^r$ term?" Each such term can be formed by selecting an x term from r of the $(x + y)$ factors; this can be done in $\binom{n}{r}$ different ways.

Example 7.2.2 In a discrete mathematics class of n students, each student is given a choice of either one of x different logic problems, or one of y different combinatorics problems. The problems are selected *with* replacement. How many different ways can the students select the problems?

For this example, we are going to give two solutions.

Solution — 1. By the Sum Principle (Theorem 5.2.6), every student has $x + y$ choices for which problem to select. Since there are n students, by the Multiplication Principle (Theorem 5.2.3), we have

$$\underbrace{(x + y)(x + y) \cdots (x + y)}_{n \text{ students}} = (x + y)^n$$

different ways the students can select the problems. ◄

Solution — 2. Suppose there are r students out of n students who wish to select a combinatorial problem; the number of ways we can select these r students is $\binom{n}{r}$. Since these problems are selected with replacement, the number of ways for r students to select a combinatorial problem (Theorem 5.6.1) is

$$\underbrace{y \times y \times \cdots \times y}_{r \text{ students}} = y^r.$$

By the Multiplication Principle, the number of ways we can select r students and that each student picks a combinatorial problem is $\binom{n}{r} y^r$.

Then, we need to consider the remaining $n - r$ students who pick a logic problem; the number of logic problems selected with replacement is

$$\underbrace{x \times x \times x \times \cdots \times x}_{n - r \text{ students}} = x^{n-r}.$$

Again, by the Multiplication Principle, the number of ways r students can select a combinatorial problem and $n - r$ students can select a logic problem is $\binom{n}{r} x^{n-r} y^r$.

Since r could be $0, 1, 2, 3, \ldots, n$, we consider each case, and add them together; that is,

$$\sum_{r=0}^{n} \binom{n}{r} x^{n-r} y^r.$$

◄

Both of these solutions give an answer to Example 7.2.2, and they are both equivalent; that is,

$$(x + y)^n = \sum_{r=0}^{n} \binom{n}{r} x^{n-r} y^r.$$

Corollary 7.2.3

$$\sum_{r=0}^{n} \binom{n}{r} = 2^n \tag{7.6}$$

We can prove Corollary 7.2.3 with an application of the Binomial Theorem.

Proof — Algebraic approach. Let $x = y = 1$. By the Binomial Theorem, we have

$$2^n = (1+1)^n = \sum_{r=0}^{n} \binom{n}{r} 1^{n-r} 1^r$$

$$= \sum_{r=0}^{n} \binom{n}{r} \qquad \text{Since } 1^{n-r} 1^r = 1.$$

◀

Another way to prove Corollary 7.2.3 is to use a combinatorial approach.

Proof — Combinatorial approach. Suppose you have n toppings to choose from while you are making a sandwich. In how many ways can you make a sandwich?

(a) For your sandwich, you can select no toppings, *or* 1 topping, *or* 2 toppings, *or* ..., *or* n toppings. Therefore, by the Sum Principle (Theorem 5.2.6), you can make a sandwich in

$$\binom{n}{0} + \binom{n}{1} + \binom{n}{2} + \cdots + \binom{n}{n} = \sum_{r=0}^{n} \binom{n}{r}$$

ways.

(b) For every topping, you can decide if you want it on your sandwich; this makes 2 choices for every topping. Therefore, the number of ways to make a sandwich is

$$\underbrace{2}_{\text{Topping 1}} \times \underbrace{2}_{\text{Topping 2}} \times \cdots \times \underbrace{2}_{\text{Topping } n} = 2^n.$$

Since parts (a) and (b) are different (yet equivalent) ways of thinking about the problem, we have shown that

$$\sum_{r=0}^{n} \binom{n}{r} = 2^n$$

as required.   ◀

By proving Corollary 7.2.3 with two different approaches, we have satisfied Pólya's step in looking back at the problem. You can decide which approach you find more appealing.

Example 7.2.4 Using the Binomial Theorem, expand $(x + y)^5$.

Solution

$$(x + y)^5 = \binom{5}{0}x^5 + \binom{5}{1}x^{5-1}y + \binom{5}{2}x^{5-2}y^2 + \binom{5}{3}x^{5-3}y^3$$

$$+ \binom{5}{4}x^{5-4}y^4 + \binom{5}{5}x^{5-5}y^5$$

$$= \binom{5}{0}x^5 + \binom{5}{1}x^4y + \binom{5}{2}x^3y^2 + \binom{5}{3}x^2y^3 + \binom{5}{4}xy^4$$

$$+ \binom{5}{5}y^5$$

$$= x^5 + 5x^4y + 10x^3y^2 + 10x^2y^3 + 5xy^4 + y^5.$$

◀

Example 7.2.5 An urn contains 4 white balls and 4 black balls. A ball is selected from the urn, recorded, and then it is returned to the urn. How many ways can 2 white balls and 3 black balls be selected in 5 drawings?

Solution Expanding $(W + B)^5$ using the Binomial Theorem, like in Example 7.2.4, we have

$$(W + B)^5 = W^5 + 5W^4B + 10W^3B^2 + 10W^2B^3 + 5WB^4 + B^5.$$

The result is W^2B^3 since this represents 2 white balls and 3 black balls. There are 10 such possible results since the coefficient of W^2B^3 is 10. ◀

Example 7.2.6 Using the Binomial Theorem, expand $(x - 7)^4$.

Solution

$$(x - 7)^4 = \binom{4}{0}x^4 + \binom{4}{1}x^{4-1}(-7) + \binom{4}{2}x^{4-2}(-7)^2$$

$$+ \binom{4}{3}x^{4-3}(-7)^3 + \binom{4}{4}x^{4-4}(-7)^4$$

$$= \binom{4}{0}x^4 + \binom{4}{1}x^3(-7) + \binom{4}{2}x^2(-7)^2 + \binom{4}{3}x^1(-7)^3$$

$$+ \binom{4}{4}(-7)^4$$

$$= x^4 + 4x^3(-7) + 6x^2(-7)^2 + 4x^1(-7)^3 + (-7)^4$$

$$= x^4 - 28x^3 + 294x^2 - 1372x + 2401.$$

Notice how the signs in the expansion alternate between $-$ and $+$. ◀

In the expansion of $(x + y)^n$,

$$(x+y)^n = \underbrace{\binom{n}{0}x^n}_{t_1} + \underbrace{\binom{n}{1}x^{n-1}y}_{t_2} + \underbrace{\binom{n}{2}x^{n-2}y^2}_{t_3} + \cdots + \underbrace{\binom{n}{r}x^{n-r}y^r}_{t_{r+1}} + \cdots + \underbrace{\binom{n}{n}y^n}_{t_{n+1}}$$

there are $n + 1$ terms.

> **Definition 7.2.7 — General term.** In the expansion of $(x + y)^n$,
>
> $$t_{r+1} = \binom{n}{r}x^{n-r}y^r \qquad (7.7)$$
>
> is the **general term** of the Binomial Theorem. The general term is the $(r + 1)^{\text{st}}$ term.

Example 7.2.8 Find the general term in the expansion of $(x + y)^5$.

Solution $t_{r+1} = \binom{5}{r}x^{5-r}y^r$ ◀

Example 7.2.9 Find the general term in the expansion of $\left(2x^2 - \dfrac{3}{x}\right)^6$, and the term containing x^3.

Solution Since $-\dfrac{3}{x} = 3x^{-1}$,

$$t_{r+1} = \binom{6}{r}(2x^2)^{6-r}(-3x^{-1})^r$$

$$= \binom{6}{r}2^{6-r}x^{12-2r}(-3)^r x^{-r}$$

$$= \binom{6}{r}(-3)^r 2^{6-r}x^{12-3r}.$$

Once we have found the general term, it is straightforward to find a particular term. For the term containing x^3,

$$12 - 3r = 3$$
$$-3r = 3 - 12$$
$$-3r = -9$$
$$r = 3.$$

Therefore, the term containing x^3 is

$$t_4 = \binom{6}{3}(-3)^3 2^{6-3} x^{12-3(3)}$$

$$= 20(-27)8x^3$$

$$= -4320x^3.$$

◀

Example 7.2.10 What is the coefficient of $x^{12} y^{13}$ in the expansion of $(x + y)^{25}$?

Solution In this case, $n = 25$, and $r = 13$ where the required term is $\binom{n}{r} x^{n-r} y^r = \binom{25}{13} x^{12} y^{13}$. Then, the required coefficient is

$$\binom{n}{r} = \binom{25}{13} = 5\,200\,300.$$

◀

Example 7.2.11 What is the fifth term in the expansion of $(x + y)^{100}$?

Solution For the fifth term in the expansion of $(x + y)^{100}$, it is when $r = 4$ and $n = 100$ in $\binom{n}{r} x^{n-r} y^r$. So,

$$\binom{100}{4} x^{100-4} y^4 = 3\,921\,225 x^{96} y^4$$

is the fifth term in the expansion of $(x + y)^{100}$.

◀

Exercises

7.2.1. Express each of the following in the form $(x + y)^n$.
  (a) $\binom{5}{0}x^5 + \binom{5}{1}x^4 y + \binom{5}{2}x^3 y^2 + \binom{5}{3}x^2 y^3 + \binom{5}{4}xy^4 + \binom{5}{5}y^5$
  (b) $\binom{4}{0}(x^2)^4 + \binom{4}{1}(x^2)^3 \left(\frac{1}{x}\right) + \binom{4}{2}(x^2)^2 \left(\frac{1}{x}\right)^2 + \binom{4}{3}(x^2) \left(\frac{1}{x}\right)^3 + \binom{4}{4} \left(\frac{1}{x}\right)^4$
  (c) $\binom{6}{0}(-3)^6 + \binom{6}{1}(-3)^5(2) + \binom{6}{2}(-3)^4(2)^2 + \binom{6}{3}(-3)^3(2)^3 + \binom{6}{4}(-3)^2(2)^4 + \binom{6}{5}(-3)(2)^5 + \binom{6}{6}(2)^6$

7.2.2. In the expansion of $(x + y)^{12}$, what is the value of k in the term that contains
  (a) $x^5 y^k$ (b) $x^k y^9$ (c) $x^k y^{2k}$ (d) $x^{k+2} y^{4k}$.

7.2.3. Evaluate the following:
  (a) $\binom{6}{0} \left(\frac{1}{4}\right)^6 + \binom{6}{1} \left(\frac{1}{4}\right)^5 \left(\frac{3}{4}\right) + \binom{6}{2} \left(\frac{1}{4}\right)^4 \left(\frac{3}{4}\right)^2 + \binom{6}{3} \left(\frac{1}{4}\right)^3 \left(\frac{3}{4}\right)^3 + \binom{6}{4} \left(\frac{1}{4}\right)^2 \left(\frac{3}{4}\right)^4 + \binom{6}{5} \left(\frac{1}{4}\right) \left(\frac{3}{4}\right)^5 + \binom{6}{6} \left(\frac{3}{4}\right)^6$
  (b) $(0.6)^7 + 7(0.6)^6(0.4) + 21(0.6)^5(0.4)^2 + 35(0.6)^4(0.4)^3 + 35(0.6)^3(0.4)^4 + 21(0.6)^2(0.4)^5 + 7(0.6)(0.4)^6 + (0.4)^7$
  (c) $(0.8)^5 + 5(0.8)^4(3) + 10(0.8)^3(3)^2 + 10(0.8)^2(3)^3 + 5(0.8)(3)^4 + (3)^5$

7.2.4. Using the Binomial Theorem, expand and simplify the following:
(a) $(x + y)^6$
(b) $(x + y)^7$
(c) $(x + y)^8$
(d) $(2x + y)^3$
(e) $(x - 2y)^4$
(f) $(1 - x)^6$

7.2.5. Expand and simplify each of the following:

(a) $\left(1 + \dfrac{1}{x}\right)^4$

(b) $\left(x - \dfrac{1}{x}\right)^6$

(c) $\left(\sqrt{x} - \dfrac{2}{\sqrt{x}}\right)^6$

(d) $\left(2x^3 + \dfrac{\sqrt{y}}{3}\right)^4$

(e) $\left(x^2 + \dfrac{3y}{x}\right)^4$

(f) $\left(3x^2 + \dfrac{1}{x}\right)^3$

7.2.6. Expand $(x + y + z)^3$. (*Hint*: Write $(x + y + z)$ in the form $(x + (y + z))$.)

7.2.7. An urn contains 10 white balls and 16 black balls. A ball is selected from the urn, recorded, and then it is returned to the urn. How many ways can 5 white balls and 3 black balls be selected in 8 drawings?

7.2.8. Suppose that a coin is tossed 8 times. How many outcomes will there be with 3 heads and 5 tails?

7.2.9. What is the coefficient of $x^6 y^3$ in the expansion of $(x + y)^9$?

7.2.10. What is the coefficient of the term containing $x^2 y^3$ in the expansion of $(2x + 3y)^5$?

7.2.11. Using the expansion of $(1 - x)^{10}$, prove that

$$1 - \binom{10}{1} + \binom{10}{2} - \binom{10}{3} + \cdots - \binom{10}{9} + \binom{10}{10} = 0.$$

7.2.12. Using the expansion of $(1 + x)^8 + (1 - x)^8$, verify that

$$\binom{8}{2} + \binom{8}{4} + \binom{8}{6} + \binom{8}{8} = 2^7 - 1.$$

7.2.13. How many ways can you choose an even number of objects from 10 objects? (Is there a fast way to calculate this?)

$$\binom{10}{0} + \binom{10}{2} + \binom{10}{4} + \binom{10}{6} + \binom{10}{8} + \binom{10}{10}$$

7.2.14. Using $\binom{n}{r} = \binom{n}{n-r}$ (Theorem 5.7.8), show that

$$\sum_{r=0}^{n} \binom{n}{r} x^{n-r} y^r = \sum_{r=0}^{n} \binom{n}{r} x^r y^{n-r}.$$

7.2.15. Evaluate the sum $\displaystyle\sum_{i=0}^{20} \binom{20}{i} 4^i - \sum_{i=0}^{20} \binom{20}{i} 3^i \cdot 2^{20-i}$.

7.2.16. Using the Binomial Theorem, show that

$$\binom{n}{0} + 2\binom{n}{1} + 2^2\binom{n}{2} + \cdots + 2^n\binom{n}{n} = 3^n.$$

(*Hint*: Look at the algebraic proof for Corollary 7.2.3, and see how you can adapt the idea to this equation.)

7.2.17. Using a combinatorial proof, show that

$$\binom{n}{0} + 2\binom{n}{1} + 2^2\binom{n}{2} + \cdots + 2^n\binom{n}{n} = 3^n.$$

(*Hint*: Look at the combinatorial proof for Corollary 7.2.3, and see how you can adapt the idea to this equation.)

7.2.18. Using the Binomial Theorem, show that

$$\binom{n}{0} + m\binom{n}{1} + m^2\binom{n}{2} + \cdots + m^n\binom{n}{n} = (m+1)^n.$$

(*Hint*: Look at your proof from Exercise 7.2.16 and extend the idea.)

7.2.19. Using a combinatorial proof, show that

$$\binom{n}{0} + m\binom{n}{1} + m^2\binom{n}{2} + \cdots + m^n\binom{n}{n} = (m+1)^n.$$

(*Hint*: Look at your proof from Exercise 7.2.17 and extend the idea.)

7.2.20 (†). The function $1 - x + x^2 - x^3 + \cdots + x^8 - x^9$ may be rewritten as $a_0 + a_1 y + \cdots + a_9 y^9$ where $y = x + 1$, and $a_0, a_1, \ldots, a_9$ are constants. Find the values of the constants a_2 and a_4. (*Hint*: Use the geometric series formula, and then use the Binomial Theorem.)

7.2.21 (†). Prove the Binomial Theorem using the Principle of Mathematical Induction.

7.3 Review

Summary

- The properties that we investigated in Pascal's Triangle are summarized in Table 7.3.
- The Binomial Theorem is

$$(x+y)^n = \binom{n}{0}x^n + \binom{n}{1}x^{n-1}y + \binom{n}{2}x^{n-2}y^2 + \cdots$$

$$+ \binom{n}{r}x^{n-r}y^r + \cdots + \binom{n}{n}y^n$$

$$= \sum_{r=0}^{n}\binom{n}{r}x^{n-r}y^r$$

- The **general term** of the Binomial Theorem is

$$t_{r+1} = \binom{n}{r}x^{n-r}y^r$$

which is the $(r+1)^{\text{st}}$ term.

Exercises

7.3.1. Using a property of Pascal's Triangle, simplify and evaluate each of the following.

(a) $\binom{20}{4} + \binom{20}{3}$

(b) $\binom{12}{0} - \binom{12}{1} + \binom{12}{2} - \binom{12}{3} + \cdots + \binom{12}{12}$

(c) $\sum_{r=0}^{16}\binom{16}{r}$

(d) $\binom{7}{0}^2 + \binom{7}{1}^2 + \binom{7}{2}^2 + \cdots + \binom{7}{7}^2$

(e) $\binom{9}{0} + \binom{9}{1} + \binom{9}{2} + \cdots + \binom{9}{9}$

7.3.2. If $\binom{n}{0} + \binom{n}{1} + \binom{n}{2} + \cdots + \binom{n}{n} = 512$, then what is the value of n?

Table 7.3   Some Properties of Pascal's Triangle

Name	Formula	Example
Pascal's Theorem	$\binom{n}{r} + \binom{n}{r+1} = \binom{n+1}{r+1}$	$4 \quad 6$ → 10 $\binom{4}{1} \quad \binom{4}{2}$ → $\binom{5}{2}$
Symmetry	$\binom{n}{r} = \binom{n}{n-r}$	1 $1 \quad 1$ $1 \quad 2 \quad 1$ $1 \quad 3 \quad 3 \quad 1$ $\binom{0}{0}$ $\binom{1}{0} \quad \binom{1}{1}$ $\binom{2}{0} \quad \binom{2}{1} \quad \binom{2}{2}$ $\binom{3}{0} \quad \binom{3}{1} \quad \binom{3}{2} \quad \binom{3}{3}$ $\binom{3}{1} = \binom{3}{2}$
Sum of Rows	$\displaystyle\sum_{i=0}^{n}\binom{n}{i} = 2^{n}$	1 $1 + 1$ $1 + 2 + 1$ $1 + 3 + 3 + 1 = 2^{3} = 8$ $\binom{0}{0}$ $\binom{1}{0} + \binom{1}{1}$ $\binom{2}{0} + \binom{2}{1} + \binom{2}{2}$ $\binom{3}{0} + \binom{3}{1} + \binom{3}{2} + \binom{3}{3} = 2^{3} = 8$
Sum of Squares	$\displaystyle\sum_{i=0}^{n}\binom{n}{i}^{2} = \binom{2n}{n}$	1^{2} $1^{2} + 1^{2}$ $1^{2} + 2^{2} + 1^{2}$ $1^{2} + 3^{2} + 3^{2} + 1^{2} = \binom{6}{3} = 20$ $\binom{0}{0}^{2}$ $\binom{1}{0}^{2} + \binom{1}{1}^{2}$ $\binom{2}{0}^{2} + \binom{2}{1}^{2} + \binom{2}{2}^{2}$ $\binom{3}{0}^{2} + \binom{3}{1}^{2} + \binom{3}{2}^{2} + \binom{3}{3}^{2} = \binom{6}{3} = 20$
Sum of Alternating Sign	$\displaystyle\sum_{i=0}^{n}(-1)^{n}\binom{n}{i} = 0,\ \text{for } n > 0$	1 $1 - 1$ $1 - 2 + 1$ $1 - 3 + 3 - 1 = 0$ $\binom{0}{0}$ $\binom{1}{0} - \binom{1}{1}$ $\binom{2}{0} - \binom{2}{1} + \binom{2}{2}$ $\binom{3}{0} - \binom{3}{1} + \binom{3}{2} - \binom{3}{3} = 0$

7.3.3. Find n and r such that each of the following is true.

(a) $\dbinom{90}{50} + \dbinom{90}{49} = \dbinom{n}{r}$ (b) $\dbinom{x-1}{y} + \dbinom{x-1}{y-1} = \dbinom{n}{r}$

7.3.4. How many different paths will spell each of the following, starting from the top and proceeding to the next row below by moving diagonally to the left or the right? (What do you notice about the words algorithm and logarithm?)

(a) ALGORITHM (b) LOGARITHM

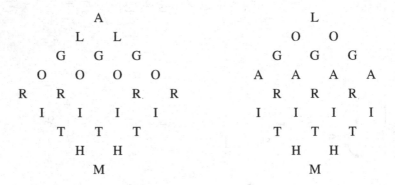

7.3.5. Using Pascal's Theorem, show that

$$\dbinom{n+3}{r+3} = \dbinom{n}{r} + 3\dbinom{n}{r+1} + 3\dbinom{n}{r+2} + \dbinom{n}{r+3}$$

for $r = 3, 4, \ldots, n$.

7.3.6. What is the coefficient of a^5b^2 in the expansion of $(a+b)^7$?

7.3.7. Express the following expression in the form $(x+y)^n$.

$$\dbinom{6}{0}x^6 + \dbinom{6}{1}x^5y + \dbinom{6}{2}x^4y^2 + \dbinom{6}{3}x^3y^3 + \dbinom{6}{4}x^2y^4 + \dbinom{6}{5}xy^5 + \dbinom{6}{6}y^6$$

7.3.8. How many terms are in the expansion of

(a) $(x+y)^{10}$?
(b) $(x+y)^{12}$?

(c) $\left(2x - \dfrac{1}{x}\right)^{12}$

7.3.9. Write the binomial expansion for

(a) $(1+x)^7$
(b) $(3-x)^6$

(c) $(4-3x)^5$
(d) $(2y-3)^8$

7.3.10. Expand and simplify.

(a) $\left(\dfrac{x}{y} + \dfrac{y}{x}\right)^3$

(b) $(\sqrt{x} - \sqrt{y})^6$

(c) $(3x + y)^5$

7.3.11. Find the coefficient of x^6 in the expansion of

(a) $\left(\dfrac{1}{3} + x\right)^{15}$

(b) $\left(3 + \dfrac{\sqrt{x}}{3}\right)^{15}$

7.4 Bibliographic Remarks

The presentation of Pascal's triangle is based on the approach presented by Pólya *et al.* (1983). This provides a glimpse into how mathematics should be taught: through exploration of patterns, generalizing results, and then proving these results. In this way, it puts the scientific method back into mathematics.

The book by Kasner and Newman (1949) is a wonderful gem that explains the beauty of mathematics to a wide audience. I am unsure why I looked at the book in the first place, but I stumbled upon a digital copy available through the Internet Archive, `archive.org`. Kasner and Newman are masters of prose; not only do Kasner and Newman explain it well, they explain it in a way that is entertaining. The example of using the Binomial Theorem to represent compactly all the outcomes of selecting balls from an urn comes from Kasner and Newman (1949, pp. 248–251). In essence, it is a thinly veiled attempt to introduce what are called **generating functions** to students without explicitly mentioning them.

I avoided giving an algebraic proof of the Binomial Theorem. Although it is possible to give it, I felt that it was not as instructive as giving a combinatorial proof that was given by Benjamin and Quinn (2003, p. 67). By asking, "How many ways can we create an $x^{n-r}y^r$ term in the expansion of $(x + y)^n$?" the theorem makes sense.

> Yes, I had seen proofs of the Binomial Theorem before, but they had seemed awkward and I wondered how anyone in his or her right mind would create such a result. But now it seemed very natural. It became a result I would never forget.
>
> —Benjamin and Quinn (2003, p. ix)

8

Introduction to Probability

> It is remarkable that a science which began with the consideration of
> games of chance should have become the most important object of
> human knowledge.
>
> —Pierre-Simon Laplace (1749–1827)
> *Théorie Analytique des Probabilités*, 1812

8.1 Introduction

We have an intuitive idea of the definition of probability, "for we were born by
mere chance."[1] Statements are made everyday such as someone has a good chance
at winning a poker tournament, or a meteorologist might say there will be 70%
chance of rain today. When we introduce and assign a numerical value to these
estimates of something happening, we are introducing the concept of probability.
The development of probability was motivated principally by games of chance:
dicing, card games, and lotteries. Out of curiosity and economic interests, it was
no wonder that the great minds of the past led to mathematical investigation of
these games.

8.2 Experiments, Sample Spaces, Sample Points, and Events

In this section, we will introduce terminology to help facilitate our discussion about
probability.

The practical value of probability theory comes when it is applied either to
real or theoretical experiments: tossing a coin, throwing two dice, selecting from a
deck of playing cards, playing roulette, playing poker, observing the lifespan of a
human being or a radioactive isotope, observing how many newborns are female
at a hospital, the number of dropped calls on a cellular (mobile) network, noise
(static) in a communications line, frequency of fatal highway accidents, frequency

---

[1] Wisdom of Solomon 2:2 (NRSV).

of crime in a neighborhood, quality control of a manufacturing process, and the chance a person is infected with a deadly disease.

All of these previous examples are vague, but in order to have a meaningful theory, we must agree upon certain conventions; we must agree upon the possible observations or outcomes of an experiment.

Definition 8.2.1 — Experiment. An experiment is a well-defined process with observable outcomes.

For example, if we toss a coin, then from our experiences (at least in my experience), we have witnessed the coin landing on its edge and rolling away (often under the refrigerator), and thus the outcome was not observed. Nonetheless, in this experiment of tossing a coin, we regard the only outcomes as *heads* or *tails*. This idealized situation simplifies the theory without overly restricting its applicability. Thus, tossing a coin and landing heads is a possible outcome, and we call this outcome a **sample point**.

Definition 8.2.2 — Sample point. A sample point is an observed outcome of an experiment.

Then, when we look at all of the possible outcomes or sample points of the experiment, we can write them down as a set. For the experiment of tossing a coin, we can denote the outcomes as $S = \{\text{heads}, \text{tails}\} = \{H, T\}$. We call the set S the **sample space**.

Definition 8.2.3 — Sample space. The sample space is the set consisting of all possible sample points, or outcomes, of an experiment. We usually denote the sample space by S.

Now, let us consider another classic experiment—rolling a die. The outcome of rolling a die is 1, or 2, or 3, ..., or 6, and the sample space is $S = \{1, 2, 3, \ldots, 6\}$. Sometimes, we may be interested in a subset of the sample space such as what are all the sample points in our die rolling experiment that are odd. This is called an **event**.

Definition 8.2.4 — Event. An event is a subset of a sample space of an experiment.

In our example of rolling a die, we can define the event E that the number on the uppermost face of the die is odd. In this case, $E = \{1, 3, 5\}$.

From Chap. 3, we can use the concepts of sets and extend them to events. If we treat events like sets, then we can perform operations such as union and intersection to create new events.

Definition 8.2.5 — Union of Two Events. The union of two events E and F is the event $E \cup F$.

$$E \cup F = \{x \in \mathcal{S} \mid x \in E \lor x \in F\} \tag{8.1}$$

where x represents a sample point, or outcome, in the sample space $\mathcal{S}$.

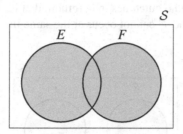

Fig. 8.1 $E \cup F$

That is, the event $E \cup F$ consists of all outcomes that is either in E or in F, or in both; the Venn diagram in Fig. 8.1 illustrates this. For example, in our die rolling example, if the event E is the outcome of the die is odd and the event F is the outcome of the die is even, then the event

$$\begin{aligned} E \cup F &= \{1, 3, 5\} \cup \{2, 4, 6\} \\ &= \{1, 2, 3, 4, 5, 6\} \end{aligned}$$

The event $E \cup F$ is all the outcomes where the number the die displays is odd or even.

Definition 8.2.6 — Intersection of Two Events. The intersection of two events E and F is the event $E \cap F$.

$$E \cap F = \{x \in \mathcal{S} \mid x \in E \land x \in F\} \tag{8.2}$$

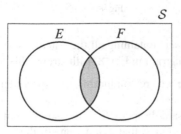

Fig. 8.2 $E \cap F$

The event $E \cap F$ consists of all the outcomes which are in both E and F. That is, the event $E \cap F$ will occur only if both E and F occur; see Fig. 8.2 for the Venn diagram. In our example of tossing the coin, if $E = \{H\}$ and $F = \{T\}$, then $E \cap F$ would not consist of any outcomes, and hence could not occur. In other words, $E \cap F = \varnothing$; in this case, the Venn diagram is given in Fig. 8.3.

> **Definition 8.2.7 — Mutually Exclusive.** Two events E and F are **mutually exclusive** if they share no outcomes in common; that is, $E \cap F = \varnothing$. Another way to say this is E and F cannot occur at the same time.

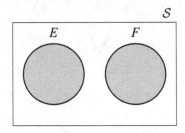

Fig. 8.3 $E \cap F = \varnothing$

> **Definition 8.2.8** The complement of an event E is the event E'.

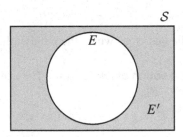

Fig. 8.4 E'

Thus, the event E' is the set containing all the outcomes in the sample space S that are not in E; the Venn diagram in Fig. 8.4 illustrates this.

Example 8.2.9 Suppose we are conducting an experiment which consists of flipping two coins.

    (a) Determine the sample space S of the experiment.
    (b) Determine the event E that at least one of the coins is heads.
    (c) Determine the event F where both coins are tails.

Solution Let H denote heads, and T denote tails.

(a) We can draw a tree diagram to illustrate the possibilities of flipping two coins; see Fig. 8.5.

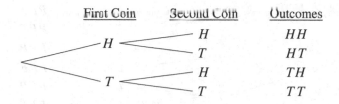

Fig. 8.5   Tree diagram for the outcomes of two coins

From the tree diagram, we can write the sample space

$$S = \{HH, TH, HT, TT\}.$$

(b) The event E is set of all outcomes where at least one of the coins is heads. This means that the first coin, or second coin, or both coins, could be heads. Thus, the event

$$E = \{HH, TH, HT\}.$$

(c) There is only one possibility in the sample space S where both coins are tails. Therefore, $F = \{TT\}$.

◀

Example 8.2.10 An experiment consists of a blood test repeated three times and observing the resulting sequence of "pass" and "fail."

(a) Describe the sample space S of the experiments.
(b) Determine the event E that exactly two passes appears.
(c) Determine the event G that at least one fail appears.

Solution

(a) The sample points may be obtained with the aid of a tree diagram; see Fig. 8.6.
The required sample space S is given by

$$S = \{PPP, PPF, PFP, PFF, FPP, FPF, FFP, FFF\}.$$

(b) By scanning the sample space S obtained in part (a), we see that the outcomes in which exactly two passes appear are given by

$$E = \{PPF, PFP, FPP\}.$$

(c) Proceeding as in part (b), we find that

$$G = \{PPF, PFP, FPP, FPF, PFF, FFP, FFF\}.$$

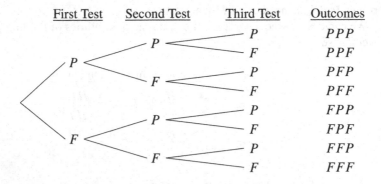

Fig. 8.6   Tree diagram for Example 8.2.10.

◀

In the previous example, the tree diagram was small and simple, and illustrated the ideas of a sample space, the outcomes or sample points, and how to find the event sets. However, for more complex problems, tree diagrams can become cumbersome. We can bypass writing out all the sample space and event sets by enumerating the sets using the counting techniques that we learned in Chap. 5. We discuss this further in Sec. 8.7.

Example 8.2.11 Consider an experiment of rolling two dice, one at a time.
  (a) Describe the sample space S of the experiment.
  (b) Determine the event E that the first die is 2.
  (c) Determine the event F that the second die is 4.
  (d) Determine the event $E \cup F$.
  (e) Determine the event $E \cap F$.
  (f) Determine the event $G \cap H$ where G is the event such that the dice are both even numbers, and H is the event that both dice are odd numbers.

Solution
  (a) We use Table 8.1 to list all the possible outcomes from rolling two dice. We can also write the sample space as

$$S = \{(1, 1), (1, 2), (1, 3), (1, 4), (1, 5), (1, 6),$$
$$(2, 1), (2, 2), (2, 3), (2, 4), (2, 5), (2, 6),$$
$$(3, 1), (3, 2), (3, 3), (3, 4), (3, 5), (3, 6),$$
$$(4, 1), (4, 2), (4, 3), (4, 4), (4, 5), (4, 6),$$
$$(5, 1), (5, 2), (5, 3), (5, 4), (5, 5), (5, 6),$$
$$(6, 1), (6, 2), (6, 3), (6, 4), (6, 5), (6, 6)\}.$$

  (b) The event $E = \{(2, 1), (2, 2), (2, 3), (2, 4), (2, 5), (2, 6)\}$.

Table 8.1 The outcomes of two dice. Note that when two dice are the same this is considered one outcome since the dice are indistinguishable.

	•	•.•	•.•	•• ••	•.• •.•	••• •••
•	$(1,1)$	$(1,2)$	$(1,3)$	$(1,4)$	$(1,5)$	$(1,6)$
••	$(2,1)$	$(2,2)$	$(2,3)$	$(2,4)$	$(2,5)$	$(2,6)$
•.•	$(3,1)$	$(3,2)$	$(3,3)$	$(3,4)$	$(3,5)$	$(3,6)$
•• ••	$(4,1)$	$(4,2)$	$(4,3)$	$(4,4)$	$(4,5)$	$(4,6)$
•.• •.•	$(5,1)$	$(5,2)$	$(5,3)$	$(5,4)$	$(5,5)$	$(5,6)$
••• •••	$(6,1)$	$(6,2)$	$(6,3)$	$(6,4)$	$(6,5)$	$(6,6)$

(c) The event $F = \{(1,4),(2,4),(3,4),(4,4),(5,4),(6,4)\}$.

(d) The event $E \cup F$ is of all the sample points where either the first die is 2 *or* the second die is 4, or both. Therefore,

$$E \cup F = \{(1,4),(2,1),(2,2),(2,3),(2,4),(2,5),$$
$$(2,6),(3,4),(4,4),(5,4),(6,4)\}$$

(e) The event $E \cap F$ is of all the sample points where the first die is a 2 *and* the second die is a 4. Therefore,

$$E \cap F = \{(2,4)\}.$$

(f) First we will need to determine the two events G and H.

$$G = \{(2,2),(4,4),(6,6)\}$$

and

$$H = \{(1,1),(3,3),(5,5)\}.$$

Then, for the event $G \cap H$, we look for the outcomes that appear in both of the events G and H, but we are unable to find any. Thus,

$$G \cap H = \{\,\} = \varnothing.$$

In this case, the events G and H are mutually exclusive.

◄

Exercises

8.2.1. List three probability statements, like those given in the introduction to this chapter, that you have found on social media. For example, these statements could be found in material from links that your friends have posted about news stories.

8.2.2. Let the event E be selecting a king from a standard deck of 52 cards, and the event F be selecting a heart. Are events E and F mutually exclusive?

8.2.3. A die is rolled and the uppermost number is observed. Let the event E be the number is even, and the event F be the number is odd. Are the events E and F mutually exclusive?

8.2.4. A game involves rolling a 12-sided die (with numbers 1 to 12). The desired outcomes in the game is to roll an even number or a multiple of 3. Are these events mutually exclusive?

8.2.5. What are *all* the possible events of the sample space $S = \{H, T\}$? (*Hint:* What is the power set of S?)

8.2.6. How many possible events are there with the sample space $S = \{1, 2, \ldots, 6\}$? (You can think of S as the sample space for rolling a die.)

8.2.7. An urn contains four balls: one red, one blue, one green, and one yellow.
   (a) Consider an experiment where you select randomly one ball from the urn, then replace it in the urn, and then draw a second ball from the urn. (This is called a **Pólya urn model**.) Draw a tree diagram and determine the sample space for this experiment.
   (b) Repeat part (a) when the second ball is drawn *without* replacing the first selected ball.

8.2.8. Suppose you are selecting six numbers from the set $\{1, 2, \ldots, 49\}$, without repetition. If the sample space is the possible combinations of these chosen numbers, how many elements are in the sample space?

8.2.9. An experiment consists of throwing two dice and observing each die's uppermost number. (*Hint:* Determine the sample space first.)
   (a) Determine the event E that the second die's number is even.
   (b) Determine the event F that the second die's number is odd.
   (c) Determine the event G that the product of the two dice's number is 15.

8.3 Probability of an Event in a Uniform Sample Space

> I filled out a thing online where I became a minister and can marry people in 40 of the 50 states. I don't know which 40, but chances are this is one of them because that's how chances work.
>
> —Eugene Mirman (1974–)
>
> from his TV Special, *Vegan on his way to the Complain Store* (2015)

All the outcomes in a **uniform sample space** have the same probability, because each outcome is **equally likely**. Here are some examples:

(a) rolling a fair die, since each side of the die is equally likely to occur

(b) flipping a fair two-sided coin

(c) selecting a card from a deck of 52 cards.

> **Definition 8.3.1 — Probability of an event in a uniform sample space.** The probability of an event E, denoted by $P(E)$, in a uniform sample space S is given by
>
> $$P(E) = \frac{n(E)}{n(S)} = \frac{\text{number of sample points in } E}{\text{number of sample points in } S}. \qquad (8.3)$$

Example 8.3.2 — Flipping a Coin. Suppose that we have a fair two-sided coin; that is, the coin that is uniformly weighted.

(a) What is the probability that the coin will land on heads?

(b) What is the probability that the coin will land on tails?

Solution Let H denote heads, and T denote tails. The sample space $S = \{H, T\}$ is a uniform sample space, where the probability of every outcome is the same. This means $n(S) = 2$. Because we are working in a uniform sample space, the coin should land on heads with the same probability as tails. Thus,

$$P(H) = \frac{n(H)}{n(S)} = \frac{1}{2}$$

and

$$P(T) = \frac{n(T)}{n(S)} = \frac{1}{2}.$$

◀

It is very important to note that Eq. (8.3) is only true when all the sample points are **equally likely**. To be more explicit, this means that no outcomes in the sample space are more likely to occur than another. This would exclude experiments with a non-uniformly weighted coin so that heads (or tails) would happen more frequently, or a die that is weighted to favor a particular number more often.

Example 8.3.3 — Picking a card from a deck. From a well-shuffled deck of 52 cards, one card is drawn. Determine the sample space and the probability that an ace is drawn from the deck.

Solution The sample space is $S = \{A\diamondsuit, 2\diamondsuit, \ldots, Q\diamondsuit, K\diamondsuit, A\heartsuit, 2\heartsuit, \ldots, Q\heartsuit,$ $K\heartsuit, A\clubsuit, 2\clubsuit, \ldots, Q\clubsuit, K\clubsuit, A\spadesuit, 2\spadesuit, \ldots, Q\spadesuit, K\spadesuit\}$. Thus, $n(S) = 52$. Let the event $E = \{A\diamondsuit, A\heartsuit, A\clubsuit, A\spadesuit\}$. Thus, $n(E) = 4$. Then,

$$P(E) = \frac{n(E)}{n(S)} = \frac{4}{52} = \frac{1}{13}.$$

Therefore, the probability of drawing an ace from a well-shuffled deck of cards is $\frac{1}{13}$ or 7.69%. ◀

The next example is an old problem whose solution has been known for 300 years. It was considered by the French mathematician Pierre Rémond de Montmort (1678–1719) in his 1708 book, *Essay d'Analyse sur les Jeux de Hazard.* It is also known as the **Montmort matching problem** in his honor.

Example 8.3.4 — Coat Check Problem. At a restaurant, Montmort and his three friends check their coats, but the incompetent attendant did not affix a ticket number to track their four coats. So when Montmort and his friends come back to collect their coats, their coat-check stubs are useless. If the attendant hands back the four coats randomly to the four friends, then what is the probability that

(a) at least one person will get his or her coat back?

(b) no person gets his or her coat back?

Solution We denote person A, B, C and D to have their respective coats A, B, C, and D. Without any loss of generality, we say that person A will select first, person B will select second, and so forth.

Draw a diagram. We can create a tree diagram to enumerate all the possible outcomes of how the selections are made by each person, such as the one seen in Fig. 8.7.

From Fig. 8.7, we see that the sample space $S = \{ABCD, ABDC, \ldots,$ $DCBA\}$; $n(S) = 24$.

(a) Let E be the event that at least one person will receive his or her coat back. Then in the sample space S, we look for any of the outcomes with A, B, C, or D since this denotes that person A, B, C, or D received the correct coat, respectively. Thus, $E = \{ABCD, ABDC, ACBD, ACDB,$ $ADBC, ADCB, BACD, BCAD, BDCA, CABD, CBAD, CBDA,$ $DACB, DBCA, DBCA\}$; $n(E) = 15$. Therefore, the probability that at least one person will receive his or her coat back is

$$P(E) = \frac{n(E)}{n(S)} = \frac{15}{24} = \frac{5}{8} = 0.625.$$

(b) Let F be the event that no one gets his or her coat back. Then, in the sample space S, we look for all outcomes in which there are no A, B, C,

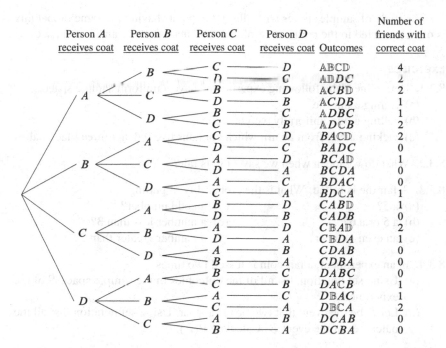

			Number of
Person A	Person B	Person C	Person D	Outcomes	friends with
receives coat	receives coat	receives coat	receives coat		correct coat

Fig. 8.7   Enumerating all of the outcomes of person A, B, C, and D selecting a coat using a tree diagram. In the outcomes, we have used $\mathbb{A}$, $\mathbb{B}$, $\mathbb{C}$, and $\mathbb{D}$ to indicate that person A, B, C, or D received the correct coat, respectively.

and $\mathbb{D}$. Thus, $F = \{BADC, BCDA, BDAC, CADB, CDAB, CDBA, DABC, DCAB, DCBA\}$; $n(F) = 9$. Therefore, the probability that no one gets his or her coat back is

$$P(F) = \frac{n(F)}{n(S)} = \frac{9}{24} = \frac{3}{8} = 0.375.$$

◀

We have investigated so far the simple case where the sample space has a finite number, N, of sample points each having probability of $\frac{1}{N}$. In this case, the probability of any event E equals the number of sample points in E divided by N.

Students may be tempted (as even mathematicians in older literature had done) to say that *every* finite sample space is a uniform probability space. This is simply not so. For example, if we look at a coin that is not uniformly weighted, the sample space still contains only the two sample points: heads and tails. The probability of heads could be p, the probability of tails is q, and $p + q = 1$, but $p \neq q$. For a particular coin, it could land on tails with probability 0.8 while landing on heads with probability 0.2.

The study of sample spaces with all sample points having the same probability is often restricted to the realm of combinatorial analysis and games of chance.

Exercises

8.3.1. Determine if the following experiments have a uniform sample space.
  (a) rolling a fair die
  (b) rolling a non-uniformly weighted die
  (c) picking a ball from an urn which contains five red and three black balls

8.3.2. What do we mean when we say a die is fair?

8.3.3. A fair die is rolled. What is the probability of getting
  (a) a 5?   (d) an odd number?
  (b) a 5 or a 6?   (e) a number less than 3?
  (c) an even number?   (f) a number greater than 2?

8.3.4. In an experiment, a fair coin is tossed two times.
  (a) Using set notation, list all of the outcomes in the sample space S of the experiment.
  (b) Let E be the event that two heads appear. Using set notation, list all the outcomes of the event E. Calculate P(E).

8.3.5. In an experiment, a fair coin is tossed three times.
  (a) Using set notation, list all of the outcomes in the sample space S of the experiment.
  (b) Let E be the event that at least two tails appear. Using set notation, list all the outcomes of the event E. Calculate P(E).

8.3.6. A single card is drawn from a well-shuffled deck of 52 cards. What is the probability that it is
  (a) a 10?   (c) a black suit?
  (b) a face card?   (d) a club or a heart?

8.3.7. In Example 8.3.4, what is the probability that exactly three people will receive the correct coat?

8.3.8. A person in born in June. What is the probability that
  (a) the person's birthday is 13 June?
  (b) the person was born in the first half of June?

8.3.9. What is the probability of drawing either a 7 or a heart face (a jack, queen, or king) card from a standard deck of 52 cards?

8.3.10. Suppose that there is one winning ticket in a large lottery. It is reasonable to believe that if you randomly selected a lottery ticket that it is not the winning ticket, since the probability that it is the winner is so very small. Is it reasonable to believe that no lottery ticket will win?

8.3.11. A letter is chosen at random from the word PROBABILITY. What is the probability that it is a B? That it is a vowel?

8.3.12. A couple decides to have four children. The probability of having children of either sex is fixed at 0.5. What is the probability that they will have more girls than boys?

8.3.13. What is the probability of picking a king of hearts in a deck of 52 well-shuffled cards?

8.3.14. Three numbers are chosen at random from the set $\{1, 2, 3, \ldots, 20\}$. Compute the probability that
    (a) their sum is odd   (b) their product is odd.

8.3.15 ([Feller (1968)], p. 24). Amongst the digits $1, 2, 3, 4, 5$, a digit is selected, and then a second selection is made amongst the remaining four digits. Assume that all 20 possible results have the same probability. Find the probability that an odd digit will be selected
    (a) the first time   (b) the second time   (c) both times.

8.3.16. An urn contains four balls: one red, one blue, one green, and one yellow.
    (a) Consider an experiment where you select randomly one ball from the urn, then replace it in the urn, and then draw a second ball from the urn. (This is called a **Pólya urn model**.)
        (i) Draw a tree diagram for this experiment.
        (ii) What is the probability that at least one green ball is drawn?
        (iii) What is the probability that two green balls are drawn?
    (b) Repeat part (a) when the second ball is drawn *without* replacing the first selected ball.
        (i) Draw a tree diagram for this experiment.
        (ii) What is the probability that at least one green ball is drawn?
        (iii) What is the probability that two green balls are drawn?

8.3.17. A traffic light at a major intersection is green for 80 seconds, yellow (amber) for 5 seconds, and red for 30 seconds. What is the probability of a vehicle arriving at the traffic light and finding the traffic light is
    (a) green?   (c) red?
    (b) yellow?   (d) either yellow or green?

8.3.18. In the game of *Dungeons and Dragons*, a fair dodecahedral die (a die with 12 faces) is used. What is the probability of rolling a number greater than 9?

8.3.19. A target that consists of three circles with radii of 5, 9, and 12 inches as seen in Fig. 8.8. If a dart lands randomly anywhere in the target, then what is the probability of it landing in

    (a) circle A?   (b) ring B?   (c) ring C?

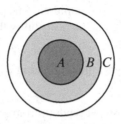

Fig. 8.8   The target for Exercise 8.3.19

(*Hint*: The formulas for a circle are in Sec. B.2 on p. 483.)

8.4 Definition and Axioms of Probability

> There is no point in using the word "impossible" to describe something that has clearly happened.

> —Douglas Adams (1952–2001)
>
> *Dirk Gently's Holistic Detective Agency*, 1987

Suppose we consider an experiment whose sample space is S. For every event $E \subseteq S$, there is an associated number $P(E)$, the **probability of an event** E, which satisfies three axioms:[2]

Axiom 1 The probability of the entire sample space, S, is 1.

$$P(S) = 1 \tag{8.4}$$

In set theory, we used Venn diagrams to visualize sets and their derived sets such as unions and intersections. We now introduce a new visual diagram for probability— the **eikosogram** or "probability picture" from the Greek words *eikos* (probability) and *gramma* (writing). The eikosogram is a square with an area equal to one (with side lengths equal to one). An area in an eikosogram represents a probability. In the eikosogram, Axiom 1 says the area of the sample space is equal to one, and thus the sample space is the entire square.

---
[2] In mathematics, we use the term *axiom* to mean a statement that is regarded as being an irreducible fact that is self-evidently true.

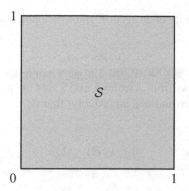

Fig. 8.9   Eikosogram for Axiom 1

Example 8.4.1 A coin flip experiment has the sample space $S = \{H, T\}$, where H is heads, and T is tails. From Example 8.3.2 (p. 315), we know $P(H) = \frac{1}{2}$ and $P(T) = \frac{1}{2}$. We can construct an eikosogram, which can be seen in Fig. 8.10.

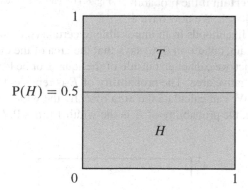

Fig. 8.10   Eikosogram for a coin flip experiment

From the eikosogram, we see that the probabilities of the two outcomes, which is the entire sample space S, sums to one;

$$P(H) + P(T) = \frac{1}{2} + \frac{1}{2} = 1.$$

Axiom 2 The probability of an event E in the sample space S is a non-negative number.

$$P(E) \geq 0 \qquad (8.5)$$

It immediately follows from Axioms 1 and 2 that the probability of an event E occurring is between zero and one inclusively; that is,

$$0 \leq P(E) \leq 1. \qquad (8.6)$$

This says that the probability assigned to an event E is a number between zero and one inclusively, and this is a measure of the chance or likelihood of E occurring. An event E is called **impossible** if the probability of E is 0; $P(E) = 0$. An event E is called **certain** if the probability of E is 1; $P(E) = 1$. We now have a numerical scale of likelihoods from impossible to certain (see Fig. 8.11).

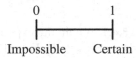

Fig. 8.11   Numerical scale of probability

In the eikosogram, condition (8.6) says that the area of the event E cannot be greater than one since we cannot go outside of the square, or be less than zero since we cannot have negative area. The probability of E is represented by the shaded area in Fig. 8.12. We can calculate the area of E by the width multiplied by the length. In this case, the probability of E is the width 1 times $P(E)$.

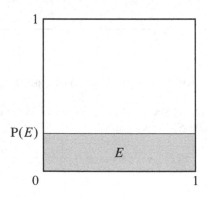

Fig. 8.12   Eikosogram of Axiom 2: an event E

Axiom 3 If the events are mutually exclusive, then the probability of the union of the events is equal to the probability of each individual event occurring.

Mathematically, for any mutually exclusive events $E_1, E_2, \ldots, E_n$, then

$$P(F_1 \sqcup F_2 \sqcup \quad \cup E_n) = P(E_1) + P(E_2) + \cdots + P(E_n). \tag{8.7}$$

In the eikosogram in Fig. 8.13, Axiom 3 says if the event is a union of events whose areas do not overlap one another, the areas can be simply added together.

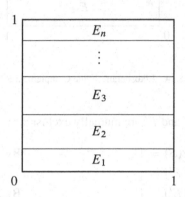

Fig. 8.13   Eikosogram for Axiom 3

Exercises

8.4.1. In an experiment, a fair coin is tossed.
   (a) What are all the possible events of the sample space $S = \{H, T\}$?
   (b) What are the probabilities of each event found in part (a)? Do these probabilities satisfy the three axioms of probability?

8.5 Simple Probability Rules

Theorem 8.5.1 — Complement Rule. The probability of an event not occurring is equal to the probability of the complement of the event. Mathematically, for an event E,

$$P(E') = 1 - P(E). \tag{8.8}$$

The two events E and E' are **complementary events**.

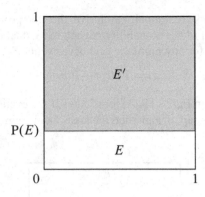

Fig. 8.14   Eikosogram for the Complement Rule

Proof The two events E and E' are mutually exclusive; $E \cup E' = S$.

$$P(S) = 1 \qquad \text{By Axiom 1}$$
$$P(E \cup E') = 1$$
$$P(E) + P(E') = 1 \qquad \text{By Axiom 3}$$
$$P(E') = 1 - P(E)$$

◀

We can ask what the eikosogram would look like if two events E and F were not mutually exclusive. Given below, two events E and F share some common area.

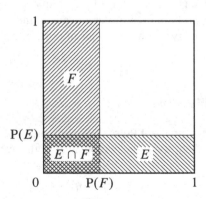

Fig. 8.15   Eikosogram with two events overlapping

This common area is the intersection of E and F, $P(E \cap F)$. If we were to find the union of $P(E)$ and $P(F)$, $P(E \cup F)$, and added the two areas of $P(E)$ and $P(F)$ together, we would count the overlapping area twice. This gives a geometrical description of how to add two event probabilities if they have a non-empty intersection. Therefore, we have the following rule.

Theorem 8.5.2 — Addition Rule.

$$P(E \cup F) = P(E) + P(F) - P(E \cap F) \qquad (8.9)$$

In words, $P(E \cup F) =$

$$\begin{array}{ll} & \text{the sum of the probabilities of individual events} \\ \textit{minus} & \text{the probability of the two-event intersection.} \end{array}$$

Notice that the addition rule is similar to the Principle of Inclusion and Exclusion that we saw in Sec. 3.4 on p. 99.

Proof

$$\begin{aligned} P(E \cup F) &= \frac{n(E \cup F)}{n(S)} && \text{By (8.3)} \\ &= \frac{n(E) + n(F) - n(E \cap F)}{n(S)} && \text{By (3.5)} \\ &= \frac{n(E)}{n(S)} + \frac{n(F)}{n(S)} - \frac{n(E \cap F)}{n(S)} \\ &= P(E) + P(F) - P(E \cap F) && \text{By (8.3)} \end{aligned}$$

◀

Example 8.5.3 What is the probability of turning up an odd number or a number greater than 2 when rolling a fair die?

Solution The sample space is $S = \{1, 2, 3, 4, 5, 6\}$; $n(S) = 6$. The event sets are $E = \{1, 3, 5\}$ and $F = \{3, 4, 5, 6\}$. So, $n(E) = 3$, and $n(F) = 4$. The set of common elements is $E \cap F = \{3, 5\}$; $n(E \cap F) = 2$. Thus, the required probability is $P(E \cup F)$.

$$\begin{aligned} P(E \cup F) &= P(E) + P(F) - P(E \cap F) \\ &= \frac{n(E)}{n(S)} + \frac{n(F)}{n(S)} - \frac{n(E \cap F)}{n(S)} \\ &= \frac{3}{6} + \frac{4}{6} - \frac{2}{6} \\ &= \frac{5}{6} \end{aligned}$$

Therefore, the probability of rolling an odd number or a number greater than 2 is $\frac{5}{6}$ or 83%. ◀

To ensure (8.9) is correct, if the intersection is empty, $P(E \cap F) = 0$, then E and F are mutually exclusive events, and

$$P(E \cup F) = P(E) + P(F) - P(E \cap F)$$
$$= P(E) + P(F) - 0$$
$$= P(E) + P(F)$$

which satisfies Axiom 3.

Example 8.5.4 What is the probability of selecting a queen or a king from a well-shuffled deck of 52 cards?

Solution The sample space is $\mathcal{S} = \{A\diamond, 2\diamond, \ldots, Q\diamond, K\diamond, A\heartsuit, 2\heartsuit, \ldots, Q\heartsuit, K\heartsuit, A\clubsuit, 2\clubsuit, \ldots, Q\clubsuit, K\clubsuit, A\spadesuit, 2\spadesuit, \ldots, Q\spadesuit, K\spadesuit\}$. Thus, $n(\mathcal{S}) = 52$. Let event $E = \{Q\diamond, Q\heartsuit, Q\clubsuit, Q\spadesuit\}$, and $F = \{K\diamond, K\heartsuit, K\clubsuit, K\spadesuit\}$, Thus, $n(E) = 4$ and $n(F) = 4$.

Since E and F are mutually exclusive, $n(E \cap F) = 0$,

$$P(E \cup F) = P(E) + P(F)$$
$$= \frac{n(E)}{n(\mathcal{S})} + \frac{n(F)}{n(\mathcal{S})}$$
$$= \frac{4}{52} + \frac{4}{52}$$
$$= \frac{8}{52}$$
$$= \frac{2}{13}.$$

The probability of selecting a queen or a king from a deck is $\frac{2}{13}$. ◄

In general, we can extend the principle of inclusion and exclusion to any finite number of events. If $E_1, E_2, E_3, \ldots E_n$ are events, then
$$P(E_1 \cup E_2 \cup \cdots \cup E_n) =$$

	the sum of the probabilities of the individual events,
minus	the probabilities of the two-event intersections,
plus	the probabilities of the three-event intersections,
minus	the probabilities of the four-event intersections,
plus	the probabilities of the five-event intersections,
⋮	⋮

and so on until you reach the probability of the n-event intersection.

Exercises

8.5.1. Grumio buys a lottery ticket to support Caesar. When Grumio was asked by his friend Markus what he thought his chances of winning the million denari jackpot, he responded with probability $\frac{1}{2}$. Grumio reasons that he will either win the jackpot or he won't.[3] Why is this incorrect reasoning?

8.5.2. Let E and F be two events of an experiment. If $P(E) = 0.3, P(F) = 0.75$, and $P(E \cap F) = 0.25$, find

    (a) $P(E \cup F)$ (c) $P\left(F'\right)$ (e) $P\left(E' \cap F'\right)$

    (b) $P\left(E'\right)$ (d) $P\left(E' \cup F'\right)$

8.5.3. Find $P(E \cup F)$ given the following information:

    (a) $P(E) = 0.85, P(F) = 0.35$, and $P(E \cap F) = 0.25$

    (b) $P(E) = \frac{3}{6}, P(F) = \frac{3}{6}$, and $P(E \cap F) = \frac{2}{6}$.

8.5.4. If E and F are events, and $P(E) = 0.2, P(F) = 0.5$, and $P(E \cup F) = 0.7$, what is

    (a) $P(E \cap F)$ (b) $P\left(E'\right)$ (c) $P\left((E \cup F)'\right)$

8.5.5. If the sample space is $S = E \cup F$, and if $P(E) = 0.75$ and $P(F) = 0.55$, find $P(E \cap F)$.

8.5.6. If E and F are mutually exclusive events with $P(E) = 0.25$ and $P(F) = 0.6$, find the probability of each of the following:

    (a) E' (c) $E \cup F$ (e) $E' \cup F'$

    (b) F' (d) $E \cap F$ (f) $E' \cap F'$

8.5.7. If E and F are mutually exclusive events, and $P(E) = 0.6$ and $P(F) = 0.2$, find

    (a) $P(E \cap F)$ (c) $P\left(E'\right)$

    (b) $P(E \cup F)$ (d) $P\left(E' \cap F\right)$

8.5.8. Let $S = \{s_1, s_2, s_3, s_4, s_5\}$ be a sample space with the following specified probabilities:

$$P(\{s_1\}) = \frac{1}{10}, P(\{s_2\}) = \frac{3}{10}, P(\{s_3\}) = P(\{s_4\}) = P(\{s_5\}) = \frac{2}{10}.$$

Given the events $E = \{s_1, s_2, s_3\}$ and $F = \{s_1, s_2, s_5\}$, calculate

    (a) $P(E)$ (b) $P(E' \cup F)$

8.5.9. Suppose a coin is weighted so that tails is twice as likely to appear as heads. What is the probability of flipping the coin and it landing

    (a) tails?   (b) heads?

8.5.10. Four fair dice are rolled. What is the probability that at least one or more 3s appear?

---

[3]From the TV series *Plebs*, Season 1, Episode 4: The Herpes Cat.

8.5.11. Suppose for tomorrow there is a 30% chance of freezing rain, 10% chance of snow and freezing rain, and 75% chance of snow or freezing rain. What is the chance of snow tomorrow?

8.5.12. Prove that

$$P(E \cup F \cup G) = P(E) + P(F) + P(G) - P(E \cap F) - P(E \cap G)$$
$$- P(F \cap G) + P(E \cap F \cap G)$$

(*Hint:* Look at the proof of Theorem 8.5.2).

8.5.13. A total of 36 students in a class play tennis, 28 play squash, and 18 play badminton. Furthermore, 22 of the students play both tennis and squash, 12 play both tennis and badminton, 9 play both squash and badminton, and 4 play all three sports. What is the probability of randomly selecting a student that plays at least one of these three sports if there are in the class
    (a) 100 students?   (b) N students?

8.5.14. An urn contains seven white, five black, and eight red balls. You randomly select a ball from the urn. What is the probability that
    (a) the ball is white?   (b) the ball is not black?

8.5.15. What is the probability of drawing either a 7 or a heart face (a jack, queen, or king) card from a standard deck of 52 cards? Are these two events mutually exclusive?

8.5.16. The four main blood types are A, B, AB, and O. The letters A and B indicate whether the two factors are present. Thus, type AB blood has both factors while type O has neither. In Canada, 42% of Canadians have type A blood, 9% of Canadians have type B blood, 3% of Canadians have type AB blood, and 46% of Canadians have type O blood. Determine the probability that a Canadian citizen
    (a) has blood factor A?   (b) doesn't have blood factor B?

8.5.17. Show that $P\left(E' \cap F'\right) = 1 - P(E \cup F)$.

8.6 Risk: Applications of Probability

I once started a lecture with the following statement: you are all going to die. This sets my students aback. What is that supposed to even mean? They have their whole life in front of them: finish university, get married, have children, and live happily ever after. It may be this invincible sense of self[4] that may be the

---

[4]"…this [adolescent] belief in personal uniqueness becomes a conviction that he will not die, that death will happen to others but not to him. This complex of beliefs in the uniqueness of his feelings and of his immortality might be called a *personal fable*…" (Elkind, 1967, p. 1031).

explanation for the leading cause of death in Americans ages 15 to 24: motor-vehicle accidents[5]—with males more than twice as likely to be involved than females.[6] Homicide and suicide follow behind in second and third place. The lights in some of my students eyes started to flicker; they understood now why their car insurance is higher than their parents. It seems like the world is out to get my students after I tell them this. Should they just stay in their dorm room and wait till it is safe? That is not necessary. If we look at them in the context of everyone else dying, the probability of dying between 15 to 24 is about 1.1%.[7] This is not high enough to become an agoraphobic—by leaving their dorm or residence, they accept the risk. In fact, every activity has some amount of risk: some people will be comfortable with the risk for a particular activity and perform it, while some are not comfortable at all with the risk for the activity and will abstain.

> **Definition 8.6.1 — Risk.** Risk is the probability of an *unwanted* event that may or may not occur.

Our mortality is a fact of life; death eventually comes, but we can delay it. With airbags, food inspection, and antibiotics, to name a few modern safety nets, these have enabled people to live long enough to die from other causes such as cancer or heart disease.

We may take this notion to be obvious now—the older you get, the higher the probability of death. Everyone knows this, just as everyone knows that the earth revolves around the sun. Nevertheless, similar to people thinking that the sun revolved around the earth,[8] people believed something different about death as well. In 17th century Europe, people believed that death was lurking around the corner, and that at any time, it could pounce on you. Death could strike down any person with equal probability. It was not until the middle of that century that scientists like Edmond Halley (1656–1742) began to investigate death. Halley noticed one day as he was walking through his parish's cemetery that each tombstone included the age of the person when they died. He started to jot down this data, examined it, and low-and-behold, he found that as you got older, you were more likely to die.

---

[5]Teen driver error was the most common reason for serious accidents involving teenagers. Inadequate surveillance, distracted driving, following too closely or too fast for conditions accounted for approximately half of all crashes (Curry *et al.*, 2011). Curry *et al.* (2011) did not specifically study cases with fatalities, but they were more interested in analyzing the causes of serious motor vehicle crashes with teen drivers.

[6]In 2013, 4636 males and 1874 females age 15 to 24 died in motor vehicle accidents in the United States. Source: Centers for Disease Control and Prevention WISQARS database.

[7]In 2013, there were 28 486 deaths for American people age 15 to 24. The total number of deaths in the United States in 2013 is 2 596 993. Thus, $\frac{28\,486}{2\,596\,993} \approx 0.010\,97$ of total deaths were people age 15 to 24. Source: Deaths—Final Data for 2013. NVSR Volume 64, Number 2 (Table 6).

[8]Contrary to popular belief, Christian scholars during medieval times did not believe the Earth was flat, and in fact knew its approximate circumference (Lindberg and Numbers, 1986, p. 342). They had adapted Aristotle's cosmological view that included a spherical Earth, but the Sun revolved around it.

Table 8.2   Abridged life table for the United States population (both sexes), 2010. Source: Murphy *et al.* (2013, Table 6, p. 29).

Age Bracket	Probability of dying in age bracket	Number of survivors at beginning of age bracket	Number of who died during age bracket	Expected remaining life at beginning of age bracket
0–1	0.006 123	100 000	612	78.7
1–5	0.001 071	99 388	106	78.1
5–10	0.000 573	99 281	57	74.2
10–15	0.000 708	99 224	70	69.3
15–20	0.002 463	99 154	244	64.3
20–25	0.004 317	98 910	427	59.5
25–30	0.004 791	98 483	472	54.7
30–35	0.005 500	98 011	539	50.0
35–40	0.006 913	97 472	674	45.2
40–45	0.009 979	96 798	966	40.5
45–50	0.016 044	95 832	1 538	35.9
50–55	0.024 343	94 295	2 295	31.4
55–60	0.035 106	91 999	3 230	27.2
60–65	0.049 847	88 770	4 425	23.1
65–70	0.074 412	84 345	6 276	19.1
70–75	0.112 312	78 068	8 768	15.5
75–80	0.174 772	69 300	12 112	12.1
80–85	0.274 365	57 189	15 691	9.1
85–90	0.430 887	41 498	17 881	6.5
90–95	0.615 150	23 617	14 528	4.6
95–100	0.783 137	9 089	7 118	3.2
> 100	1.000 000	1 971	1 971	2.3

We live differently than our ancestors did; we also die differently than our ancestors did. Large risks that were killing us have been eliminated. For example, in 1900s America, you expected to live 50.70 years if you were a woman, and 47.88 years if you were a man (Arias, 2006, Table 11, p. 30), dying mostly from pneumonia, influenza, diphtheria, tuberculosis, and diarrhea (Linder and Grove, 1947, Table 12, pp. 210–215). With advances in medicine and public health, American males born in 2015 expect to live 76.4 years and females 81.2, with only pneumonia and influenza remaining on the top 10 list of leading causes of death. If you lived further up north in Canada, you increase your life expectancy almost by 3 years to 79.33 if you are male, and if you are female almost $2\frac{1}{2}$ years to 83.60 (Statistics Canada, 2013b, Table 1, pp. 4, 6).

After eliminating large risks, such as polio, typhoid fever, diphtheria, and so on, we are left with small risks, and with these small risks, we need a way to conceptualize them. So Richard Wilson (1979) came up with a winsome little unit of risk called the micromort (from the words *micro* meaning 10^{-6} and *mort* for mortality).

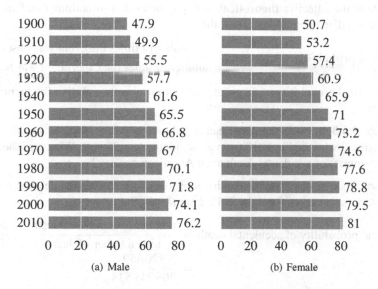

Fig. 8.16   Life Expectancy (in years) for Americans. Source: Arias (2006, Table 10, p. 26) for 1900 to 1990, Minino *et al.* (2002, p. 6) for 2000 data, and Murphy *et al.* (2013, p. 6) for 2010 data.

▌ Definition 8.6.2 — Micromort. A micromort is a one-in-a-million probability of death. It is a measure of risk of death.

To give you an idea of what kind of activities will incur a cost of one micromort against your life, consider the following examples. Suppose you are a smoker, and you like to go outside and enjoy a cigarette: smoking 1.4 cigarettes is one micromort, and the likely cause of death is either cancer or heart disease. We also know that second-hand smoke is also risky: living with a smoker for two months is one micromort. If during dinner, you consume $\frac{1}{2}$ L of wine that's another micromort (liver disease), and eating 100 charcoal broiled steaks is another micromort (cancer from benzopyrene). Even living in particular cities can incur one micromort: being in Boston or New York for 2 days (air pollution), or Denver for 2 months (cancer from cosmic radiation) (Wilson, 1979, p. 45). To a large extent, our day to day activities involve some level of risk, and this information helps us to make informed decisions. Considering risk is not restricted to people performing everyday activities, but risk is considered by regulators, engineers, and scientists to ensure the safety of the general public.

You might be wondering after seeing the different amounts of micromorts for each activity and ask, how were these calculated? In the next example we show how to calculate the number of micromorts for an activity, but it based on the *empirical probability* of an event occurring. The adjective **empirical** denotes that something has been collected through observation or experimentation. This is the opposite of the adjective **theoretical**, which denotes that something that happens under an idealized conditions. Thus, the

$$\textbf{empirical probability} \text{ of an event} = \frac{\text{number of times the event was observed}}{\text{number of times the experiment was repeated}}.$$

The empirical probability of an event is also known as the **relative frequency** of an event.

Example 8.6.3 In 2010, the American population was 308 745 538. The number of Americans who died in 2010 from an accident was 120 859. What was the risk (in micromorts) of accidental death to an American per day?

Solution First, we calculate the empirical probability of accidental death in a year for an American citizen from the 2010 data; that is, the

$$
\begin{aligned}
\text{empirical probability of accidental death} &= \frac{\text{number of accidental deaths in 2010}}{\text{total number of deaths in 2010}} \\
&= \frac{120\,859}{308\,745\,538} \\
&= 0.000\,391.
\end{aligned}
$$

To convert this to a micromort, we multiply by 10^6; that is, $0.000\,391 \times 10^6 = 391$. This means that the risk of accidental death is 391 micromorts per year. Dividing by 365 (for the number of days in a year), we have $\frac{391}{365} = 1.07$. Therefore, the estimate of accidental death for an American is 1.07 micromorts per day based on 2010 data.   ◄

Table 8.3   Risk of death by cause in Canada, the United States, and England and Wales in 2011, 2010, and 2012 respectively. Calculations are based on 2011 Canadian population of 33 476 688, 2010 American population of 308 745 538, and 2012 England and Wales population of 56 567 800. Source: Statistics Canada (2014), Murphy *et al.* (2013), and Office for National Statistics (2013).

Death from	Micromorts per day		
	Canada	US	England & Wales
All causes	19.81	21.90	24.18
Cancer	5.93	5.10	7.04
Heart disease	3.90	5.30	5.63
Cerebrovascular diseases	1.09	1.15	1.74
Chronic lower respiratory diseases	0.92	1.23	1.38
Accidents (unintentional injuries)	0.88	1.07	0.55
Diabetes	0.59	0.61	0.24
Alzheimer's disease	0.52	0.74	0.43
Suicide	0.31	0.34	0.18
Influenza and pneumonia	0.47	0.44	1.27

This type of calculation can be repeated for other causes of death, and the results are given in Table 8.3.

We do not know the true theoretical probability of someone dying from an accident, but we can estimate the theoretical probability with an empirical probability. Empirical probabilities may differ from the theoretical probabilities when the experiment is repeated only a few times. Such differences or fluctuations can result in an event occurring more frequently or less frequently than the theoretical probability. When the experiment is repeated a large number of times, however, these fluctuations tend to cancel out, and the empirical probability usually approaches the theoretical probability.

Example 8.6.4 — Rolling a die. Consider an experiment in which you are repeatedly rolling a die.

Perhaps in this experiment, the goal is to determine if the die is fair; that is, a die is fair if the probability of each outcome is equally likely.

To do this, you could record how many times you observe the number 3 in a certain number of rolls, and see how close it is to the theoretical probability of $\frac{1}{6}$. You ob-

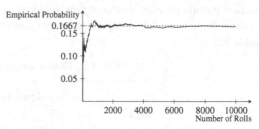

Fig. 8.17 The empirical probability of rolling a 3 with a die calculated after each roll for 10 000 rolls.

serve from Table 8.4 that the empirical probability differs considerably when the number of rolls is small, but as the number of rolls becomes large, the empirical probability approaches the number $\frac{1}{6} \approx 0.1667$; you feel this is enough evidence to support the claim that the die is fair (see Exercise 8.6.2). In Fig. 8.17, the calculation of the empirical probability after each roll for all 10 000 rolls of the die is presented. Table 8.4 shows a small portion of the data that was collected. We

Table 8.4   Results from a computer simulation of rolling a die. As the number of rolls increases, the empirical probability approaches the theoretical probability of $\frac{1}{6} \approx 0.1667$ for rolling a 3 with a die.

Number of Rolls, n	Number of 3s observed, m	Empirical Probability, $\frac{m}{n}$
1	0	$\frac{0}{1} = 0$
10	0	$\frac{0}{10} = 0$
100	12	$\frac{12}{100} = 0.12$
1 000	167	$\frac{167}{1000} = 0.167$
10 000	1651	$\frac{1651}{1000} = 0.1651$

see that there are fluctuations, but as the number of rolls increases, these fluctuations get smaller and smaller allowing the empirical probability to approach the theoretical probability.   ◄

Exercises

8.6.1. Explain the difference between empirical probability and theoretical probability.

8.6.2. In Example 8.6.4, is inductive reasoning or deductive reasoning used when the statement "you feel this is enough evidence to support the claim that the die is fair" is made? Why?

8.6.3. An insurance company offers customers coverage against accidental damage to their mobile phones. The company selected randomly 1000 customers in a city to see if there is a relationship between age and accidents, and collected the data in Table 8.5.

Table 8.5   1000 customers in a city

Age	Accidents in a year				
	0	1	2	3	over 4
under 15	43	50	40	28	20
16–20	20	36	62	14	24
21–25	82	68	93	35	3
26–30	38	50	32	35	4
over 30	67	64	53	32	7

Calculate the (empirical) probability of randomly selecting a customer in the city

(a) with no accidents in a year
(b) being under 15 years old and having no accidents in a year
(c) being 21–25 years old and having two or more accidents in a year
(d) being 16–20 years old and having only one accident in a year
(e) being 16–20 years old or having only one accident in a year
(f) being over 30 years old
(g) not being over 30 years old

8.7 Using Counting Techniques to Calculate Probability

It becomes clear that enumerating all possible cases becomes tedious and unwieldy as the number of elements in the sample space increases. It is for this reason that we use the results we learned from combinatorial analysis in that renders direct enumeration unnecessary.

While the problems of Chap. 5 dealing with permutations and combinations may seem to have a dry and dreary look, like "how many ways the letters of Mississauga can be rearranged," those theorems form an important foundation of probability. We need to know how to calculate the total number of ways an event can happen before aspiring to predict the likelihood of it happening.

8.7.1 Permutations

Example 8.7.1 — A monkey at the typewriter. A monkey is sitting at a typewriter and begins to hit keys at random. If all the characters typed were uppercase letters of the alphabet or spaces, what is the probability that the monkey wrote "TO BE" in the first five characters?

Fig. 8.18   A monkey at a typewriter

Solution The experiment is randomly selecting (or typing) five uppercase letters or spaces. The sample space S consists of all the possible permutations of letters and spaces (with replacement). Then, the number of sample points in S is

$$n(S) = 27 \times 27 \times 27 \times 27 \times 27 = 27^5 = 14\,348\,907.$$

Let E be the event that the monkey has typed "TO BE." There is only one possible way to do this; $n(E) = 1$.

Thus, the probability of a monkey writing "TO BE" is

$$P(E) = \frac{n(E)}{n(S)} = \frac{1}{14\,348\,907}.$$

◀

8.7.2 Combinations

Example 8.7.2 — Winning the Lottery. In a national lottery, a person selects six numbers, in which order does not matter, from $\{1, 2, 3, \ldots, 49\}$ without replacement. What is the probability that the person's selected six numbers will win him the prize?

Solution The experiment is randomly selecting six numbers from the set $\{1, 2, 3, \ldots, 49\}$ without replacement. So,

$$
\begin{aligned}
n(S) &= \binom{49}{6} \\
&= \frac{49!}{(49 - 6!)6!} \\
&= \frac{49 \times 48 \times 47 \times 46 \times 45 \times 44 \times \cancel{43!}}{\cancel{43!} \times 6!} \\
&= \frac{10\,068\,347\,520}{720} \\
&= 13\,983\,816.
\end{aligned}
$$

Let E be the event of picking the winning six numbers. There is only one way to pick the winning six numbers; that is, $n(E) = \binom{6}{6} = 1$.

Therefore, the probability of winning the lottery is

$$
P(E) = \frac{n(E)}{n(S)} = \frac{1}{13\,983\,816}.
$$

◄

Comparing this to Example 8.7.1, we see that winning the lottery has a higher probability than a monkey typing the first two words of Shakespeare's greatest soliloquy. Now remember when the problem stated that the monkeys began to hit the keys *at random*. Well, this is an idealized behavior, or an idealized monkey. There was an art experiment conducted in 2002, where a computer with a keyboard was put inside of the enclosure of six Sulawesi Macaque monkeys at the Paignton Zoo in England. Unfortunately, the monkeys mostly kept hitting the letter S before they partially destroyed it by using it as a toilet. After conducting the experiments, the conclusion was that real monkeys are not good random letter generators.

With virtual monkeys that really can hit keys randomly, they can recreate any written work. American software engineer Jesse Anderson in October 2011 successfully reproduced the entire works of William Shakespeare with one million virtual monkeys in 46 days. Each virtual monkey generated nine letters at a time; if it matched a segment for any of the plays, the segment was kept, otherwise it was discarded (Anderson, 2011b).

> All the world's a stage,
> And all the monkeys merely players;
> They have their typos and their hits,
> And one monkey in his time plays many parts,
> His acts being 38 works of Shakespeare.

—Anderson (2011a)

Example 8.7.3 — A Flush in Poker. In a poker hand, you are dealt five cards from a standard well-shuffled deck. What is the probability that you have a flush, where all five cards are from the same suit?

Solution The experiment is to select five cards at random from a deck of 52 cards. The sample space S consists of all the possible ways to select five cards from 52 cards (without replacement). Then, the number of sample points in S is

$$
\begin{aligned}
n(S) &= \binom{52}{5} \\
&= \frac{52!}{(52-5)!\,5!} \\
&= \frac{52!}{47!\,5!} \\
&= \frac{52 \times 51 \times 50 \times 49 \times 48 \times \cancel{47!}}{\cancel{47!}\,5!} \\
&= \frac{311\,875\,200}{120} \\
&= 2\,598\,960.
\end{aligned}
$$

Let E be the event that all of the cards in the poker hand are the same suit. There are $\binom{4}{1} = 4$ ways to pick a suit, and there are $\binom{13}{5} = 1287$ ways to pick five cards from that suit. Then by the Multiplication Principle,

$$
n(E) = \underbrace{\binom{4}{1}}_{\text{suit}} \times \underbrace{\binom{13}{5}}_{\text{rank}} = 4 \times 1287 = 5148.
$$

Therefore, the desired probability is

$$
P(E) = \frac{n(E)}{n(S)} = \frac{5148}{2\,598\,960} = \frac{33}{16\,660} \approx 0.002.
$$

◀

Example 8.7.4 — The Lazy Professor Problem. An assignment consists of 15 questions. In order to save time, the professor corrects only three of them, which he selects at random. If in your assignment only one problem is wrong or incomplete, what is the probability that it will be amongst those selected for correction?

Solution The experiment is to select three questions at random, where order does not matter, from a total of 15 questions. The sample space S consists of all the possible ways of selecting three from the 15 problems (without replacement). Then, the number of sample points in S is

$$n(S) = \binom{15}{3} = 455.$$

Let E be the event that the instructor has selected three problems which includes the incorrect problem. Then,

$$n(E) = \underbrace{\binom{14}{2}}_{\text{Correct}} \times \underbrace{\binom{1}{1}}_{\text{Incorrect}} = 91.$$

The number of ways to select a correct problem is $\binom{14}{2}$, since there are 14 problems that are correct, and we select two, so that we have three problems selected in total. The number of ways to select one incorrect problem is $\binom{1}{1}$ since there is only one incorrect problem to choose. By the Multiplication Principle, we multiply these binomial coefficients together to get the number of ways to select two correct problems and one incorrect problem.

The desired probability is therefore

$$P(E) = \frac{n(E)}{n(S)} = \frac{\binom{14}{2} \times \binom{1}{1}}{\binom{15}{3}} = \frac{91}{455} = \frac{1}{5} = 0.2.$$

So there is a 20% chance that the incorrect/incomplete problem will be selected, and you will receive $\frac{2}{3} \approx 67\%$, and there is a 80% chance that you will receive $\frac{3}{3} = 100\%$.   ◄

Exercises

8.7.1. A fresh pack of 52 cards is opened, and then is well-shuffled. What is the probability, after shuffling, that all of the cards are in their original initial order?

8.7.2. Two monkeys are sitting at typewriters and begin to hit keys at random. If all the characters typed were uppercase letters of the alphabet or spaces, what is the probability that both monkeys wrote "TO BE."

8.7.3. Anne, Marilla, Matthew, Diana, and Gilbert are going to a party. What is the probability that the friends will arrive in order of ascending age?

8.7.4. Given that you have 6 lieutenants, 5 captains, and 2 majors, find

(a) the probability that you form a committee of six people containing two lieutenants, two captains, and two majors.

(b) the probability that you form a committee of three people containing one major as president, one captain as vice president, and one lieutenant as secretary.

8.7.5. A committee of 6 members to plan a new mathematics tutoring drop-in center is to be selected from amongst 5 professors and 9 students. If the committee is randomly selected, what is the probability that it will include at least 4 professors?

8.7.6. A box contains 23 regular screws and 2 defective screws. If 4 screws are used, what is the probability that none of the screws selected are defective?

8.7.7. A student is able to solve 7 of the 15 problems of the homework problems. The professor selects 5 questions out of the 15 for a quiz. What is the probability that the student can solve all 5 problems the professor has selected?

8.7.8. A group of art experts are testing their ability of spotting a forgery. There are 20 paintings in which there are 19 legitimate paintings and one forgery. The judges of the competition allow each expert to select three paintings as a possible forgery. What is the probability that a novice—who knows absolutely nothing about art—picks the forgery if he randomly selects three paintings?

8.7.9. A professor surveys his discrete mathematics class of 29 students and finds that 15 students read the relevant section in the textbook before class, 12 students had attempted the homework, and 7 students did both.

(a) How many students did not read the relevant sections nor attempt the homework?

(b) The professor decides to randomly put students into groups of 3 to discuss the material. What is the probability that all 3 students in a group did not do the reading nor the homework?

8.7.10. A hockey team has two goalies, six defenders, eight wingers, and four centers. If the team randomly selects four players to attend a charity function, what is the chance that they are all wingers?

8.7.11. Given a deck of 52 well-shuffled cards, you are dealt three cards. What is the probability of getting three cards

(a) of the same rank?

(b) of the same suit?

(c) that are in sequential order (*e.g.*, ace-2-3, 2-3-4, ...)?

8.7.12. In five-card poker, a straight flush is a hand that contains five cards in sequence, all of the same suit, such as $8\heartsuit$, $9\heartsuit$, $10\heartsuit$, $J\heartsuit$, $Q\heartsuit$. Aces can play low, such as $A\clubsuit$, $2\clubsuit$, $3\clubsuit$, $4\clubsuit$, $5\clubsuit$, or aces can play high, such as $10\clubsuit$, $J\clubsuit$, $Q\clubsuit$, $K\clubsuit$, $A\clubsuit$. What is the probability of getting a straight flush?

8.7.13. In a game of Crazy Eights (a card game), what is the probability that a seven-card hand will contain at most one 8 card?

8.7.14. If four cards are drawn from a well-shuffled deck of playing cards, what is the probability that there will be
    (a) one of each suit?   (b) at least two of the same suit?

8.7.15. A child wants to steal some candy from a bowl, but the bowl is sitting on the kitchen counter that is high enough so that he cannot see into the bowl, but he still can reach into it. If the bowl contains 5 small chocolates and 3 suckers, and the kid steals 2 pieces of candy, what is the probability that
    (a) he grabbed 2 chocolates?   (b) he didn't grab 2 suckers?

8.7.16. Powerball is a major multistate lottery in the United States, which selects five balls at random from a set of balls numbered $\{1, 2, \ldots, 59\}$. Then the Powerball number is selected from a second set of balls numbered $\{1, 2, \ldots, 35\}$. When a person buys a ticket, each ticket has on it imprinted the five selected numbers plus the Powerball number. What is the probability of having all five winning numbers
    (a) and the Powerball number?
    (b) but not the Powerball number?

8.7.17. A Euchre deck consists of 9's, 10's, jacks, queens, kings and aces for each of the four suits. A Euchre hand consists of five cards. What is the chance of being dealt exactly two jacks and any three other cards?

8.7.18. An urn contains seven white, five black, and eight red balls. From this urn, you select three balls at random without replacement.
    (a) Are all possible outcomes equally likely? Why?
    (b) What is the probability that all selected balls are red?
    (c) What is the probability that none of the balls are red?

8.7.19 (The Secretary Problem). The Head of the Mathematics Department seeks to hire a secretary; the former secretary retired. There are N candidates for the position that has been supplied by the university's human resources department. He is unable to select outside this pool of candidates. The candidates are ordered randomly, and then the Head interviews the candidates sequentially according to that order. It is assumed that he can rank all of the candidates from best to worst without ties. The decision to offer the job to the candidate must be based only on the relative ranks of those applicants interview so far. Once the Head rejects the candidate, the candidate will seek employment elsewhere, and will no longer be under consideration. Suppose after the Head has already interviewed $t - 1$ candidates, he ranks the current candidate he is interviewing as the best so far. What is the probability that the best candidate is in the first t candidates he has interviewed?

8.8 Conditional Probability

We begin this section with the following motivating example.

Example 8.8.1
 (a) What is the probability of rolling a 6 with a fair die?
 (b) What is the probability of rolling a 6 with a fair die if you know that the outcome is greater than 3?

Solution — 1. Let E be the event of rolling a 6; $E = \{6\}$.
 (a) The sample space is $S = \{1, 2, \ldots, 6\}$. Therefore, the required probability is

$$P(E) = \frac{n(E)}{n(S)} = \frac{1}{6}.$$

 (b) Now that we know that the outcome is greater than 3, the sample space is $S = \{4, 5, 6\}$. Therefore, the required probability is

$$P(E) = \frac{n(E)}{n(S)} = \frac{1}{3}.$$

◀

S

S

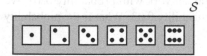

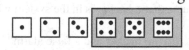

(a) The sample space S of rolling a die with no additional information.

(b) The sample space S of rolling a die given that the outcome is greater than 3.

Fig. 8.19 A visualization of the sample space from Example 8.8.1 without and with the additional information. With the additional information, it restricts the possible outcomes in the sample space.

We can see from Example 8.8.1 that the probability of rolling a 6 is different when we have the new given information that the outcome is greater than 3. When we have new given information like this affecting the probability, we are calculating what is called the **conditional probability**.

> **Definition 8.8.2 — Conditional Probability.** For two events E and F, $P(F \mid E)$, which we read as "the probability of F given E" is the probability that the event F will occur given that the event E has occurred. Then
>
> $$P(F \mid E) = \frac{P(E \cap F)}{P(E)} = \frac{P(F \cap F)}{P(E)}, \quad \text{if } P(E) \neq 0. \tag{8.10}$$

Let us repeat our analysis of Example 8.8.1 using this new definition of conditional probability.

Solution — 2. Let E be the event of rolling a 6; $E = \{6\}$. Let F be the event that the outcome is greater than 3; $F = \{4, 5, 6\}$. The intersection of E and F is $E \cap F = \{6\}$. Therefore, the required probability is

$$P(F \mid E) = \frac{P(E \cap F)}{P(E)} = \frac{\frac{n(E \cap F)}{n(S)}}{\frac{n(E)}{n(S)}} = \frac{\frac{1}{6}}{\frac{3}{6}} = \frac{1}{3}.$$

◄

The equation for conditional probability

$$P(F \mid E) = \frac{P(E \cap F)}{P(E)}$$

can be rearranged to give

$$P(E \cap F) = P(E) \times P(F \mid E). \tag{8.11}$$

This is helpful when we know the conditional probability, but we need to determine the compound probability $P(E \cap F)$. (We will also use this again in Sec. 8.11 on Bayes' Theorem.)

Example 8.8.3 — Computer Network Security. If a (black hat) hacker has gained access to a computer network, the network's software alerts the administrator with probability 0.95. The software sometimes generates a false alarm; that is, there is an alert when a hacker is not present in the network with probability 0.10. We assume that a hacker is in the system with probability 0.03. What is the probability of

(a) no hacker and a false alarm?

(b) the software not detecting a hacker even though there is a hacker in the network?

Solution Let E be the event that there is a hacker in the network, and F be the event that the software alerts the administrator. Then their complements are the following: E' denotes a hacker is not in the network, and F' denotes the software does not alert the administrator.

Draw a diagram. A diagram will help us organize the data we are given, and it helps us to spot which data are missing. We can create a tree diagram (Fig. 8.20) with the different possible outcomes, but we can modify the diagram to include the probabilities on the branches; we call this a **probability tree**. Note that the probabilities on the second set of branches of the probability tree are the conditional probabilities based on the previous outcome.

From the given information, we know that a hacker is in the network with probability 0.03; that is,

$$P(E) = 0.03.$$

Then, by the Complement Rule,

$$P(E') = 1 - 0.03 = 0.97.$$

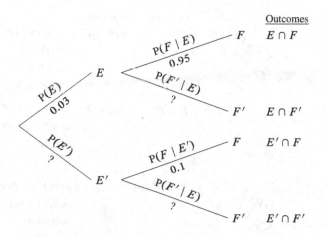

Fig. 8.20   Probability tree diagram for Example 8.8.3 with the given data

We are given two further pieces of information. We are also told the conditional probability of the software sounding an alert to the administrator given there is a hacker is 0.95; that is,

$$P(F \mid E) = 0.95.$$

Then, the complement is

$$P(F' \mid E) = 1 - 0.95 = 0.05.$$

Furthermore, the conditional probability of an alert given that there is no hacker is 0.10; that is,

$$P(F' \mid E) = 0.10.$$

We can add this information to our probability tree diagram; see Fig. 8.21.

(a) The required probability is

$$P(\text{no hacker and a false alarm}) = P(E' \cap F)$$
$$= P(E') P(F \mid E')$$
$$= (0.97)(0.10)$$
$$= 0.097.$$

Therefore, the probability that there is no hacker and a false alarm is 9.7%.

(b) The required probability is

$$P(\text{hacker and no alarm}) = P(E \cap F')$$
$$= P(E) P(F' \mid E)$$
$$= (0.03)(0.05)$$
$$= 0.0015.$$

Therefore, the probability that there is a hacker but no alarm is 0.15%.

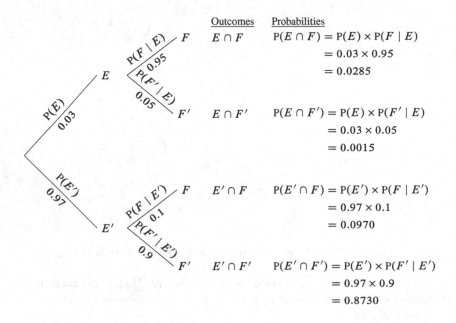

Fig. 8.21   The completed probability tree diagram for Example 8.8.3

◀

Example 8.8.4 — Royal Flush. A royal flush in poker is when the cards are the ten, jack, queen, king, and ace of one suit. Given that a poker hand is a flush, what is the probability that it is also a royal flush?

Solution Let F be the event that a hand is a royal flush, and let the event E be that the hand is a flush. Since a royal flush is a flush, this means that $F \subset E$. Then, $E \cap F = F$, which implies that $P(E \cap F) = P(F)$; see Fig. 8.22.

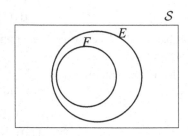

Fig. 8.22 $F \subset E$ and $E \cap F = F$

From Example 8.7.3 (on p. 337), you know that the number of flushes is 5148; that is, $n(F) = 5148$. The number of royal flushes is 4: you select one of the four suits, and the ranks are specified; that is, $n(E) = 4$. The total number of possible five card poker hands is $n(S) = \binom{52}{5} = 2\,598\,960$.

Therefore, the probability that the poker hand is a royal flush, given that you know it is a flush, is

$$P(F \mid E) = \frac{P(E \cap F)}{P(E)} = \frac{P(F)}{P(E)} = \frac{\frac{n(F)}{n(S)}}{\frac{n(E)}{n(S)}} = \frac{\frac{4}{2\,598\,960}}{\frac{5148}{2\,598\,960}} = \frac{4}{5148} = \frac{1}{1287}.$$

◀

Example 8.8.5 What is the probability of drawing, without replacement, 2 kings in a row from a well-shuffled deck of 52 cards?

Solution Let E be the event that a king is drawn on the first card, and F is a king on the second card. For the first card, there are 4 kings in the 52 cards, so

$$P(E) = \frac{4}{52}.$$

Then, we draw the next card without replacing the first card. So, there are 3 kings left in the 51 remaining cards; that is,

$$P(F \mid E) = \frac{3}{51}.$$

We require the compound probability of a king drawn on the first *and* second card; that is, we want $P(E \cap F)$. Therefore, the required probability, using Eq. (8.11), is

$$P(E \cap F) = P(E) \times P(F \mid E)$$

$$= \frac{4}{52} \times \frac{3}{51}$$

$$= \frac{12}{2652}$$

$$= \frac{1}{221}.$$

◀

We leave the probability diagram tree to you as Exercise 8.8.2.

Let us revisit Montmort's matching problem (Example 8.3.4 on p. 316). In this example, we will use the extended version of

$$P(E \cap F) = P(E) \times P(F \mid E);$$

that is,

$$P(E_1 \cap E_2 \cap \cdots \cap E_n) = P(E_1) \times P(E_2 \mid E_1) \times P(E_3 \mid E_1 \cap E_2) \times \cdots$$
$$\times P(E_n \mid E_1 \cap E_2 \cap \cdots \cap E_n) \qquad (8.12)$$

$$= P(E_1) \times \prod_{i=2}^{n} P(E_i \mid E_1 \cap \cdots \cap E_{i-1}) \qquad (8.13)$$

Example 8.8.6 — Coat Check Problem Revisited. At a restaurant, Montmort and his three friends check their coats, but the incompetent attendant did not affix a ticket number to track the coats. So when Montmort and his friends come back to collect their coats, their coat-check stubs are useless. If the attendant hands back the coats randomly to the customers, then what is the probability that

(a) at least one person will get his or her coat back?

(b) no person gets his or her coat back?

Solution *Understand the problem.* Let us introduce some suitable notation. Let E_1 be the event that person A received the correct coat, ..., and E_4 be the event that person D received the correct coat. The missing data is the probability that at least one person receives the correct coat—this is what we are required to find. Then, what we need to find is $P(E_1 \cup E_2 \cup E_3 \cup E_4)$, which is the probability that at least one person receives the correct coat, and we also need to find $1 - P(E_1 \cup E_2 \cup E_3 \cup E_4)$, which is the probability that no one receives the correct coat.

Devise a plan. We have seen before how to find the probabilities of the union of events before, namely by using the Principle of Inclusion and Exclusion (Sec. 8.5); that is, in the case for four events $P(E_1 \cup E_2 \cup E_3 \cup E_4)$ is

	the sum of the probabilities of the individual events,
minus	the probabilities of the two-event intersections,
plus	the probabilities of the three-event intersections,
minus	the probability of the four-event intersection.

or in mathematical notation,

$$
\begin{aligned}
P(E_1 \cup E_2 \cup E_3 \cup E_4) = {} & P(E_1) + P(E_2) + P(E_3) + P(E_4) \\
& - P(E_1 \cap E_2) - P(E_1 \cap E_3) - P(E_1 \cap E_4) \\
& - P(E_2 \cap E_4) - P(E_2 \cap E_3) - P(E_3 \cap E_4) \\
& + P(E_1 \cap E_2 \cap E_3) + P(E_1 \cap E_2 \cap E_4) \\
& + P(E_2 \cap E_3 \cap E_4) + P(E_1 \cap E_3 \cap E_4) \\
& - P(E_1 \cap E_2 \cap E_3 \cap E_4)
\end{aligned}
$$

The plan is now is to find all those probabilities required to calculate $P(E_1 \cup E_2 \cup E_3 \cup E_4)$.

Carry out the plan. The probability that a person would get his or her correct coat is $\frac{1}{4}$. Thus,

$$
P(\text{person gets the correct coat}) = P(E_1) = P(E_2) = P(E_3) = P(E_4) = \frac{1}{4}.
$$

Now, we consider the probability of two people getting their correct coat, which is

$$P\begin{pmatrix} 2 \text{ people get} \\ \text{their correct} \\ \text{coats} \end{pmatrix} = P\begin{pmatrix} 1^{st} \text{ person gets} \\ \text{the } \quad \text{correct} \\ \text{coat} \end{pmatrix} \times P\begin{pmatrix} 2^{nd} \text{ person gets} \\ \text{the correct coat} \end{pmatrix} \begin{vmatrix} 1^{st} \text{ person gets} \\ \text{the correct coat} \end{vmatrix}$$

$$= \frac{1}{4} \times \frac{1}{3}$$

$$= \frac{1}{4 \times 3}$$

$$= \frac{1}{P(4,2)}$$

$$= \frac{1}{12}.$$

Thus,

$$P(E_1 \cap E_2) = P(E_1) \times P(E_2 \mid E_1) = \frac{1}{4} \times \frac{1}{3} = \frac{1}{4 \times 3} = \frac{1}{P(4,3)} = \frac{1}{12}$$

$$P(E_1 \cap E_3) = P(E_1) \times P(E_3 \mid E_1) = \frac{1}{4} \times \frac{1}{3} = \frac{1}{4 \times 3} = \frac{1}{P(4,3)} = \frac{1}{12}$$

$$\vdots$$

$$P(E_3 \cap E_4) = P(E_3) \times P(E_4 \mid E_3) = \frac{1}{4} \times \frac{1}{3} = \frac{1}{4 \times 3} = \frac{1}{P(4,3)} = \frac{1}{12}.$$

For the probability of three people getting their correct coats, it is

$$P\begin{pmatrix} 3 \text{ people get} \\ \text{their correct} \\ \text{coats} \end{pmatrix} = P\begin{pmatrix} 1^{st} \text{ person gets} \\ \text{the } \quad \text{correct} \\ \text{coat} \end{pmatrix} \times P\begin{pmatrix} 2^{nd} \text{ person gets} \\ \text{the correct coat} \end{pmatrix} \begin{vmatrix} 1^{st} \text{ person gets} \\ \text{the correct coat} \end{vmatrix}$$

$$\times P\begin{pmatrix} 3^{rd} \text{ person gets} \\ \text{the correct coat} \end{pmatrix} \begin{vmatrix} 1^{st} \text{ and } 2^{nd} \text{ get the} \\ \text{correct coat} \end{vmatrix}$$

$$= \frac{1}{4} \times \frac{1}{3} \times \frac{1}{2}$$

$$= \frac{1}{4 \times 3 \times 2}$$

$$= \frac{1}{P(4,3)}$$

$$= \frac{1}{24}.$$

Thus,

$$P(E_1 \cap E_2 \cap E_3) = P(E_1) \times P(E_2 \mid E_1) \times P(E_3 \mid E_1 \cap E_2) = \frac{1}{4} \times \frac{1}{3} \times \frac{1}{2} = \frac{1}{P(4,3)} = \frac{1}{24}$$

$$P(E_1 \cap E_2 \cap E_4) = P(E_1) \times P(E_2 \mid E_1) \times P(E_4 \mid E_1 \cap E_2) = \frac{1}{4} \times \frac{1}{3} \times \frac{1}{2} = \frac{1}{P(4,3)} = \frac{1}{24}$$

$$P(E_2 \cap E_3 \cap E_4) = P(E_2) \times P(E_3 \mid E_2) \times P(E_4 \mid E_2 \cap E_3) = \frac{1}{4} \times \frac{1}{3} \times \frac{1}{2} = \frac{1}{P(4,3)} = \frac{1}{24}$$

$$P(E_1 \cap E_3 \cap E_4) = P(E_1) \times P(E_3 \mid E_1) \times P(E_4 \mid E_1 \cap E_3) = \frac{1}{4} \times \frac{1}{3} \times \frac{1}{2} = \frac{1}{P(4,3)} = \frac{1}{24}.$$

Lastly, the probability that all four people get their correct coats is

$$P\begin{pmatrix}\text{4 people get}\\\text{their correct}\\\text{coats}\end{pmatrix} = P\begin{pmatrix}1^{\text{st}}\text{ person gets}\\\text{the} \qquad \text{correct}\\\text{coat}\end{pmatrix} \times P\begin{pmatrix}2^{\text{nd}}\text{ person gets}\mid 1^{\text{st}}\text{ person gets}\\\text{the correct coat}\mid\text{the correct coat}\end{pmatrix}$$

$$\times P\begin{pmatrix}3^{\text{rd}}\text{ person gets}\mid 1^{\text{st}}\text{ and }2^{\text{nd}}\text{ get the}\\\text{the correct coat}\mid\text{correct coat}\end{pmatrix}$$

$$\times P\begin{pmatrix}4^{\text{th}}\text{ person gets}\mid 1^{\text{st}},2^{\text{nd}},\text{ and }3^{\text{rd}}\text{ get the}\\\text{the correct coat}\mid\text{correct coat}\end{pmatrix}$$

$$= \frac{1}{4} \times \frac{1}{3} \times \frac{1}{2} \times \frac{1}{1}$$

$$= \frac{1}{P(4,4)}$$

$$= \frac{1}{24}.$$

Thus,

$$P(E_1 \cap E_2 \cap E_3 \cap E_4) = \frac{1}{4} \times \frac{1}{3} \times \frac{1}{2} \times \frac{1}{1} = \frac{1}{P(4,4)} = \frac{1}{24}.$$

(a) By the Principle of Inclusion and Exclusion, the probability that at least one person gets his or her correct coat is

$$P(E_1 \cup E_2 \cup E_3 \cup E_4) = \frac{1}{4} + \frac{1}{4} + \frac{1}{4} + \frac{1}{4}$$

$$-\frac{1}{12} - \frac{1}{12} - \frac{1}{12} - \frac{1}{12} - \frac{1}{12} - \frac{1}{12}$$

$$+\frac{1}{24} + \frac{1}{24} + \frac{1}{24} + \frac{1}{24}$$

$$-\frac{1}{24}$$

$$= 4\left(\frac{1}{4}\right) - 6\left(\frac{1}{12}\right) + 4\left(\frac{1}{24}\right) - \frac{1}{24}$$

$$= 1 - \frac{1}{2} + \frac{1}{6} - \frac{1}{24}$$

$$= \frac{5}{8}.$$

(b) The probability that no one get his or her correct coat is

$$1 - P(E_1 \cup E_2 \cup E_3 \cup E_4) = 1 - \frac{5}{8} = \frac{3}{8}.$$

Looking back. Notice that we could have rewritten it as

$$P(E_1 \cup E_2 \cup E_3 \cup E_4) = 4\left(\frac{1}{P(4,1)}\right) - 6\left(\frac{1}{P(4,2)}\right) + 4\left(\frac{1}{P(4,3)}\right)$$

$$- \frac{1}{P(4,4)}$$

$$= \binom{4}{1}\frac{1}{P(4,1)} - \binom{4}{2}\frac{1}{P(4,2)} + \binom{4}{3}\frac{1}{P(4,3)}$$

$$+ \binom{4}{4}\frac{1}{P(4,4)}.$$

Why can we use binomial coefficients? Consider, for example, the probabilities of the two-people getting their correct coats. Since the probability is $\frac{1}{P(4,2)} = \frac{1}{12}$ for each pair of persons getting their correct coats, we just need to count the number of terms of the two-event intersections and multiply it by $\frac{1}{12}$: the number of ways to choose 2 people from 4 people is $\binom{4}{2}$. Similarly, this can be done for each triple, and all four people getting their correct coats.

By the definition of $\binom{n}{r}$,

$$\binom{n}{r} = \frac{n!}{r!(n-r!)}$$

$$\binom{n}{r} = \frac{P(n,r)}{r!}$$

$$r! = \frac{P(n,r)}{\binom{n}{r}}$$

$$\frac{1}{r!} = \frac{\binom{n}{r}}{P(n,r)} = \binom{n}{r}\frac{1}{P(n,r)}.$$

By substituting $\binom{n}{r}\frac{1}{P(n,r)}$ with $\frac{1}{r!}$ for each term, we have

$$P(E_1 \cup E_2 \cup E_3 \cup E_4) = 1 - \frac{1}{2!} + \frac{1}{3!} - \frac{1}{4!} = \frac{5}{8}.$$

We can extend this for n people:

$$P(E_1 \cup E_2 \cup \cdots \cup E_n) = \binom{n}{1}\frac{1}{P(n,1)} - \binom{n}{2}\frac{1}{P(n,2)} + \cdots$$

$$+ (-1)^n \binom{n}{n}\frac{1}{P(n,n)}$$

$$= \frac{1}{1!} - \frac{1}{2!} + \frac{1}{3!} - \frac{1}{4!} + \cdots + (-1)^n\frac{1}{n!}. \qquad (8.14)$$

◀

Exercises

8.8.1. Determine the following:

(a) Given: $P(E \cap F) = 0.5$ (c) Given: $P(F \mid E) = 0.4$
           $P(E) = 0.8$ $P(E) = 0.2$
     Find: $P(F \mid E)$ Find: $P(E \cap F)$

(b) Given: $P(E \mid F) = 0.3$ (d) Given: $P(E \cap F) = 0.2$
           $P(F \mid E) = 0.4$ $P(F \cap G) = 0.3$
           $P(E \cap F) = 0.2$ $P(F) = 0.6$
     Find: $P(E)$ Find: $P(E \mid F)$
           $P(F)$ $P(G \mid F)$

8.8.2. Draw the probability tree for Example 8.8.5. What is the probability of
  (a) a king is drawn on the first card, but not on the second card?
  (b) a king is not drawn on either the first or second card?

8.8.3. A fair die is rolled. Consider the events $A = \{2, 4, 6\}$, $B = \{3, 4\}$, and
$C = \{1, 2, 3, 4, 5\}$. Find
  (a) $P(A \cap B)$ (d) $P(A \mid B)$ (g) $P(B \mid C)$
  (b) $P(A \cap C)$ (e) $P(B \mid A)$ (h) $P(C \mid B)$
  (c) $P(A \mid C)$ (f) $P(C \mid A)$

8.8.4. Suppose you throw a pair of fair dice. What is the probability that doubles
are thrown given that the sum of the dice is even?

8.8.5. Suppose you draw three cards without replacement from a well-shuffled
deck of cards. What is the probability that all three cards is not a king?

8.8.6. Two cards are drawn one after the other and without replacement from a
well-shuffled deck of 52 playing cards. What is the probability that
  (a) the first card is an ace?
  (b) the second card is black if the first was red?
  (c) the second card is 10 (of any suit) knowing that the first card is an ace?
  (d) the first card is a heart given that the second card is a diamond?

8.8.7. The data in Table 8.6 are from the Life Table for males in Canada from 2009 to 2011. The Life Tables are prepared by Statistics Canada from census records and observed deaths. Starting with a 100 000 babies, as they age the number of survivors decreases. What is the probability that a 20-year-old Canadian male will survive

    (a) to be 70 years old?   (b) to be 100 years old?

Table 8.6 Life Table for Canadian Males from 2009 to 2011

Age	Number Surviving
0	100 000
20	99 047
40	97 399
70	80 301
100	2 049

Table 8.7 Life Table for Canadian Females from 2009 to 2011

Age	Number Surviving
0	100 000
20	99 282
40	98 434
70	87 202
100	4 708

8.8.8. The data in Table 8.7 are from the Life Table for females in Canada from 2009 to 2011. What is the probability that a 40-year-old Canadian female will survive

    (a) to be 70 years old?   (b) to be 100 years old?

8.8.9. In 2011, 4.3% of the population in Canada had an Aboriginal identity (Statistics Canada, 2013a). Also, 18.2% of the total Aboriginal population were youths aged 15 to 24, 6% of the total Aboriginal population were seniors aged 65 and over. Twenty-eight percent of the total aboriginal population are children aged 14 and under, of which 49.6% were living in a family with both their parents. What is the (empirical) probability that a randomly selected person in Canada at that time was

    (a) an Aboriginal youth?
    (b) an Aboriginal senior?
    (c) not an Aboriginal senior or Aboriginal youth?
    (d) an Aboriginal child not living with both their parents?

8.8.10. A family has two children. What is the conditional probability that both are boys given that at least one of them is a boy? Assume that the sample space S is given by $S = \{(b,b), (b,g), (g,b), (g,g)\}$, and all outcomes are equally likely. Note that (b,g), for example, means that the older child is a boy and that the younger child is a girl.

8.8.11. An administration assistant prepares six letters and envelopes to send to six different people, but then stuffs randomly (or absentmindedly) the letters into envelopes. Realizing the mistake after sending the letters, what is the probability that the administration assistant had put all the letters into the correct corresponding envelopes?

8.8.12. Suppose an urn contains six red and nine black balls. We draw two balls from the urn without replacement. Assuming that each ball in the urn is equally likely to be drawn, what is the probability that both drawn balls are black?

Fig. 8.23   xkcd.com/1374/

8.8.13. Suppose that your employer decides to have a Secret Santa in the office. Secret Santa is a Western tradition in which everyone in the office is randomly assigned a person to give a gift anonymously. Everyone in the office puts his or her name on a card, and then puts it into a hat. Then, each person takes turns selecting names. There are 10 employees in the office, including you. What is the probability that

    (a) no one selects his or her own name?

    (b) at least one person selects his or her own name?

    (c) you select your own name?

8.8.14. A maze is a tool used in psychological laboratory experiments to measure spatial learning and memory. These types of experiments can help identify cognitive defects that are responsible for diseases such as Alzheimer's disease. A reward, such as cheese, is placed at the end of the maze as seen in Fig. 8.24, and a mouse is placed at the beginning. The mouse's goal is to complete the maze to receive the reward. The paths are narrow enough such that the mouse is not able to backtrack if it takes a wrong turn.

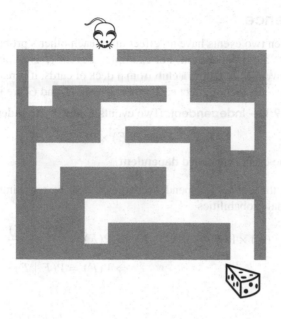

Fig. 8.24   The mouse maze

(a) If the mouse selects randomly with equal probability which direction to turn at a junction, then what is the probability the mouse will complete the maze on the first try?

(b) Suppose the mouse uses its sense of smell. We estimate the rat's probability of selecting correctly which direction to be 0.5 on the first junction, and increasing by turn by 0.1 on the subsequent turns to a maximum of 0.9.

  (i) What is the probability of the mouse successfully completes the maze on this first trial?

  (ii) What is the probability the mouse successfully completes the maze on the second trial?

  (iii) What is the probability the mouse fails the first trial, but completes it successfully on the second trial?

8.8.15 (Bertrand's Box). Three boxes are identical in appearance. Each has two drawers, each drawer contains a medal. The medals in the first box are gold; those in the second box are silver; the third box contains a gold medal and a silver medal. A box is randomly selected.

  (a) What is the probability to find in its drawers, a gold and a silver medal?

  (b) A drawer is randomly selected, and the medal in the drawer is gold. What is the probability that the third box was selected?

8.9 Independence

Informally, when two events have no effect upon each other's probability, we say that these events are **independent**. For example, if event H is a coin flip lands on heads, and the event C is getting a club from a deck of cards, it is reasonable to say that these two events do not affect each other; events H and C are independent.

> **Definition 8.9.1 — Independent.** Two events E and F are **independent**, if
>
> $$P(E \cap F) = P(E) \times P(F); \qquad (8.15)$$
>
> otherwise, the events are called **dependent**.

To see why this defines independence is made clear by rewriting the definition using conditional probabilities

$$P(E) \times P(F) = P(E \cap F) \Leftrightarrow P(F) = \frac{P(E \cap F)}{P(E)}$$

$$\Leftrightarrow P(F) = P(F \mid E)$$

and similarly,

$$P(E) \times P(F) = P(E \cap F) \Leftrightarrow P(E) = \frac{P(E \cap F)}{P(F)}$$

$$\Leftrightarrow P(E) = P(E \mid F).$$

Hence, if events E and F are independent, we see that the occurrence of F does not affect the probability of E, and vice versa:

$$P(E) = P(E \mid F) \text{ and } P(F) = P(F \mid E). \qquad \bullet$$

Although the derived expressions may seem more intuitive, they are not the preferred definition, as the conditional probabilities may be undefined if $P(E)$ or $P(F)$ are zero. Furthermore, the preferred definition makes clear by symmetry that when E is independent of F, F is also independent of E.

Example 8.9.2 What is the probability of getting heads in a coin flip *and* a club card from a deck of cards?

Solution — 1. We define the following events events: H denotes the event getting heads in a coin flip, and C denotes the event of getting a club card from a deck of cards.

We are required to find $P(H \cap C)$. These two events are independent since the two events have no effect upon each other; thus, we can use Eq. (8.15) to find $P(E \cap F)$. The probability of getting heads in a coin flip is

$$P(H) = \frac{1}{2}.$$

The probability of selecting a club from a deck of cards is

$$P(C) = \frac{13}{52} = \frac{1}{4}.$$

By Eq. (8.15),

$$P(H \cap C) = P(H) \times P(C) = \frac{1}{2} \times \frac{1}{4} = \frac{1}{8}.$$

◄

Here is a second solution with the use of an eikosogram.

Solution — 2. We define the events like we do in the previous solution: H is the event getting heads in a coin flip, and C is the event getting a club card from a deck of cards. The probability of getting heads in a coin flip is

$$P(H) = \frac{1}{2}.$$

The probability of getting a club from a deck of cards is

$$P(C) = \frac{13}{52} = \frac{1}{4}.$$

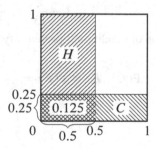

Fig. 8.25   Eikosogram of two independent events H and C

We know that these two events are independent, so we can draw the eikosogram in the following way (Fig. 8.25):

- We place a rectangle along the vertical axis, with a width of 0.5 (and a height of 1) to represent the probability of H; $P(H) = 0.5$ and so the area of $0.5 \times 1 = 0.5$ on the diagram represents the probability of the event H.
- We place a rectangle along the horizontal axis, overlaying the previous rectangle, with a height of 0.25 (and width 1) to represent the probability of C; $P(C) = 0.25$ and so the area of $0.25 \times 1 = 0.25$ on the diagram represents the probability of the event C.

Therefore, the probability of getting heads in a coin flip *and* a club card from a deck of cards is the area of the overlapping events in the eikosogram

$$P(H) \times P(C) = 0.5 \times 0.25 = 0.125 = \frac{1}{8}.$$

◀

Example 8.9.3 Suppose we draw a single card from a deck of 52 cards.
    (a) What is the probability that a heart or spade is drawn?
    (b) What is the probability that the drawn card is not a 2?
    (c) Given that a red card has been drawn, what is the probability that it is a diamond? Are the events E = red card is drawn, and F = diamond is drawn, independent events?

Solution
    (a) Define the events. Let E be heart is drawn, and F be spade is drawn. The events E and F are mutually exclusive events with $P(E) = P(F) = \frac{1}{4}$. We seek $P(E \cup F)$. By the Addition Rule for mutually exclusive events,

$$P(E \cup F) = P(E) + P(F) = \frac{1}{4} + \frac{1}{4} = \frac{1}{2}.$$

    (b) Define the event E be a 2 is shown. Then $P(E) = \frac{4}{52} = \frac{1}{13}$. We are required to find $P(E')$. By the Complement Rule (Theorem 8.5.1),

$$P(E') = 1 - \frac{1}{13} = \frac{12}{13}.$$

    (c) From the definition of conditional probability, $P(F \mid E) = \frac{P(E \cap F)}{P(E)}$, we have

$$P(E \cap F) = P(F) = \frac{13}{52} = \frac{1}{4}$$

and

$$P(E) = \frac{26}{52} = \frac{1}{2}.$$

Thus,

$$P(F \mid E) = \frac{1/4}{1/2} = \frac{1}{2}.$$

Since $P(F) = \frac{1}{4}$, we see that $P(F \mid E) \neq P(F)$. Thus, E and F are not independent events. (This is because knowing that a red card was drawn increases the probability that a diamond was drawn.)

◀

Example 8.9.4 — Guessing a Password. A password for a computer network is made up of a string of 6 lowercase letters. Let the event A be selecting the letter "a" correctly, and the event B be selecting the letter "b" correctly. Assuming that each letter in the password is independent, what is the probability of guessing the correct sequence *abbaab*?

Solution We first set up suitable notation to solve this problem. If we let $A = \{a\}$, and $B = \{b\}$, then the problem is asking us to find $P(A \cap B \cap B \cap A \cap A \cap B)$. Since the events are independent, we have

$$P(A \cap B \cap B \cap A \cap A \cap B) = P(A) \times P(B) \times P(B) \times P(A)$$
$$\times P(A) \times P(B)$$

$$= \frac{1}{26} \times \frac{1}{26} \times \frac{1}{26} \times \frac{1}{26} \times \frac{1}{26} \times \frac{1}{26}$$

$$= \left(\frac{1}{26}\right)^6$$

$$= \frac{1}{26^6}$$

$$= \frac{1}{308\,915\,776}.$$

Looking back, we could have solved this a different way using counting techniques. The sample space S is the number of ways that we could form a string of 6 lowercase letters; hence, $n(S) = 26^6 = 308\,915\,776$ by Theorem 5.6.1. Let E be the event that the correct password *abbaab* has been guessed; there is only one way to do this, and so $n(E) = 1$. Therefore, the probability of correctly guessing the password is

$$P(E) = \frac{n(E)}{n(S)} = \frac{1}{308\,915\,776}.$$

We have arrived at the same answer by two different methods, and so we can now feel confident in our answer. ◄

Example 8.9.5 — Sex Distribution. Consider a family with three children. What is the probability that the family has children of male and female sexes and there is at most one girl? Are these events independent? (Feller, 1968, p. 126)

Solution We assume that each of the eight possibilities $\{bbb, bbg, bgb, bgg, gbb, gbg, gbb, ggg\}$ has equal probability of occurring, namely $\frac{1}{8}$. Let H be the event "the family has children of male and female sexes," and A the event "there is at most one girl." Then $P(H) = \frac{6}{8}$ and $P(A) = \frac{4}{8}$. For both events A and H to occur means one of the possibilities bbg, bgb, gbb, and therefore $P(A \cap H) = \frac{3}{8} = P(A) \cdot P(H)$. Thus in families with three children the two events are independent. ◄

Example 8.9.6 — The Birthday Problem. Suppose you are in a room with n people. What is the probability that at least two people will have the same birthday? What is the smallest number of people, n, such that the probability is greater than or equal to $\frac{1}{2}$ that at least two people have the same birthday.

Solution We make the following assumptions:

- There are 365 possible birthdays in a year. (We ignore the people who are born on a leap year.)
- A person is equally likely to be born on each day. (We ignore people[9] like twins who are born on the same day.)
- A person's birthday is independent of every other person's birthday.

With all these stated assumptions, the probability of a person born on a particular day is $\frac{1}{365}$.

Let the event E be defined as no two people having the same birthday. We will use *indirect reasoning*; it is easier to analyze the complement of the event. So,

$$P(\underbrace{2 \text{ or more have same birthday}}_{E'}) = 1 - P(E)$$

since E and E' are the only two possibilities, they are also mutually exclusive. By our assumption at the beginning that each birthday is independent, we can multiply the probabilities together.

For two people, the probability that they do not share a birthday is

$$\underbrace{\frac{365}{365}}_{\text{Person 1}} \times \underbrace{\frac{364}{365}}_{\text{Person 2}} = \frac{365 \times 364}{365^2} = \frac{P(365, 2)}{365^2} = 0.997\,26$$

because if one person is born on a particular day, then the second person can only be born on the remaining 364 days in order for the birthdays to be different.

For three people, the probability that they do not share a birthday is

$$\underbrace{\frac{365}{365}}_{\text{Person 1}} \times \underbrace{\frac{364}{365}}_{\text{Person 2}} \times \underbrace{\frac{363}{365}}_{\text{Person 3}} = \frac{365 \times 364 \times 363}{365^3}$$

$$= \frac{P(365, 3)}{365^3}$$

$$= 0.991\,80$$

Continuing in this manner, for n people (where $n \le 365$)

$$\underbrace{\frac{365}{365}}_{\text{Person 1}} \times \underbrace{\frac{364}{365}}_{\text{Person 2}} \times \cdots \times \underbrace{\frac{365 - n + 1}{365}}_{\text{Person } n}$$

$$= \frac{365 \times 364 \times \cdots \times (365 - n + 1)}{365^n}$$

$$= \frac{P(365, n)}{365^n}$$

---

[9] You may be thinking by now that we are ignoring a lot of people. You might actually be one of those people we are ignoring. If so, I apologize. But, by excluding these cases from the analysis, it helps simplify the problem so we can solve it. Other readers might not have thought about these people and just assumed that everyone was equally likely with probability $\frac{1}{365}$. Since mathematics is science, we need to state such assumptions since these are considered limitations to our solution; we need to know when our solution will work and when it will fail in the real world.

Therefore, the probability of at least two people out of n people having the same birthday is

$$1 - \frac{P(365, n)}{365^n}. \tag{8.16}$$

By looking at Fig. 8.26 or Table 8.8 on the following page, we see that the smallest number of people such that the probability of at least two people having the same birthday is greater than or equal to $\frac{1}{2}$ is $n = 23$. We should remark that by the Pigeonhole Principle, the chance that two people will share a birthday is 100% when there are $n = 366$ people, but if we look at the table, 99.9% probability is reached with just 70 people! This is very counter-intuitive for many people, and hence this problem is often called the **Birthday Paradox** (even though this is not a logical paradox).

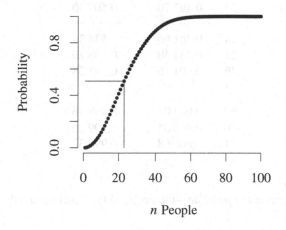

Fig. 8.26 The probability of at least two people out of n people having the same birthday

◀

8.9.1 Independence versus Mutually Exclusive Events

The two concepts of independent events and mutually exclusive events are not related at all. Consider the following situation. Suppose that all men were either "bald" or "not bald." These two types of men are mutually exclusive; you cannot be both bald and not bald. The events "bald" and "not bald" are parallel to each other in an eikosogram (Fig. 8.27(a)).

Then, let us say that picking a bald man from a population has a probability of 0.6, and the probability of picking a non-bald man with probability of 0.4. If these two events were independent (the events on an eikosogram are perpendicular), then the probability of picking a bald and non-bald man would be $0.6 \times 0.4 = 0.24$ (Fig. 8.27(b)).

Table 8.8 Twenty-three people is the smallest number of people where the probability
is greater than 50% that at least two people
share the same birthday.

n	$\dfrac{P(365,n)}{365^n}$	$1 - \dfrac{P(365,n)}{365^n}$
⋮	⋮	⋮
20	0.588 56	0.411 44
21	0.556 31	0.443 69
22	0.524 30	0.475 70
23	0.492 70	0.507 30
24	0.461 66	0.538 34
25	0.431 30	0.568 70
26	0.401 76	0.598 24
⋮	⋮	⋮
69	0.001 04	0.998 96
70	0.000 84	0.999 16
71	0.000 68	0.999 32

But in fact this is not possible—the probability of picking a bald and non-bald man is zero.

In summary, on an eikosogram

   (a) if the two events are **mutually exclusive**, then the events are parallel, with no overlapping areas.

   (b) if the two events are **independent**, then the events are perpendicular.

8.9.2 The Sally Clark Case

> It is notorious that the theory of probability has often been applied falla-
> ciously. The most common mistake is to neglect the interdependence of
> two or more probabilities and combine them by formulas which apply
> only to independent probabilities.

> —Eddington (1935, p. 123)

There was a famous case in Britain that happened at the turn of the century that involved the conviction of a mother who was accused of killing her two children. This has been aptly named the Sally Clark case, and since then, it has

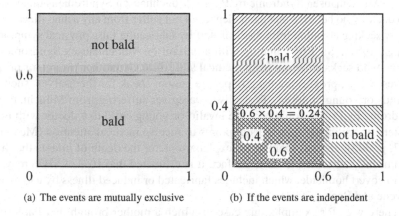

(a) The events are mutually exclusive   (b) If the events are independent

Fig. 8.27   Eikosograms of bald and non-bald men

been discussed in many books and scholarly articles concerning the misuse of probability by forensic experts in the courts.[10]

Sally Clark was a young lawyer working in London in 1996 when she gave birth to a son, Christopher. Sally became worried about her son when he developed the sniffles, and was reassured by doctors that it was a bad cold and not to worry. One night Christopher stopped breathing, and died suddenly. An infection of his lungs was discovered during autopsy, but the death was ruled as SIDS—Sudden Infant Death Syndrome (Schneps and Colmez, 2013, p. 2). A death is ruled as SIDS if the pathologist finds no specific cause of death, but believes it was natural.

In the fall of 1997, Sally had a second son, but he suffered the same fate. The family rushed to the hospital, and for the second time, the doctors were unable to save Sally's son.

An autopsy was performed on the second son gave surprising and shocking evidence. The pathologist claimed to have seen retinal hemorrhaging, and he could feel a broken rib; both indications of smothering. Harry also had large amounts of bacteria in his nose, throat, lungs, and stomach, but although the pathologist noted it, he did not find it relevant. After the pathologist's findings, an investigation of abuse began (Schneps and Colmez, 2013, p. 3).

Sally and her husband were both charged with the murder of their two children. Sally's husband was exonerated from any wrongdoing; Sally was charged with double murder.

During the trial, the eminent pediatrician Sir Roy Meadow took the stand for the prosecution. Meadow (1977) was famous for coining the term that led him to

---

[10]For example, see Watkins (2000), Aitken and Taroni (2004, pp. 211–215), Dawid (2005), Hill (2005), Buchanan (2007), Sesardic (2007), Wyatt *et al.* (2011, p. 553), and Schneps and Colmez (2013) to name a few.

fame—Münchhausen Syndrome by Proxy. Münchhausen Syndrome is diagnosed by a doctor who believes that the patient does not suffer from any actual illness, but simply seeking attention from medical staff by fabricating false physical symptoms (Asher, 1951). Sir Roy came up with a variant on Münchhausen Syndrome: a person who seeks attention from medical staff through another person, or proxy. In these cases, proxies are usually people cannot speak for themselves, such as invalids or young children. Typically, a caregiver suffering from Münchhausen Syndrome by Proxy would take the invalid or young child to a doctor with non-existent or artificially induced symptoms to receive medical attention (Meadow, 1977, p. 343). He was looking for explanations for the death of infants that had been usually attributed to SIDS. In fact, it is estimated that 10% of SIDS may be due to covert homicide, which includes fabricated or induced illness by caregivers (Levene and Bacon, 2004).

There was, for example, one case in which a mother brought her baby of 6 weeks of age to the hospital at least once a month with salt poisoning. The doctors noticed that once the child was kept in the hospital and became healthy, he would immediately relapse when his mother visited. Before social services could remove the boy from his mother, the boy was brought to the hospital, collapsed with extreme hypernatremia (elevated sodium level in the blood) and died (Meadow, 1977, p. 344).

Thus, Sally's visits to the doctors with her sick sons suggested a similar pattern of Münchhausen Syndrome by Proxy. "None of the pathologists or clinicians had described or classified the death of either child as an example of sudden infant death syndrome" (Meadow, 2002). Meadow testified that, in his opinion, "neither child's death was an example of sudden infant death syndrome" (Meadow, 2002, p. 42), and in fact, the second child had signs of physical abuse, and the first child had been smothered (Dyer, 2005b). When considering the possibility that the children had died from SIDS, Sir Roy presented the probability of a child dying from SIDS in a family with a mother who was over 27 years of age and a non-smoker, both conditions of which Sally satisfied, was $1/8543$. Continuing, he said that the probability of two children dying suddenly in that family is the square of $1/8543$, or approximately 1 in 73 million (Meadow, 2002, p. 42). The underlying assumption in his calculation, which he neglected to state in court, was that the events were assumed to be independent (Wyatt *et al.*, 2011, p. 553). "He added that the chance of two genuine deaths from the syndrome in one family arose 'once in every hundred years' and that the odds could be compared to that of four different horses winning the Grand National in consecutive years at odds of 80 to one" (Dyer, 2005a).

Independence is a very strong assumption, which should be verified empirically. There could be other causes which are unknown for SIDS such as environmental factors and/or genetic risk factors. If one child is exposed to one of these factors, it is likely that the second child will be exposed to the same factors;

thus, the probability of a second death is more likely, not less (Wyatt *et al.*, 2011, p. 553). For example, people who have a family history of heart disease are at a higher risk of having heart disease themselves compared to those without a family history.

The jury convicted Sally. The extremely low probability of 1 in 73 million was, as Lord Justice Kay stated when he quashed the conviction in appeal, "dramatic evidence that one could confidently expect to have a dramatic impact on the jury" (Dyer, 2003). This is also a fault in the logic of the jury: a low probability of an event occurring does not mean that it cannot happen. If a person wins the lottery[11] we (normally) do not accuse the person of cheating.[12]

While Sally was incarcerated, Sally's husband and a team of legal experts finally got hold of medical reports that were not made available to them during the trial, and discovered that eight different colonies of lethal bacterium *Staphylococcus aureus* had been found in Harry's body. Armed with these reports, independent medical experts agreed that Harry could have died from the serious infection—his death should not been considered unexplained or else as sudden infant syndrome (Schneps and Colmez, 2013, p. 18). At the beginning of 2003, the Criminal Cases Review Commission quashed her conviction, and Sally returned home after spending three years in prison.

In 2005, Sir Roy was struck off the medical register when the General Medical Council found him "guilty of serious professional misconduct over evidence he gave at the trial of the solicitor Sally Clark" (Dyer, 2005c), but was reinstated by the High Court since he had acted in good faith (Dyer, 2010). The pathologist, who did not disclose the *S. aureus*, was removed from the Home Office pathologists list (Dyer, 2005d), only to be reinstated upon appeal (Dyer, 2007). Experts believed that *S. aureus* was a postmortem contaminant, even though other experts for the Clark defense believed it to be the cause of death (Dyer, 2007).

Sally never recovered from this catastrophic experience; she died in her home of acute alcohol poisoning on 16 March 2007 at age 42 (Schneps and Colmez, 2013, p. 19).

Exercises

8.9.1. What is the difference between mutually exclusive events and independent events? How do you show that two events are mutually exclusive? How do you show that two events are independent?

---

[11] A person winning a lottery where they need to pick six correct numbers from the set $\{1, 2, 3, \ldots, 49\}$ has the low probability of approximately 1 in 14 million. See Example 8.7.2 on p. 336.

[12] There has been fraud in the past. For example, see how Rosenthal (2014) used probability and statistics to expose the Ontario Lottery Retailer scandal. Even more impressive is in June 2003, Mohan Srivastava, a geological statistician, spotted a pattern that could allow him to identify a winning Ontario scratch lottery ticket. He notified the Ontario Lottery and Gaming Corp., and within a couple of days they recalled the game; see Lehrer (2011).

8.9.2. Are the events E and F independent given the following
  (a) $P(E) = 0.6, P(F) = 0.3$, and $P(E \cap F) = 0.18$
  (b) $P(E) = 0.2, P(F) = 0.3$, and $P(E \cap F) = 0.05$
  (c) $P(E) = 0.7, P(F) = 0.2$, and $P(E \cup F) = 0.76$

8.9.3. Determine if the following are independent or dependent events.

	First Event	Second Event
(a)	Rolling a die and getting a 6	Rolling a die and getting a 4
(b)	Not looking where you are walking	Bumping into a wall
(c)	Spinning a Roulette wheel (see Fig. 8.29 on p. 366) and landing on black	Spinning a Roulette wheel again and landing on 26
(d)	Going to the bookshop	Buying a new book
(e)	Studying for a final exam	Passing the final exam
(f)	A coin landing on tails	A coin landing on heads

8.9.4. A fair die is rolled. Consider the events $A = \{2, 4, 6\}$, $B = \{3, 4\}$, and $C = \{1, 2, 3, 4, 5\}$. Determine if the following pairs of events are independent:
  (a) A and B (b) A and C (c) B and C

8.9.5. Let A and B be two independent events. Using eikosograms, show that the following events are also independent:
  (a) A' and B (b) A and B' (c) A' and B'

8.9.6. Let A and B be two independent events. Using an algebraic approach, show that the following events are also independent:
  (a) A' and B (b) A and B' (c) A' and B'

8.9.7. What is the probability of throwing a 6 on the first and second throw of a fair die?

8.9.8. Suppose a submarine distress flare will work with probability 0.75. If two flares are fired in quick succession, what is the probability that at least one will work?

8.9.9. Brandan can hit a target with his rifle once in 3 shots. If each shot he fires is independent, and he fires 3 shots in succession, what is the probability that he will hit his target?

8.9.10. First used in 2005 during the War in Afghanistan, the M777 howitzer is a towed 155 mm artillery piece used by the United States Marine Corps and United

States Army. Testing at the Yuma Proving Ground by the United States Army placed 13 of 14 rounds, fired from up to 24 kilometers away, within 10 meters of their target.[13] Assuming that three more rounds will be fired from this distance (of 24 kilometers), and the rounds are independent, what is the probability that at least one round will hit within 10 meters of their target?

8.9.11. Determine the probability that at least two people in a group of ten will have the same birthday.

8.9.12. Determine the probability that at least two friends in a group of five will have the same birth month.

8.9.13. Five friends go to a restaurant, which has 14 entrées on the menu. If the friends are equally likely to choose any of the entrées, what is the chance that at least two of them will order the same entrée?

8.9.14. Suppose you have 100 employees, and every quarter (3 months), 25 employees are randomly selected for drug testing. The names are replaced before every draw. The probability of an employee being selected in a quarter is 25%. What is the probability that an employee will be selected at least once in a year?

8.9.15. Suppose I wrote down a number between 1 and 10 (inclusively), and you wrote down a number between 1 and 10 (inclusively) as well. What is the chance that my number is higher than yours?

8.9.16. During an election, political parties hire research companies that will poll constituents by telephone. Research companies will phone constituents between 5 pm and 9 pm. If you are a member of this constituency being polled, there is a 20% chance that you will receive such a call from a research company. If you eat dinner from 6:30 pm to 7:30 pm, what is the probability that your dinner will be interrupted?

8.9.17. Four ants are sitting on the corners of a square (see Fig. 8.28). From where it is sitting, each ant will randomly pick an edge of the square and travel along that edge. What is the probability that none of the ants collide?

Fig. 8.28   Four ants sitting on the corners of a square.

---

[13]Source: http://en.wikipedia.org/wiki/M777_howitzer

8.9.18. The American version of Roulette revolves around a wheel divided into 38 compartments that is mounted on a heavy rotor (see Fig. 8.29). The 38 compartments are numbered 1 through 36 (with red and black colors) plus 0 and 00 (colored green). As the wheel revolves, a *tourneur* launches a ball along the track in the direction counter to that of the wheel's rotation. When the ball's angular momentum drops below that is necessary to hold it in the track, it crosses onto the rotor, and eventually settles into one of the 38 compartments.

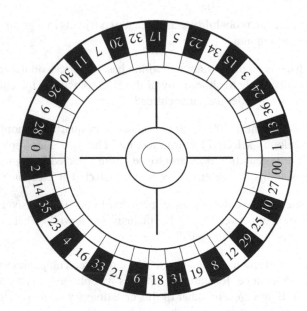

Fig. 8.29 An American roulette wheel. The green numbers, 0 and 00, are shown in gray, the red numbers are shown in white, and the black numbers are black.

(a) What is the probability of winning a bet on a single-number (*en plein*) on the Roulette wheel?

(b) What is the probability of winning if you bet on two numbers (*à cheval*)?

(c) What is the probability of having a black number appear ten times in a row?

(d) Suppose that you place a bet 12 times on the same eight numbers. What is the probability of not winning a single time?

(e) After hitting three green, five red, and two black, what is the probability that each of the next three consecutive spins comes up 26 black?

(f) A recent invention at some casinos—the 14th wager—allows players to bet that a particular color will appear in the next three consecutive spins. What is the probability of red appearing three times in a row?

(g) Do we have a uniform sample space if the sample space is all the possible number outcomes?

(h) Is it a uniform sample space if the sample space is all the possible color outcomes?

8.9.19 (Testing for the independence of three events). In this exercise, we will see that it is *not always true* that the events E, F, and G are all **pairwise independent** if and only if

$$P(E \cap F \cap G) = P(E) \times P(F) \times P(G).$$

Consider the Venn diagram in Fig. 8.30 with the probabilities labeled.

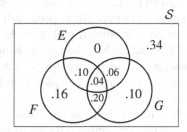

Fig. 8.30   Venn diagram for Exercise 8.9.19

Show that

(a) $P(E \cap F \cap G) = P(E) \times P(F) \times P(G)$

(b) $P(E \cap F) \neq P(E) \times P(F)$

(c) $P(E \cap G) \neq P(E) \times P(G)$

(d) $P(F \cap G) \neq P(F) \times P(G)$

8.10 The Origins of Probability: Gambling

After examining some of the basics of probability, we should ask ourselves, "Why would someone think of this?" Or, "What made mathematicians to start thinking about these kinds of questions?" In the history of mathematics, it is a recurring theme that real-world problems or applications along with curious people asking the right questions will lead to new mathematical theory.[14] Many incorrectly attribute the creation of probability theory to Pascal and Fermat in 1654, but as we will see, probability theory may have existed even earlier.

Dice existed in one form or another since ancient times. For example, in Egypt during the First Dynasty (3500 BC), dice were used in board games. When such

---

[14]The converse is true: new mathematical theory can inspire real-world applications. In fact, it is this interplay between the real-world and mathematical theory that makes mathematics a thriving subject to study.

a game was found by archæologists, they called it "Hounds and Jackals" which based on the depictions of these animals, resembles our modern game of "Snakes and Ladders" (David, 1962, p. 5). Another example is Palamedes, during the Trojan Wars, invented games of chance to stifle boredom and to boost morale of the soldiers (David, 1962, p. 6). Until playing cards were introduced around 1350, dice remained the common instrument for gambling for nearly 1000 years (Kendall, 1956, p. 1).

There were many attempts by Church and State to try and outlaw gambling. The clergy and lawmakers were not so much concerned about the games of chance itself, but with the drinking, swearing, crime, and even chess (!) that were common vices amongst gamblers (Kendall, 1956, p. 1). In England, beginning with the great military leader King Edward III (1312–1377), a series of laws prohibiting certain games was enacted to promote more masculine sports such as archery to ready the population for time of war. This continued with King Henry VIII (1491–1547) who specifically outlawed dice and cards; although, the King set a bad example for his subjects. There is, however, no evidence that suggests that any class—educated, middle, or poor—ever stopped playing dice games from Roman times until the Renaissance (Kendall, 1956, p. 2).

Since dice and gambling have existed in one form or another since ancient times, why did it take until the 13th century for us to arrive at the concept of equally-likely outcomes of a die throw? There are many possible reasons as to why this took so long:

(a) *The material of the die.* A die may have been made out of bone or other materials which does not have uniform density. If someone throws a bone die, then it could land more often on one side (David, 1962, p. 22).

(b) *Divinity.* Dice were used by priests to express divine intent. They may have manipulated the outcome by how they threw the dice. Another reason is that they may have believed that attempting to penetrate the mysteries of the divine would bring retribution (David, 1962, pp. 22–23). We must remember that at this time, people were superstitious; all speculation about future events belonged only to the omniscient and omnipotent deities.

(c) *Difficulty with numbers.* In ancient times, it was difficult to write numbers and so it hindered tallying the outcomes of throws required to find the (empirical) probability (David, 1962, p. 22).

(d) *The calculations were performed*; however these may have been lost or not divulged (David, 1962, p. 23). It may also be that someone with enough intelligence worked the elements of a theory, but kept it secret on account of its monetary value or power (Kendall, 1956, p. 3).

None of these reasons provides any definitive answer for the lack of probability theory development.

The first hint at enumerating the possibilities of dice is found in a Latin poem *De Vetula* written in France. There have been many candidates as to who wrote it, but the current contender is Richard de Fournival (1200–1259), which means that the poem is likely to have been written between 1220 and 1250. The poem enumerates the 216 ways to throw three dice (Kendall, 1956, pp. 4–5). It was circulated widely in the late Middle Ages and some authors quoted it in their work Bellhouse (2000) believes this provides evidence that elementary probability theory existed in Europe from 1250.

In the 16th century, one of the most colorful characters in the history of mathematics was born. The Italian mathematician and inveterate gambler Gerolamo Cardano (1501–1576) is infamous for his feud with another mathematician, Niccolò Tartaglia (1499–1557), over who had first solved the cubic equation; that is, in modern notation, $Ax^3 + Bx + C = 0$ (Burton, 2011, pp. 321–324). Tartaglia had disclosed his method to Cardano under the promise that Cardano would keep it a secret, but Cardano reneged and published the solution anyway in his textbook on algebra in 1545 (Hald, 1990, p. 37). As with Galileo,[15] Cardano got himself in a spot of trouble with the Roman Catholic Church for casting the horoscope of Jesus Christ. Cardano was imprisoned for a time, but was later released and received an annuity from the Pope. As a father, Cardano failed miserably: his first son was executed for poisoning his wife, and Cardano personally clipped his second son's ears for attempting to commit the same crime.

During his life, Cardano was constantly short of funds, and resorted to gambling and playing chess for money. He must have been a dubious player since his book, *Liber de ludo aleæ* published posthumously in 1663, had a section dedicated to cheating with false die, marked cards, palming cards, tilted gaming tables, and accomplices (Bellhouse, 2005, p. 181). Although his book would not be printed for approximately 100 years, Cardano likely spread these ideas through his popular lectures, and they would have been familiar to his friends and pupils (Burton, 2011, p. 445). The book has been viewed by historians as either a dizzying mathematical work, or a gambling manual. "It is so badly printed as to be scarcely intelligible" (Todhunter, 1865, p. 2). It contains many paragraphs concerned with gambling, such as the game descriptions and precautions one should take to avoid being cheated. There is only a small section dedicated to the discussion of chance, in which Cardano gave a primitive definition of probability: the probability of a particular outcome was some of the possible ways of achieving the outcome divided by the number of all the possibilities (Todhunter, 1865, p. 2). Bellhouse (2005) argues that, in fact, some of Cardano's calculations with dice are based on his reading and understanding of *De Vetula*.

When Cardano investigated the probabilities of rolling a die, he states (in modern terms) that if the die is fair, then each outcome is equally likely to occur. This is the first time we see the transition from empirical probability (probability

---

[15]See Sec. 4.2.

estimated by experiment of a die) to theoretical probability (the probability that an outcome should have if the die is fair) (Burton, 2011, p. 445).

It was approximately ten years before Cardano's book appeared in print that Blaise Pascal (1623–1662) and Pierre de Fermat (1601–1665)[16] exchanged letters in the summer of 1654. Pascal sent a letter to Fermat about two problems that his friend had posed. This exchange is the point where the formal study of probability started.

Pascal made friends with one of France's social elite, the distinguished Antoine Gombaud (1607–1684), better known as the Chevalier de Méré,[17] a title he used for his writings. At that time, it was desirable and fashionable to attend salons and discuss philosophy in witty banter amongst attendees. Gombaud was a common fixture at these gatherings and often expressed contempt for mathematicians since they "do not know how to entertain us except by numbers and figures" (Ore, 1960, p. 409). When Gombaud recounted how he met Pascal, he recalled him as the nerdy mathematician who carried scraps of paper with him which he brought out from time to time to take notes. Continuing, he said, "this man, who had neither taste nor sentiment, could not refrain from mingling into all we said, but he almost always surprised us and often made us laugh" (Ore, 1960, p. 409). It appears that through Pascal's comedic talent, Pascal and Gombaud became friends, and more than likely ended up gambling together (Ore, 1960, p. 410).

Bonding over their common interest, Gombaud must have posed gambling problems to Pascal. Many letters of correspondence are unfortunately lost. Amongst the missing is the first letter from Pascal formulating problems and his solutions to Pierre de Fermat, and also Fermat's response.

In a letter from Pascal to Fermat dated 29 July 1654, he recounts the proposition from Gombaud that if one throws a die, the chance of getting a six in four throws is 51.77%.[18] This is correct; see Exercise 8.10.2. But, what happens when one throws two dice?

> **Problem 8.10.1 — Dice Problem.** When one throws two dice, how many throws must one be allowed in order to have a better than 50% chance of getting double-sixes at least once?

We can apply indirect reasoning. First, we will calculate the probability of not obtaining double-sixes. From Table 8.1, we see that there are 35 out of the 36 outcomes which do not yield double-sixes; hence, the probability of not getting double-sixes is $\frac{35}{36}$.

---

[16] It was this exchange between these two mathematicians who are popularly, but erroneously, supposed to have founded the study of probability theory (Kendall, 1956, p. 1).

[17] In French, *chevalier* means knight, and Méré is a small village in France where he was educated.

[18] It was stated (in French, of course) as the advantage in getting it in 4 throws is as 671 is to 625 (Fermat, 1894, p. 296).

If we throw the dice multiple times, the probability of no double-sixes

in 2 throws is $\underbrace{\dfrac{35}{36}}_{\text{Throw 1}} \times \underbrace{\dfrac{35}{36}}_{\text{Throw 2}}$ $= \left(\dfrac{35}{36}\right)^2$

in 3 throws is $\underbrace{\dfrac{35}{36}}_{\text{Throw 1}} \times \underbrace{\dfrac{35}{36}}_{\text{Throw 2}} \times \underbrace{\dfrac{35}{36}}_{\text{Throw 3}}$ $= \left(\dfrac{35}{36}\right)^3$

$\vdots$

in n throws is $\underbrace{\dfrac{35}{36}}_{\text{Throw 1}} \times \underbrace{\dfrac{35}{36}}_{\text{Throw 2}} \times \underbrace{\dfrac{35}{36}}_{\text{Throw 3}} \times \cdots \times \underbrace{\dfrac{35}{36}}_{\text{Throw } n}$ $= \left(\dfrac{35}{36}\right)^n$

The probability of the complement event, getting double-sixes at least once in n throws, is

$$p_n = 1 - \left(\dfrac{35}{36}\right)^n \tag{8.17}$$

To have a better than even chance (greater than 50% probability), $p_n > \frac{1}{2}$. If we let $n = 24$, then $p_{24} = 1 - \left(\dfrac{35}{36}\right)^{24} = 0.4914$. If we let $n = 25$, then

$p_{25} = 1 - \left(\dfrac{35}{36}\right)^{25} = 0.5055$. Thus, a gambler is in an advantageous position if a gambler is allowed 25 or more throws to get at least one double-sixes.

Exercises

8.10.1. How many outcomes are there in throwing three dice?

8.10.2. Calculate the probability of getting a six in four throws or less.

8.10.3. Given $1 - \left(\dfrac{35}{36}\right)^n = \dfrac{1}{2}$, solve for n. (*Hint*: Check out Example B.6.3 on p. 492 on how to solve equations that have a variable in the exponent.)

8.11 Bayes' Theorem

Traditionally, it has been thought that Bayes' Theorem was first penned by the dissenting English minister Thomas Bayes (c. 1702–1761), and posthumously published in 1763, three years after his death. There are some questions that exist on who first discovered the theorem.[19]

---

[19] Stigler (1983) attempts to answer the question of who was the first person to discover Bayes' Theorem.

Theorem 8.11.1 — Bayes' Theorem. Let E and F be events in the sample space S. Then the conditional probability, $P(E \mid F)$, is computed by

$$P(E \mid F) = \frac{P(E)\,P(F \mid E)}{P(E)\,P(F \mid E) + P(E')\,P(F \mid E')}. \qquad (8.18)$$

Proof

$$P(E \mid F) = \frac{P(E \cap F)}{P(F)} \qquad \text{By (8.10)}$$

$$= \frac{P(E)\,P(F \mid E)}{P(F)} \qquad \text{By (8.11)}$$

$$= \frac{P(E)\,P(F \mid E)}{P(E \cap F) + P(E' \cap F)}$$

$$= \frac{P(E)\,P(F \mid E)}{P(E)\,P(F \mid E) + P(E')\,P(F \mid E')} \qquad \text{By (8.11)}$$

◀

Bayes' Theorem is often used for inference. There are many causes that may lead to an effect, and often we see only the effect. But, we may perhaps like to *infer* what caused the effect. Bayes' Theorem is a simple, yet powerful formula that allows to calculate the probability of a cause for a given effect. We will see that Bayesian reasoning can be applied to a breadth of problems including:

- if a plane is observed hitting a building, what is the probability that it was caused by terrorists?
- if a patient tests positive for HIV, what is the probability that the positive test actually was caused by the presence of HIV in the patient?

In each of these questions, we want to quantify the likelihood of a particular cause inducing an observed effect.

Example 8.11.2 — The World Trade Center attacks.

Most of us would have assigned almost no probability to terrorists crashing planes into buildings in Manhattan when we woke up that morning. But we recognized that a terrorist attack was an obvious possibility once the first plane hit the World Trade Center. And we had no doubt we were being attacked once the second tower was hit. Bayes' Theorem can replicate this result.

—Silver (2012, p. 247)

We define the events E that a plane hits a building in Manhattan, and F that terrorists are attacking.

Our initial estimate of terrorists attacking Manhattan would be low, say

$$P(F) = 0.000\,05,$$

which is about a 0.005% chance. The complement probability, the probability that terrorists are *not* attacking Manhattan, would be

$$P(F') = 1 - P(F) = 1 - 0.00005 = 0.99995.$$

We then consider assigning a conditional probability for a plane hitting a building in Manhattan given that terrorists are attacking:

$$P(E \mid F) = 1.$$

What is the probability that a plane hits a building in Manhattan and it is *not* by terrorists? Prior to the attacks in Manhattan on 11 September 2001, there had only been two accidents where planes had hit buildings: a plane struck the Empire State Building in 1945, and in the following year another hit at 40 Wall Street. Thus, the empirical probability of a plane hitting a building in Manhattan (over 25 000 days) given that it was an accident (or in other words, not a terrorist attack) is

$$P(E \mid F') = \frac{2}{25\,000} = 0.000\,08.$$

Therefore, by Bayes' Theorem, the probability of a terrorist attack given that a building has been hit by a plane is

$$\begin{aligned} P(F \mid E) &= \frac{P(F)\,P(E \mid F)}{P(F)\,P(E \mid F) + P(F')\,P(E \mid F')} \\ &= \frac{(0.000\,05)(1)}{(0.000\,05)(1) + (0.999\,95)(0.000\,08)} \\ &= 0.384\,63. \end{aligned}$$

However, we can continue to update our probability estimates as we receive new information; this is the power of Bayes' theorem. Thus, our *posterior*[20] probability of a terror attack after the first plane hit, 38%, becomes our *prior*[21] probability before the second plane hit.

So, we now let our estimate of terrorist attack be $P(F) = 0.384\,63$, we keep the rest of our probabilities for the other events. Then, the probability of a terrorist attack given that a second plane hits is

$$\begin{aligned} P(F \mid E) &= \frac{P(F)\,P(E \mid F)}{P(F)\,P(E \mid F) + P(F')\,P(E \mid F')} \\ &= \frac{(0.384\,63)(1)}{(0.384\,63)(1) + (1 - 0.384\,63)(0.000\,08)} \\ &= \frac{0.384\,63}{0.384\,63 + (0.615\,37)(0.000\,08)} \\ &= 0.999\,87. \end{aligned}$$

---

[20]The posterior probability of a random event or an uncertain proposition is the conditional probability that is assigned after the relevant evidence or background is taken into account.
[21]A prior probability of a random event is the probability that is assigned before new evidence of background is taken into account.

Thus, the probability that it was a terrorist attack becomes almost certain with 99.99% probability after the second plane crash. ◄

Example 8.11.3 — Tank Drivers. Suppose that you are a leader of an armored squadron: for operating a tank, you classify that 75 percent of your tankers are "good drivers" and the rest as "risky drivers" (or better suited for other positions in the tank operation).[22] Tankers in the "risky" category allow the tank to run off the road 40 percent of the time, whereas those in the "good" category allow the tank to run off the road only 10 percent of the time. What percentage of off-road driving is done by tankers in the "risky driver" category?

Solution The question can be rephrased as the following: Calculate the probability that a tanker is a "risky driver" given that the driver has driven off the road.

Let R be the event that a tanker is a "risky driver." The probability $P(R) = 0.25$. Let R' be the event that a tanker is not a "risky driver." The probability $P(R') = 0.75$. Let OF be the event that a tanker is off-roading. $P(OD \mid R) = 0.4$ and $P(OF \mid R') = 0.1$.

We want to find the probability $P(R \mid OF)$. By Bayes' Theorem,

$$
\begin{aligned}
P(R \mid OF) &= \frac{P(R)\,P(OF \mid R)}{P(R)\,P(OF \mid R) + P(R')\,P(OF \mid R')} \\
&= \frac{(0.25)(0.4)}{(0.25)(0.4) + (0.75)(0.1)} \\
&= 0.571\,43.
\end{aligned}
$$

Therefore, there is a 57.1% chance that the off-road driving is by tankers in the "risky driver" category. ◄

Screening is a public health strategy where a large portion of the population is tested for a disease to find undiagnosed individuals so that the individual can receive treatment and/or prevent further spread of the disease in the population. Screening has been used for diseases such as cancer (cervical, breast, skin, and colon), tuberculosis, sexually transmitted infections, and heart disease. Since this affects the entire population, it is important for all of us to understand some of the probability concepts, such as Bayes' Theorem, that apply to these tests to understand our results. Economical medical tests that deliver rapid results for screening purposes are especially sensitive, and because of this sensitivity, there are occasions when the test for a disease is positive even if the person is not infected; this is called a **false positive**.[23] When a person tests positive, a second test is administered that uses a different method. If both tests are true, then the

---

[22] I was asked, "Who cares if a tank drives off the road?" The answer is me. In particular, in the city of Kingston, Ontario, it is not uncommon to see armored vehicles driving down the city roads. But, nothing strikes fear in my heart more than seeing these armored vehicles driving around, and the sign *Student Driver* hanging off the rear hitch.

[23] This is also called a **false alarm** (Fawcett, 2006, p. 862).

person is identified as being infected with the disease. False positives can lead to unnecessary stress and anxiety for the person, but with Bayes' Theorem, it can help them understand what the probability of he or she actually having the disease.

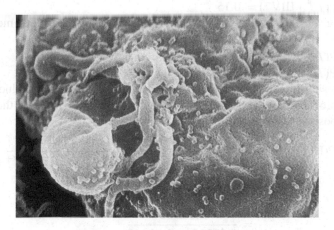

Fig. 8.31   Scanning electron micrograph of HIV-1 virions budding from a cultured lymphocyte. From the Centers for Disease Control and Prevention (CDC), Image ID: 1197, by C. Goldsmith, 1984.

Example 8.11.4 — HIV Testing. The human immunodeficiency virus (HIV), as seen in Fig. 8.31, is a lentivirus that causes acquired immunodeficiency syndrome (AIDS)—a condition that leads to the eventual failure of the immune system and allows the patient to be susceptible to other life threatening infections and cancers.

In Canada, it is estimated that 71 300 people are living with HIV, 25% of whom were unaware of their infection (due to a lack of testing and/or diagnosis).[24] In 2011, 33 476 688 people were enumerated in the census.[25]

A manufacturer claims that its at-home HIV test will detect the HIV virus (that is, test positive for a person who has HIV) 95% of the time. Further, 15% of all HIV-negative persons also test positive.

(a) If a patient has just tested positive, what is the probability that the patient does have HIV? (This is called a true positive.)

(b) If a patient has just tested positive, what is the probability that the patient does not have HIV? (This is called a false positive.)

Solution — 1. Let HIV^+ denote that the person has HIV, and HIV^- denote that the person does not have HIV. Let T^- be the event that the test is negative, and T^+ be the event that the test is positive.

---

[24]Source: http://www.phac-aspc.gc.ca/aids-sida/publication/survreport/2011/dec/index-eng.php

[25]Source: http://www12.statcan.ca/census-recensement/2011/as-sa/98-310-x/98-310-x2011001-eng.cfm

The given data are the following:
- $P(HIV^+) = \frac{71\,300}{33\,476\,688} \approx 0.002\,13$
- $P(T^+ \mid HIV^+) = 0.95$
- $P(T^+ \mid HIV^-) = 0.15$

From the given data, we can calculate the following using the Complement Rule:
- $P(HIV^-) = \frac{33\,476\,688 - 71\,300}{33\,476\,688} = \frac{33\,405\,388}{33\,476\,688} \approx 0.997\,87$
- $P(T^- \mid HIV^+) = 1 - P(T^+ \mid HIV^+) = 1 - 0.95 = 0.05$
- $P(T^- \mid HIV^-) = 1 - P(T^+ \mid HIV^-) = 1 - 0.15 = 0.85$

(a) Using Bayes' Theorem and substituting the appropriate values for the probabilities, the probability of a person who has HIV given they have a positive test is

$$
\begin{aligned}
P(HIV^+ \mid T^+) &= \frac{P(HIV^+)\,P(T^+ \mid HIV^+)}{P(HIV^+)\,P(T^+ \mid HIV^+) + P(HIV^-)\,P(T^+ \mid HIV^-)} \\
&= \frac{(0.002\,13)(0.95)}{(0.002\,13)(0.95) + (0.997\,87)(0.15)} \\
&= \frac{0.002\,02}{0.002\,02 + 0.149\,68} \\
&= 0.013\,32
\end{aligned}
$$

or 1.33%. This is also known as the true positive rate.

(b) The probability of a person who does not have HIV and have a positive test is

$$
\begin{aligned}
P(HIV^- \mid T^+) &= \frac{P(HIV^-)\,P(T^+ \mid HIV^-)}{P(HIV^-)\,P(T^+ \mid HIV^-) + P(HIV^+)\,P(T^+ \mid HIV^+)} \\
&= \frac{(0.997\,87)(0.15)}{(0.997\,87)(0.15) + (0.002\,13)(0.95)} \\
&= \frac{0.149\,68}{0.149\,68 + 0.002\,02} \\
&= 0.986\,68
\end{aligned}
$$

or 98.67%. This is also known as the false positive rate.   ◄

Solution — 2. In this solution, we will construct a modified tree diagram to help us organize the given information.

Let HIV^+ denote that the person has HIV, and HIV^- denote that the person does not have HIV. Let T^- be the event that the test is negative, and T^+ be the event that the test is positive.

In Fig. 8.32, we created a probability tree diagram. Recall that the probabilities on the second set of branches of the probability tree diagram are conditional probabilities based on the previous outcome.

Using the probability tree diagram in Fig. 8.32, and the observation that

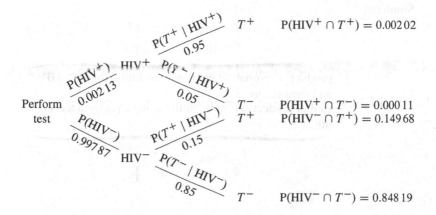

Fig. 8.32   Probability tree diagram for Example 8.11.4.

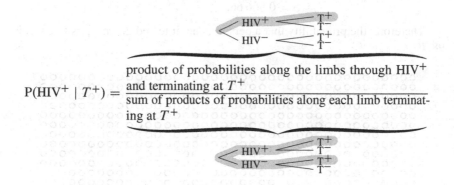

$$P(HIV^+ \mid T^+) = \frac{\text{product of probabilities along the limbs through } HIV^+ \text{ and terminating at } T^+}{\text{sum of products of probabilities along each limb terminating at } T^+}$$

Then,

$$P(HIV^+ \mid T^+) = \frac{P(HIV^+ \cap T^+)}{P(HIV^+ \cap T^+) + P(HIV^- \cap T^+)}$$

$$= \frac{0.002\,02}{0.002\,02 + 0.149\,68}$$

$$= 0.013\,32$$

Therefore, the probability that a person is infected given a positive test is 1.33%.

Similarly,

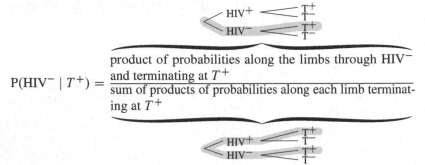

$$P(\text{HIV}^- \mid T^+) = \frac{\text{product of probabilities along the limbs through HIV}^- \text{ and terminating at } T^+}{\text{sum of products of probabilities along each limb terminating at } T^+}$$

Then,

$$
\begin{aligned}
P(\text{HIV}^- \mid T^+) &= \frac{P(\text{HIV}^- \cap T^+)}{P(\text{HIV}^+ \cap T^+) + P(\text{HIV}^- \cap T^+)} \\
&= \frac{0.149\,68}{0.000\,11 + 0.149\,68} \\
&= 0.986\,66.
\end{aligned}
$$

Therefore, the probability that a person is not infected given a positive test is 98.7%. ◀

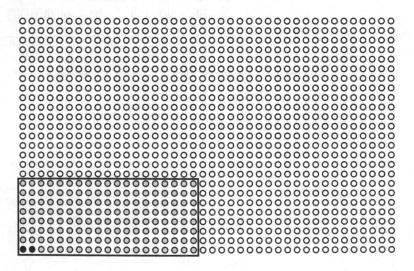

Fig. 8.33   Visualizing a sample of 1000 Canadians as dots from Example 8.11.4.

Figure 8.33 can help us visualize a sample of 1000 Canadians as dots from Example 8.11.4. The two black dots represent two people infected with HIV, and the other 150 gray dots represents people who are not infected, but tested positive. So, in total there are 152 people that will test positive. Given that a test is positive, the probability that the person has HIV is $2/152 = 0.013$ or 1.3%. To increase the probability that the person has HIV given that the test is positive, you have to reduce the number of false positives.

You might have asked yourself while reading Example 8.11.4 the following: why is there a low probability of having HIV given that the test is positive? This is known as the **false positive paradox**: when there is a low incidence (of a disease, for example) in a population, there is a higher probability of a false positive test. Thus, if you are looking for something that is incredibly rare, then the test you are using should match how rare the thing for which you are searching.

This same idea can be applied to testing for steroid or drug usage. As the number of students who uses steroids or drugs decrease, the more likely a student who is not using steroids or drugs will be identified as a user.

A **false negative**, a test yielding a negative result but the person is infected, can also occur. With HIV, the most frequent reason for a false negative is that the individual is newly infected and has not begun to produce antibodies against HIV. In this case, individuals who have a history or currently engage in behavior that puts them at risk of HIV infection are encouraged to undergo repeated testing over time.[26]

In summary, we give Table 8.9 which is the **confusion matrix** and formulas for commonly calculated probabilities related to the confusion matrix.

Table 8.9 Confusion matrix and associated formulas. Adapted from Fawcett (2006).

Test Outcome	Condition	
	Positive (P)	Negative (N)
Positive (P)	True Positive (TP)	False Positive (FP)
Negative (N)	False Negative (FN)	True Negative (TN)

$$\text{True positive probability (rate)} = \frac{TP}{TP + FN} \tag{8.19}$$

$$\text{False positive probability (rate)} = \frac{FP}{FP + TN} \tag{8.20}$$

$$\text{True negative probability (rate)} = \frac{TN}{TN + FP} \tag{8.21}$$

$$\text{False negative probability (rate)} = \frac{FN}{TP + FN} \tag{8.22}$$

$$\text{Accuracy} = \frac{TP + TN}{TP + FP + FN + TN} \tag{8.23}$$

---

[26]http://www.who.int/hiv/abouthiv/en/fact_sheet_hiv.htm

If we let the events $E_1, E_2, \ldots, E_n$ be the causes, and F be the effect, then the probability $P(E_i \mid F)$ is the chance that E_i was the cause of F. For example, a disease is the cause of symptoms:

$$\text{disease} \to \text{symptom};$$

but, the converse is not true:

$$\text{symptom} \not\to \text{disease}.$$

Since a symptom does not imply (or guarantee) that we have a certain disease, we would like to know what the probability of having a particular disease for a given symptom we are experiencing. Doctors use this kind of probabilistic inference in medical diagnostics with the help from the general form of Bayes' Theorem.

Example 8.11.5 — Medical diagnostics. If a doctor knows nothing about a patient's symptoms, then all the doctor can say is that it could be any one of many diseases that are prevalent in the population. The doctor could say the patient is more likely to have a common disease rather than a rare disease, but nothing more. In other words, the doctor would be guessing on the basis of empirical probability (or relative frequency) of a particular disease in the population. However, as the doctor begins to acquire some information about the patient such as the symptoms, then the doctor knows that some symptoms are more common in certain diseases than in others.

Suppose a patient presents to a doctor with a rash (this is event F, the effect), and the doctor would like to estimate the chance of five mutually exclusive and collectively exhaustive causes:

- cause 1 (event E_1) is the patient came into contact with an allergen,
- cause 2 (event E_2) is the patient came into contact with an irritant like poison ivy,
- cause 3 (event E_3) is the patient may have a skin disease such as dermatitis or acne,
- cause 4 (event E_4) is the patient has poor personal hygiene such as not showering, and
- cause 5 (event E_5) is due to reasons other than the aforementioned causes.

If the doctor knows the probabilities of each cause, $P(E_1), \ldots, P(E_5)$, and the conditional probabilities of a rash appearing given a particular cause $P(F \mid E_1), \ldots, P(F \mid E_5)$, then the doctor can calculate the probability that E_i caused the rash; that is,

$$P(E_i \mid F) = \frac{P(E_i) \, P(F \mid E_i)}{P(E_1) \, P(F \mid E_1) + P(E_2) \, P(F \mid E_2) + \cdots + P(E_5) \, P(F \mid E_5)}.$$

In this example, Bayes' Theorem represents the combining of medical knowledge and knowledge of disease incidence to enable the doctor to assign a numerical value to the probability of a correct diagnosis. ◄

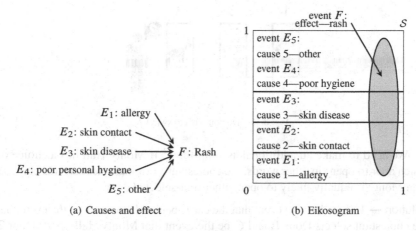

Fig. 8.34 Underlying causes of a rash

Theorem 8.11.6 — General Form of Bayes' Theorem. Let $E_1, E_2, \ldots, E_n$ be mutually exclusive events of a sample space S, and let F be an event of the experiment such that $P(F) \neq 0$. Then the posterior probability of event E_i occurring given F is

$$P(E_i \mid F) = \frac{P(E_i)\,P(F \mid E_i)}{P(E_1)\,P(F \mid E_1) + \cdots + P(E_n)\,P(F \mid E_n)}$$

$$= \frac{P(E_i)\,P(F \mid E_i)}{\sum_{j=1}^{n} P(E_j)\,P(F \mid E_j)}$$

To help in remembering the Bayes' Theorem formula, you can first construct a probability tree diagram. Then the required conditional probability $P(E_i \mid F)$ is found by

$$P(E_i \mid F) = \frac{\text{product of probabilities along the limbs through } E_i \text{ and terminating at } F}{\text{sum of products of probabilities along each limb terminating at } F}.$$

Example 8.11.7 — The Monty Hall problem. Suppose a contestant is on the game show "Let's Make a Deal" with Monty Hall. Behind one of three doors is a new sports car, and behind the remaining doors are goats. The contestant picks Door 1. Monty Hall, who knows which door has the prize, and so he opens Door 3 to reveal a goat. Then, Monty offers the contestant a choice to stay with Door 1, or switch to Door 2. Is it to the advantage of the contestant to switch from Door 1 to Door 2, or stay with Door 1?

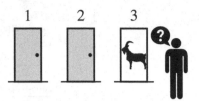

Fig. 8.35   Do you wish to switch?

We need to make an additional assumption:[27] if Monty Hall has a choice of which door to open (in other words, the contestant's original selection was correct), then Monty is equally likely to open either non-selected door.

Solution — 1. Let A be the event that the car is behind Door 1, B be the event that the contestant selects Door 1, and C be the event that Monty Hall opens Door 2. Then,

$$P\left(\begin{matrix} \text{Car behind} \\ \text{Door 1} \end{matrix}\middle| \begin{matrix} \text{Contestant selects Door 1} \\ and\ \text{Monty opens Door 2} \end{matrix}\right) = P(A \mid B \cap C)$$

$$= \frac{P(A \cap B \cap C)}{P(B \cap C)}$$

$$= \frac{P(C \mid A \cap B) \cdot P(B \mid A) \cdot P(A)}{P(C \mid B) \cdot P(B)}$$

$$= \frac{\frac{1}{2} \cdot \frac{1}{3} \cdot \frac{1}{3}}{\frac{1}{2} \cdot \frac{1}{3}}$$

$$= \frac{1}{3}.$$

If the contestant remains with Door 1, then the probability of winning the car is $\frac{1}{3}$. If the contestant switches to Door 3, then his probability of winning the car is $\frac{2}{3}$. ◀

If the first solution did not convince you, consider the following solution where we use a probability tree diagram.

Solution — 2. Using the probability tree diagram in Fig. 8.36 on the following page, the probability of switching and winning is $\frac{1}{3} + \frac{1}{3} = \frac{2}{3}$, and the probability of not switching and winning is $\frac{1}{6} + \frac{1}{6} = \frac{1}{3}$.

◀

There were many mathematicians who did not believe the solution to the problem. If you are still unconvinced, this is what Monty Hall had to say about the problem...

---

[27]Rosenthal (2008) points out that we need this assumption to do the following analysis. Many people make this assumption without explicitly saying so.

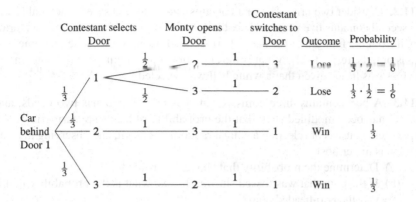

Fig. 8.36 Probability tree diagram for the Monty Hall problem.

After one [door] is seen to be empty, his chances are no longer 50/50 but remain what they were in the first place, one out of three. It just seems to the contestant that one [door] having been eliminated, he stands a better chance. Not so. (Selvin, 1975)

I couldn't have said it better myself.

Exercises

8.11.1. Given the probability tree diagram in Fig. 8.37,

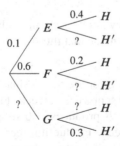

Fig. 8.37 Probability tree for Exercise 8.11.1

(a) fill in in the missing probabilities
(b) find

(i) $P(H \mid E)$	(iv) $P(E \mid H)$	(vii) $P(E \mid H')$
(ii) $P(H \mid F)$	(v) $P(F \mid H)$	(viii) $P(F \mid H')$
(iii) $P(H' \mid G)$	(vi) $P(G \mid H)$	(ix) $P(G \mid H')$

8.11.2. Consider two urns. The first contains two white and seven black balls, and the second contains five white and six black balls. We flip a fair coin and then draw a ball from the first urn or the second urn depending on whether the outcome was heads or tails, respectively. What is the conditional probability that the outcome of the toss was heads given that a white ball was selected?

8.11.3. A box contains three coins. Coin A is fair, coin B has two heads, and coin C has been modified such that the probability of heads appearing is $\frac{1}{3}$. An experiment consists of selecting a coin at random, tossing it, and observing the side that lands uppermost.
  (a) Determine the probability that "heads" appeared.
  (b) If the coin that was tossed showed heads, what is the probability that it was the two-headed coin?

8.11.4. In a survey, 700 undergraduates and 300 graduate students were asked the following question: do you support the idea of having a café in the library? The results of the survey are shown in Table 8.10.

Table 8.10   Results of student survey for Exercise 8.11.4

Response	Undergraduate (%)	Graduate (%)
Support	77	59
Oppose	14	31
Undecided	9	10

If a participant in the survey is selected randomly, what is the probability that he or she is a graduate student given that the student supports the idea of a café in the library?

8.11.5 (Defective sport cars). Suppose that a small premium electric sports car manufacturer, produces a 1000 sports cars a week at three plants. Plant A produces 350 sports cars a week, plant B produces 250 sports cars a week, and plant C produces 400 sports cars a week. Production records indicate that 5 percent of the sports cars at plant A will be defective, 3 percent of those produced at plant B will be defective, and 7 percent of those produced at plant C will be defective. All sports cars are shipped to a central car dealership in Southern California. If a sports car at the dealer is found to be defective, what is the probability that it was produced at
  (a) at plant A?
  (b) at plant B?
  (c) at plant C?

8.11.6. A computer program is attempting to predict whether a message is an email or junk mail, otherwise known as *spam*. The goal of the program is to filter out spam before it clogs the user's mailbox and waste limited storage space. Table 8.11 gives the data about the program.

Table 8.11   Data about the program in Exercise 8.11.6. Source: Hastie *et al.* (2009, p. 313)

True Class	Predicted Class	
	Email	Spam
Email	57.3%	4.0%
Spam	5.3%	33.4%

What is the probability that the program predicts the message is
  (a) spam given that the message is actually spam?
  (b) not spam given that the message is an email?

8.11.7. Game Console Russian Roulette is played with a 6-shot revolver with one bullet (and five empty chambers). The cylinder is spun at the beginning of each round and the player points it at his or her game console and pulls the trigger.
  (a) What is the probability of the player's game console survival on the first round?
  (b) Draw a probability tree for three rounds.
  (c) What is the probability of the player's game console survival on the third round?
  (d) What is the probability of the player's game console will not survive by the third round?

8.11.8. "Let's play a little game of Russian roulette," the interviewer of an investment company says who is considering you for a position with them. The company uses these types of questions to see how well applicants can handle probability concepts. "I have a gun, with six chambers, all empty, and I get up and you watch me place two bullets into two adjacent chambers. Closing the barrel, and spinning it, I put the muzzle up to your head and pull the trigger. Click! You're still alive." (What is the probability of you living?) "Now before we discuss your qualifications," she says laying your résumé on the desk, "I'm going to pull the trigger one more time. Now here is the interesting question that I'll let you decide: would you prefer if I just pull the trigger again, or spin the barrel first?"

8.11.9 (Bank robbery and polygraph tests). Three robbers hold up a bank, and then blend in with the other 27 customers and 10 staff. The police hold everyone for questioning, and have each person submit to a polygraph test. The polygraph test can identify a robber 95% of the time. The polygraph test also identifies a non-robber as a robber 5% of the time. If a person is identified by the polygraph test to be a robber, what is the probability that the person is actually a robber?

8.11.10 (The host forgets). You have selected one of three doors in which a prize is behind one of them. The host has forgotten accidentally which door the prize is behind, and he randomly picks either of the unselected doors. It just happens that the prize is not behind the door the host opened. *Now*, what are the probabilities that you will win the prize if you stick with your original selection compared to switching to the remaining door?

8.11.11 (Three prisoners problem). Three convicted felons—A, B, and C—received the death penalty and were kept in separate cells. The governor decides to pardon only one of the felons. He puts their names in a hat, and randomly selects a name. He phones the Warden, and requests the name of the lucky felon be withheld for several days. Rumors in the prison begin to circulate, and it reaches Felon A. When the Warden passes by A's cell, he begs to know which one is saved, but the Warden refuses and continues to walk.

"Wait! Come back!" A called after him. "I've something important to say!"

This sounded promising, certainly: the Warden turned and came back again.

"Well since you cannot tell me who exactly will die," said A, "then tell me this. If B is to be set free, give me C's name. If C is to be set free, give me B's name. If I'm the one to be set free, flip a coin to decide either to give me B or C's name."

The Warden thought for a moment. "But if you see me flip a coin, wouldn't you know immediately you're the one to be set free? On the other hand, if I don't flip a coin, then you'll know it's either you or the person I don't name."

"You must tell me some other time then," A murmured. "Tell me in the morning Sir."

The Warden thought it over, and the next morning he came back to A's cell, and told him that B was going to be executed.

What is the probability that
    (a) A will live?   (b) B will live?   (c) C will live?

8.11.12. What is the accuracy of the at-home HIV test in Example 8.11.4?

8.11.13 (Singing in the shower). Either Emily or Madison are equally likely to be in the shower. Emily always sings in the shower while Madison only sings a quarter of the time. If you hear singing in the shower, what is the chance that the singer is Emily?

8.11.14. Consider a woman who presents with painful urination to her general practitioner (doctor). The prevalence of urinary tract infection (UTI) is 55% of the female patients who present to their general practitioner. Two dipstick tests are performed; one detects nitrites, and the other test detects leukocyte esterase which are indicators of a UTI. The true positive rate for a UTI if a dipstick test is positive for either nitrites or leukocyte esterase is 90%. The true negative rate for a UTI if both nitrites and leukocyte esterase are negative is 60%. What is the probability that the woman has a UTI after at least one of the dipstick tests is positive? What is the accuracy of the test? (*Hint*: Create a confusion matrix with a total of 1000 women.)

8.11.15 (Counterfeit money). A bank is interested in determining whether a note (paper money) is counterfeit or not. From previous experience, 0.5% of all notes accepted for deposit at the bank are counterfeit. A non-destructive test by holding the notes under an ultraviolet light to reveal a hidden watermark has two outcomes: the watermark is not seen and the tester identifies the note as a counterfeit (a positive test result), or the watermark is present and the tester concludes it is a legitimate note (negative test result). The accuracy of this test is such that there is a 90% chance of detecting a counterfeit note, but 15% chance of a positive result when the note is legitimate (the watermark might have been obscured by dirt and the detector did not see it). Suppose that two tests are conducted independently on the same note, and both tests gave a positive result. What is the probability that the note is indeed counterfeit?

8.11.16 (†). A box initially contains 5 white beads and 5 red ones. An experiment consists of the following steps:
- Roll a fair die twice in succession, and observe the sum of the two upper face numbers.
- If the sum is less than or equal to 4 (Sum $\leq$ 4), randomly select 1 bead from the box.
- If the sum is greater than or equal to 10 (Sum $\geq$ 10), randomly select 2 beads from the box without replacement.
- If the sum is any other value (that is, 5, 6, 7, 8 or 9), do not select any beads.
(a) Draw the probability tree for this experiment showing the probabilities associated with each branch.
(b) Calculate the probability that at least one red bead is withdrawn from the box.
(c) Calculate the conditional probability that Sum $\geq$ 10 given that at least one red bead is withdrawn from the box.

8.11.17 (†). If person A, B, C, and D each speak the truth once in three times (independently), and A affirms that B denies that C declares that D is a liar, what is the probability that D was speaking the truth?

8.12 Review

Summary

- The **probability** of an event E in a uniform sample space S is

$$P(E) = \frac{n(E)}{n(S)}.$$

- The three probability axioms:
  (a) $P(\varnothing) = 0$.
  (b) If S is the sample space, then $P(S) = 1$.
  (c) If E and F are mutually exclusive events, then

$$P(E \cup F) = P(E) + P(F).$$

- **Complement Rule**: If E and E' are complementary events, then

$$P(E) = 1 - P(E').$$

- **Addition Rule**: If E and F are two events, then

$$P(E \cup F) = P(E) + P(F) - P(E \cap F).$$

- **Conditional Probability**: The probability of an event F given an event E is

$$P(F \mid E) = \frac{P(E \cap F)}{P(E)} = \frac{P(F \cap E)}{P(E)} \qquad \text{if } P(E) \neq 0.$$

- Two events, E and F, are **independent** if and only if

$$P(E \cap F) = P(E) \times P(F).$$

- **Bayes' Theorem**: Let E and F be mutually exclusive events in the sample space S. Then the conditional probability, $P(E \mid F)$, is computed by

$$P(E \mid F) = \frac{P(E)\,P(F \mid E)}{P(E)\,P(F \mid E) + P(E')\,P(F \mid E')}.$$

Exercises

8.12.1. There are 38 numbers on a Roulette wheel (see Fig. 8.29 on p. 366). On any spin, what is the probability of the ball landing on
  (a) an odd number?   (b) a prime number?   (c) a multiple of five?

8.12.2. Consider the following data for the events A and B:

$$P(A) = 0.4, \ P(B) = 0.75, \ \text{and } P(A \cap B) = 0.25.$$

Calculate

(a) $P(A')$ (c) $P(A' \cap B)$ (e) $P(A \mid B)$

(b) $P(A \cup B)$ (d) $P\big((A \cup B)'\big)$ (f) $P(B \mid A')$.

8.12.3. Let E and F be two events with

$$P(E) = 0.2, P(F) = 0.6, \text{ and } P(E \mid F) = 0.1.$$

Determine the following:

(a) $P(E \cap F)$ (c) $P(F \mid E)$

(b) $P(E \cup F)$ (d) $P(E' \cup F)$.

8.12.4. A sample space contains two events, A and B, and

$$P(A) = 0.38, \ P(B) = 0.25, \text{ and } P(A \cap B) = 0.16.$$

Calculate the following:

(a) $P(A')$ (c) $P(A \cup B)$ (e) $P(A' \cup B')$

(b) $P(B')$ (d) $P(A \mid B)$ (f) $P(A' \cap B')$.

8.12.5. Blackjack is a popular casino game. Suppose you are playing Blackjack against a dealer. The dealer will deal you two cards face up, and he will deal for himself one card face down and one card face up. Thus, you are unable to see one of the dealer's cards. Kings, queens, and jacks are counted as ten points. You and the dealer can count his or her own ace as 1 point or 11 points. All the other cards are counted as the numeric value shown on the card. Given that you have a jack of spades and queen of hearts (a hand that equals to 20 points in Blackjack), what is the probability that the dealer has 21 points given that the one card is showing is king of hearts?

8.12.6. In the game of Yahtzee, five dice are shaken in a cup and then rolled all at once. What is the probability that you got

(a) a Yahtzee (all five dice are the same)?

(b) four of a kind (four dice are the same and one is different)?

(c) a full house (a three-of-a-kind and a pair)?

8.12.7. There are six people in a room. What is the probability that two or more were born in the same month (but not necessarily the same year)?

8.12.8. At a party, there are seven female-male couples. For a dance, each male is paired randomly with a female. What is the probability that at least one of the pairs is a couple?

8.12.9. The DJ of a local radio show conducts a contest in which the listeners phone in the answer to a question. If the DJ estimates the probability of the caller being right is 30%, what is the probability that the fifth caller will win?

8.12.10. Two cards are taken from a well-shuffled deck of 52 cards. What is the probability that they are both kings?

8.12.11. Events E and F are independent with $P(E) = 0.4$ and $P(F) = 0.3$. Find $P(E \cup F)$.

8.12.12. Aaron and Trent are sometimes late for work. Let A be the event that Aaron is late for work, and T be the event that Trent is late for work with the following probabilities: $P(A) = 0.2$, $P(A \cap T) = 0.1$, and $P(A' \cap T') = 0.5$.
  (a) What is the probability that on a randomly selected day that
     (i) Aaron, or Trent, or both are late for work?
     (ii) Trent is late for work?
  (b) Given that Trent is late, what is the chance that Aaron is also late?
  (c) Aaron and Trent's boss suspect that their lateness is linked. Check to see if the probability of Aaron being late, and Trent being late is independent.

8.12.13. Nine graduate students, five male and four female, work in a psychology laboratory at a university. Their supervisor wants to send them to a conference in Paris, but has only enough funds to send four of them. The supervisor said that he randomly selected the students, and all the students selected were female. Some of the male students protested, and voiced their concern that they were being discriminated by the supervisor. What is the probability that all four selected students were female? Do you believe that the supervisor was not discriminating against the men?

8.12.14. A committee comprising four freshman Arts students will be formed to advise the Faculty on the growing importance and relevance of Mathematics in the Arts curriculum. There are 10 equally-qualified candidates who have put their names forward to serve on this committee. Seven of them are male; the remaining three are female.
  (a) Suppose James is one of the candidates. Calculate the probability that he will be selected for this committee, assuming that names will be chosen at random from the candidate list.
  (b) What is the probability that James is one of the candidates and that the committee will comprise an equal number of males and females?

8.12.15. An experiment consists of drawing two cards from a well-shuffled deck of 52 playing cards.
  (a) What is the probability that the two cards drawn are of the same suit?
  (b) Calculate the probability that the first card drawn is a heart, given the fact that the second card drawn is a heart.

8.13 Bibliographic Remarks

The creation of probability theory is quite interesting and unclear. For an interesting (and readable) brief account, see Kendall (1956). The authoritative expert on the history of the mathematical development of probability theory is Todhunter (1865), but many who read it lament that it is "just about as dull as any book on probability could be" (Kendall, 1963). A more recent work by Hald (1990) gives a more livelier account by placing developments in a historical context.

The idea of using the eikosograms came from Cherry and Oldford (2002). Cherry and Oldford argue that John Venn's intention of using his circle-based diagrams was to demonstrate logic and set theory. Venn diagrams used in a probability setting are lacking an important key piece of information—the amount of probability that is associated with the event. To remedy this situation, most people will simply label the events with a probability, with sets displayed as all the same size; thus, the Venn diagram becomes a fancier tableau. By switching from a circle to a square, we are now able to calculate easily the area which now we have a visual representation of how much probability is associated with a probability and can perform quick calculations. It appears that Kemeny *et al.* (1956, pp. 133–134) had also thought of expressing probability in terms of areas in a diagram.

The sad case of Sally Clark who was wrongfully convicted for the death of her child is a summary of the account given in Schneps and Colmez (2013). Schneps and Colmez provide more detail the case in Chapter 1 of their book, and the remaining chapters cover other cases in which probability and statistics were used wrongfully in a court of law.

In the Bayes' Theorem section, the example of the World Trade Center attacks came from Silver (2012, pp. 247–248). The rest of Silver's book tries to explain why so many analysts had failed to predict the 2008 financial market crisis. As Hurley (2007) points out, not many probability and statistics textbooks emphasize the use of probability trees in aiding in probability calculations. In particular, the way I remember Bayes' Theorem is to write out the formula as I am looking at the limbs of a probability tree. It seems that only in Finite Mathematics textbooks that this visual aid has caught on with the ground work laid by Kemeny *et al.* (1956). The basis for Example 8.11.5 on medical diagnostics was based off of Hall (1967).

In the exercises of Sec. 8.11, many of these problems are notorious teasers in probability theory. The common ingredient to all these teasers is that they should be solved using conditional probability, but the difficulty lays in identifying the correct conditioning event (Bar-Hillel and Falk, 1982). The Prisoner's paradox was first described by Gardner (1959, pp. 180, 182) and discussed further in Bar-Hillel and Falk (1982, pp. 118–119). Nonetheless, Rosenthal (2009) points out that the Prisoner's paradox and the Monty Hall problem (Example 8.11.7) are just different versions of Bertrand's Box (Bertrand, 1889, p. 2) in Exercise 8.8.15. Exercise 8.11.10 where the host forgets which door the prize is behind was adapted from Rosenthal (2008). Exercise 8.11.17 was originally posed by Eddington (1935, p. 121) who said that the question came from a mock examination paper.

9

Random Variables

We also know there are known unknowns; that is to say we know there are some things we do not know.[1]

—Donald H. Rumsfeld (1932–)

9.1 Introduction

In algebra, you may have seen an equation such as $x^2 - 4 = 0$, and asked to solve for x. In this case, $x = 2$ or $x = -2$. We see that a variable used in algebra can have more than one possible value which solves the equation. Now, if we think about an experiment such as flipping a coin twice, we can let X represent the number of times the coin lands on heads. Thus, it could be that $X = 0$ (no heads appeared in the two flips), $X = 1$ (one head appeared in the two flips), or $X = 2$ (two heads appeared in the two flips). In this context of probability, we call X a *random* variable since we do not know what the value of X will be until we perform the experiment. Since we are not certain what the value of X will be before the experiment, we can assign probabilities to each values of X to give us the likelihood of 0, 1, or 2 heads appearing; that is, we can find values for $P(X = 0)$, $P(X = 1)$, and $P(X = 2)$. Perhaps, when the Secretary of Defense Donald Rumsfeld gave a news briefing in 2002, he could have said, "we also know there are [random variables], that is to say we know there are some things we do not know." In this chapter, we will explore different kinds of random variables, and how to calculate their probabilities.

9.2 Random Variables & Probability Distributions

When we perform an experiment, we may not be so interested in the details of the outcome, but only on some numerical quantity determined by the result. For example, a person playing craps—a dice game played by throwing two dice—may

---

[1]Source: http://www.defense.gov/transcripts/transcript.aspx?transcriptid=2636.

not be interested in the numbers that appear on each die, but only the sum of the two dice. It is only natural to refer to this variable quantity of interest as a **random variable**.

> **Definition 9.2.1 — Random variable.** A random variable assigns a numerical value to the outcome from the sample space. We usually denote random variables by a capital letter such as X, Y, etc.

A random variable is **continuous** if its value can be any real number. Usually, a continuous random variable will measure the amount of something. For example, a continuous random variable could be the height of a person, the length of time of a phone call, or the amount of coffee poured by a barista into a cup.

A random variable is **discrete** if its value can be from any set of countable numbers such as the set of integers $\mathbb{Z} = \{\ldots, -3, -2, -1, 0, 1, 2, 3, \ldots\}$. Usually, a discrete random variable will count something. For example, a discrete random variable could be the number of children in a family, the number of lattes sold at a coffee shop, or the number of highway accidents in a day. In this chapter, we will focus on discrete random variables.

Consider a gambler who is interested in the sum of two dice, the discrete random variable X takes the outcome, say $(6, 2)$, and adds them together to give 8. Notice that there can be multiple points in the sample space $\mathcal{S}$, in which $X = 8$: $(6, 2)$, $(5, 3)$, $(4, 4)$, $(3, 5)$, $(2, 6)$. Thus, there is a set of points in $\mathcal{S}$ in which $X = 8$. We can now calculate the probability of this random variable:

$$P(X = 8) = \frac{n\{(6, 2), (5, 3), (4, 4), (3, 5), (2, 6)\}}{n(\mathcal{S})} = \frac{5}{36}.$$

With rolling two dice, we know the sum of two dice can be $2, 3, 4, \ldots, 12$; see Table 9.1.

Table 9.1   The sum of two dice

+	⚀	⚁	⚂	⚃	⚄	⚅
⚀	2	3	4	5	6	7
⚁	3	4	5	6	7	8
⚂	4	5	6	7	8	9
⚃	5	6	7	8	9	10
⚄	6	7	8	9	10	11
⚅	7	8	9	10	11	12

This means that the random variable X can have a value of $\{2, 3, 4, \ldots, 12\}$.

We have already calculated the probability of when two dice sum to 8, $P(X = 8) = \frac{5}{36}$, but we can also do this for $P(X = 2)$, $P(X = 3)$, *etc.* When we start to calculate and list in a table, like we do in Table 9.2, the probabilities of each value the random variable can be; this is creating a **probability distribution**.

Table 9.2   Probability distribution of the sum of two dice

x	2	3	4	5	6	7	8	9	10	11	12
$P(X = x)$	$\frac{1}{36}$	$\frac{2}{36}$	$\frac{3}{36}$	$\frac{4}{36}$	$\frac{5}{36}$	$\frac{6}{36}$	$\frac{5}{36}$	$\frac{4}{36}$	$\frac{3}{36}$	$\frac{2}{36}$	$\frac{1}{36}$

It may be difficult to visualize the probability distribution function in a table, so we can also generate a graph of the probability distribution, which we have done in Fig. 9.1.

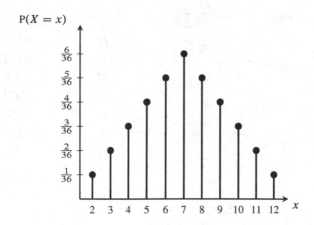

Fig. 9.1   Probability distribution of the sum of two dice

From the table and the graph, we see that the sum of 7 (lucky 7) has the highest probability of appearing, and the sum of 2 (snake eyes) and 12 (boxcars) have the lowest probability.

All probability distributions must sum up to one:

$$\sum_{\text{all } x} P(X = x) = 1. \tag{9.1}$$

In the case with the sum of two dice, the probability distribution function adds up to one:

$$\sum_{x=2}^{12} P(X = x) = P(X = 2) + P(X = 3) + P(X = 4) + \cdots + P(X = 12)$$

$$= \frac{1}{36} + \frac{2}{36} + \frac{3}{36} + \cdots + \frac{1}{36}$$

$$= 1.$$

Example 9.2.2 — Flipping a Coin. Let X be the number of heads in three coin flips.

    (a) Draw a tree diagram.
    (b) Determine $P(X = 0)$, $P(X = 1)$, $P(X = 2)$, and $P(X = 3)$.
    (c) What is the probability that in the three coin flips there was more than zero but less than 3.

Solution   (a) The tree diagram is given in Fig. 9.2.

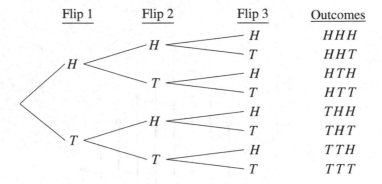

Fig. 9.2   Tree diagram of three coin flips

    (b)

$$P(X = 0) = \frac{n(\{TTT\})}{n(\mathcal{S})} = \frac{1}{8}$$

$$P(X = 1) = \frac{n(\{TTH, THT, HTT\})}{n(\mathcal{S})} = \frac{3}{8}$$

$$P(X = 2) = \frac{n(\{HHT, HTH, THH\})}{n(\mathcal{S})} = \frac{3}{8}$$

$$P(X = 3) = \frac{n(\{HHH\})}{n(\mathcal{S})} = \frac{1}{8}$$

(c) In the notation for random variables, this translates to $0 < X < 3$. Thus, we need to determine $P(0 < X < 3)$.

$$\begin{aligned} P(0 < X < 3) &= P(X = 1 \cup X = 2) \\ &= P(X = 1) + P(X = 2) \\ &= \frac{3}{8} + \frac{3}{8} \\ &= \frac{3}{4} \end{aligned}$$

◀

Example 9.2.3 — From a Frequency Distribution to a Probability Distribution.
Peter Norvig (1956–) is an American computer scientist who is currently the Director of Research at Google. He is known primarily for his research in artificial intelligence and natural language processing. Google Books is an ambitious project that is attempting to digitize all of the world's books.

Table 9.3   Number of Occurrences by Word Length in the English language. Data from `http://norvig.com/mayzner.html`.

Word Length	Frequency (in millions)	Word Length	Frequency (in millions)
1	22 301.22	13	3 850.58
2	131 293.85	14	1 653.08
3	152 568.38	15	565.24
4	109 988.33	16	151.22
5	79 589.32	17	72.81
6	62 391.21	18	28.62
7	59 052.66	19	8.51
8	44 207.29	20	6.35
9	33 006.93	21	0.13
10	22 883.84	22	0.81
11	13 098.06	23	0.32
12	7 124.15	Σ	743 842.91

As of October 2015, it has scanned more than 25 million books, which means that approximately 10 billion pages have been digitized (Heyman, 2015). To give you an idea of the magnitude of this project, the United States Library of Congress has 32 million books (Raymond, 2009)—so if Google Books had started with the Library of Congress it would be about 78% done. In 2012, while going through the massive 23 gigabyte Google Books data set, Dr Norvig compiled several tables including the frequency of word lengths in the English language. Using his data in Table 9.3,

(a) what is the probability distribution for the length of an English word?

(b) graph the probability distribution.

Solution The probability distribution is given in Table 9.4, and the graph of it is in Fig. 9.3.

Table 9.4　Probability distribution of the word length, x, in the English language

x	$P(X = x)$		x	$P(X = x)$
1	$\frac{22\,301.22}{743\,842.91} \approx 0.029\,981\,09$		13	$\frac{3850.58}{743\,842.91} \approx 0.005\,176\,60$
2	$\frac{131\,293.85}{743\,842.91} \approx 0.176\,507\,50$		14	$\frac{1653.08}{743\,842.91} \approx 0.002\,222\,35$
3	$\frac{152\,568.38}{743\,842.91} \approx 0.205\,108\,33$		15	$\frac{565.24}{743\,842.91} \approx 0.000\,759\,89$
4	$\frac{109\,988.33}{743\,842.91} \approx 0.147\,865\,00$		16	$\frac{151.22}{743\,842.91} \approx 0.000\,203\,30$
5	$\frac{79\,589.32}{743\,842.91} \approx 0.106\,997\,48$		17	$\frac{72.81}{743\,842.91} \approx 0.000\,097\,88$
6	$\frac{62\,391.21}{743\,842.91} \approx 0.083\,876\,86$		18	$\frac{28.62}{743\,842.91} \approx 0.000\,038\,48$
7	$\frac{59\,052.66}{743\,842.91} \approx 0.079\,388\,62$		19	$\frac{8.51}{743\,842.91} \approx 0.000\,011\,44$
8	$\frac{44\,207.29}{743\,842.91} \approx 0.059\,430\,95$		20	$\frac{6.35}{743\,842.91} \approx 0.000\,008\,54$
9	$\frac{33\,006.93}{743\,842.91} \approx 0.044\,373\,52$		21	$\frac{0.13}{743\,842.91} \approx 0.000\,000\,17$
10	$\frac{22\,883.84}{743\,842.91} \approx 0.030\,764\,35$		22	$\frac{0.81}{743\,842.91} \approx 0.000\,001\,09$
11	$\frac{13\,098.06}{743\,842.91} \approx 0.017\,608\,64$		23	$\frac{0.32}{743\,842.91} \approx 0.000\,000\,43$
12	$\frac{7124.15}{743\,842.91} \approx 0.009\,577\,49$		$\sum$	$1.000\,000\,00$

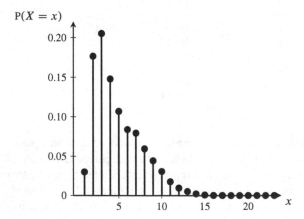

Fig. 9.3　Graph of the probability distribution for the length of an English word

Remark: The most common word in English is "the" with it accounting for 7.14% of all of the words in the Google Books data.　◀

Exercises

9.2.1. Classify the random variable as either discrete or continuous.

    (a) X = the number of phone calls a person makes in a day

    (b) Y = the temperature in a kitchen

    (c) Y = the number of girls in five-child family

    (d) X = the number of times a student takes a driver's test before passing

    (e) Y = the distance a person walks in a day

    (f) Y = the number of steps a person takes in a day

9.2.2. Classify each of the following random variables as discrete or continuous:

    (a) Q = the number of questions that are assigned for homework

    (b) T = the length of time you take to do your homework

    (c) P = the number of reams of paper ordered for a university department

    (d) E = the number of eggs laid each year by a chicken

    (e) A = the number of accidents per year on a stretch of highway

    (f) T = the length of time to play a game of chess

9.2.3. Given $P(X = x) = \frac{1}{6}$, where $x \in \{1, 2, 3, 4, 5, 6\}$,

    (a) state the probability distribution in tabular form, and give its graph

    (b) verify that it satisfies the condition that $\sum_{\text{all } x} P(X = x) = 1$

9.2.4. Recall the coat check problem from Example 8.3.4 (on p. 316). State the probability distribution of 0, 1, 2, 3, or 4 people correctly receiving their coats in tabular form, and give its graph.

9.2.5. Given the following probability distribution, what is the value of a?

x	1	2	3	4	5	6	7
$P(X = x)$	$\frac{1}{21}$	$\frac{2}{21}$	$\frac{3}{21}$	a	$\frac{4}{21}$	$\frac{2}{21}$	$\frac{1}{21}$

9.2.6. Find the probability distribution for the total number of tails obtained in four tosses of a fair coin.

9.2.7. If the random variable X is the number of sixes that appear when an ordinary die is thrown four times, what is

    (a) $P(X = 2)$ (c) $P(X > 2)$

    (b) $P(X \leq 2)$ (d) $P(1 < X < 4)$

9.2.8. Determine if the following function is a probability distribution.

$$P(X = x) = \begin{cases} \dfrac{x + 2}{26} & \text{for } x = 3, 4, 5, 6 \\ 0 & \text{otherwise} \end{cases}$$

9.2.9. Let N be a random variable that has the following probability distribution.

n	2	6	12	60	120
$P(N = n)$	0.15	0.05	0.1	0.3	0.4

Find $P(N > 6 \mid N \leq 60)$.

9.2.10. The probability distribution for a random variable X is given as a graph in Fig. 9.4.

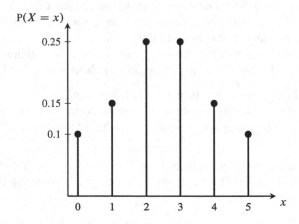

Fig. 9.4   Graph of the probability distribution for Exercise 9.2.10.

Calculate
(a) $P(X = 2)$ (c) $P(X$ is odd)
(b) $P(X \geq 3)$ (d) $P(1 \leq X \leq 4)$

9.2.11. Suppose in an office that the number of accidents per month is modeled by a random variable N with

$$P(N = n) = \frac{1}{(n + 1)(n + 2)}, \qquad \text{for } n \geq 0.$$

What is the probability that there is at least one accident in a particular month given that there is at most 4 accidents in that month?

9.2.12. The Autobahn is Germany's highway system where more than half of it does not have a speed limit. On speed-unrestricted stretches, an advisory speed limit (German: *Richtgeschwindigkeit*) of 130 kph (81 mph) applies. The speed of 2300 vehicles on a speed-unrestricted stretch was observed with 97 vehicles going less than 130 kph, and 2179 vehicles going less than 150 kph. Let X be the speed of a vehicle on the highway.

(a) Find the (empirical) probability distribution of X.

(b) What is the probability that a randomly selected vehicle on this stretch of Autobahn will be going at least 150 kph?

9.2.13. Five hundred participants with mobile phones responded to a survey conducted to determine the age of RMC students' mobile phones. The survey results giving the ages in months are summarized in Table 9.5.

Table 9.5   RMC Student Mobile Phone Survey Results

Months Phone is Kept, x	Respondents
$0 \leq x < 2$	4
$2 \leq x < 4$	30
$4 \leq x < 6$	100
$6 \leq x < 8$	132
$8 \leq x < 10$	121
$x \geq 10$	113

What is the (empirical) probability that a student that is randomly selected has a mobile phone

(a) six months or older?

(b) less than four months old?

(c) greater than or equal to four months but less than ten months?

9.2.14. How would you design an experiment to show that a Roulette wheel is fair? (For a description of the American Roulette wheel, see Example 8.9.18.)

9.3 Expected Value

On average, one in every 200 men alive today is descended from Genghis Khan.

Source: Zerjal *et al.* (2003)

The **expected value** is an important measure that is often used to summarize information contained in a random variable's probability distribution. The expected value of a random variable X, written $E(X)$, is a measure of central location for the random variable.

People, such as gamblers, are quite concerned about the expected value. For instance, if a gambler is playing a game where he expects to lose 25 cents on every dollar he bets, he would not consider the game to be fair enough to play. Others, such as experimenters, might want to know what the most probable outcome of a given experiment might be.

The expected value of X is

$$E(X) = \sum_{\text{all } x} x\, P(X = x). \tag{9.2}$$

The notation $\sum\limits_{\text{all } x}$ means the summation of all the possible values that the random variable X may take. For (9.2), we do not know the lower and upper limits of summation and hence cannot write them. Moreover, the random variable may be defined such that it does not take consecutive integer values; for example, $x = 2, 4, 6, 10$.

Example 9.3.1 What is the expected value of a die? Consider the random variable X having $P(X = x) = \frac{1}{6}$ for $x = 1, 2, \ldots, 6$. Find $E(X)$.

Solution Let X be the random variable whose value is the outcome of the random die. Thus, $P(X = x) = \frac{1}{6}$ for $x = 1, 2, \ldots, 6$.

Table 9.6　Probability distribution of a dice, and expectation

x	$P(X = x)$	$x\, P(X = x)$
1	$\frac{1}{6}$	$\frac{1}{6}$
2	$\frac{1}{6}$	$\frac{2}{6}$
3	$\frac{1}{6}$	$\frac{3}{6}$
4	$\frac{1}{6}$	$\frac{4}{6}$
5	$\frac{1}{6}$	$\frac{5}{6}$
6	$\frac{1}{6}$	$\frac{6}{6}$
$\sum$	1	$\frac{21}{6}$

To find the expected value, we can create a table like Table 9.6, or we use Eq. (9.2).

$$E(X) = \sum_{x=1}^{6} x\, P(X = x)$$

$$= 1\left(\frac{1}{6}\right) + 2\left(\frac{1}{6}\right) + 3\left(\frac{1}{6}\right) + 4\left(\frac{1}{6}\right) + 5\left(\frac{1}{6}\right) + 6\left(\frac{1}{6}\right)$$

$$= \frac{1}{6} + \frac{2}{6} + \frac{3}{6} + \frac{4}{6} + \frac{5}{6} + \frac{6}{6}$$

$$= \frac{21}{6}$$

$$= \frac{7}{2}$$

$$= 3.5$$

◀

Example 9.3.2 What is the expected value of the sum of two fair dice?

Solution Let X be the random variable whose values is the sum of two fair dice.

Table 9.7 Probability distribution of the sum of two dice, and expectation

x	$P(X = x)$	$x\,P(X = x)$
2	$\frac{1}{36}$	$\frac{2}{36}$
3	$\frac{2}{36}$	$\frac{6}{36}$
4	$\frac{3}{36}$	$\frac{12}{36}$
5	$\frac{4}{36}$	$\frac{20}{36}$
6	$\frac{5}{36}$	$\frac{30}{36}$
7	$\frac{6}{36}$	$\frac{42}{36}$
8	$\frac{5}{36}$	$\frac{40}{36}$
9	$\frac{4}{36}$	$\frac{36}{36}$
10	$\frac{3}{36}$	$\frac{30}{36}$
11	$\frac{2}{36}$	$\frac{22}{36}$
12	$\frac{1}{36}$	$\frac{12}{36}$
$\sum$	1	$\frac{252}{36} = 7$

Therefore, the expected value of the sum of two fair dice is

$$E(X) = \sum_{\text{all } x} x\,P(X = x)$$

$$= 2\,P(X = 2) + 3\,P(X = 3) + \cdots + 12\,P(X = 12)$$

$$= 2\left(\frac{2}{36}\right) + 3\left(\frac{2}{36}\right) + 4\left(\frac{3}{36}\right) + \cdots + 12\left(\frac{1}{36}\right)$$

$$= \frac{2}{36} + \frac{6}{36} + \frac{12}{36} + \frac{20}{36} + \frac{30}{36} + \frac{42}{36} + \frac{40}{36} + \frac{36}{36} + \frac{30}{36} + \frac{22}{36} + \frac{12}{36}$$

$$= \frac{252}{36}$$

$$= 7.$$

◀

We can apply Example 9.3.2 to a gambling application. Suppose that a gambler throws two fair die where the sum of the two dice is $\{2, 3, \ldots, 12\}$. After the throw, the casino will pay as many dollars as the sum of the two dice. If he plays many times, the gambler would expect to win one dollar about $\frac{2}{36}$ of the time, and two dollars about $\frac{6}{36}$, *etc.* Therefore, the expected winnings per game would be $7.00.

Based on the expected value, Blaise Pascal (1623–1662) argued in his book *Pensées* (English: Thoughts) that the value of a game is the prize to be won multiplied by the probability of winning it, and that it should be compared to the risked amount multiplied the probability of losing it (Ore, 1960); that is,

$$\text{Game Value} = \underbrace{\text{Prize Value} \times \text{P(Win)}}_{\text{Expected Winnings}} - \underbrace{\text{Cost to play} \times \text{P(Lose)}}_{\text{Expected Losses}}. \qquad (9.3)$$

A **fair price** to play the game in Example 9.3.2 would be the same amount that the gambler expects to win. He would pay $7.00 per game and he would expect to win $7.00 per game, so his net expected winnings per game is zero; this would be called a **fair game**. If either the gambler or the casino can expect to make more money than the other, then the game is called **unfair**.

Later in Pascal's book, he introduces the idea of the wager that every human makes: to decide if God exists or not; if one decides to believe in God then an infinite reward is gained; if one decides not to believe, then nothing is gained and nothing is lost. Pascal's logic created such an uproar that "one must admit that this is hardly a field for applied mathematics" (Ore, 1960). Nonetheless, this is one of the leading philosophers and founders of probability theory in 17th century France, and so it is understandable that Pascal's two passions began to overlap in his mind.

Example 9.3.3 — Roulette. In Roulette, there are 38 equal compartments labeled with the numbers 0 through 36 and 00 along the circumference of the wheel; see Fig. 8.29 on p. 366. The wheel is spun and a ball comes to rest in one of these compartments. If a player wagers $1 on a single number, and the ball comes to rest in the compartment corresponding to the player's selected number, the player wins $35; otherwise, the wager is lost. What is the expected winnings for the player?

Solution Let X be the amount of the player's winnings.

The probability of the ball landing on a single particular number on the roulette wheel is $\frac{1}{38}$, and the probability that he loses his bet is $\frac{37}{38}$.

Therefore, the expected player's winnings is

$$E(X) = \underbrace{35[P(X = 35)]}_{\text{Win}} + \underbrace{(-1)[P(X = -1)]}_{\text{Loss}}$$

$$= 35\left(\frac{1}{38}\right) - 1\left(\frac{37}{38}\right)$$

$$= -0.052\,63.$$

This can be interpreted that in the long run the player can expect to *lose* about 5 cents for every dollar the player bets on a single number. ◄

Example 9.3.4 — Life Insurance Policy. A Canadian insurance company charges $1000 for a life insurance policy for Canadian males who are 19 years old that pays out $150 000 upon death. The probability that a 19 year old Canadian male dies before his 20th birthday is 0.0007 (Statistics Canada, 2013b). What is the expected gain per policy for the insurance company?

Solution We use Table 9.8 to organize the information about the two possible outcomes.

Table 9.8   Insurance Policy Outcomes for Example 9.3.4

Outcome	Company Gain	Probability
No death	$1000	$1 - 0.0007 = 0.9993$
Death	$-\$150\,000 + \$1000 = -\$149\,000$	0.0007

Let X be the expected gain for the insurance company. Then,

$$E(X) = \underbrace{(\$1000 \times 0.9993)}_{\text{Gain}} - \underbrace{(\$149\,000 \times 0.0007)}_{\text{Loss}} = \$895.00.$$

This says that on average, the insurance company expects to gain $895 per policy. Our intuition is confirmed: the expected gain is positive because an insurance company is in business to make a profit. If the company is not satisfied with its gain, it can raise the price of the policy. ◄

Example 9.3.5 — The Lazy Professor, continued. Recall Example 8.7.4 (on p. 337). What is the expected value of the student's grade if each question is worth only one point?

Solution The expected value of the student's grade is $(\frac{1}{5} \times \frac{2}{3}) + (\frac{4}{5} \times \frac{3}{3}) = 93.3\%$. In fact, the expected value matches the same grade as if the instructor marked all of the questions: $\frac{14}{15} = 93.3\%$. ◄

Exercises

9.3.1. In Example 9.3.3 with Roulette, what is the expected amount that the casino will make on each dollar wagered by the player? (This is also known as the *house average* or *house edge*.)

9.3.2. If you bet on the ball landing on red or black in Roulette and win, then you win the same amount as your bet, otherwise you lose your bet. If you bet $1 on red and $1 on black in a single play in Roulette (see Fig. 8.29 on p. 366), then what is the expected value of your winnings?

9.3.3. Suppose a casino is offering the following game: The gambler rolls a fair die. After the throw, the casino will pay as many dollars as there are dots on the uppermost face, except for when the 6 face appears. If the 6 face appears uppermost, then the gambler must pay $6.00. How much should the gambler expect to pay if this is a fair game?

9.3.4. In the city of Kingston, 21% of the population have unlisted phone numbers, and none of the residences have more than one phone number. A mathematician rides into town and selects 2100 names at random from the phone book. How many of the selected phone numbers would the mathematician expect to be unlisted phone numbers?

9.3.5. Recall the coat check problem from Example 8.3.4 (on p. 316). What is the expected number of people who will receive the correct coat?

9.3.6. Suppose a Canadian insurance company charges $800 for a life insurance policy for Canadian females who are 19 years old that pays out $300 000 upon death. Calculate the expected gain per policy for the insurance company if
- (a) the probability that a 19 year old Canadian female dies before her 20th birthday is 0.000 29 (Statistics Canada, 2013b).
- (b) the female is from Québec who has a smaller probability of dying before her 20th birthday of 0.000 24 (Statistics Canada, 2013b).

9.3.7. Mathematically, why do casinos and life insurance companies make a profit?

9.3.8. A mathematics department is holding a lottery to raise funds for a local charity. The first prize in a lottery is $1000, second prize is $500 and the third prize is $250. Tickets cost $2 each. If the department sells 5000 tickets, what is the expected return for an individual who bought five tickets?

9.3.9. From the probability distribution of the random variable X given as a graph in Fig. 9.4 (on p. 400), find the expected value.

9.3.10. Given the following probability distribution

x	1	2	3	4	5
$P(X = x)$	$6a$	$5a$	$4a$	$2a$	a

(a) Find a.

(b) Find the expected value of X, $E(X)$.

9.3.11. Given the following probability distribution

x	2	3	5	7	11
$P(X = x)$	$\frac{1}{6}$	$\frac{1}{3}$	$\frac{1}{4}$	y	z

and $E(X) = 4\frac{2}{3}$, find y and z.

9.3.12. The number of highway accidents that occurs along a stretch of road during the rush hours of 4 P.M. to 7 P.M. are shown in Table 9.9. Find the expected number of accidents during the rush hours.

Table 9.9   Probability distribution of accidents in Example 9.3.12

Accidents, x	0	1	2	3	4	5
$P(X = x)$	0.10	0.40	0.30	0.10	0.05	0.05

9.3.13. Harry wishes to purchase a life insurance policy that pays out $100 000 upon death. Harry is Canadian and just turned 23 years old; the probability he will survive another year is 0.999 24 (Statistics Canada, 2013b). What is the minimum amount Harry would expect to pay for the policy for the year?

9.3.14. What is the expected length of a word in the English language? (*Hint:* Look at Table 9.4 on p. 398.)

9.3.15. Consider a vehicle owner who has a 90% chance of no accidents in a year, a 10% chance of one accident in a year, and no chance of being in more than two accidents in a year. Assume that if the owner has an accident, then there is a 60% chance that the car will need a $1500 repair, 30% chance that the car will need a $5000 repair, and a 10% chance the car will need to be replaced at a cost of $15 000. What is the owner's expected loss in a year?

9.3.16. In order to set a reasonable price for a cup with unlimited refills at a convenience store, the manager recorded the number of times customers refilled their cup. The frequency distribution is given in Table 9.10.

Table 9.10   Frequency Distribution of Soda Refills for Exercise 9.3.16. The frequency is the number of customers that get x refills.

x	0	1	2	3	4
Frequency	3	6	4	2	1

(a) What is the expected number of refills per customer?
(b) If it costs the manager $1.50 for the cup and the first fill, and 50 cents for each subsequent refill, what is the fair price for the manager to charge?
(c) Why is it unlikely the manager will set his price for the cup as the fair price as determined in part (b)?

9.4 Variance & Standard Deviation

In this section, we look at another measure to characterize the amount of spread or dispersion of a random variable. To motivate the need to quantify dispersion, consider the following random variables X and Y with the following probability distributions given in Table 9.11.

Table 9.11   The probability distribution of random variables X and Y

k	$P(X = k)$	$P(Y = k)$
3	0	$\frac{1}{4}$
4	$\frac{1}{2}$	$\frac{1}{4}$
5	$\frac{1}{2}$	$\frac{1}{4}$
6	0	$\frac{1}{4}$

The expected value of X is

$$E(X) = 3(0) + 4\left(\frac{1}{2}\right) + 5\left(\frac{1}{2}\right) + 6(0) = \frac{4+5}{2} = \frac{9}{2} = 4.5$$

and similarly, the expected value of Y is

$$E(Y) = 3\left(\frac{1}{4}\right) + 4\left(\frac{1}{4}\right) + 5\left(\frac{1}{4}\right) + 6\left(\frac{1}{4}\right) = \frac{3+4+5+6}{4} = \frac{18}{4} = 4.5.$$

We see that the expected value of X and Y are the same.

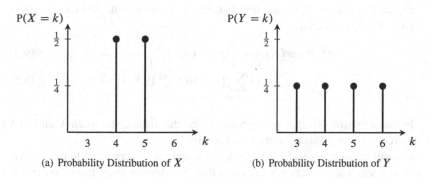

(a) Probability Distribution of X (b) Probability Distribution of Y

Fig. 9.5   Probability distributions of two random variables X and Y

The graphs of the probability distributions are plotted in Fig. 9.5. Even though the expected value of X and Y are the same, both probability distributions differ considerably in how the probability is dispersed about the expected value. The number of possible values of Y is larger than the number of possible values of X. We say that there is more variability in Y than X.

To describe the dispersion of X, we calculate the expected value of the distance from $E(X)$ squared; this is called the **variance** of X, which we denote by var(X). We can calculate the variance of X,

$$E\left((X - E(X))^2\right) = (4 - 4.5)^2\left(\frac{1}{2}\right) + (5 - 4.5)^2\left(\frac{1}{2}\right)$$

$$= 0.125 + 0.125$$

$$= 0.25$$

and similarly for Y,

$$E\left((Y - E(Y))\right) = (3 - 4.5)^2\left(\frac{1}{4}\right) + (4 - 4.5)^2\left(\frac{1}{4}\right)$$

$$+ (5 - 4.5)^2\left(\frac{1}{4}\right) + (6 - 4.5)^2\left(\frac{1}{4}\right)$$

$$= 0.5625 + 0.0625 + 0.0625 + 0.5625$$

$$= 1.25.$$

Since $E\left((Y - E(Y))^2\right) > E\left((X - E(X))^2\right)$, it reflects the fact that Y has more variability than X.

We expect a random variable X to take on values around its expected value $E(X)$. Moreover, we need a way to measure how far away X is from its expected value, on average.

> **Definition 9.4.1 — Variance.** The **variance** of a random variable X is denoted by $\text{var}(X)$. The other notation is σ^2, which is read as "sigma-squared." The variance is, on average, the squared distance X is away from the expected value $E(X)$. Mathematically,
>
> $$\sigma^2 = \text{var}(X) = E\left[(X - E(X))^2\right] \qquad (9.4)$$
>
> $$= \sum_{\text{all } x} [x - E(X)]^2 \, P(X = x). \qquad (9.5)$$

In the definition of variance, the reason why the distance between X and $E(X)$ is squared is to keep this quantity non-negative.

A small variance indicates that random variable X is very close to the expected value, whereas a larger variance indicates that X is farther away from the expected value.

Example 9.4.2 Consider the random variable X having $P(X = x) = \frac{1}{6}$ for $x = 1, 2, \ldots, 6$. Find $\text{var}(X)$.

Solution — 1. To find the variance, we use Eq. (9.5).

$$
\begin{aligned}
\text{var}(X) =& \frac{1}{6}\left(1 - \frac{7}{2}\right)^2 + \frac{1}{6}\left(2 - \frac{7}{2}\right)^2 + \frac{1}{6}\left(3 - \frac{7}{2}\right)^2 \\
&+ \frac{1}{6}\left(4 - \frac{7}{2}\right)^2 + \left(5 - \frac{7}{2}\right)^2 + \left(6 - \frac{7}{2}\right)^2 \\
=& \frac{1}{6}\left(-\frac{5}{2}\right)^2 + \frac{1}{6}\left(-\frac{3}{2}\right)^2 + \frac{1}{6}\left(-\frac{1}{2}\right)^2 + \frac{1}{6}\left(\frac{1}{2}\right)^2 + \frac{1}{6}\left(\frac{3}{2}\right)^2 \\
&+ \frac{1}{6}\left(\frac{5}{2}\right)^2 \\
=& \frac{25}{24} + \frac{9}{24} + \frac{1}{24} + \frac{1}{24} + \frac{9}{24} + \frac{25}{24} \\
=& \frac{70}{24} \\
=& \frac{35}{12} \\
\approx& 2.92
\end{aligned}
$$

◀

Solution — 2. Another approach is to use a table, like Table 9.12, to calculate the required values.

Table 9.12   Probability distribution, expectation, and variance of a die in Example 9.4.2

x	$P(X = x)$	$x\,P(X = x)$	$(x - E(X))^2\,P(X = x)$
1	$\dfrac{1}{6}$	$\dfrac{1}{6}$	$\left(1 - \dfrac{21}{6}\right)^2\left(\dfrac{1}{6}\right) = \dfrac{25}{24}$
2	$\dfrac{1}{6}$	$\dfrac{2}{6}$	$\left(2 - \dfrac{21}{6}\right)^2\left(\dfrac{1}{6}\right) = \dfrac{9}{24}$
3	$\dfrac{1}{6}$	$\dfrac{3}{6}$	$\left(3 - \dfrac{21}{6}\right)^2\left(\dfrac{1}{6}\right) = \dfrac{1}{24}$
4	$\dfrac{1}{6}$	$\dfrac{4}{6}$	$\left(4 - \dfrac{21}{6}\right)^2\left(\dfrac{1}{6}\right) = \dfrac{1}{24}$
5	$\dfrac{1}{6}$	$\dfrac{5}{6}$	$\left(5 - \dfrac{21}{6}\right)^2\left(\dfrac{1}{6}\right) = \dfrac{9}{24}$
6	$\dfrac{1}{6}$	$\dfrac{6}{6}$	$\left(6 - \dfrac{21}{6}\right)^2\left(\dfrac{1}{6}\right) = \dfrac{25}{24}$
$\sum$ 1		$\dfrac{21}{6}$	$\dfrac{35}{12}$

◀

Theorem 9.4.3

$$\text{var}(X) = E(X^2) - [E(X)]^2 \qquad (9.6)$$

Proof First, we start with the definition of variance.

$$\text{var}(X) = E\left[(X - E(X))^2\right]$$
$$= \sum_{\text{all } x} (x - E(X))^2\,P(X = x)$$

We expand and simplify

$$(x - E(X))^2 = (x - E(X))(x - E(X))$$
$$= x^2 - x\,E(X) - E(X)x + [E(X)]^2$$
$$= x^2 - 2x\,E(X) + [E(X)]^2$$

Hence,

$$\text{var}(X) = \sum_{\text{all } x} \left(x^2 - 2\,E(X) + [E(X)]^2\right) P(X = x)$$

Next, we distribute $P(X = x)$ over $\left(x^2 - 2\,E(X) + [E(X)]^2\right)$.

$$= \sum_{\text{all } x} \left(x^2\,P(X = x) - 2\,E(X)\,P(X = x) + [E(X)]^2\,P(X = x)\right)$$

Using Theorem 6.5.5, on p. 229, we can break up the summation into three smaller summations.

$$= \sum_{\text{all } x} x^2\,P(X = x) - \sum_{\text{all } x} 2x\,E(X)\,P(X = x) + \sum_{\text{all } x} [E(X)]^2\,P(X = x)$$

Using Theorem 6.5.3 (on p. 228), we can bring quantities that do not depend on the summation index in front of the summation. In the second summation, 2 and $E(X)$ are constants, so we can bring it in front of the summation. The same with the third summation: we can bring $[E(X)]^2$ in front of the summation.

$$= \underbrace{\sum_{\text{all } x} x^2\,P(X = x)}_{E(X^2)} - 2\,E(X) \underbrace{\sum_{\text{all } x} x\,P(X = x)}_{E(X)} + [E(X)]^2 \underbrace{\sum_{\text{all } x} P(X = x)}_{1}$$

Recall that $\sum_{\text{all } x} P(X = x)$ is the sum of all the possible values that the random variable X can be. The probability of all these possible values must sum to 1.

$$= E(X^2) - 2\underbrace{E(X)\,E(X)}_{[E(X)]^2} + [E(X)]^2$$

$$= E(X^2) - 2[E(X)]^2 + [E(X)]^2$$

$$= E(X^2) - [E(X)]^2$$

◀

To calculate the variance of a random variable X with $P(X = x) = \frac{1}{6}$ for $x = 1, 2, \ldots, 6$, we will use Theorem 9.4.3.

Solution — 3. From Example 9.3.1, we know that $E(X) = \frac{7}{2}$. Next,

$$E(X^2) = \sum_{\text{all } x} x^2\,P(X = x)$$

$$= \sum_{i=1}^{6} x^2\left(\frac{1}{6}\right)$$

$$= (1)^2\left(\frac{1}{6}\right) + (2)^2\left(\frac{1}{6}\right) + (3)^2\left(\frac{1}{6}\right) + (4)^2\left(\frac{1}{6}\right)$$

$$+ (5)^2\left(\frac{1}{6}\right) + (6)^2\left(\frac{1}{6}\right)$$

$$= \frac{1}{6} + \frac{4}{6} + \frac{9}{6} + \frac{16}{6} + \frac{25}{6} + \frac{36}{6}$$

$$= \frac{1 + 4 + 9 + 16 + 25 + 36}{6}$$

$$= \frac{91}{6}.$$

Then,

$$\mathrm{var}(X) = \mathrm{E}(X^2) - [\mathrm{E}(X)]^2 = \frac{91}{6} - \left(\frac{7}{2}\right)^2 = \frac{91}{6} - \frac{49}{4} = \frac{35}{12} \approx 2.92.$$

◄

If we would like to know the distance between the random variable X and the expected value $\mathrm{E}(X)$, then we can take the square root of the variance; this is called the **standard deviation**.

> **Definition 9.4.4 — Standard Deviation.** The **standard deviation** of a random variable X is
>
> $$\sigma = \sqrt{\mathrm{var}(X)} = \sqrt{\sigma^2}. \tag{9.7}$$
>
> The standard deviation is the positive square root of the variance, and is a measure of the distance from X to the expected value on average.

Example 9.4.5 — The Standard Deviation of a Die. What is the standard deviation of a single die?

Solution Let X be the random variable of the die's outcome. From Example 9.4.2, $\mathrm{var}(X) = \frac{35}{12}$. Then, the standard deviation of X is therefore

$$\sigma = \sqrt{\mathrm{var}(X)} = \sqrt{\frac{35}{12}} \approx 1.707\,83.$$

◄

Exercises

9.4.1. For the sum of two dice, calculate the
  (a) variance   (b) standard deviation

9.4.2. From the probability distribution of the random variable X given as a graph in Fig. 9.4 (on p. 400), find the
  (a) variance of X.   (b) standard deviation of X.

9.4.3. Referring to Exercise 9.3.15 (on p. 407), what is the variance of the vehicle owner's loss? What is the standard deviation of the vehicle owner's loss?

9.4.4. A discrete mathematics class wrote a quiz worth 10 points. Let X be the score of a student's quiz. On the next page, Table 9.13 gives the (incomplete) probability distribution of X.

Table 9.13   Discrete Mathematics Class Marks Distribution

x	1	2	3	4	5	6	7	8	9	10
$P(X = x)$	0.06	0.08	0.12	0.04	0.14	0.12	0.16	0.14	a	0.04

    (a) What is the value of a?

    (b) Find the expected value, $E(X)$.

    (c) Find the standard deviation.

9.4.5 (†). Let Y be a random variable such that

$$Y = aX + b$$

where a and b are constants, and X is a random variable with expected value $E(X)$ and variance $var(X) = E\left((X - E(X))^2\right)$.

    (a) Show that

        (i) $E(Y) = a\,E(X) + b$ (ii) $var(Y) = a^2\,var(X)$

    (b) Let $Y = 2X + 3$, where X is a random variable with $E(X) = 3$ and $var(X) = 4$. Calculate $E(Y)$ and $var(Y)$.

9.5 Review

Summary

- A **random variable** assigns a numerical value to the outcome from the sample space.
- All probability distributions must sum up to one:

$$\sum_{\text{all } x} P(X = x) = 1.$$

- Continuous versus discrete random variables.
- The **expected value** of a random variable X is

$$E(X) = \sum_{\text{all } x} x\, P(X = x).$$

- A game is **fair** if the expected value is zero.
- The **variance** of a random variable X is

$$\sigma^2 = var(X) = E\left[(X - E(X))^2\right]$$

$$= \sum_{\text{all } x} [x - E(X)]^2\, P(X = x)$$

$$= E(X^2) - [E(X)]^2.$$

- The **standard deviation** of a random variable X is

$$\sigma = \sqrt{\sigma^2} = \sqrt{var(X)}.$$

Exercises

9.5.1. If $P(X < 5) = 0.7564$, then what is the value of $P(X \geq 5)$?

9.5.2. Suppose that X is a random variable with a probability distribution given in Table 9.14. What is the expected value and variance of X?

Table 9.14   Probability Distribution for X

x	0	1	2
$P(X = x)$	0.25	0.40	0.35

9.5.3. In a game of craps, a player rolls two dice. They win at once if the total is 7 or 11, and lose at once if the total is 2, 3, or 12. Calculate the probability of
  (a) winning on the first roll?
  (b) losing on the first roll?
  (c) neither winning nor losing on the first roll?

9.5.4. Given the following probability distribution of X in Table 9.15,
  (a) what should the value of a be?
  (b) state the formula for the expected value (using sigma notation), and then find the expected value of X.
  (c) state the formula for the variance (using sigma notation), and then find the variance of X.
  (d) what is the standard deviation of X?

Table 9.15   Probability Distribution for Exercise 9.15

x	1	2	3	4
$P(X = x)$	0.1	0.2	0.3	a

9.5.5. Consider an experiment or game where three coins are flipped at the same time. Coin 1 is biased with $P(\text{Head}) = 0.8$ and $P(\text{Tail}) = 0.2$. Coins 2 and 3 are both fair; that is, $P(\text{Head}) = P(\text{Tail}) = 0.5$.
  (a) What is the sample space for this experiment?
  (b) Is this a uniform sample space? Explain.
  (c) Let Y denote the number of Tails obtained in one repetition of this game. Determine the probability distribution of Y.

(d) Suppose that you win \$10 if $Y = 3$, \$2 if $Y = 2$, and \$5 (*i.e.,* you lose \$5) otherwise. Should you play this game? Support your answer by the appropriate calculation.

10

Probability Distributions

In this chapter, we will look at probability distributions that arise frequently when investigating certain kinds of experiments.

10.1 Uniform Distribution

> **Definition 10.1.1 — The Uniform Distribution.** A random variable X has a **uniform distribution** if
>
> $$P(X = x) = \frac{1}{n} \tag{10.1}$$
>
> where n is the number of outcomes in the experiment.

The uniform distribution also satisfies the condition that the sum of all of its probability sums to one; that is,

$$\sum_{\text{all } x} P(X = x) = 1.$$

Proof Let $x \in \{x_1, x_2, \ldots, x_n\}$. Then,

$$\sum_{\text{all } x} P(X = x) = P(X = x_1) + P(X = x_2) + P(X = x_3) + \cdots + P(X = x_n)$$

$$= \overbrace{\frac{1}{n} + \frac{1}{n} + \frac{1}{n} + \cdots + \frac{1}{n}}^{n \text{ times}}$$

$$= \frac{\overbrace{1 + 1 + 1 + \cdots + 1}^{n \text{ times}}}{n}$$

$$= \frac{n}{n}$$

$$= 1.$$

◀

Example 10.1.2 A random variable X takes on the values $x \in \{1, 2, 3, \ldots, 10\}$, and $P(X = x)$ is constant. Find
- (a) the probability distribution
- (b) $P(X \leq 4)$
- (c) graph the probability distribution.

Solution

(a) Let $P(X = x) = c$, where c is some constant for $x \in \{1, 2, 3, \ldots, 10\}$. Then, since $P(X = x)$ for all x must sum to one,

$$P(X = 1) + P(X = 2) + \cdots + P(X = 10) = 1$$
$$10c = 1$$
$$c = \frac{1}{10}.$$

Therefore, $P(X = x) = \frac{1}{10}$ for $x = 1, 2, 3, \ldots, 10$.

(b)

$$P(X \leq 4) = P(X = 1) + P(X = 2) + P(X = 3) + P(X = 4)$$
$$= \frac{1}{10} + \frac{1}{10} + \frac{1}{10} + \frac{1}{10}$$
$$= \frac{4}{10}$$
$$= \frac{2}{5}$$

(c) The probability distribution of X is plotted in Fig. 10.1.

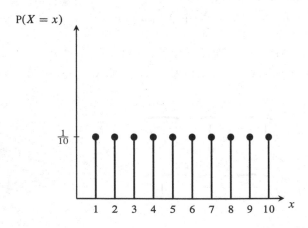

Fig. 10.1   Graph of the probability distribution for Example 10.1.2

◄

A random variable that has the same probability for each outcome, such as the random variable in Example 10.1.2, is called a **uniform random variable**. Because the probability is evenly spread across the possible values of X, we also say that X is a **uniform distribution**.

Theorem 10.1.3 The expected value of a random variable X with a uniform distribution is

$$E(X) = \frac{1}{n} \sum_{\text{all } x} x. \tag{10.2}$$

Proof Let $x \in \{x_1, x_2, \ldots, x_n\}$. Then,

$$E(X) = \sum_{\text{all } x} x \, P(X = x)$$

$$= x_1 \, P(X = x_1) + x_2 \, P(X = x_2) + \cdots + x_n \, P(X = x_n)$$

$$= x_1 \left(\frac{1}{n}\right) + x_2 \left(\frac{1}{n}\right) + \cdots + x_n \left(\frac{1}{n}\right)$$

$$= \frac{1}{n}(x_1 + x_2 + \cdots + x_n)$$

$$= \frac{1}{n} \sum_{\text{all } x} x.$$

◀

Example 10.1.4 What is the expected value of a die?

Solution Since a die has a uniform probability distribution, $P(X = x) = \frac{1}{6}$ for $x \in \{1, 2, 3, 4, 5, 6\}$, the expected value is

$$E(X) = \frac{1}{6}(1 + 2 + 3 + 4 + 5 + 6) = \frac{21}{6} = \frac{7}{2} = 3.5.$$

Looking back at Example 9.3.1, we see that we have the same answer. ◀

Theorem 10.1.5 The variance of a random variable X with a uniform distribution is

$$\text{var}(X) = \frac{1}{n} \sum_{\text{all } x} x^2 - \left[\frac{1}{n} \sum_{\text{all } x} x\right]^2. \tag{10.3}$$

Proof Recall that the variance can be found by

$$\text{var}(X) = E(X^2) - [E(X)]^2.$$

We will first find $E(X^2)$.

$$E(X^2) = \sum_{\text{all } x} x^2 \, P(X = x)$$

$$= \sum_{\text{all } x} x^2 \left(\frac{1}{n}\right)$$

$$= \frac{1}{n} \sum_{\text{all } x} x^2$$

Then,

$$\text{var}(X) = E(X^2) - [E(X)]^2$$

$$= \frac{1}{n} \sum_{\text{all } x} x^2 - \left[\frac{1}{n} \sum_{\text{all } x} x\right]^2.$$

◄

Example 10.1.6 What is the variance of a die?

Solution A fair die follows a uniform probability distribution, $P(X = x) = \frac{1}{6}$. Hence, we can use Theorem 10.1.5 to find the variance. Therefore,

$$\text{var}(X) = \frac{1}{n} \sum_{\text{all } x} x^2 - \left[\frac{1}{n} \sum_{\text{all } x} x\right]^2$$

$$= \frac{1}{6} \sum_{x=1}^{6} x^2 - \left[\frac{1}{n} \sum_{x=1}^{6} x\right]^2$$

$$= \frac{1}{6}(1^2 + 2^2 + \cdots + 6^2) - \left[\frac{1}{6}(1 + 2 + \cdots + 6)\right]^2$$

$$= \frac{91}{6} - \left[\frac{21}{6}\right]^2$$

$$= \frac{91}{6} - \frac{441}{36}$$

$$= \frac{546}{36} - \frac{441}{36}$$

$$= \frac{105}{36}$$

$$= \frac{35}{12} \approx 2.92.$$

Looking back at Example 9.4.2, we see that our answer matches.   ◄

Exercises

10.1.1. Which of the following experiments would you consider to follow a uniform distribution?

    (a) selecting a card from a well shuffled deck of cards

    (b) drawing a name from a hat

    (c) the color of shirts students are wearing in a lecture hall

    (d) winning a lottery

10.1.2. What is the probability distribution of randomly selecting a number from the set $\{10, 11, \ldots, 15\}$? Graph the probability distribution.

10.1.3. What is the probability distribution of randomly selecting a number from the set $\{i, i + 1, \ldots, j - 1, j\}$? Graph the probability distribution.

10.1.4. In an experiment of throwing two fair dice, why is taking the sum of two dice not an example of a uniform probability distribution?

10.1.5. A charity lottery sells $2\,000\,000$ tickets, and each ticket costs \$1.00. The prizes and the number of tickets that will receive each prize is stated in Table 10.1.

Table 10.1   Charity Lottery for Example 10.1.5

Prize (\$)	Number of Tickets
500 000	2
50 000	2
5 000	10
500	20
50	2 000
5	20 000

    (a) What is the expected value of each ticket? What is the variance?

    (b) Is this a fair game?

    (c) What is the amount that the charity should receive from this lottery?

10.1.6 (†). What is the expected value and the variance of a uniform distribution whose random variable can take on the values $1, 2, 3, \ldots, n$?

0.2 Binomial Distribution

The prefix *bi* at the beginning of a word generally denotes the fact that it involves two things. A random variable that follows a **binomial distribution** has exactly two outcomes for each trial: a *success* and a *failure*.

Example 10.2.1 When Sterling pulls the trigger on his revolver, the bullet hits the target 90% of the time. What is the probability that Sterling hits his target 0, 1, 2, 3, or 4 times?

Solution Let X be the random variable of the number of bullets that hit the target. In this case, we consider a *success* to be Sterling hitting his target, whereas a *failure* is him missing his target. The probability of a bullet hitting the target is $\frac{9}{10}$; the probability of a bullet not hitting the target is $\frac{1}{10}$.

The probabilities of 0, 1, 2, 3, or 4 bullets hitting the target can be found by using a probability tree, like in Fig. 10.2. By using a probability tree this way and looking at the sequences of hits and misses in the four shots, we are assuming that the events are *independent*; that is, the outcome of a shot does not affect the outcome of the next shot.

From the probability tree in Fig. 10.2,

$$P(X = 4) = \left(\frac{9}{10}\right)^4$$

$$P(X = 3) = \left(\frac{9}{10}\right)^3 \left(\frac{1}{10}\right) + \left(\frac{9}{10}\right)^3 \left(\frac{1}{10}\right)$$
$$+ \left(\frac{9}{10}\right)^3 \left(\frac{1}{10}\right) + \left(\frac{9}{10}\right)^3 \left(\frac{1}{10}\right)$$
$$= 4\left(\frac{9}{10}\right)^3 \left(\frac{1}{10}\right)$$

$$P(X = 2) = \left(\frac{9}{10}\right)^2 \left(\frac{1}{10}\right)^2 + \left(\frac{9}{10}\right)^2 \left(\frac{1}{10}\right)^2$$
$$+ \left(\frac{9}{10}\right)^2 \left(\frac{1}{10}\right)^2 + \left(\frac{9}{10}\right)^2 \left(\frac{1}{10}\right)^2$$
$$+ \left(\frac{9}{10}\right)^2 \left(\frac{1}{10}\right)^2 + \left(\frac{9}{10}\right)^2 \left(\frac{1}{10}\right)^2$$
$$= 6\left(\frac{9}{10}\right)^2 \left(\frac{1}{10}\right)^2$$

$$P(X = 1) = \left(\frac{9}{10}\right)^1 \left(\frac{1}{10}\right)^3 + \left(\frac{9}{10}\right)^1 \left(\frac{1}{10}\right)^3$$
$$+ \left(\frac{9}{10}\right)^1 \left(\frac{1}{10}\right)^3 + \left(\frac{9}{10}\right)^1 \left(\frac{1}{10}\right)^3$$
$$= 4\left(\frac{9}{10}\right)^1 \left(\frac{1}{10}\right)^3$$

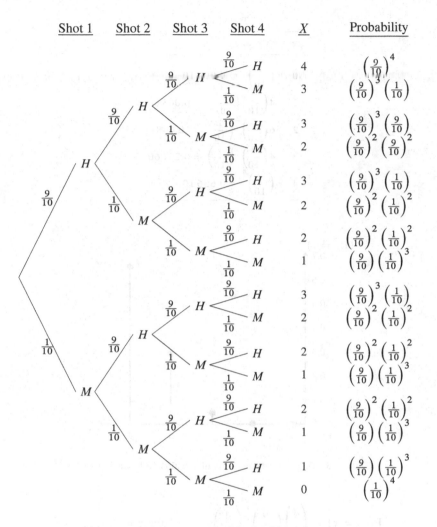

Fig. 10.2   Probability tree of Sterling hitting a target

$$P(X = 0) = \left(\frac{1}{10}\right)^4$$

Thus, the probability distribution is in tabular form in Table 10.2. The probability distribution can be plotted as well; see Fig. 10.3.

There is something interesting to note about the coefficients in the probability distribution: $1, 4, 6, 4, 1$. These are the same numbers that is in row 4 of Pascal's triangle. Thus, we can rewrite the coefficient's using binomial coefficients such as the following Another way of writing this distribution is

Table 10.2

x	$P(X = x)$
0	$\left(\dfrac{1}{10}\right)^{4} \approx 0.000\,10$
1	$4\left(\dfrac{9}{10}\right)\left(\dfrac{1}{10}\right)^{3} = 0.003\,60$
2	$6\left(\dfrac{9}{10}\right)^{2}\left(\dfrac{1}{10}\right)^{2} \approx 0.048\,60$
3	$4\left(\dfrac{9}{10}\right)^{3}\left(\dfrac{1}{10}\right) \approx 0.291\,60$
4	$\left(\dfrac{9}{10}\right)^{4} \approx 0.656\,10$

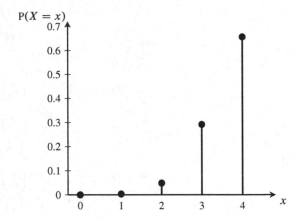

Fig. 10.3   Graph of the probability distribution for Example 10.2.1

$$P(X = x) = \binom{4}{x}\left(\frac{9}{10}\right)^{x}\left(\frac{1}{10}\right)^{4-x} \qquad \text{for } x = 0, 1, 2, 3, 4.$$

◀

The previous example is an example of a binomial distribution. If the following three conditions are met, then the number of successes in repeated trials will follow a binomial distribution. For a series of repeated trials,

   (a) there are exactly two outcomes of each trial (which may be labeled as
        either a *success* or a *failure*)
   (b) the trials are independent
   (c) the trials are identical.

Table 10.3

x	$P(X = x)$
0	$\binom{4}{0}\left(\frac{1}{10}\right)^4$
1	$\binom{4}{1}\left(\frac{9}{10}\right)^1\left(\frac{1}{10}\right)^3$
2	$\binom{4}{2}\left(\frac{9}{10}\right)^2\left(\frac{1}{10}\right)^2$
3	$\binom{4}{3}\left(\frac{9}{10}\right)^3\left(\frac{1}{10}\right)$
4	$\binom{4}{4}\left(\frac{9}{10}\right)^4$

Definition 10.2.2 — The Binomial Distribution. A random variable X has a **binomial distribution** if

$$P(X = x) = \binom{n}{x}p^x(1 - p)^{n-x} \qquad \text{for } x = 0, 1, 2, \ldots, n \qquad (10.4)$$

where

- x represents the number of *successes*
- n is the number of trials
- p is the probability of *success*
- $1 - p$ is the probability of *failure*.

A binomial distribution is denoted by the symbol $\mathcal{B}(n, p)$ where constants n (the number of trials) and p (the probability of a success) are the **parameters**. A random variable X that is binomial-distributed is denoted by $X \sim \mathcal{B}(n, p)$.

The binomial distribution also satisfies condition (9.1) that the sum of all of its probability sums to one; that is,

$$\sum_{\text{all } x} P(X = x) = 1.$$

Proof

$$\sum_{\text{all } x} P(X = x) = \sum_{x=0}^{n}\binom{n}{x}p^x(1 - p)^{n-x}$$

$$= \left(p + (1 - p)\right)^n \qquad \text{By Binomial Theorem 7.2.1}$$

$$= 1^n$$

$$= 1$$

◀

Basic Discrete Mathematics

Example 10.2.3 Given $X \sim \mathcal{B}(20, \frac{1}{3})$, find $P(X = x)$.

Solution We know that this is a binomial distribution where the number of trials $n = 20$, and the probability of success on each trial $p = \frac{1}{3}$. Thus,

$$P(X = x) = \binom{20}{x}\left(\frac{1}{3}\right)^x\left(\frac{2}{3}\right)^{20-x} \qquad \text{for } x = 0, 1, 2, \ldots, 20.$$

◄

Example 10.2.4 Suppose you toss a uniformly weighted coin seven times. What is the probability distribution of the number of heads that appear in the seven tosses?

Solution Let X be the number of heads (considered a *success* in this experiment) in $n = 7$ independent tosses of a uniformly weighted coin (both sides are weighted the same so that the probability of landing on heads is $\frac{1}{2}$). The probability distribution function of X is

$$P(X = x) = \binom{n}{x}\left(\frac{1}{2}\right)^x\left(\frac{1}{2}\right)^{7-x} \qquad \text{for } x = 0, 1, 2, \ldots, 7.$$

We can calculate for the probability of x heads landing in 7 tosses. The probability distribution function can be either given as Table 10.4, or as a graph in Fig. 10.4.

Table 10.4 Probability Distribution for Example 10.2.4

x	$P(X = x)$
0	$\binom{7}{0}\left(\frac{1}{2}\right)^0\left(\frac{1}{2}\right)^7 = \frac{1}{128}$
1	$\binom{7}{1}\left(\frac{1}{2}\right)^1\left(\frac{1}{2}\right)^6 = \frac{7}{128}$
2	$\binom{7}{2}\left(\frac{1}{2}\right)^2\left(\frac{1}{2}\right)^5 = \frac{21}{128}$
3	$\binom{7}{3}\left(\frac{1}{2}\right)^3\left(\frac{1}{2}\right)^4 = \frac{35}{128}$
4	$\binom{7}{4}\left(\frac{1}{2}\right)^4\left(\frac{1}{2}\right)^3 = \frac{35}{128}$
5	$\binom{7}{5}\left(\frac{1}{2}\right)^5\left(\frac{1}{2}\right)^2 = \frac{21}{128}$
6	$\binom{7}{6}\left(\frac{1}{2}\right)^6\left(\frac{1}{2}\right)^1 = \frac{7}{128}$
7	$\binom{7}{7}\left(\frac{1}{2}\right)^7\left(\frac{1}{2}\right)^0 = \frac{1}{128}$

◄

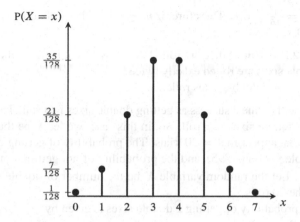

Fig. 10.4   Graph of the probability distribution for Example 10.2.4

Example 10.2.5 What is the probability that in a family of five children, exactly three will be girls?

Solution Let G be the event the child is a girl (and considered a *success*) and the event B is the child is a boy (and considered a *failure*), then we assume that a boy or a girl born is equally likely; $p = \frac{1}{2}$. Then,

$$P(3 \text{ girls}) = \binom{5}{3}\left(\frac{1}{2}\right)^3\left(\frac{1}{2}\right)^2 = 10\left(\frac{1}{2}\right)^5 = \frac{5}{16}.$$

◀

Example 10.2.6 Assume that the probability of a boy or a girl is equally likely. How many children must a family plan on so that the probability of at least one girl will be greater than or equal to 0.9?

Solution Let the event E denote "at least one girl," and the event E' denote "no girl." Therefore,

$$P(E) = 1 - P\left(E'\right)$$
$$= 1 - \binom{8}{0}\left(\frac{1}{2}\right)^n$$
$$= 1 - \frac{1}{2^n}.$$

We wish to determine n so that

$$1 - \frac{1}{2^n} \geq 0.9$$
$$\frac{1}{2^n} \leq 0.1.$$

Note that $\frac{1}{2^4} = \frac{1}{16} \leq 0.1$. Therefore, if $n \geq 4$ then the probability of E will be greater than 0.9. ◀

Example 10.2.7 A pair of dice are rolled 20 times. What is the probability that
  (a) double sixes are rolled exactly twice?
  (b) at least two double sixes rolled?

Solution We will define a success as getting double sixes in a roll. Then, a failure is not getting double sixes in a roll. So, in this case, we let X be the number of double sixes that appear in $n = 20$ trials. The probability of getting double sixes is $\frac{1}{36}$ (see Table 9.2 on p. 395), and the probability of not getting double sixes is $1 - \frac{1}{36} = \frac{35}{36}$. Let the random variable X be the number of double sixes that is observed in the 20 rolls.

  (a) The probability of getting 2 double sixes is given by

$$P(X = 2) = \binom{20}{2}\left(\frac{1}{36}\right)^2\left(\frac{35}{36}\right)^{20-2}$$

$$= \frac{20!}{2!18!} \times \left(\frac{1}{36}\right)^2 \times \left(\frac{35}{36}\right)^{18}$$

$$\approx 0.088.$$

Therefore, the probability of rolling exactly 2 double sixes in 20 rolls of a pair of dice is approximately 8.8%.

  (b) The probability of getting at least 2 double sixes is given by

$$P(X \geq 2) = P(X = 2) + P(X = 3) + \cdots + P(X = 20)$$

$$= 1 - P(X = 1) - P(X = 0)$$

$$= 1 - \left[\binom{20}{1}\left(\frac{1}{36}\right)^1\left(\frac{35}{36}\right)^{20-1}\right]$$

$$- \left[\binom{20}{0}\left(\frac{1}{36}\right)^0\left(\frac{35}{36}\right)^{20-0}\right]$$

$$\approx 1 - 0.569 - 0.325$$

$$= 0.106.$$

Therefore, the probability of at least 2 double sixes in 20 rolls is 10.6%. ◀

Example 10.2.8 — Continued from Example 10.2.1.
  (a) How many shots would Sterling expect to hit the target?
  (b) What is the variance of Sterling's shots?

Solution

(a) To calculate the number of shots that Sterling expects to hit the target, we use Eq. (9.2).

$$\sum_{x=0}^{4} x\,P(X = x) = 0\binom{4}{0}\left(\frac{1}{10}\right)^{4} + 1\binom{4}{1}\left(\frac{9}{10}\right)^{1}\left(\frac{1}{10}\right)^{3}$$

$$+ 2\binom{4}{2}\left(\frac{9}{10}\right)^{2}\left(\frac{1}{10}\right)^{2} + 3\binom{4}{3}\left(\frac{9}{10}\right)^{3}\left(\frac{1}{10}\right)$$

$$+ 4\binom{4}{4}\left(\frac{9}{10}\right)^{4}$$

$$= 0 + 0.0036 + 0.0972 + 0.8748 + 2.6244$$

$$= 3.6$$

Therefore, the number of shots that Sterling expects to hit the target is 3.6 shots out of 4.

(b) To calculate the variance of Sterling's shots, we use Eq. (9.5).

$$\sigma^{2} = \sum_{x=0}^{4} [x - E(X)]^{2}\,P(X = x)$$

$$= (0 - 3.6)^{2}\binom{4}{0}\left(\frac{1}{10}\right)^{4} + (1 - 3.6)^{2}\binom{4}{1}\left(\frac{9}{10}\right)^{1}\left(\frac{1}{10}\right)^{3}$$

$$+ (2 - 3.6)^{2}\binom{4}{2}\left(\frac{9}{10}\right)^{2}\left(\frac{1}{10}\right)^{2}$$

$$+ (3 - 3.6)^{2}\binom{4}{3}\left(\frac{9}{10}\right)^{3}\left(\frac{1}{10}\right)$$

$$+ (4 - 3.6)^{2}\binom{4}{4}\left(\frac{9}{10}\right)^{4}$$

$$= 0.0013 + 0.0243 + 0.1244 + 0.1050 + 0.1050$$

$$= 0.36$$

◄

Theorem 10.2.9 The expected value of a random variable X that follows a binomial distribution with n independent trials and probability of a success p on each trial is

$$E(X) = np. \tag{10.5}$$

The variance of X is

$$\text{var}(X) = np(1 - p). \tag{10.6}$$

Proof For this proof, we will require the combinatorial identity

$$x\binom{n}{x} = n\binom{n-1}{x-1}. \tag{10.7}$$

For the proof of this identity, please refer to Exercise 5.7.4 (c).

By definition, the expected value is $E(X) = \sum_{\text{all } x} x\, P(X = x)$. So, with the binomial distribution,

$$E(X) = \sum_{x=0}^{n} x\binom{n}{x} p^x(1-p)^{n-x} \qquad \text{Set index } x = 1. \text{ See Exercise 10.2.3.}$$

$$= \sum_{x=1}^{n} \underbrace{x\binom{n}{x}}_{n\binom{n-1}{x-1}} p^x(1-p)^{n-x} \qquad \text{By identity (10.7).}$$

$$= \sum_{x=1}^{n} n\binom{n-1}{x-1} p\, p^{x-1}(1-p)^{n-x} \qquad \text{By Theorem 6.5.3.}$$

$$= np \sum_{x=1}^{n} \binom{n-1}{x-1} p^{x-1}(1-p)^{n-x} \qquad \text{Shift } down \text{ the index } x \text{ by 1.}$$

$$= np \underbrace{\sum_{x=0}^{n-1} \binom{n-1}{x} p^x(1-p)^{n-1-x}}_{\text{Sum of } \mathcal{B}(n-1, p)}$$

$$= np(1)$$

$$= np$$

$$E(X^2) = \sum_{\text{all } x} x^2\, P(X = x)$$

$$= \sum_{x=0}^{n} x^2\binom{n}{x} p^x(1-p)^{n-x} \qquad \text{Set index } x = 1. \text{ See Exercise 10.2.3.}$$

$$= \sum_{x=1}^{n} x^2\binom{n}{x} p^x(1-p)^{n-x} \qquad x^2 = x \cdot x$$

$$= \sum_{x=1}^{n} x \, x \underbrace{\binom{n}{x}}_{n\binom{n-1}{x-1}} p^x (1-p)^{n-x} \qquad \text{By identity (10.7).}$$

$$= \sum_{x=1}^{n} x \cdot n \binom{n-1}{x-1} pp^{x-1}(1-p)^{n-x} \qquad \text{By Theorem 6.5.3.}$$

$$= np \sum_{x=1}^{n} x \binom{n-1}{x-1} p^{x-1}(1-p)^{n-x} \qquad \text{Shift } down \text{ the index } x \text{ by 1.}$$

$$= np \sum_{x=0}^{n-1} (x+1) \binom{n-1}{x} p^x (1-p)^{n-1-x} \qquad \text{Distribute } x+1.$$

$$= np \left[\underbrace{\sum_{x=0}^{n-1} x \binom{n-1}{x} p^x (1-p)^{n-1-x}}_{\text{Expected value of } \mathcal{B}(n-1, p)} \right.$$

$$\left. + \underbrace{\sum_{x=0}^{n-1} \binom{n-1}{i} p^i (1-p)^{n-1-i}}_{\text{Sum of } \mathcal{B}(n-1, p)} \right]$$

$$= np[(n-1)p + 1]$$

Then,

$$\text{var}(X) = \text{E}(X^2) - [\text{E}(X)]^2$$
$$= np[(n-1)p + 1] - (np)^2$$
$$= np[np - p + 1] - (np)^2$$
$$= np[\cancel{np} - p + 1 - \cancel{np}]$$
$$= np(1 - p).$$

◄

Exercises

10.2.1. Given the random variable $X \sim \mathcal{B}(10, 0.7)$, calculate

(a) $P(X = 0)$ (g) $P(0 < X < 2)$
(b) $P(X = 1)$ (h) $P(0 < X \leq 2)$
(c) $P(X = 2)$ (i) $P(X < 2)$
(d) $P(X > 2)$ (j) $P(X \leq 2)$
(e) $P(X \geq 2)$ (k) $E(X)$
(f) $P(0 \leq X \leq 2)$ (l) $\text{var}(X)$

10.2.2. Given the random variable $Y \sim \mathcal{B}(15, 0.3)$, calculate

(a) $P(Y = 2)$ (e) $P(Y \leq 1)$
(b) $P(Y = 1)$ (f) $P(1 \leq Y \leq 2)$
(c) $P(Y > 1)$ (g) $E(Y)$
(d) $P(Y < 1)$ (h) $\text{var}(Y)$

10.2.3. Explain why the following is true.

(a) $\displaystyle\sum_{x=0}^{n} x \binom{n}{x} p^x (1-p)^{n-x} = \sum_{x=1}^{n} x \binom{n}{x} p^x (1-p)^{n-x}$

(b) $\displaystyle\sum_{x=0}^{n} x^2 \binom{n}{x} p^x (1-p)^{n-x} = \sum_{x=1}^{n} x^2 \binom{n}{x} p^x (1-p)^{n-x}$

10.2.4. The probability of a business person canceling a reservation at the Kingston Hotel is estimated to be 20%.

(a) Generate and graph the probability distribution for the random variable that represents the number of business people canceling when there are five reservations.

(b) Use the probability distribution to determine the probability of at least four of the five reservations being canceled.

10.2.5. The probability that a marksman misses his target on any given shot is 0.10.

(a) If he fires six shots, what is the probability that the target is missed two or more times?

(b) If he fires 20 times, what is the expected number of misses?

10.2.6. It is known that 7% of computer chips produced by a machine are defective.

(a) Find the probability that out of 30 chips produced
  (i) exactly 3 will be defective   (ii) 3 or more will be defective

(b) What is the expected number of defective chips out of 30?

10.2.7. A local newspaper in Kingston has 35 000 subscribers, of whom 15 300 are women and the rest are men. Forty percent of men read the sports section while 33% of the women read the sports section. A random sample of 100 subscribers is picked. What is the expected number of subscribers in the random sample who read the sports section?

10.2.8. Reliability is defined as the proportion of non-defective items produced over the long term. A ballpoint pen manufacturer claims that 99% of their pens are reliable. If we assume the claim to be true, then in a random sample of 1000 pens, what is the expected number of defective pens?

10.2.9. In Canada, 46% of Canadians have type O blood.[1] Suppose we have randomly selected four people. Let X be the random number of persons out of the four people who have type O blood.

    (a) What is the probability that none of the four people have type O blood?

    (b) What is the probability that two or fewer have type O blood?

    (c) Out of the four persons, what is the expected number of them having type O blood?

10.2.10. A light-bulb manufacturer estimates that 0.5% of the bulbs manufactured are defective. What is the probability that two out of a set of five bulbs are found to be defective.

10.2.11. If the probability of hitting a target is $\frac{1}{7}$ and ten shots are fired independently, what is the probability of the target being hit at least three times?

10.2.12. A discrete mathematics student forgot to study for the multiple choice examination that has 20 questions. For each question there are five choices with only one being the correct answer. Suppose the student is able to eliminate one of the choices because it is obviously false. The student then makes a random guess from the remaining choices. Find the expected value and variance of the student's correct answers.

10.2.13. Consider a 12 member jury, where each member acts independently of each other. If each member has a probability 0.7 of correctly determining the guilt or innocence of a defendant, then what is the probability that the majority of the jurors will make the correct choice?

10.2.14. A military recruitment center is looking at evaluating recruit candidates. Each candidate must meet a minimum score which is less than an average score estimate which, in turn, is less than the above average score. (The score benchmarks are a secret, so we cannot tell you.) The outcome for any candidate is independent of the outcomes for all other candidates. The recruitment center has found, from past records, the following probabilities from the outcomes of candidates.

---

[1] Source: http://www.blood.ca/

Outcome	Probability
Candidate does not achieve the minimum score	0.15
Candidate achieves at least the minimum score	0.85
Candidate achieves at least the average score estimate	0.50
Candidate achieves at least the above average score	0.175

For example, the probability that a candidate achieves at least the average estimate but not the above average estimate is 0.325. A particular month includes exactly 40 candidates that may be assumed to be random sample of such candidates. Use binomial distributions to find the probability that

(a) at most 10 candidates do not achieve the minimum score?

(b) exactly 2 candidates achieve at least the above average score estimate?

(c) more than 10 but fewer than 15 candidates achieve at least the minimum score but not the average score estimate?

10.2.15. A secretary typing at a constant speed of 120 words per minute makes a mistake on any particular word with probability 0.02 independently from word to word. Each incorrect word must be corrected, and requires five seconds per correction. For a 500-word memo,

(a) how many seconds will the secretary spend on average making corrections?

(b) how many seconds does it take the secretary to complete the memo on average?

10.2.16 (Online advertising). A business wishes to purchase an advertisement link on a particular social media app. The probability that a user clicks on the advertisement link while browsing on the app is 0.1. Each user's probability of clicking on the link is independent from every other user.

(a) If there are 50 000 unique users browsing the app per day, how many people are expected to click on the business' link?

(b) If the business pays 10 cents per link click, how much should the business expect to pay per day for the advertisement link?

10.2.17 (Overbooking). A small restaurant has 10 tables. From past experience, the restaurant's manager estimates the chance that a party is a no-show for their reservation is 20%, assuming that each party/reservation is independent. Since it is expected that some people will cancel, the manager instated an overbooking policy—a policy that the restaurant will take more reservations than actual tables.

(a) If 11 reservations were made, what is the chance that all of 11 parties show up?

(b) If there is overbooking, the manager is comfortable with risking at most a 10% chance that all of the parties show up for their reservation. What is the maximum number of reservations the restaurant can make while staying within the manager's risk?

10.2.18. Amélie isn't suppose to eat candy before dinner, but she reaches inside of a candy jar anyway that contains 5 caramel squares and 10 pieces of brittle, and scoops out 3 pieces of candy. She discovers that she got 3 caramel squares—her favorite! Amélie is just about to eat them when her father, a cruel mathematician, catches her in the act. He says to her, "If you can repeat that performance 2 out of 3 times Amélie, I'll let you eat those delicious caramel squares. But if you fail, you'll not have any dessert or candy for the whole week!" What is the probability that Amélie will be enjoying those candies?

10.2.19. The United States Postal Service employs machines that helps to sort the mail based on ZIP code. Consider a computer program that examines a scan of a single handwritten digit. The program is able to correctly read the handwritten number 98.4% of the time.[2]
(a) If the computer program's correctness at predicting a handwritten digit is independent, what is the probability that the program reads all 5 digits in a ZIP code correctly?
(b) In 100 pieces of mail, what is the expected number of incorrectly read ZIP codes by the computer program?

10.3 The Poisson Distribution

In this section, we will see a probability distribution that allows us to determine the probability of a given number of events occurring in a period of time with a known average rate. We will be able to answer questions such as the following:
(a) What is the probability of a certain number of misprints on a page?
(b) What is the probability of a certain number of telephone calls were answered during a day?
(c) What is the probability of a certain number of alpha particles emitted by a radioactive source?
(d) What is the probability of a certain number of people in a city that will live to be 100 years old?

The probability distribution that enables us to answer such questions is the **Poisson distribution**. Siméon Denis Poisson (1781–1840) introduced this distribution in 1837 while he was investigating the probability of verdicts in criminal and civil matters.

---

[2] Source: LeCun (1989), Hastie *et al.* (2009, p. 407).

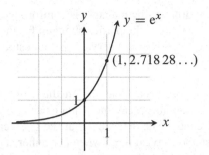

Fig. 10.5   The exponential function

The Poisson distribution relies on a special number in mathematics called $e \approx 2.718\,28$. It is a transcendental number that was investigated by the Swiss mathematician Leonhard Euler (1707–1783). Euler may have selected the letter e because of its relationship to the exponential function (see Fig. 10.5), but because of him, many mathematicians call it **Euler's number**.

The exponential function, as shown in Fig. 10.5, can be represented by an infinite series

$$e^x = \sum_{n=0}^{\infty} \frac{x^n}{n!} = 1 + \frac{x}{1!} + \frac{x^2}{2!} + \frac{x^3}{3!} + \frac{x^4}{4!} + \cdots \tag{10.8}$$

By setting $x = 1$, we get **Euler's number** e:

$$e = e^1 = \sum_{n=0}^{\infty} \frac{1}{n!} = 1 + \frac{1}{1!} + \frac{1}{2!} + \frac{1}{3!} + \frac{1}{4!} + \cdots \tag{10.9}$$

or approximately,

$$e \approx 2.718\,28. \tag{10.10}$$

Definition 10.3.1 — Poisson distribution. A random variable X has a **Poisson distribution** if

$$P(X = x) = \frac{\lambda^x e^{-\lambda}}{x!} \tag{10.11}$$

where
- $x = 0, 1, 2, \ldots$ is the number of events that will occur,
- λ is the average rate of event occurrence in a time period or space, and
- $e \approx 2.718\,28$ is Euler's number.

A random variable X that is Poisson-distributed is denoted by $X \sim \mathrm{Poi}(\lambda)$, where λ is the parameter.

The Poisson distribution arises when the event occurrences are

(a) independent of each other
(b) the probability that two or more events occur simultaneously is zero
(c) happen randomly in time or space
(d) uniformly (that is, the expected number of event occurrence in an interval is directly proportional to the length of the interval).

Example 10.3.2 A secretary works an eight-hour shift in a marketing department answering calls on a single telephone line with no call waiting feature. On average, she answers 25 phone calls during her shift. If the number of calls she answers follows a Poisson distribution, what is the probability that she will answer
    (a) 20 calls during her shift?   (b) x calls during her shift?

Besides the fact that in Example 10.3.2 we were told that the number of calls answered by the secretary follows a Poisson distribution, it makes sense that the calls answered follow the event conditions from which the Poisson distribution arises:

(a) the calls arriving are independent; the previous call arrival does not affect the arrival of the next call;
(b) the secretary will answer the call, and since it is a single telephone line with no call waiting feature, she can only answer one call at a time;
(c) the calls happen randomly in time; and
(d) the secretary expects to answer 25 calls during her shift, so for example, in 4 shifts she would expect to answer $25 \times 4 = 100$ calls.

Solution Let the random variable X represent the number of phone calls the secretary answers during her shift. Given that X follows a Poisson distribution, we set $\lambda = 25$, which is the given average number of phone calls she will answer in her shift.

(a) We are required to calculate $P(X = 20)$.

$$P(X = 20) = \frac{(25)^{20}e^{-25}}{20!} \approx 0.051\,92.$$

Therefore, the probability of the secretary answering exactly 20 calls in her eight-hour shift is approximately 5.19%.

(b) To calculate the probability of x number of calls answered in her shift,

$$P(X = x) = \frac{(25)^{x}e^{-25}}{x!}.$$

The graph of $P(X = x)$ for $x = 0, 1, 2, \ldots$ is in Fig. 10.6 on the next page.

◄

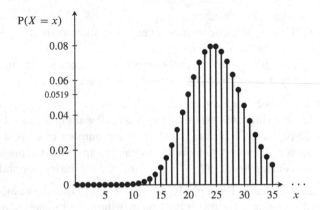

Fig. 10.6   Graph of the probability of the number of calls, x, the secretary will answer in her eight-hour shift. Note that from part (a), that $P(X = 20) \approx 0.0519$.

Theorem 10.3.3 The Poisson distribution satisfies condition (9.1) that the sum of all of its probability sums to one; that is,

$$\sum_{\text{all } x} P(X = x) = 1.$$

Proof

$$\sum_{\text{all } x} P(X = x) = \sum_{x=0}^{\infty} P(X = x)$$

$$= \sum_{x=0}^{\infty} \frac{\lambda^x e^{-\lambda}}{x!}$$

$$= e^{-\lambda} \underbrace{\sum_{x=0}^{\infty} \frac{\lambda^x}{x!}}_{e^{\lambda}} \qquad \text{By Theorem 6.5.3}$$

$$= e^{-\lambda} e^{\lambda} \qquad\qquad \text{By (10.8)}$$

$$= e^{-\lambda + \lambda}$$

$$= e^{0}$$

$$= 1$$

◄

The key parameter of the Poisson distribution is λ since it is the value of the expected value and the variance.

Theorem 10.3.4 The expected value of a random variable X that follows a Poisson distribution is

$$E(X) = \lambda \qquad (10.12)$$

and the variance is

$$\text{var}(X) = \lambda. \qquad (10.13)$$

Proof We will prove that the expected value is $E(X) = \lambda$.

$$E(X) = \sum_{\text{all } x} x\, P(X = x)$$

$$= \sum_{x=0}^{\infty} x\, \frac{e^{-\lambda}\lambda^x}{x!}$$

$$= \sum_{x=0}^{\infty} \frac{xe^{-\lambda}\lambda\lambda^{x-1}}{x!}$$

$$= \lambda e^{-\lambda} \sum_{x=0}^{\infty} \frac{x\lambda^{x-1}}{x!}$$

$$= \lambda e^{-\lambda} \sum_{x=0}^{\infty} \frac{\lambda^{x-1}}{(x-1)!}$$

$$= \lambda e^{-\lambda} \underbrace{\sum_{i=0}^{\infty} \frac{\lambda^i}{i!}}_{e^\lambda} \qquad i = x - 1$$

$$= \lambda e^{-\lambda} e^\lambda$$

$$= \lambda e^{-\lambda+\lambda}$$

$$= \lambda e^0$$

$$= \lambda$$

Now, we need to prove the variance, $\text{var}(X)$. Recall, the formula for variance is $\text{var}(X) = E(X^2) - [E(X)]^2$, but we need to determine $E(X^2)$.

$$E(X^2) = \sum_{\text{all } x} x^2\, P(X = x)$$

$$= \sum_{x=0}^{\infty} x^2 \frac{e^{-\lambda}\lambda^x}{x!}$$

$$= \sum_{x=0}^{\infty} \frac{xxe^{-\lambda}\lambda\lambda^{x-1}}{x!}$$

$$= \lambda \sum_{x=0}^{\infty} \frac{xe^{-\lambda}\lambda^{x-1}}{(x-1)!} \qquad\qquad i = x - 1$$

$$= \lambda \sum_{i=0}^{\infty} \frac{(i+1)e^{-\lambda}\lambda^{i}}{i!}$$

$$= \lambda \sum_{i=0}^{\infty} \frac{ie^{-\lambda}\lambda^{i} + e^{-\lambda}\lambda^{i}}{i!}$$

$$= \lambda \left(\underbrace{\sum_{i=0}^{\infty} \frac{ie^{-\lambda}\lambda^{i}}{i!}}_{E(X)=\lambda} + \underbrace{\sum_{i=0}^{\infty} \frac{e^{-\lambda}\lambda^{i}}{i!}}_{\sum_{\text{all } x} P(X=x)=1} \right)$$

$$= \lambda(\lambda + 1)$$

Then,

$$\mathrm{var}(X) = E(X^2) - [E(X)]^2$$
$$= \lambda(\lambda + 1) - \lambda^2$$
$$= \lambda^2 + \lambda - \lambda^2$$
$$= \lambda.$$

◄

The next example is one of the first applications of the Poisson distribution applied to real data.

Example 10.3.5 — Killed by a Horse. Ladislaus von Bortkewitsch[3] (1868–1931), a Polish-born Russian economist and statistician, published a book in 1898 entitled *Das Gesetz der kleinen Zahlen* (English: The Law of Small Numbers). In one of his examples, he studied the data of Prussian soldiers killed by a horse over a 20 year span, from 1875–1894. During that time, cavalry (horse) units were in use until the development and transition to tank warfare in World War I. The death of a soldier by a horse was a relatively rare event, so von Bortkewitsch believed that he could use the Poisson distribution to calculate the probability of the number of accidents in an Army corps per year. Von Bortkewitsch calculated the average number of accidents per corps per year and the variance, and found that the mean and the variance was the same. Thus, it was reasonable to assume that the number of accidents follows a Poisson distribution. Using the data collected by von Bortkewitsch in Table 10.5, develop a Poisson distribution model

    (a) to determine what is the probability of

---

[3]Polish name: Władysław Bortkiewicz

Table 10.5 Number of Prussian soldiers killed by horse kick by year and corps. Data from von Bortkewitsch (1898, p. 24).

Year	II	III	IV	V	VII	VIII	IX	X	XIV	XV	Total
					Army Corps						
1875	0	0	0	0	1	1	0	0	1	0	3
1876	0	0	1	0	0	0	0	0	1	1	3
1877	0	0	0	0	1	0	0	1	2	0	4
1878	2	1	1	0	0	0	0	1	1	0	6
1879	0	1	1	2	0	1	0	0	1	0	6
1880	2	1	1	1	0	0	2	1	3	0	11
1881	0	2	1	0	1	0	1	0	0	0	5
1882	0	0	0	0	0	1	1	2	4	1	9
1883	1	2	0	1	1	0	1	0	0	0	6
1884	1	0	0	0	1	0	0	2	1	1	6
1885	0	0	0	0	0	0	2	0	0	1	3
1886	0	0	1	1	0	0	1	0	3	0	6
1887	2	1	0	0	2	1	1	0	2	0	9
1888	1	0	0	1	0	0	0	0	1	0	3
1889	1	1	0	1	0	0	1	2	0	2	8
1890	0	2	0	1	2	0	2	1	2	2	12
1891	0	1	1	1	1	1	0	3	1	0	9
1892	2	0	1	1	0	1	1	0	1	0	7
1893	0	0	0	1	2	0	0	1	0	0	4
1894	0	0	0	0	0	1	0	1	0	0	2
Total	12	12	8	11	12	7	13	15	24	8	122

  (i) no deaths in a corps in a year?
  (ii) there is at least one death in a corps in a year?
  (iii) more than one death in a corps in a year?
  (iv) no deaths in a corps over 5 years?
 (b) to graph the probability distribution of the number of deaths in a corps in a year; and
 (c) to compare the actual number of deaths to the expected number of deaths.

Solution Let the random variable X be the number of accidents in a corps for one year. We begin by calculating the average, which we will use as the expected value for X. Using Table 10.5, there were 122 accidents out of 200 data points, so the

Basic Discrete Mathematics

average is

$$\frac{122}{200} = 0.61.$$

Hence, we let the expected value of X be $E(X) = \lambda = 0.61$. Thus, the probability distribution is

$$P(X = x) = \frac{(0.61)^x e^{-0.61}}{x!}.$$

(a) (i) The probability of no deaths in a corps in a year is

$$P(X = 0) = \frac{(0.61)^0 e^{-0.61}}{0!}$$

$$= e^{-0.61}$$

$$= 0.543\,35.$$

(ii) The probability that there is at least one death in a corps in a year is

$$P(X \geq 1) = 1 - P(X < 1)$$

$$= 1 - P(X = 0)$$

$$= 1 - 0.543\,35$$

$$= 0.456\,65.$$

(iii) The probability of more than one death in a corps in a year is

$$P(X > 1) = 1 - P(X \leq 1)$$

$$= 1 - [P(X = 0) + P(X = 1)]$$

$$= 1 - \left[\frac{(0.61)^0 e^{-0.61}}{0!} + \frac{(0.61)^1 e^{-0.61}}{1!} \right]$$

$$= 1 - [0.543\,35 + 0.331\,44]$$

$$= 1 - 0.874\,79$$

$$= 0.125\,21.$$

(iv) This is slightly different because we are asked not about the number of accidents in *one* year, but *five* years. So, we introduce another random variable Y to be the number of accidents in a corps over *five* years. We make the assumption that the expected value of Y would be the expected accidents in a year in a corps multiplied by 5 years; that is

$$E(Y) = \lambda = 0.61 \times 5 = 3.05.$$

Hence, the probability distribution of Y is

$$P(Y = y) = \frac{(3.05)^y e^{-3.05}}{y!}.$$

Therefore, the probability of no deaths in a corps over five years is

$$P(Y = 0) = \frac{(3.05)^0 e^{-3.05}}{0!}$$

$$= e^{-3.05}$$

$$= 0.047\,36.$$

(b) The probability distribution is given in Table 10.6, and the graph is in Fig. 10.7 is on the next page.

Table 10.6   Probability distribution of x number of deaths of soldiers in Example 10.3.5

x	$P(X = x)$	
0	$\dfrac{(0.61)^0 e^{-0.61}}{0!}$	$\approx 0.543\,35$
1	$\dfrac{(0.61)^1 e^{-0.61}}{1!}$	$\approx 0.331\,44$
2	$\dfrac{(0.61)^2 e^{-0.61}}{2!}$	$\approx 0.101\,09$
3	$\dfrac{(0.61)^3 e^{-0.61}}{3!}$	$\approx 0.020\,56$
4	$\dfrac{(0.61)^4 e^{-0.61}}{4!}$	$\approx 0.003\,13$
5	$\dfrac{(0.61)^5 e^{-0.61}}{5!}$	$\approx 0.000\,38$

(c) The comparison of actual number of deaths to the expected number of deaths is given in Table 10.7.

Table 10.7   The actual number of deaths compared to the expected number of deaths in the Prussian Army.

Number of deaths in a corps in a year, x	Actual number of corps with x deaths in a year	Expected number of 200 corps with x deaths in a year, $200 \times P(X = x)$
0	109	$200(0.543\,35) = 108.67$
1	65	$200(0.331\,44) = 66.29$
2	22	$200(0.101\,09) = 20.22$
3	3	$200(0.020\,56) = 4.11$
4	1	$200(0.003\,13) = 0.63$
5 or more	0	$200(0.000\,44) = 0.09$

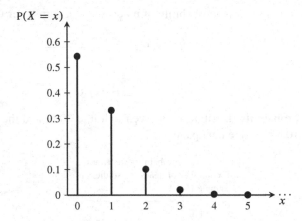

Fig. 10.7   Graph of the probability distribution of *x* number of deaths of soldiers in Example 10.3.5

◀

Consider the two patterns in Fig. 10.8. Which of the two scatterplots are random? Before, reading on, decide for yourself. Is the scatterplot on the left more random than the one on the right? Or, is the scatterplot on the right more random than the one on left?

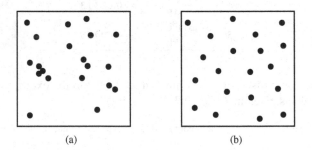

(a)   (b)

Fig. 10.8   Two plots of "random numbers." Which plot is more random?

The pattern on the left is in fact more random; this surprises most people. The scatterplot on the left was generated by atmospheric noise,[4] while the scatterplot on the right was created by me placing points on a grid with a couple of variations. This is known as the **clustering illusion**, which appears as streaks or clusters in small samples from a random distribution. Humans are so accustomed to finding patterns (see Sec. 6.1) that we can sometimes see patterns where there are no patterns to be found.

---

[4]The data was generated from www.random.org.

Example 10.3.6 — V-1 Flying Bomb Attacks on London. During the doodlebug raids on London, England in World War II, it was believed that the German V-1 flying-bombs (German: *Vergeltungswaffe 1*), an early form of cruise missile, landed in clusters—that is, the Germans were using some sort of guidance systems to bomb specific targets.[5] To disprove this, the British statistician R. D. Clarke (1946) examined an area of South London comprised of 144 square kilometers. Then, the area was subdivided into 576 squares of $\frac{1}{4}$ square kilometers each, and the number of bombs per square was recorded; see Table 10.8. In the 576 squares, there were 537 bomb strikes yielding an average of $\frac{537}{576} \approx 0.932$ bomb strikes per square. Clarke believed that the bombings were not clustering, but instead the number of bombs that landed in a square followed a Poisson distribution and hence, they were randomly distributed.

If we let the random variable X follow a Poisson distribution with $\lambda = 0.932$ (which means that there is on average about 1 bomb per square), we can calculate the theoretical probability of the number of bomb strikes in a square. The probability distribution of X is

$$P(X = x) = \frac{(0.932)^x e^{-0.932}}{x!} \tag{10.14}$$

which is given in Fig. 10.9.

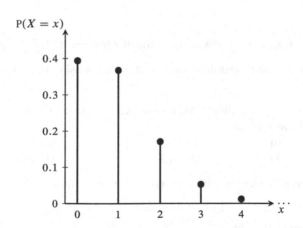

Fig. 10.9   Graph of the probability distribution of the number of bombs landing in a square.

We can use $P(X = x)$ as a proportion of the 576 squares that have x bomb hits; see Table 10.8.

---

[5]The V-1's guidance system used a mechanical autopilot to regulate altitude (barometric device), yaw and pitch (a pair of gyroscopes), and azimuth control (Zaloga, 2005, p. 8). The V-1 was pointed in the general direction of the target, launched, and the guidance system controlled the flight.

Table 10.8   Number of German V1 flying-bomb strikes in 576 quarter square-kilometer squares in London during World War II. Data from Clarke (1946).

Number of bomb strikes in a square, x	Actual number of squares with x bomb strikes	Expected number of 576 squares with x bomb strikes, $576 \times \mathrm{P}(X = x)$
0	229	$576(0.393\,77) = 226.812$
1	211	$576(0.366\,99) = 211.386$
2	93	$576(0.171\,02) = 98.508$
3	35	$576(0.053\,13) = 30.603$
4	7	$576(0.012\,38) = 7.131$
5 or more	1	$576(0.002\,71) = 1.561$
Total	537	536.560

Table 10.8 indicates the bombs landed randomly in London, and were not targeting any particular building or landmark. They killed indiscriminately like a devastating game of Russian roulette played upon the entire population of London. To see an interactive map of all the World War II bomb strikes in London, check out www.bombsight.org. ◀

Exercises

10.3.1. Using Eq. (10.9), estimate the value of e for $n = 1, 2, 3, 4, 5, 6$.

10.3.2. Calculate the probabilities of $x = 0, 1, 2, 3, 4$ using Eq. (10.14) in Example 10.3.6.

10.3.3. If X is a Poisson-distributed random variable with an expected value of 4, find (to five decimal places):
   (a) $\mathrm{P}(X = 0)$ (c) $\mathrm{P}(X = 2)$ (e) $\mathrm{P}(X < 2)$
   (b) $\mathrm{P}(X = 1)$ (d) $\mathrm{P}(X \geq 2)$ (f) $\mathrm{P}(X < 3)$

10.3.4. If X is a Poisson-distributed random variable, and

$$\mathrm{P}(X = 4) = 3\,\mathrm{P}(X = 3),$$

then find λ and $\mathrm{P}(X = 5)$.

10.3.5. Assuming the weekly demand for a laptop is a Poisson random variable with expectation of 3. What is the probability that the store will sell
   (a) at least 2 in a week?   (c) less than 5 in a week?
   (b) at most 5 in a week?

10.3.6. Incoming requests to a library catalogue database arrive at random times. On Wednesday afternoons, during the academic year, there is a constant average rate of 10 requests per hour. What is the probability of receiving 4 or more requests in a particular hour?

10.3.7 (Errors on a Page). Suppose that the number of typographical errors on a single page of this book follows a Poisson distribution with parameter $\lambda = 0.2$. What is the probability that this particular page has

(a) only one error? (b) at least two errors?

10.3.8. A page in a technical report is expected to have two grammatical errors. What is the probability that a particular page has

(a) no errors? (c) two errors?
(b) one error? (d) more than two errors?

10.3.9 (Living to 100). On average in Canada, 2049 males out of 100 000 live to be 100 years old. What is the probability that there will be more than one male who lives to be 100 years old?

10.3.10. A grocery store rents rug cleaners to the general public at $50 per day. The average number of rentals is 3.1 per day.

(a) Calculate the expected daily income from the rentals (assuming that there is an unlimited supply of rug cleaners available).
(b) The demand for rug cleaners follows a Poisson distribution. On a particular day, what is the probability that
  (i) no rug cleaners are rented? (iv) 3 rug cleaners or more are
  (ii) 1 rug cleaner is rented? rented?
  (iii) 2 rug cleaners are rented?
(c) If one of the rug cleaners is broken, then what is the expected daily income from the rentals?

10.3.11. Potterat *et al.* (1990) analyzed two decades (1970–1990) of data in Colorado Springs, CO, USA. They found that the number of female prostitutes to be about 23 per 100 000 people, or about 0.023% of the population.

(a) In 2014, Colorado Springs had an estimated population of 445 830 people. Estimate the number of female prostitutes in the city.
(b) Colorado Springs has an area of 194.7 square miles. Estimate the number of female prostitutes per square mile.
(c) Suppose that female prostitutes in Colorado Springs are distributed at random in the city. If a sexologist, a person who engages in the scientific study of human sexuality, selects an area of 4 square miles at random in the city for further study, then what is the probability of the sexologist will find at least 3 female prostitutes in this area?

10.3.12. A Petri dish, named after the German microbiologist Julius Richard Petri (1852–1921), is a shallow plastic lidded dish that is used to culture (grow) cells such as bacteria. A Petri dish with bacterial colonies is shown with a grid in Fig. 10.10.

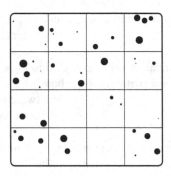

Fig. 10.10   A sample of bacterial colonies in a Petri dish. The size of each colony can vary, but each dot represents one bacterial colony.

   (a) What is the average number of bacterial colonies in a square of the grid in the Petri dish?
   (b) If the number of bacterial colonies in a square follows a Poisson distribution, then what is the probability of there being in a square

(i) no bacterial colony?	(v) 4 bacterial colonies?
(ii) 1 bacterial colony?	(vi) 5 bacterial colonies?
(iii) 2 bacterial colonies?	(vii) 6 bacterial colonies?
(iv) 3 bacterial colonies?	(viii) 6 or more bacterial colonies?

   (c) Fill in Table 10.9.

Table 10.9   Number of bacterial colonies in a square for Exercise 10.3.12.

Number of bacterial colonies in a square, x	Actual number of squares with x bacterial colonies	Expected number of 16 squares with x bacterial colonies, $16\,P(X = x)$
0		
1		
2		
3		
4		
5		
6 or more		

10.3.13. The number of telephone calls to a help desk for technical issues often follows a Poisson distribution. If there is an average of 7.2 calls per hour, what is the probability that there are

(a) exactly 5 calls in one hour?

(b) 4 or less calls in one hour?

(c) exactly 5 calls in *two* hours?

(d) exactly 2 calls in *half* an hour?

10.3.14 (†). Show that $\sum_{n=1}^{\infty} \frac{n^2}{n!} = 2e$. (*Hint*: $n^2 = n + n(n-1)$.)

10.4 The Geometric Distribution

The next probability distribution that we will investigate answers questions about the probability of how many trials of an experiment we will have to perform in order to see a particular outcome for the first time.

Example 10.4.1 In the board game, Monopoly, in order to get out of jail you must either roll doubles or pay $50 to the bank. What is the probability that you will get out of jail on the third roll (without paying)?

Solution The probability of getting doubles is $\frac{6}{36} = \frac{1}{6}$ (see Table 8.1 on p. 313), and by the Complement Rule, the probability of not getting doubles is $1 - \frac{1}{6} = \frac{5}{6}$. Thus,

- the probability of not getting doubles on the 1st roll is $\frac{5}{6}$,
- the probability of not getting doubles on the 2nd roll is $\frac{5}{6}$, and
- the probability of getting doubles on the 3rd roll is $\frac{1}{6}$.

Then, since these rolls are independent, we can multiply these probabilities together; that is,

$$\left(\frac{5}{6}\right)\left(\frac{5}{6}\right)\left(\frac{1}{6}\right) = \left(\frac{5}{6}\right)^2 \left(\frac{1}{6}\right) = \frac{25}{216}.$$

Therefore, the probability of getting out of jail on the 3rd roll is $\frac{25}{216}$. ◀

Example 10.4.1 consisted of rolling dice repeatedly. These rolls can be viewed as identical, independent trials which result in either a success or a failure. The example was concerned about when we would first see a success (seeing doubles). The general probability distribution that describes these kinds of experiments is the **geometric distribution**.

> **Definition 10.4.2 — The Geometric Distribution.** A random variable X has a **geometric distribution** if
>
> $$P(X = x) = (1-p)^{x-1}p \qquad (10.15)$$
>
> where
> - p is the probability of success on each trial, and

- $x = 1, 2, 3, \ldots$.

The random variable X is the number of completed trials when the first success occurs.

Example 10.4.3 Calculate the probability distribution for getting out of jail in Monopoly in x rolls of the dice.

Solution Let X be the random variable of the number of rolls when you first observe doubles. Then, the probability distribution of X is

$$P(X = x) = \left(\frac{5}{6}\right)^{x-1} \left(\frac{1}{6}\right).$$

The probability distribution can be either given as Table 10.10, or as a graph in Fig. 10.11.

Table 10.10 The probability distribution
for rolling doubles on the xth roll

x	$P(X = x)$
1	$\left(\frac{5}{6}\right)^{0} \left(\frac{1}{6}\right) = \frac{1}{6} \approx 0.166\,67$
2	$\left(\frac{5}{6}\right)^{1} \left(\frac{1}{6}\right) = \frac{5}{36} \approx 0.138\,89$
3	$\left(\frac{5}{6}\right)^{2} \left(\frac{1}{6}\right) = \frac{25}{216} \approx 0.115\,74$
4	$\left(\frac{5}{6}\right)^{3} \left(\frac{1}{6}\right) = \frac{125}{1296} \approx 0.096\,45$
5	$\left(\frac{5}{6}\right)^{4} \left(\frac{1}{6}\right) = \frac{625}{7776} \approx 0.080\,38$
$\vdots$	$\vdots$

The distribution will continue on forever since you can theoretically continue to roll the dice and not observe doubles. However, the probability of not observing doubles for the xth roll converges to zero since the probabilities must all sum to one.   ◄

The geometric distribution also satisfies condition (9.1) that the sum of all of its probability sums to one; that is,

$$\sum_{\text{all } x} P(X = x) = 1.$$

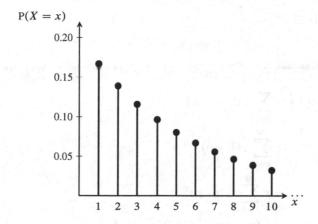

Fig. 10.11 Graph of the probability distribution for Example 10.4.3

Proof

$$\sum_{\text{all }x} P(X = x) = \sum_{x=1}^{\infty} (1 - p)^{x-1} p$$

$$= \sum_{x=1}^{\infty} \underbrace{p(1 - p)^{x-1}}_{a=p,\ r=1-p} \qquad \text{Infinite geometric series (Theorem 6.9.2)}$$

$$= \frac{p}{1 - (1 - p)}$$

$$= \frac{p}{p}$$

$$= 1$$

◄

The next question that we can answer with the geometric distribution is when we *expect* to see a success. This can be answered by calculating the expected value.

Theorem 10.4.4 The expected value of a random variable X with a geometric distribution is

$$E(X) = \frac{1}{p} \qquad (10.16)$$

and the variance is

$$\text{var}(X) = \frac{1-p}{p^2}. \tag{10.17}$$

Proof

$$E(X) = \sum_{\text{all } x} x\,P(X = x)$$

$$= \sum_{x=1}^{\infty} x(1-p)^{x-1}p$$

$$= p\sum_{x=1}^{\infty} x(1-p)^{x-1}$$

$$= p\left[1 + 2(1-p) + 3(1-p)^2 + 4(1-p)^3 + \cdots\right]$$

Next, we use the same style of argument that we used to prove the infinite geometric series (see p. 237).

$E(X)$	$=$	$p[1$	$+$	$2(1-p)$	$+$	$3(1-p)^2$	$+$	$4(1-p)^3$ $+\cdots]$
$-(1-p)\,E(X)$	$=$	$-p[$		$(1-p)$	$+$	$2(1-p)^2$	$+$	$3(1-p)^3$ $+\cdots]$
$E(X) - (1-p)\,E(X)$	$=$	$p[1$	$+$	$(1-p)$	$+$	$(1-p)^2$	$+$	$(1-p)^3$ $+\cdots]$

Then, we solve for $E(X)$

$$E(X) - (1-p)\,E(X) = p\underbrace{[1 + (1-p) + (1-p)^2 + (1-p)^3 + \cdots]}_{\text{Geometric series, } a = 1, r = 1 - p}$$

$$E(X) - (1-p)\,E(X) = p\frac{1}{1-(1-p)}$$

$$E(X)[1 - (1-p)] = \frac{p}{1-(1-p)}$$

$$E(X) = \frac{p}{[1-(1-p)]^2} = \frac{p}{(1-1+p)^2} = \frac{p}{p^2} = \frac{1}{p}.$$

◀

Example 10.4.5 In Monopoly, how many times will you expect to roll the dice in order to get out of jail?

Solution Let X be a random variable that follows a geometric distribution with probability of success $\frac{1}{6}$. Then,

$$E(X) = \frac{1}{p} = \frac{1}{\frac{1}{6}} = 6.$$

Therefore, on average we expect to see doubles appear on the 6th roll of the dice; that is, you expect to get out of jail on the 6th roll. Perhaps, you may want to consider paying the $50 to the bank.   ◀

Example 10.4.6 You are standing at a particular corner waiting to cross a busy street. To cross the street, it takes you 7 seconds. If there is traffic 80% of the 7-second intervals, then how long do you expect to wait before you can cross the street?

Solution Each 7-second interval can be viewed as identical, independent trials in which the probability of a success (no traffic in a 7-second interval) is 0.2. Let X be a random variable that follows a geometric distribution with $p = 0.2$. Then,

$$E(X) = \frac{1}{p} = \frac{1}{0.2} = 5.$$

Thus, you expect to wait for 4 intervals, and then to cross the street on the 5th interval. Therefore, you will expect to wait $4 \times 7 = 28$ seconds. ◀

In general, if you would like to calculate the expected waiting time until a success occurs, then

$$E(X) - 1 = \frac{1}{p} - 1 = \frac{1}{p} - \frac{p}{p} = \frac{1-p}{p}. \tag{10.18}$$

Exercises

10.4.1. If X is a geometrically distributed random variable with expected value 5, find

(a) $P(X = 1)$ (c) $P(X = 3)$ (e) $P(X \geq 3)$

(b) $P(X = 2)$ (d) $P(X < 3)$ (f) $\text{var}(X)$

10.4.2. Create a table and a graph of the geometric distribution for $x = 1, 2, 3, 4$, where the probability of success on each trial is

(a) $p = 0.2$ (b) $p = 0.5$ (c) $p = 0.7$

10.4.3. In an experiment of repeatedly rolling a single die, what is the probability that you will roll a 3 for the first time on the fourth roll?

10.4.4. In an experiment of repeatedly rolling a single die, what is the probability that you will roll a 3 or a 4 for the first time on the fifth roll?

10.4.5. There are 38 numbers on a Roulette wheel (see Fig. 8.29 on p. 366). If you bet on the same number for each spin, what is the expected number of spins required for you to win?

10.4.6. A professor wants to open her door. She has ten keys of which one fits her office door. She randomly keeps trying keys so that each try each key has probability $\frac{1}{10}$ of being tried. What is the probability the professor will succeed within three tries?

10.4.7. A local coffee shop has a promotion where 1 in 12 cups has a hidden message on the bottom of the cup indicating to the customer that they have won a free doughnut.

(a) What is your chance of winning a free doughnut
    (i) on your first cup?   (iii) within 4 cups purchased?
    (ii) on your second cup?
(b) How many cups do you expect to have purchased when you win a free doughnut?
(c) Your friend Sally bought 14 cups of coffee from the shop already, and she still has not won. Should she complain to a consumer watch group?

10.4.8. On a game show, the finalist randomly selects one of three doors in which the grand prize is behind one of them. What is the chance that for three consecutive shows, no finalist wins the grand prize?

10.4.9. On a multiple-choice examination with four possible answers per question, how many questions will a student who is randomly guessing expect to answer incorrectly before the student gets one correct?

10.4.10. What is the expected number of people you would have to ask in order to find someone who has the same birthday as you?

10.4.11. At a call center, Casey randomly selects people in a city until someone answers. The chance that someone will answer the phone is 45% of the time.
(a) What is the probability that Casey will
    (i) talk to someone in 3 phone calls?
    (ii) have to make more than 2 phone calls?
(b) What is the variance of the number of calls he will have to make?
(c) What is the standard deviation of the number of calls he will have to make?

10.4.12. Céline has a success rate of 78% for scoring a goal in soccer. How many goals will she expect to score before she misses?

10.4.13. Another form of the geometric distribution is

$$P(X = x) = (1 - p)^x p \qquad (10.19)$$

where
- p is the probability of success on each trial, and
- $x = 0, 1, 2, 3, \ldots$.

The random variable X is the number of trials *before* the first success occurs with expectation $E(X) = \frac{1-p}{p}$. Thus, X can be the number of trials you must wait, or the **waiting period**. Why is (10.19) equivalent to (10.15)?

10.4.14 (†). Suppose that a particular production line has a 10% chance of producing a defective computer memory chip. The probability that a quality control inspector will have to check at least x chips to find the first defective chip is 0.05. What is the value of x?

10.5 The Hypergeometric Distribution

A new probability distribution arises when we try to describe an experiment whose repeated trials are *not independent*. Drawing cards without returning them to the deck or balls from an urn without putting them back in the urn are examples in which the probability of the outcomes change from trial to trial. This kind of experiment that exhibits random selection without replacement follows a **hypergeometric distribution**. We will provide the following example, and then we will generalize in order to state the distribution.

Example 10.5.1 — Quality Control. Suppose we have a batch of 50 discs of which 12% are defective. An inspector randomly selects 15 discs without replacement to test. These 15 discs are called a **sample** from the batch of 50 discs.

    (a) What is the probability that the sample contains
        (i) 3 defective discs?   (ii) x defective discs?
    (b) State the probability distribution for the number of defective discs as a table.
    (c) Graph the probability distribution.

Solution First, it is important to note that in the batch of 50 discs, there are $0.12 \times 50 = 6$ defective discs.

    (a)   (i) Let E be the event that the inspector finds 3 defective discs in the sample. We would like to know $n(E)$; that is, we need to calculate how many samples include exactly 3 defective discs. The number of ways 3 defective discs can be chosen from 6 defective discs is $\binom{6}{3} = 20$. Then the remaining $15 - 3 = 12$ discs of the sample are selected from the $50 - 6 = 44$ non-defective discs; the number of ways to do is $\binom{50-6}{15-3} = \binom{44}{12}$. So, by the Multiplication Principle,

$$n(E) = \overbrace{\binom{6}{3}}^{\text{Defective}} \times \overbrace{\binom{50-6}{15-3}}^{\text{Non-defective}}$$

$$= \binom{6}{3} \times \binom{44}{12}$$

$$= 20 \times 21\,090\,682\,613$$

$$= 421\,813\,652\,260.$$

The sample space S is the set of all possible samples of 15 discs from the batch of 50. Thus,

$$n(S) = \binom{50}{15} = 2\,250\,829\,575\,120.$$

Hence, by the fundamental definition of probability,

$$P(E) = \frac{n(E)}{n(S)}$$

$$= \frac{\binom{6}{3}\binom{50-6}{15-3}}{\binom{50}{15}}$$

$$= \frac{421\,813\,652\,260}{2\,250\,829\,575\,120}$$

$$\approx 0.18740.$$

Therefore, the probability of 3 defective discs in the sample is 18.7%.

(ii) Let X be the number of defective discs that the inspector finds in the sample. We can use similar reasoning in part (a), and simply replace 3 defective discs with x defective discs. Therefore,

$$P(X = x) = \frac{\binom{6}{x}\binom{50-6}{15-x}}{\binom{50}{15}}$$

where $x = 0, 1, 2, \ldots, 6$.

(b) The values of $P(X = x)$, for $x = 0, 1, 2, \ldots, 6$ is given as a probability distribution in Table 10.11.

(c) The probability distribution of X is plotted in Fig. 10.12.

◄

In general, if there is a batch of N items with r defective items, and an inspector samples n items, then the probability that the inspector will have x defective items in the sample is

$$P(X = x) = \frac{\binom{r}{x}\binom{N-r}{n-x}}{\binom{N}{n}} \tag{10.20}$$

and this gives us the hypergeometric distribution.

Table 10.11 Probability distribution for Example 10.5.1

x	$P(X = x)$
0	$\dfrac{\binom{6}{0}\binom{50-6}{15-0}}{\binom{50}{15}} = \dfrac{\binom{6}{0}\binom{44}{15}}{\binom{50}{15}} \approx 0.102\,15$
1	$\dfrac{\binom{6}{1}\binom{50-6}{15-1}}{\binom{50}{15}} = \dfrac{\binom{6}{1}\binom{44}{14}}{\binom{50}{15}} \approx 0.306\,44$
2	$\dfrac{\binom{6}{2}\binom{50-6}{15-2}}{\binom{50}{15}} = \dfrac{\binom{6}{2}\binom{44}{13}}{\binom{50}{15}} \approx 0.345\,98$
3	$\dfrac{\binom{6}{3}\binom{50-6}{15-3}}{\binom{50}{15}} = \dfrac{\binom{6}{3}\binom{44}{12}}{\binom{50}{15}} \approx 0.187\,40$
4	$\dfrac{\binom{6}{4}\binom{50-6}{15-4}}{\binom{50}{15}} = \dfrac{\binom{6}{4}\binom{44}{11}}{\binom{50}{15}} \approx 0.051\,11$
5	$\dfrac{\binom{6}{5}\binom{50-6}{15-5}}{\binom{50}{15}} = \dfrac{\binom{6}{5}\binom{44}{10}}{\binom{50}{15}} \approx 0.006\,61$
6	$\dfrac{\binom{6}{6}\binom{50-6}{15-6}}{\binom{50}{15}} = \dfrac{\binom{6}{6}\binom{44}{9}}{\binom{50}{15}} \approx 0.000\,31$

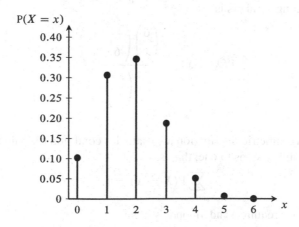

Fig. 10.12 Graph of the probability distribution for Example 10.5.1

Definition 10.5.2 — Hypergeometric distribution. A random variable X follows a **hypergeometric distribution** if

$$P(X = x) = \frac{\binom{r}{x}\binom{N-r}{n-x}}{\binom{N}{n}} \tag{10.21}$$

where

- n is the number of items in the sample
- x is the number of successes in the sample
- N is the number of items in the batch (or the population)
- r is the number of successes in the batch (or the population).

Example 10.5.3 Consider a national lottery in which the six winning numbers are drawn from the set $\{1, 2, \ldots, 49\}$ without replacement. A player wins the jackpot if all six numbers on their ticket matches the winning numbers. What is the probability that the ticket has x of the 6 winning numbers?

Solution Let X be the number of winning numbers on the ticket. Then X follows a hypergeometric distribution with parameters $N = 49$ (we are sampling from 49 numbers), $n = r = 6$ (the sample consists of 6 numbers, and the number of winning numbers is 6 as well). Therefore, the probability of the player's ticket having x winning numbers is

$$P(X = x) = \frac{\binom{6}{x}\binom{49-x}{6-x}}{\binom{49}{6}}.$$

◄

The hypergeometric distribution also satisfies condition (9.1) that the sum of all of its probability sums to one; that is,

$$\sum_{\text{all x}} P(X = x) = 1.$$

To show this, we require Vandermonde's identity.

Theorem 10.5.4 — Vandermonde's identity. For $r \geq 0$, and $s \geq 0$, then

$$\sum_{x=0}^{n} \binom{r}{x}\binom{s}{n-x} = \binom{r+s}{n}. \tag{10.22}$$

Proof — Combinatorial approach. We consider the following motivating question. From a discrete mathematics class consisting of r males and s females, how many ways can we form a committee with n students?

(a) We first select x males for the committee from r males in the class; there are $\binom{r}{x}$ ways to accomplish this. Next, we select the remaining $n - x$ committee members from the group of s female students; there are $\binom{s}{n-x}$ ways to do this. By the Multiplication Principle, $\binom{r}{x}\binom{s}{n-x}$. Since x could be $0, 1, 2, \ldots, n$, these cases can be combined by the Sum Principle. Therefore, there are altogether $\sum_{x=0}^{n} \binom{r}{x}\binom{s}{n-x}$ ways to form such committees.

(b) There are $r + s$ students, and we need to select n of them; there are $\binom{r+s}{n}$ ways to do this.

Since (a) and (b) are equivalent ways of answering the question,

$$\sum_{x=0}^{n} \binom{r}{x}\binom{s}{n-x} = \binom{r+s}{n}.$$

◀

With Vandermonde's identity proved, we can use it to show that the sum of a hypergeometric distribution is one.

Proof

$$\sum_{\text{all } x} P(X = x) = \sum_{x=0}^{n} \frac{\binom{r}{x}\binom{N-r}{n-x}}{\binom{N}{n}}$$

$$= \frac{1}{\binom{N}{n}} \sum_{x=0}^{n} \binom{r}{x}\binom{N-r}{n-x} \qquad \text{By Vandermonde's identity}$$

$$= \frac{1}{\binom{N}{n}} \binom{r + N - r}{n} \qquad r + N - r = N$$

$$= \frac{\binom{N}{n}}{\binom{N}{n}}$$

$$= 1$$

◀

Theorem 10.5.5 The expected value of a random variable X that follows a hypergeometric distribution is

$$E(X) = \frac{rn}{N} \qquad (10.23)$$

and the variance is

$$\text{var}(X) = \frac{rn}{N}\left[\frac{(r-1)(n-1)}{N-1} + 1 - \frac{rn}{N}\right]. \tag{10.24}$$

Proof For this proof we will require two combinatorial identities:

$$x\binom{r}{x} = r\binom{r-1}{x-1} \tag{10.25}$$

and

$$\binom{N}{n} = \frac{N}{n}\binom{N-1}{n-1}. \tag{10.26}$$

For proofs of these identities, please refer to Exercise 5.7.4 (c).

$$
\begin{aligned}
E(X) &= \sum_{\text{all } x} x\,P(X = x) \\
&= \sum_{x=0}^{n} x\,\frac{\binom{r}{x}\binom{N-r}{n-x}}{\binom{N}{n}} &&\text{Why? See Exercise 10.5.5} \\
&= \sum_{x=1}^{n} x\,\frac{\binom{r}{x}\binom{N-r}{n-x}}{\binom{N}{n}} &&\text{By (10.25)} \\
&= \sum_{x=1}^{n} \frac{r\binom{r-1}{x-1}\binom{N-r}{n-x}}{\binom{N}{n}} &&\text{By (10.26)} \\
&= \sum_{x=1}^{n} \frac{r\binom{r-1}{x-1}\binom{N-r}{n-x}}{\frac{N}{n}\binom{N-1}{n-1}} &&\text{By Theorem 6.5.3} \\
&= \frac{rn}{N} \sum_{x=1}^{n} \frac{\binom{r-1}{x-1}\binom{N-r}{n-x}}{\binom{N-1}{n-1}} &&\text{Shift } \textit{down} \text{ the index} \\
&= \frac{rn}{N} \sum_{x=0}^{n-1} \frac{\binom{r-1}{x}\binom{N-r}{n-(x+1)}}{\binom{N-1}{n-1}} &&N - r = (N-1) - (r-1) \\
&= \frac{rn}{N} \underbrace{\sum_{x=0}^{n-1} \frac{\binom{r-1}{x}\binom{(N-1)-(r-1)}{(n-1)-x}}{\binom{N-1}{n-1}}}_{\substack{1 \\ \text{Hypergeometric distribution} \\ \text{with parameters } n-1, r-1, \\ \text{and } N-1, \text{ which sums to 1.}}} \\
&= \frac{rn}{N}
\end{aligned}
$$

Next,

$$E(X^2) = \sum_{\text{all } x} x^2\,P(X = x)$$

$$= \sum_{x=0}^{n} x^2 \frac{\binom{r}{x}\binom{N-r}{n-x}}{\binom{N}{n}} \qquad \text{See Exercise 10.5.5}$$

$$= \sum_{x=1}^{n} x^2 \cdot \frac{\binom{r}{x}\binom{N-r}{n-x}}{\binom{N}{n}}$$

$$= \sum_{x=1}^{n} x \frac{x\binom{r}{x}\binom{N-r}{n-x}}{\binom{N}{n}} \qquad \text{By (10.25)}$$

$$= \sum_{x=1}^{n} x \frac{r\binom{r-1}{x-1}\binom{N-r}{n-x}}{\binom{N}{n}} \qquad \text{By (10.26)}$$

$$= \sum_{x=1}^{n} x \frac{r\binom{r-1}{x-1}\binom{N-r}{n-x}}{\frac{N}{n}\binom{N-1}{n-1}}$$

$$= \sum_{x=1}^{n} \frac{rn}{N} x \frac{\binom{r-1}{x-1}\binom{N-r}{n-x}}{\binom{N-1}{n-1}} \qquad \text{By Theorem 6.5.3.}$$

$$= \frac{rn}{N} \sum_{x=1}^{n} x \frac{\binom{r-1}{x-1}\binom{N-r}{n-x}}{\binom{N-1}{n-1}} \qquad \text{Shift } down \text{ the index.}$$

$$= \frac{rn}{N} \sum_{x=0}^{n-1} (x+1) \frac{\binom{r-1}{x}\binom{N-r}{n-(x+1)}}{\binom{N-1}{n-1}} \qquad N-r=(N-1)-(r-1)$$

$$= \frac{rn}{N} \sum_{x=0}^{n-1} (x+1) \frac{\binom{r-1}{x}\binom{(N-1)-(r-1)}{(n-1)-x}}{\binom{N-1}{n-1}}$$

$$= \frac{rn}{N} \sum_{x=0}^{n-1} \left[x \frac{\binom{r-1}{x}\binom{(N-1)-(r-1)}{(n-1)-x}}{\binom{N-1}{n-1}} + \frac{\binom{r-1}{x}\binom{(N-1)-(r-1)}{(n-1)-x}}{\binom{N-1}{n-1}} \right]$$

$$= \frac{rn}{N} \left[\underbrace{\sum_{x=0}^{n-1} x \frac{\binom{r-1}{x}\binom{(N-1)-(r-1)}{(n-1)-x}}{\binom{N-1}{n-1}}}_{\substack{\frac{(r-1)(n-1)}{N-1} \\ \text{Expectation of Hypergeometric} \\ \text{distribution with parameters} \\ n-1, r-1, \text{ and } N-1}} + \underbrace{\sum_{x=0}^{n-1} \frac{\binom{r-1}{x}\binom{(N-1)-(r-1)}{(n-1)-x}}{\binom{N-1}{n-1}}}_{\substack{1 \\ \text{Hypergeometric distribution} \\ \text{with parameters } n-1, r-1, \\ \text{and } N-1, \text{ which sums to 1.}}} \right]$$

$$= \frac{rn}{N} \left[\frac{(r-1)(n-1)}{N-1} + 1 \right].$$

Therefore, the variance is

$$\text{var}(X) = \text{E}(X^2) - [\text{E}(X)]^2$$

$$= \frac{rn}{N} \left[\frac{(r-1)(n-1)}{N-1} + 1 \right] - \left(\frac{rn}{N} \right)^2$$

$$= \frac{rn}{N} \left[\frac{(r-1)(n-1)}{N-1} + 1 - \frac{rn}{N} \right].$$

◄

Example 10.5.6 — Estimating an animal population size. A hundred deer in a conservation area were caught and tagged by the conservation authority at the beginning of the spring. Near the end of the summer, 25 deer were observed of which only 10 of them were tagged. What is the estimate of the current size of the deer population?

Solution Let the random variable X be the number of tagged deer caught. Then X follows a hypergeometric distribution, where

$$r = \text{the number of tagged deer in the conservation area}$$
$$= 100$$
$$N - r = \text{the number of untagged deer in the conservation area}$$
$$= ?$$
$$n = \text{the number of deer observed}$$
$$= 25$$

The 10 deer that were observed and tagged provide an estimate of $E(X)$. Thus, we need to solve for N in the equation of the expected value.

$$E(X) = \frac{nr}{N}$$
$$10 = \frac{25 \times 100}{N}$$
$$N = \frac{25 \times 100}{10}$$
$$= 250$$

Therefore, there is an estimated 250 deer in the conservation area at the end of the summer.   ◄

The problem of estimating an animal population size can be reformulated as an urn model problem, which is concerned with drawing black and white balls from an urn; we highlight the similarities in Table 10.12. The problem of calculating the probability of exactly x tagged animals in a sample of r animals is equivalent to finding the probability of drawing exactly x white balls from an urn (that contains white and black balls) in a sample of r balls. This illustrates an important problem solving technique: try to spot the similarities between two problems. *Find something familiar.*[6] When I am faced with a new problem, one of the first things I try is to fit or to adapt a solution that I know from another problem; there is no point in reinventing the wheel. In Table 10.12, the underlying counting problem is the same, but the objects of interest are different—balls, deer, bottles of beer, or other gear.

---

[6]For other problem solving strategies, see Pólya's approach, which is summarized in the Preface starting on p. xviii.

Table 10.12   Comparing the animal population problem to the urn model problem

Parameters and Variables	Animal problem	Urn problem
N	Population size	Number of balls in the urn
n	Number of animals on the first catch that one tagged	Number of white balls
r	The size of the sample	The number of balls drawn for a sample.
x	The number of tagged animals in the sample of r animals	The number of white balls in the sample of r balls.
$P(X = x)$	The probability of exactly x tagged animals in the sample of r animals.	The probability of exactly x white balls in the sample of r balls.

Exercises

10.5.1. Create a table and a graph for the probability distribution of X in Example 10.5.3.

10.5.2. Explain why each situation is not best described with a hypergeometric distribution, and how you could modify the situation so that it does follow a hypergeometric distribution.
   (a) Thirteen cards are dealt from a well-shuffled deck of 52 cards. The number of cards of each suit is recorded.
   (b) In a telephone survey, 1000 people were asked whether or not they would vote for a particular candidate. None of the pollsters coordinated to ensure they did not phone the same person twice. The number of people who said they would vote for the particular candidate was recorded.
   (c) Students are randomly selected one at a time (without replacement) to form a committee from 5 males and 9 females. The selection process continues until a female has been selected for the committee.

10.5.3. Use Vandermonde's identity to show that $\displaystyle\sum_{x=0}^{n} \binom{n}{x}^2 = \binom{2n}{n}$.

10.5.4. Show that
$$\binom{k-1}{0}\binom{r}{1} + \binom{k-1}{1}\binom{r}{2} + \cdots + \binom{k-1}{r-1}\binom{r}{r} = \binom{r+k-1}{r-1}$$
   (a) using an algebraic proof. (*Hint:* Vandermonde's Identity.)

(b) using a combinatorial proof. (*Hint:* Suppose there are $k - 1$ females and r males. How many ways can you pick $r - 1$ people to form a committee?)

10.5.5. Show that

(a) $\displaystyle\sum_{x=0}^{n} x \frac{\binom{r}{x}\binom{N-r}{n-x}}{\binom{N}{n}} = \sum_{x=1}^{n} x \frac{\binom{r}{x}\binom{N-r}{n-x}}{\binom{N}{n}}.$

(b) $\displaystyle\sum_{x=0}^{n} x^2 \frac{\binom{r}{x}\binom{N-r}{n-x}}{\binom{N}{n}} = \sum_{x=1}^{n} x^2 \frac{\binom{r}{x}\binom{N-r}{n-x}}{\binom{N}{n}}.$

10.5.6. In poker, 5 cards are dealt from a well-shuffled deck of 52 cards. Let X be the number of kings in the 5-card hand.
   (a) Find the probability distribution of X. State it as a table.
   (b) Graph the probability distribution of X.
   (c) $P(X \geq 2)$ (d) $P(X \neq 1)$ (e) $E(X)$ (f) $\text{var}(X)$

10.5.7. In a batch of 20 tablets, there are 3 defective. A technician randomly selects 5 without replacement for a sample. What is the probability that the sample contains
   (a) exactly one defective tablet?
   (b) less than 3 defective tablets? First solve the problem using indirect reasoning, and then re-solve the problem using direct reasoning.

10.5.8. An urn contains 3 red marbles, 5 yellow marbles, and 4 blue marbles. Five marbles are randomly selected without replacement. What is the probability of
   (a) all 4 blue marbles are in the sample?
   (b) none of the marbles are yellow in the sample?

10.5.9. From a well-shuffled deck of 52 cards, 13 cards are dealt. What is the probability that the hand contains
   (a) no hearts?   (c) less than 3 hearts?
   (b) exactly 3 hearts?   (d) more than 10 hearts?

10.5.10. In a small town with 2000 people, 667 of them have an asymptomatic nasal staph infection. If a company has 10 employees, how many of the employees are expected to have the infection?

10.5.11. A company has 35 female and 20 male employees. Five employees are randomly selected by drawing their names from a hat to go on a conference to Bermuda. What is the probability that
   (a) all five selected are male?   (c) there is more women than men
   (b) all five selected are female?   selected?

10.5.12. Four friendly jet fighters and eight enemy aircraft engage in combat. This is clearly not a fair fight, so the commander decided to launch four surface-to-air missiles (SAMs). The SAMs are heat seeking and targets randomly any of the aircraft. What is the probability that

    (a) none of the SAMs strike the friendly jet fighters?

    (b) at least one of the SAMs will strike a friendly jet fighter?

10.6 The Normal (or Gaussian) Distribution: A Bridge from Discrete to Continuous Mathematics

We have to recall Problem 7.1.1 that we saw earlier when we were discovering Pascal's triangle (see p. 277). The problem we solved was to count the number of paths through a grid. We were able to calculate this by starting at the top of the grid, and working our way to the bottom, to arrive at the total number of possible paths. We apply the same idea to the following situation. Suppose we drop a bead at the top of a **Galton box** (see Fig. 10.13), and it falls downward. Once it hits one of the hexagons (or you can think of them as pegs), it will travel to the left with probability $\frac{1}{2}$, or to the right with probability $\frac{1}{2}$. This repeats each time the bead encounters a hexagon until it eventually falls into one of the bins at the bottom.

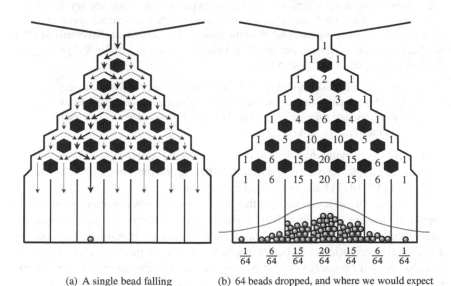

(a) A single bead falling   (b) 64 beads dropped, and where we would expect them to land

Fig. 10.13 An experiment where beads are dropped down a grid and land into bins. This is a **Galton box** named after Sir Francis Galton (1822–1911). In his book on *Natural Inheritance*, Galton (1889, p. 63) used it to demonstrate visually that the normal distribution is approximate to the binomial distribution.

While we cannot control exactly in which box the bead will land, we *expect* more beads in the bins that have more paths leading to them. Thus, the bins in the middle have more paths, and therefore, we expect more beads in the middle bins than in the bins further away from the middle.

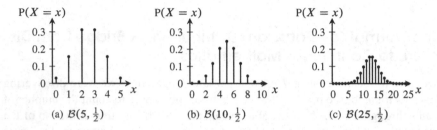

Fig. 10.14   As we let the number of trials n increase in the binomial distribution $\mathcal{B}(n, p)$ with parameter $p = \frac{1}{2}$, we see that it is becoming more and more like the normal distribution.

As we can see in Fig. 10.14, as the number of trials increases in a binomial probability distribution with parameter $p = \frac{1}{2}$, it begins to take on the appearance of a bell shape. Time and time again, this bell shape in a probability distribution keeps appearing in applications—trying to explain how if we are trying to aim for a target value, that there is some random error that we cannot predict. This distribution is called the **normal distribution** or the **Gaussian distribution** after the German mathematician Johann Carl Friedrich Gauss (1777–1855) who gave the first derivation of the distribution.

> Everyone believes [in the normal distribution]—once expressed to me by Dr Lippmann—since experimenters imagine that it is a mathematical theorem, and the mathematicians because it is an experimental fact (Poincaré, 1912, p. 171).[7]

There were many mathematicians who had discovered it before, but Gauss was the first to provide the derivation, and it would have gone unnoticed if Pierre-Simon Laplace (1749–1827) had not realized its usefulness.

From Fig. 10.14, it can be seen that the normal distribution can be used as an approximation of the binomial distribution with parameter $p = \frac{1}{2}$ for a sufficiently large number of trials. Since calculating the exactly probabilities of success of a binomial distributions can become tedious, a general rule is that the binomial distribution can be approximated by the normal distribution whenever np and $n(1 - p)$ are both at least 5 or more.

---

[7]The original French reads: *Tout le monde y croit cependant, me disait un jour M. Lippmann, car les expérimentateurs s'imaginent que c'est un théorème de mathématiques, et les mathématiciens que c'est un fait expérimental* (Poincaré, 1912, p. 171). Gabriel Lippmann (1845–1921) that Poincaré refers to in his statement was a French physicist who won the 1908 Nobel Prize in Physics for his method of color photography based on the interference phenomenon of light.

Definition 10.6.1 — Normal Distribution. A continuous random variable X has a **normal distribution** if

$$f(x) = \frac{1}{\sigma\sqrt{2\pi}} e^{-\frac{(x-\mu)^2}{2\sigma^2}} \qquad (10.27)$$

where

- $x \in \mathbb{R}$ (for any real number x),
- $e \approx 2.718\,28$ is Euler's number,
- μ is the expected value (or mean), and
- σ is the standard deviation (and σ^2 is the variance).

A normal distribution is denoted by the symbol $\mathcal{N}(\mu, \sigma^2)$, where constants μ (the expected value) and σ (the standard deviation) are the parameters. A random variable X that is normally-distributed is denoted by $X \sim \mathcal{N}(\mu, \sigma^2)$.

This can help us to understand that we do not have complete control over randomness. Consider a coach of a football team. On average, the performance of the team is good. But, if the performance of the team was terrible in a particular game, the coach may yell at the team, and in the subsequent game the team will have a good performance (which appears to be an improvement).[8] If the performance of the team was outstanding, the coach may praise them, and in the subsequent game the team will play as usual (which appears as a deterioration). From this observation, the coach may incorrectly deduce that yelling at the team results in improvement, and praising them results in deterioration, when really his yelling or praising has no affect on the outcome that is due to chance.[9] This illustrates an important point: *we need to observe the experiment long enough to see the overall probability distribution before we can start to make changes to the experiment.* How could the coach improve the team's performance? The coach would have to improve the team's average (or expected) performance.

The normal distribution is shown in Fig. 10.15. If $\mu = 0$ and $\sigma = 1$, the distribution is called the **standard normal distribution**, denoted by $\mathcal{N}(0, 1)$.

A normal distribution has the following three properties:

(a) It is symmetric about the expected value, μ (read as the Greek letter "mu"). (If you fold the distribution along the dashed line of the expected value, the two halves are mirror images of each other.)

(b) The standard deviation, σ (read as the Greek letter "sigma"), is the distance from the expected value to the points of inflection of the curve (a point of inflection is where a "hill" begins to turn into a "valley").

(c) The area under the curve equals to one.

---

[8]This is known as **regression toward the mean**. It is a phenomenon that occurs when the first observation of an experiment is extreme, and subsequent observations are closer to the expected value.

[9]When people believe their actions affect the outcome, when in fact their action have no effect at all, this is called a **superstition**.

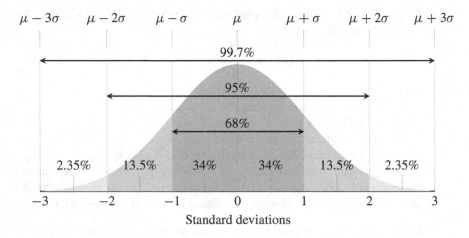

$\mu - 3\sigma$ $\quad\mu - 2\sigma$ $\quad\mu - \sigma$ $\quad\quad\mu$ $\quad\quad\mu + \sigma$ $\quad\mu + 2\sigma$ $\quad\mu + 3\sigma$

Fig. 10.15   The normal distribution

Of the observations, 68.26%, 95.44%, and 99.74% will lie within one, two, and three standard deviations from the expected value, respectively; that is,

$$P(\mu - \sigma \leq X \leq \mu + \sigma) = 0.6826$$
$$P(\mu - 2\sigma \leq X \leq \mu + 2\sigma) = 0.9544$$
$$P(\mu - 3\sigma \leq X \leq \mu + 3\sigma) = 0.9974.$$

This has been called the **68-95-99.7 rule**. (See Exercise 10.6.3.)

Example 10.6.2 — Bell curve grading system. Students also know the normal distribution for grading as the **bell curve**, since the curve in Fig. 10.15 resembles the shape of a bell. There are some professors who use the following system to distribute grades based on standard deviations away from the expected value:

(a) within one standard deviation from the expected value represents 68% of the class; these students would earn a C grade;

(b) within one and two standard deviations *above* the expected value represents 13.5% of the class; these students would receive a B grade. Within one and two standard deviations *below* the expected value (13.5% of the class), these students would get a D; and

(c) two standard deviations *above* the expected value represents 2.35% of the class, and these students would get the coveted A grade. Meanwhile, two standard deviations *below* the expected value (2.35% of the class), these students would receiving the failing F grade.

If we consider a class of 50 students,

- 34 students would get a C grade

- 6 or 7 students would get a B, while 6 or 7 receive a D
- 1 or 2 students receive an A, while 1 or 2 students receives an F.

The goal of using a bell curve is to minimize the variation between instructors by grading students relative to their peers in that class. I find it curious that students often ask if it is possible for the class to have the insurance policy of the bell curve; either intentional or not, they wish to sacrifice one of their own for the security of the majority to receive an average grade. ◄

The area under the standard normal distribution cannot be found using a formula; the area under the curve must be calculated using numerical methods that are executed on a computer. Because of this, we provide a table of probability values for each Z-value up to two decimal places in Appendix A.1 (on pp. 480–481).

Example 10.6.3 If $X \sim \mathcal{N}(0, 1)$, find the following probabilities using the standard normal distribution table in Appendix A.1.

    (a) $P(X < 1.5)$ (b) $P(X > -1)$ (c) $P(-1 < X < 2)$

Solution

  (a) The area under the normal curve $\mathcal{N}(0, 1)$ to determine $P(X < 1.5)$ is given below.

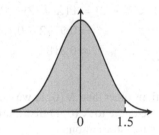

    From the table, $P(X < 1.5) = 0.9332$.

  (b) The area under the normal curve $\mathcal{N}(0, 1)$ to determine $P(X > -1)$ is given below.

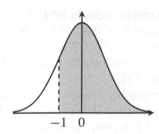

From the tables, $P(X < -1) = 0.1587$. The total area under the curve is 1 (see property (c) of the standard normal curve), so

$$P(X > -1) = 1 - P(X < -1)$$
$$= 1 - 0.1587$$
$$= 0.8413.$$

(c) The area under the normal curve $\mathcal{N}(0, 1)$ to determine $P(-1 < X < 2)$ is given below.

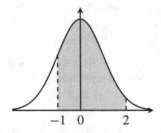

To determine this area we subtract areas that can be found from the table. So

$$P(-1 < X < 2) = P(X < 2) - P(X < -1)$$
$$= 0.9772 - 0.1587$$
$$= 0.8185.$$

◀

For normal distributions other than $\mathcal{N}(0, 1)$ we can perform a transformation on the variables to reduce the curve to $\mathcal{N}(0, 1)$ distribution. Suppose $X \sim \mathcal{N}(\mu, \sigma^2)$. We can do the following transformation:

$$Z = \frac{X - \mu}{\sigma}. \tag{10.28}$$

This new random variable Z now has expected value $\mu = 0$ and $\sigma^2 = 1$ which is a standard normally distributed random variable. This will allow us now to look up values in the Z table.

Example 10.6.4 — Canadian male height. The height of adult Canadian males is approximately normally distributed with an expected value of 176 cm (5' 9") and a standard deviation of 7.1 cm (2.8").

(a) What is the probability a randomly selected adult Canadian male is taller than 185 cm (6' 1")?

(b) What is the probability a randomly selected adult Canadian male has a height between 180 and 190 cm (5' 11" and 6' 3")?

(c) What is the 90th percentile of the height of Canadian male?

Solution

(a) We want $P(X > 185)$.

$$P(X > 185) = P\left(\frac{X - \mu}{\sigma} > \frac{185 - \mu}{\sigma}\right)$$

$$= P\left(Z > \frac{185 - 176}{7.1}\right)$$

$$= P(Z > 1.27)$$

Now, we go to our Z table (on p. 481), and we look up the value of $P(Z < 1.27) = 0.8980$.

Therefore,

$$P(X > 185) = P(Z > 1.27)$$

$$= 1 - P(Z < 1.27)$$

$$= 1 - 0.8980$$

$$= 0.1020.$$

So, the probability that a randomly selected Canadian male that is taller than 185 cm is 10.2%.

(b)

$$P(180 < X < 190) = P\left(\frac{180 - 176}{7.1} < Z < \frac{190 - 176}{7.1}\right)$$

$$= P(0.56 < Z < 1.97)$$

$$= P(Z < 1.97) - P(Z < 0.56)$$

$$= 0.9756 - 0.7134$$

$$= 0.2622$$

So, the probability that a randomly selected adult Canadian male is between 180 and 190 cm is 26.22%.

(c) The nth **percentile** is the value or score in which n percent of the observations are below this value. For example, a male who is the 90th percentile in height means that 90% of the population is shorter than him. We want to find what value of X (or Z) will give us the area under the curve equal to 0.9 (or 90%). Rearranging our equation of Z, we get

$$Z = \frac{X - \mu}{\sigma}$$

$$X = \mu + \sigma Z.$$

To find the value for Z, we look in the table for the area under the curve that will give us 0.9. This corresponds to the value of $Z = 1.28$. Substituting $Z = 1.28$ into our formula above we have

$$X = 176 + 7.1(1.28) = 185.$$

So, the 90th percentile adult Canadian male is 185 cm. This means that a person who is 185 cm is taller than 90% of the adult male population in Canada.

◀

Example 10.6.5 The annual precipitation for the city of Toronto, Ontario, Canada is normally distributed with an expected value of 846.61 mm and a variance of 14 512.97 mm.[10] What is the probability of annual precipitation in Toronto exceeding 1000 mm?

Solution *Step 1*: Calculate the Z-value.

The Z-value corresponding to 1000 mm of precipitation where $\mu = 846.61$ and $\sigma^2 = 14\,512.97$ is

$$Z = \frac{1000 - 846.61}{\sqrt{14\,512.97}} = 1.27.$$

Therefore, having more than 1000 mm of annual precipitation is 1.27 standard deviations above the expected annual precipitation of 846.61 mm.

Step 2: Using the Z-value, look up the probability value in the standard normal distribution table (in Appendix A.1).

From the standard normal distribution table, the Z-value of 1.27 corresponds to a probability of 0.8980. This means, that in approximately 90 years out of a 100, the annual precipitation in Toronto should fall between 0 and 1000 mm.

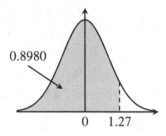

Step 3: Evaluate the probability value.

The probability value that the standard normal distribution table gives is the area of the curve to the left of the Z-value. In fact, what we require is the area under the curve to the *right* of the Z-value. Since the total area under the curve is one (see property (c) of the standard normal curve), the correct answer is found by subtracting the probability calculated in Step 2. Thus,

$$1 - 0.8980 = 0.1020$$

is the probability that the annual precipitation in Toronto will exceed 1000 mm.

---

[10]Source: http://climate.weather.gc.ca based upon the annual precipitation records for Toronto, Ontario, Canada from 1987 to 2011.

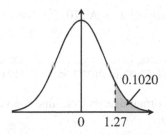

This means, that in approximately 10 years out of 100, the annual precipitation in Toronto will be more than 1000 mm. ◄

Exercises

10.6.1. For the normal distribution in Fig. 10.16, what is the
    (a) expected value?   (c) variance?
    (b) standard deviation?

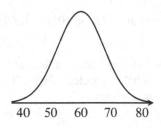

Fig. 10.16   A normal distribution for Exercise 10.6.1

10.6.2. True or false.
    (a) The normal distribution is symmetric about its expected value μ.
    (b) The normal distribution is always symmetric about zero (the origin).
    (c) The area under the normal distribution within one standard deviation of the expected value is approximately 50%.

10.6.3. Given a random variable $Z \sim \mathcal{N}(0, 1)$, calculate the following using the standard normal distribution table (in Appendix A.1):
    (a) $P(-1 \le Z \le 1)$ (b) $P(-2 \le Z \le 2)$ (c) $P(-3 \le Z \le 3)$.

10.6.4. Given a random variable $Z \sim \mathcal{N}(0, 1)$, find
    (a) $P(Z < 0.87)$ (b) $P(Z > 1.49)$.

10.6.5. Given a random variable $X \sim \mathcal{N}(70, 5^2)$, find
    (a) $P(X < 65)$ (b) $P(X > 75)$.

10.6.6. A normally distributed random variable X has expected value $\mu = 322$ and standard deviation $\sigma = 45$. Find $P(277 \leq X \leq 412)$.

10.6.7. For the standard normal distribution, find the value of a which satisfies $P(-a \leq Z \leq a) = 0.5160$.

10.6.8. Let X be normally distributed with expected value $\mu = 1.5$ and variance $\sigma^2 = 4.41$.
    (a) Find $P(X > 4)$.
    (b) Find $P(-3 < X < 1.7)$.
    (c) Determine the value of x such that $P(X \leq x) = 0.7517$.

10.6.9. Let $X \sim \mathcal{N}(\mu, 5^2)$ and $P(X < 20) = 0.8413$.
    (a) Find the value of μ.
    (b) What is the value of $P(\mu < X < 20)$?

10.6.10. Let $X \sim \mathcal{N}(100, \sigma^2)$ and $P(X < 105) = 0.6915$. What is the standard deviation σ?

10.6.11. The marks on a discrete mathematics final examination can be modeled with a normal distribution, which the expected value is 70 and the standard deviation is 5. What is the probability that a mark is
    (a) greater than 75?   (c) between 65 and 70?
    (b) less than 60?

10.6.12. Developed by German psychologist William Stern (1871–1938), an intelligence quotient, or IQ, is a score derived from a test designed to access human intelligence. A person with an IQ score of 100 has average intelligence. IQ scores follow a normal distribution where the expected value is 100, and the standard deviation is 15. If a person is randomly selected, what is the probability that the person's IQ score is
    (a) between 85 and 115?   (c) below 50?
    (b) above 125?

10.6.13. The weight, X milligrams, of protein in a protein shake in a bottle can be modeled by a normal distribution with expected value 507.5 and standard deviation 4.0.
    (a) Find
        (i) $P(X < 515)$ (iii) $P(500 < X < 515)$
        (ii) $P(X \geq 515)$ (iv) $P(X \neq 507.5)$.
    (b) Determine the value of x such that $P(X < x) = 0.96$.

10.6.14. The volume of soda in a can may be modeled by a normal distribution with expected value $\mu = 355$ mL and standard deviation $\sigma = 2.5$ mL. Determine the probability that a randomly selected can of soda is

(a) less than 360 mL? (c) between 350 and 360 mL?
(b) more than 348 mL? (d) not exactly 355 mL?

10.6.15. The heights of American women (20 years and over) approximately follows a normal distribution with an expected value of 63.8 inches and a standard deviation of 0.16.[11] If Layla is in the 95th percentile in height for American women, then what is her height?

10.6.16. The weight of a bag of frozen peas, X, may be modeled by a normal distribution with an expected value of 500 g and a standard deviation of 10 g.

(a) Determine the probability that a randomly selected bag is
  (i) exactly 500 g? (iii) between 495 and 505 g?
  (ii) less than 510 g?
(b) If $P(X < x) = 0.98$, what is the value of x?

10.6.17. Health insurance claims for a particular insurance company follows a normal distribution with expected value $10\,000 and a standard deviation of $7400. What is the probability that a claim exceeds $15\,000?

10.6.18. The daily demand for blood (in liters) at a local hospital is $\mathcal{N}(50, 100)$. How many liters of blood should be stocked at the beginning of the day to ensure that only a 5% chance of running out by the end of the day?

10.6.19. At a university, the scores on a discrete mathematics final examination follow a normal distribution with an expected value of 70 and a standard deviation of 10. The scores on a calculus final also follow a normal distribution with an expected value of 80 and a standard deviation of 8. If a student scored 79 on the discrete mathematics final, and 85 on the calculus final, then relative to the students in each respective class, which subject did the student perform better?

0.7 Review

Summary

- A random variable follows a distribution if it satisfies certain attributes.
- **Uniform distribution**: Every outcome of the experiment is equally likely.
- **Binomial distribution**: The number of successes in n identical and independent trials in which the probability of success at each trial is p. Each trial has only two outcomes called a success or a failure.

---

[11] Source: http://www.cdc.gov/nchs/data/nhsr/nhsr010.pdf National Health Statistics Reports, 2008. (Table 10)

- **Poisson distribution**: The number of occurrences of a specific event type in a period of time when the events occur randomly at an expected rate λ per unit time.
- **Geometric distribution**: The number of trials required to obtain the first success in a sequence of identical and independent trials in which the probability of success at each trial is p.
- **Hypergeometric distribution**: There are r objects that have a specific property out of a total of N objects. In a sample of n objects, we want to know the probability of x objects in the sample that have the specific property.
- The formula and parameters for each distribution is summarized in Table 10.13.

Exercises

10.7.1. Graph the following probability distributions, and identify the type of distribution.

(a) $P(X = x) = \dfrac{1}{n}$ where $x \in \{1, 2, 3, \dots, n\}$.

(b) $P(X = x) = \dbinom{n}{x} p^x (1 - p)^{n-x}$ where $n = 5$ and $p = \dfrac{1}{5}$.

(c) $P(X = x) = (1 - p)^{x-1} p$ where $p = \dfrac{1}{4}$ and $x \in \{1, 2, 3, \dots\}$.

10.7.2. Identify the appropriate probability distribution to use in each situation, and then explain your reasoning for the selection.

(a) Seven balls are drawn from an urn, which contains 5 red balls and 7 white balls. What is the chance that 4 white balls are drawn?

(b) A fair die is rolled 6 times. What is the chance that a two appears exactly 4 of the 6 rolls?

(c) Six employees in a company of 25 are asked whether they took discrete mathematics in university or not. Fifteen of the employees did not. What is the probability that none of the six took discrete mathematics?

(d) Ten tickets numbered from 1 to 10 are placed in an urn. If a ticket is randomly selected, then what is the chance the ticket numbered 4 will be selected?

(e) In a lottery, 6 balls are selected from 49 numbered balls. What is probability of selecting the winning ticket?

(f) Calculate the waiting time before winning the lottery.

(g) What is the probability of rolling a die five times before a 6 appears for the first time?

(h) A car starts in cold weather 75% of the time. What is the probability that the car will start all five times during the cold weather?

Table 10.13 Discrete probability distributions

Distribution Name	Probability Distribution	Parameter values	$E(X)$	$var(X)$
Uniform	$P(X = x) = \dfrac{1}{n}$	$n \in \mathbb{N}$	$\dfrac{1}{n}\sum_{\text{all }x} x$	$\dfrac{1}{n}\left(\sum_{\text{all }x} x^2 - \sum_{\text{all }x} x\right)$
Binomial	$P(X = x) = \dbinom{n}{x} p^x (1 - p)^{n-x}$ for $x = 0, 1, 2, \ldots, n$	$n > 0$ $0 \leq p \leq 1$	np	$np(1 - p)$
Poisson	$P(X = x) = \dfrac{\lambda^x e^{-\lambda}}{x!}$ for $x = 0, 1, \ldots$	$\lambda > 0$	λ	λ
Geometric	$P(X = x) = (1 - p)^{x-1} p$ for $x = 1, 2, 3, \ldots$	$0 < p \leq 1$	$\dfrac{1}{p}$	$\dfrac{1 - p}{p^2}$
Hyper-geometric	$P(X = x) = \dfrac{\binom{r}{x}\binom{N-r}{n-x}}{\binom{N}{n}}$	$N \in \{0, 1, 2, \ldots\}$ $r \in \{0, 1, 2, \ldots, N\}$ $n \in \{0, 1, 2, \ldots, N\}$	$\dfrac{nr}{N}$	$\dfrac{rn}{N}\left[\dfrac{(r-1)(n-1)}{N-1} + 1 - \dfrac{rn}{N}\right]$

10.7.3. Suppose X is a random variable where

$$P(X = x) = \binom{20}{x}(0.25)^x(0.75)^{20-x}$$

for $x = 0, 1, \ldots, 20$. What is the expected value and variance of X?

10.7.4. A large lot of ammunition that has been in storage for 10 years is known to have a defective rate of 20%. Five rounds of this ammunition are selected at random and test fired.
  (a) If X denotes the number of misfires that occur during the test, explain why X has a binomial distribution.
  (b) What is the probability that exactly two of the five rounds misfire?
  (c) What is the probability that at least one round misfires?
  (d) What is the expected number of rounds that will misfire?
  (e) What is the standard deviation of the number of rounds that will misfire?

10.7.5. The waiting period is associated with which probability distribution?

10.7.6. In a particular county, the number of tornadoes in a five-day period follows a Poisson distribution with $\lambda = 2$. The number of tornadoes in any five-day period is independent. What is the probability that less than 3 tornadoes occur in a 15-day period?

10.7.7. An industrial plant releases on average 2.50 tons of waste water into a lake every week. If the standard deviation is 0.5 tons, what is the probability that in a week, less than 2.75 tons of waste water was released?

10.7.8. On a Discrete Mathematics final examination of several hundred students, the professor observed the expected mark to be 68% with a standard deviation of 12%. If the marks are assumed to be normally distributed, what percentage of students
  (a) failed the course, if the pass mark is 50%?
  (b) got an A+ (92% or higher)?
  (c) obtained a mark between 62% and 74%?

10.8 Bibliographic Remarks

The probability distributions that have been discussed in this chapter have been long established. The books of Feller (1968), Stewart *et al.* (1988), and Ross (2010) have heavily influenced how I have presented this material.

Appendix A

Probability Distribution Tables

A.1 Standard Normal Distribution Table

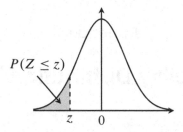

$P(Z \leq z)$

z	0.00	0.01	0.02	0.03	0.04	0.05	0.06	0.07	0.08	0.09
−3.5	.0002	.0002	.0002	.0002	.0002	.0002	.0002	.0002	.0002	.0002
−3.4	.0003	.0003	.0003	.0003	.0003	.0003	.0003	.0003	.0003	.0002
−3.3	.0005	.0005	.0005	.0004	.0004	.0004	.0004	.0004	.0004	.0003
−3.2	.0007	.0007	.0006	.0006	.0006	.0006	.0006	.0005	.0005	.0005
−3.1	.0010	.0009	.0009	.0009	.0008	.0008	.0008	.0008	.0007	.0007
−3.0	.0013	.0013	.0013	.0012	.0012	.0011	.0011	.0011	.0010	.0010
−2.9	.0019	.0018	.0018	.0017	.0016	.0016	.0015	.0015	.0014	.0014
−2.8	.0026	.0025	.0024	.0023	.0023	.0022	.0021	.0021	.0020	.0019
−2.7	.0035	.0034	.0033	.0032	.0031	.0030	.0029	.0028	.0027	.0026
−2.6	.0047	.0045	.0044	.0043	.0041	.0040	.0039	.0038	.0037	.0036
−2.5	.0062	.0060	.0059	.0057	.0055	.0054	.0052	.0051	.0049	.0048
−2.4	.0082	.0080	.0078	.0075	.0073	.0071	.0069	.0068	.0066	.0064
−2.3	.0107	.0104	.0102	.0099	.0096	.0094	.0091	.0089	.0087	.0084
−2.2	.0139	.0136	.0132	.0129	.0125	.0122	.0119	.0116	.0113	.0110
−2.1	.0179	.0174	.0170	.0166	.0162	.0158	.0154	.0150	.0146	.0143
−2.0	.0228	.0222	.0217	.0212	.0207	.0202	.0197	.0192	.0188	.0183
−1.9	.0287	.0281	.0274	.0268	.0262	.0256	.0250	.0244	.0239	.0233
−1.8	.0359	.0351	.0344	.0336	.0329	.0322	.0314	.0307	.0301	.0294
−1.7	.0446	.0436	.0427	.0418	.0409	.0401	.0392	.0384	.0375	.0367
−1.6	.0548	.0537	.0526	.0516	.0505	.0495	.0485	.0475	.0465	.0455
−1.5	.0668	.0655	.0643	.0630	.0618	.0606	.0594	.0582	.0571	.0559
−1.4	.0808	.0793	.0778	.0764	.0749	.0735	.0721	.0708	.0694	.0681
−1.3	.0968	.0951	.0934	.0918	.0901	.0885	.0869	.0853	.0838	.0823
−1.2	.1151	.1131	.1112	.1093	.1075	.1056	.1038	.1020	.1003	.0985
−1.1	.1357	.1335	.1314	.1292	.1271	.1251	.1230	.1210	.1190	.1170
−1.0	.1587	.1562	.1539	.1515	.1492	.1469	.1446	.1423	.1401	.1379
−0.9	.1841	.1814	.1788	.1762	.1736	.1711	.1685	.1660	.1635	.1611
−0.8	.2119	.2090	.2061	.2033	.2005	.1977	.1949	.1922	.1894	.1867
−0.7	.2420	.2389	.2358	.2327	.2296	.2266	.2236	.2206	.2177	.2148
−0.6	.2743	.2709	.2676	.2643	.2611	.2578	.2546	.2514	.2483	.2451
−0.5	.3085	.3050	.3015	.2981	.2946	.2912	.2877	.2843	.2810	.2776
−0.4	.3446	.3409	.3372	.3336	.3300	.3264	.3228	.3192	.3156	.3121
−0.3	.3821	.3783	.3745	.3707	.3669	.3632	.3594	.3557	.3520	.3483
−0.2	.4207	.4168	.4129	.4090	.4052	.4013	.3974	.3936	.3897	.3859
−0.1	.4602	.4562	.4522	.4483	.4443	.4404	.4364	.4325	.4286	.4247
−0.0	.5000	.4960	.4920	.4880	.4840	.4801	.4761	.4721	.4681	.4641

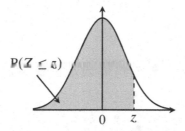

z	0.00	0.01	0.02	0.03	0.04	0.05	0.06	0.07	0.08	0.09
0.0	.5000	.5040	.5080	.5120	.5160	.5199	.5239	.5279	.5319	.5359
0.1	.5398	.5438	.5478	.5517	.5557	.5596	.5636	.5675	.5714	.5753
0.2	.5793	.5832	.5871	.5910	.5948	.5987	.6026	.6064	.6103	.6141
0.3	.6179	.6217	.6255	.6293	.6331	.6368	.6406	.6443	.6480	.6517
0.4	.6554	.6591	.6628	.6664	.6700	.6736	.6772	.6808	.6844	.6879
0.5	.6915	.6950	.6985	.7019	.7054	.7088	.7123	.7157	.7190	.7224
0.6	.7257	.7291	.7324	.7357	.7389	.7422	.7454	.7486	.7517	.7549
0.7	.7580	.7611	.7642	.7673	.7704	.7734	.7764	.7794	.7823	.7852
0.8	.7881	.7910	.7939	.7967	.7995	.8023	.8051	.8078	.8106	.8133
0.9	.8159	.8186	.8212	.8238	.8264	.8289	.8315	.8340	.8365	.8389
1.0	.8413	.8438	.8461	.8485	.8508	.8531	.8554	.8577	.8599	.8621
1.1	.8643	.8665	.8686	.8708	.8729	.8749	.8770	.8790	.8810	.8830
1.2	.8849	.8869	.8888	.8907	.8925	.8944	.8962	.8980	.8997	.9015
1.3	.9032	.9049	.9066	.9082	.9099	.9115	.9131	.9147	.9162	.9177
1.4	.9192	.9207	.9222	.9236	.9251	.9265	.9279	.9292	.9306	.9319
1.5	.9332	.9345	.9357	.9370	.9382	.9394	.9406	.9418	.9429	.9441
1.6	.9452	.9463	.9474	.9484	.9495	.9505	.9515	.9525	.9535	.9545
1.7	.9554	.9564	.9573	.9582	.9591	.9599	.9608	.9616	.9625	.9633
1.8	.9641	.9649	.9656	.9664	.9671	.9678	.9686	.9693	.9699	.9706
1.9	.9713	.9719	.9726	.9732	.9738	.9744	.9750	.9756	.9761	.9767
2.0	.9772	.9778	.9783	.9788	.9793	.9798	.9803	.9808	.9812	.9817
2.1	.9821	.9826	.9830	.9834	.9838	.9842	.9846	.9850	.9854	.9857
2.2	.9861	.9864	.9868	.9871	.9875	.9878	.9881	.9884	.9887	.9890
2.3	.9893	.9896	.9898	.9901	.9904	.9906	.9909	.9911	.9913	.9916
2.4	.9918	.9920	.9922	.9925	.9927	.9929	.9931	.9932	.9934	.9936
2.5	.9938	.9940	.9941	.9943	.9945	.9946	.9948	.9949	.9951	.9952
2.6	.9953	.9955	.9956	.9957	.9959	.9960	.9961	.9962	.9963	.9964
2.7	.9965	.9966	.9967	.9968	.9969	.9970	.9971	.9972	.9973	.9974
2.8	.9974	.9975	.9976	.9977	.9977	.9978	.9979	.9979	.9980	.9981
2.9	.9981	.9982	.9982	.9983	.9984	.9984	.9985	.9985	.9986	.9986
3.0	.9987	.9987	.9987	.9988	.9988	.9989	.9989	.9989	.9990	.9990
3.1	.9990	.9991	.9991	.9991	.9992	.9992	.9992	.9992	.9993	.9993
3.2	.9993	.9993	.9994	.9994	.9994	.9994	.9994	.9995	.9995	.9995
3.3	.9995	.9995	.9995	.9996	.9996	.9996	.9996	.9996	.9996	.9997
3.4	.9997	.9997	.9997	.9997	.9997	.9997	.9997	.9997	.9997	.9998
3.5	.9998	.9998	.9998	.9998	.9998	.9998	.9998	.9998	.9998	.9998

Appendix B

Prerequisite Knowledge

In this appendix, we give a quick review of concepts that you would have seen in previous courses that we will need in this book.

B.1 Number Systems

The following are some special sets of numbers or number systems that we commonly use in mathematics:

- $\mathbb{N} = \{1, 2, 3, 4, \ldots\}$ is the set of **natural numbers**.
- $\mathbb{W} = \{0, 1, 2, 3, \ldots\}$ is the set of **whole numbers**. It is the set of natural numbers and zero.
- $\mathbb{Z} = \{\ldots, -2, -1, 0, 1, 2, \ldots\}$ is the set of **integers**. It is the set of negative and positive whole numbers (including zero).
- $\mathbb{Q} = \{x \mid x = \frac{m}{n} \text{ for integers } m \text{ and } n, \text{ and } n \neq 0\}$ is the set of **rationals**. It is the set of all numbers that can be expressed as a fraction.
- $\mathbb{Q}' = \{x \mid x \neq \frac{m}{n} \text{ for any integers } m \text{ and } n, \text{ and } n \neq 0\}$ is the set of **irrationals**. It is the set of all numbers that cannot be expressed as a fraction such as $\sqrt{2}$, and π.
- $\mathbb{R} = \mathbb{Q} \cup \mathbb{Q}'$ is the set of **reals**. A real number can be either a rational or irrational number.

Exercises

B.1.1. Does the number $\frac{3}{4}$ belong to
  (a) $\mathbb{Q}$?   (b) $\mathbb{Q}'$?   (c) $\mathbb{Z}$?   (d) $\mathbb{R}$?   (e) $\mathbb{N}$?

B.2 Geometric Formulas

The geometric formulas that we will require in this book are given in Fig. B.1 on p. 484.

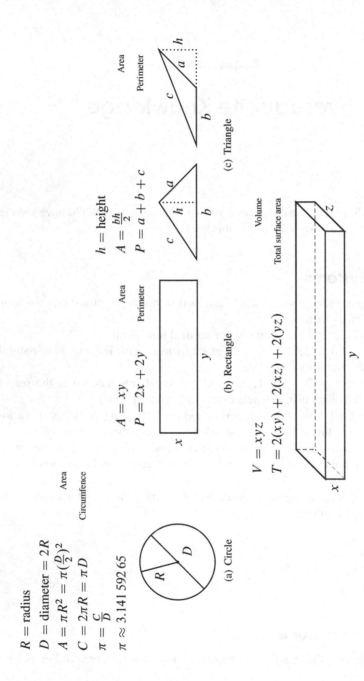

Fig. B.1 Geometric Formulas

B.3 Basic Properties of Real Numbers

There are five basic properties that hold for the operations of addition and multiplication of real numbers.

> **Definition B.3.1 — Associativity.** If x, y, and z are real numbers, then
>
> $$(x + y) + z = x + (y + z)$$
>
> and
>
> $$(x \cdot y) \cdot z = x \cdot (y \cdot z).$$

Example B.3.2 If you add three real numbers or you multiply three real numbers, the associativity property says the placement of the parentheses does not matter.

(a) $(3 + 9) + 11 = 12 + 11 = 23$ or $3 + (9 + 11) = 3 + 20 = 23$
(b) $(3 \cdot 4) \cdot 5 = 12 \cdot 5 = 60$ or $3 \cdot (4 \cdot 5) = 3 \cdot 20 = 60$

◀

In view of this, many people will forgo the use of parentheses in such a simple context, and they will write simply $x + y + z$ in the case of the addition operation, or $x \cdot y \cdot z$ in the case of the multiplication operation.

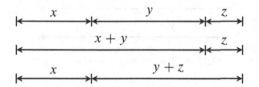

Fig. B.2   The associativity of addition

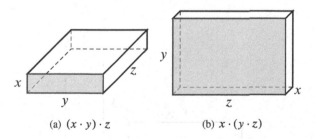

(a) $(x \cdot y) \cdot z$ (b) $x \cdot (y \cdot z)$

Fig. B.3   The associativity of multiplication. The volume of the rectangular prism in (a) is the same as (b).

Definition B.3.3 — Commutativity. If x and y are real numbers, then

$$x + y = y + x$$

and

$$x \cdot y = y \cdot x$$

Example B.3.4 If you add two real numbers or you multiply two real numbers, the commutative property says that order of the numbers does not matter.

(a) $3 + 9 = 12 = 9 + 3$ (b) $3 \cdot 4 = 12 = 4 \cdot 3$

◄

For the commutativity of addition, we see in Fig. B.4, $x + y$ and $y + x$ represent different expressions of the length for the same line segment; therefore, $x + y = y + x$.

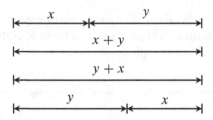

Fig. B.4   The commutativity of addition.

For the commutativity of multiplication, we see in Fig. B.5(a), the area is $x \cdot y$, and in Fig. B.5(b) the area is $y \cdot x$. Since Fig. B.5(a) and Fig. B.5(b) represent different expressions of area for the same rectangle, therefore, $x \cdot y = y \cdot x$.

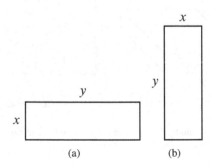

Fig. B.5   The commutativity of multiplication

Definition B.3.5 — Distributivity. If x, y, and z are real numbers, then

$$x \cdot (y + z) = x \cdot y + x \cdot z$$

and

$$(y + z) \cdot x = y \cdot x + z \cdot x.$$

Example B.3.6

(a) $3 \cdot (4 + 5) = 3 \cdot 9 = 27$ or $3 \cdot (4 + 5) = 3 \cdot 4 + 3 \cdot 5 = 12 + 15 = 27$
(b) $(4 + 5) \cdot 3 = 9 \cdot 3 = 27$ or $(4 + 5) \cdot 3 = 4 \cdot 3 + 5 \cdot 3 = 12 + 15 = 27$

◄

For the case of $x \cdot (y + z) = x \cdot y + x \cdot z$, we can measure the area of the rectangle in Fig. B.6(a) in two different ways: $x(y + z)$ and $x \cdot y + x \cdot z$ represent the area of the same rectangle. For the case of $(y + z) \cdot x = y \cdot x + z \cdot x$ in Fig. B.6(b), we can again measure the area of the rectangle in two different ways: $(y + z) \cdot x$ and $y \cdot x + z \cdot x$ represent the area of the same rectangle.

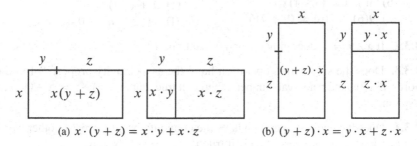

(a) $x \cdot (y + z) = x \cdot y + x \cdot z$ (b) $(y + z) \cdot x = y \cdot x + z \cdot x$

Fig. B.6   The distributive property

Definition B.3.7 — Identity. For every real number x, there is a unique real number 0, called the **additive identity**, such that

$$x + 0 = x \quad \text{and} \quad 0 + x = x.$$

For every real number x, there is a unique real number 1, called the **multiplicative identity**, such that

$$x \cdot 1 = x \quad \text{and} \quad 1 \cdot x = x.$$

Example B.3.8

(a) $3 + 0 = 3$ and $0 + 3 = 3$.   (b) $3 \cdot 1 = 3$ and $1 \cdot 3 = 3$.

◄

Definition B.3.9 — Inverse. For every real number x, there is a unique real number $-x$, called the **additive inverse** of x, such that

$$x + (-x) = 0 \quad \text{and} \quad (-x) + x = 0.$$

For every nonzero real number x, there is a (unique) real number $\frac{1}{x}$, called the **multiplicative inverse** of x, such that

$$x \cdot \frac{1}{x} = 1 \quad \text{and} \quad \frac{1}{x} \cdot x = 1.$$

Example B.3.10
  (a) $3 + (-3) = 0$ and $(-3) + 3 = 0$.   (b) $3 \cdot \frac{1}{3} = 1$ and $\frac{1}{3} \cdot 3 = 1$.

◀

Exercises

B.3.1. Determine what basic property each of the following statements is illustrating.

  (a) $9 + 3(6) = 9 + 6(3)$
  (b) $3(4x) = (3 \times 4)x$
  (c) $9(6) + 7(6) = (9 + 7)6$

  (d) $(-3)(-\frac{1}{3}) = 1$
  (e) $a + (-a) = 0$
  (f) $x(yz + a) + 0 = x(yz + a)$

B.3.2. If $x \cdot y = 1$, does either x or y have to be 1?

B.3.3. Does the subtraction operation have the associativity property? In other words, if x, y, and z are real numbers, then is it true that $(x - y) - z = x - (y - z)$? Explain.

B.3.4. Does the division operation have the commutative property? In other words, if x and y are real numbers, then is it true that $\frac{x}{y} = \frac{y}{x}$? Explain.

B.3.5. Use a rectangle to show the extended distributive property

$$w \cdot (x + y + z) = w \cdot x + w \cdot y + w \cdot z.$$

B.3.6. Do the natural numbers $\mathbb{N} = \{1, 2, 3, \dots\}$ have for every $x \in \mathbb{N}$
  (a) an additive inverse?   (c) an additive identity?
  (b) a multiplicative inverse?   (d) a multiplicative identity?

B.3.7. Do the whole numbers $\mathbb{W} = \{0, 1, 2, \dots\}$ have for every $x \in \mathbb{W}$
  (a) an additive inverse?   (c) an additive identity?
  (b) a multiplicative inverse?   (d) a multiplicative identity?

B.3.8. Do the integers $\mathbb{Z} = \{\dots, -2, -1, 0, 1, 2, \dots\}$ have for every $x \in \mathbb{Z}$

(a) an additive inverse? (c) an additive identity?
(b) a multiplicative inverse? (d) a multiplicative identity?

B.3.9. Do the rationals $\mathbb{Q} = \{x \mid x = \frac{m}{n}$ for integers m and n, and $n \neq 0\}$ have for every $x \in \mathbb{Q}$

(a) an additive inverse? (c) an additive identity?
(b) a multiplicative inverse? (d) a multiplicative identity?

B.4 Solving Quadratic Equations

> **Theorem B.4.1 — Quadratic Formula.** For the equation $ax^2 + bx + c = 0$, where a, b, and c are constants, and $a \neq 0$, then
>
> $$x = \frac{-b \pm \sqrt{b^2 - 4ac}}{2a}.$$ (B.1)

Proof — From Hoehn (1975). For $a \neq 0$,

$$ax^2 + bx + c = 0$$

$$4a(ax^2 + bx + c) = 4a(0) \qquad \text{Multiply by } 4a \text{ on both sides.}$$

$$4a^2x^2 + 4abx + 4ac = 0$$

$$4a^2x + 4abx = -4ac \qquad \text{Add } b^2 \text{ on both sides}$$

$$4a^2x + 4abx + b^2 = b^2 - 4ac \qquad \text{Factor}$$

$$(2ax + b)^2 = b^2 - 4ac$$

$$2ax + b = \pm\sqrt{b^2 - 4ac}$$

$$2ax = -b \pm \sqrt{b^2 - 4ac}$$

$$x = \frac{-b \pm \sqrt{b^2 - 4ac}}{2a}.$$

◀

Exercises

B.4.1. Solve for x using the quadratic formula.
(a) $x^2 + 2x - 48 = 0$ (c) $x^2 - 4x = 12$ (e) $6x^2 - 7x + 1 = 0$
(b) $x^2 - 4x - 21 = 0$ (d) $3x^2 + x - 30 = 0$ (f) $12x^2 + 11x = 15$

B.4.2. In the proof of the quadratic formula, we factored $4a^2x + 4abx + b^2$ to get $(2ax + b)^2$. Show by expanding that $(2ax + b)^2 = 4a^2x + 4abx + b^2$.

B.5 Exponent Laws

Exponential notation, such as 3^3 to express $3 \times 3 \times 3$, was developed by the French mathematician René Descartes (1596–1650) in 1637 to express large and small numbers in a more compact way. It is more convenient for representing powers, but the following laws allow to us to manipulate them as well.

Table B.1   Exponent Laws

Name of Law	Exponent Law	Example
Law of Multiplication	$x^m \times x^n = x^{m+n}$	$x^3 \times x^4 = x^7$
Law of Division	$\dfrac{x^m}{x^n} = x^{m-n}, x \neq 0$	$\dfrac{x^7}{x^5} = x^2$
Law of Powers	$(x^m)^n = x^{m \times n}$	$(x^4)^5 = x^{20}$
Power of a Product	$(xy)^m = x^m \cdot y^m$	$(xy)^4 = x^4 \cdot y^4$
Power of a Quotient	$\left(\dfrac{x}{y}\right)^m = \dfrac{x^m}{y^m}, y \neq 0$	$\left(\dfrac{x}{y}\right)^3 = \dfrac{x^3}{y^3}$
Power of Zero	$x^0 = 1$	$(xy)^0 = 1$
Power of a Fraction	$x^{\frac{p}{q}} = \sqrt[q]{x^p}, x \geq 0, q \neq 0$	$x^{\frac{3}{5}} = \sqrt[5]{x^3}$
Power of a Negative	$x^{-m} = \dfrac{1}{x^m}, x \neq 0, m > 0$	$x^{-3} = \dfrac{1}{x^3}$

Exercises

B.5.1. Evaluate the expression.

(a) $4^{-3} \times 4^4$

(b) $\dfrac{8^{\frac{5}{3}}}{8^{-\frac{1}{3}}}$

(c) $\left(64^{\frac{4}{3}}\right)^{-0.5}$

(d) $(16 \times 81)^{-\frac{1}{4}}$

(e) $\left(\dfrac{3^{\frac{1}{2}}}{2^{\frac{1}{4}}}\right)^4$

B.5.2. Simplify the expression.

(a) $y^{-3} \times y^2$

(b) $\dfrac{3 \times 3^3 \times 2^3}{2^{-3} \times (3^3)^2}$

(c) $\left(x^{\frac{r}{s}}\right)^{\frac{s}{r}}$

(d) $\dfrac{(a^m \times a^{-n})^{-2}}{(a^{m+n})^2}$

B.5.3. Simplify the expression.

(a) $c^2 \cdot c^3 \cdot c$

(b) $y^2 \cdot y^3 \cdot y^4$

(c) $x \cdot x^6 \cdot x^7$

(d) $\dfrac{x^4}{x}$

(e) $\dfrac{x^9}{x^5}$

(f) $\dfrac{w^4}{w^3}$

(g) $(x^5)^3$

(h) $x^6 \cdot x^4$

(i) $\dfrac{a^9}{a^2}$

(j) $(x^2)^3$

(k) $(x^2 y^4)^3$

(l) $\left(\dfrac{a^3}{b^4}\right)^5$

(m) $\dfrac{(a^2b^3)^4}{(ab^2)^3}$

(n) $\left(\dfrac{x^2y^3}{z^4}\right)^4$

(o) $(x^2y^5)^2$

(p) $(x^3)^2(3x)^2$

(q) $(2x^3)^4(5x^3)^4$

(r) $(4m^3n^5)^3(-3m^2n^6)^3$

(s) $\left(\dfrac{12x^3y^4}{9x^2y}\right)\left(\dfrac{18x^5y^4}{4x^2y^2}\right)$

B.5.4. Express as the same base.

(a) $x^2\sqrt{x}$

(b) $y^3\sqrt[3]{y}$

(c) $(z^2+2)(z^2+2)^{-\frac{1}{2}}$

(d) $\sqrt{(x+5)^2}\cdot(x+5)^2$

B.5.5. Solve the following equations for x.

(a) $6^{2x}=6^4$

(b) $5^{-x}=5^3$

(c) $8^x=\left(\frac{1}{32}\right)^{x-2}$

(d) $4^{x^2-x}=2^{12}$

(e) $(1.3)^{x-2}=(1.3)^{2x+1}$

(f) $5^{2x}-2\times5^x+1=0$

B.6 Logarithms

Definition B.6.1 — Exponential and Logarithm form.

$$\underbrace{b^y=x}_{\text{Exponential Form}}\quad\Leftrightarrow\quad\underbrace{y=\log_b(x)}_{\text{Logarithmic Form}},\qquad b>0\text{ and }b\neq1\qquad\text{(B.2)}$$

You can convert from an exponential form to a logarithmic form and *vice versa* using the above definition.

The four properties of logarithms are the following:

$$b^0=1\Leftrightarrow\log_b(1)=0\qquad\text{(B.3)}$$
$$b^x=b^x\Leftrightarrow\log_b(b^x)=x\qquad\text{(B.4)}$$
$$b^1=b\Leftrightarrow\log_b(b)=1\qquad\text{(B.5)}$$
$$b^{\log_b(x)}=x\Leftrightarrow\log_b(x)=\log_b(x).\qquad\text{(B.6)}$$

Table B.2   Logarithm Laws. For $x>0$, $y>0$, and $r\in\mathbb{R}$

Name of Law	Logarithm Law	Example
Law of Multiplication	$\log_b(xy)=\log_b(x)+\log_b(y)$	$\log_b(10\times100)=\log_b(10)+\log_b(100)$
Law of Division	$\log_b\left(\dfrac{x}{y}\right)=\log_b(x)-\log_b(y)$	$\log_b\left(\dfrac{100}{10}\right)=\log_b(100)-\log_b(10)$
Law of Exponent	$\log_b(x^r)=r\log_b(x)$	$\log_b(3^4)=4\log_b(3)$
Law of Square Root	$\log_b(\sqrt[y]{x})=\log_b\left(x^{\frac{1}{y}}\right)$	$\log_b\left(\sqrt[4]{3}\right)=\log_b\left(3^{\frac{1}{4}}\right)$

Note that log (without any base specified) is called the **common logarithm**, and it is often implied that the base is 10. This is the function that is often found on hand-held calculators. Another base that is used frequently is the **natural logarithm**. The natural logarithm uses e $= 2.71828\ldots$ as its base, and it is denoted by ln; that is, $\ln x = \log_e x$. If you do not have 10 as your base in your logarithm, you can still use your calculator by using the **Change of Base** formula: for any positive real numbers b and x (with $b \neq 1$)

$$\log_b(x) = \frac{\log(x)}{\log(b)}. \tag{B.7}$$

Proof We know that

$$b^{\log_b(x)} = x \qquad\qquad\qquad \text{Apply log on both sides.}$$
$$\log\left(b^{\log_b(x)}\right) = \log(x) \qquad\qquad \text{Law of Exponent for log}$$
$$\log_b(x) \cdot \log(b) = \log(x)$$
$$\log_b(x) = \frac{\log(x)}{\log(b)}$$

◄

Example B.6.2 Calculate $\log_4(256)$.

Solution With a calculator, we find that

$$\log_4(256) = \frac{\log(256)}{\log(4)} \approx \frac{2.40823997}{0.602059991}$$
$$= 4.$$

◄

These laws of logarithms can help us solve equations where the unknown variable x is in the exponent.

Example B.6.3 Given $2^x = 128$, solve for x.

Solution — 1.

$$2^x = 128 \qquad\qquad\qquad \text{Apply log on both sides.}$$
$$\log\left(2^x\right) = \log(128) \qquad\qquad\qquad \text{Law of exponent}$$
$$x\log(2) = \log(128)$$
$$x = \frac{\log(128)}{\log(2)} \approx \frac{2.10721}{0.30103}$$
$$x = 7$$

Therefore, $2^7 = 128$.   ◄

Solution — 2.

$$2^x = 128$$
$$x = \log_2(128) \qquad \text{By (B.2)}$$
$$= \frac{\log(128)}{\log(2)} \qquad \text{By (B.7)}$$
$$= 7$$

Therefore, $2^7 = 128$. ◀

Exercises

B.6.1. Evaluate the following:

(a) $32^{\frac{4}{5}}$

(b) $7^{\log_7 24}$

(c) $\log_c(1)$

(d) $\log_4(2)$

(e) $\log_6(18) + \log_6(12)$

(f) $\log 10^{1000}$

B.6.2. Simplify.

(a) $\dfrac{(-6x^4y^3)(9x^{-8}y^2)^{\frac{1}{2}}}{3x^{-2}y^7}$

(b) $\dfrac{(r^{3x+4})(r^{2x-5})}{(r^{x-7})(r^{2x-3})}$

B.6.3. Write $\log_a p = r$ in exponential form.

B.6.4. Calculate $2\log(100) - \log(1000) + \log(10)$.

B.6.5. Rewrite the following as a single logarithm:

(a) $2\log(a) - \log(b) + \log(c)$ (b) $\log(a) + 5\log(b) - \log(c)$

B.6.6. Evaluate the expressions without a calculator. Use the following approximations for $\log(3) \approx 0.4771$ and $\log(4) \approx 0.6021$.

(a) $\log(12)$

(b) $\log(\frac{3}{4})$

(c) $\log(\sqrt{3})$

B.6.7. Evaluate without a calculator. (Give exact answers.)

(a) $\log_2(\sqrt[5]{16})$

(b) $\log_3(9\sqrt{27})$

(c) $\frac{1}{2}\log_8(16) + \log_8(48) - \log_8(3)$

(d) $\log_8\left(\frac{1}{4}\right)$

B.6.8. Express the given equations in logarithmic form.

(a) $2^6 = 64$

(b) $10^{-3} = 0.001$

(c) $32^{3/5} = 8$

B.6.9. Simplify the expression.

(a) $\log_2 8^{-x}$

(b) $\log_2 5 - \log_2 90 + 2\log_2 3$

(c) $\log_b \sqrt[3]{\dfrac{x^2\sqrt{y}}{z^3}}$

B.6.10. Solve for x.

(a) $2^x = 5$

(b) $5^x = 7$

(c) $6^x = 30$

(d) $3^x + 1 = 10$

(e) $7 = 12 - 4^x$

(f) $1 - \left(\frac{35}{36}\right)^x = \frac{1}{2}$

B.6.11. Solve for x.

(a) $x = \log_4 8$

(b) $\log_2(x - 3) + \log_2(x - 4) = 1$

(c) $e^{3x-4} = 2$

(d) $2^x = 3$

B.6.12. Show that $\log_b(x) = \dfrac{1}{\log_x(b)}$.

B.6.13. Show that $\log_b\left(\sqrt[y]{x}\right) = \frac{1}{y}\log_b(x)$.

B.6.14. A car depreciates at 19% per year. How many years will it take for the car to be worth half of its original value?

B.6.15. A swarm of locusts doubles every three weeks. If there are 4500 locusts now, how many locusts were there 9 weeks ago (to the nearest locust)?

B.6.16. In Springfield, the population is declining by 7.5% per annum. If the current population is 15 000, how many people will there be in $3\frac{1}{2}$ years?

B.6.17. Show that the following statement is true: $\dfrac{3}{\log_2(a)} = \dfrac{1}{\log_2(a)}$.

B.7 Scientific Notation

In this book, we will encounter some very large numbers like

$$5\,973\,000\,000\,000\,000\,000\,000$$

(which is the mass of the earth in kilograms), or very small numbers like

$$0.000\,000\,000\,000\,000\,000\,000\,000\,000\,000\,910\,9$$

(which is the mass of an electron in kilograms). For these kinds of numbers, it becomes tiresome to write out the placeholding zeros. Instead, we can express these kinds of numbers as a product of a number and an integer power of 10:

$$5\,973\,000\,000\,000\,000\,000\,000 = 5.9736 \times 10^{24}$$

and

$$0.000\,000\,000\,000\,000\,000\,000\,000\,000\,000\,910\,9 = 9.109 \times 10^{-31}.$$

Expressing these numbers in this fashion is called **scientific notation**.

Definition B.7.1 — Scientific notation. In scientific notation, all numbers are written in the form

$$m \times 10^n, \qquad \text{where } m \text{ is any real number, and } n \text{ is an integer.}$$

Basic operations such as multiplication or division can be easily done with scientific notation with the use of exponent laws. Given two numbers in scientific notation,

$$x = m \times 10^n$$

and

$$y = p \times 10^q$$

then

(a) to multiply: $x \times y = (m \times 10^n) \times (p \times 10^q) = mp \times 10^{n+q}$;

(b) to divide: $\dfrac{x}{y} = \dfrac{m \times 10^n}{p \times 10^q} = \dfrac{m}{p} \times 10^{n-q}$.

Example B.7.2

(a) $(1.23 \times 10^6) \times (4.56 \times 10^4) = 5.6088 \times 10^{10}$

(b) $\dfrac{1.23 \times 10^6}{4.56 \times 10^4} = \dfrac{1.23}{4.56} \times 10^{6-4} \approx 0.26974 \times 10^2 = 2.6974 \times 10^1$

◄

Exercises

B.7.1. Express the following in scientific notation.

(a) 2

(b) 300

(c) 4321.768

(d) −53 000

(e) 6 720 000 000

(f) 0.000 000 007 51

Appendix C

Solutions to Exercises

1.1.1. Logic is the study of methods and principles used in distinguishing correct from incorrect arguments.

1.1.2. He was a pioneer in pointing out that the process of analysis does not depend on the interpretation of the symbols which are employed, but solely upon the laws of their combination.

1.1.3. Boole died from pneumonia, which he contracted after walking two miles in drenching rain to give a class.

1.1.4. De Morgan's *Formal Logic* was unsuccessful owing to his complicated notation and Boole's more readable *Mathematical Analysis of Logic*.

1.1.5. De Morgan was a unique figure since he made no startling discoveries. His chief contributions to logic is his applications of mathematical methods and the subsequent development of symbolic logic in the form of five papers he published.

1.1.6. De Morgan refused to submit to the Church of England religious test, which prevented him from proceeding to the master's degree or continuing as a fellow at Cambridge or Oxford.

1.1.7. The book was the first to demonstrate the close ties between mathematics and formal logic.

1.2.1. (a) This is a proposition. In fact, Bonaparte did lose the battle against the Seventh Coalition.
   (b) This is a proposition. In fact, Beethoven had originally dedicated the symphony to Bonaparte, but when Bonaparte declared himself Emperor, Beethoven scratched-out the name so furiously, that it almost left a hole in the title page!
   (c) This is not a proposition; it is a question.
   (d) No, this is not a proposition. It is either a command or exclamation in which we cannot assign a truth value.
   (e) This is a proposition; in fact, it is a true proposition.
   (f) This is a proposition. In fact, it is a false proposition; it says that four is greater than ten, which is clearly false.
   (g) This is not a proposition, because the pronoun 'we' is not defined. (Unless, you knew this was from *Casablanca* and you knew 'we' refers to Rick and Ilsa.)
   (h) This is a proposition. In fact, it is a false proposition since Mexico is in North America.
   (i) This is not a proposition. It is a command.
   (j) This is a proposition. No further comment.

1.2.2. (a) Yes, it is a proposition. You can assign a value of true or false. (In fact, he was the president of the USA.)

   (b) This is a proposition.

   (c) This is not a proposition. It's a question.

   (d) This is a proposition.

   (e) This is a question. So, it is not a proposition.

   (f) This is a demand or an exclamation. There is no truth to determine, so it is not a proposition.

1.2.3. The prime proposition in Exercise 1.2.2 is (a).

1.2.4. The compound propositions in Exercise 1.2.2 are (b) and (d).

1.2.5. This is a paradox. It is sometimes called the **Socratic paradox**, and it is a well-known saying that is derived from Plato's account of the Greek philosopher Socrates.

1.3.1. (a) It is snowing.

   (b) Fish do have gills.

   (c) The longest bridge in Great Britain is not the Forth Bridge.

   (d) New York City is the capital of New York.

   (e) I never gave or took any excuse.

   (f) Not every child is an artist.

   (g) It is not the case that no baseball team wins two games in a row.

   (h) It rains and the wind blows.

   (i) It is not the case that the market is not growing, and the market is being measured correctly.

   (j) It is not the case that the national debt is not decreasing rapidly enough or our government spending policies are helping.

   (k) It is not the case that if sales taxes are lowered, then consumer spending will increase. (An alternative: Sales taxes are lowered and consumer spending will not increase.)

1.3.2. (a) $p \wedge q$: House prices are high, and houses are in demand.

   (b) $p \wedge \neg q$: House prices are high, but houses are not in demand.

   (c) $\neg p \wedge \neg q$: House prices are not high and houses are not in demand.

   (d) $p \vee \neg q$: House prices are high or houses are not in demand, or both.

   (e) $\neg(p \wedge q)$: It is not the case that both house prices are high and houses are in demand.

   (f) $\neg(\neg p \vee \neg q)$: It is not the case that house prices are not high, or houses are not in demand, or both.

1.3.3. (a) $q \to r$ (c) $q \to r$

      (b) $p \to q$ (d) $\neg p \to \neg r$

1.3.4. (a) F   (b) F   (c) T   (d) T   (e) T   (f) F

1.3.5. (a) T   (b) F   (c) F   (d) T   (e) F   (f) F

1.3.6. (a) T   (b) F   (c) T   (d) F

1.3.7. (a) Let p be "Personnel should be promoted," and q be "They are qualified." Then symbolically, $q \to p$.

   (b) Let p be "Bacon wrote Shakespeare's plays," q be "Marlowe wrote Shakespeare's plays." Then symbolically, $p \veebar q$.

   (c) Let p be "The child is admitted to school immediately," and q be "The child is sent home and notified of his admission." Then symbolically, $p \veebar q$.

   (d) Let p be "You talk to Marian," and q be "Marian talks to you." Then symbolically, $\neg p \to \neg q$.

1.3.8. (a) $r \vee ((\neg p) \wedge q)$

(b) $q \rightarrow \left(\neg(\neg(\neg p)) \wedge r\right)$

1.3.9. (a) If Darcy practices her serve daily, then she will have a good chance of winning.

(b) If the computer is not brand-new, then I won't buy it.

1.3.10. The antecedent is false: the moon is not made of cheese. The consequent is also false: Socrates has been long dead before even the creation of Canada. Since both the antecedent and the consequent are both false, then *the implication is true*.

1.3.11. The truth table is given below.

p	q	$p \uparrow q$
T	T	F
T	F	T
F	T	T
F	F	T

1.4.1. The completed truth table is given below.

p	q	$\neg p$	$\neg q$	$p \wedge \neg q$	$\neg p \vee q$	$\neg p \rightarrow q$
T	T	F	F	F	T	T
T	F	F	T	T	F	T
F	T	T	F	F	T	T
F	F	T	T	F	T	F

1.4.2. (a) $\neg(p \vee q)$

p	q	$p \vee q$	$\neg(p \vee q)$
T	T	T	F
T	F	T	F
F	T	T	F
F	F	F	T

(b) $\neg(p \wedge q)$

p	q	$p \wedge q$	$\neg(p \wedge q)$
T	T	T	F
T	F	F	T
F	T	F	T
F	F	F	T

(c) $(\neg p) \to q$

p	q	$\neg p$	$(\neg p) \to q$
T	T	F	T
T	F	F	T
F	T	T	T
F	F	T	F

(d) $[(p \to q) \wedge (\neg p)] \to (\neg q)$

p	q	$p \to q$	$\neg p$	$(p \to q) \wedge (\neg p)$	$\neg q$	$[(p \to q) \wedge (\neg p)] \to (\neg q)$
T	T	T	F	F	F	T
T	F	F	F	F	T	T
F	T	T	T	T	F	F
F	F	T	T	T	T	T

(e) $p \vee (q \wedge r)$

p	q	r	$q \wedge r$	$p \vee (q \wedge r)$
T	T	T	T	T
T	T	F	F	T
T	F	T	F	T
T	F	F	F	T
F	T	T	T	T
F	T	F	F	F
F	F	T	F	F
F	F	F	F	F

(f) $(p \vee q) \wedge (p \vee r)$

p	q	r	$p \vee q$	$p \vee r$	$(p \vee q) \wedge (p \vee r)$
T	T	T	T	T	T
T	T	F	T	T	T
T	F	T	T	T	T
T	F	F	T	T	T
F	T	T	T	T	T
F	T	F	T	F	F
F	F	T	F	T	F
F	F	F	F	F	F

(g) $(p \wedge q) \vee (p \wedge \neg r)$

p	q	r	$\neg r$	$p \wedge q$	$p \wedge \neg r$	$(p \wedge q) \vee (p \wedge \neg r)$
T	T	T	F	T	F	T
T	T	F	T	T	T	T
T	F	T	F	F	F	F
T	F	F	T	F	T	T
F	T	T	F	F	F	F
F	T	F	T	F	F	F
F	F	T	F	F	F	F
F	F	F	T	F	F	F

1.4.3. (a) $q \to p$

p	q	$q \to p$
T	T	T
T	F	T
F	T	F
F	F	T

(b) $(r \wedge q) \to r$

r	q	$r \wedge q$	$(r \wedge q) \to r$
T	T	T	T
T	F	F	T
F	T	F	T
F	F	F	T

(c) $(r \to s) \to r$

r	s	$r \to s$	$(r \to s) \to r$
T	T	T	T
T	F	F	T
F	T	T	F
F	F	T	F

(d) $(p \vee q) \to p$

p	q	$p \vee q$	$(p \vee q) \to p$
T	T	T	T
T	F	T	T
F	T	T	F
F	F	F	T

(e) $(p \to q) \to q$

p	q	$p \to q$	$(p \to q) \to q$
T	T	T	T
T	F	F	T
F	T	T	T
F	F	T	F

(f) $(p \wedge q) \to (p \vee q)$

p	q	$p \wedge q$	$p \vee q$	$(p \wedge q) \to (p \vee q)$
T	T	T	T	T
T	F	F	T	T
F	T	F	T	T
F	F	F	F	T

(g) $(p \vee q) \to [(\neg p) \vee q]$

p	q	$p \vee q$	$\neg p$	$\neg p \vee q$	$(p \vee q) \to (\neg p \vee q)$
T	T	T	F	T	T
T	F	T	F	F	F
F	T	T	T	T	T
F	F	F	T	T	T

(h) $(p \to r) \to [(\neg p) \vee r]$

p	r	$p \to r$	$\neg p$	$\neg p \vee r$	$(p \to r) \to (\neg p \vee r)$
T	T	T	F	T	T
T	F	F	F	F	T
F	T	T	T	T	T
F	F	T	T	T	T

(i) $(v \wedge w) \to [(\neg w) \to v]$

v	w	$v \wedge w$	$\neg w$	$\neg w \to v$	$(v \wedge w) \to (\neg w \to v)$
T	T	T	F	T	T
T	F	F	T	T	T
F	T	F	F	T	T
F	F	F	T	F	T

(j) $(p \vee q) \rightarrow (p \rightarrow q)$

p	q	$p \vee q$	$p \rightarrow q$	$(p \vee q) \rightarrow (p \rightarrow q)$
T	T	T	T	T
T	F	T	F	F
F	T	T	T	T
F	F	F	T	T

(k) $(p \wedge q) \rightarrow [(\neg q) \rightarrow (\neg p)]$

p	q	$p \wedge q$	$\neg q$	$\neg p$	$\neg q \rightarrow \neg p$	$(p \wedge q) \rightarrow (\neg q \rightarrow \neg p)$
T	T	T	F	F	T	T
T	F	F	T	F	F	T
F	T	F	F	T	T	T
F	F	F	T	T	T	T

(l) $(\ell \rightarrow k) \rightarrow [(\neg k) \rightarrow (\neg \ell)]$

ℓ	k	$\ell \rightarrow k$	$\neg k$	$\neg \ell$	$\neg k \rightarrow \neg \ell$	$(\ell \rightarrow k) \rightarrow (\neg k \rightarrow \neg \ell)$
T	T	T	F	F	T	T
T	F	F	T	F	F	T
F	T	T	F	T	T	T
F	F	T	T	T	T	T

(m) $(p \vee q) \rightarrow [p \wedge (\neg q)]$

p	q	$p \vee q$	$\neg q$	$p \wedge \neg q$	$(p \vee q) \rightarrow (p \wedge \neg q)$
T	T	T	F	F	F
T	F	T	T	T	T
F	T	T	F	F	F
F	F	F	T	F	T

(n) $(p \rightarrow q) \rightarrow r$

p	q	r	$p \rightarrow q$	$(p \rightarrow q) \rightarrow r$
T	T	T	T	T
T	T	F	T	F
T	F	T	F	T
T	F	F	F	T
F	T	T	T	T
F	T	F	T	F
F	F	T	T	T
F	F	F	T	F

1.5.1. The truth table is given below.

p	q	r	$q \vee r$	$p \wedge (q \vee r)$
T	T	T	T	T
T	T	F	T	T
T	F	T	T	T
T	F	F	F	F
F	T	T	T	F
F	T	F	T	F
F	F	T	T	F
F	F	F	F	F

1.5.2. (a) Switches p, q, and r are in series. Thus, the switching circuit can be written as $(p \wedge q) \wedge r$.

   (b) The switches p and $\neg p$ are in parallel. This can be written as $p \vee \neg p$.

   (c) At the bottom of the switching circuit, the switches q and r are in series, which can be written as $(q \wedge r)$. Then p and $(q \wedge r)$ are in parallel. This can be written as $p \vee (q \wedge r)$.

   (d) At the top of the switching circuit, the switches p and q are in series; this can be written as $p \wedge q$. At the bottom of the switching circuits, the switches p and r are also in series; this can be written as $p \wedge r$. Then $(p \wedge q)$ and $(p \wedge r)$ are in parallel. Thus, the switching circuit can be written as $(p \wedge q) \vee (p \wedge r)$.

1.5.3. (a)

(b)

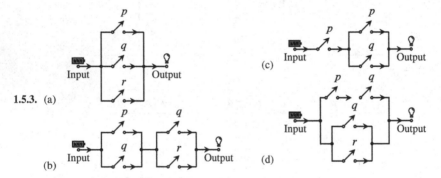

(c)

(d)

1.6.1. A tautology must have all true truth values; a contradiction must have all false truth values. A contingency has neither all true or all false truth values.

1.6.2. (a) $(p \wedge \neg p) \rightarrow q$

p	q	$\neg p$	$p \wedge \neg p$	$(p \wedge \neg p) \rightarrow q$
T	T	F	F	**T**
T	F	F	F	**T**
F	T	T	F	**T**
F	F	T	F	**T**
				T

Therefore, it is a tautology.

(b) $\neg(p \wedge \neg p)$

p	$\neg p$	$p \wedge \neg p$	$\neg(p \wedge \neg p)$
T	F	F	**T**
F	T	F	**T**
			T

Therefore, it is a tautology.

(c) $(\neg p \to p) \to p$

p	$\neg p$	$\neg p \to p$	$(\neg p \to p) \to p$
T	F	T	**T**
F	T	F	**T**
			T

Therefore, it is a tautology.

1.6.3. (a) $p \to (q \to r)$

p	q	r	$q \to r$	$p \to (q \to r)$
T	T	T	T	**T**
T	T	F	F	**F**
T	F	T	T	**T**
T	F	F	T	**T**
F	T	T	T	**T**
F	T	F	F	**T**
F	F	T	T	**T**
F	F	F	T	**T**
				$\neq$T

Therefore, it is not a tautology. (It is a contingency.)

(b) $(q \vee r) \to (\neg r \to q)$

q	r	$\neg r$	$q \vee r$	$(\neg r \to q)$	$(q \vee r) \to (\neg r \to q)$
T	T	F	T	T	**T**
T	F	T	T	T	**T**
F	T	F	T	T	**T**
F	F	T	F	F	**T**
					T

This is a tautology.

(c) $[(p \vee q) \wedge (\neg q)] \to p$

p	q	$p \vee q$	$\neg q$	$(p \vee q) \wedge (\neg q)$	$(p \vee q) \wedge (\neg q) \to p$
T	T	T	F	F	**T**
T	F	T	T	T	**T**
F	T	T	F	F	**T**
F	F	F	T	F	**T**

$$\underbrace{}_{\text{T}}$$

Therefore, this is a tautology.

(d) $[(p \wedge \neg q) \vee (q \wedge \neg r)] \vee (r \wedge \neg p)$

p	q	r	$\neg p$	$\neg q$	$\neg r$	$[(p \wedge \neg q)$	$\vee$	$(q \wedge \neg r)]$	$\vee$	$(r \wedge \neg p)$
T	T	T	F	F	F	F	F	F	**F**	F
T	T	F	F	F	T	F	T	T	**T**	F
T	F	T	F	T	F	T	T	F	**T**	F
T	F	F	F	T	T	T	T	F	**T**	F
F	T	T	T	F	F	F	F	F	**T**	T
F	T	F	T	F	T	F	T	T	**T**	F
F	F	T	T	T	F	F	F	F	**T**	T
F	F	F	T	T	T	F	F	F	**F**	F

$$\underbrace{}_{\neq \text{T}}$$

Therefore, it is not a tautology. (It is a contingency.)

1.6.4. (a) $(p \wedge q) \vee (\neg p \wedge \neg q)$

p	q	$p \wedge q$	$\neg p$	$\neg q$	$\neg p \wedge \neg q$	$(p \wedge q) \vee (\neg p \wedge \neg q)$
T	T	T	F	F	F	**T**
T	F	F	F	T	F	**F**
F	T	F	T	F	F	**F**
F	F	F	T	T	T	**T**

Therefore, it is a contingency.

(b) $(p \vee \neg q) \wedge (q \vee \neg p)$

p	q	$\neg p$	$\neg q$	$p \vee \neg q$	$q \vee \neg p$	$(p \vee \neg q) \wedge (q \vee \neg p)$
T	T	F	F	T	T	**T**
T	F	F	T	T	F	**F**
F	T	T	F	F	T	**F**
F	F	T	T	T	T	**T**

Therefore, it is a contingency.

(c) $\mathbb{T} \to (\neg p \wedge p)$

$\mathbb{T}$	p	$\neg p$	$\neg p \wedge p$	$\mathbb{T} \to (\neg p \wedge p)$
T	T	F	F	F
T	F	T	F	F

F

Therefore, it is a contradiction.

(d) $\mathbb{F} \to (p \to r)$

$\mathbb{F}$	p	r	$p \to r$	$\mathbb{F} \to (p \to r)$
F	T	T	T	T
F	T	F	F	T
F	F	T	T	T
F	F	F	T	T

T

Therefore, it is a tautology.

1.7.1. (a) If you are a monkey's uncle, then you love bananas.
(b) If your nephew is not a chimp, then you do not love bananas.
(c) You are a monkey's uncle if and only if your nephew is a chimp.
(d) You are a monkey's uncle and your nephew is a chimp.

1.7.2. (a) $\neg p \leftrightarrow q$

p	q	$\neg p$	$\neg p \leftrightarrow q$
T	T	F	F
T	F	F	T
F	T	T	T
F	F	T	F

(b) $\neg p \leftrightarrow (p \to \neg q)$

p	q	$\neg p$	$\neg q$	$p \to \neg q$	$\neg p \leftrightarrow (p \to \neg q)$
T	T	F	F	F	T
T	F	F	T	T	F
F	T	T	F	T	T
F	F	T	T	T	T

(c) $(p \vee q) \leftrightarrow (q \vee p)$

p	q	$p \vee q$	$q \vee p$	$(p \vee q) \leftrightarrow (q \vee p)$
T	T	T	T	T
T	F	T	T	T
F	T	T	T	T
F	F	F	F	T

(d) $(p \rightarrow q) \leftrightarrow (\neg p \vee q)$

p	q	$p \rightarrow q$	$\neg p$	$\neg p \vee q$	$(p \rightarrow q) \leftrightarrow (\neg p \vee q)$
T	T	T	F	T	T
T	F	F	F	F	T
F	T	T	T	T	T
F	F	T	T	T	T

(e) $(p \vee q) \leftrightarrow (\neg r \wedge \neg s)$

p	q	r	s	$p \vee q$	$\neg r$	$\neg s$	$\neg r \wedge \neg s$	$(p \vee q) \leftrightarrow (\neg r \wedge \neg s)$
T	T	T	T	T	F	F	F	F
T	T	T	F	T	F	T	F	F
T	T	F	T	T	T	F	F	F
T	T	F	F	T	T	T	T	T
T	F	T	T	T	F	F	F	F
T	F	T	F	T	F	T	F	F
T	F	F	T	T	T	F	F	F
T	F	F	F	T	T	T	T	T
F	T	T	T	T	F	F	F	F
F	T	T	F	T	F	T	F	F
F	T	F	T	T	T	F	F	F
F	T	F	F	T	T	T	T	T
F	F	T	T	F	F	F	F	T
F	F	T	F	F	F	T	F	T
F	F	F	T	F	T	F	F	T
F	F	F	F	F	T	T	T	F

(f) $(p \vee q) \leftrightarrow (p \wedge q)$

p	q	$p \vee q$	$p \wedge q$	$(p \vee q) \leftrightarrow (p \wedge q)$
T	T	T	T	T
T	F	T	F	F
F	T	T	F	F
F	F	F	F	T

(g) $[p \rightarrow (q \vee r)] \wedge [p \leftrightarrow \neg r]$

p	q	r	$q \vee r$	$p \rightarrow (q \vee r)$	$\neg r$	$p \leftrightarrow \neg r$	$[p \rightarrow (q \vee r)] \wedge [p \leftrightarrow \neg r]$
T	T	T	T	T	F	F	F
T	T	F	T	T	T	T	T
T	F	T	T	T	F	F	F
T	F	F	F	F	T	T	F
F	T	T	T	T	F	T	T
F	T	F	T	T	T	F	F
F	F	T	T	T	F	T	T
F	F	F	F	T	T	F	F

1.7.3. (a) $p \to q$

(b) $q \to p$

(c) $p \to q$

(d) $p \leftrightarrow q$

(e) $p \leftrightarrow q$

(f) $q \to p$

1.7.4. (a) $\neg(p \lor q) \Leftrightarrow (\neg p \land \neg q)$

p	q	$p \lor q$	$\neg(p \lor q)$	$\neg p$	$\neg q$	$\neg p \land \neg q$
T	T	T	F	F	F	F
T	F	T	F	F	T	F
F	T	T	F	T	F	F
F	F	F	T	T	T	T

The columns of $\neg(p \lor q)$ and $\neg p \land \neg q$ have the same truth values. Therefore, they are logically equivalent.

(b) $\neg(p \land q) \Leftrightarrow (\neg p \lor \neg q)$

p	q	$p \land q$	$\neg(p \land q)$	$\neg p$	$\neg q$	$\neg p \lor \neg q$
T	T	T	F	F	F	F
T	F	F	T	F	T	T
F	T	F	T	T	F	T
F	F	F	T	T	T	T

The columns $\neg(p \land q)$ and $(\neg p \lor \neg q)$ have the same truth values. Therefore, they are logically equivalent.

(c) $\neg(\neg p \lor q) \Leftrightarrow (p \land \neg q)$

p	q	$\neg p$	$\neg q$	$\neg p \lor q$	$\neg(\neg p \lor q)$	$p \land \neg q$
T	T	F	F	T	F	F
T	F	F	T	F	T	T
F	T	T	F	T	F	F
F	F	T	T	T	F	F

The columns of $\neg(\neg p \lor q)$ and $(p \land \neg q)$ have the same truth values. Therefore, they are logically equivalent.

1.7.5. (a)   (i) Converse: If cars must stop, then the traffic light is red.

(ii) Inverse: If the traffic light is not red, then cars must not stop.

(iii) Contrapositive: If cars must not stop, then the traffic light is not red.

(b)   (i) Converse: If it is an apple, then the fruit is red.

(ii) Inverse: If the fruit is not red, then it is not an apple.

(iii) Contrapositive: If it is not an apple, then the fruit is not red.

(c)   (i) Converse: Change is sufficient for unrest.

(ii) Inverse: No unrest is sufficient for no change.

(iii) Contrapositive: No change is sufficient for no unrest.

(d)   (i) Converse: A responsibility implies a right.

(ii) Inverse: No right implies no responsibility.

(iii) Contrapositive: No responsibility implies no right.
(e)   (i)   Converse: A duty implies also the right to search for the truth.
       (ii) Inverse: No right to search for the truth implies no duty.
       (iii) Contrapositive: No duty implies also no right to search for the truth.
(f)   (i)   Converse: If it improves upon the silence, then speak.
       (ii) Inverse: Do not speak only if it does not improve the silence.
       (iii) Contrapositive: If it does not improve upon the silence, then do not speak.

1.7.6. The truth table is given below.

p	q	$p \wedge q$	$\neg(p \wedge q)$	$\neg p$	$\neg q$	$(\neg p \wedge \neg q)$
T	T	T	**F**	F	F	**F**
T	F	F	**T**	F	T	**F**
F	T	F	**T**	T	F	**F**
F	F	F	**T**	T	T	**T**

The columns of $\neg(p \wedge q)$ and $(\neg p \wedge \neg q)$ do not have the same truth values. Therefore, they are not logically equivalent.

1.7.7. The truth table is given below.

p	q	$p \rightarrow q$	$\neg q$	$\neg p$	$\neg q \rightarrow \neg p$
T	T	**T**	F	F	**T**
T	F	**F**	T	F	**F**
F	T	**T**	F	T	**T**
F	F	**T**	T	T	**T**

The columns of $p \rightarrow q$ and $\neg q \rightarrow \neg p$ have the same truth values. Therefore, they are logically equivalent.

1.7.8. The truth table is given below.

p	q	$q \rightarrow p$	$\neg p$	$\neg q$	$\neg p \rightarrow \neg q$
T	T	**T**	F	F	**T**
T	F	**T**	F	T	**T**
F	T	**F**	T	F	**F**
F	F	**T**	T	T	**T**

The columns of $q \rightarrow p$ and $\neg p \rightarrow \neg q$ have the same truth values. Therefore, they are logically equivalent.

1.7.9. (a) $q \rightarrow p$
           (i)   Contrapositive: $\neg p \rightarrow \neg q$.
           (ii) Converse: $p \rightarrow q$.
           (iii) Inverse: $\neg q \rightarrow \neg p$.
       (b) $\neg p \rightarrow \neg q$
           (i)   Contrapositive: $\neg(\neg q) \rightarrow \neg(\neg p) \Leftrightarrow q \rightarrow p$.

(ii) Converse: $\neg q \rightarrow \neg p$.

(iii) Inverse: $\neg(\neg p) \rightarrow \neg(\neg q) \Leftrightarrow p \rightarrow q$.

1.7.10. (a) (i) Contrapositive: If I am not human, then I cannot think

(ii) Converse: If I am human, then I can think.

(iii) Inverse: If I cannot think, then I am not human.

(b) I can think if and only if I am human.

(c) The two conditions are:

(i) I can think, but I am not human.

(ii) I cannot think but I am not human.

1.7.11. (a) $p \rightarrow q$: "If an integer is divisible by 9, then it is divisible by 3." This is true.

(b) $q \rightarrow p$: "If an integer is divisible by 3, then it is divisible by 9." This is false.

(c) $p \leftrightarrow q$ is therefore false since it is not true for $q \rightarrow p$.

1.7.12. True.

1.7.13. (a) False   (b) True   (c) False

1.7.14. (a) The biconditional statement is "the date is the 29th of the month if and only if it is not February." False. It can still be February if it is a leap year.

(b) The biconditional statement is "the computer is unplugged if and only if the computer will not run." False. The computer may not be switched on.

(c) The biconditional statement is "you download 12 songs for $11.88 if and only if each song costs $0.99." False. You could have downloaded 6 songs for $0.69 each, and the other 6 songs at $1.29 per song.

(d) The biconditional statement is "Jake lives in Las Vegas if and only if he lives in Nevada." False. Jake can live in Nevada, but he could live in any city in the state.

(e) The biconditional statement is "London is not in England if and only if $3 + 6 = 11$." True. Both propositions "London is not in England" and "$3 + 6 = 11$" are false, and hence the biconditional proposition is true.

1.7.15. (a) $p \wedge (q \vee r) \Leftrightarrow (p \wedge q) \vee (p \wedge r)$

p	q	r	$q \vee r$	$p \wedge (q \vee r)$	$p \wedge q$	$p \wedge r$	$(p \wedge q) \vee (p \wedge r)$
T	T	T	T	T	T	T	T
T	T	F	T	T	T	F	T
T	F	T	T	T	F	T	T
T	F	F	F	F	F	F	F
F	T	T	T	F	F	F	F
F	T	F	T	F	F	F	F
F	F	T	T	F	F	F	F
F	F	F	F	F	F	F	F

The columns of $p \wedge (q \vee r)$ and $(p \wedge q) \vee (p \wedge r)$ have the same truth values. Therefore, they are logically equivalent.

(b) $p \vee (q \wedge r) \Leftrightarrow (p \vee q) \wedge (p \vee r)$

p	q	r	$q \wedge r$	$p \vee (q \wedge r)$	$p \vee q$	$p \vee r$	$(p \wedge q) \wedge (p \wedge r)$
T	T	T	T	**T**	T	T	**T**
T	T	F	F	**T**	T	T	**T**
T	F	T	F	**T**	T	T	**T**
T	F	F	F	**T**	T	T	**T**
F	T	T	T	**T**	T	T	**T**
F	T	F	F	**F**	T	F	**F**
F	F	T	F	**F**	F	T	**F**
F	F	F	F	**F**	F	F	**F**

The columns of $p \vee (q \wedge r)$ and $(p \wedge q) \wedge (p \wedge r)$ have the same truth values. Therefore, they are logically equivalent.

1.7.16. The truth table is given below.

p	q	$p \leftrightarrow q$	$\neg(p \leftrightarrow q)$	$\neg q$	$p \wedge \neg q$	$\neg p$	$q \wedge \neg p$	$(p \wedge \neg q) \vee (q \wedge \neg p)$
T	T	T	**F**	F	F	F	F	**F**
T	F	F	**T**	T	T	F	F	**T**
F	T	F	**T**	F	F	T	T	**T**
F	F	T	**F**	T	F	T	F	**F**

The columns of $\neg(p \leftrightarrow q)$ and $(p \wedge \neg q) \vee (q \wedge \neg p)$ have the same truth values. Therefore, they are logically equivalent.

1.7.17. (a) $\neg p \vee q$ (b) $\neg(p \wedge \neg q)$

1.8.1. (a) No, it is not a proposition. Who is "she" in this case?.

(b) No, it is not a proposition; who is "he" in this case?

(c) Yes, it is a proposition.

(d) Yes, it is a proposition.

(e) Yes, it is a proposition.

(f) Yes, it is a proposition.

(g) No, it is not a proposition; who is "she" in this case?

(h) No, it is not a proposition; it is a question.

(i) No, it is not a proposition; it is a question.

(j) Yes, it is a proposition.

1.8.2. This is called the **Crocodile's dilemma**, and it is indeed a paradox that was known to the Ancient Greeks. This was reintroduced by Walker (1847, p. 159) and Prantl (1855, p. 493), and it could be restated instead of a crocodile with pirates who kidnaps the child. If the crocodile keeps the child, then the father is correct, and by the terms of the agreement, the crocodile should give the child back. But, if the crocodile does not keep the child, then the father is not correct, and by the term of the agreement, the crocodile should keep the child.

1.8.3. (a) $4 + 5 \neg 15$

(b) -3 is not a positive number.

(c) Either Alexander the Great was not a Roman leader, or $2 + 1 \neq 3$.

1.8.4. (a) T   (c) T   (e) T

(b) F   (d) F   (f) T

1.8.5. If you can do it today, then you should not put it off until tomorrow.

1.8.6. (a)   (i) Converse: If I won't wear my coat, then it's cold outside and it's not snowing.

  (ii) Contrapositive: If I wear my coat, then it is not cold outside or it is not snowing.

  (iii) Negation: It is cold outside and it's not snowing, and I will wear my coat.

(b) (i) Converse: If you'll need a snorkel, then you live in Atlantis.

  (ii) Contrapositive: If you will not need a snorkel, then you do not live in Atlantis.

  (iii) Negation. You live in Atlantis and you'll not need a snorkel.

1.8.7. (a) $\neg p \wedge (p \rightarrow q)$

p	q	$\neg p$	$p \rightarrow q$	$\neg p \wedge (p \rightarrow q)$
T	T	F	T	F
T	F	T	F	F
F	T	F	T	F
F	F	T	T	T

It is a contingency.

(b) $(p \rightarrow q) \vee (q \rightarrow p)$

p	q	$p \rightarrow q$	$q \rightarrow p$	$(p \rightarrow q) \vee (q \rightarrow p)$
T	T	T	T	T
T	F	F	T	T
F	T	T	F	T
F	F	T	T	T
				T

It is a tautology.

(c) $(p \vee q) \wedge (p \vee \neg r)$

p	q	r	$p \vee q$	$\neg r$	$p \vee \neg r$	$(p \vee q) \wedge (p \vee \neg r)$
T	T	T	T	F	T	T
T	T	F	T	T	T	T
T	F	T	T	F	T	T
T	F	F	T	T	T	T
F	T	T	T	F	F	F
F	T	F	T	T	T	T
F	F	T	F	F	F	F
F	F	F	F	T	T	T

It is a contingency.

(d) $(p \vee q) \wedge (\neg p \wedge \neg q)$

p	q	$p \vee q$	$\neg p$	$\neg q$	$\neg p \wedge \neg q$	$(p \vee q) \wedge (\neg p \wedge \neg q)$
T	T	T	F	F	F	F
T	F	T	F	T	F	F
F	T	T	T	F	F	F
F	F	F	T	T	T	F
						F

This is a contradiction.

(e) $\neg(p \wedge q) \leftrightarrow (q \wedge p)$

p	q	$p \wedge q$	$\neg(p \wedge q)$	$q \wedge p$	$\neg(p \wedge q) \leftrightarrow (q \wedge p)$
T	T	T	F	T	F
T	F	F	T	F	F
F	T	F	T	F	F
F	F	F	T	F	F
					F

This is a contradiction.

1.8.8. The truth table is given below.

p	q	$p \leftrightarrow q$	$\neg p$	$\neg q$	$p \wedge \neg q$	$q \wedge \neg p$	$(p \wedge \neg q) \vee (q \wedge \neg p)$	$\neg((p \wedge \neg q) \vee (q \wedge \neg p))$
T	T	**T**	F	F	F	F	F	**T**
T	F	**F**	F	T	T	F	T	**F**
F	T	**F**	T	F	F	T	T	**F**
F	F	**T**	T	T	F	F	F	**T**

The columns of $p \leftrightarrow q$ and $\neg((p \wedge \neg q) \vee (q \wedge \neg p))$ are the same. Therefore, they are logically equivalent.

2.2.1. (a) Commutative   (c) Distributive
     (b) Commutative   (d) Associative

2.2.2. (a) Idempotence   (d) Distributivity
     (b) Commutativity   (e) Distributivity
     (c) Associativity   (f) Commutativity

2.2.3. (a) The truth table is given below.

p	q	r	$q \vee r$	$p \wedge (q \vee r)$	$p \wedge q$	$p \wedge r$	$(p \wedge q) \vee (p \wedge r)$
T	T	T	T	**T**	T	T	**T**
T	T	F	T	**T**	T	F	**T**
T	F	T	T	**T**	F	T	**T**
T	F	F	F	**F**	F	F	**F**
F	T	T	T	**F**	F	F	**F**
F	T	F	T	**F**	F	F	**F**
F	F	T	T	**F**	F	F	**F**
F	F	F	F	**F**	F	F	**F**

The columns of $p \wedge (q \vee r)$ and $(p \wedge q) \vee (p \wedge r)$ are the same. Therefore, they are logically equivalent.
(b) The truth table is given below.

p	q	$p \vee q$	$\neg(p \vee q)$	$\neg p$	$\neg q$	$\neg p \wedge \neg q$
T	T	T	F	F	F	F
T	F	T	F	F	T	F
F	T	T	F	T	F	F
F	F	F	T	T	T	T

The columns of $\neg(p \vee q)$ and $(\neg p \wedge \neg q)$ have the same truth values. Therefore, they are logically equivalent.

(c) The truth table is given below.

p	q	$p \wedge q$	$p \vee (p \wedge q)$
T	T	T	T
T	F	F	T
F	T	F	F
F	F	F	F

The columns of p and $p \vee (p \wedge q)$ have the same truth values. Therefore, they are logically equivalent.

2.2.4. (a) $q \to p \Leftrightarrow \neg q \vee p$
(b) $(p \wedge q) \to q \Leftrightarrow \neg(p \wedge q) \vee q$
(c) $((p \wedge q) \wedge r) \to s \Leftrightarrow \neg((p \wedge q) \wedge r) \vee s$
(d) $\neg p \vee (r \leftrightarrow s) \Leftrightarrow p \to (r \leftrightarrow s)$

2.2.5. $(p \to q) \to r \Leftrightarrow \neg(p \to q) \vee r$
$$\Leftrightarrow \neg(\neg p \vee q) \vee r$$

2.2.6. (a) $((p \to q) \vee r) \vee v \Leftrightarrow (p \to q) \vee (r \vee s)$
(b) $u \wedge (v \wedge w) \Leftrightarrow (u \wedge v) \wedge w$
(c) $(p \to q) \wedge (r \wedge s) \Leftrightarrow ((p \to q) \wedge r) \wedge s$

2.2.7. (a) $j \wedge k \Leftrightarrow k \wedge j$
(b) $(p \to q) \wedge (q \to p) \Leftrightarrow (q \to p) \wedge (p \to q)$
(c) $(p \to q) \leftrightarrow (q \to p) \Leftrightarrow (q \to p) \leftrightarrow (p \to q)$

2.2.8. (a) $\neg((p \to q) \vee (q \to r)) \Leftrightarrow \neg(p \to q) \wedge \neg(q \to r)$
(b) $\neg((p \wedge q) \wedge (q \wedge p)) \Leftrightarrow \neg(p \wedge q) \vee \neg(q \wedge p)$
(c) Solution 1: $\neg(p \vee q) \wedge \neg(q \vee p) \Leftrightarrow (\neg p \wedge \neg q) \wedge (\neg q \wedge \neg p)$.
Solution 2: $\neg(p \vee q) \wedge \neg(q \vee p) \Leftrightarrow \neg((p \vee q) \vee (q \vee p))$.
(d) $\neg(p \to q) \vee \neg(q \to r) \Leftrightarrow \neg((p \to q) \wedge (q \to r))$

2.2.9. $p \to q \Leftrightarrow \neg p \vee q$ By (2.24)
$$\Leftrightarrow q \vee \neg p \qquad \text{By (2.6)}$$
$$\Leftrightarrow \neg q \to \neg p \qquad \text{By (2.24)}$$

2.2.10. $\neg p \to \neg q \Leftrightarrow \neg(\neg p) \vee \neg q$ By (2.24)

$\qquad\qquad\quad \Leftrightarrow p \vee \neg q$ By (2.22)

$\qquad\qquad\quad \Leftrightarrow \neg q \vee p$ By (2.6)

$\qquad\qquad\quad \Leftrightarrow \neg(\neg q) \to p$ By (2.24)

$\qquad\qquad\quad \Leftrightarrow q \to p$ By (2.22)

2.2.11. $(\neg p) \wedge \big(\neg(p \vee q)\big) \Leftrightarrow \neg(p \vee (p \vee q))$ By (2.10)

$\qquad\qquad\qquad\qquad \Leftrightarrow \neg((p \vee p) \vee q)$ By (2.4)

$\qquad\qquad\qquad\qquad \Leftrightarrow \neg(p \vee q)$ By (2.2)

2.2.12. $(p \wedge \neg q) \vee q \Leftrightarrow q \vee (p \wedge \neg q)$ By (2.6)

$\qquad\qquad\quad \Leftrightarrow (q \vee p) \wedge (q \vee \neg q)$ By (2.9)

$\qquad\qquad\quad \Leftrightarrow (q \vee p) \wedge \mathbb{T}$ By (2.18)

$\qquad\qquad\quad \Leftrightarrow q \vee p$ By (2.14)

$\qquad\qquad\quad \Leftrightarrow p \vee q$ By (2.6)

2.2.13. $p \vee (q \vee r) \Leftrightarrow p \vee (r \vee q)$ By (2.6)

$\qquad\qquad\quad \Leftrightarrow (p \vee r) \vee q$ By (2.4)

$\qquad\qquad\quad \Leftrightarrow (r \vee p) \vee q$ By (2.6)

$\qquad\qquad\quad \Leftrightarrow r \vee (p \vee q)$ By (2.4)

$\qquad\qquad\quad \Leftrightarrow r \vee (q \vee p)$ By (2.6)

2.2.14. $p \to (p \vee q) \Leftrightarrow \neg p \vee (p \vee q)$ By (2.24)

$\qquad\qquad\quad \Leftrightarrow (\neg p \vee p) \vee q$ By (2.4)

$\qquad\qquad\quad \Leftrightarrow (p \vee \neg p) \vee q$ By (2.6)

$\qquad\qquad\quad \Leftrightarrow \mathbb{T} \vee q$ By (2.18)

$\qquad\qquad\quad \Leftrightarrow q \vee \mathbb{T}$ By (2.6)

$\qquad\qquad\quad \Leftrightarrow \mathbb{T}$ By (2.14)

2.2.15. $(p \wedge q) \to p \Leftrightarrow \neg(p \wedge q) \vee p$ By (2.24)

$\qquad\qquad\quad \Leftrightarrow (\neg p \vee \neg q) \vee p$ By (2.11)

$\qquad\qquad\quad \Leftrightarrow (\neg q \vee \neg p) \vee p$ By (2.6)

$\qquad\qquad\quad \Leftrightarrow \neg q \vee (\neg p \vee p)$ By (2.4)

$\qquad\qquad\quad \Leftrightarrow \neg q \vee (p \vee \neg p)$ By (2.6)

$\qquad\qquad\quad \Leftrightarrow \neg q \vee \mathbb{T}$ By (2.18)

$\qquad\qquad\quad \Leftrightarrow \mathbb{T}$ By (2.14)

2.2.16. $\neg p \leftrightarrow q \Leftrightarrow (\neg p \to q) \wedge (q \to \neg p)$ By (2.26)

$\Leftrightarrow (\neg\neg p \vee q) \wedge (\neg q \vee \neg p)$ By (2.24)(×2)

$\Leftrightarrow (p \vee q) \wedge (\neg q \vee \neg p)$ By (2.22)

$\Leftrightarrow (q \vee p) \wedge (\neg p \vee \neg q)$ By (2.6)(×2)

$\Leftrightarrow (\neg\neg q \vee p) \wedge (\neg p \vee \neg q)$ By (2.22)

$\Leftrightarrow (\neg q \to p) \wedge (p \to \neg q)$ By (2.24)(×2)

$\Leftrightarrow (p \to \neg q) \wedge (\neg q \to p)$ By (2.5)

$\Leftrightarrow p \leftrightarrow \neg q$ By (2.26)

2.2.17. $\neg(p \vee \neg(p \wedge q)) \Leftrightarrow \neg p \wedge \neg(\neg(p \wedge q))$ By (2.10)

$\Leftrightarrow \neg p \wedge (p \wedge q)$ By (2.22)

$\Leftrightarrow (\neg p \wedge p) \wedge q$ By (2.3)

$\Leftrightarrow (p \wedge \neg p) \wedge q$ By (2.5)

$\Leftrightarrow \mathbb{F} \wedge q$ By (2.19)

$\Leftrightarrow q \wedge \mathbb{F}$ By (2.5)

$\Leftrightarrow \mathbb{F}$ By (2.15)

2.2.18. $\Big[(p \wedge \neg(\neg p \vee q)) \vee (p \wedge q)\Big] \to p$

$\Leftrightarrow \Big[\Big(p \wedge (\neg(\neg p) \wedge \neg q)\Big) \vee (p \wedge q)\Big] \to p$ By (2.10)

$\Leftrightarrow \Big[(p \wedge (p \wedge \neg q)) \vee (p \wedge q)\Big] \to p$ By (2.22)

$\Leftrightarrow \Big[((p \wedge p) \wedge \neg q) \vee (p \wedge q)\Big] \to p$ By (2.3)

$\Leftrightarrow [(p \wedge \neg q) \vee (p \wedge q)] \to p$ By (2.1)

$\Leftrightarrow [p \wedge (\neg q \wedge q)] \to p$ By (2.8)

$\Leftrightarrow [p \wedge (q \wedge \neg q)] \to p$ By (2.5)

$\Leftrightarrow (p \wedge \mathbb{F}) \to p$ By (2.15)

$\Leftrightarrow \mathbb{T}$ By (2.16)

2.2.19. $\neg(p \to q) \to p \Leftrightarrow \neg(\neg p \vee q) \to p$ By (2.24)

$\Leftrightarrow (\neg\neg p \wedge \neg q) \to p$ By (2.10)

$\Leftrightarrow (p \wedge \neg q) \to p$ By (2.22)

$\Leftrightarrow \neg(p \wedge \neg q) \vee p$ By (2.24)

$\Leftrightarrow (\neg p \vee \neg\neg q) \vee p$ By (2.11)

$\Leftrightarrow (\neg p \vee q) \vee p$ By (2.22)

$\Leftrightarrow (q \vee \neg p) \vee p$ By (2.6)

$\Leftrightarrow q \vee (\neg p \vee p)$ By (2.4)

$\Leftrightarrow q \vee (p \vee \neg p)$ By (2.6)

$\Leftrightarrow q \vee \mathbb{T}$ By (2.18)

$\Leftrightarrow \mathbb{T}$ By (2.14)

2.2.20. Let p be "You are from the state of New York," and q be "You will receive a scholarship."

$$\neg(p \to q) \Leftrightarrow \neg(\neg p \vee q) \qquad \text{By (2.24)}$$
$$\Leftrightarrow \neg\neg p \wedge \neg q \qquad \text{By (2.10)}$$
$$\Leftrightarrow p \wedge \neg q \qquad \text{By (2.22)}$$

Thus, the negation of the implication is the following: you are from the state of New York, and you will not receive a scholarship.

2.3.1. (a) Let p denote "The TSX fails to increase this period," and q denote "The market will suffer for several periods later." Then symbolically, the argument is $[(p \to q) \wedge \neg p] \to \neg q$. It is invalid argument since $[(p \to q) \wedge \neg p] \to \neg q$ is not a tautology.

p	q	$[(p \to q) \wedge \neg p] \to \neg q$
T	T	T
T	F	T
F	T	F
F	F	T

(b) Let p denote "A country is developing," q denote "A country cannot devote much of its financial resources to technological development," and r denote "The country's economy will not grow." Then symbolically, the argument is $[(p \to q) \wedge (q \to r)] \to (p \to r)$. It is a valid argument since $[(p \to q) \wedge \neg p] \to \neg q$ is a tautology.

p	q	$[(p \to q) \wedge (q \to r)] \to (p \to r)$
T	T	T
T	F	T
F	T	T
F	F	T
		T

2.4.1. (a) Let p denote "Wei earns more than \$1000," and q denote "Wei will go to France." Then it has the same argument as *modus ponens*.

(b) Let p denote "Gabriel was not in school when the fire alarm was pulled," and q denote "Gabriel could not be the culprit." The statement that he was at the doctor's office at 2 P.M. is another way of saying that he was not a school, hence p. So this is the same argument as *modus ponens*.

2.4.2. (a) Therefore, I will study.

(b) Therefore, I'll catch a cold.

(c) This can be rewritten as the following: If I need to speak the truth, then I improve upon the silence. I cannot restrain my tongue; I must speak. Therefore, by *modus ponens*, I am improving upon the silence.

2.7.1. If we let c denote coffee, k denote kerosene, and e energy, then we can represent the argument symbolically as

$c \to e$	Premise
$k \to e$	Premise
$\therefore k \to c$	Conclusion

To determine the validity of the argument, we can construct a truth table.

c	k	e	$c \to e$	$k \to e$	$(c \to e) \land (c \to k)$	$k \to c$	$[(c \to e) \land (c \to k)] \to (k \to c)$
T	T	T	T	T	T	T	T
T	T	F	F	F	F	T	T
T	F	T	T	T	T	T	T
T	F	F	F	T	F	T	T
F	T	T	T	T	T	F	F
F	T	F	T	F	F	F	T
F	F	T	T	T	T	T	T
F	F	F	T	T	T	T	T

Since the last column is not a tautology, the argument is a fallacy.

2.7.2. The argument can be written as follows:

If I'm in Kingston, then I'm in Ontario.	Major Premise
I'm in Kingston.	Minor Premise
Therefore, I'm in Ontario.	Conclusion

2.7.3. Let p be "An animal meows," and q be "The animal is a cat." Then the argument has the form in Table 2.16. Therefore, the argument is denying the antecedent.

2.7.4. Let p be "It is snowing," and q "It is winter." Then the argument has the form in Table 2.14. Thus, the argument is affirming the consequent.

2.7.5. Let p be "We're improving things," and q be "Things must change." Then the argument has the form in Table 2.14. Thus, the argument is affirming the consequent.

2.8.1. (a) It is not true. If $n = 16$, then $16^2 + 16 + 16 = 289 = 17 \times 17$, so it is not prime.
(b) It is not true. If $n = 40$, then $40^2 + 40 + 40 = 1681 = 41 \times 41$, so it is not prime.
(c) True.
(d) True.

2.8.2. Let p denote "Aspasia," q denote "a woman," and r denote "mortal." Table C.1 shows that the argument is invalid.

2.9.1. We create Table C.2.
Since $[((p \to q) \land (\neg p \to r)) \land (\neg p \lor p)] \to (q \lor r)$ is a tautology, the argument is valid.

2.9.2. Let p denote the proposition that "I study mathematics." Let q denote the proposition that "I study economics." Let r denote the proposition that "I take English." Then, the argument has the following form:

Table C.1   Inductive argument

p	q	r	(p → q)	∧	p → r)	→	(q → r)
T	T	T	T	T	T	**T**	T
T	T	F	T	F	F	**T**	F
T	F	T	F	F	T	**T**	T
T	F	F	F	F	F	**T**	T
F	T	T	T	T	T	**T**	T
F	T	F	T	T	T	**F**	F
F	F	T	T	T	T	**T**	T
F	F	F	T	T	T	**T**	T

Table C.2   Protagoras' and Euathlus' argument

p	q	r	¬p	[((p → q)	∧	(¬p → r))	∧	(¬p ∨ p)]	→	(q ∨ r)
T	T	T	F	T	T	T	T	T	**T**	T
T	T	F	F	T	T	T	T	T	**T**	T
T	F	T	F	F	F	T	F	T	**T**	T
T	F	F	F	F	F	T	F	T	**T**	F
F	T	T	T	T	T	T	T	T	**T**	T
F	T	F	T	T	F	F	F	T	**T**	T
F	F	T	T	T	T	T	T	T	**T**	T
F	F	F	T	T	F	F	F	T	**T**	F

p ∨ q	Premise
r → q	Premise
p	Premise
∴ ¬r	Conclusion

Using a truth table, we have

p	q	r	[((p ∨ q)	∧	(r → q))	∧	p]	→	¬r
T	T	T	T	T	T	T	T	**F**	F
T	T	F	T	T	T	T	T	**T**	T
T	F	T	T	F	F	F	T	**T**	F
T	F	F	T	T	T	T	T	**T**	T
F	T	T	T	T	T	F	F	**T**	F
F	T	F	T	T	T	F	F	**T**	T
F	F	T	F	F	F	F	F	**T**	F
F	F	F	F	F	T	F	F	**T**	T

It is not a tautology, and therefore, it is not valid.

2.9.3. $\quad [(p \wedge (q \to \neg s)) \wedge (p \to s)] \to \neg q$

$\Leftrightarrow \neg [(p \wedge (\neg q \vee \neg \neg s)) \wedge (\neg p \vee s)] \vee \neg q$ $\qquad$ By (2.24)

$\Leftrightarrow \neg [(p \wedge (\neg q \vee s)) \wedge (\neg p \vee s)] \vee \neg q$ $\qquad$ By (2.22)

$\Leftrightarrow [\neg (p \wedge (\neg q \vee s)) \vee \neg (\neg p \vee s)] \vee \neg q$ $\qquad$ By (2.11)

$\Leftrightarrow [(\neg p \vee \neg(\neg q \vee s)) \vee \neg(\neg p \vee s)] \vee \neg q$ $\qquad$ By (2.11)

$\Leftrightarrow [(\neg p \vee (\neg \neg q \wedge s)) \vee (\neg \neg p \wedge \neg s)] \vee \neg q$ $\qquad$ By (2.10)

$\Leftrightarrow [(\neg p \vee (q \wedge s)) \vee (p \wedge \neg s)] \vee \neg q$ $\qquad$ By (2.22)

$\Leftrightarrow [((\neg p \vee q) \wedge (\neg p \vee s)) \vee (p \wedge \neg s)] \vee \neg q$ $\qquad$ By (2.9)

$\Leftrightarrow [(p \wedge \neg s) \vee ((\neg p \vee q) \wedge (\neg p \vee s))] \vee \neg q$ $\qquad$ By (2.6)

$\Leftrightarrow [((p \wedge \neg s) \vee (\neg p \vee q)) \wedge ((p \wedge \neg s) \vee (\neg p \vee s))] \vee \neg q$ $\qquad$ By (2.6)

$\Leftrightarrow [((p \wedge \neg s) \vee (\neg p \vee q)) \wedge ((p \wedge \neg s) \vee (\neg p \vee \neg \neg s))] \vee \neg q$ $\qquad$ By (2.22)

$\Leftrightarrow [((p \wedge \neg s) \vee (\neg p \vee q)) \wedge ((p \wedge \neg s) \vee \neg (p \wedge \neg s))] \vee \neg q$ $\qquad$ By (2.22)

$\Leftrightarrow [((p \wedge \neg s) \vee (\neg p \vee q)) \wedge \mathbb{T}] \vee \neg q$ $\qquad$ By (2.18)

$\Leftrightarrow [(p \wedge \neg s) \vee (\neg p \vee q)] \vee \neg q$ $\qquad$ By (2.12)

$\Leftrightarrow [((p \wedge \neg s) \vee \neg p) \vee q] \vee \neg q$ $\qquad$ By (2.4)

$\Leftrightarrow ((p \wedge \neg s) \vee \neg p) \vee [q \vee \neg q]$ $\qquad$ By (2.4)

$\Leftrightarrow ((p \wedge \neg s) \vee \neg p) \vee \mathbb{T}$ $\qquad$ By (2.18)

$\Leftrightarrow \mathbb{T}$ $\qquad$ By (2.14)

2.9.4. (a) $\quad [((p \to r) \wedge (q \to r)) \wedge (p \vee q)] \to r$

$\Leftrightarrow \neg [((\neg p \vee r) \wedge (\neg q \vee r)) \wedge (p \vee q)] \vee r$ $\qquad$ By (2.24)

$\Leftrightarrow [\neg((\neg p \vee r) \wedge (\neg q \vee r)) \vee \neg(p \vee q)] \vee r$ $\qquad$ By (2.11)

$\Leftrightarrow [(\neg(\neg p \vee r) \vee \neg(\neg q \vee r)) \vee \neg(p \vee q)] \vee r$ $\qquad$ By (2.11)

$\Leftrightarrow [((\neg \neg p \wedge \neg r) \vee (\neg \neg q \wedge \neg r)) \vee \neg(p \vee q)] \vee r$ $\qquad$ By (2.10)

$\Leftrightarrow [((p \wedge \neg r) \vee (q \wedge \neg r)) \vee \neg(p \vee q)] \vee r$ $\qquad$ By (2.22)

$\Leftrightarrow [((\neg r \wedge p) \vee (\neg r \wedge q)) \vee \neg(p \vee q)] \vee r$ $\qquad$ By (2.5)

$\Leftrightarrow [(\neg r \wedge (p \vee q)) \vee \neg(p \vee q)] \vee r$ $\qquad$ By (2.8)

$\Leftrightarrow (\neg r \wedge (p \vee q)) \vee [\neg(p \vee q) \vee r]$ $\qquad$ By (2.4)

$\Leftrightarrow (\neg r \wedge (p \vee q)) \vee [r \vee \neg(p \vee q)]$ $\qquad$ By (2.6)

$\Leftrightarrow (\neg r \wedge (p \vee q)) \vee [\neg \neg r \vee \neg(p \vee q)]$ $\qquad$ By (2.22)

$\Leftrightarrow (\neg r \wedge (p \vee q)) \vee \neg [\neg r \wedge \neg \neg(p \vee q)]$ $\qquad$ By (2.11)

$\Leftrightarrow \underbrace{(\neg r \wedge (p \vee q))}_{s} \vee \neg \underbrace{[\neg r \wedge (p \vee q)]}_{s}$ $\qquad$ By (2.22)

$\Leftrightarrow s \vee \neg s$

$\Leftrightarrow \mathbb{T}$ $\qquad$ By (2.18)

2.9.5. Let p denote the proposition, "The accused lives at a small rate." Let q denote the proposition, "The accused maintains a large household." Let r denote the proposition, "The accused is rich." The argument can be represented symbolically as

$p \to r$	Premise
$q \to r$	Premise
$p \vee q$	Premise
$\therefore r$	Conclusion

From Exercise 2.9.4(a), we know that this is a simple constructive dilemma, which is a valid argument.

2.9.6. Let p denote the proposition, "The accused lives at a small rate." Let q denote the proposition, "The accused maintains a large household." Let r denote the proposition, "The accused is poor." The argument can be represented symbolically as

$p \to r$	Premise
$q \to r$	Premise
$p \vee q$	Premise
$\therefore r$	Conclusion

From Exercise 2.9.4(a), we know that this is a simple constructive dilemma, which is a valid argument.

2.9.7. Let p denote the proposition, "Christians have committed crimes." Let q denote the proposition, "Christians have not committed crimes." Let r denote the proposition, "Refusal to permit a public enquiry is irrational." Let s denote the proposition, "It is unjust to punish them." The argument can be represented symbolically as

$p \to r$	Premise
$q \to s$	Premise
$p \vee q$	Premise
$\therefore r \vee s$	Conclusion

From Exercise 2.9.4(b), we know that this is a complex constructive dilemma, which is a valid argument.

2.10.1. Solutions will vary.

2.10.2. The argument is valid. It follows the hypothetical syllogism. However, it is not sound since the premise—time is money—is not true; hence leading to a conclusion that may be true or false.

2.11.1. See Table 2.26.

2.11.2. You conclude the following:
   (a) by *modus ponens*, I will buy some milk.
   (b) by *modus tollens*, the animal does not eat bananas.
   (c) by hypothetical syllogism, Christine is nice.
   (d) by disjunctive syllogism, Renata bicycled to work.

2.11.3. (a) It is valid, since it is *modus tollens*.

(b) It is valid, since it is hypothetical syllogism.

(c) Let w denote "women," m denote "moral," and p denote "penguins." Then in symbolic form,

$$
\begin{array}{ll}
w \to m & \text{Premise} \\
p \to m & \text{Premise} \\
\hline
\therefore p \to w & \text{Conclusion}
\end{array}
$$

Using a truth table, we can check to see if $[(w \to m) \land (p \to m)] \to (p \to w)$ is a tautology.

w	m	p	$[(w \to m)$	$\land$	$(p \to m)]$	$\to$	$p \to w$
T	T	T	T	T	T	T	T
T	T	F	T	T	T	T	T
T	F	T	F	F	F	T	T
T	F	F	F	F	T	T	T
F	T	T	T	T	T	F	F
F	T	F	T	T	T	T	T
F	F	T	T	F	F	T	F
F	F	F	T	T	T	T	T
						$\neq$T	

Hence, it is not a tautology. Therefore, the argument is not valid.

(d) It is valid, since it is a disjunctive syllogism.

(e) It is not valid, since it is the denying the antecedent fallacy.

(f) It is valid since it is *modus tollens*.

(g) Let q denote "Lieselotte is successful" and p denote "Lieselotte studies." The way it is written, the first premise is q, if p. Thus, the argument in symbolic form is

$$
\begin{array}{ll}
p \to q & \text{Premise} \\
\neg p & \text{Premise} \\
\hline
\therefore \neg q & \text{Conclusion}
\end{array}
$$

This is an invalid argument, since it is denying the antecedent.

(h) Let p be "Michelle studies," q be "Michelle is successful," and r be "Michelle is happy." Then, the argument has the form

$$
\begin{array}{ll}
p \to q & \text{Premise} \\
q \to r & \text{Premise} \\
\hline
\therefore r \to p & \text{Conclusion}
\end{array}
$$

Then using a truth table, we can check if $[(p \to q) \land (q \to r)] \to (r \to p)$ is a tautology.

p	q	r	$[(p \to q)$	$\wedge$	$(q \to r)]$	$\to$	$(r \to p)$
T	T	T	T	T	T	T	T
T	T	F	T	F	F	T	T
T	F	T	F	F	T	T	T
T	F	F	F	F	T	T	T
F	T	T	T	T	T	F	F
F	T	F	T	F	F	T	T
F	F	T	T	T	T	F	F
F	F	F	T	T	T	T	T

$$\neq T$$

Since it is not a tautology, it is an invalid argument.

3.1.1. If the two sets have the same cardinality, they do not have to be equal. For example, consider the sets $A = \{1, 2, 3\}$ and $B = \{2, 3, 4\}$. $n(A) = n(B) = 3$, but $A \neq B$.

3.1.2. No. By definition of a proper subset, there must be at least one more element in A than in the set B. Therefore, $n(B) < n(A)$.

3.1.3. (a) T. The order in which the elements are listed in the set does not matter.
(b) F. The number 2 is not in the set.
(c) F. The number 2 is not in the set.
(d) F. The set $\{3, 4, 8\}$ is not an element of $\{1, 2, 3, \ldots, 8\}$.
(e) T. The set $\{3, 4, 8\}$ is a subset of $\{1, 2, 3, \ldots, 8\}$.
(f) F. The null set, $\varnothing$, is not in $\{1, 2, 3, \ldots, 8\}$.
(g) T. The null set, $\varnothing$, is a subset of every set.

3.1.4. (a) Finite   (c) Finite   (e) Finite
(b) Finite   (d) Infinite   (f) Infinite

3.1.5. (a) $\{1, 2, 3, 4, 5, 6, 7, 8, 9\}$
(b) $\{0, 1, 2, 3, 4, 5, 6, 7, 8, 9, 10\}$
(c) $\{1, 3, 5, 7, 9\}$
(d) $\{A, E, I, O, U\}$

3.1.6. (a) $A = \{4, 5, 6, 7, 8, 9\}$
(b) $B = \{2, 4, 6\}$
(c) There are no natural numbers that satisfy $5 + x = 3$. Therefore, C contains no elements; that is, $C = \varnothing = \{\ \}$.
(d) $D = \{3, 6, 9, 12, \ldots\}$.

3.1.7. $\varnothing, \{1\}, \{2\}, \{3\}, \{1, 2\}, \{1, 3\}, \{2, 3\}, \{1, 2, 3\}$.

3.1.8. To show that A is not a subset of B, it is necessary to find at least one element in A that does not belong to B. The element $5 \in A$ is not even, and hence $5 \notin B$. Therefore, A is not a subset of B.

3.1.9. Each element of A is in B, hence $A \subseteq B$. However, $1 \in B$ but $1 \notin A$. Therefore, $A \neq B$. Therefore, $A \subset B$ (A is a proper subset of B).

3.1.10. (a) Yes.
(b) Yes.
(c) No. $\{1, 2, 3, 4\} \not\subset \{1, 2, 3, 4\}$.

(d) Yes.
(e) Yes.
(f) Yes.

3.1.11 They are all equal. The order or repetition of elements does not change the set.

3.3.1. (a) $A \cup B = \{1, 2, 3, 4, 5, 6\}$ (b) $A \cap B = \{3, 4, 5\}$

3.3.2. (a)   (i) $A \cup B = \{$Rob, Rich, River, Rami, Rico$\}$
        (ii) $A \cap B = \{$Rob$\}$
      (iii) $A \smallsetminus B = \{$Rich, Rami$\}$
      (iv) $B \cup C = \{$Rob, River, Rico, Renée$\}$
       (v) $B \cap C = \{$River, Rico$\}$
      (vi) $B \smallsetminus C = \{$Rob$\}$
     (vii) $A \cup C = \{$Rob, Rich, Rami, River, Rico, Renée$\}$
    (viii) $A \cap C = \{\,\} = \varnothing$
      (ix) $A \smallsetminus C = \{$Rob, Rich, Rami$\}$
  (b)   (i) $A' = \{$River, Rico, Renée$\}$
       (ii) $B' = \{$Rich, Rami, Renée$\}$
      (iii) $C' = \{$Rob, Rich, Rami$\}$
      (iv) $A' \cup B' = \{$River, Rico, Renée, Rich, Rami$\}$
       (v) $B' \cup C' = \{$Rich, Rami, Renée, Rob$\}$
      (vi) $A' \cup C' = \{$River, Rico, Renée, Rob, Rich, Rami$\}$
     (vii) $A' \cap B' = \{$Renée$\}$
    (viii) $B' \cap C' = \{$Rich, Rami$\}$
      (ix) $A' \cap C' = \{\,\} = \varnothing$
       (x) $A' \smallsetminus B' = \{$River, Rico$\}$
      (xi) $B' \smallsetminus C' = \{$Renée$\}$
     (xii) $A' \smallsetminus C' = \{$River, Rico, Renée$\}$

3.3.3. The table is provided below.

$\cap$	$\varnothing$	$\{1\}$	$\{2\}$	$\{1, 2\}$
$\varnothing$	$\varnothing$	$\varnothing$	$\varnothing$	$\varnothing$
$\{1\}$	$\varnothing$	$\{1\}$	$\varnothing$	$\{1\}$
$\{2\}$	$\varnothing$	$\varnothing$	$\{2\}$	$\{2\}$
$\{1, 2\}$	$\varnothing$	$\{1\}$	$\{2\}$	$\{1, 2\}$

3.3.4. The table is provided below.

$\cup$	$\varnothing$	$\{1\}$	$\{2\}$	$\{1, 2\}$
$\varnothing$	$\varnothing$	$\{1\}$	$\{2\}$	$\{1, 2\}$
$\{1\}$	$\{1\}$	$\{1\}$	$\{1, 2\}$	$\{1, 2\}$
$\{2\}$	$\{2\}$	$\{1, 2\}$	$\{2\}$	$\{1, 2\}$
$\{1, 2\}$	$\{1, 2\}$	$\{1, 2\}$	$\{1, 2\}$	$\{1, 2\}$

3.3.5. (a) $A \cap B = \{1, 2, 3, 4\} = A$
(b) $A \cap B' = \{1, 2, 3, 4\} \cap \{6, 7, 8, 9, 10\} = \varnothing$
(c) $A' \cap B' = \{5, 6, 7, 8, 9, 10\} \cap \{6, 7, 8, 9, 10\} = \{6, 7, 8, 9, 10\} = B'$

3.3.6. See Fig. 3.33 on p. 111. No, it is not the same as Fig. 3.11.

3.3.7. The Venn diagram is given in Fig. C.1.

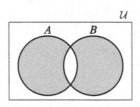

Fig. C.1 $A \veebar B$

3.3.8. (a)   (i) $A \cap B \cap C$
(ii) $C \smallsetminus (A \cup B)$
(iii) $((A \cap B) \cup (A \cap C)) \smallsetminus (A \cap B \cap C)$
(iv) $(A \cap B) \cup (B \cap C)$
(v) $(A \cup B \cup C) \smallsetminus (A \cap B \cap C)$

(b)   (i) $1, 2, 4, 5, 6$ (xi) $4, 7$ (xxi) $2, 3, 5, 6, 7, 8$
(ii) $1, 2, 3, 4, 6, 7$ (xii) $3, 7$ (xxii) $1, 2, 4, 5, 7, 8$
(iii) $1, 2, 3, 4, 5, 7$ (xiii) $4, 5, 7, 8$ (xxiii) $1, 3, 4, 5, 7, 8$
(iv) $1, 2$ (xiv) $3, 6, 7, 8$ (xxiv) $1, 3, 4, 6, 7, 8$
(v) $1, 3$ (xv) $2, 5, 6, 8$ (xxv) $1, 2, 3, 6, 7, 8$
(vi) $1, 4$ (xvi) $7, 8$ (xxvi) $1, 2, 3, 5, 6, 8$
(vii) $3, 6$ (xvii) $5, 8$ (xxvii) $1, 2, 4, 5, 6, 8$
(viii) $2, 6$ (xviii) $6, 8$ (xxviii) 8
(ix) $2, 5$ (xix) $3, 4, 5, 6, 7, 8$ (xxix) $5, 6, 8$
(x) $4, 5$ (xx) $2, 4, 5, 6, 7, 8$

3.3.9. (a)   (i) $A \cap B$ (iii) $A \cup B$
(ii) A (iv) $(A \cup B \cup C \cup D)'$

(b)   (i) $1, 2, 3, 4, 5, 6, 7, 11, 12, 13, 14, 15$ (vi) $6, 12, 15, 16$
(ii) $1, 2, 3, 4, 5, 7, 8, 9, 10, 11, 13, 14$ (vii) 16
(iii) $1, 3, 4, 13$ (viii) $2, 7$
(iv) $5, 6, 10, 11, 12, 14, 15, 16$ (ix) 1
(v) $2, 3, 4, 5, 6, 7, 8, 9, 10, 11, 12, 16$

3.3.10. The Venn diagram is given in Fig. C.2.

3.3.12. (a) Let A be the set of all relationship types where there was more than 60 homicides in both 2012 and 2013. Then $A = \{$Family, Acquaintance$\}$.
(b) Let B be the set of all relationship types where there was less than 50 homicides in 2012 or 2013. Then $B = \{$Intimate, Criminal, Stranger, Unknown$\}$.
(c) Let C be the set of all relationship types where the number of homicides decreased from 2012 to 2013. Then $C = \{$Family, Acquaintance, Stranger, Unknown$\}$.

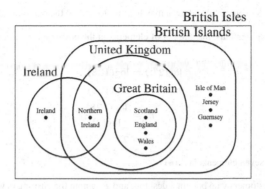

Fig. C.2 Venn diagram of British political and geographical names

(d) Let D be the set of all relationship types where the number of homicides increased from 2012 to 2013. Then $D = \{$Intimate, Criminal$\}$.

3.3.13. The Venn diagrams are in Fig. C.3.

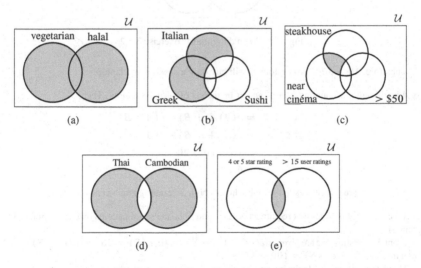

Fig. C.3 Venn diagrams for Exercise 3.3.13

3.3.14. $\emptyset$, $\{1\}$, $\{2\}$, $\{3\}$, $\{4\}$, $\{1,2\}$, $\{1,3\}$, $\{1,4\}$, $\{2,3\}$, $\{2,4\}$, $\{3,4\}$, $\{1,2,3\}$, $\{1,2,4\}$, $\{1,3,4\}$, $\{2,3,4\}$, $\{1,2,3,4\}$.

3.3.15. (a) $2^5 = 32$
(b) $2^{10} = 1024$
(c) $2^{230} \approx 1.725 \times 10^{69}$

3.3.16. $2 \times 2 \times 2 = 2^3 = 8$. You can also draw a tree diagram to find all the outcomes and then count them.

3.3.17. Given a set A with m elements, the number of elements in the power set of A is $n(\mathcal{P}(A)) = 2^m = 128$. Thus,

$$\log(2^m) = \log(128)$$
$$m \log(2) = \log(128)$$
$$m = \frac{\log(128)}{\log(2)}$$
$$m = 7.$$

Therefore, there are seven elements in set A.

3.4.2. Let D denote the customers who bought a desktop, and L denote the customers who bought a laptop. Then the corresponding Venn diagram is Fig. C.4.

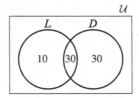

Fig. C.4   Venn diagram for Exercise 3.4.2

Therefore, $10 + 30 = 40$ customers will buy only a desktop or a laptop.

3.4.3. Let A be the set of accounting majors, and B be the set of business majors. Then,

$$n(A \cup B) = n(A) + n(B) - n(A \cap B)$$
$$n(A \cap B) = n(A) + n(B) - n(A \cup B)$$
$$= 16 + 10 - 20$$
$$= 6.$$

Therefore, there are 6 members in the club who are both accounting and business majors.

3.4.4. Let X be the set of patients that require an X-rays, and S be the set of patients that need a referral to a specialist.

From the question, we are given $n(\mathcal{U}) = 100, n(X) = 40, n(S) = 36$, and $n((X \cup S)') = 34$, which means that $n(X \cup S) = 100 - 34 = 66$.

Using the Principle of Inclusion and Exclusion for two sets,

$$n(X \cup S) = n(X) + n(S) - n(X \cap S)$$
$$n(X \cap S) = n(X) + n(S) - n(X \cup S)$$
$$= 40 + 36 - 66$$
$$= 10.$$

The Venn diagram is given in Fig. C.5.

Therefore, 10 patients needed an X-ray and a referral to a specialist.

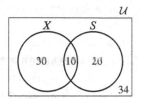

Fig. C.5   Venn diagram for Exercise 3.4.4

3.4.5. Let A denote the student answered the first question correctly; $n(A) = 90$. Let B denote the student answered the second question correctly; $n(B) = 60$. The number of students who correctly answered both questions is $n(A \cap B) = 50$. The Venn diagram is given in Fig. C.6.

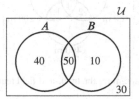

Fig. C.6   Venn diagram for Exercise 3.4.5

(a) $n(A \cup B) = n(A) + n(B) - n(A \cap B) = 90 + 60 - 50 = 100$
(b) $n((A \cup B)') = n(\mathcal{U}) - n(A \cup B) = 130 - 100 = 30$
(c) $40 + 10 = 50$
(d) 10
(e) $30 + 40 = 70$

3.4.6. (a) From the question, we have the following information.

$$n(100 \text{ m}) = 15$$
$$n(200 \text{ m}) = 17$$
$$n(400 \text{ m}) = 8$$
$$n(100 \text{ m} \cap 200 \text{ m}) = 7$$
$$n(200 \text{ m} \cap 400 \text{ m}) = 5$$
$$n(100 \text{ m} \cap 400 \text{ m}) = 6$$
$$n(100 \text{ m} \cap 200 \text{ m} \cap 400 \text{ m}) = 4$$

The number of runners is equal to $n(100 \text{ m} \cup 200 \text{ m} \cup 400 \text{ m})$. Using the Principle of Inclusion and Exclusion,

$$
\begin{aligned}
n(100 \text{ m} \cup 200 \text{ m} \cup 400 \text{ m}) =& n(100 \text{ m}) + n(200 \text{ m}) + n(400 \text{ m}) \\
& - n(100 \text{ m} \cap 200 \text{ m}) \\
& - n(200 \text{ m} \cap 400 \text{ m}) \\
& - n(100 \text{ m} \cap 400 \text{ m}) \\
& + n(100 \text{ m} \cap 200 \text{ m} \cap 400 \text{ m}) \\
=& 15 + 17 + 8 - 7 - 5 - 6 + 4 \\
=& 26.
\end{aligned}
$$

Therefore, there are 26 runners.

(b) See Fig. C.7 for the Venn diagram. From the diagram, we see that the total number of runners is $6 + 3 + 9 + 2 + 4 + 1 + 1 = 26$ (which is the same answer that we got in part (a)).

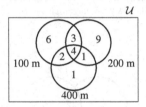

Fig. C.7   Venn diagram for the track-and-field team in Exercise 3.4.6

3.4.7. (a) The Venn diagram is given in Fig. C.8.

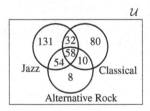

Fig. C.8   Venn diagram for Exercise 3.4.7

(b)   (i) at least two of the genres: $32 + 58 + 54 + 10 = 154$
   (ii) exactly one genre: $131 + 80 + 8 = 219$
   (iii) only jazz: 131
   (iv) exactly two genres: $32 + 54 + 10 = 96$.

3.4.8. The Venn diagram is given in Fig. C.9.

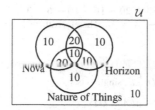

Fig. C.9   Venn diagram of television shows watched by students in Exercise 3.4.8

(a) 90   (b) 10   (c) 20   (d) 50

3.4.9. The Venn diagram is in Fig. C.10.

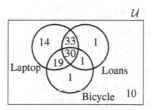

Fig. C.10   Venn diagram for newspaper survey in Exercise 3.4.9

(a) 14%   (b) 1%   (c) 1%   (d) 1%

3.4.10. (a) $n(A \cup B) = n(A) + n(B)$ (b) $n(A \cap B) = \varnothing$

3.4.11. Let H denote a player with a hurt hip; $n(H) = 8$. Let I denote a player with a hurt hand; $n(I) = 6$. Let K denote a player with a hurt knee; $n(K) = 5$. The rest of the data are $n(H \cap I) = 3$, $n(H \cap K) = 2$, $n(I \cap K) = 2$, and $n(H \cap I \cap K) = 0$. Then, $n(H \cup I \cup K) = n(H) + n(I) + n(K) - n(H \cap I) - n(H \cap K) - n(I \cap K) + n(H \cap I \cap K) = 8 + 6 + 5 - 3 - 2 - 2 + 0 = 12$. There is a total of 12 players with injuries, but there is suppose to be a maximum of 10 players. (The story doesn't add up.) The reporter was fired for this inaccuracy.

3.4.12. Let A be the set of numbers that are divisible by 2, B be the set of numbers that are divisible by 3, and C be the set of numbers that are divisible by 5. Then, $n(A) = 50$, $n(B) = 33$, and $n(C) = 20$. We require $n((A \cup B \cup C)') = n(\mathcal{U}) - n(A \cup B \cup C)$, but we also need $n(A \cup B \cup C)$. Then, $n(A \cup B \cup C) = n(A) + n(B) + n(C) - n(A \cap B) - n(A \cap C) - n(B \cap C) + n(A \cap B \cap C)$, where

(a) $A \cap B$ corresponds to numbers divisible by 2 and 3, which is equivalent to numbers divisible by 6. So, $A \cap B = \{6, 12, 18, 24, 30, 36, 42, 48, 54, 60, 66, 72, 78, 84, 90, 96\}$, and thus, $n(A \cap B) = 16$.

(b) $A \cap C$ corresponds to numbers divisible by 2 and 5, which is equivalent to numbers divisible by 10. So $A \cap C = \{10, 20, 30, 40, 50, 60, 70, 80, 90, 100\}$, and thus, $n(A \cap C) = 10$.

(c) $B \cap C$ corresponds to numbers divisible by 3 and 5, or equivalently numbers divisible by 15. So $B \cap C = \{15, 30, 45, 60, 75, 90\}$, and thus, $n(B \cap C) = 6$.

(d) $A \cap B \cap C$ corresponds to numbers that are divisible by 2, 3, and 5, which is equivalent to numbers divisible by 30. So $A \cap B \cap C = \{30, 60, 90\}$, and thus, $n(A \cap B \cap C) = 3$.

Thus, $n(A \cup B \cup C) = 50 + 33 + 20 - 16 - 10 - 6 + 3 = 74$, and so, $n\left((A \cup B \cup C)'\right) = n(\mathcal{U}) - n(A \cup B \cup C) = 100 - 74 = 26$. Therefore, there are 26 numbers that are not divisible by 2, 3, or 5.

3.4.13. $n(A \cup B \cup C \cup D)$

$\left. = n(A) + n(B) + n(C) + n(D) \right\}$ all single sets

$- n(A \cap B) - n(A \cap C) - n(A \cap D)$

$\left. - n(B \cap C) - n(B \cap D) - n(C \cap D) \right\}$ all two-way intersections

$+ n(A \cap B \cap C) + n(A \cap B \cap D) + n(A \cap C \cap D)$

$\left. + n(B \cap C \cap D) \right\}$ all three-way intersections

$\left. - n(A \cap B \cap C \cap D) \right\}$ all four-way intersections

3.5.1. Let x be an arbitrary element in $(A \cap B)'$. Then,

$$
\begin{aligned}
x \in (A \cap B)' &\Leftrightarrow x \notin (A \cap B) && \text{By (3.3)} \\
&\Leftrightarrow \neg(x \in (A \cap B)) && \text{By (3.28)} \\
&\Leftrightarrow \neg(x \in A \wedge x \in B) && \text{By (3.2)} \\
&\Leftrightarrow \neg(x \in A) \vee \neg(x \in B) && \text{By (2.11)} \\
&\Leftrightarrow (x \notin A) \vee (x \notin B) && \text{By (3.28)} \\
&\Leftrightarrow (x \in A') \vee (x \in B') && \text{By (3.3)} \\
&\Leftrightarrow x \in (A' \cup B') && \text{By (3.2)}
\end{aligned}
$$

Since every element of $(A \cap B)'$ is an element of $(A' \cup B')$, and vice versa, the two sets are equal.

3.5.2. Let x be an arbitrary element in $A \smallsetminus B$. Then,

$$
\begin{aligned}
x \in (A \smallsetminus B) &\Leftrightarrow (x \in A) \wedge (x \notin B) \\
&\Leftrightarrow (x \in A) \wedge (x \in B') \\
&\Leftrightarrow x \in (A \cap B')
\end{aligned}
$$

Since every element of $(A \smallsetminus B)$ is an element of $(A \cap B')$, and vice versa, the two sets are equal.

3.5.3.   (a) $\begin{aligned}[t] A \cap (A' \cup B) &= (A \cap A') \cup (A \cap B) && \text{By (3.15)} \\
&= \varnothing \cup (A \cap B) && \text{By (3.24)} \\
&= (A \cap B) \cup \varnothing && \text{By (3.14)} \\
&= A \cap B && \text{By (3.20)}
\end{aligned}$

(b) $\begin{aligned}[t] (A' \cup B') \cap (A \cap B) &= (A \cap B)' \cap (A \cap B) && \text{By (3.18)} \\
&= (A \cap B) \cap (A \cap B)' && \text{By (3.13)} \\
&= \varnothing && \text{By (3.24)}
\end{aligned}$

(c) $\begin{aligned}[t] (A \cup B) \cap (A \cup B') &= A \cup (B \cap B') && \text{By (3.16)} \\
&= A \cup \varnothing && \text{By (3.24)} \\
&= A && \text{By (3.20)}
\end{aligned}$

(d) $\begin{aligned}[t] (A' \cap B') \cup (A \cup B) &= (A \cup B)' \cup (A \cup B) && \text{By (3.17)} \\
&= (A \cup B) \cup (A \cup B)' && \text{By (3.14)} \\
&= \mathcal{U} && \text{By (3.23)}
\end{aligned}$

3.5.4. $(A \cap B)' \cup (A' \cap B' \cap C)' \cup A = (A' \cup B') \cup (A'' \cup B'' \cup C') \cup A$

$$= (A' \cup B') \cup (A \cup B \cup C') \cup A$$

$$= A' \cup (A \cup A) \cup (B \cup B') \cup C'$$

$$= A' \cup A \cup U \cup C'$$

$$= U$$

3.6.1. This is from Carroll (1886, p. 34).

3.6.3. This was adapted from Carroll (1886).

3.6.5. This exercise is from Carroll (1886, p. 33).

3.7.1. (a) The set A is finite since there are 10 provinces in Canada; $n(A) = 10$.

(b) The set B is finite since there are 12 months in a year; $n(B) = 12$.

(c) There are no natural numbers less than 1; hence $C = \{\} = \varnothing$. Therefore, $n(C) = 0$, and C is finite.

(d) The set D is infinite. $D = \{\ldots, -3, -1, 1, 3, 5, \ldots\}$.

(e) The set $E = \{1, 2, 7\}$. Hence, $n(E) = 3$, and therefore E is finite.

(f) While this may be difficult to count all of the dogs living in Germany, there is still a finite number of them at any given point in time. Therefore, F is finite.

3.7.2. This is known as the **Barber's Paradox**.

3.7.3. True.

3.7.4. (a) $n(C \cup D) = n(C) + n(D) - n(C \cap D)$

(b) $n(C \cup D) = n(C) + n(D) - n(\varnothing) = n(C) + n(D) - 0 = n(C) + n(D)$

3.7.5. (a) $A \cap B = \{c\}$

(b) $A \cup B = \{a, c, d, e, f, g\}$

(c) $A \smallsetminus C = \{e, f\}$

(d) $n(B) = 3$

(e) $A' = U \smallsetminus A = \{b, d, g, h\}$

(f) $\mathcal{P}(B) = \{\varnothing, \{c\}, \{d\}, \{g\}, \{c, d\}, \{c, g\}, \{d, g\}, \{c, d, g\}\}$

3.7.6. (a) Let A be the set of cities where the number of homicides increased from 2012 to 2013. Then $A = \{\text{Vancouver, Hamilton, Ottawa, Kingston}\}$.

(b) Let B be the set of cities that had more than 30 homicides in 2012 or 2013. Then $B = \{\text{Toronto, Montréal, Winnipeg, Vancouver}\}$.

(c) Let C be set of cities that had less than 9 homicides in 2012 and 2013. Then $C = \{\text{Kingston}\}$.

(d) Let D be set of cities that had a decrease of 3 homicides or more from 2012 to 2013. Then $D = \{\text{Montréal, Winnipeg}\}$.

3.7.7. First, we define the following sets: D be the set of doctors, F be the set of females, and S be the set of supervisors. Then $n(D) = 25$, $n(F) = 35$, $n(S) = 19$, $n(F \cap D) = 10$, $n(F \cap S) = 11$, $n(F \cap D \cap S) = 3$, and $n(D \cap S') = 15$. The Venn diagram is given in Fig. C.11.

(a) The number of male supervisors is $7 + 1 = 8$.

(b) The percentage of female supervisors is $\frac{8+3}{1+7+3+8} = \frac{11}{19} \approx 57.89\%$.

(c) $n(D \cup S) = n(D) + n(S) - n(D \cap S) = 25 + 19 - 10 = 34$

3.7.8. $250 + 100 - 40 = 310$

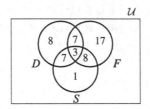

Fig. C.11   Venn diagram for Exercise 3.7.7

3.7.9. Let E be the set of even numbers from 1 to 100. Let F be the set of multiples of 3 from 1 to 100. Let G be the set of multiples of 5 from 1 to 100. $E = \{2, 4, 6, \ldots, 100\}$; $n(E) = 50$. $F = \{3, 6, 9, \ldots, 99\}$; $n(F) = 33$. $G = \{5, 10, 15, \ldots, 100\}$; $n(G) = 20$. $E \cap F = \{6, 12, 18, \ldots, 96\}$; $n(E \cap F) = 16$. $E \cap G = \{10, 20, 30, \ldots, 100\}$; $n(E \cap G) = 10$. $F \cap G = \{15, 30, 45, \ldots, 90\}$; $n(F \cap G) = 6$. $E \cap F \cap G = \{30, 60, 90\}$; $n(E \cap F \cap G) = 3$. The Venn diagram is given in Fig. C.12.

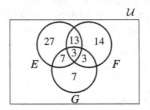

Fig. C.12   Venn diagram for Exercise 3.7.9

(a) $27 + 14 + 7 = 48$ (b) $7 + 3 + 13 = 23$

4.1.1. Suppose in order to derive a contradiction, we will assume that $\sqrt{2}$ is rational. Then, we can express $\sqrt{2}$ as a fraction

$$\frac{a}{b} = \sqrt{2} \qquad \text{for } a, b \in \mathbb{N}, \text{ and } b \neq 0.$$

We can always write a and b in such a way that a and b are both not even (meaning we can write the fraction in a reduced, simplified form). Taking the square of both sides, we have

$$\left(\frac{a}{b}\right)^2 = (\sqrt{2})^2$$

$$\frac{a^2}{b^2} = 2$$

$$a^2 = 2b^2.$$

This shows that a^2 is even; hence a is even. This means that $a = 2k$ for some natural number k. Making this substitution,

$$(2k)^2 = 2b^2$$

$$4k^2 = 2b^2$$

$$2k^2 = b^2$$

we see that b^2 is also even; hence, b is even. This contradicts our assumption that a and b are both not even. Hence, we cannot express $\sqrt{2}$ as a fraction. Therefore, $\sqrt{2}$ is irrational.

4.1.2. Suppose in order to derive a contradiction, we will assume that $\sqrt{3}$ is a rational number. Then, we can express $\sqrt{3}$ as a fraction

$$\frac{a}{b} = \sqrt{3} \qquad \text{for } a, b \in \mathbb{N}, \text{ and } b \neq 0.$$

We can always write a and b in such a way that a and b have no common factors; we can express the fraction in a reduced, simplified form. Taking the squares of both sides, we have

$$\left(\frac{a}{b}\right)^2 = (\sqrt{3})^2$$

$$\frac{a^2}{b^2} = 3$$

$$a^2 = 3b^2.$$

This means that a^2 is divisible by 3; a is also divisible by 3. Then, $a = 3k$ for some natural number k. Making this substitution,

$$(3k)^2 = 3b^2$$

$$9k^2 = 3b^2$$

$$3k^2 = b^2.$$

Hence, b^2 is divisible by 3, and this implies that b is divisible by 3. As we can see, a and b are both divisible by 3. However, this contradicts our assumption that a and b do not have common factors. Hence, we cannot express $\sqrt{3}$ as a fraction. Therefore, $\sqrt{3}$ is irrational.

4.1.3. Suppose in order to derive a contradiction that $\sqrt{n}$ is rational. Then we can write $\sqrt{n}$ as a ratio of two integers p and q such that $\frac{p}{q}$ is irreducible (has no common factors). Then

$$\sqrt{n} = \frac{p}{q}$$

$$n = \frac{p^2}{q^2}$$

$$q^2 n = p^2.$$

From this, p should be divisible by q. But we have chosen p and q where they don't have a common factor (irreducible). So we have a contradiction. So there is not a p and q to satisfy $\sqrt{n} = p/q$ so $\sqrt{n}$ is irrational.

4.2.3.

$$\mathbb{N} = \{\ 1,\ 2,\ 3,\ 4,\ 5,\ 6,\ 7,\ 8,\ 9,\ \dots\}$$
$$\updownarrow\ \ \updownarrow\ \ \updownarrow\ \ \updownarrow\ \ \updownarrow\ \ \updownarrow\ \ \updownarrow\ \ \updownarrow\ \ \updownarrow$$
$$\mathbb{W} = \{\ 0,\ 1,\ 2,\ 3,\ 4,\ 5,\ 6,\ 7,\ 8,\ \dots\}$$

5.1.1. There needs to be 51 or more students to ensure that at least two students are from the same American state.

5.1.2. You need to pick 11 numbers from the numbers 1 to 20.

5.1.3. If each box contains at most one object, then there are at most n objects.

5.1.4. In 2013, there were 4.1 million males living in London.[12] In comparison, a human head can contain upwards of 500 000. By the Pigeonhole Principle, there must be at least two males in London who have the same number of hairs.

5.1.5. Four. Since there are only three colors, the Pigeonhole Principle asserts that at least two of the socks must be the same color.

5.1.6. If you have 50 pieces of sushi, and 20 mouths, then there can fill each mouth with 2 pieces of sushi. This leaves $50 - 2(20) = 10$ pieces. Each of the 20 mouths already have 2 pieces, so we must place a piece of sushi in the mouth of someone that already has 2 pieces. Therefore, there is a person who will get at least three or more pieces.

5.1.7. Let there be n seats labeled $0, 1, 2, \ldots, n - 1$. If a person knows no people, then we put them in seat 0, if they know one person then they go into seat 1, *etc.* We see that the first and last seats may not simultaneously be occupied; if some person knows nobody, no-one else can know everyone, and vice versa. Thus, at most $n - 1$ chairs are occupied. Since there are n seats, by the Pigeonhole Principle, some seat must have more than one person!

5.1.8. Say we have a lady named Alice. Of the five other people at the party, there are either three whom Alice knows or three whom she does not. In the first case, the three are either complete strangers, or two of them are acquaintances and form a trio with Alice. The other case is similar.

Remark: This problem was posed by Bostwick (1958) and solved by Bostwick *et al.* (1959).

5.2.1. We immediately spot the word *or* in the exercise. This tells us that we need to apply the Sum Principle. Using the Sum Principle, there are $10 + 5 = 15$ choices.

5.2.2. You can choose either a main dish with meat *or* without meat. Using the Sum Principle, there are $5 + 6 = 11$ choices.

5.2.3. We immediately spot the word *and* in the exercise. This tells us that we need to apply the Multiplication Principle somehow. The dime can land in two ways, and the quarter in two ways, and $2 \times 2 = 4$, therefore they can fall in four ways.

5.2.4. This problem could have been re-worded to say, "If one die is thrown *and* another die is thrown, how many ways can they land?" As you can see, the wording implied *and*; we use the Multiplication Principle to solve this exercise. The first die can land in six ways, and the second in six ways, and $6 \times 6 = 36$; therefore there are 36 ways in which the two dice can land.

5.2.5. The first prize can be given to any of the 10 students, and the second prize can be given to any of the remaining 9 students, and $10 \times 9 = 90$; therefore, we have the choice of 90 ways of giving the two prizes.

5.2.6. He can select one of the four doors to enter the classroom. Once inside the classroom, he will have the choice of four doors to exit. Therefore, by the Multiplication Principle, Chris can enter and leave the room in $4 \times 4 = 16$ different ways.

5.2.7. By the Multiplication Principle, there are $3 \times 4 \times 3 = 36$ different ways to order your meal.

5.2.8. The tree diagram is given in Fig. C.13. Therefore, there are $3 \times 2 = 6$ possible itineraries.

---

[12] Source: http://www.neighbourhood.statistics.gov.uk/HTMLDocs/dvc134_a/index.html

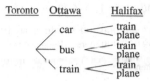

Fig. C.13   Possible itineraries for Sterling

5.2.9. For the first digit, it can be $1, 2, \ldots, 9$; that is, there are 9 choices for the first digit. In order for the number to be divisible by 2 or 5, the last digit must be $2, 4, 5, 6, 8$, or 0. Thus, for the second digit there are 3 choices; that is, you can select from 1, 3, or 7. Therefore, there are

$$\underbrace{9}_{1^{st}\ digit} \times \underbrace{3}_{2^{nd}\ digit} = 27$$

numbers that are not divisible by 2 or 5.

5.2.10. (a) $\underbrace{6}_{1,2,3,4,5,6} \times \underbrace{7}_{0\ to\ 6} \times \underbrace{7}_{0\ to\ 6} = 294.$

(b) *Take cases.* If the last digit is a 1, then

$$\underbrace{5}_{2,3,4,5,6} \times \underbrace{4}_{One\ less\ choice} \times \underbrace{1}_{a\ '1'} = 20.$$

If the last digit is a 3, then

$$\underbrace{5}_{1,2,4,5,6} \times \underbrace{4}_{One\ less\ choice} \times \underbrace{1}_{a\ '3'} = 20.$$

If the last digit is a 5, then

$$\underbrace{5}_{1,2,3,4,6} \times \underbrace{4}_{One\ less\ choice} \times \underbrace{1}_{a\ '5'} = 20.$$

By the Sum Principle, there are $20 + 20 + 20 = 60$ ways to do this.

5.2.11. *Solution 1*: There is only one single-digit number that ends in 9, and that is 9. For a double-digit number, the number of choices for the first digit is 9, and the second digit there is only one choice. By the Multiplication Principle, there are $9 \times 1 = 9$ double-digit numbers that end in the digit 9. Therefore, by the Sum Principle there are $1 + 9 = 10$ single or double-digit numbers that end in 9.
Solution 2: You could just list all the possibilities. They are 9, 19, 29, 39, 49, 59, 69, 79, 89, and 99; there are 10 possibilities.

5.2.12. By the Multiplication Principle, the number of different possible outcomes is $2 \times 2 \times 2 \times 2 = 2^4 = 16$.

5.2.13. By the Multiplication Principle, the number of different possible outcomes is $6 \times 6 \times 6 \times 6 \times 6 \times 6 = 6^6 = 46\,656$.

5.2.14. (a)   For Canada, the number of possible postal codes is

$$\underbrace{26}_{1st\ Char} \times \underbrace{10}_{2nd\ Char} \times \underbrace{26}_{3rd\ Char} \times \underbrace{10}_{4th\ Char} \times \underbrace{26}_{5th\ Char} \times \underbrace{10}_{6th\ Char} = 17\,576\,000.$$

For the United States, the number of possible zip codes is $10 \times 10 \times 10 \times 10 \times 10 = 10^5 = 100\,000$.

(b) There are $17\,576\,000 - 100\,000 = 17\,476\,000$ more postal codes than ZIP codes.

5.2.15. (a) There are $10^5 \times 10^4 = 10^9 = 1\,000\,000\,000$ different ZIP+4 codes.

(b) There are $10^9 - 10^5 = 999\,900\,000$ more ZIP+4 codes than ZIP codes.

5.2.16. (a) If there are no restrictions, then

$$\underbrace{7}_{\text{Driver Seat}} \times \underbrace{6}_{\text{Seat 2}} \times \underbrace{5}_{\text{Seat 3}} \times \underbrace{4}_{\text{Seat 4}} \times \underbrace{3}_{\text{Seat 5}} \times \underbrace{2}_{\text{Seat 6}} \times \underbrace{1}_{\text{Seat 7}} = 5040$$

different seating arrangements.

(b) If there are only three people can drive, then

$$\underbrace{3}_{\text{Driver Seat}} \times \underbrace{6}_{\text{Seat 2}} \times \underbrace{5}_{\text{Seat 3}} \times \underbrace{4}_{\text{Seat 4}} \times \underbrace{3}_{\text{Seat 5}} \times \underbrace{2}_{\text{Seat 6}} \times \underbrace{1}_{\text{Seat 7}} = 2160$$

seating arrangements.

5.2.17. Yes. There are $26 \times 26 = 676$ possible first and last initials. Consequently, this is less than the number of patrons at the concert. So, by the Pigeonhole Principle, we can conclude that there are at least two people with the same first and last initial.

5.2.18. The number of possible Ontario license plates is $26 \times 26 \times 26 \times 10 \times 10 \times 10 = 26^3 \times 10^3 = 17\,576\,000$.

5.2.19. The number of possible Ontario license plates after 1997 is $26 \times 26 \times 26 \times 26 \times 10 \times 10 \times 10 = 26^4 \times 10^3 = 456\,976\,000$. Thus, there are $456\,976\,000 - 17\,576\,000 = 439\,400\,000$ more license plates possible than before 1997.

5.2.20. We assume that two related people cannot be married.

Solution 1: Assuming that a marriage is between a lady and a gentlemen, then if they were all unrelated, we might make the match in $12 \times 10 = 120$ ways; but this will include the $3 \times 2 = 6$ ways in which the selected lady and gentlemen are sister and brother. Therefore, the number of eligible ways is $120 - 6 = 114$.

Solution 2: If we remove the two-sex assumption, then if they are all unrelated, we might make the match in $(12 + 10) \times (12 + 10 - 1) = 22 \times 21 = 462$ ways; but this will still include the $(3 + 2) \times (3 + 2 - 1) = 5 \times 4 = 20$ ways in which the couple is related. Therefore, the number of eligible ways is $462 - 20 = 442$.

5.2.21. The Camaro can be ordered $4 \times 2 \times 9 \times 3 = 216$ different ways according to the Multiplication Principle.

5.2.22.
$$\left[\underbrace{(\overset{\text{or}}{\overbrace{\underset{\text{skirts}}{6} + \underset{\text{pants}}{7}})} \times \underset{\text{shirts}}{5}}_{} \right] + \underset{\text{dresses}}{\overset{\text{or}}{\overbrace{8}}} = 13(5) + 8 = 65 + 8 = 73 \text{ possible outfits.}$$

5.2.23. (a) The number of passwords that can be created with 3 digits is $26 \times 26 \times 26 \times 26 \times 10 \times 10 \times 10 = 456\,976\,000$, and the number of passwords that can be created with 4 digits is $26 \times 26 \times 26 \times 26 \times 10 \times 10 \times 10 \times 10 = 4\,569\,760\,000$. So the total number of passwords that can be created is $456\,976\,000 + 4\,569\,760\,000 = 5\,026\,736\,000$.

(b) The number of passwords that can be created with 3 non-repeating digits is $26 \times 26 \times 26 \times 26 \times 10 \times 9 \times 8 = 329\,022\,720$, and the number of passwords that can be created with 4 non-repeating digits is $26 \times 26 \times 26 \times 26 \times 10 \times 9 \times 8 \times 7 = 2\,303\,159\,040$. Thus, the total number of passwords that can be created with no repeating digits is $329\,022\,720 + 2\,303\,159\,040 = 2\,632\,181\,760$.

(c) The number of passwords that can be created with 4 non-repeating letters and 3 digits is $26 \times 25 \times 24 \times 23 \times 10 \times 10 \times 10 = 358\,800\,000$, and the number of passwords that can be created with 4 non-repeating letters and 4 digits is $26 \times 25 \times 24 \times 23 \times 10 \times 10 \times 10 \times 10 = 3\,588\,000\,000$. Thus, the total number of passwords that can be created is $358\,800\,000 + 3\,588\,000\,000 = 3\,946\,800\,000$.

5.2.24. We consider four cases. *Case 1*: 1 symbol; 2. *Case 2*: 2 symbols; 2^2. *Case 3*: 3 symbols; 2^3. *Case 4*: 4 symbols; 2^4. Therefore, there are $2 + 2^2 + 2^3 + 2^4 = 30$ different sequences of one to four dots or dashes.

5.2.25. See Solution 1 of Example 7.2.2 on p. 295.

5.3.1. (a) $\frac{5!}{3!} = \frac{5 \times 4 \times 3!}{3!} = 5 \times 4 = 20$

  (b) $\frac{20!}{17!} = \frac{20 \times 19 \times 18 \times 17!}{17!} = 20 \times 19 \times 18 = 6840$

  (c) $\frac{500!}{499!} = \frac{500 \times 499!}{499!} = 500$

5.3.2. (a) $6!$ (b) $\frac{10!}{7!}$ (c) $\frac{20!}{17!4!}$ (d) $\frac{13!}{8!}$

5.3.3. (a) $(n+3)(n+2)! = (n+3)!$

  (b) $(n^2 + 5n + 6)(n+1)! = (n+3)(n+2)(n+1)! = (n+3)!$

  (c) $\frac{(n+2)!}{n+2} = \frac{(n+2)(n+1)!}{n+2} = (n+1)!$

5.3.4. (a) $\dfrac{n!}{(n-1)!} = \dfrac{n \times (n-1)!}{(n-1)!} = n$

  (b) $n(n-1)! = n!$

  (c) $\dfrac{n!}{(n-2)!} = \dfrac{n \times (n-1) \times (n-2)!}{(n-2)!} = n \times (n-1)$

  (d) $(n+1)n! = (n+1)!$

  (e) $\dfrac{(3n)!}{(3n-1)!} = \dfrac{(3n) \times (3n-1)!}{(3n-1)!} = 3n$

  (f) $n[n! + (n-1)!] = nn! + n(n-1)! = nn! + n! = (n+1)n! = (n+1)!$

5.3.5. $A = \{1, 2, 3, 4, 5, 6\}$

5.3.6. The factorial $n!$ is used to calculate the number of permutations of n objects. So, the number of ways to order zero objects is only one way; hence, $0! = 1$.

5.3.7. $5! = 120$. For the first position, there are 5 people from which you can choose. In the second position, there are 4 people remaining from which to choose. In the third position, there are 3 people remaining. In the fourth position, there are 2 people remaining. Lastly, in the fifth position, there is only 1 person remaining. Thus, by the Multiplication Principle, the number of different ways to arrange five people is $5 \times 4 \times 3 \times 2 \times 1 = 120 = 5!$.

5.3.8. $11! = 39\,916\,800$. For the first lady, she can choose a gentleman in 11 ways; then the second lady has a choice of the remaining 10; the third has choice of 9, *etc.* Therefore, the ladies can take gentlemen partners altogether in

$$\underbrace{11}_{\text{Lady 1}} \times \underbrace{10}_{\text{Lady 2}} \times \underbrace{9}_{\text{Lady 3}} \times \underbrace{8}_{\text{Lady 4}} \times \underbrace{7}_{\text{Lady 5}} \times \underbrace{6}_{\text{Lady 6}} \times \underbrace{5}_{\text{Lady 7}} \times \underbrace{4}_{\text{Lady 8}} \times \underbrace{3}_{\text{Lady 9}} \times \underbrace{2}_{\text{Lady 10}} \times \underbrace{1}_{\text{Lady 11}} = 39\,916\,800$$

ways.

5.3.9. $6 \times 5 \times 4 \times 3 \times 2 \times 1 = 6! = 720$

5.3.10. $8! = 8 \times 7 \times 6 \times 5 \times 4 \times 3 \times 2 \times 1 = 40\,320$

5.3.11. There are $12! = 12 \times 11 \times 10 \times \cdots \times 2 \times 1 = 479\,001\,600$ possible tone rows.

5.3.12. $\underbrace{26}_{\text{Letter 1}} \times \underbrace{25}_{\text{Letter 2}} \times \underbrace{24}_{\text{Letter 3}} \times \cdots \times \underbrace{2}_{\text{Letter 25}} \times \underbrace{1}_{\text{Letter 26}} = 26! \approx 4.033 \times 10^{26}$. Therefore, there are 26!

different substitution ciphers. The number of *useful* substitution ciphers is $26! - 1$, since we want to remove one possible substitution cipher were the plaintext and the ciphertext are the same letters.

5.4.1. (a) $P(6,3) = \dfrac{6!}{(6-3)!} = \dfrac{6!}{3!} = \dfrac{6 \times 5 \times 4 \times \cancel{3!}}{\cancel{3!}} = 6 \times 5 \times 4 = 120$

(b) $P(8,0) = \dfrac{8!}{(8-0)!} = \dfrac{8!}{8!} = 1$

(c) $P(3,2) = \dfrac{3!}{(3-2)!} = \dfrac{3!}{1!} = 3! = 3 \times 2 \times 1 = 6$

(d) $P(6,4) = \dfrac{6!}{(6-4)!} = \dfrac{6!}{2!} = \dfrac{6 \times 5 \times 4 \times 3 \times 2 \times 1}{2 \times 1} = \dfrac{720}{2} = 360$

(e) $P(7,6) = \dfrac{7!}{(7-6)!} = \dfrac{7!}{1!} = 7! = 5040$

(f) $P(12,6) = \underbrace{12 \times 11 \times 10 \times 9 \times 8 \times 7}_{P(12,6)} = 665\,280$

(g) $P(4,3) \times P(3,1) = \underbrace{(4 \times 3 \times 2)}_{P(4,3)} \times \underbrace{3}_{P(3,1)} = 24 \times 3 = 72$

(h) $P(10,2) \times P(5,3) = \underbrace{(10 \times 9)}_{P(10,2)} \times \underbrace{(5 \times 4 \times 3)}_{P(5,3)} = 90 \times 60 = 5400$

(i) $P(n, n-2) = \dfrac{n!}{(n - (n-2))!}$

$\qquad\qquad = \dfrac{n!}{(n - n + 2)!}$

$\qquad\qquad = \dfrac{n!}{2}$

(j) $\dfrac{9!}{5!4!} \times \dfrac{5 \times 4!}{3!2!} = 126 \times 10 = 1260$

5.4.2. (a) $\qquad \dfrac{n!}{(n-1)!} = 6$

$\qquad \dfrac{n \times \cancel{(n-1)!}}{\cancel{(n-1)!}} = 6$

$\qquad\qquad n = 6$

(b) $\qquad \dfrac{n!}{(n-2)!} = 90$

$\qquad \dfrac{n \times (n-1) \times \cancel{(n-2)!}}{\cancel{(n-2)!}} = 90$

$\qquad\qquad n^2 - n = 90$

$\qquad\qquad n^2 - n - 90 = 0$

$\qquad\qquad (n-10)(n+9) = 0$

So $n = 10, -9$. Since $n \in \mathbb{N}, n = 10$.

Another way we can solve $n^2 - n - 90 = 0$ is by using the quadratic formula (Theorem B.4.1 on p. 489), where $a = 1$, $b = -1$, and $c = -90$. Then,

$$n = \frac{-(-1) \pm \sqrt{(-1)^2 - 4(1)(-90)}}{2(1)} = \frac{1 \pm \sqrt{1 + 360}}{2} = \frac{1 \pm \sqrt{361}}{2} = \frac{1 \pm 19}{2}.$$

Thus, $n = \frac{1+19}{2} = \frac{20}{2} = 10$ or $n = \frac{1-19}{2} = -\frac{18}{2} = -9$, and since $n \in \mathbb{N}$, therefore $n = 10$.

(c)
$$P(n, 5) = 14 \times P(n, 4)$$
$$\cancel{n} \times \cancel{(n-1)} \times \cancel{(n-2)} \times \cancel{(n-3)} \times (n-4) = 14 \times \cancel{n} \times \cancel{(n-1)} \times \cancel{(n-2)} \times \cancel{(n-3)}$$
$$n - 4 = 14$$
$$n = 18$$

(d)
$$P(n, 3) = 17 \times P(n, 2)$$
$$\cancel{n} \times \cancel{(n-1)} \times (n-2) = 17 \times \cancel{n} \times \cancel{(n-1)}$$
$$n - 2 = 17$$
$$n = 19$$

(e)
$$\frac{n!}{(n-1)!} = 7$$
$$\frac{n\cancel{(n-1)!}}{\cancel{(n-1)!}} = 7$$
$$n = 7$$

(f)
$$\frac{n!}{(n-2)!} = 12$$
$$\frac{n(n-1)\cancel{(n-2)!}}{\cancel{(n-2)!}} = 12$$
$$n^2 - n = 12$$
$$n^2 - n - 12 = 0$$
$$(n-4)(n+1) = 0$$
Thus, $n = 4, -1$. Since $n \in \mathbb{N}$, $n = 4$.

(g)
$$8P(n, 3) = 7P(n + 1, 3)$$
$$8\cancel{n}\cancel{(n-1)}(n-2) = 7(n+1)\cancel{n}\cancel{(n-1)}$$
$$8(n-2) = 7(n+1)$$
$$8n - 16 = 7n + 7$$
$$n = 7 + 16$$
$$n = 23$$

(h)
$$3P(n, 4) = P(n-1, 5)$$
$$3n\cancel{(n-1)}\cancel{(n-2)}\cancel{(n-3)} = \cancel{(n-1)}\cancel{(n-2)}\cancel{(n-3)}(n-4)(n-5)$$
$$3n = (n-4)(n-5)$$
$$3n = n^2 - 9n + 20$$
$$0 = n^2 - 12n + 20$$
$$0 = (n-2)(n-10)$$
So n could be 2 or 10. But $P(2, 4)$ is not allowed. Therefore, $n = 10$.

5.4.3. $P(6, 6) = 6! = 6 \times 5 \times 4 \times 3 \times 2 \times 1 = 720$

5.4.4. $P(16, 3) = 16 \times 15 \times 14 = 3360$

5.4.5. $\underbrace{9}_{\text{1}^{\text{st}} \text{ digit}} \times \underbrace{9}_{\text{2}^{\text{nd}} \text{ digit}} \times \underbrace{8}_{\text{3}^{\text{rd}} \text{ digit}} \times \underbrace{7}_{\text{4}^{\text{th}} \text{ digit}} = 9 \times P(9, 3) = 9 \times 504 = 4536$

5.4.6. $P(10, 5) = 10 \times 9 \times 8 \times 7 \times 6 = 30\,240$

5.4.7. $P(8, 3) = 8 \times 7 \times 6 = 336$

5.4.8. (a) $P(5, 4) = 5 \times 4 \times 3 \times 2 = 120$ (b) $P(8, 4) = 8 \times 7 \times 6 \times 5 = 1680$

5.4.9. (a) *By direct reasoning.* The other 8 cadets and Cadet James may be arranged in 9! ways; then, two out of the potential positions are not available to Cadet Stewart, hence the total number of arrangements is $9! \times 8 = 2\,903\,040$ (Finucan, 1970).

     (b) *By indirect reasoning.* The total possible ways for 10 cadets to line up is 10!. We treat James and Stewart temporarily as a single cadet (so they are side-by-side in the arrangement); thus, there are 9! to arrange the cadets. By themselves, the James-Stewart pair can be arranged 2! ways. Thus, by the Multiplication Principle the number of arrangements to exclude is $9! \times 2!$. Therefore, the number of permissible arrangements is $10! - (9! \times 2!) = 2\,903\,040$.

     With both types of reasoning, we see that they yielded the same answer.

5.4.10. Use *indirect reasoning.* There are 7! possible arrangements of the 7-volume encyclopedia. There is only one correct way to put the 7-volume on the shelf: in ascending order by the volume number. Therefore, there are $7! - 1 = 5040 - 1 = 5039$ incorrect arrangements.

5.4.11. (a) There are 19 books in total. Therefore, there are $19! \approx 1.2165 \times 10^{17}$ ways the books can be arranged.

     (b) Amongst the German books, there are 5! ways they can be arranged. For the French books, there are 6! ways, and 8! for the English books. Also, when each set is then prepared, the three sets can be placed on the shelves in 3! ways. Therefore, by the Multiplication Principle, the permissible number of ways is $5! \times 6! \times 8! \times 3! = 20\,901\,888\,000$.

5.4.12. (a) $11! = 39\,916\,800$

     (b) $11 \times 10 \times 9 \times 8 \times 7 \times 1 \times 5 \times 4 \times 3 \times 2 \times 1 = P(11, 5) \times 5! = 6\,652\,800$

     (c) $\underbrace{1}_{\text{Rhea}} \times 9 \times 8 \times 7 \times 6 \times 5 \times 4 \times 3 \times 2 \times 1 \times \underbrace{1}_{\text{Jerilyn}} = 9! = 362\,880$

     (d) You can group Brian and Stewart as one pair. But remember that there are two possible permutations of Brian and Stewart. So, $2 \times 10! = 7\,257\,600$.

5.4.13. (a) $P(25, 4) = 303\,600$

     (b) $25 \times 25 \times 25 \times 25 = 25^4 = 390\,625$

5.5.1. By Theorem 5.5.2, the number of ways of arranging the players in a circle is $(5 - 1)! = 4! = 24$.

5.5.2. The number of ways of arranging beads in a necklace is at most $(20 - 1)! = 19!$. Since a necklace can be turned over without changing the arrangement of the beads, the total number of necklaces is $\frac{19!}{2} \approx 6.0823 \times 10^{16}$.

5.5.3. Using Theorem 5.5.2, we need to solve $(n - 1)! = 24$. Therefore, $n = 5$, since $(5 - 1)! = 4! = 24$.

5.5.4. We wish to count the number of circular permutations or the number of ways to seat A, B, C, D, E and F where perhaps A and B don't get along. Seat A at the head of the table, and B cannot be placed to the left or right of A. There are 4 choices for the person to the left of A and 3 choices for the person to the right of A, and the remaining seats can be filled in 2! ways. Therefore, the number of seating arrangements in which A and B are not seated together is $4 \times 3 \times 2! = 24$.

5.5.5. There are 4! circular permutations of the gentlemen, and there are 4! circular permutations of the ladies. A circular setting of 10 people which alternate between gentlemen and ladies corresponds to interweaving of a circular permutation of the gentlemen with a circular permutation of the ladies; there are 5 ways to do this. Therefore, the total number of ways that the gentlemen and ladies can sit around a circular table while alternating is $4! \times 4! \times 5 = 2880$.

5.6.1. If repetition is not allowed, then there is 26 choices for the first letter, 25 choices for the second letter, 24 choices for the third letter, *etc.* Therefore, $26 \times 25 \times 24 \times 23 \times 22 \times 21 \times 20 \times 19 = P(26, 8) = 62\,990\,928\,000$ possible passwords. By imposing this restriction, there are $26^8 - P(26, 8) = 145\,836\,136\,576$ less passwords.

5.6.2. $6^{12} = 2\,176\,782\,336$

5.6.3. $2^6 = 64$

5.6.4. $2^{10} = 1024$

5.6.5. $4^3 = 64$

5.6.6. There are six employees from which to choose for the first task. There are six employees from which to choose for the second task, and so on and so forth. Therefore, there are

$$\underbrace{6}_{\text{Task 1}} \times \underbrace{6}_{\text{Task 2}} \times \underbrace{6}_{\text{Task 3}} \times \underbrace{6}_{\text{Task 4}} \times \underbrace{6}_{\text{Task 5}} = 6^5 = 7776$$

ways for the manager to assign five different tasks to six employees.

5.6.7. (a) $10^4 = 10\,000$
(b) $9 \times 10^3 = 9000$

5.6.8. $2^6 = 64$. Yes, there are enough for all alpha-numeric symbols.

5.6.9. (a)   (i) 7 wives
     (ii) $7 \times 7 = 7^2 = 49$ sacks
     (iii) $7 \times 7 \times 7 = 7^3 = 343$ cats
     (iv) $7 \times 7 \times 7 \times 7 = 7^4 = 2401$ kits
(b) Remember we also met one man with whom he had the wives, sacks, cats, and kits. Adding them together, we have $1 + 7 + 49 + 343 + 2401 = 2801$.
(c) One—just the narrator.

5.6.10. The committee could choose $1, 2, 3, \ldots, 6$ people. Since each person can be either on or off the committee, there are $2^6 = 64$ possibilities. Finally, we remove the single case where there is no one on the committee. Therefore, there are $2^6 - 1 = 63$ committees that contain at least one member.

5.6.11. Use *indirect reasoning*. There are 10^4 possible combinations. There are 9^4 possible combinations that do not contain the number 2. Therefore, the number of ways, at most, she will have to try is $10^4 - 9^4 = 3439$.

5.6.12. Toronto had a population of $2\,615\,000$ in 2011 (making it the fourth most populous city in North America after Mexico City, New York City, and Los Angeles). There are only $26^3 = 17\,576$ unique initials. Since there are more people than unique initials, then by the Pigeonhole principle, there are at least two people with the same initials.

5.6.13. By the Multiplication Principle and Theorem 5.6.1, there are $11^{16} \times 6^{16} = 1.296 \times 10^{29}$ possible compositions.

5.6.14. For each pixel, there are 256^3 possibilities. Then,

$$\underbrace{256^3 \times 256^3 \times \cdots \times 256^3}_{m \times n \text{ pixels}} = (256^3)^{mn} = 256^{3mn}.$$

5.7.1. (a) $\dbinom{7}{3} = \dfrac{7!}{3!(7-3)!} = \dfrac{7!}{3!4!} = \dfrac{7 \times 6 \times 5 \times \cancel{4!}}{3! \cancel{4!}} = \dfrac{7 \times 6 \times 5}{3 \times 2 \times 1} = \dfrac{210}{6} = 35$

(b) $\dbinom{8}{8} = \dfrac{8!}{8!(8-8)!} = \dfrac{8!}{8!0!} = \dfrac{8!}{8! \times 1} = \dfrac{8!}{8!} = 1$

(c) $\dbinom{7}{4} = \dfrac{7!}{4!(7-4)!} = \dfrac{7!}{4!3!} = \dfrac{7 \times 6 \times 5 \times \cancel{4!}}{\cancel{4!}3!} = \dfrac{7 \times 6 \times 5}{3 \times 2 \times 1} = 35$

(d) $\dbinom{10}{2} = \dfrac{10!}{2!(10-2)!} = \dfrac{10!}{2!8!} = \dfrac{10 \times 9 \times \cancel{8!}}{2!\cancel{8!}} = \dfrac{10 \times 9}{2} = \dfrac{90}{2} = 45$

(e) $\dbinom{n}{2} = \dfrac{n!}{2!(n-2)!} = \dfrac{n \times (n-1) \times \cancel{(n-2)!}}{2!\cancel{(n-2)!}} = \dfrac{n \times (n-1)}{2}$

(f) $\dbinom{7}{r} = \dfrac{7!}{r!(7-r)!}$

(g) $\dbinom{7}{0} = \dfrac{7!}{0!(7-0)!} = \dfrac{7!}{7!} = 1$

(h) $\dbinom{n}{n-2} = \dfrac{n!}{(n-2)!(n-(n-2))!} = \dfrac{n!}{(n-2)!(n-n+2)!} = \dfrac{n!}{(n-2)!2!}$

5.7.2. $\dbinom{10}{5} = 252$, $\dbinom{11}{7} = 330$, and $\dbinom{15}{2} = 105$. Thus, $\dbinom{15}{2}$ is the smallest value.

5.7.3. (a) $\dbinom{9}{3} = \dbinom{9}{9-3} = \dbinom{9}{6}$ (c) $\dbinom{8}{r} = \dbinom{8}{8-r}$

(b) $\dbinom{43}{19} = \dbinom{43}{43-19} = \dbinom{43}{24}$ (d) $\dbinom{n}{3} = \dbinom{n}{n-3}$

5.7.4. (a) On the left side, we have $\dbinom{10}{5} = \dfrac{10!}{5!(10-5)!} = \dfrac{10!}{5!5!} = 252$. On the right side, we have

$$\dbinom{9}{4} + \dbinom{9}{5} = \dfrac{9!}{4!(9-4)!} + \dfrac{9!}{5!(9-5)!} = \dfrac{9!}{4!5!} + \dfrac{9!}{5!4!} = 2\left(\dfrac{9!}{4!5!}\right) = 252.$$

Thus, we have shown that the left side is equal to the right side.

(b) $5\dbinom{n}{5} = 5\dfrac{n!}{5!(n-5)!} = \cancel{5}\dfrac{n!}{\cancel{5} \times 4!(n-5)!}$

$\qquad\quad = \dfrac{n(n-1)!}{4!(n-5)!} = n\dfrac{(n-1)!}{4!((n-1)-4)!} = n\dbinom{n-1}{4}$

(c) $n\binom{n-1}{x-1} = n\dfrac{(n-1)!}{(x-1)!(n-1-(x-1))!}$

$= n\dfrac{(n-1)!}{(x-1)!(n-1-x+1)!}$

$= n\dfrac{(n-1)!}{(x-1)!(n-x)!}$

$= \dfrac{n!}{(x-1)!(n-x)!}$

$= \dfrac{x}{x}\dfrac{n!}{(x-1)!(n-x)!}$

$= x\dfrac{n!}{x!(n-x)!}$

$= x\binom{n}{x}$

(d) $2\binom{2n-1}{n-1} = 2\dfrac{(2n-1)!}{(n-1)!(2n-1-(n-1))!}$

$= 2\dfrac{(2n-1)!}{(n-1)!(2n-1-n+1)!}$

$= 2\dfrac{(2n-1)!}{(n-1)!(2n-n)!}$

$= 2\dfrac{n}{n}\dfrac{(2n-1)!}{(n-1)!(2n-n)!}$

$= \dfrac{2n!}{n!(2n-n)!}$

$= \binom{2n}{n}$

5.7.5. This statement is true. To see this, recall the formula for permutations is $P(n,r) = \frac{n!}{(n-r)!}$. Then,

$$r!\binom{n}{r} = r!\frac{n!}{(n-r)!r!} = r!\frac{n!}{(n-r)!r!} = \frac{n!}{(n-r)!} = P(n,r).$$

5.7.6.

$$\binom{n}{n-2} = 28$$

$$\frac{n!}{(n-2)!(n-(n-2))!} = 28$$

$$\frac{n!}{(n-2)!2!} = 28$$

$$\frac{n!}{(n-2)!} = 28 \times 2!$$

$$\frac{n \times (n-1) \times (n-2)!}{(n-2)!} = 28 \times 2$$

$$n \times (n-1) = 56$$

$$n^2 - n = 56$$

$$n^2 - n - 56 = 0$$

$$(n + 7)(n - 8) = 0$$

So, $n = -7$ or $n = 8$, and since $n \in \mathbb{N}$, therefore $n = 8$.

Another way we can solve $n^2 - n - 56 = 0$ is by using the quadratic formula (Theorem B.4.1 on p. 489), where $a = 1$, $b = -1$, and $c = -56$. Then,

$$n = \frac{-(-1) \pm \sqrt{(-1)^2 - 4(1)(-56)}}{2(1)} = \frac{1 \pm \sqrt{1 + 224}}{2} = \frac{1 \pm \sqrt{225}}{2} = \frac{1 \pm 15}{2}.$$

Thus, $n = \frac{1+15}{2} = \frac{16}{2} = 8$ or $n = \frac{1-15}{2} = -\frac{14}{2} = -7$, and since $n \in \mathbb{N}$, therefore $n = 8$.

5.7.7. There are 8 points to choose from, in which you need to select 2 points; that is,

$$\binom{8}{2} = \frac{8!}{2!(8-2)!} = \frac{8!}{2!6!} = \frac{8 \times 7 \times \cancel{6!}}{2!\cancel{6!}} = \frac{8 \times 7}{2} = 28.$$

5.7.8. There are $\binom{5}{3} = \frac{5!}{3!2!} = 10$ subsets containing exactly three elements. They are $\{a, b, c\}$, $\{a, b, d\}$, $\{a, b, e\}$, $\{a, c, d\}$, $\{a, c, e\}$, $\{a, d, e\}$, $\{b, c, d\}$, $\{b, c, e\}$, $\{b, d, e\}$, $\{c, d, e\}$.

5.7.9. (a) $\binom{26}{5} = 65\,780$ (c) $\binom{13}{5} = 1287$

(b) $\binom{12}{5} = 792$

(d) The even ranks are $2, 4, 6, 8, 10$ and there are four suits. Thus, there are 20 cards from which to choose. Therefore, $\binom{20}{5} = 15\,504$.

5.7.10. $\binom{13}{1} \times \binom{13}{1} \times \binom{13}{1} \times \binom{13}{1} \times \binom{52-4}{1} = 13 \times 13 \times 13 \times 13 \times 48 = 1\,370\,928$

5.7.11. (a) $\binom{32}{15} = 565\,722\,720$ (b) $\binom{32-15}{15} = \binom{17}{15} = 136$

5.7.12. The order in which he puts his books into his knapsack is not important; the order does not matter. There are three cases to consider since he can carry *at most* three books. This means that he could carry $0, 1, 2$ *or* 3 books; the *or* suggests we need to apply the Sum Principle.

(a) Case 1: No books. $\binom{10}{0} = 1$ (c) Case 3: Two books. $\binom{10}{2} = 45$

(b) Case 2: One book. $\binom{10}{1} = 10$ (d) Case 4: Three books. $\binom{10}{3} = 120$

Adding up all of the cases, there are $1 + 10 + 45 + 120 = 176$ different combinations of books that he can select.

5.7.13. (a) $\binom{20}{3} = 1140$

(b) $\binom{19}{2} = 171$

(c) $\binom{19}{3} = 969$. Another way is $\binom{20}{3} - \binom{19}{2} = 969$.

5.7.14. For the first round, the number of games that must be played is

$$\binom{10}{2} = \frac{10!}{2!(10-2)!} = \frac{10!}{2!8!} = \frac{10 \times 9 \times 8\!\!\!/}{2!8\!\!\!/} = \frac{10 \times 9}{2!} = \frac{10 \times 9}{2} = \frac{90}{2} = 45.$$

Similarly for the second round, the number of games is $\binom{10}{2} = 45$. Therefore, the total number of games that must be played is $45 + 45 = 90$.

5.7.15. $\binom{161}{3} \times \binom{97}{3} \times \binom{37}{3} \times \binom{7}{1} = 682\,640 \times 147\,440 \times 7770 \times 7 = 5\,474\,268\,738\,624\,000$

5.7.16. $\underbrace{\binom{10}{10}}_{A} \times \underbrace{\binom{10}{2}}_{B} \times \underbrace{\binom{10}{3}}_{C} = 1 \times 45 \times 120 = 5400$

5.7.17. (a) $\binom{12}{4} = 495$

(b) $\binom{12-4}{4} = \binom{8}{4} = 70$

(c) $\underbrace{\binom{4}{2}}_{\text{2 defective}} \times \underbrace{\binom{12-4}{2}}_{\text{2 non-defective}} = 6 \times 28 = 168$

5.7.18. The number of handshakes is

$$\binom{30}{2} = \frac{30!}{2!(30-2)!} = \frac{30!}{2!28!} = \frac{30 \times 29 \times 28\!\!\!/}{2!28\!\!\!/} = \frac{30 \times 29}{2} = 435.$$

5.7.19. We are given that there are 120 handshakes and $\binom{n}{2} = 120$. We are required to find n. Then, we solve:

$$\binom{n}{2} = 120$$

$$\frac{n!}{2!(n-2)!} = 120$$

$$\frac{n!}{(n-2)!} = 120 \times 2!$$

$$\frac{n \times (n-1) \times (n-2)\!\!\!/}{(n-2)\!\!\!/} = 120 \times 2$$

$$n(n-1) = 240$$

$$n^2 - n - 240 = 0$$

$$(n-16)(n+15) = 0.$$

Therefore, the solutions to $n^2 - n - 240 = 0$ is $n = -15$ or $n = 16$. Since $n \in \mathbb{N}$, therefore $n = 16$.

Another way we can solve $n^2 - n - 240 = 0$ is by using the quadratic formula (Theorem B.4.1 on p. 489), where $a = 1$, $b = -1$, and $c = -240$. Then,

$$n = \frac{-(-1) \pm \sqrt{(-1)^2 - 4(1)(-240)}}{2(1)} = \frac{1 \pm \sqrt{1 + 960}}{2} = \frac{1 \pm \sqrt{961}}{2} = \frac{1 \pm 31}{2}.$$

Thus, $n = \frac{1+31}{2} = \frac{32}{2} = 16$ or $n = \frac{1-31}{2} = -\frac{30}{2} = -15$, and since $n \in \mathbb{N}$, therefore $n = 16$.

5.7.20. (a) Using *direct reasoning*. The first way is to *take cases*. The manager can assign one employee to the task, *or* two employees, *or* ..., *or* six employees. The number of ways we can assign an employee to the task is $\binom{6}{1}$. The number of ways we can assign 2 employees to the task is $\binom{6}{2}$, and so on and so forth. By the Sum Principle, we can add up all of the cases. Therefore, the number of ways to assign at least one employee is

$$\underbrace{\binom{6}{1}}_{\substack{\text{Assign} \\ \text{1 employee}}} + \underbrace{\binom{6}{2}}_{\substack{\text{Assign} \\ \text{2 employees}}} + \underbrace{\binom{6}{3}}_{\substack{\text{Assign} \\ \text{3 employees}}} + \underbrace{\binom{6}{4}}_{\substack{\text{Assign} \\ \text{4 employees}}} + \underbrace{\binom{6}{5}}_{\substack{\text{Assign} \\ \text{5 employees}}} + \underbrace{\binom{6}{6}}_{\substack{\text{Assign} \\ \text{6 employees}}}$$

$$= 6 + 15 + 20 + 15 + 6 + 1$$
$$= 63.$$

(b) Using *indirect reasoning*. The second way to solve this problem is to think of the total number of subsets, $2^6 = 64$, and subtract the empty set (the case where the manager assigns no employee to the task. Thus, there are $64 - 1 = 63$ ways.

5.7.21. There are 12 face cards in a standard deck of playing cards.
(a) Using *direct reasoning* (or take cases). In the hand, there could be 3 face cards, *or* 4 face cards, *or* 5 face cards. By the Sum Principle, we can add up all of these cases together. Therefore, the number of 5-card hands that contain at least 3 face cards is

$$\underbrace{\binom{12}{3}\binom{52-12}{2}}_{\text{3 face cards}} + \underbrace{\binom{12}{4}\binom{52-12}{1}}_{\text{4 face cards}} + \underbrace{\binom{12}{5}\binom{52-12}{0}}_{\text{5 face cards}}$$

$$= (220)(780) + (495)(40) + (792)(1)$$
$$= 192\,192.$$

(b) Using *indirect reasoning*.

$$\underbrace{\binom{52}{5}}_{\text{all possible hands}} - \underbrace{\binom{12}{0}\binom{52-12}{5}}_{\text{no face cards}} - \underbrace{\binom{12}{1}\binom{52-12}{4}}_{\text{1 face card}} - \underbrace{\binom{12}{2}\binom{52-12}{3}}_{\text{2 face cards}}$$

$$= 2\,598\,960 - (1)(658\,008) - (12)(91\,390) - (66)(9880)$$
$$= 192\,192$$

5.7.22.

$$\binom{n}{4} = \frac{n!}{4!(n-4)!} = 330$$

$$\frac{n!}{(n-4)!} = 330 \cdot 4!$$

$$\frac{n(n-1)(n-2)(n-3)(n-4)!}{(n-4)!} = 330 \cdot 24$$

$$n(n-1)(n-2)(n-3) = 7920$$

Now, we have to figure out what n is by trial and error; see Table C.3.

Table C.3 Finding n by trial and error

n	1	2	3	4	5	6	7	8	9	10	11
$n(n-1)(n-2)(n-3)$	0	0	0	24	120	360	840	1680	3024	5040	7920

We see that $n = 11$ satisfies the equation. We can easily check that $\binom{11}{4} = 330$. Therefore, there are 11 items on the menu that will allow 330 different combination plates.

Alternatively, it might have been easier just to try different values of n in $\binom{n}{4}$; see Table C.4.

Table C.4 Calculating different values of n in $\binom{n}{4}$

n	4	5	6	7	8	9	10	11
$\binom{n}{4}$	1	5	15	35	70	126	210	330

5.7.23. (a) $\binom{10}{8} = \frac{10!}{8!2!} = \frac{10 \times 9 \times 8!}{8!2!} = \frac{90}{2} = 45$

(b) $\binom{10}{10} = \frac{10!}{10!0!} = 1$

(c) $\binom{10}{8} + \binom{10}{9} + \binom{10}{10} = 45 + 10 + 1 = 56$

5.7.24. We need to pick 6 men from 15 possible choices, $\binom{15}{6}$, and pick 6 women from 12 possible choices $\binom{12}{6}$. By the Multiplication Principle, $\binom{15}{6} \times \binom{12}{6} = 5005 \times 924 = 4\,624\,620$.

5.7.25. (a) Pick 20 as the smallest element, and we select the remaining 6 elements from $\{21, 22, \ldots, 50\}$. Therefore, there are $\binom{30}{6} = 593\,775$ subsets with 7-elements and 20 as the smallest element.

(b) Pick 40 as the largest element, and we select the remaining 6 elements from $\{1, 2, \ldots, 39\}$. Therefore, there are $\binom{39}{6} = 3\,262\,623$ subsets with 7-elements and 40 as the largest element.

(c) Pick 20 as the smallest element and 40 as the largest, and we need to select the remaining 5 elements from $\{21, 22, \ldots, 39\}$. Therefore, there are $\binom{19}{5} = 11\,628$ subsets.

(d) Let 25 be the middle element. This means that the first 3 elements must be selected from $\{1, 2, \ldots, 24\}$; that is, $\binom{24}{3}$. The last 3 elements must be selected from $\{26, 27, \ldots, 50\}$; that is, $\binom{25}{3}$. Therefore, the number of subsets is $\binom{24}{3} \times \binom{25}{3} = 2024 \times 2030 = 4\,655\,200$.

5.7.26. Using Theorem 5.7.8, $\binom{7}{6} = \binom{7}{1}$, $\binom{7}{4} = \binom{7}{3}$, $\binom{7}{2} = \binom{7}{5}$, and $\binom{7}{0} = \binom{7}{7}$. Therefore, $\binom{7}{0} + \binom{7}{2} + \binom{7}{4} + \binom{7}{6} = \binom{7}{1} + \binom{7}{3} + \binom{7}{5} + \binom{7}{7}$ as required.

5.7.27. (a) One of the 5 elements must be special. Then, the remaining four elements are chosen from the seven non-special elements; that is, $\binom{7}{4} = 35$.

(b) The five elements must be chosen amongst the seven non-special elements; that is, $\binom{7}{5} = 21$.

(c) Since we consider all cases where either the special element is included in the subset or not, this is equivalent to considering all possible 5-element subsets of A. Thus, $\binom{7}{4} + \binom{7}{5} = 35 + 21 = 56 = \binom{8}{5}$.

5.7.28. There are $\binom{n}{r}$ ways to pick and paint r homes red. Then, the contractor would have $n - r$ remaining homes painted blue: $\binom{n-r}{b}$. The rest of the homes are painted yellow; there is only one way to accomplish this. Therefore, by the Multiplication Principle, there are $\binom{n}{r} \times \binom{n-r}{b}$ ways to paint the homes.

Remark: This exercise was adapted from Pólya *et al.* (1983, p. 10).

5.7.29. Other answers are possible, but here is one possible answer: how many ways can a committee of 5 women and 2 men be formed by selecting from 8 women and 9 men?

5.7.30. (a) On the right side, the number of ways to pick 7 students to form a team from a total of 14 is $\binom{14}{7}$. On the left side,
- the number of ways you can pick no boys and all girls is $\binom{7}{0}\binom{7}{7}$. By Theorem 5.7.8, $\binom{7}{0} = \binom{7}{7}$. Therefore, $\binom{7}{0}\binom{7}{7} = \binom{7}{0}\binom{7}{0} = \binom{7}{0}^2$.
- the number of ways you can pick 1 boy and 6 girls is $\binom{7}{1}\binom{7}{6}$. By Theorem 5.7.8, $\binom{7}{1} = \binom{7}{6}$. Therefore, $\binom{7}{1}\binom{7}{6} = \binom{7}{1}\binom{7}{1} = \binom{7}{1}^2$.
- this continues up to the number of ways you can pick all boys and no girls, which is $\binom{7}{7}\binom{7}{0}$. By Theorem 5.7.8, $\binom{7}{7} = \binom{7}{0}$. Therefore, $\binom{7}{7}\binom{7}{0} = \binom{7}{7}\binom{7}{7} = \binom{7}{7}^2$.

Therefore, with all of the above cases by the Sum Principle, we have $\binom{7}{0}^2 + \binom{7}{1}^2 + \binom{7}{2}^2 + \cdots + \binom{7}{7}^2$.
With the two different (yet equivalent) approaches, we have proved the equation as required.

(b) We can extend this reasoning for the general case. On the right side, the number of ways to pick n students to form a team from a total of $2n$ students is $\binom{2n}{n}$. On the left side,
- the number of ways you can pick no boys and all girls is $\binom{n}{0}\binom{n}{n}$. By Theorem 5.7.8, $\binom{n}{0} = \binom{n}{n}$. Therefore, $\binom{n}{0}\binom{n}{n} = \binom{n}{0}\binom{n}{0} = \binom{n}{0}^2$.
- the number of ways you can pick 1 boy and $n - 1$ girls is $\binom{n}{1}\binom{n}{n-1}$. By Theorem 5.7.8, $\binom{n}{1}\binom{n}{n-1} = \binom{n}{1}\binom{n}{1} = \binom{n}{1}^2$.
- this continues up to the number of ways you can pick all boys and no girls, which is $\binom{n}{n}\binom{n}{0}$. By Theorem 5.7.8, $\binom{n}{n} = \binom{n}{0}$. Therefore, $\binom{n}{n}\binom{n}{0} = \binom{n}{n}\binom{n}{n} = \binom{n}{n}^2$.

Therefore, with all of the above cases, by the Sum Principle we have $\binom{n}{0}^2 + \binom{n}{1}^2 + \cdots + \binom{n}{n}^2$.
With the two different (yet equivalent) approaches, we have proved the equation for the general case.

5.8.1. Example 5.8.1 uses direct reasoning. We count all the possibilities by taking cases.

5.8.2. The tree diagram is in Fig. C.14.

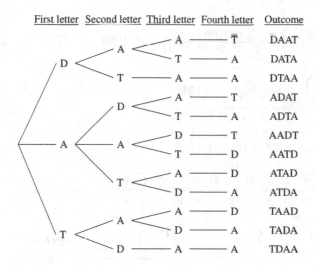

Fig. C.14   Tree diagram for the permutations of DATA in Exercise 5.8.2

It is still direct reasoning because we are only counting the possibilities in which we are interested.

5.8.3. (a) $\dbinom{3}{2,1} = \dfrac{3!}{2!1!} = \dfrac{6}{2} = 3$

(b) $\dbinom{4}{2,0,2} = \dfrac{4!}{2!0!2!} = \dfrac{24}{2 \times 1 \times 2} = 6$

(c) $\dbinom{15}{11,4} = \dfrac{15!}{11!4!} = \dfrac{15 \times 14 \times 13 \times 12 \times \cancel{11!}}{\cancel{11!} \times 4!} = \dfrac{32\,760}{24} = 1365$

(d) $\dbinom{20}{15,2,3} = \dfrac{20!}{15!2!3!}$

$$= \dfrac{20 \times 19 \times 18 \times 17 \times 16 \times \cancel{15!}}{\cancel{15!} \times 2! \times 3!}$$

$$= \dfrac{1\,860\,480}{12}$$

$$= 155\,040$$

5.8.4. It is not a valid multinomial coefficient since $10 + 2 + 2 + 7 = 21 \neq 20$.

5.8.5. To account for repetition, we divide by the number of ways each repeated letter can be arranged.

    (a) For the letter A, there are 2!. For the letter C, there are 2! ways. For the letter E, there are 2! ways. Therefore, the number of distinguishable permutations that can be made from the letters ACCEPTABLE is $\dbinom{10}{2,2,2,1,1,1} = \dfrac{10!}{2!2!2!} = 453\,600$.

(b) $\begin{pmatrix} 6 \\ 2,1,1,2 \end{pmatrix} = \dfrac{6!}{\underbrace{2!}_{B} \ \underbrace{1!}_{A} \ \underbrace{1!}_{M} \ \underbrace{2!}_{O}} = \dfrac{720}{4} = 180$

(c) $\begin{pmatrix} 10 \\ 1,2,2,3,1,1 \end{pmatrix} = \dfrac{10!}{\underbrace{1!}_{B} \ \underbrace{2!}_{O} \ \underbrace{2!}_{K} \ \underbrace{3!}_{E} \ \underbrace{1!}_{P} \ \underbrace{1!}_{R}} = 151\,200$

(d) $\begin{pmatrix} 6 \\ 1,3,1,1 \end{pmatrix} = \dfrac{6!}{\underbrace{1!}_{C} \ \underbrace{3!}_{A} \ \underbrace{1!}_{N} \ \underbrace{1!}_{D}} = 120$

(e) $\begin{pmatrix} 18 \\ 1,1,1,2,2,2,1,3,1,1,1,1,1 \end{pmatrix}$

$= \dfrac{18!}{\underbrace{1!}_{C} \ \underbrace{1!}_{O} \ \underbrace{1!}_{N} \ \underbrace{2!}_{T} \ \underbrace{2!}_{R} \ \underbrace{2!}_{A} \ \underbrace{1!}_{F} \ \underbrace{3!}_{I} \ \underbrace{1!}_{B} \ \underbrace{1!}_{U} \ \underbrace{1!}_{L} \ \underbrace{1!}_{E} \ \underbrace{1!}_{S}}$

$= 1.333 \times 10^{14}$

(f) $\begin{pmatrix} 11 \\ 2,2,2,1,1,1,1,1 \end{pmatrix} = \dfrac{11!}{\underbrace{2!}_{M} \ \underbrace{2!}_{A} \ \underbrace{2!}_{T} \ \underbrace{1!}_{H} \ \underbrace{1!}_{E} \ \underbrace{1!}_{I} \ \underbrace{1!}_{C}, \ \underbrace{1!}_{S}} = 4\,989\,600$

(g) $\begin{pmatrix} 10 \\ 1,1,2,1,2,3 \end{pmatrix} = \dfrac{10!}{\underbrace{1!}_{M} \ \underbrace{1!}_{A} \ \underbrace{2!}_{T} \ \underbrace{1!}_{R} \ \underbrace{2!}_{E} \ \underbrace{3!}_{S}} = 151\,200$

(h) $\begin{pmatrix} 11 \\ 1,4,4,2 \end{pmatrix} = \dfrac{11!}{\underbrace{1!}_{M} \ \underbrace{4!}_{I} \ \underbrace{4!}_{S} \ \underbrace{2!}_{P}} = \dfrac{11!}{1!4!4!2!} = 34\,650.$

(i) $\begin{pmatrix} 5 \\ 1,1,1,2 \end{pmatrix} = \dfrac{5!}{\underbrace{1!}_{P} \ \underbrace{1!}_{R} \ \underbrace{1!}_{E} \ \underbrace{2!}_{S}} = 60$

(j) $\begin{pmatrix} 10 \\ 1,1,1,1,1,2,3 \end{pmatrix} = \left(\dfrac{10!}{\underbrace{1!}_{P} \ \underbrace{1!}_{R} \ \underbrace{1!}_{I} \ \underbrace{1!}_{N} \ \underbrace{1!}_{C} \ \underbrace{2!}_{E} \ \underbrace{3!}_{S}} \right) = 302\,400$

(k) $\begin{pmatrix} 12 \\ 2,3,2,3,1,1 \end{pmatrix} = \dfrac{12!}{\underbrace{2!}_{Q} \ \underbrace{3!}_{U} \ \underbrace{2!}_{I} \ \underbrace{3!}_{N} \ \underbrace{1!}_{E} \ \underbrace{1!}_{M}} = 3\,326\,400$

(l) $\begin{pmatrix} 10 \\ 3,3,1,2,1 \end{pmatrix} = \dfrac{10!}{\underbrace{3!}_{S} \ \underbrace{3!}_{T} \ \underbrace{1!}_{A} \ \underbrace{2!}_{I} \ \underbrace{1!}_{C}} = 5040$

(m) $\begin{pmatrix} 9 \\ 2,1,1,2,1,2 \end{pmatrix} = \dfrac{9!}{\underbrace{2!}_{S} \ \underbrace{1!}_{U} \ \underbrace{1!}_{P} \ \underbrace{2!}_{E} \ \underbrace{1!}_{N} \ \underbrace{2!}_{D}} = 45\,360$

(n) $\begin{pmatrix} 7 \\ 2,3,1,1 \end{pmatrix} = \dfrac{7!}{\underbrace{2!}_{T} \underbrace{3!}_{O} \underbrace{1!}_{R} \underbrace{1!}_{N}} = 420$

(o) $\begin{pmatrix} 8 \\ 1,1,1,1,1,1,2 \end{pmatrix} = \dfrac{8!}{\underbrace{1!}_{W} \underbrace{1!}_{A} \underbrace{1!}_{T} \underbrace{1!}_{E} \underbrace{1!}_{R} \underbrace{1!}_{L} \underbrace{2!}_{O}} = 20\,160$

(p) $\begin{pmatrix} 12 \\ 1,3,2,1,1,2,1,1 \end{pmatrix} = \dfrac{12!}{\underbrace{1!}_{S} \underbrace{3!}_{A} \underbrace{2!}_{L} \underbrace{1!}_{D} \underbrace{1!}_{W} \underbrace{2!}_{R} \underbrace{1!}_{B} \underbrace{1!}_{E}} = 19\,958\,400$

5.8.6. $\begin{pmatrix} 12 \\ 4,4,4 \end{pmatrix} = 34\,650$

5.8.7. Using Theorem 5.8.3, the total number of arrangements is $\begin{pmatrix} 10 \\ 1,2,4,3 \end{pmatrix} = \frac{10!}{1!2!4!3!} = 12\,600$.

5.8.8. $\begin{pmatrix} 11 \\ 6,5 \end{pmatrix} = \dfrac{11!}{6!5!} = 462$

5.8.9. $\begin{pmatrix} 8 \\ 3 \end{pmatrix}\begin{pmatrix} 8-3 \\ 2 \end{pmatrix}\begin{pmatrix} 8-3-2 \\ 3 \end{pmatrix} = \begin{pmatrix} 8 \\ 3 \end{pmatrix}\begin{pmatrix} 5 \\ 2 \end{pmatrix}\begin{pmatrix} 3 \\ 3 \end{pmatrix} = \dfrac{8!}{3!2!3!} = \begin{pmatrix} 8 \\ 3,2,3 \end{pmatrix} = 560$

5.8.10. Let us glue the two 2s together. Then we have to count the permutations of the set $\{1,1,1,2,3,4,5\}$. By Theorem 5.8.3, there are $\begin{pmatrix} 7 \\ 3,1,1,1,1 \end{pmatrix} = \frac{7!}{3!} = \frac{5040}{6} = 840$ permutations.

5.8.11. $\begin{pmatrix} 10 \\ 6,4 \end{pmatrix} = \dfrac{10!}{6!4!} = 210$

5.8.12. $\begin{pmatrix} 9 \\ 4,5 \end{pmatrix} = \dfrac{9!}{4!5!} = 126$

5.8.13. (a) $\begin{pmatrix} 7 \\ 3,1,1,1,1 \end{pmatrix} = \dfrac{7!}{\underbrace{3!}_{M} \underbrace{1!}_{A} \underbrace{1!}_{X} \underbrace{1!}_{I} \underbrace{1!}_{U}} = \dfrac{7!}{3!} = 840$

(b) $\begin{pmatrix} 6 \\ 3,1,1,1 \end{pmatrix} = \dfrac{6!}{\underbrace{3!}_{M} \underbrace{1!}_{A} \underbrace{1!}_{I} \underbrace{1!}_{U}} = \dfrac{6!}{3!} = 120$

(c) $\begin{pmatrix} 6 \\ 2,1,1,1,1 \end{pmatrix} = \dfrac{6!}{\underbrace{2!}_{M} \underbrace{1!}_{A} \underbrace{1!}_{X} \underbrace{1!}_{I} \underbrace{1!}_{U}} = \dfrac{6!}{2!} = 360$

5.8.14. (a) $\begin{pmatrix} 8 \\ 2,2,2,2 \end{pmatrix} = \dfrac{8!}{2!2!2!2!} = 2520$ (b) $\begin{pmatrix} 8 \\ 4,4 \end{pmatrix} = \dfrac{8!}{4!4!} = 70$

5.8.15. As we can see in Fig. 5.15, there is a total of 9 notes in the opening bar, but some of these notes are repeated: g4 is repeated 3 times, b4 is repeated 3 times, and d5 is repeated twice. Using Theorem 5.8.3, there are

$$
\binom{9}{3,3,2,1} = \frac{\overbrace{9!}^{\text{9 notes in total}}}{\underbrace{3!}_{g4}\ \underbrace{3!}_{b4}\ \underbrace{2!}_{d5}\ \underbrace{1!}_{g5}} = 5040
$$

distinguishable permutations of Bach's opening notes.

5.8.16. As we can see in Fig. 5.16, there is total of 15 notes in the opening bar, but some of these notes are repeated: e5 is repeated 4 times, a5 is repeated 6 times, c6 is repeated twice, and g♯5 is repeated twice. Using Theorem 5.8.3, there are

$$
\binom{15}{4,6,2,1,2} = \frac{\overbrace{15!}^{\text{15 notes in total}}}{\underbrace{4!}_{e5}\ \underbrace{6!}_{a5}\ \underbrace{2!}_{c6}\ \underbrace{2!}_{g\sharp5}\ \underbrace{1!}_{a4}} = 18\,918\,900
$$

distinguishable permutations of Bach's opening notes.

5.8.17. (a) From the hotel, Lauren can go 3 blocks east, and 3 blocks south to the MoMA. Therefore,
$$
\binom{3+3}{3,3} = \binom{6}{3,3} = \frac{6!}{3!3!} = \frac{720}{6 \times 6} = 20.
$$

(b) From the hotel, Lauren can go 3 blocks east, and 7 blocks south to the Rockefeller Center. Therefore, $\binom{3+7}{3,7} = \binom{10}{3,7} = \frac{10!}{3!7!} = \frac{3\,628\,800}{6 \times 5040} = 120.$

(c) From the hotel to Times Square, Lauren can go 2 blocks east and 10 blocks south. Therefore,
$$
\binom{2+10}{2,10} = \binom{12}{2,10} = \frac{12!}{2!10!} = \frac{479\,001\,600}{2 \times 3\,628\,800} = 66.
$$

(d) From Carnegie Hall to St. Patrick's Cathedral, Lauren can go 2 blocks east and 7 blocks south. Therefore, $\binom{2+7}{2,7} = \binom{9}{2,7} = \frac{9!}{2!7!} = \frac{9 \times 8 \times \cancel{7!}}{2!\cancel{7!}} = \frac{9 \times 8}{2} = 36.$

(e) From the Apple Store to St. Patrick's Cathedral, there is only one path. She can only walk down 5th Avenue.

5.8.18. (a) (i) $\dfrac{(6+6)!}{6!6!} = 924$ (ii) $\dfrac{(3+6)!}{3!6!} = 84$

(b) (i) $\dfrac{(10+10)!}{10!10!} = 184\,756$ (iii) $\dfrac{(7+14)!}{7!14!} = 116\,280$

(ii) $\dfrac{(x+x)!}{x!x!} = \dfrac{(2x)!}{x!x!}$ (iv) $\dfrac{(x+y)!}{x!y!}$

5.8.19. $\binom{5}{3,2} \times 2 \times \binom{6}{4,2} = \frac{5!}{3!2!} \times 2 \times \frac{6!}{4!2!} = 300$

5.8.20. (a) (i) $\dfrac{(1+1+1)!}{1!1!1!} = \dfrac{3!}{1!1!1!} = 3! = 6$ (iv) $\dfrac{(2+2+2)!}{2!2!2!} = \dfrac{6!}{2!2!2!} = 90$

(ii) $\dfrac{(2+1+1)!}{2!1!1!} = \dfrac{4!}{2!} = 12$ (v) $\dfrac{(3+2+2)!}{3!2!2!} = \dfrac{7!}{3!2!2!} = 210$

(iii) $\dfrac{(2+2+1)!}{2!2!1!} = \dfrac{5!}{2!2!} = 30$

(b) (i) $\dfrac{(10+10+10)!}{10!10!10!} = \dfrac{30!}{10!10!10!} = 5\,550\,996\,791\,340$

(ii) $\dfrac{(x+x+x)!}{x!x!x!} = \dfrac{(3x)!}{x!x!x!}$

(iii) $\dfrac{(10+8+12)!}{10!8!12!} = \dfrac{30!}{10!8!12!} = 3\,784\,770\,539\,550$

(iv) $\dfrac{(x+y+z)!}{x!y!z!}$

5.8.21. $\dfrac{(2+2+2)!}{2!2!2!} \times \dfrac{(1+1+1)!}{1!1!1!} \times \dfrac{(3+2+2)!}{3!2!2!} = \dfrac{6!}{2!2!2!} \times \dfrac{3!}{1!1!1!} \times \dfrac{7!}{3!2!2!}$

$$= 90 \times 6 \times 210$$

$$= 113\,400$$

5.8.22. (a) There are 9 squares in which 3 squares are red, 3 squares are blue, 2 squares are yellow, and 1 square is black. Since all the squares must be painted, $3+3+2+1 = 9$, then the number of permutations of these 9 squares is given by the multinomial coefficient $\dbinom{9}{3,3,2,1} = \dfrac{9!}{3!3!2!1!} = 5040$.

(b) Since there are only three red squares, there can only be one red square in each row. (We do not have to worry about having more than one red square in a row.)

Beginning with row 1, we make the first square red. Then there are 6 (non-red) choices for the second square, and 5 (non-red) remaining choices for the third square. Now, there are 3 ways that we can have the red square (in the first position, the second position, or the third position) for this row. Therefore, there is $6 \times 5 \times 3 = 90$ ways to paint row 1. Now, we only have 4 non-red choices left since we used up two of them in row 1.

For row 2, we make the first square red, and in the second square there are 4 (non-red) choices for the second square, and 3 (non-red) choices for the third square. There are 3! permutations for this row. Therefore, there is $4 \times 3 \times 3 = 36$ ways to paint row 2. Now, we only have 2 non-red choices left since we used up four of them in rows 1 and 2.

For row 3, we make the first square red, and in the second square there are two choices, and in the last square there is only one choice left. There are 3 permutations for this row. Therefore, there are $2 \times 1 \times 3 = 6$ ways to paint row 3.

Adding up the possibilities for each row, we have $90 + 36 + 6 = 132$ ways to paint the canvas with a red square in every row.

(c) There are 3! ways to ensure that there is only one red square in each row and in each column. Then, the remaining squares are filled with the remaining 6 non-red colors. Therefore, the number of ways to paint the canvas with a red square in every row and in every column is $3! \times 6! = 4320$.

5.9.1. By symmetry of combinations (Theorem 5.7.8),

$$\binom{n+r-1}{r} = \binom{n+r-1}{n+\not{r}-1-\not{r}} = \binom{n+r-1}{n-1}.$$

5.9.2. There are $n = 5$ distinct flavors and taken $r = 2$ at a time. Then,

$$\binom{5+2-1}{2} = \binom{6}{2} = \dfrac{6!}{(6-2)!2!} = \dfrac{6!}{4!2!} = \dfrac{6 \times 5 \times \not{4!}}{\not{4!}2!} = \dfrac{30}{2} = 15.$$

5.9.3. $\binom{5+20-1}{5} = \binom{24}{5} = 42\,504$

5.9.4. (a) (i) $\binom{25+5-1}{5} = \binom{29}{5} = 118\,755$

(ii) $\binom{25}{5} = 53\,130$

(b) There are $118\,755 - 53\,130 = 65\,625$ more choices if you allow repetition.

5.9.5. (a) If we let $n = 5$, and $r = 12$, then

$$\binom{5 + 12 - 1}{12} = \frac{(5 + 12 - 1)!}{12!(5 - 1)!} = \frac{16!}{12!4!} = 1820.$$

Therefore, there are 1820 ways.

(b) If we let $n = 5$, and $r = 12 - 5 = 7$.

$$\binom{5 + 7 - 1}{7} = \binom{11}{7} = \frac{11!}{7!4!} = 330$$

Therefore, there are 330 ways.

(c) For each uniquely colored marble, there are 5 different containers in which it can be placed. So there are

$$5 \times 5 \times 5 \times 5 \times 5 \times 5 \times 5 \times 5 \times 5 \times 5 \times 5 \times 5 = 5^{12} = 244\,140\,625.$$

5.9.6. Set $r = 4$ and $n = 2$. There are $\binom{2+4-1}{4} = \binom{5}{4} = 5$ such solutions. The solutions, in the form (a, b), are $(0, 4)$, $(1, 3)$, $(2, 2)$, $(3, 1)$, $(4, 0)$.

5.9.7. Let $r = 16$, and $n = 4$. Then, by Theorem 5.9.2, there are

$$\binom{4 + 16 - 1}{16} = \binom{19}{16} = 969$$

solutions to the equation.

5.9.8. We know by Theorem 5.9.2 that the number of distinct solutions to $r_1 + r_2 + r_3 + r_4 = 20$ is equal to $\binom{4+20-1}{20}$ when $r_1, r_2, r_3, r_4 \in \{0, 1, 2, \ldots\}$. We need to convert the restrictions in the question to match this. Given that $r_1 \in \{1, 2, 3, \ldots\}$, let us define a new variable $x_1 = r_1 - 1$, then $x_1 \in \{0, 1, 2, \ldots\}$. Similarly, for $r_2 \in \{3, 4, 5, \ldots\}$, we can define a new variable $x_2 = r_2 - 3$, then $x_2 \in \{0, 1, 2, \ldots\}$. Substituting x_1 and x_2 into the equation, we have

$$r_1 + r_2 + r_3 + r_4 = 20$$
$$(x_1 + 1) + (x_2 + 3) + r_3 + r_4 = 20$$
$$x_1 + x_2 + r_3 + r_4 = 20 - 1 - 3$$
$$x_1 + x_2 + r_3 + r_4 = 16$$

where $x_1, x_2, r_3, r_4 \in \{0, 1, 2, \ldots\}$. Therefore, the number of distinct solutions to the equation is $\binom{4+16-1}{16} = \binom{19}{16} = 969$.

Remark: This problem involved finding something familiar. We reduced the problem down to a problem whose solution we already know. It is important to try to reduce the problem to similar problem that we have seen; this is one of the strategies in Pólya's approach to problem solving.

5.9.9. By Theorem 5.9.2, $\binom{3+60-1}{60} = \binom{62}{60} = 1891$ ways to create a secret success formula.

5.9.10. Since Curtis has specified that he will be donating \$20 to the CNIB, we have three charities left in which to allocate the remaining \$80 he wishes to donate. Since the donation must be made in increments of \$10, the problem reduces to finding the number of whole number solutions to $r_1 + r_2 + r_3 = 8$; $r = 8$ and $n = 3$. By Theorem 5.9.2, there are $\binom{3+8-1}{8} = \binom{10}{8} = 45$ ways.

5.9.11. Any 4-digit number n can be represented by $n_1 n_2 n_3 n_4$ where n_1, n_2, n_3, and n_4 are the digits of n. We wish to know how many of n there are, given that $n_1 + n_2 + n_3 + n_4 = 8$. The number n can be expressed in the stars and bars notation. For example, the number 2231 can be represented by $\star\star|\star\star|\star\star\star|\star$. Since n_1 must be a 1 or higher (otherwise it is not a 4-digit number), one star must be located in the first location. We have $r = 7$ remaining stars and $n = 4$ locations. Thus, by Theorem 5.9.2, there are $\binom{4+7-1}{7} = \binom{10}{7} = 120$ 4-digit numbers whose sum of their digits is 8.

5.9.13. (a) $\binom{6+12-1}{12} = \binom{17}{12} = 6188$

(b) $\binom{6+13-1}{13} = \binom{18}{13} = 8568$

(c) Pick two of each flavor that satisfy the criteria. We made a select then of 12 bagels. This means that we still need to pick the remaining 12 bagels, so $\binom{6+12-1}{12} = 6188$.

(d) *Take cases.* Case 1: Buy no blueberry bagels; the number of ways to do this is $\binom{5+24-1}{24} = \binom{28}{24} = 20\,475$.

Case 2: Buy 1 blueberry bagel; the number of ways to do this is $\binom{5+23-1}{23} = \binom{27}{23} = 17\,550$.

Case 3: Buy 2 blueberry bagels; the number of ways to do this is $\binom{5+22-1}{22} = \binom{26}{22} = 14\,950$.

By the Sum Principle, there are $20\,475 + 17\,550 + 14\,950 = 52\,975$ ways to select the bagels.

(e) First, we select the 5 blueberry and the 4 French toast bagels. Then, to select the remaining $24 - 5 - 4 = 15$ bagels, the number of ways to do is $\binom{6+15-1}{15} = \binom{20}{15} = 15\,504$.

5.10.1. (a) $2^7 = 128$

(b) $\binom{7}{6} = 7$

5.10.2. (a) 2^n

(b) $\binom{n}{n-1} = \dfrac{n!}{(n-n-1)!(n-1)!} = \dfrac{n!}{(n-1)!} = \dfrac{n \times (n-1)!}{(n-1)!} = n$

5.10.3. Refer to Fig. 5.23 on p. 193.

5.10.4. We will show that $P(n,r) - P(n-1,r) = r \times P(n-1,r-1)$. Then,

$$P(n,r) - P(n-1,r) = \frac{n!}{(n-r)!} - \frac{(n-1)!}{((n-1)-r)!}$$

$$= \frac{n \times (n-1)!}{(n-r)!} - \frac{(n-1)!}{((n-1)-r)!}\frac{(n-r)}{(n-r)}$$

$$= \frac{n \times (n-1)!}{(n-r)!} - \frac{(n-1)! \times (n-r)}{(n-r)!}$$

$$= \frac{n \times (n-1)! - (n-1)!(n-r)}{(n-r)!}$$

$$= (n-n+r)\frac{(n-1)!}{(n-r)!}$$

$$= r\frac{(n-1)!}{(n-r)!}$$

$$= r\frac{(n-1)!}{((n-1)-(r-1))!}$$

$$= r \times P(n-1,r-1)$$

as required.

5.10.5. For signals, the order matters. So, $P(6,2) + P(6,3) + P(6,4) + P(6,5) + P(6,6) = 30 + 120 + 360 + 720 + 720 = 1950$.

5.10.6. Since Daniel and Mark are required to attend, this removes 2 spots. Thus, there are $11 - 2 = 9$ remaining students from which to choose for the remaining $6 - 2 = 4$ spots. Therefore, the number of ways 6 people from the lab, with the restriction, is $\binom{9}{4} = 126$.

5.10.7. When it says that in 5 playing cards there are at least 4 hearts, this means that there could be 4 hearts *or* 5 hearts; the *or* suggests we need to apply the Sum Principle.

Therefore,

$$\underbrace{\binom{13}{4}\binom{52-13}{1}}_{\text{4 hearts}} \overset{\text{Sum Principle}}{+} \underbrace{\binom{13}{5}\binom{52-13}{0}}_{\text{5 hearts}} = (715)(39) + (1287)(1)$$

$$= 27\,885 + 1287$$
$$= 29\,172.$$

5.10.8. $\underbrace{(1+1)}_{2} \times \underbrace{(2+1)}_{3} - \underbrace{1}_{\text{empty set}} = 5$

5.10.9. (a) $\underbrace{(2+1)}_{2} \times \underbrace{(1+1)}_{7} \times \underbrace{(1+1)}_{71} - \underbrace{1}_{\text{empty set}} = 11$

(b) $\underbrace{1}_{2} \times \underbrace{(1+1)}_{2} \times \underbrace{(1+1)}_{7} \times \underbrace{(1+1)}_{71} = 8$

5.10.10. Counting the number of ways a vehicle is sold each day is equivalent to the problem of counting the number of *distinguishable* permutations of a word that has some letters repeating (like Example 5.8.4 on p. 176). For example, we could have a string of characters that represents a possible way that the sales representative sold each car on each day: $TTTSTTCCSS$ (where T is truck, C is crossover, and S is sedan). We also have the restriction that 3 trucks and 1 sedan were sold in the first 4 days *and* then she sold the remaining vehicles in the remaining 6 days; the *and* gives a hint that we should use the multiplication principle.

Vehicles sold in first 4 days Vehicles sold in remaining 6 days

$$\underbrace{\frac{4!}{\underbrace{3!}_{\text{3 trucks}}\underbrace{1!}_{\text{1 sedan}}}}_{} \overset{\text{Multiplication Principle}}{\times} \underbrace{\frac{6!}{\underbrace{2!}_{\text{5 − 3 trucks}}\underbrace{2!}_{\text{3 − 1 crossovers}}\underbrace{2!}_{\text{2 sedans}}}}_{} = \frac{4 \times \cancel{3!}}{\cancel{3!}} \times \frac{6!}{2!2!2!}$$

$$= 4 \times \frac{720}{8}$$
$$= \frac{720}{2}$$
$$= 360$$

5.10.11. *Take cases* by counting the number of ways to place each book on the bookshelves.

- *Book 1.* This first book can be placed on any of the r shelves.
- *Book 2.* We can place this book either to the left or to the right of book 1, *or* on any of the remaining shelves. By the Sum Principle, there are $2 + (r-1) = r + 1$ ways to do this.
- *Book 3.* There are two subcases to consider:
  (a) The first 2 books are on the same shelf, and so the third book can be placed to the left, the middle, or to the right. The third book can also be placed on any of the remaining $r - 1$ empty shelves. By the Sum Principle, there are $3 + (r-1) = r + 2$ ways to do this.
  (b) The first 2 books are not on the same shelf. The third book can be placed to the left or the right of the first book; there are 2 ways. Or, the third book can be placed to the left or the right of the second book; there are 2 ways. The third way is that the third book can be placed on any of the

remaining $r - 2$ shelves. Thus, by the Sum Principle, there are $2 + 2 + (r - 2) = r + 2$ ways.

We see a pattern begin to emerge.

- For book 4, there are $r + 3$ ways.
- For book 5, there are $r + 4$ ways, ...
- For book n, there are $r + n - 1$ ways.

By the Multiplication Principle, the total number of ways is

$$r(r + 1)(r + 2)(r + 3) \cdots (r + n - 1) = \frac{(n + k - 1)!}{(k - 1)!}.$$

5.10.12. (a) We can *take cases* (direct reasoning).

Case 1: no couples. We choose 10 of the 25 couples, and then from each pick one person: $\binom{25}{10} \times 2^{10} = 3\,347\,210\,240$.

Case 2: exactly one couple. We choose 1 of the 25 couples, and then we choose the remaining 8 members by choosing 8 of the 24 couples and picking one person: $\binom{25}{1} \times \binom{24}{8} \times 2^8 = 4\,707\,014\,400$.

Case 3: exactly two couples. We choose 2 of the 25 couples, and then we choose the remaining 6 members by choosing 6 of the 24 members and picking one person: $\binom{25}{2} \times \binom{23}{6} \times 2^6 = 1\,938\,182\,400$.

Adding together all of the cases, the number of ways we can form a committee of 10 members that include at most 2 married couples is $\binom{25}{10} \times 2^{10} + \binom{25}{1} \times \binom{24}{8} \times 2^8 + \binom{25}{2} \times \binom{23}{6} \times 2^6 = 9\,992\,407\,040$.

(b) We pick a couple, and then there two ways to pick the person from the selected couple. Then, we pick the remaining 9 people from the group excluding the selected couple. That is, $\binom{25}{1} \times 2 \times \binom{48}{9} = 83\,855\,332\,000$.

(c) The number of ways to do this is zero. By the Pigeonhole Principle, we cannot pick 26 members from amongst 25 couples and not include at least one married couple.

5.10.13.

$$\underbrace{5}_{\text{Bun}} \times \underbrace{5}_{\text{Meat}} \times \underbrace{(8 + \overbrace{1}^{\text{No cheese}})}_{\text{Cheese}} \times \underbrace{\left[\binom{n}{0} + \binom{n}{1} + \cdots + \binom{n}{n} \right]}_{\text{Toppings}} = 5 \times 5 \times 9 \times 2^n$$

$$= 225 \times 2^n$$

Then,

$$225 \times 2^n \geq 10^9$$

$$2^n \geq \frac{10^9}{225}$$

$$\log(2^n) \geq \log\left(\frac{10^9}{225}\right)$$

$$n \geq \frac{\log\left(\frac{10^9}{225}\right)}{\log(2)} \approx 22.08$$

So Fat King Burger needs to have at least 23 toppings or more to support their claim.

5.11.1. There are 6 different appetizers, 7 main courses, and 4 beverage choices. Using the Multiplication Principle, there are $6 \times 7 \times 4 = 168$ ways a customer can order a meal.

5.11.2. You never order more than one type of gin and one type of vermouth. If you assumed (correctly) that you need a garnish, then there are $4 \times 3 \times 3 = 36$ ways to order a martini. If you assumed (incorrectly) that you could order a martini without a garnish (4 options—nothing, olive, onion, or lemon), then there are $4 \times 3 \times 4 = 48$. I don't know why you would skip the garnish; the olive is the best part!

5.11.5. (a) Permutation   (b) Combination   (c) Combination

5.11.6. $2^8 - 1 = 256 - 1 = 255$

5.11.7. $3 \times 2 \times 5 \times \left(\binom{10}{0} + \binom{10}{1} + \binom{10}{2} + \cdots + \binom{10}{9} + \binom{10}{10} \right)$

$= 3 \times 2 \times 5 \times (1 + 10 + 45 + 120 + 210 + 252 + 210 + 120 + 45 + 10 + 1)$

$= 3 \times 2 \times 5 \times 1024$

$= 30\,720$

Another way of solving the problem is by using the Multiplication Principle with the toppings. For each topping, you have two outcomes, either you have the selected topping or you don't. Thus, there are $2^{10} = 1024$ different combinations of topping for the burger. Thus, there are $3 \times 2 \times 5 \times 1024 = 30\,720$ ways of ordering a burger.

5.11.8. (a) $\underbrace{\binom{5}{3}}_{\text{males}} \times \underbrace{\binom{7}{3}}_{\text{female}} = 10 \times 35 = 350$

(c) Use *indirect reasoning.*

$\binom{12}{6} - \underbrace{\binom{7}{6}}_{\text{all females}} = 924 - 7 = 917$

(b) $\underbrace{\binom{5}{2}}_{\text{males}} \times \underbrace{\binom{7}{4}}_{\text{females}} = 10 \times 35 = 350$

5.11.9. $P(20, 5) = 20 \times 19 \times 18 \times 17 \times 16 = 1\,860\,480$

5.11.10. $\binom{7}{3, 4} = \frac{7!}{3!4!} = \frac{7 \times 6 \times 5 \times \cancel{4!}}{3!\cancel{4!}} = \frac{210}{6} = 35$

5.11.11. (a) For the first digit, we have 9 choices (since it cannot be zero), then 10 choices for the second digit. At the third digit, we have no more choices, since the third digit must be the same as the second digit. The last digit must also match the first digit. Therefore, there are $9 \cdot 10 = 90$ 4-digit palindromes.

(b) $9 \cdot 10 \cdot 10 = 900$ 5-digit palindromes.

5.11.12. (a) $\underbrace{2}_{1^{\text{st}} \text{ digit}} \times \underbrace{3}_{2^{\text{nd}} \text{ digit}} \times \underbrace{2}_{3^{\text{rd}} \text{ digit}} \times \underbrace{1}_{4^{\text{th}} \text{ digit}} = 12$

(c) $\underbrace{2}_{1^{\text{st}} \text{ digit}} \times \underbrace{2}_{2^{\text{nd}} \text{ digit}} \times \underbrace{1}_{3^{\text{rd}} \text{ digit}} \times \underbrace{1}_{4^{\text{th}} \text{ digit}} = 4$

(b) $\underbrace{2}_{1^{\text{st}} \text{ digit}} \times \underbrace{2}_{2^{\text{nd}} \text{ digit}} \times \underbrace{1}_{3^{\text{rd}} \text{ digit}} \times \underbrace{2}_{4^{\text{th}} \text{ digit}} = 8$

5.11.13. (a) $\underbrace{5}_{1^{\text{st}} \text{ symbol}} \times \underbrace{15}_{2^{\text{nd}} \text{ symbol}} \times \underbrace{15}_{3^{\text{rd}} \text{ symbol}} \times \underbrace{15}_{4^{\text{th}} \text{ symbol}} = 5 \times 15^3 = 16\,875$

(b) $\underbrace{14}_{1^{\text{st}} \text{ symbol}} \times \underbrace{14}_{2^{\text{nd}} \text{ symbol}} \times \underbrace{13}_{3^{\text{rd}} \text{ symbol}} \times \underbrace{12}_{4^{\text{th}} \text{ symbol}} = 14 \times P(14, 3) = 30\,576$

(c) $\underbrace{5}_{1^{\text{st}} \text{ symbol}} \times \underbrace{14}_{2^{\text{nd}} \text{ symbol}} \times \underbrace{13}_{3^{\text{rd}} \text{ symbol}} \times \underbrace{12}_{4^{\text{th}} \text{ symbol}} = 5 \times P(14, 3) = 10\,920$

5.11.14. $\dbinom{n}{r} = \dfrac{P(n,r)}{r!}$

5.11.15. The cards in Fig. 5.24 are considered different even though they have the same numbers, the numbers are in a different order; hence, order matters. This means we should use a permutation, not a combination.

    (a)   (i) In the first row of B, there are 15 choices of numbers. For the second row, we have used up one of the numbers are left with 14 choices of numbers, and so on. Thus, for column B, the number of possible arrangements is

$$15 \times 14 \times 13 \times 12 \times 11 = P(15,5) = 360\,360.$$

        (ii) For column I, the number of possible arrangements is

$$15 \times 14 \times 13 \times 12 \times 11 = P(15,5) = 360\,360.$$

       (iii) For column N, there is one less row due to the "free" space. Thus, the number of possible arrangements for column N is

$$15 \times 14 \times 13 \times 12 = P(15,4) = 32\,760.$$

    (b) By the Multiplication Principle, the total possible Bingo cards is

$$\underbrace{(15 \times 14 \times 13 \times 12 \times 11)^4}_{\text{For B, I, G, and O columns}} \times \underbrace{15 \times 14 \times 13 \times 12}_{\text{For column N}} = (P(15,5))^4 \times P(15,4)$$

$$= (360\,360)^4 \times 32\,720$$

$$\approx 5.52 \times 10^{26}.$$

5.11.16. $\dbinom{300}{5} = \dfrac{300!}{5!(300-5)!} = \dfrac{300!}{5!295!} = \dfrac{300 \times 299 \times 298 \times 297 \times 296 \times \cancel{295!}}{5!\cancel{295!}}$

$$= \dfrac{300 \times 299 \times 298 \times 297 \times 296}{5 \times 4 \times 3 \times 2 \times 1}$$

$$= 19\,582\,837\,560$$

5.11.17. $\underbrace{\dbinom{10}{5}}_{\text{Employee 1}} \times \underbrace{\dbinom{10-5}{5}}_{\text{Employee 2}} = \dbinom{10}{5} \times \dbinom{5}{5} = 252 \times 1 = 252$

5.11.18. There are nine letters in the word ALGORITHM. The important word in the question is *or*; this means that the sum rule will be used.

3-letter words or 4-letter words

$$= \Big(\underbrace{9}_{\text{Letter 1}} \times \underbrace{8}_{\text{Letter 2}} \times \underbrace{7}_{\text{Letter 3}} \Big) + \Big(\underbrace{9}_{\text{Letter 1}} \times \underbrace{8}_{\text{Letter 2}} \times \underbrace{7}_{\text{Letter 3}} \times \underbrace{6}_{\text{Letter 4}} \Big)$$

$$\underbrace{}_{\text{3-letter words}} \qquad \underbrace{}_{\text{4-letter words}}$$

$$= P(9,3) + P(9,4)$$

$$= 504 + 3024 = 3528$$

5.11.19. $\dbinom{50}{4} \times \dbinom{42}{4} = 230\,300 \times 111\,930 = 25\,777\,479\,000$

5.11.20. There are 6 players available. We can use *indirect reasoning*, and count all the possible ways by calculating all the possible ways that the members can line up and remove the ways where 2 defensemen are together.

$$\underbrace{6!}_{\text{number of ways to line up}} \quad - \quad \underbrace{(5! \times 2)}_{\text{number of ways defensemen together}} \quad = 720 - 240 = 480$$

Therefore, there are 480 ways to line up the players with 2 defensemen not together.

5.11.21. When it says that in 5 playing cards there are at least 3 kings, this means that there could be 3 kings *or* 4 kings; the *or* suggests we need to apply the Sum Principle. Note that you cannot have more than 4 kings.

$$\underbrace{\binom{4}{3}\binom{48}{2}}_{\text{3 kings}} \overset{\text{Sum Principle}}{+} \underbrace{\binom{4}{4}\binom{48}{1}}_{\text{4 kings}} = 4(1128) + 1(48) = 4560$$

Therefore, there are 4560 ways in which you can select 5 playing cards with at least 3 kings.

5.11.22. (a) $\binom{24}{5} = 42\,504$ (c) $\binom{24}{5} - \binom{20}{5} = 27\,000$

(b) $\binom{20}{5} = 15\,504$

5.11.23. (a) $\binom{5}{2} \times \binom{3}{1} \times \binom{3}{1} = 10 \times 3 \times 3 = 90$

(b) $5 \times 4 \times 3 \times 3 = 180$

5.11.24.

$$\overset{\text{First 5 games}}{\underbrace{\frac{5!}{\underbrace{3!}_{\text{3 wins}} \underbrace{2!}_{\text{2 losses}}}}} \overset{\text{Multiplication Principle}}{\times} \overset{\text{Last 7 games}}{\underbrace{\frac{7!}{\underbrace{4!}_{7-3 \text{ wins}} \underbrace{2!}_{\text{2 ties}} \underbrace{1!}_{3-2 \text{ losses}}}}} = \frac{5 \times 4 \times \cancel{3!}}{\cancel{3!}2} \times \frac{7 \times 6 \times 5 \times \cancel{4!}}{\cancel{4!}2}$$

$$= \frac{5 \times \cancel{4} \times 7 \times 6 \times 5}{\cancel{4}}$$

$$= 1050$$

5.11.25. $\binom{10}{2,2,2,2,2} = \frac{10!}{2!2!2!2!2!} = 113\,400$

5.11.26.

$$P(n,r) = \frac{n!}{(n-r)!} = 6720$$

and

$$\binom{n}{r} = \frac{n!}{(n-r)!r!} = 56$$

$$\frac{n!}{(n-r)!} = 56 \cdot r!$$

Therefore,

$$56 \cdot r! = 6720$$
$$r! = \frac{6720}{56}$$
$$r! = 120$$
$$r = 5.$$

5.11.27.

$$P(n,r) = \frac{n!}{(n-r)!} = 720$$

and

$$\binom{n}{r} = \frac{n!}{(n-r)!r!} = 120$$

$$\frac{n!}{(n-r)!} = 120 \cdot r!$$

Therefore,

$$120 \cdot r! = 720$$
$$r! = \frac{720}{120}$$
$$r! = 6$$
$$r = 3.$$

5.11.28.
$$\frac{n-r+1}{r}\binom{n}{r-1} = \frac{n-r+1}{r} \cdot \frac{n!}{(n-(r-1))!(r-1)!}$$

$$= \frac{n-r+1}{r} \cdot \frac{n!}{(n-r+1)!(r-1)!}$$

$$= \frac{\cancel{n-r+1}}{r} \cdot \frac{n!}{\cancel{(n-r+1)}(n-r)!(r-1)!}$$

$$= \frac{n!}{(n-r)!r!}$$

$$= \binom{n}{r}$$

5.11.29. She can make a selection of $r = 3$ bills, and there are $n = 4$ kinds of bills; therefore, by Theorem 5.9.2, the number of assortments of bills Emma can select is $\binom{4+3+1}{3} = \binom{8}{3} = 56$.

6.1.1. (a) If Fibonacci started with two rabbits, then his sequence would be 2, 2, 4, 6, 10, 16, 26, 42, 68, 110, 178, 288. Thus, there would be 288 rabbits in one year.

(b) If Fibonacci started with three rabbits, then his sequence would be 3, 3, 6, 9, 15, 24, 39, 63, 102, 165, 267, 432. Thus, there would be 432 rabbits in one year.

6.1.2. In modern usage, it is more common to express the Fibonacci sequence as 0, 1, 1, 2, 3, 5, ... The difference between the modern usage and the expression we found in Example 6.1.4 is that it does not include zero.

6.1.3. (a) 0, 2, 0, 2, 0, 2

(b) $-3, 9, -27, 81, -243, 729$

(c) $0, 1, 3, 6, 10, 15$

(d) $-2, 4, -6, 8, -10, 12$

(e) $-1, \frac{1}{2}, -\frac{1}{3}, \frac{1}{4}, -\frac{1}{5}, \frac{1}{6}$

(f) $\frac{1}{2}, \frac{1}{6}, \frac{1}{12}, \frac{1}{20}, \frac{1}{30}, \frac{1}{42}$

(g) $\frac{1}{2}, \frac{1}{6}, \frac{1}{12}, \frac{1}{20}, \frac{1}{30}, \frac{1}{42}$

(h) $1, \frac{1}{2}, 3, \frac{1}{4}, 5, \frac{1}{6}$

6.1.4. (a) $t_1 = 5, t_2 = 8, t_3 = 11$.

(b) $t_{11} = 35, t_{12} = 38$

(c) $t_n = 5 + (n - 1)3$

6.1.5. (a) $t_1 = 20, t_2 = 10, t_3 = 5$

(b) $t_n = 20(\frac{1}{2})^{n-1}$

6.1.6. (a) $t_n = 5 + (n - 1)2$

(b) $t_n = 1 + (n - 1)3$

(c) $t_n = \dfrac{n}{n + 1}$

(d) $t_n = n^2$

(e) $t_n = n^3$

(f) $t_n = (-1)^{n+1} \dfrac{1}{n}$

6.1.7. One, Two, Three, Four, Five, Six, Seven, Eight, ...

6.1.8. (a) $1, 5, 21, 85, 341, 1365$

(b) $2, 1, \frac{3}{2}, \frac{7}{6}, \frac{19}{14}, \frac{47}{38}$

(c) $1, 2, 3, 5, 8, 13$

(d) $3, 5, 4, -9, \frac{5}{13}, -\frac{56}{61}$

6.1.10. (a) geometric sequence: $t_n = 10(0.1)^{n-1}$

(b) geometric sequence: $t_n = -\frac{5}{6}(-\frac{2}{3})^{n-1}$

(c) arithmetic sequence: $t_n = 2k + (n - 1)2$

(d) arithmetic sequence: $t_n = \frac{a}{b} + (n - 1)(-\frac{2a}{b})$

(e) geometric sequence: $t_n = \frac{m^2}{n}(-\frac{n}{m})^{n-1}$

6.1.11. The first term is $a = x$, and the common ratio $r = \frac{y}{x}$. Therefore, the eighth term is

$$t_{10} = ar^9 = x\left(\frac{y}{x}\right)^9 = \frac{y^9}{x^8}.$$

6.1.12. $t_n = \begin{cases} a & \text{if } n = 1, \\ t_{n-1} + d & \text{if } n = 2, 3, 4, \ldots . \end{cases}$

6.1.13. $t_n = \begin{cases} a & \text{if } n = 1, \\ t_{n-1} \times r & \text{if } n = 2, 3, 4, \ldots . \end{cases}$

6.1.14. This sequence is a geometric sequence, since each cell will divide into two in each generation: $b_1 = 4, b_2 = 8, b_3 = 16$. The common ratio is $\frac{t_n}{t_{n-1}} = 2$. Therefore, the geometric sequence $t_1, t_2, t_3, \ldots$ is a geometric sequence with $a = 4$ and common ratio 2; that is $t_n = (4)2^{n-1}$ for $n = 1, 2, 3, \ldots$.

6.1.15. (a) $100\% - 95\% = 5\%$ remains.

(b) $a = 1, r = 0.05, n = 10$. Then, $t_{10} = ar^{n-1} = 1(0.05)^{10-1} = (0.05)^9 \approx 1.953\,13 \times 10^{-12}$.

(c)
$$0.1 = (0.05)^{n-1}$$
$$\log(0.1) = \log\left((0.05)^{n-1}\right)$$
$$\log(0.1) = (n-1)\log(0.05)$$
$$n - 1 = \frac{\log(0.1)}{\log(0.05)}$$
$$n \approx 0.7686 + 1$$
$$n = 1.7686$$

Since $n \in \mathbb{N}$, we will round up to 2. Therefore, the glass needs to be rinsed at least twice to ensure 1% or less soap residue remains.

6.1.16. The next diagram in the sequence is given below.

6.1.17. In the sequence, there are 5♣s and 2◇s. Thus, there are $5-2 = 3$ more ♣s written than ◇s. When the sequence is copied 75 times, in total there are $75 \times 3 = 225$ ♣s written and a total of $75 \times 2 = 150$ ◇s written. Therefore, there are $225 - 150 = 75$ more ♣s written than ◇. (This was adapted from the 2015 CEMC Pascal Contest problem 11.)

6.1.18. (a) True. Let $a = 2c$ for some $c \in \mathbb{Z}$, and $b = 2d$ for some $d \in \mathbb{Z}$. Then, $a + b = 2c + 2d = \underbrace{2(c+d)}_{e} = 2e$, and by definition is also an even number.

(b) False. Let $a = 2c + 1$ for some $c \in \mathbb{Z}$, and $b = 2d + 1$ for some $d \in \mathbb{Z}$. Then, $a + b = 2c + 1 + 2d + 1 = \underbrace{2(c+d)}_{\text{even}} + \underbrace{2}_{\text{even}}$, and by part (a) is also even.

6.1.19. Since $t_1, t_2, t_3, \ldots$ is an arithmetic sequence, by definition, $t_n = a + (n-1)d$ in which $a = t_1$ and d is the common difference. Let u_n be the nth term for the sequence $t_2, t_4, t_6, \ldots$. Then, $u_n = t_{2n} = a + (2n-1)d$. Now, we will show that the difference between the consecutive terms, $t_2, t_4, t_6, \ldots$ is constant.

$$u_n - u_{n-1} = a + (2n-1)d - \{a + [2(n-1) - 1]d\}$$
$$= (2n-1)d - (2n-3)d$$
$$= 2d$$

Since the difference between the consecutive terms is constant, $t_2, t_4, t_6, \ldots$ is an arithmetic sequence.

6.2.1. In each case, look for the value of the largest exponent with a variable (not on a constant).

(a) 2

(b) 2

(c) 3

(d) 4

(e) 13

(f) 20

(g) 2
(h) 3

6.2.2. First, you need to construct the table of finite differences for the terms that are provided. Then, you just have to extend the table to find the next three terms.

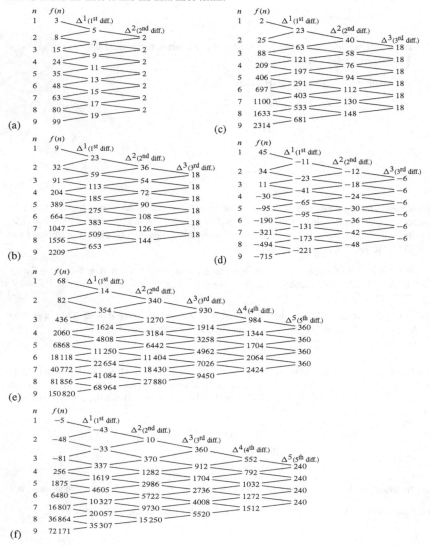

6.3.1. (a) $1^3 + 2^3 + 3^3 + 4^3 + 5^3$
     (b) $1 + \frac{1}{2} + \frac{1}{3} + \frac{1}{4} + \frac{1}{5}$

(c)

$$\frac{2-1}{4+3} + \frac{3-1}{9+3} + \frac{4-1}{16+3} + \frac{5-1}{25+3} + \frac{6-1}{36+3}$$

$$= \frac{1}{7} + \frac{2}{12} + \frac{3}{19} + \frac{4}{28} + \frac{5}{39}$$

$$= \frac{1}{7} + \frac{1}{6} + \frac{3}{19} + \frac{1}{7} + \frac{5}{39}$$

(d) $\binom{20}{4} + \binom{20}{5} + \binom{20}{6} + \binom{20}{7} + \binom{20}{8}$

6.3.2. (a) $t_6 = 8$
(b) $t_6 = 6! + 1 = 720 + 1 = 721$
(c) $t_6 = \frac{1}{8}$

6.3.3. (a) $2 + 2 + 2 = 6$
(b) $\frac{1}{2} + \frac{1}{3} + \frac{1}{4} = \frac{13}{12}$
(c) $1! + 2! + 3! = 1 + 2 + 6 = 9$
(d) $\binom{7}{3} + \binom{7}{4} + \binom{7}{5} + \binom{7}{6} = 35 + 35 + 21 + 7 = 98$

6.3.4. (a) 10   (d) 2016   (g) 225
(b) 62   (e) 28   (h) 98
(c) 121   (f) 5   (i) 45

6.3.5. $\displaystyle\sum_{i=0}^{20} x^i = x^0 + x^1 + x^2 + \cdots + x^{20} = 1 + x + x^2 + \cdots + x^{20}$

6.3.6. (a) $\displaystyle\sum_{i=1}^{8} i$ (d) $\displaystyle\sum_{i=3}^{n} i^3$ (g) $\displaystyle\sum_{i=1}^{n} (2i-1)$

(b) $\displaystyle\sum_{i=3}^{7} i$ (e) $\displaystyle\sum_{i=2}^{n} 2n$ (h) $\displaystyle\sum_{i=1}^{n} \frac{1}{k+i}$

(c) $\displaystyle\sum_{i=4}^{20} i^2$ (f) $\displaystyle\sum_{i=1}^{n} 3^i$ (i) $\displaystyle\sum_{i=1}^{n} i x_i$

6.3.7. (a) $\displaystyle\sum_{i=1}^{5} \frac{i}{i+1}$ (b) $\displaystyle\sum_{i=0}^{8} i \binom{8}{i}$

6.3.8. (a) $\displaystyle\sum_{i=1}^{n} (2i-1)(3i)(7 + (n-1)4)$ (d) $\displaystyle\sum_{i=1}^{r} t_i (x_i)^{i+1}$

(b) $\displaystyle\sum_{i=1}^{n} \frac{1}{i(i+1)}$ (e) $\displaystyle\sum_{i=1}^{k-1} i f(ix)$

(c) $\displaystyle\sum_{i=1}^{n} (2i-1)x^i$ (f) $\displaystyle\sum_{i=1}^{n-1} (2i-1)x^i$

6.3.9. (a) $\displaystyle\sum_{i=1}^{n} x_i y_i$ (b) $\displaystyle\sum_{i=1}^{3} x_i^i$ (c) $\displaystyle\sum_{i=1}^{n} i t_i$ (d) $\displaystyle\sum_{i=1}^{n} (-1)^{i+1} i t_i$

6.3.10. $\displaystyle\sum_{i=1}^{n}(t_i - t_{i-1}) = (t_1 - t_0) + (t_2 - t_1) + (t_3 - t_2) + \cdots + (t_n - t_{n-1})$

$$= -t_0 + \cancel{t_1} - \cancel{t_1} + \cancel{t_2} - \cancel{t_2} + \cdots + \cancel{t_{n-1}} - \cancel{t_{n-1}} + t_n$$

$$= t_n - t_0$$

6.4.1. (a) We use Gauss' method.

S	$=$	2	$+$	4	$+$	$\cdots$	$+$	198	$+$	200
$+S$	$=$	200	$+$	198	$+$	$\cdots$	$+$	4	$+$	2
$2S$	$=$	202	$+$	202	$+$	$\cdots$	$+$	202	$+$	202

$$\underbrace{\qquad\qquad\qquad\qquad\qquad\qquad\qquad\qquad}_{100 \text{ terms}}$$

So,

$$2S = 202 \times 100$$
$$S = \frac{202 \times 100}{2}$$
$$S = 10\,100$$

(b) $\displaystyle\sum_{i=1}^{100} = 2i.$

6.4.2. $a = 7, d = 4, n = 20$

$$S_{20} = \frac{20}{2}[2(7) + (20-1)4] = 10[14 + 19(4)] = 900$$

6.4.3. *Method 1*: Using the Gaussian summation formula. If we multiply on both sides the Gaussian summation by 2, we have

$$2\sum_{i=1}^{n} i = \cancel{2}\frac{n(n+1)}{\cancel{2}}$$

$$2(1 + 2 + 3 + \cdots + n) = n(n+1)$$
$$2 + 4 + 6 + \cdots + 2n = n(n+1).$$

Method 2: Using Eq. (6.13). $a = 2, d = 2$, and n. Then,

$$S_n = \frac{n}{2}[2(2) + (n-1)2] = \frac{n}{2}(4 + 2n - 2) = \frac{n}{2}(2n + 2) = \frac{n}{\cancel{2}} \cdot \cancel{2}(n+1) = n(n+1).$$

6.4.4. (a) $a = 6, d = 3, t_n = 54, n = ?$

$$t_n = 54 = a + (n-1)d$$
$$54 = 6 + (n-1)3$$
$$3n = 51$$
$$n = 17$$

Therefore, $6 + 9 + 12 + \cdots + 54 = S_{17} = \frac{17}{2}[2(6) + (17-1)3] = 510.$

(b) $a = 11, d = 11, t_n = 143, n = ?$

$$t_n = 143 = a + (n-1)d$$
$$143 = 11 + (n-1)11$$
$$143 = \cancel{11} + 11n - 11$$
$$n = \frac{143}{11} = 13$$

Therefore, $11 + 22 + 33 + \cdots + 143 = S_{13} = \frac{13}{2}[2(11) + (13-1)11] = 1001$.

(c) $a = 9, d = -5, t_n = -101, n = ?$

$$t_n = -101 = a + (n-1)d$$
$$-101 = 9 + (n-1)(-5)$$
$$-101 = 9 - 5n + 5$$
$$-5n = -101 - 9 - 5$$
$$n = \frac{-115}{-5} = 23$$

Therefore, $9 + 4 + (-1) + \cdots + (-101) = S_{23} = \frac{23}{2}[2(9) + (23-1)(-5)] = -1058$.

(d) $a = \frac{1}{7}, d = \frac{3}{7}, t_n = \frac{37}{7}, n = ?$

$$t_n = \frac{37}{7} = a + (n-1)d$$
$$\frac{37}{7} = \frac{1}{7} + (n-1)(\frac{3}{7})$$
$$\frac{37}{7} = \frac{1}{7} + \frac{3}{7}n - \frac{3}{7}$$
$$\frac{3}{7}n = \frac{39}{7}$$
$$n = \frac{39}{\cancel{7}} \cdot \frac{\cancel{7}}{3} = \frac{39}{3} = 13$$

Therefore, $\frac{1}{7} + \frac{4}{7} + \frac{7}{7} + \cdots + \frac{37}{7} = S_{13} = \frac{13}{2}[2(\frac{1}{7}) + (13-1)(\frac{3}{7})] = \frac{247}{7}$.

6.4.5. (a) The cost of drilling the last meter of the well is the xth term of the series. Therefore, the cost of drilling the last meter is $t_{30} = 50 + (30-1)5 = \$195$.

(b) The cost to drill the entire well constitutes an arithmetic series whose terms are the cost for each meter. Therefore, taking $a = 50$, $d = 5$, and $n = 30$ in Eq. (6.13) gives the total cost of $S_{30} = \frac{30}{2}[2(50) + (30-1)5] = \$3675$.

6.4.6. We are required to solve

$$1 + 2 + 3 + \cdots + n = 120$$
$$\frac{n(n+1)}{2} = 120$$
$$n(n+1) = 240$$
$$n^2 + n = 240$$
$$n^2 + n - 240 = 0.$$

Using the quadratic formula, $n = \frac{-1 \pm \sqrt{(1)^2 - 4(1)(-240)}}{2(1)} = \frac{-1 \pm \sqrt{961}}{2} = \frac{-1 \pm 31}{2} = 15$ or $\cancel{-16}$. Therefore, you will need 15 children to sell the 120 chocolate bars.

6.4.7. (a) $n = 12$

(b) $n = 15$

(c) $n = 12$

(d) $n = 13$

6.4.8. We want to show that $\sum_{i=1}^{n} i = \dfrac{n(n+1)}{2} = \dbinom{n+1}{2}$. Thus, we need to show that $\dbinom{n+1}{2} = \dfrac{n(n+1)}{2}$.

$$\binom{n+1}{2} = \frac{(n+1)!}{2!(n+1-2)!} = \frac{(n+1)!}{2!(n-1)!} = \frac{(n+1)n\cancel{(n-1)!}}{2!\cancel{(n-1)!}} = \frac{n(n+1)}{2!} = \frac{n(n+1)}{2}$$

6.4.9. (a) $\displaystyle\sum_{i=0}^{100} i = \binom{101}{2} = 5050$

(b) $\displaystyle\sum_{j=45}^{100} j = \binom{101}{2} - \binom{45}{2} = 5050 - 990 = 4060$

(c) $\displaystyle\sum_{k=30}^{45} k = \binom{46}{2} - \binom{30}{2} = 1035 - 435 = 600$

6.4.10. (a)   (i) See Table C.5.

Table C.5   The sum of odd numbers

1	$=$	$1 = 1^2$
$1 + 3$	$=$	$4 = 2^2$
$1 + 3 + 5$	$=$	$9 = 3^2$
$1 + 3 + 5 + 7$	$=$	$16 = 4^2$
$1 + 3 + 5 + 7 + 9$	$=$	$25 = 5^2$
$1 + 3 + 5 + 7 + 9 + 11$	$=$	$36 = 6^2$

(ii) The pattern is that the sum the odd numbers appear to be square numbers.

(iii) $\underbrace{1 + 3 + 5 + 7 + 9 + 11 + 13}_{\text{The first 7 odd numbers}} = 7^2 = 49$.

(iv) The sum of the first n odd numbers $1 + 3 + 5 + \cdots + (2n - 1) = n^2$.

(b)   (i)   i. $1 + 3 = 4$

      ii. $1 + 3 + 5 = 9$

      iii. $1 + 3 + 5 + 7 = 16$

(ii) In Fig. 6.13, the square has $4 \times 4 = 16$ circles, but if we start in the top-left corner, there is one circle that represents the number 1. It is then surrounded by 3 circles (representing the number 3), 5 circles (representing the number 5), and then 7 circles (representing the number 7). We were able to arrange each odd number of circles into the formation of a square! Thus, $1 + 3 + 5 + 7 = 4^2 = 16$. Therefore, the sum of the first n odd numbers must be a square number as well. If we continue the pattern, then we would surround the circles in Fig. 6.13 with an additional 9 circles, like in Fig. C.15.

Thus, the sum of the first 6 odd numbers is $1 + 3 + 5 + 7 + 9 = 6^2 = 36$.

(iii) In general, the sum of the first n odd numbers is $1 + 3 + 5 + \cdots + (2n - 1) = n^2$.

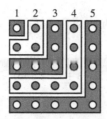

Fig. C.15   Extending the diagram in Exercise 6.4.10

6.4.11. (a) $(1 + 2 + 3 + 4)^2 = \left[\dfrac{4(5)}{2}\right]^2 = \left(\dfrac{20}{2}\right)^2 = 10^2 = 100.$

(b) $(1 + 2 + \cdots + n)^2 = \left[\dfrac{n(n+1)}{2}\right]^2$

6.4.12. (a) $3 + 6 + 9 + \cdots + 3n = 3(1 + 2 + 3 + \cdots + n) = 3\left(\displaystyle\sum_{i=1}^{n} i\right) = 3\left[\dfrac{n(n+1)}{2}\right]$

(b) $4 + 8 + 12 + \cdots + 4n = 4(1 + 2 + 3 + \cdots + n) = 4\left(\displaystyle\sum_{i=1}^{n} i\right) = 4\left[\dfrac{n(n+1)}{2}\right]$

(c) $n + 2n + 3n + \cdots + n^2 = n(1 + 2 + 3 + \cdots + n)$

$$= n\left(\sum_{i=1}^{n} i\right)$$

$$= n\left[\frac{n(n+1)}{2}\right]$$

$$= \frac{n^2(n+1)}{2}$$

(d) $(1 + 2 + 3 + \cdots + 25)^2 = \left[\dfrac{25(26)}{2}\right]^2 = (325)^2 = 105\,625$

(e) $(1 + 2 + 3 + \cdots + n)^2 = \left(\displaystyle\sum_{i=1}^{n} i\right)^2 = \left[\dfrac{n(n+1)}{2}\right]^2$

(f) $(n + 2n + 3n + \cdots + n^2)^2 = [n(1+2+3+\cdots+n)]^2 = \left[n\left(\displaystyle\sum_{i=1}^{n} i\right)\right]^2 = \left[\dfrac{n^2(n+1)}{2}\right]^2$

(g) $(1 + 2 + 3 + \cdots + 25)^3 = \left[\dfrac{25(26)}{2}\right]^3 = (325)^3 = 34\,328\,125$

(h) $(1 + 2 + 3 + \cdots + n)^3 = \left(\displaystyle\sum_{i=1}^{n} i\right)^3 = \left[\dfrac{n(n+1)}{2}\right]^3$

(i) $(n + 2n + 3n + \cdots + n^2)^3 = [n(1+2+3+\cdots+n)]^3 = \left[n\left(\displaystyle\sum_{i=1}^{n} i\right)\right]^3 = \left[\dfrac{n^2(n+1)}{2}\right]^3$

(j) $(1 + 2 + 3 + \cdots + 25)^k = (325)^k$

(k) $(1 + 2 + 3 + \cdots + n)^k = \left[\dfrac{n(n+1)}{2}\right]^k$

(l) $(n + 2n + 3n + \cdots + n^2)^k = [n(1 + 2 + 3 + \cdots + n)]^k = \left[n \left(\sum_{i=1}^{n} i \right) \right]^k = \left[\dfrac{n^2(n+1)}{2} \right]^k$

6.4.13. $(1 + 2 + 3 + \cdots + 500) - (3 + 6 + 9 + \cdots + 498)$

$= (1 + 2 + 3 + \cdots + 500) - 3(1 + 2 + 3 + \cdots + 166)$

$= \dfrac{500(501)}{2} - \dfrac{3(166)(167)}{2}$

$= 125\,250 - 41\,583$

$= 83\,667$

6.4.14. For this series, $S_2 = 19$ and $S_4 = 50$. Substitute into formula (6.13) for both sums.

For S_2:   For S_4:

$$S_n = \frac{n}{2}[2a + (n-1)d]$$ $$S_n = \frac{n}{2}[2a + (n-1)d]$$

$$19 = \frac{2}{2}[2a + (2-1)d]$$ $$50 = \frac{4}{2}[2a + (4-1)d]$$

$$19 = 1[2a + 1d]$$ $$50 = 2[2a + 3d]$$

$$19 = 2a + d.$$ $$25 = 2a + 3d.$$

Solve the system of two equations:

$$19 = 2a + d$$

$$25 = 2a + 3d$$

$$-6 = -2d$$

$$d = \frac{-6}{-2} = 3.$$

Substitute $d = 3$ into one of the equations

$$19 = 2a + d$$

$$19 = 2a + 3$$

$$16 = 2a$$

$$a = \frac{16}{2} = 8.$$

With $a = 8$ and $d = 3$, the first six terms of the series are $8 + 11 + 14 + 17 + 20 + 23$. The sum of the first six terms is $S_6 = \frac{6}{6}[2(8) + (6-1)(3)] = 3[16 + 5(3)] = 93$, and the first 15 terms is $S_{15} = \frac{15}{2}[2(8) + (15-1)(3)] = 435$.

6.5.1. (a) By Theorem 6.5.1, $\displaystyle\sum_{i=1}^{9} 4 = 4(9) = 36.$

(b) By Theorem 6.5.1, $\displaystyle\sum_{j=1}^{n} 5 = 5n.$

(c) By Theorem 6.5.1, $\displaystyle\sum_{x=1}^{n-1} 6 = 6(n-1) = 6n - 6.$

(d) $\displaystyle\sum_{x=20}^{45} 6 = \sum_{x=1}^{45} 6 - \sum_{x=1}^{19} 6 = \underbrace{6(45)}_{\text{Theorem 6.5.1}} - \underbrace{6(19)}_{\text{Theorem 6.5.1}} = 270 - 114 = 156$

(e) $\displaystyle\sum_{i=1}^{17} 7i = \underbrace{7\sum_{i=1}^{17} i}_{\text{Theorem 6.5.3}} = 7\,\underbrace{\frac{17(18)}{2}}_{\text{Theorem 6.4.1}} = 1\,071$

(f) $\displaystyle\sum_{j=6}^{17} 7j = 7\sum_{j=6}^{17} j$ By Theorem 6.5.3

$$= 7\left[\sum_{j=1}^{17} j - \sum_{j=1}^{5} j\right]$$

$$= 7\left[\frac{17(18)}{2} - \frac{5(6)}{2}\right] \quad \text{By Theorem 6.4.1}$$

$$= 7(153 - 15) = 966$$

(g) $\displaystyle\sum_{j=6}^{n} 7j = 7\sum_{j=6}^{n} j$ By Theorem 6.5.3

$$= 7\left[\sum_{j=1}^{n} j - \sum_{j=1}^{5} j\right]$$

$$= 7\left[\frac{n(n+1)}{2} - \frac{5(6)}{2}\right] \quad \text{By Theorem 6.4.1}$$

$$= 7\left[\frac{n(n+1)}{2} - 15\right]$$

(h) $\displaystyle\sum_{i=1}^{15}(5i+3) = \sum_{i=1}^{15} 5i + \sum_{i=1}^{15} 3 = 5\sum_{i=1}^{15} i + 3(15) = 5\frac{15(16)}{2} + 45 = 600 + 45 = 645$

(i) $\displaystyle\sum_{k=1}^{100}(10k+3) = \sum_{k=1}^{100} 10k + \sum_{k=1}^{100} 3$

$$= 10\sum_{k=1}^{100} k + 3(100)$$

$$= 10\frac{100(101)}{2} + 300$$

$$= 50\,800$$

(j) $\displaystyle\sum_{j=1}^{30}(9j+6) = \sum_{j=1}^{30} 9j + \sum_{j=1}^{30} 6 = 9\sum_{j=1}^{30} j + 6(30) = 9\frac{30(31)}{2} + 180 = 4\,365$

(k) $\displaystyle\sum_{j=5}^{30}(9j+6) = \sum_{j=1}^{30}(9j+6) - \sum_{j=1}^{4}(9j+6) = 4365 - \left(9\sum_{j=1}^{4} j + \sum_{j=1}^{4} 6\right)$

$$= 4365 - \left(9\frac{4(5)}{2} + 6(4)\right) = 4365 - (90 + 24)$$

$$= 4\,251$$

(l) $\displaystyle\sum_{j=5}^{n}(9j+6)=\sum_{j=1}^{n}(9j+6)-\sum_{j=1}^{4}(9j+6)$

$$=9\sum_{j=1}^{n}j+\sum_{j=1}^{n}6-\left(9\sum_{j=1}^{4}j+\sum_{j=1}^{4}6\right)$$

$$=9\frac{n(n+1)}{2}+6n-\left(9\frac{4(5)}{2}+6(4)\right)$$

$$=\frac{9n(n+1)}{2}+6n-114$$

6.5.2. (a)

$$\sum_{i=1}^{n}i=91$$

$$\frac{n(n+1)}{2}=91$$

$$n(n+1)=182$$

$$n^2+n-182=0$$

$$(n-13)(n+14)=0$$

Thus, $n=13$ or $n=-14$. Since $n\in\mathbb{W}$, then $n=13$.

(b)

$$\sum_{i=1}^{n}(4i-1)=1830$$

$$\sum_{i=1}^{n}4i-\sum_{i=1}^{n}1=1830$$

$$4\sum_{i=1}^{n}i-n=1830$$

$$\frac{4n(n+1)}{2}-n=1830$$

$$2n(n+1)-n=1830$$

$$2n^2+2n-n=1830$$

$$2n^2+n-1830=0$$

$$n=30,-30.5$$

Since $n\in\mathbb{W}$, then $n=30$.

6.5.3. (a) $\displaystyle\sum_{i=1}^{n}(x_i-\bar{x})=\sum_{i=1}^{n}x_i-n\bar{x}$

$$=\sum_{i=1}^{n}x_i-n\frac{1}{n}\sum_{i=1}^{n}x_i$$

$$=\sum_{i=1}^{n}x_i-\sum_{i=1}^{n}x_i$$

$$=0$$

(b) $\displaystyle\sum_{i=1}^{n}(x_i-\bar{x})^2=\sum_{i=1}^{n}(x_i-\bar{x})(x_i-\bar{x})$

$$=\sum_{i=1}^{n}(x_i^2-2x_i\bar{x}+\bar{x}^2)$$

$$=\sum_{i=1}^{n}x_i^2-2\bar{x}\sum_{i=1}^{n}x_i+\sum_{i=1}^{n}\bar{x}^2$$

Notice that $\sum_{i=1}^{n} x_i = n\frac{1}{n} \sum_{i=1}^{n} x_i = n\bar{x}$. Then,

$$\sum_{i=1}^{n} (x_i - \bar{x})^2 = \sum_{i=1}^{n} x_i^2 - 2\bar{x}n\bar{x} + n\bar{x}^2$$

$$= \sum_{i=1}^{n} x_i^2 - 2n\bar{x}^2 + n\bar{x}^2$$

$$= \sum_{i=1}^{n} x_i^2 - \bar{x}^2(2n - n)$$

$$= \sum_{i=1}^{n} x_i^2 - n\bar{x}^2.$$

6.5.4. On the left side, we have $\sum_{i=1}^{4} i^2 = 1^2 + 2^2 + 3^2 + 4^2 = 30$ and on the right side, we have

$$\left(\sum_{i=1}^{4} i\right)^2 = \left(\frac{4(5)}{2}\right)^2 = 10^2 = 100. \text{ Since } 30 \neq 100, \sum_{i=1}^{4} i^2 \neq \left(\sum_{i=1}^{4} i\right)^2.$$

6.5.5. On the left side, we have $\sum_{i=1}^{4} = i(i+1) = 1(2) + 2(3) + 3(4) + 4(5) = 40$. On the right side,

we have $\left(\sum_{i=1}^{4} i\right)\left(\sum_{i=1}^{4} (i+1)\right) = (1 + 2 + 3 + 4)(2 + 3 + 4 + 5) = 10(14) = 140$. Since

$40 \neq 140$, then $\sum_{i=1}^{4} i(i+1) \neq \left(\sum_{i=1}^{4} i\right)\left(\sum_{i=1}^{4} (i+1)\right)$.

6.5.6. $\sum_{i=1}^{n} [af(i) + bg(i)] = \underbrace{\sum_{i=1}^{n} af(i) + \sum_{i=1}^{n} bg(i)}_{\text{Theorem 6.5.5}} = \underbrace{a \sum_{i=1}^{n} f(i)}_{\text{Theorem 6.5.3}} + \underbrace{b \sum_{i=1}^{n} g(i)}_{\text{Theorem 6.5.3}}$

6.5.7. $2^1 \times 2^2 \times 2^3 \times \cdots \times 2^n = 2^{1+2+3+\cdots+n} = 2^{\frac{n(n+1)}{2}} = 2^{\frac{1}{2}(n)(n+1)} = (\sqrt{2})^{n(n+1)}$

6.5.8.

$$\sum_{i=0}^{49} (2i + 1) = 1 + \sum_{i=1}^{49} (2i + 1)$$

$$= 1 + \sum_{i=1}^{49} 2i + \sum_{i=1}^{49} 1$$

$$= 1 + 2 \sum_{i=1}^{49} i + 49$$

$$= 1 + 2\frac{49 \times 50}{2} + 49$$

$$= 1 + 49 \times 50 + 49$$
$$= 2500$$

6.5.9. $\displaystyle\sum_{i=1}^{50} 2i = 2\sum_{i=1}^{50} i = \cancel{2}\frac{50(51)}{\cancel{2}} = 50(51) = 2550$

6.5.10. $\displaystyle\sum_{i=1}^{20} 5i = 5\sum_{i=1}^{20} i = 5\left[\frac{20(21)}{2}\right] = 1050$

6.5.11. (a) Shift *up* the index by 1, and shift *down* the term by 1; that is,

$$\sum_{i=1}^{n} \frac{3i-1}{i+1} = \sum_{i=2}^{n+1} \frac{3(i-1)-1}{(i-1)+1} = \sum_{i=2}^{n+1} \frac{3i-3-1}{i-1+1} = \sum_{i=2}^{n+1} \frac{3i-4}{i}.$$

(b) Shift *down* the index by 3, and shift *up* the term by 3; that is,

$$\sum_{j=6}^{17} \frac{j^2}{2j+1} = \sum_{j=6-3}^{17-3} \frac{(j+3)^2}{2(j+3)+1} = \sum_{j=3}^{14} \frac{(j+3)^2}{2j+6+1} = \sum_{j=3}^{14} \frac{(j+3)^2}{2j+7} = \sum_{k=3}^{14} \frac{(k+3)^2}{2k+7}.$$

6.5.12. The change of base formula is

$$\log_i 1000! = \frac{\log 1000!}{\log i} \quad\Leftrightarrow\quad \frac{1}{\log_i 1000!} = \frac{\log i}{\log 1000!}.$$

Then,

$$\sum_{i=2}^{1000} \frac{1}{\log_i 1000!} = \sum_{i=2}^{1000} \frac{\log i}{\log 1000!}$$

$$= \frac{1}{\log 1000!} \sum_{i=2}^{1000} \log i$$

$$= \frac{1}{\log 1000!} (\log 2 + \log 3 + \cdots + \log 9999 + \log 1000) \qquad \text{Law of Multiplication, Table B.2}$$

$$= \frac{1}{\log 1000!} \log(2 \times 3 \times \cdots \times 9999 \times 1000)$$

$$= \frac{1}{\log 1000!} \log(1 \times 2 \times 3 \times \cdots \times 9999 \times 1000)$$

$$= \frac{1}{\cancel{\log 1000!}} \cancel{\log 1000!}$$

$$= 1$$

6.6.1. $n! = \begin{cases} \displaystyle\prod_{i=1}^{n} i & \text{if } n \geq 1 \\ 1 & \text{if } n = 0 \end{cases}$

6.6.2. $P(n,r) = \begin{cases} \displaystyle\prod_{i=0}^{r-1} (n-i) & \text{if } r \geq 0 \\ 1 & \text{if } r = 0 \end{cases}$

6.6.3. (a) $2 \times 2 \times 2 = 2^3 = 8$ (c) $2(3) \times 2(4) \times 2(5) \times 2(6) = 5760$

      (b) $\frac{1}{2} \times \frac{1}{3} \times \frac{1}{4} = \frac{1}{24}$ (d) $\binom{4}{2} \times \binom{4}{3} \times \binom{4}{4} = 24$

6.6.4. (a) $\displaystyle\prod_{i=1}^{n} c = \underbrace{c \times c \times \cdots \times c}_{n \text{ times}} = c^n$

      (b) $\displaystyle\prod_{i=1}^{n} ct_i = ct_1 \times ct_2 \times \cdots \times ct_n$

$$= \underbrace{c \times c \times c \times \cdots \times c}_{n \text{ times}} \times t_1 \times t_2 \times \cdots \times t_n$$

$$= c^n (t_1 \times t_2 \times \cdots \times t_n)$$

$$= c^n \prod_{i=1}^{n} t_i$$

      (c) $\displaystyle\prod_{i=1}^{n} (a_i \times b_i) = (a_1 \times b_1) \times (a_2 \times b_2) \times \cdots \times (a_n \times b_n)$

$$= (a_1 \times a_2 \times \cdots \times a_n) \times (b_1 \times b_2 \times \cdots \times b_n)$$

$$= \prod_{i=1}^{n} a_i \times \prod_{i=1}^{n} b_i$$

6.7.1. (a) $a = 12, r = 3, n = 10.$ $S_{10} = \frac{12(1-3^{10})}{1-3} = 354\,288$

      (b) $a = 7, r = \frac{1}{4}, n = 8.$ $S_8 = \frac{7(1-(\frac{1}{4})^8)}{1-\frac{1}{4}} = \frac{152\,915}{16\,384} \approx 9.333\,19$

      (c) $a = \frac{1}{5}, r = -3, n = 12.$ $S_{12} = \frac{\frac{1}{5}(1-(-3)^{12})}{1-(-3)} = -26\,572$

      (d) $a = 3, r = 1, n = 6.$ $S_6 = 3(6) = 18$

6.7.2. (a) $a = 27, r = \frac{1}{3}, t_n = \frac{1}{243}.$

$$t_n = \frac{1}{243} = ar^{n-1}$$

$$\frac{1}{243} = 27\left(\frac{1}{3}\right)^{n-1}$$

$$\frac{1}{243 \cdot 27} = \left(\frac{1}{3}\right)^{n-1}$$

$$\log\left(\left(\frac{1}{3}\right)^{n-1}\right) = \log\left(\frac{1}{6561}\right)$$

$$(n-1)\log\left(\frac{1}{3}\right) = \log\left(\frac{1}{6561}\right)$$

$$n = \frac{\log\left(\frac{1}{6561}\right)}{\log\left(\frac{1}{3}\right)} + 1 = 9$$

Then, $S_9 = \frac{27(1-(\frac{1}{3})^9)}{1-(\frac{1}{3})} = \frac{9841}{243} \approx 40.497\,94.$

(b) $a = \frac{1}{3}, r = \frac{2}{3}, t_n = \frac{128}{6561}, n = ?$

$$t_n = \frac{128}{6561} = ar^{n-1}$$

$$\frac{128}{6561} = \left(\frac{1}{3}\right)\left(\frac{2}{3}\right)^{n-1}$$

$$\frac{128}{2187} = \left(\frac{2}{3}\right)^{n-1}$$

$$\log\left(\frac{128}{2187}\right) = \log\left(\left(\frac{2}{3}\right)^{n-1}\right)$$

$$\log\left(\frac{128}{2187}\right) = (n-1)\log\left(\frac{2}{3}\right)$$

$$n - 1 = \frac{\log\left(\dfrac{128}{2187}\right)}{\log\left(\dfrac{2}{3}\right)}$$

$$n = 7 + 1 = 8$$

Then, $S_8 = \frac{\frac{1}{3}[1-(\frac{2}{3})^8]}{1-\frac{2}{3}} = \frac{6305}{6561} \approx 0.960\,98.$

6.7.3. (a) $S_{12} = \displaystyle\sum_{i=1}^{12} 4^i = \sum_{i=1}^{12} \underbrace{4}_{a}\underbrace{(4}_{r})^{i-1} = \frac{4(1-4^{12})}{1-4} = 22\,369\,620$

(b) $S_{10} = \displaystyle\sum_{i=1}^{10} 3^{-i} = \sum_{i=1}^{10}\left(\frac{1}{3}\right)^i = \sum_{i=1}^{10} \underbrace{\left(\frac{1}{3}\right)}_{a}\underbrace{\left(\frac{1}{3}\right)}_{r}^{i-1} = \frac{\frac{1}{3}[1-(\frac{1}{3})^{10}]}{1-\frac{1}{3}} = \frac{29\,524}{59\,049} \approx 0.5$

(c) $\displaystyle\sum_{i=10}^{20} 3^i = \sum_{i=10}^{20} \underbrace{3}_{a}\underbrace{(3}_{r})^{i-1}$

$= S_{20} - S_9$

$= \dfrac{3[1-(3)^{20}]}{1-3} - \dfrac{3[1-(3)^9]}{1-3}$

$= 5\,230\,176\,600 - 29\,523$

$= 5\,230\,147\,077$

6.7.4. The required sum is $2 + 2^2 + 2^3 + \cdots + 2^{12}$. If we set $a = 1, r = 2$, and $n = 13$, in the geometric series formula, then $1 + (1)2 + (1)2^2 + (1)2^3 + \cdots + (1)2^{12}$, and so

$$1 + 2 + 2^2 + 2^3 + \cdots + 2^{12} = \sum_{i=1}^{13} 2^{i-1}$$

$$2 + 2^2 + 2^3 + \cdots + 2^{12} = \sum_{i=1}^{13} 2^{i-1} - 1$$

$$2 + 2^2 + 2^3 + \cdots + 2^{12} = \frac{2^{13} - 1}{2-1} - 1 = 2^{13} - 1 - 1 = 2^{13} - 2 = 8192 - 2 = 8190.$$

6.7.5. $a = 3, r = 3, S_n = 3279, n = ?$

$$S_n = 3279 = \frac{3(1 - 3^n)}{1 - 3}$$

$$3279(-2) = 3 - 3(3^n)$$

$$-3 - 6558 = -3(3^n)$$

$$\frac{-6561}{-3} = 3^n$$

$$3^n = 2187$$

$$n = \log_3(2187) = \frac{\log(2187)}{\log(3)} = 7$$

6.7.6.

$$\sum_{i=1}^{10} t_i = 244 \sum_{i=1}^{5} t_i$$

$$\frac{\cancel{a}(1 - r^{10})}{\cancel{1-r}} = 244 \frac{\cancel{a}(1 - r^5)}{\cancel{1-r}}$$

$$1 - r^{10} = 244(1 - r^5)$$

$$1 - r^{10} = 244 - 244r^5$$

$$244r^5 - r^{10} = 243$$

$$r^5(244 - r^5) = 243$$

Then, $r^5 = 243 \Leftrightarrow r = \sqrt[5]{243} = 3.$

6.7.7. (a) $1(4^0) + 1(4^1) + 1(4^2) + \cdots = \displaystyle\sum_{i=1}^{n} 4^{i-1}$

(b) $S_{10} = \displaystyle\sum_{i=1}^{10} 4^{i-1} = \frac{1 - 4^{10}}{1 - 4} = 349\,525.$ Therefore, after 10 levels of this infection, there are 349 525 infected phones.

6.7.8. (a) $\displaystyle\sum_{i=1}^{500}(3^i - 3^{i-1}) = (\cancel{3} - 1) + (3^2 - \cancel{3}) + (\cancel{3^3} - \cancel{3^2}) + \cdots + (\cancel{3^{499}} - \cancel{3^{498}})$

$$+ (3^{500} - \cancel{3^{499}})$$

$$= -1 + 3^{500}$$

$$\approx 3.636 \times 10^{238}$$

(b) $\displaystyle\sum_{i=1}^{100}\left(\frac{1}{i} - \frac{1}{i+1}\right) = \left(\frac{1}{1} - \frac{1}{2}\right) + \left(\frac{1}{2} - \frac{1}{3}\right) + \left(\frac{1}{3} - \frac{1}{4}\right) + \cdots + \left(\frac{1}{100} - \frac{1}{101}\right)$

$$= 1 + \left(-\frac{1}{2} + \frac{1}{2}\right) + \left(-\frac{1}{3} + \frac{1}{3}\right) + \left(-\frac{1}{4} + \frac{1}{4}\right) + \cdots$$

$$+ \left(-\frac{1}{100} + \frac{1}{100}\right) - \frac{1}{101}$$

$$= 1 - \frac{1}{101}$$

$$= \frac{100}{101}$$

(c) $\displaystyle\sum_{i=1}^{n}[(i+1)^3 - i^3] = (2^3 - 1^3) + (3^3 - 2^3) + (4^3 - 3^3) + \cdots + [(n+1)^3 - n^3]$

$$= (n+1)^3 - 1^3$$
$$= n^3 + 3n^2 + 3n$$

6.7.9. $\displaystyle\sum_{i=1}^{n}\frac{1}{i(i+1)} = \sum_{i=1}^{n}\left(\frac{1}{i} - \frac{1}{i+1}\right)$

$$= \left(1 - \frac{1}{2}\right) + \left(\frac{1}{2} - \frac{1}{3}\right) + \cdots + \left(\frac{1}{n} - \frac{1}{n+1}\right)$$

$$= 1 + \left(-\frac{1}{2} + \frac{1}{2}\right) + \left(-\frac{1}{3} + \frac{1}{3}\right) + \cdots + \left(-\frac{1}{n} + \frac{1}{n+1}\right) - \frac{1}{n+1}$$

$$= 1 - \frac{1}{n+1}$$

6.7.10. $\displaystyle\sum_{i=1}^{n}\log\left(\frac{i+1}{i}\right) = \sum_{i=1}^{n}[\log(i+1) - \log i]$

$$= (\log 2 - \log 1) + (\log 3 - \log 2) + (\log 4 - \log 3) + \cdots$$
$$\qquad + (\log(n+1) - \log n)$$
$$= (-\log 1 + \log 2) + (-\log 2 + \log 3) + (-\log 3 + \log 4) + \cdots$$
$$\qquad + (-\log n + \log(n+1))$$
$$= -\log 1 + (\log 2 - \log 2) + (\log 3 - \log 3) + \cdots + (\log n - \log n)$$
$$\qquad + \log(n+1)$$
$$= \log(n+1) - \log 1$$
$$= \log\left(\frac{n+1}{1}\right)$$
$$= \log(n+1)$$

6.7.11. Let $S_n = \displaystyle\sum_{i=1}^{n} i\,r^{i-1}$. We write the series S_n explicitly, multiply each term by the common ratio r, and subtract:

Hence,

$$S_n - rS_n = 1\frac{(1-r^n)}{1-r} - r^n$$

$$S_n(1-r) = \frac{(1-r^n)}{1-r} - r^n$$

$$S_n = \frac{\frac{(1-r^n)}{1-r} - r^n}{1-r}$$

$$= \frac{(1-r^n)}{(1-r)^2} - \frac{r^n}{1-r}.$$

6.8.1. (a) $P = \$100, i = 0.014, n = 3$
(b) $P = \$200, i = 0.02, n = 6$
(c) $P = \$500, i = 0.18, n = 10$
(d) $P = \$1000, i = 0.0325, n = 20$
(e) $P = \$15\,000, i = 0.03, n = 9$
(f) $P = \$20\,000, i = 0.0425, n = 14$

6.8.2. (a) $P = \$100, i = 0.014, n = 3$.

$$A = 100(1 + 0.014)^3 = 100(1.014)^3 = \$104.26 \text{ to the nearest cent}$$

(b) $P = \$200, i = 0.02, n = 5$.

$$A = 200(1 + 0.02)^5 = 200(1.02)^5 = \$220.82 \text{ to the nearest cent}$$

(c) $P = \$500, i = \frac{0.02}{2} = 0.01, n = 5 \times 2 = 10$.

$$A = 500(1 + 0.01)^{10} = 500(1.01)^{10} = \$552.31 \text{ to the nearest cent}$$

(d) $P = \$1000, i = \frac{0.0325}{4} = 0.008\,13, n = 10 \times 4 = 40$

$$A = 1000(1 + 0.008\,13)^{40} = 1000(1.008\,13)^{40} = \$1382.49 \text{ to the nearest cent}$$

(e) $P = \$1500, i = \frac{0.12}{12} = 0.01, n = 7 \times 12 = 84$.

$$A = 1500(1 + 0.01)^{84} = 1500(1.01)^{84} = \$3460.08 \text{ to the nearest cent}$$

(f) $P = \$20\,000, i = \frac{0.0425}{12} = 0.003\,54, n = 5 \times 12 = 60$.

$$A = 20\,000(1 + 0.003\,54)^{60} = 20\,000(1.003\,54)^{60} = \$24\,723.57 \text{ to the nearest cent}$$

6.8.3. $P = \$200, i = \frac{0.04}{12} = \frac{4}{1200} = \frac{1}{300}, n = 40 \times 12 = 480$.

$$A = 200\left(1 + \frac{1}{300}\right)^{480} = \$987.97 \text{ to the nearest cent}$$

6.8.4. (a) $P = \$500, i = 0.05, n = 5. \ A = 500(1 + 0.05)^5 = \$638.14 \text{ to the nearest cent.}$
(b) $P = \$500, i = \frac{0.045}{12} = 0.003\,75, n = 5 \times 12 = 60$.

$$P = 500(1 + 0.003\,75)^{60} = \$625.90 \text{ to the nearest cent}$$

Therefore, investment (a) is a better investment than investment (b).

6.8.5. Present value $= \dfrac{1500}{(1.03)^5} \times \dfrac{[(1.03)^5 - 1]}{1.03 - 1} = \$6869.56 \text{ to the nearest cent}$

6.8.6. $i = \frac{0.0325}{4} = 0.008\,125, n = 4 \times 4 = 16$.

$$50\,000 = \frac{a[(1.008\,125)^{16} - 1]}{1.008\,125 - 1}$$

$$50\,000(1.008\,125 - 1) = a[(1.008\,125)^{16} - 1]$$

$$a = \frac{50\,000(1.008\,125 - 1)}{(1.008\,125)^{16} - 1}$$

$$= \$2938.94 \text{ to the nearest cent}$$

6.8.7. $i = \frac{0.025}{2} = 0.0125, n = 6$. He needs $\$500 - \$75 = \$425$.

$$425 = \frac{a[(1.0125)^6 - 1]}{1.0125 - 1}$$

$$a = \frac{425(1.0125 - 1)}{(1.0125)^6 - 1}$$

$$= \$68.65 \text{ to the nearest cent}$$

6.8.8. $198\,402(1 + 0.0101)^{n} = 200\,000$

$$198\,402(1.0101)^{n} = 200\,000$$

$$(1.0101)^{n} = \frac{200\,000}{198\,402}$$

$$\log\left((1.0101)^{n}\right) = \log\left(\frac{200\,000}{198\,402}\right)$$

$$n\log(1.0101) = \log\left(\frac{200\,000}{198\,402}\right)$$

$$n = \frac{\log\left(\dfrac{200\,000}{198\,402}\right)}{\log(1.0101)}$$

$$n \approx 0.7983 \text{ years}$$

Then to figure out how many months we multiply by 12: $0.7983 \times 12 = 9.5796$. Therefore, the city will grow to $200\,000$ in approximately 9.6 months.

6.8.9. (a) 2020: $6.884(1.01)^{10} = 7.604$ billion. 2030: $6.884(1.01)^{20} = 8.400$ billion.

(b) 2020: $6.884(1.02)^{10} = 8.392$ billion. 2030: $6.884(1.02)^{20} = 10.229$ billion.

(c)

Rate of Growth	2010	2020	2030	2040	2050	2060	2070
1%	6.884	7.604	8.400	9.279	10.249	11.322	12.506
2%	6.884	8.392	10.229	12.469	15.200	18.529	22.587

(d) The graph is given in Fig. C.16.

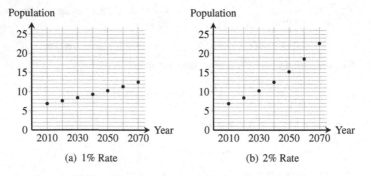

(a) 1% Rate   (b) 2% Rate

Fig. C.16   The predicted world population growth

6.8.10. $6.490(1 + i)^{5} = 6.884$

$$(1 + i)^{5} = \frac{6.884}{6.490}$$

$$(1 + i)^{5} = 1.060\,71$$

$$1 + i = (1.06071)^{\frac{1}{5}}$$

$$1 + i = 1.01186$$

$$i = 0.011\,86$$

Therefore, there was a 1.186% increase per year from 2005 to 2010.

6.8.11. $1(0.9)^5 - (0.9)^5 = 0.59049$ or 59% of original story remains.

6.8.12. See the solution given in Bussey and Hartwell (1914).

6.9.1. Yes, since

$$\sum_{i=0}^{\infty} ar^i = \underbrace{a}_{i=0} + \underbrace{ar}_{i=1} + \underbrace{ar^2}_{i=2} + \cdots = \frac{a}{1-r}$$

and

$$\sum_{i=1}^{\infty} ar^{i-1} = \underbrace{a}_{i=1} + \underbrace{ar}_{i=2} + \underbrace{ar^2}_{i=3} + \cdots = \frac{a}{1-r}$$

are equivalent.

6.9.2. (a) $\displaystyle\sum_{i=1}^{\infty} 2\left(\frac{1}{3}\right)^{i-1} = \frac{2}{1-\frac{1}{3}} = \frac{2}{\frac{2}{3}} = 2 \times \frac{3}{2} = 3$

(b) $\displaystyle\sum_{i=1}^{\infty} 2(3)^{-i} = \sum_{i=1}^{\infty} 2\left(\frac{1}{3}\right)^i$

$$= \sum_{i=1}^{\infty} 2\left(\frac{1}{3}\right)\left(\frac{1}{3}\right)^{i-1}$$

$$= \frac{1}{3}\sum_{i=1}^{\infty} 2\left(\frac{1}{3}\right)^{i-1}$$

$$= \frac{1}{3}\left(\frac{2}{1-\frac{1}{3}}\right)$$

$$= \frac{1}{3}\left(\frac{2}{\frac{2}{3}}\right)$$

$$= \frac{1}{3}(3)$$

$$= 1$$

(c) $\displaystyle\sum_{i=1}^{\infty} \left(-\frac{1}{2}\right)^{i-1} = \frac{1}{1-\left(-\frac{1}{2}\right)} = \frac{1}{1+\frac{1}{2}} = \frac{1}{\frac{3}{2}} = \frac{2}{3}$

(d) $\displaystyle\sum_{i=1}^{\infty} 4(3)^{-i-1} = \sum_{i=1}^{\infty} 4(3)^{-(i+1)}$

$$= \sum_{i=1}^{\infty} 4\left(\frac{1}{3}\right)^{i+1}$$

$$= \sum_{i=1}^{\infty} 4\left(\frac{1}{3}\right)^{2}\left(\frac{1}{3}\right)^{i-1}$$

$$= \left(\frac{1}{3}\right)^{2} \sum_{i=1}^{\infty} 4\left(\frac{1}{3}\right)^{i-1}$$

$$= \frac{1}{3}\left[\frac{4}{1-\frac{1}{3}}\right]$$

$$= \frac{1}{3}\left[\frac{4}{\frac{2}{3}}\right]$$

$$= \frac{1}{3}\left[4 \times \frac{3}{2}\right]$$

$$= \frac{1}{3}(6)$$

$$= 2$$

(e) $a = 7$ and $r = \frac{3.5}{7} = \frac{1}{2}$. Thus, $7 + 3.5 + 1.75 + \cdots = \frac{7}{1-\frac{1}{2}} = \frac{7}{\frac{1}{2}} = 7 \times 2 = 14$.

(f) $a = 5, r = \frac{1}{2}$. Thus, $5 + 2.5 + 1.25 + 0.625 + \cdots = \frac{5}{1-\frac{1}{2}} = \frac{5}{\frac{1}{2}} = 10$

(g) $a = \frac{1}{4}, r = \frac{1}{4}$. Thus, $\frac{1}{4} + \frac{1}{16} + \frac{1}{64} + \frac{1}{256} + \cdots = \frac{\frac{1}{4}}{1-\frac{1}{4}} = \frac{\frac{1}{4}}{\frac{3}{4}} = \frac{1}{3}$.

(h) $a = \frac{1}{2}, r = -\frac{1}{2}$. Thus, $\frac{1}{2} - \frac{1}{4} + \frac{1}{8} - \frac{1}{16} + \cdots = \frac{\frac{1}{2}}{1-(-\frac{1}{2})} = \frac{1}{3}$.

Remark: This is one of the first infinite series that was calculated in mathematics; this was done by Archimedes.

(i) $a = \frac{1}{3}, r = \frac{3}{4}$. Thus, $\frac{1}{3} + \frac{1}{4} + \frac{3}{16} + \cdots = \frac{\frac{1}{3}}{1-\frac{3}{4}} = \frac{\frac{1}{3}}{\frac{1}{4}} = \frac{1}{3} \times 4 = \frac{4}{3}$.

(j) $a = \frac{1}{5}, r = \frac{5}{6}$. Thus, $\frac{1}{5} + \frac{1}{6} + \frac{5}{36} + \frac{25}{216} + \cdots = \frac{\frac{1}{5}}{1-\frac{5}{6}} = \frac{\frac{1}{5}}{\frac{1}{6}} = 1$.

6.9.3. $a = 55\,000$ and $r = 0.17$. Then, $\frac{55\,000}{1-0.17} = \frac{55\,000}{0.83} = 66\,265.06$ to the nearest cent. Therefore, Uncle Vladimir should give you an extra $66\,265.06 - 55\,000 = 11\,265.06$ to cover the taxes.

6.9.4. (a) $\underbrace{\left(\frac{1}{3} + \frac{1}{9} + \frac{1}{27} + \cdots\right)}_{a=\frac{1}{3},\ r=\frac{1}{3}} + \underbrace{\left(\frac{1}{4} + \frac{1}{16} + \frac{1}{64} + \cdots\right)}_{a=\frac{1}{4},\ r=\frac{1}{4}} = \frac{\frac{1}{3}}{1-\frac{1}{3}} + \frac{\frac{1}{4}}{1-\frac{1}{4}}$

$$= \frac{\frac{1}{3}}{\frac{2}{3}} + \frac{\frac{1}{4}}{\frac{3}{4}}$$

$$= \frac{1}{2} + \frac{1}{3}$$

$$= \frac{5}{6}$$

(b) Rearrange so that $\left(\dfrac{1}{3} + \dfrac{1}{9} + \dfrac{1}{27} + \cdots\right) + \left(\dfrac{1}{4} + \dfrac{1}{16} + \dfrac{1}{64} + \cdots\right)$, and you get the same answer as the previous part.

6.9.5. (a) $q = 0.19 \mid 0.0019 \mid \cdots = \underbrace{\dfrac{19}{10^2} + \dfrac{19}{10^4} + \dfrac{19}{10^6} + \cdots}_{a=\frac{19}{10^2},\; r=\frac{1}{10^2}} = \dfrac{\frac{19}{10^2}}{1 - \frac{1}{10^2}} = \dfrac{\frac{19}{100}}{\frac{99}{100}} = \dfrac{19}{99}$

(b) $q = 0.19 + 0.005\,3 + 0.000\,053 + \cdots$

$= \dfrac{19}{10^2} + \underbrace{\dfrac{53}{10^4} + \dfrac{53}{10^6} + \cdots}_{a=\frac{53}{10^4},\; r=\frac{1}{10^2}}$

$= \dfrac{19}{10^2} + \dfrac{\frac{53}{10^4}}{1 - \frac{1}{10^2}}$

$= \dfrac{19}{10^2} + \dfrac{\frac{53}{10^4}}{\frac{10^2-1}{10^2}}$

$= \dfrac{19}{10^2} + \dfrac{\frac{53}{10^4}}{\frac{99}{10^2}}$

$= \dfrac{19}{10^2} + \dfrac{53}{10^4} \times \dfrac{10^2}{99}$

$= \dfrac{19}{10^2} + \dfrac{53}{99 \times 10^2}$

$= \dfrac{19}{100} + \dfrac{53}{9900}$

$= \dfrac{782}{4003}$

6.9.6. $q = 0.\overline{9} = 0.9 + 0.09 + 0.009 + \cdots = \underbrace{\dfrac{9}{10} + \dfrac{9}{10^2} + \dfrac{9}{10^3} + \cdots}_{a=\frac{9}{10},\; r=\frac{1}{10}} = \dfrac{\frac{9}{10}}{1 - \frac{1}{10}} = \dfrac{\frac{9}{10}}{\frac{9}{10}} = 1$

6.9.7.

$$\sum_{i=1}^{\infty}(1+x)^{-i} = \sum_{i=1}^{\infty}\left(\dfrac{1}{1+x}\right)^{i}$$

$$= \sum_{i=1}^{\infty}\left(\dfrac{1}{1+x}\right)\left(\dfrac{1}{1+x}\right)^{i-1}$$

$$= \dfrac{1}{1+x}\sum_{i=1}^{\infty}\left(\dfrac{1}{1+x}\right)^{i-1}$$

$$= \dfrac{1}{1+x}\left(\dfrac{1}{1 - \frac{1}{1+x}}\right)$$

$$= \dfrac{1}{1+x}\left(\dfrac{1}{\frac{1+x-1}{1+x}}\right)$$

$$= \frac{1}{1+x}\left(\frac{1+x}{x}\right)$$

$$= \frac{1}{x}$$

6.9.8. The infinite geometric series is $\underbrace{\dfrac{1}{4} \times \dfrac{1}{16} \times \dfrac{1}{64} \times \cdots}_{a=\frac{1}{4},\; r=\frac{1}{4}} = \frac{\frac{1}{4}}{1-\frac{1}{4}} = \frac{\frac{1}{4}}{\frac{3}{4}} = \frac{1}{3}$. Then,

$$\frac{1}{4} \times \frac{1}{16} \times \frac{1}{64} \times \cdots = \frac{1}{3}$$

$$\frac{1}{4}\log 2 \times \frac{1}{16}\log 2 \times \frac{1}{64}\log 2 \times \cdots = \frac{1}{3}\log 2$$

$$\log 2^{\frac{1}{4}} \times \log 2^{\frac{1}{16}} \times \log 2^{\frac{1}{64}} \times \cdots = \log 2^{\frac{1}{3}}$$

$$2^{\frac{1}{4}} \times 2^{\frac{1}{16}} \times 2^{\frac{1}{64}} \times \cdots = 2^{\frac{1}{3}}.$$

6.9.9. The total distance, both up and down, the basketball traveled is

$$\underbrace{10}_{\text{down}} + \underbrace{10(0.3)}_{\text{up}} + \underbrace{10(0.3)}_{\text{down}} + \underbrace{10(0.3)^2}_{\text{up}} + \underbrace{10(0.3)^2}_{\text{down}} + \cdots = 10 + 2[10(0.3)] + 2[10(0.3)^2] + \cdots$$

$$= 10 + 20(0.3) + 20(0.3)^2 + 20(0.3)^3 + \cdots = 10 + \frac{20(0.3)}{1-0.3} = 10 + \frac{6}{0.7} \approx 18.571 \text{ feet.}$$

6.9.11. (a) Using Theorem 6.9.2, $a = 1$ and $r = -1$ (and ignoring the restriction on r), the sum is $\frac{1}{1-\frac{1}{2}} = \frac{1}{2}$.

(b) $\underbrace{(1-1)}_{0} + \underbrace{(1-1)}_{0} + \underbrace{(1-1)}_{0} + \cdots = 0$

(c) Theorem 6.9.2 is applied incorrectly in part (a). The theorem requires $-1 < r < 1$. In this case, $r = 1$.

6.9.12. It is a geometric series with $a = \frac{1}{2}$ and $r = \frac{1}{2}$. Thus, $\frac{1}{2} + \frac{1}{4} + \frac{1}{8} + \frac{1}{16} + \cdots = \frac{\frac{1}{2}}{1-\frac{1}{2}} = \frac{\frac{1}{2}}{\frac{1}{2}} = 1$.

6.9.13. Using Gauss' method, we have the following:

$$
\begin{array}{rcccccccccc}
S & = & 1 & + & 2x & + & 3x^2 & + & 4x^3 & + & \cdots \\
-xS & = & & - & x & - & 2x^2 & - & 3x^3 & - & \cdots \\
\hline
S - xS & = & 1 & + & x & + & x^2 & + & x^3 & + & \cdots
\end{array}
$$

which is a geometric series (with $a = 1$). Hence,

$$S - xS = \frac{1}{1-x}$$

$$(1-x)S = \frac{1}{1-x}$$

$$S = \frac{1}{(1-x)(1-x)} = \frac{1}{(1-x)^2}$$

6.10.1. (a) $3(k+1) = 3k + 3$

(b) $3(k+1) + 1 = 3k + 3 + 1 = 3k + 4$

(c) $\frac{1}{k+1}$

(d) 3^k

6.10.2. (a) $\displaystyle\sum_{i=1}^{n+1} 2^{n-1} = 2^{n+1} - 1$

(b) $\displaystyle\sum_{i=1}^{n+1} (4n+1) = (n+1)(2(n+1)+3) \qquad (n+1)(2n+2+3) = (n+1)(2n+3)$

(c) $\displaystyle\sum_{i=1}^{n+1} (2n-1)^2 = \frac{1}{3}(n+1)(4(n+1)^2 - 1)$

6.10.3. We will prove this using the Principle of Mathematical Induction.

(a) *Base step.* For $n = 1$, the proposition is true since on the left side is 1, and for the right side is $1^2 = 1$.

(b) *Inductive step.*

(i) We *assume* that the proposition is true for $n = k$, $k \in \mathbb{N}$. That is, we assume that $1 + 3 + 5 + \cdots + (2k - 1) = k^2$ is true; this is the inductive hypothesis.

(ii) We need to *prove* that the proposition is true for $n = k + 1$. That is, we need to show that

$$1 + 3 + 5 + \cdots + (2k - 1) + (2(k - 1) + 1) = (k + 1)^2$$

is also true. Starting on the left side,

$$\underbrace{1 + 3 + 5 + \cdots + 2k - 1}_{k^2 \text{ by inductive hypothesis}} + 2k + 1 = k^2 + 2k + 1 = (k + 1)^2,$$

which is the same as the right side. Hence, we have proved that the proposition is true for $n = k + 1$, and this concludes the inductive step.

Therefore, by the Principle of Mathematical Induction, $1 + 3 + 5 + \cdots + (2n - 1) = n^2$ is true for every natural number n.

6.10.4. We will prove this using the Principle of Mathematical Induction.

(a) *Base step.* For $n = 1$, the proposition states that $4(1) - 2 = 4 - 2 = 2(1)^2$, which is true.

(b) *Inductive step.*

(i) We *assume* that the proposition is true for $n = k$, $k \in \mathbb{N}$. That is, we assume that $2 + 6 + 10 + \cdots + (4k - 2) = 2k^2$ is true; this is the inductive hypothesis.

(ii) The $k + 1$ term is $4(k + 1) - 2 = 4k + 4 - 2 = 4k + 2$. We need to *prove* that

$$2 + 6 + 10 + \cdots + (4k - 2) + 4k + 2 = 2(k + 1)^2.$$

Then,

$$\underbrace{2 + 6 + 10 + \cdots + (4k - 2)}_{2k^2 \text{ by the inductive hypothesis}} + 4k + 2 = 2k^2 + 4k + 2$$

$$= 2(k^2 + 2k + 1)$$

$$= 2(k + 1)^2$$

Hence, we have proven that the proposition is true for $n = k + 1$, and this concludes the inductive step.

Therefore, $2 + 6 + 10 + \cdots + (4n - 2) = 2n^2$ is true for all natural numbers n by the Principle of Mathematical Induction.

6.10.5. We will prove this using the Principle of Mathematical Induction.

(a) *Base step.* For $n = 1$, the proposition states that $3(1) - 2 = 1 = \frac{1(3(1)-1)}{2}$, which is true.

(b) *Inductive step.*

(i) We *assume* that the proposition is true for $n = k, k \in \mathbb{N}$. That is, we assume that

$$1 + 4 + 7 + \cdots + (3k - 2) = \frac{k(3k - 1)}{2}.$$

This is the inductive hypothesis.

(ii) The $k + 1$ term is $3(k + 1) - 2 = 3k + 3 - 2 = 3k + 1$. We need to prove that the proposition is true for $n = k + 1$. That is, we need to show that

$$1 + 4 + 7 + \cdots + (3k - 2) + (3k + 1) = \frac{(k + 1)(3k + 2)}{2}.$$

So,

$$\underbrace{1 + 4 + 7 + \cdots + (3k - 2)}_{\frac{k(3k-1)}{2}} + (3k + 1) = \frac{k(3k - 1)}{2} + (3k + 1)$$

$$= \frac{k(3k - 1)}{2} + \frac{2(3k + 1)}{2}$$

$$= \frac{k(3k - 1) + 2(3k + 1)}{2}$$

$$= \frac{3k^2 - k + 6k + 2}{2}$$

$$= \frac{3k^2 + 5k + 2}{2}$$

$$= \frac{(k + 1)(3k + 2)}{2}.$$

Hence, we have proved the proposition is true for $n = k + 1$, and this concludes the inductive step.

Therefore, $1 + 4 + 7 + \cdots + (3n - 2) = \frac{n(3n-1)}{2}$ is true for all natural numbers n by the Principle of Mathematical Induction.

6.10.6. We will prove this using the Principle of Mathematical Induction.

(a) *Base step.* For $n = 1$, the proposition states that $(1)2^1 = 2 = 2 + 0(2)^2 = 2 + (1 - 1)2^{1+1}$, which is true.

(b) *Inductive step.*

(i) We *assume* that the proposition is true for $n = k, k \in \mathbb{N}$. That is, we assume

$$1 \cdot 2 + 2 \cdot 2^2 + 3 \cdot 2^3 + \cdots + k2^k = 2 + (k - 1)2^{k+1}.$$

This is the inductive hypothesis.

(ii) We need to prove that the proposition is true for $n = k + 1$. That is, we need to show that

$$1 \cdot 2 + 2 \cdot 2^2 + 3 \cdot 2^3 + \cdots + k2^k + (k + 1)2^{k+1} = 2 + k2^{k+2}$$

So,

$$\underbrace{1 \cdot 2 + 2 \cdot 2^2 + 3 \cdot 2^3 + \cdots + k2^k}_{2+(k-1)2^{k+1} \text{ by the inductive hypothesis}} + (k + 1)2^{k+1}$$

$$= 2 + (k - 1)2^{k+1} + (k + 1)2^{k+1}$$

$$= 2 + 2^{k+1}(k - 1 + k + 1)$$

$$= 2 + 2^{k+1}(2k)$$

$$= 2 + k2^{k+2}$$

Hence, we have proved the proposition is true for $n = k + 1$, and this concludes the inductive step.

Therefore, $1 \cdot 2 + 2 \cdot 2^2 + 3 \cdot 2^3 + \cdots + n2^n = 2 + (n-1)2^{n+1}$ is true for all natural numbers n by using the Principle of Mathematical Induction.

6.10.7. We will prove this using the Principle of Mathematical Induction.

(a) *Base step.* For $n = 1$, the proposition states that $1(2(1) + 1) = 1(2 + 1) = 1(3) = 3 = \frac{18}{6} = \frac{1(2)(9)}{6} = \frac{1\big((1)+1\big)\big(4(1)+5\big)}{6}$, which is true.

(b) *Inductive step.*

    (i) We will *assume* that the proposition is true for $n = k$, $k \in \mathbb{N}$. That is, we assume

$$1 \cdot 3 + 2 \cdot 5 + 3 \cdot 7 + \cdots + k(2k + 1) = \frac{k(k + 1)(4k + 5)}{6}.$$

    (ii) The $k + 1$ term is $(k + 1)(2(k + 1) + 1) = (k + 1)(2k + 3)$. We need to *prove*

$$1 \cdot 3 + 2 \cdot 5 + 3 \cdot 7 + \cdots + k(2k + 1) + (k + 1)(2k + 3) = \frac{(k + 1)(k + 2)(4k + 9)}{6}.$$

So,

$$\underbrace{1 \cdot 3 + 2 \cdot 5 + 3 \cdot 7 + \cdots + k(2k + 1)}_{\frac{k(k+1)(4k+5)}{6}} + (k + 1)(2k + 3)$$

$$= \frac{k(k + 1)(4k + 5)}{6} + (k + 1)(2k + 3)$$

$$= \frac{k(k + 1)(4k + 5)}{6} + \frac{6(k + 1)(2k + 3)}{6}$$

$$= \frac{(k + 1)[k(4k + 5) + 6(2k + 3)]}{6}$$

$$= \frac{(k + 1)[4k^2 + 5k + 12k + 18]}{6}$$

$$= \frac{(k + 1)[4k^2 + 17k + 18]}{6}$$

$$= \frac{(k + 1)(k + 2)(4k + 9)}{6}$$

Hence, we have proven that the proposition is true for $n = k + 1$, and this concludes the inductive step.

Therefore, $1 \cdot 3 + 2 \cdot 5 + \cdots + n(2n + 1) = \frac{n(n+1)(4n+5)}{6}$ is true for all natural numbers n by using the Principle of Mathematical Induction.

6.10.8. We will prove this using the Principle of Mathematical Induction.

(a) *Base step.* For $n = 1$, the proposition states $(1 + \frac{3}{1}) = 4 = 2^2 = (1 + 1)^2$, which is true.

(b) *Inductive step.*

    (i) We *assume* that the proposition is true for $n = k$, $k \in \mathbb{N}$. That is, we assume $(1 + \frac{3}{1})(1 + \frac{5}{4}) \cdots (1 + \frac{2k+1}{k^2}) = (k + 1)^2$; this is the inductive hypothesis.

(ii) We need to *prove* that the proposition is true for $n = k + 1$. That is, we need to show that $(1 + \frac{3}{1})(1 + \frac{5}{4}) \cdots (1 + \frac{2k+1}{k^2})(1 + \frac{2k+3}{(k+1)^2}) = (k + 2)^2$. So,

$$\underbrace{\left(1 + \frac{3}{1}\right)\left(1 + \frac{5}{4}\right) \cdots \left(1 + \frac{2k+1}{k^2}\right)}_{(k+1)^2 \text{ by the inductive hypothesis}} \left(1 + \frac{2k+3}{(k+1)^2}\right)$$

$$=(k + 1)^2\left(1 + \frac{2k+3}{(k+1)^2}\right)$$

$$=(k + 1)^2 + \frac{\cancel{(k+1)^2}(2k+3)}{\cancel{(k+1)^2}}$$

$$=k^2 + 2k + 1 + 2k + 3$$

$$=k^2 + 4k + 4$$

$$=(k + 2)^2$$

and this concludes the inductive step.
Therefore $(1 + \frac{3}{1})(1 + \frac{5}{4})(1 + \frac{7}{9}) \cdots (1 + \frac{2n+1}{n^2}) = (n + 1)^2$ is true for every natural number n by the Principle of Mathematical Induction.

6.10.9. We will prove this using the Principle of Mathematical Induction.

(a) *Base step.* For $n = 1$, the proposition $\sum\limits_{i=1}^{n} i^3 = \left[\frac{n(n + 1)}{2}\right]^2$ is true, since on the left side is

$\sum\limits_{i=1}^{1} i^3 = 1^3 = 1$, and on the right side of $\left[\frac{1(2)}{2}\right]^2 = 1^2 = 1$.

(b) *Inductive step.*

(i) We *assume* that the proposition is true for $n = k$, $k \in \mathbb{N}$. That is, we assume that $\sum\limits_{i=1}^{k} i^3 = \left[\frac{k(k + 1)}{2}\right]^2$ is true; this is the inductive hypothesis.

(ii) We need to *prove* that the proposition is true for $n = k + 1$. That is, we need to show that $\sum\limits_{i=1}^{k+1} i^3 = \left[\frac{(k + 1)(k + 2)}{2}\right]^2$. Starting on the left side,

$$\sum_{i=1}^{k+1} i^3 = \sum_{i=1}^{k} i^3 + (i + 1)^3$$

$$= \left[\frac{k(k + 1)}{2}\right]^2 + (k + 1)^3 \qquad \text{By inductive hypothesis}$$

$$= \frac{k^2(k + 1)^2}{4} + \frac{4(k + 1)^3}{4}$$

$$= \frac{k^2(k + 1)^2 + 4(k + 1)^3}{4}$$

$$= \frac{(k + 1)^2[k^2 + 4(k + 1)]}{4}$$

$$= \frac{(k + 1)^2(k^2 + 4k + 4)}{4}$$

$$= \frac{(k + 1)^2(k + 2)^2}{4}$$

$$= \left[\frac{(k+1)(k+2)}{2}\right]^2$$

which is the same as the right side. Hence, we have shown that the proposition is true for $n = k + 1$ and this concludes the inductive step.

Therefore, by the Principle of Mathematical Induction, $\sum_{i=1}^{n} i^3 = \left[\frac{n(n+1)}{2}\right]^2$ is true for every natural number n.

6.10.10. We will prove this using the Principle of Mathematical Induction.

(a) *Base step.* For $n = 1$, the proposition is true since on the left side is $\sum_{i=1}^{1}(2i-1) = 1$ and $1^2 = 1$ is true.

(b) *Inductive step.*

(i) We *assume* that the proposition is true for $n = k, k \in \mathbb{N}$. That is, $\sum_{i=1}^{k}(2i-1) = k^2$; this is the inductive hypothesis.

(ii) We need to *prove* that the proposition is true for $n = k + 1$. That is, we need to prove that

$$\sum_{i=1}^{k+1}(2i-1) = (k+1)^2$$

is also true. So, on the left side

$$\sum_{i=1}^{k+1}(2i-1) = \sum_{i=1}^{k}(2i-1) + (2(k+1)-1)$$

$$= \underbrace{\sum_{i=1}^{k}(2i-1)}_{k^2} + (2k+1)$$

$$= k^2 + 2k + 1$$

$$= (k+1)^2$$

which is the same as the right side. This concludes the inductive step.

Therefore, by the Principle of Mathematical Induction, $\sum_{i=1}^{n}(2i-1) = n^2 = (k+1)^2$ is true for every natural number n.

6.10.11. We will prove this using the Principle of Mathematical Induction.

(a) *Base step.* For $n = 1$, the proposition is true since on the left side is $\sum_{i=1}^{1}(2i-1)^2 =$ $(2(1)-1)^2 = (2-1)^2 = 1^2 = 1$, and on the right side is $\frac{1[2(1)-1][2(1)+1]}{3} = \frac{1(3)}{3} = 1$.

(b) *Inductive step.*

(i) We *assume* that the proposition is true for $n = k, k \in \mathbb{N}$. That is, $\sum_{i=1}^{k}(2k-1)^2 =$ $\frac{k(2k-1)(2k+1)}{3}$ is true; this is the inductive hypothesis.

(ii) We need to *prove* that the proposition is true for $n = k + 1$. That is, we need to prove that

$$\sum_{i=1}^{k+1} (2i - 1)^2 = \frac{(k+1)[2(k+1) - 1][2(k+1) + 1]}{3}$$

$$= \frac{(k+1)(2k+1)(2k+3)}{3}$$

is also true. Then, on the left side

$$\sum_{i=1}^{k+1} (2k - 1)^2 = \sum_{i=1}^{k} (2i - 1)^2 + (2(k+1) - 1)^2$$

$$= \frac{k(2k-1)(2k+1)}{3} + (2k+1)^2 \qquad \text{By the inductive hypothesis}$$

$$= \frac{k(2k-1)(2k+1)}{3} + \frac{3(2k+1)^2}{3}$$

$$= \frac{k(2k-1)(2k+1) + 3(2k+1)^2}{3}$$

$$= \frac{(2k+1)[k(2k-1) + 3(2k+1)]}{3}$$

$$= \frac{(2k+1)(2k^2 - k + 6k + 3)}{3}$$

$$= \frac{(2k+1)(2k^2 + 5k + 3)}{3}$$

$$= \frac{(2k+1)(2k+3)(k+1)}{3}$$

$$= \frac{(k+1)(2k+1)(2k+3)}{3}$$

which is the same as the right side. This concludes the inductive step.

Therefore, by the Principle of Mathematical Induction, $\displaystyle\sum_{i=1}^{n} (2i - 1)^2 = \frac{n(2n-1)(2n+1)}{3}$ is true for every natural number n.

6.10.12. We will prove this using the Principle of Mathematical Induction.

(a) *Base step.* For $n = 1$, the proposition is true since it states on the left side $\frac{1}{1 \cdot 2 \cdot 3} = \frac{1}{6}$ and on the right side $\frac{1}{2}[\frac{1}{2} - \frac{1}{2 \cdot 3}] = \frac{1}{2}[\frac{1}{2} - \frac{1}{6}] = \frac{1}{6}$.

(b) *Inductive step.*

(i) We *assume* that the proposition is true for $n = k$, $k \in \mathbb{N}$. That is, we assume

$$\sum_{i=1}^{k} \frac{1}{i(i+1)(i+2)} = \frac{1}{2}\left[\frac{1}{2} - \frac{1}{(k+1)(k+2)}\right].$$

This is the inductive hypothesis.

(ii) We need to *prove* that the proposition is true since $n = k + 1$. That is, we need to show that

$$\sum_{i=1}^{k+1} \frac{1}{i(i+1)(i+2)} = \frac{1}{2}\left[\frac{1}{2} - \frac{1}{(k+2)(k+3)}\right].$$

So,

$$\sum_{i=1}^{k+1} \frac{1}{i(i+1)(i+2)} = \sum_{i=1}^{k} \frac{1}{i(i+1)(i+2)} + \frac{1}{(k+1)(k+2)(k+3)}$$

$$= \frac{1}{2}\left[\frac{1}{2} - \frac{1}{(k+1)(k+2)}\right] + \frac{1}{(k+1)(k+2)(k+3)}$$

$$= \frac{1}{4} - \frac{1}{2(k+1)(k+2)} + \frac{1}{(k+1)(k+2)(k+3)}$$

$$= \frac{1}{4} - \frac{k+3}{2(k+1)(k+2)(k+3)} + \frac{2}{2(k+1)(k+2)(k+3)}$$

$$= \frac{1}{4} - \frac{(k+1)}{2(k+1)(k+2)(k+3)}$$

$$= \frac{1}{2}\left[\frac{1}{2} - \frac{1}{(k+2)(k+3)}\right].$$

Hence, the proposition is true for $n = k + 1$, and this concludes the inductive step.

Therefore, $\sum_{i=1}^{n} \frac{1}{i(i+1)(i+2)} = \frac{1}{2}\left[\frac{1}{2} - \frac{1}{(n+1)(n+2)}\right]$ is true for every natural number n by the Principle of Mathematical Induction.

6.10.13. We will prove this using the Principle of Mathematical Induction.

(a) *Base step.* For $n = 1$, the proposition is true since on the left side $[2(1)]^3 = 2^3 = 8$ and on the right side is $2(1)^2(1+1)^2 = 2(2)^2 = 2 \cdot 4 = 8$.

(b) *Inductive step.*

(i) We *assume* that the proposition is true for $n = k, k \in \mathbb{N}$. That is, we assume that

$$\sum_{i=1}^{k} (2i)^3 = 2k^2(k+1)^2.$$

This is the inductive hypothesis.

(ii) We need to *prove* that the proposition is true for $n = k + 1$. That is, we need to show that

$$\sum_{i=1}^{k+1} (2i)^3 = 2(k+1)^2(k+2)^2.$$

So,

$$\sum_{i=1}^{k+1} (2i)^3 = \sum_{i=1}^{k} (2i)^3 + [2(k+1)]^3$$

$$= 2k^2(k+1)^2 + [2(k+1)]^3$$

$$= 2k^2(k+1)^2 + [2^3(k+1)^3]$$

$$= 2(k+1)^2[k^2 + 2^2(k+1)]$$

$$= 2(k+1)^2[k^2 + 4k + 4]$$

$$= 2(k+1)^2(k+2)^2.$$

Hence, we have proved the proposition is true for $n = k + 1$, and this concludes the inductive step.

Therefore, by the Principle of Mathematical Induction, $\sum\limits_{i=1}^{n}(2i)^3 = 2n^2(n+1)^2$ for every natural number n.

6.10.14. We will prove this using the Principle of Mathematical Induction.
    (a) *Base step.* For $n = 1$, the proposition states that $1 < 2^1 = 2$, which is true.
    (b) *Inductive step.*
        (i) We assume that the proposition is true for $n = k, k \in \mathbb{N}$. That is, we assume that $k < 2^k$; this is the inductive hypothesis.
        (ii) We need to prove that the proposition is true for $n = k + 1$; that is, we need to show that $k + 1 < 2^{k+1}$. So, multiplying both sides by 2 gives $2k < 2 \cdot 2^k$. Also, $2 \cdot 2^k = 2^{k+1}$ and $k + 1 \leq 2k$. Combining the previous statements, we have $k + 1 \leq 2k < 2 \cdot 2^k = 2^{k+1}$. Therefore, we have proved that $k + 1 < 2^{k+1}$, and this concludes the inductive step.

Therefore, $n < 2^n$ is true for every natural number n by the Principle of Mathematical Induction.

6.10.15. We will prove this using the Principle of Mathematical Induction.
    (a) *Base step.* For $n = 1$, the proposition is true since $1 = 1! \leq 1^1 = 1$.
    (b) *Inductive step.*
        (i) We *assume* that the proposition is true for $n = k$ for $k \in \mathbb{N}$. That is, we assume that $k! \leq k^k$ is true; this is the inductive hypothesis.
        (ii) We need to *prove* that the proposition is true for $n = k + 1$. That is, we need to show that $(k + 1)! \leq (k + 1)^{k+1}$ is also true.

$$(k + 1)! = (k + 1)k! \leq (k + 1)k^k$$
$$\leq (k + 1)k^k k$$
$$= (k + 1)k^{k+1}$$

Hence, $(k + 1)! \leq (k + 1)k^{k+1}$ is true, and this completes the inductive step.
Therefore, by the Principle of Mathematical Induction, $n! \leq n^n$ for every natural number n.

6.10.16. We will prove this using the Principle of Mathematical Induction.
    (a) *Base step.* For $n = 1$, the proposition states that $1^2 + 1$ is even, which is true since $1^2 + 1 = 2$.
    (b) *Inductive step.*
        (i) We *assume* that the proposition is true for $n = k, k \in \mathbb{N}$. That is, we assume that $k^2 + k$ is even; this is the inductive hypothesis.
        (ii) We need to *prove* that the proposition is true for $n = k + 1$. That is, we need to show that $(k + 1)^2 + (k + 1)$ is even. Then,

$$(k + 1)^2 + (k + 1) = k^2 + 2k + 1 + k + 1$$
$$= \underbrace{k^2 + k}_{\text{inductive hypothesis}} + \underbrace{2k + 2}_{\text{even}}$$

and the sum of two even numbers is even. Hence, $(k + 1)^2 + (k + 1)$ is even, and this concludes the inductive step.

Therefore, for every natural number n, $n^2 + n$ is even by the Principle of Mathematical Induction.

6.10.17. We will prove this using the Principle of Mathematical Induction.
    (a) *Base step.* For $n = 1$, the proposition is true, since $1(1 + 5) = 6$ is divisible by 2; $\frac{6}{2} = 3$.
    (b) *Inductive step.*

(i) We *assume* that the proposition is true for $n = k$, $k \in \mathbb{N}$. That is, we need to assume that $n(n + 5)$ is divisible by 2; this is the inductive hypothesis.

(ii) We need to prove that the proposition is true for $n = k + 1$. That is, we need to show that $(k + 1)\big((k + 1) + 5\big) = (k + 1)(k + 6)$ is also divisible by 2. Then,

$$(k + 1)(k + 6) = k^2 + 7k + 6$$
$$= (k^2 + 5k) + (2k + 6)$$
$$= k(k + 5) + 2(k + 3).$$

By the inductive hypothesis, $k(k + 5)$ is divisible by 2; that is $k(k + 5) = 2\ell$ for some $\ell \in \mathbb{N}$. Continuing, we have

$$(k + 1)(k + 6) = 2\ell + 2(k + 3)$$
$$= 2(\ell + k + 3).$$

Since we were able to factor out a 2, this means that 2 is one of the factors of $(n+1)(n+2)$ and hence $(n + 1)(n + 2)$ is divisible by 2. This concludes the inductive step.

Therefore, $n(n + 5)$ is divisible by 2 for all natural numbers n by the Principle of Mathematical Induction.

5.10.18. We will prove this using the Principle of Mathematical Induction. Let t_n denote the nth term of an arithmetic sequence in which $t_n = a + (n - 1)d$.

(a) *Base step.* For $n = 1$, the proposition states that $t_1 = a + (1 - 1)d = a$, which is true since the first term is indeed a.

(b) *Inductive step.*

(i) We assume that the proposition is true for $n = k$, $k \in \mathbb{N}$. That is, we assume that $t_k = a + (k - 1)d$; this is the inductive hypothesis.

(ii) We need to *prove* that the proposition is true for $n = k + 1$. That is, we need to show that $t_{k+1} = a + (k + 1 - 1)d = a + kd$. Then,

$$t_{k+1} = t_k + d$$
$$= a + (k - 1)d + d \qquad \text{By the inductive hypothesis}$$
$$= a + kd - d + d$$
$$= a + (k + 1 - 1)d.$$

Hence, we have proved the formula is true for t_{k+1}, and this concludes the inductive step. Therefore, $t_n = a + (n - 1)d$ is true for every natural number n by the Principle of Mathematical Induction.

5.10.19. (a) We calculate S_1, S_2, S_3, and S_4.

n	$S_n = \displaystyle\sum_{i=1}^{n} \dfrac{1}{i(i + 1)}$
1	$\frac{1}{2}$
2	$\frac{1}{2} + \frac{1}{6} = \frac{2}{3}$
3	$\frac{1}{2} + \frac{1}{6} + \frac{1}{12} = \frac{3}{4}$
4	$\frac{1}{2} + \frac{1}{6} + \frac{1}{12} + \frac{1}{20} = \frac{4}{5}$
$\vdots$	$\vdots$
n	$\frac{n}{n+1}$

(b) We make a guess that the formula should be $n/(n + 1)$. So we try proving it is true using the Principle of Mathematical Induction.

(c) We *assume* that $\displaystyle\sum_{i=1}^{k} \frac{1}{i(i + 1)} = \frac{k}{k + 1}, k \in \mathbb{N}$; this is the inductive hypothesis. We will prove

that $\displaystyle\sum_{i=1}^{k+1} \frac{1}{i(i + 1)} = \frac{k + 1}{k + 2}$. So,

$$\sum_{i=1}^{k+1} \frac{1}{i(i + 1)} = \sum_{i=1}^{k} \frac{1}{i(i + 1)} + \frac{1}{(k + 1)(k + 2)}$$

$$= \frac{k}{k + 1} + \frac{1}{(k + 1)(k + 2)}$$

$$= \frac{k(k + 2)}{(k + 1)(k + 2)} + \frac{1}{(k + 1)(k + 2)}$$

$$= \frac{k^2 + 2k + 1}{(k + 1)(k + 2)}$$

$$= \frac{(k + 1)^2}{(k + 1)(k + 2)}$$

$$= \frac{k + 1}{k + 2}.$$

Hence, we have proved the proposition is true for $n = k + 1$, and this concludes the inductive step. Therefore, $\displaystyle\sum_{i=1}^{n} \frac{1}{i(i + 1)} = \frac{n}{n + 1}$ is true for every natural number n by the Principle of Mathematical Induction.

6.10.20. (a) We calculate S_1, S_2, S_3, and S_4 in the hope of finding a pattern.

n	$S_n = \displaystyle\sum_{i=1}^{n} \frac{i}{(i + 1)!}$
1	$\frac{1}{(1+1)!} = \frac{1}{2!} = \frac{1}{2}$
2	$\frac{1}{2} + \frac{2}{(2+1)!} = \frac{1}{2} + \frac{2}{3!} = \frac{1}{2} + \frac{2}{6} = \frac{5}{6}$
3	$\frac{5}{6} + \frac{3}{4!} = \frac{5}{6} + \frac{3}{24} = \frac{23}{24}$

(b) The guess for the formula is $\displaystyle\sum_{i=1}^{n} \frac{i}{(i + 1)!} = \frac{(n + 1)! - 1}{(n + 1)!}$.

(c) We *assume* that $\displaystyle\sum_{i=1}^{k} \frac{i}{(i + 1)!} = \frac{(k + 1)! - 1}{(k + 1)!}$ is true. We need to *prove* that

$$\sum_{i=1}^{k+1} \frac{i}{(i + 1)!} = \frac{(k + 2)! - 1}{(k + 2)!}$$

Then,

$$\sum_{i=1}^{k} \frac{i}{(i+1)!} + \frac{k+1}{(k+2)!} = \frac{(k+1)!-1}{(k+1)!} + \frac{k+1}{(k+2)!}$$

$$= \frac{(k+1)!-1}{(k+1)!} + \frac{k+1}{(k+2)(k+1)!}$$

$$= \frac{[(k+1)!-1](k+2)}{(k+2)(k+1)!} + \frac{k+1}{(k+2)(k+1)!}$$

$$= \frac{[(k+1)!-1](k+2)}{(k+2)!} + \frac{k+1}{(k+2)!}$$

$$= \frac{[(k+1)!-1](k+2)+(k+1)}{(k+2)!}$$

$$= \frac{(k+2)(k+1)!-(k+2)+(k+1)}{(k+2)!}$$

$$= \frac{(k+2)!-\cancel{k}-2+\cancel{k}+1}{(k+2)!}$$

$$= \frac{(k+2)!-1}{(k+2)!}.$$

Therefore $\displaystyle\sum_{i=1}^{n} \frac{i}{(i+1)!} = \frac{(n+1)!-1}{(n+1)!}$ is true for every natural number n by the Principle of Mathematical Induction.

5.10.22. We will prove this using the Principle of Mathematical Induction.
Base step. For $n = 1$,

$$t_1 = \frac{1}{\sqrt{5}}\left[\left(\frac{1+\sqrt{5}}{2}\right)-\left(\frac{1-\sqrt{5}}{2}\right)\right] = 1.$$

For $n = 2$,

$$t_2 = \frac{1}{\sqrt{5}}\left[\left(\frac{1+\sqrt{5}}{2}\right)^2-\left(\frac{1-\sqrt{5}}{2}\right)^2\right] = 1.$$

Inductive step. Let $a = \frac{1+\sqrt{5}}{2}$ and $b = \frac{1-\sqrt{5}}{2}$. We assume that $t_n = \frac{1}{\sqrt{5}}(a^n - b^n)$ is the nth Fibonacci number, and we need to show that if we assume this, then this implies that $t_{n+1} = \frac{1}{\sqrt{5}}(a^{n+1} - b^{n+1})$ is the $(n+1)^{\text{st}}$ Fibonacci number. In our proof, we require two identities: $a^2 = a+1$ and $b^2 = b+1$. To see that $a^2 = a+1$, $a^2 = \left(\frac{1+\sqrt{5}}{2}\right)^2 = \frac{(1+\sqrt{5})^2}{2^2} = \frac{(1+\sqrt{5})(1+\sqrt{5})}{4} = \frac{1+2\sqrt{5}+5}{4} = \frac{2+2\sqrt{5}+4}{4} = \frac{2+2\sqrt{5}}{4} + \frac{4}{4} = \frac{1+\sqrt{5}}{2} + 1$. The second identity is proved in a similar fashion.

Starting with the left side,

$$t_{n+1} = t_{n-1} + t_n$$

$$= \frac{1}{\sqrt{5}}(a^{n-1} - b^{n-1}) + \frac{1}{\sqrt{5}}(a^n - b^n) \qquad \text{By the inductive hypothesis.}$$

$$= \frac{1}{\sqrt{5}}[a^{n-1} - b^{n-1} + a^n - b^n]$$

$$= \frac{1}{\sqrt{5}}[a^{n-1}(a+1) - b^{n-1}(b+1)]$$

$$= \frac{1}{\sqrt{5}}[a^{n-1}a^2 - b^{n-1}b^2]$$

$$= \frac{1}{\sqrt{5}}[a^{n-1+2} - b^{n-1+2}]$$

$$= \frac{1}{\sqrt{5}}[a^{n+1} - b^{n+1}]$$

which is the same as the right side. Hence, we have shown that Binet's formula is true for t_{n+1}, and this concludes the inductive step.

Therefore, by the Principle of Mathematical Induction, Binet's formula is true for every natural number n.

6.10.23. (a) $S_1 = 1! = 1 = 3^{1-1}$ is true. $S_2 = 1! + 2! = 3 = 3^{2-1}$ is true. $S_3 = 1! + 2! + 3! = 9 = 3^{3-1}$ is true.

(b) No. $S_4 = 1! + 2! + 3! + 4! = 33 \neq 27 = 3^{4-1}$.

6.10.24. The flaw lays in the imprecise definition of a heap, and not in the mathematics. This same kind of thinking could be applied to another conundrum: all men are not bald. If you start with a man who has a full head of hair, and you remove one hair, is he bald? If you assume that after removing k hairs, then does removing $k + 1$ hairs make him bald? The issue lays in the imprecise definition of a bald man.

6.11.1. (a) $\sum\limits_{i=0}^{7}(7 + 3i)$ (c) $\sum\limits_{i=1}^{10} i \cdot i!$ (e) $\sum\limits_{i=0}^{9} \frac{1}{(3i+4)(3i+7)}$

(b) $\sum\limits_{i=-1}^{5} \frac{1}{2^i}$ (d) $\sum\limits_{i=1}^{15} i(i+1)$ (f) $\sum\limits_{i=0}^{20} x^i$

6.11.2. $\sum\limits_{i=4}^{10} \frac{i}{i+1} = \frac{4}{5} + \frac{5}{6} + \frac{6}{7} + \frac{7}{8} + \frac{8}{9} + \frac{9}{10} + \frac{10}{11}$

6.11.3. It asking to find $100 + 102 + 104 + \cdots + 200$; this is an arithmetic series with $a = 100$, $d = 2$, $n = 51$. Then, $S_{51} = \frac{51}{2}(100 + 200) = 7650$.

6.11.4. (a) $\frac{150(151)}{2} = 11\,325$

(b) $\frac{150(151)(2(150)+1)}{6} = 1\,136\,275$

(c) $a = \frac{1}{2}, r = \frac{1}{2}, n = 10$. Then, $S_{10} = \frac{\frac{1}{2}(1-(\frac{1}{2})^{10})}{1-\frac{1}{2}} = 1 - \frac{1}{1024} = \frac{1023}{1024}$

(d) $a = 100, d = 105, n = 11$. Then, $S_{11} = \frac{11}{2}(100 + 150) = 1375$.

(e) $\displaystyle\sum_{i=1}^{150} i(i+1) = \sum_{i=1}^{150}(i^2+i)$

$$= \sum_{i=1}^{150} i^2 + \sum_{i=1}^{150} i$$

$$= \frac{150(151)(2\cdot 150+1)}{6} + \frac{150(151)}{2}$$

$$= 1\,136\,275 + 11\,325$$

$$= 1\,147\,600$$

(f) $\displaystyle\left(1-\frac{1}{2}\right)+\left(\frac{1}{2}-\frac{1}{3}\right)+\left(\frac{1}{3}-\frac{1}{4}\right)+\cdots+\left(\frac{1}{149}-\frac{1}{150}\right)$

$$=1+\left(-\frac{1}{2}+\frac{1}{2}\right)+\left(-\frac{1}{3}+\frac{1}{3}\right)+\cdots+\left(-\frac{1}{149}+\frac{1}{149}\right)-\frac{1}{150}$$

$$=1-\frac{1}{150}$$

$$=\frac{149}{150}$$

6.11.5. $\displaystyle\sum_{i=1}^{60}(17a_i + 8b_i) = 17\sum_{i=1}^{60} a_i + 8\sum_{i=1}^{60} b_i = 17(-4) + 8(7) = -68 + 56 = -12$

6.11.6. $\displaystyle\prod_{i=5}^{12} \frac{1}{i+1} = \frac{1}{5+1} \times \frac{1}{6+1} \times \frac{1}{7+1} \times \cdots \times \frac{1}{12+1} = \frac{1}{6} \times \frac{1}{7} \times \frac{1}{8} \times \cdots \times \frac{1}{13}$

6.11.8. $\displaystyle\sum_{i=1}^{10} i2^{i-1} = (1)2^0 + (2)2^1 + (3)2^2 + \cdots + (10)2^9 = 9217$

6.11.9. We are given that $a = 4$, $t_n = 25$, $n = 8$, and that it is an arithmetic series. Using the formula $S_n = \frac{n}{2}(a + t_n)$ we have $S_8 = \frac{8}{2}(4 + 25) = 4(29) = 116$.

6.11.10. We are given that $d = 7$, $a = 49$, and $t_n = 119$. We need to find the value of n. To find n, since it is an arithmetic series, the general term is $t_n = a + (n-1)d$. Then,

$$t_n = 119 = a + (n-1)d$$
$$119 = 49 + (n-1)7$$
$$119 = 49 + 7n - 7$$
$$7n = 119 - 49 + 7$$
$$n = \frac{77}{7} = 11$$

Therefore, $S_{11} = \frac{11}{2}(49 + 119) = 924$.

6.11.11. $\displaystyle\sum_{i=1}^{\infty} 3^{1-i} = \sum_{i=1}^{\infty} 3 \cdot 3^{-i}$

$$= 3 \sum_{i=1}^{\infty} 3^{-i}$$

$$= 3 \sum_{i=1}^{\infty} \left(\frac{1}{3}\right)^i$$

$$= 3 \sum_{i=1}^{\infty} \frac{1}{3}\left(\frac{1}{3}\right)^{i-1}$$

$$= \cancel{3}\,\frac{\frac{1}{\cancel{3}}}{1 - \frac{1}{3}}$$

$$= \frac{1}{1 - \frac{1}{3}} = \frac{1}{\frac{2}{3}} = \frac{3}{2}$$

6.11.12.

$$\sum_{i=1}^{n} \left(\frac{1}{i} - \frac{1}{i+2}\right) = \left(1 - \frac{\cancel{1}}{\cancel{3}}\right) + \left(\frac{1}{2} - \frac{\cancel{1}}{\cancel{4}}\right) + \left(\frac{\cancel{1}}{\cancel{3}} - \frac{\cancel{1}}{\cancel{5}}\right) + \left(\frac{\cancel{1}}{\cancel{4}} - \frac{\cancel{1}}{\cancel{6}}\right) + \cdots + \left(\frac{\cancel{1}}{\cancel{n}} - \frac{1}{n+2}\right)$$

$$= 1 + \frac{1}{2} - \frac{1}{n+1} - \frac{1}{n+2}$$

$$= \frac{3}{2} - \frac{1}{n+1} - \frac{1}{n+2}$$

6.11.13. We will prove this using the Principle of Mathematical Induction.

    (a) *Base step.* For $n = 1$, the proposition is true, since on the left side we have $2^1 = 2$, and on the right side we have $(\sqrt{2})^{1(2)} = (2)^{\frac{1}{2}(2)} = 2^1 = 2$.

    (b) *Inductive step.*

        (i) We *assume* that the proposition is true for $n = k$, $k \in \mathbb{N}$. That is, we assume that $2^1 \times 2^2 \times 2^3 \times \cdots \times 2^k = (\sqrt{2})^{k(k+1)}$ is true.

        (ii) We need to *prove* that

$$2^1 \times 2^2 \times 2^3 \times \cdots \times 2^k \times 2^{k+1} = (\sqrt{2})^{(k+1)(k+2)}$$

is also true.

$$\underbrace{2^1 \times 2^2 \times 2^3 \times \cdots \times 2^k}_{(\sqrt{2})^{k(k+1)}} \times 2^{k+1} = (\sqrt{2})^{k(k+1)} \times 2^{k+1}$$

$$= (\sqrt{2})^{k(k+1)} \times (\sqrt{2})^{2(k+1)}$$

$$= (\sqrt{2})^{k(k+1)+2(k+1)}$$

$$= (\sqrt{2})^{(k+1)(k+2)}$$

Hence, we have shown that $2^1 \times 2^2 \times 2^3 \times \cdots \times 2^k = (\sqrt{2})^{(k+1)(k+2)}$ is also true, and this concludes the inductive step.

Therefore, $2^1 \times 2^2 \times 2^3 \times \cdots \times 2^n = (\sqrt{2})^{n(n+1)}$ is true for every natural number n by the Principle of Mathematical Induction.

6.11.14. We will prove this using the Principle of Mathematical Induction.

(a) *Base step.* For $n = 1$, we have on the left side a, and on the right side is $\frac{a(1-r^1)}{1-r} = \frac{a(1-r)}{1-r} = a$. Thus, it is true for $n = 1$.

(b) *Inductive step.*

(i) We *assume* that the proposition is true for $n = k$, $k \in \mathbb{N}$. That is, we assume that
$$a + ar + ar^2 + \cdots + ar^{k-1} = \frac{a(1-r^k)}{1-r}$$ is true; this is the inductive hypothesis.

(ii) We need to *prove* that the proposition is true for $n = k + 1$. That is, we need to show that
$$a + ar + ar^2 + \cdots + ar^{k-1} + ar^k = \frac{a(1-r^{k+1})}{1-r}$$ is also true. So,

$$\underbrace{a + ar + ar^2 + \cdots + ar^{k-1}}_{\frac{a(1-r^k)}{1-r} \text{ by the inductive hypothesis}} + ar^k = \frac{a(1-r^k)}{1-r} + ar^k$$

$$= \frac{a(1-r^k)}{1-r} + \frac{(1-r)ar^k}{1-r}$$

$$= \frac{a(1-r^k) + (1-r)ar^k}{1-r}$$

$$= \frac{a - ar^k + ar^k - ar^{k+1}}{1-r}$$

$$= \frac{a - ar^{k+1}}{1-r}$$

$$= \frac{a(1-r^{k+1})}{1-r}.$$

Hence, we have shown that it is also true, and this concludes the inductive step.

Therefore, $a + ar + ar^2 + \cdots + ar^{k-1} = \frac{a(1-r^k)}{1-r}$ is true for every natural number n by the Principle of Mathematical Induction.

7.1.1. (a) $\binom{10}{8}$ (b) $\binom{20}{2}$ (c) $\binom{25}{25}$ (d) $\binom{26}{0}$

7.1.3. (a)

$$\boxed{210}\ \boxed{252}\ \ 210\ \ \boxed{120}$$
$$\boxed{462}\ \boxed{462}\ \ 330$$
$$924\ \ \ \ 792$$
$$\boxed{1716}$$

(b)

$$\boxed{3003}\ \ 5005\ \ \boxed{6435}\ \ \boxed{6435}$$
$$8008\ \ \boxed{11\,440}\ \ \boxed{12\,870}$$
$$19\,448\ \ 24\,310$$
$$\boxed{43\,758}$$

7.1.4. (a) $\binom{8}{3} + \binom{8}{4} = \binom{9}{4}$

(b) $\binom{17}{6} + \binom{17}{7} = \binom{18}{7}$

(c) $\binom{9}{5} + \binom{9}{4} = \binom{10}{5}$

(d) $\binom{15}{11} + \binom{15}{5} = \binom{15}{15-11} + \binom{15}{5} = \binom{15}{4} + \binom{15}{5} = \binom{16}{5}$

(e) $\binom{n-1}{r-1} + \binom{n-1}{r} = \binom{n}{r}$

(f) $\binom{n-1}{r-1} + \binom{n-1}{n-r-1} = \binom{n-1}{r-1} + \binom{n-1}{r} = \binom{n}{r}$

7.1.5. (a)

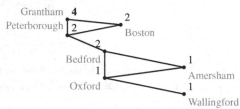

(b)

7.1.6. (a) There are 4 routes from Wallingford to Grantham.

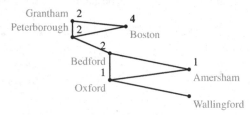

(b) There are 4 routes from Amersham to Boston.

Grantham 2
Peterborough $\boxed{2}$ 4
Boston
2
Bedford
1   1
Amersham
Oxford
Wallingford

(c) Yes.

7.1.7. From Fig. C.17, there are $20 + 35 + 34 + 14 = 103$ ways for the checker to reach the opposite side of the checkerboard.

7.1.8. By the symmetry property of Pascal's triangle, $\binom{9}{0} = \binom{9}{9}$, $\binom{9}{2} = \binom{9}{7}$, and $\binom{9}{4} = \binom{9}{5}$. Therefore, $\binom{9}{0} + \binom{9}{2} + \binom{9}{4} = \binom{9}{5} + \binom{9}{7} + \binom{9}{9}$.

7.1.9. (a) $\binom{n}{0} = \frac{n!}{0!(n-0)!} = \frac{n!}{n!} = 1 = \frac{n!}{n!0!} = \frac{n!}{n!(n-n)!} = \binom{n}{n}$

      (b) $\binom{r}{0} = \frac{r!}{0!(r-0)!} = \frac{r!}{r!} = 1 = \frac{(r+1)!}{(r+1)!} = \frac{(r+1)!}{0!(r+1-0)!} = \binom{r+1}{0}$

      (c) $\binom{r}{r} = \frac{r!}{r!(r-r)!} = \frac{r!}{r!0!} = \frac{r!}{r!} = 1 = \frac{(r+1)!}{(r+1)!} = \frac{(r+1)!}{(r+1)!(r+1-(r+1))!} = \binom{r+1}{r+1}$

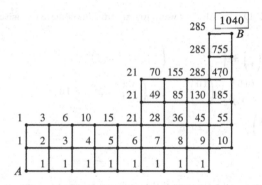

Fig. C.17   Number of paths of a checker on a checkerboard

7.1.10. There are 1040 paths from point A to point B. See Fig. C.18.

									285	**1040**
										B
									285	755
					21	70	155	285	470	
					21	49	85	130	185	
1	3	6	10	15	21	28	36	45	55	
1	2	3	4	5	6	7	8	9	10	
A	1	1	1	1	1	1	1	1		

Fig. C.18   Number of paths from point A to point B

7.1.11. There are 574 paths that Dr Algebrix can take from his home to the university. Refer to Fig. C.19.

7.1.12. (a) The table is given below.

n		Sum
0	1	1
1	$1 + 1$	2
2	$1 + 2 + 1$	4
3	$1 + 3 + 3 + 1$	8
4	$1 + 4 + 6 + 4 + 1$	16
5	$1 + 5 + 10 + 10 + 5 + 1$	32
6	$1 + 6 + 15 + 20 + 15 + 6 + 1$	64
7	$1 + 7 + 21 + 35 + 35 + 21 + 7 + 1$	128

(b) $2^8 = 256 = 1 + 8 + 28 + 56 + 70 + 56 + 28 + 8 + 1$

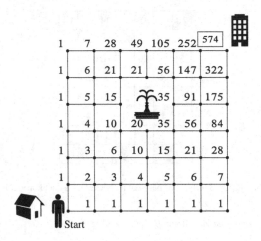

Fig. C.19   Dr. Algebrix wants to walk from his home to the university

(c)   (i) $\dbinom{9}{0} + \dbinom{9}{1} + \dbinom{9}{2} + \cdots + \dbinom{9}{9} = 2^9 = 512$

(ii) $\dbinom{10}{0} + \dbinom{10}{1} + \dbinom{10}{2} + \cdots + \dbinom{10}{10} = 2^{10} = 1024$

(iii) $\displaystyle\sum_{i=0}^{11} \dbinom{11}{i} = 2^{11} = 2048$

(d) $\displaystyle\sum_{i=0}^{n} \dbinom{n}{i} = 2^n$

(e)   The combinatorial proof is the same as the combinatorial proof of Corollary 7.2.3 on p. 296.

(f)   We will prove this using the Principle of Mathematical Induction. *Base step.* For $n = 0$, $\dbinom{0}{0} = \frac{0!}{0!0!} = 2^0$. *Inductive step.*

(i)   We *assume* that the proposition is true for $n = k$, $k \in \mathbb{W}$. That is,

$$\binom{k}{0} + \binom{k}{1} + \cdots + \binom{k}{k} = \sum_{i=0}^{k} \binom{k}{i} = 2^k.$$

This is the inductive hypothesis.

(ii)   We need to *prove* that the proposition is true for $n = k + 1$. That is, we need to prove that

$$\binom{k+1}{0} + \binom{k+1}{1} + \cdots + \binom{k+1}{k+1} = \sum_{i=0}^{k+1} \binom{k+1}{i} = 2^{k+1}.$$

We require Pascal's Theorem and two identities: $\binom{k}{0} = \binom{k+1}{0}$ and $\binom{k}{k} = \binom{k+1}{k+1}$ (see Exercise 7.1.9). On the left side,

$$\underbrace{\binom{k+1}{0} + \binom{k+1}{1}}_{\binom{k}{0}+\binom{k}{1}} + \underbrace{\binom{k+1}{2}}_{\binom{k}{1}+\binom{k}{2}} + \cdots + \underbrace{\binom{k+1}{k}}_{\binom{k}{k-1}+\binom{k}{k}} + \binom{k+1}{k+1}$$

$$= \binom{k+1}{0} + \binom{k}{0} + \binom{k}{1} + \binom{k}{1} + \binom{k}{2} + \cdots + \binom{k}{k-1} + \binom{k}{k} + \binom{k+1}{k+1}$$

$$\underbrace{\qquad}_{\binom{k}{0}} \qquad\qquad\qquad\qquad\qquad\qquad\qquad\qquad\qquad \underbrace{\qquad}_{\binom{k}{k}}$$

$$= \binom{k}{0} + \binom{k}{0} + \binom{k}{1} + \binom{k}{1} + \binom{k}{2} + \cdots + \binom{k}{k-1} + \binom{k}{k} + \binom{k}{k}$$

$$= 2 \sum_{i=0}^{k} \binom{k}{i}$$

$$= 2 \times 2^k$$

$$= 2^{k+1}$$

which is the same as the right side. This concludes the inductive step.

Therefore, by the Principle of Mathematical Induction it is true for every whole number n.

7.1.13. $32 = 2^5$. Therefore, $n = 5$.

7.1.14. $256 = 2^8$. Therefore, $n = 8$.

7.1.15. (a) The completed chart is given below.

n		Sum
0	1^2	1
1	$1^2 + 1^2$	2
2	$1^2 + 2^2 + 1^2$	6
3	$1^2 + 3^2 + 3^2 + 1^2$	20
4	$1^2 + 4^2 + 6^2 + 4^2 + 1^2$	70
5	$1^2 + 5^2 + 10^2 + 10^2 + 5^2 + 1^2$	252
6	$1^2 + 6^2 + 15^2 + 20^2 + 15^2 + 6^2 + 1^2$	924
7	$1^2 + 7^2 + 21^2 + 35^2 + 35^2 + 21^2 + 7^2 + 1^2$	3432

(b) Find these numbers in Pascal's triangle: 1, 2, 6, 20, 70, ...

(c) $\binom{2n}{n}$. For $n = 8$, $1^2 + 8^2 + 28^2 + 56^2 + 70^2 + 56^2 + 28^2 + 8^2 + 1 = 12\,870 = \binom{16}{8}$.

(d) (i) $\binom{9}{0}^2 + \binom{9}{1}^2 + \binom{9}{2}^2 + \cdots + \binom{9}{9}^2 = \binom{18}{9} = 48\,620$

   (ii) $\binom{10}{0}^2 + \binom{10}{1}^2 + \binom{10}{2}^2 + \cdots + \binom{10}{10}^2 = \binom{20}{10} = 184\,756$

(e) $\binom{n}{0}^2 + \binom{n}{1}^2 + \binom{n}{2}^2 + \cdots + \binom{n}{n}^2 = \binom{2n}{n}$

(f) Please refer to the solution of Exercise 5.7.30 (b).

7.1.16. $n = 10$

7.1.17. (a) The completed chart is given below.

n		Sum
0	1	1
1	$1 - 1$	0
2	$1 - 2 + 1$	0
3	$1 - 3 + 3 - 1$	0
4	$1 - 4 + 6 - 4 + 1$	0
5	$1 - 5 + 10 - 10 + 5 - 1$	0
6	$1 - 6 + 15 - 20 + 15 - 6 + 1$	0
7	$1 - 7 + 21 - 35 + 35 - 21 + 7 - 1$	0

(b) $1 - 8 + 28 - 56 + 70 - 56 + 28 - 8 + 1 = 0$

(c) (i) $\dbinom{9}{0} - \dbinom{9}{1} + \dbinom{9}{2} - \cdots - \dbinom{9}{9} = 0$

 (ii) $\dbinom{10}{0} - \dbinom{10}{1} + \dbinom{10}{2} - \cdots - \dbinom{10}{9} + \dbinom{10}{10} = 0$

 (iii) $\dbinom{11}{0} - \dbinom{11}{1} + \dbinom{11}{2} - \cdots - \dbinom{11}{9} + \dbinom{11}{10} - \dbinom{11}{11} = 0$

(d) $\dbinom{n}{0} - \dbinom{n}{1} + \dbinom{n}{2} - \dbinom{n}{3} + \cdots + (-1)^n \dbinom{n}{n} = 0$

7.1.18. (a) $\dbinom{16}{9} + \dbinom{16}{10} = \dbinom{17}{10} = 19\,448$

(b) $\dbinom{21}{18} + \dbinom{21}{19} = \dbinom{22}{19} = 1\,540$

(c) $\dbinom{14}{4} + \dbinom{14}{3} = \dbinom{15}{4} = 1\,365$

(d) $\displaystyle\sum_{r=11}^{12} \dbinom{17}{r} = \dbinom{17}{11} + \dbinom{17}{12} = \dbinom{18}{12} = 18\,564$

(e) $\dbinom{11}{0} + \dbinom{11}{1} + \dbinom{11}{2} + \cdots + \dbinom{11}{11} = 2^{11} = 2\,048$

(f) $\dbinom{12}{0} - \dbinom{12}{1} + \dbinom{12}{2} - \cdots - \dbinom{12}{12} = 0$

(g) $\dbinom{15}{0}^2 + \dbinom{15}{1}^2 + \dbinom{15}{2}^2 + \cdots + \dbinom{15}{15}^2 = \dbinom{30}{15} = 155\,117\,520$

7.1.19. (a) 15

(b) They are along the third diagonal of Pascal's triangle. See Fig. C.20.

(c) $\dbinom{n+1}{2}$

(a) Highlighting the triangular numbers

(b) Highlighting the triangular numbers as binomial coefficients

Fig. C.20   Triangular numbers in Pascal's triangle

(d) The formula is $\binom{1+1}{2} + \binom{2+1}{2} + \binom{3+1}{2} + \cdots + \binom{n+1}{2} = \binom{n+2}{3}$, or using

sigma notation, $\displaystyle\sum_{i=1}^{n} \binom{i+1}{2} = \binom{n+2}{3}$.

Fig. C.21   The sum of the triangular numbers follows a hockey-stick pattern

From Fig. C.21, we see that the sum of the first four triangular numbers is $1 + 3 + 6 + 10 = 20$. Notice that the highlighted shape in Fig. C.21 looks like a hockey-stick. Using our formula we get the same answer: $\binom{4+2}{3} = \binom{6}{3} = \dfrac{6!}{3!3!} = \dfrac{6 \times 5 \times 4 \times \cancel{3!}}{\cancel{3!} \times 3!} = \dfrac{\cancel{6} \times 5 \times 4}{\cancel{6}} = 20$.

(e) *Inductive step.* We assume that $\displaystyle\sum_{i=1}^{k} \binom{i+1}{2} = \binom{k+2}{3}$ is true for $n = k$. Now, we need to

show that it is true for $n = k + 1$; that is, $\displaystyle\sum_{i=1}^{k+1} \binom{i+1}{2} = \binom{k+3}{3}$. Then,

$$\sum_{i=1}^{k+1} \binom{i+1}{2} = \underbrace{\sum_{i=1}^{k} \binom{i+1}{2}}_{\binom{k+2}{3}} + \binom{k+2}{2}$$

$$= \binom{k+2}{3} + \binom{k+2}{2}$$

$$= \binom{k+2}{2} + \binom{k+2}{3} \qquad \text{By commutativity}$$

$$= \binom{k+2}{3} \qquad \text{By Pascal's Theorem 7.1.3}$$

as required.

7.1.20. (a) *Algebraic proof.*

$$\binom{n+2}{r+2} = \binom{n+1+1}{r+2}$$

$$= \binom{n+1}{r+1} + \binom{n+1}{r+2} \qquad \text{Apply Pascal's Theorem}$$

$$= \binom{n}{r} + \binom{n}{r+1} + \binom{n}{r+1} + \binom{n}{r+2} \qquad \begin{array}{l}\text{Apply Pascal's Theorem}\\\text{to each term.}\end{array}$$

$$= \binom{n}{r} + 2\binom{n}{r+1} + \binom{n}{r+2}$$

(b) *Combinatorial proof.* Suppose we have to select a committee with $(r+2)$ members from a class of $(n+2)$ students. In how many ways can we form a committee?

   (i) (Left side of the identity.) By the definition of the binomial coefficient, the number of ways to select $(r+2)$ members from $(n+2)$ students is $\binom{n+2}{r+2}$.

   (ii) (Right side of the identity.) *Take cases.* There are $2^2 = 4$ cases to consider.

      i. *Case 1*: Diane and Shelly are both not on the committee. Since they are not available to be selected, then we must chose $(r+2)$ members from amongst the remaining n students; the number of ways to do this is $\binom{n}{r+2}$.

      ii. *Case 2*: Diane is on the committee but Shelly is not. Then, we want to select $(r+1)$ other members from amongst the remaining n students; the number of ways to do this is $\binom{n}{r+1}$.

      iii. *Case 3*: Shelly is on the committee but Diane is not. This is the same as Case 2 but with the names exchanged. Therefore, the number of ways to do this is $\binom{n}{r+1}$.

      iv. *Case 4*: Diane and Shelly are both on the committee. Then, we need to select r other members from amongst the remaining n students; the number of ways to do this is $\binom{n}{r}$.

By the Sum Principle, we can add the four cases together to get

$$\binom{n}{r} + \binom{n}{r+1} + \binom{n}{r+1} + \binom{n}{r+2} = \binom{n}{r} + 2\binom{n}{r+1} + \binom{n}{r+2}$$

Since both parts (i) and (ii) are equivalent ways of thinking about the situation, we have shown that

$$\binom{n+2}{r+2} = \binom{n}{r} + 2\binom{n}{r+1} + \binom{n}{r+2}$$

as required.

7.1.21. (a) *Combinatorial proof.* Suppose we have to select a committee from n people. One of the members of the committee must be a chairperson. In how many ways can we form a committee?

   (i) (Left side of the identity). *Take cases.* Suppose we select i persons from n people. Then, there are $\binom{n}{i}$ ways to do this. From these i persons, we need to select a chairperson; there are i ways to do this. Therefore, by the Multiplication Principle, there are $i\binom{n}{i}$ ways to select a committee with i persons and a chairperson. By the Sum Principle, we can add these i cases together to get

$$1\binom{n}{1} + 2\binom{n}{2} + \cdots + n\binom{n}{n} = \sum_{i=0}^{n} i\binom{n}{i},$$

(ii) (Right side of the identity). Suppose we have to select a chairperson, then there are n choices. The rest of the $(n-1)$ persons can either be in or out of the committee; there are 2^{n-1} ways to do this. By the Multiplication Principle, there are $n2^{n-1}$ ways to form the committee.

Since parts (i) and (ii) are equivalent ways of thinking about the problem, we have shown that the identity is true.

(b) *Proof by mathematical induction. Base step*: For $n = 1$, $1\binom{1}{1} = 1 = 1(2^0)$. *Inductive step.*

(i) We *assume* that the proposition is true for $n = k$, $k \in \mathbb{N}$. That is,

$$\sum_{i=1}^{k} i\binom{k}{i} = k(2^{k-1}).$$

(ii) We need to *prove* that the proposition is true for $n = k + 1$. That is, we need to prove that

$$\sum_{i=1}^{k+1} i\binom{i+1}{i} = (k+1)2^k.$$

Then,

$$\sum_{i=1}^{k+1} i\binom{k+1}{i} = \sum_{i=1}^{k+1} (k+1)\binom{k}{i-1} \qquad \text{By the given hint}$$

$$= (k+1)\sum_{i=1}^{k+1} \binom{k}{i-1}$$

$$= (k+1)\underbrace{\sum_{i=0}^{k} \binom{k}{i}}_{2^k} \qquad \text{Shift index}$$

$$= (k+1)2^k \qquad \text{By Exercise 7.1.12}$$

This concludes the inductive step.

Therefore, by the Principle of Mathematical Induction, $\sum_{i=1}^{n} i\binom{n}{i} = n(2^{n-1})$ is true for every natural number n.

Remark: To see why the hint is true, see Exercise 5.7.4 (c).

7.2.1. (a) $(x+y)^5$ (b) $(x^2 + \frac{1}{x})^4$ (c) $(-3+2)^6$

7.2.2. The exponents in each case must sum to 12.

(a) $5 + k = 12$. Therefore, $k = 12 - 5 = 7$.

(b) $k + 9 = 12$. Therefore, $k = 12 - 9 = 3$.

(c) $k + 2k = 12$. Hence, $3k = 12$, and therefore $k = \frac{12}{3} = 4$.

(d) $k + 2 + 4k = 12$. Hence, $5k = 10$, and therefore $k = \frac{10}{5} = 2$.

7.2.3. (a) $(\frac{1}{4} + \frac{3}{4})^6 = (1)^6 = 1$

(b) $(0.6 + 0.4)^7 = 1^7 = 1$

(c) $(0.8 + 3)^5 = (3.8)^5 = 792.351\,68$

7.2.4. (a) $(x + y)^6 = \binom{6}{0}x^6 + \binom{6}{1}x^5y + \binom{6}{2}x^4y^2 + \binom{6}{3}x^3y^3 + \binom{6}{4}x^2y^4$

$$+ \binom{6}{5}xy^5 + \binom{6}{6}y^6$$

$$= x^6 + 6x^5y + 15x^4y^2 + 20x^3y^3 + 15x^2y^4 + 6xy^5 + y^6$$

(b) $(x + y)^7 = \binom{7}{0}x^7 + \binom{7}{1}x^6y + \binom{7}{2}x^5y^2 + \binom{7}{3}x^4y^3 + \binom{7}{4}x^3y^4$

$$+ \binom{7}{5}x^2y^5 + \binom{7}{6}xy^6 + \binom{7}{7}y^7$$

$$= x^7 + 7x^6y + 21x^5y^2 + 35x^4y^3 + 35x^3y^4 + 21x^2y^5$$
$$+ 7xy^6 + y^7$$

(c) $(x + y)^8 = \binom{8}{0}x^8 + \binom{8}{1}x^7y + \binom{8}{2}x^6y^2 + \binom{8}{3}x^5y^3 + \binom{8}{4}x^4y^4$

$$+ \binom{8}{5}x^3y^5 + \binom{8}{6}x^2y^6 + \binom{8}{7}xy^7 + \binom{8}{8}y^8$$

$$= x^8 + 8x^7y + 28x^6y^2 + 56x^5y^3 + 70x^4y^4 + 56x^3y^5$$
$$+ 28x^2y^6 + 8xy^7 + y^8$$

(d) $(2x + y)^3 = \binom{3}{0}(2x)^3 + \binom{3}{1}(2x)^2y + \binom{3}{2}(2x)y^2 + \binom{3}{3}y^3$

$$= (1)2^3x^3 + (3)2^2x^2y + (3)2xy^2 + (1)y^3$$

$$= 8x^3 + 12x^2y + 6xy^2 + y^3$$

(e) $(x - 2y)^4 = \binom{4}{0}x^4 + \binom{4}{1}x^3(-2y) + \binom{4}{2}x^2(-2y)^2 + \binom{4}{3}x(-2y)^3$

$$+ \binom{4}{4}(-2y)^4$$

$$= \binom{4}{0}x^4 + \binom{4}{1}x^3(-2y) + \binom{4}{2}x^2(-2)^2y^2 + \binom{4}{3}x(-2)^3y^3$$

$$+ \binom{4}{4}(-2)^4y^4$$

$$= (1)x^4 + (4)x^3(-2y) + (6)x^2(-2)^2y^2 + (4)x(-2)^3y^3$$
$$+ (1)(-2)^4y^4$$

$$= x^4 - 8x^3y + 24x^2y^2 - 32xy^3 + 16y^4$$

(f) $(1-x)^6 = \binom{6}{0}(1)^6 + \binom{6}{1}(1)^5(-x) + \binom{6}{2}(1)^4(-x)^2 + \binom{6}{3}(1)^3(-x)^3$

$\qquad + \binom{6}{4}(1)^2(-x)^4 + \binom{6}{5}(1)(-x)^5 + \binom{6}{6}(-x)^6$

$\qquad = (1)(1)^6 + (6)(1)^5(-x) + (15)(1)^4(-x)^2 + (20)(1)^3(-x)^3$

$\qquad\qquad + (15)(1)^2(-x)^4 + (6)(1)(-x)^5 + (1)(-x)^6$

$\qquad = 1 - 6x + 15x^2 - 20x^3 + 15x^4 - 6x^5 + x^6$

7.2.5. (a) $\left(1+\dfrac{1}{x}\right)^4 = \binom{4}{0}(1)^4 + \binom{4}{1}(1)^3\left(\dfrac{1}{x}\right) + \binom{4}{2}(1)^2\left(\dfrac{1}{x}\right)^2 + \binom{4}{3}(1)\left(\dfrac{1}{x}\right)^3$

$\qquad\qquad + \binom{4}{4}\left(\dfrac{1}{x}\right)^4$

$\qquad\quad = (1)(1)^4 + (4)(1)^3\left(\dfrac{1}{x}\right) + (6)(1)^2\left(\dfrac{1}{x}\right)^2 + (4)(1)\left(\dfrac{1}{x}\right)^3$

$\qquad\qquad + (1)\left(\dfrac{1}{x}\right)^4$

$\qquad\quad = 1 + \dfrac{4}{x} + \dfrac{6}{x^2} + \dfrac{4}{x^3} + \dfrac{1}{x^4}$

(b) $\left(x-\dfrac{1}{x}\right)^6 = \binom{6}{0}x^6 + \binom{6}{1}x^5\left(-\dfrac{1}{x}\right) + \binom{6}{2}x^4\left(-\dfrac{1}{x}\right)^2 + \binom{6}{3}x^3\left(-\dfrac{1}{x}\right)^3$

$\qquad\qquad + \binom{6}{4}x^2\left(-\dfrac{1}{x}\right)^4 + \binom{6}{5}x\left(-\dfrac{1}{x}\right)^5 + \binom{6}{6}\left(-\dfrac{1}{x}\right)^6$

$\qquad\quad = (1)x^6 + 6x^5\left(-\dfrac{1}{x}\right) + 15x^4\left(-\dfrac{1}{x}\right)^2 + 20x^3\left(-\dfrac{1}{x}\right)^3$

$\qquad\qquad + 15x^2\left(-\dfrac{1}{x}\right)^4 + 6x\left(-\dfrac{1}{x}\right)^5 + 1\left(-\dfrac{1}{x}\right)^6$

$\qquad\quad = x^6 - 6x^4 + 15x^2 - 20 + \dfrac{15}{x^2} - \dfrac{6}{x^4} + \dfrac{1}{x^6}$

(c) $\left(\sqrt{x}-\dfrac{2}{\sqrt{x}}\right)^6 = \binom{6}{0}(\sqrt{x})^6 + \binom{6}{1}(\sqrt{x})^5\left(-\dfrac{2}{\sqrt{x}}\right) + \binom{6}{2}(\sqrt{x})^4\left(-\dfrac{2}{\sqrt{x}}\right)^2$

$\qquad\qquad + \binom{6}{3}(\sqrt{x})^3\left(-\dfrac{2}{\sqrt{x}}\right)^3 + \binom{6}{4}(\sqrt{x})^2\left(-\dfrac{2}{\sqrt{x}}\right)^4$

$\qquad\qquad + \binom{6}{5}(\sqrt{x})\left(-\dfrac{2}{\sqrt{x}}\right)^5 + \binom{6}{6}\left(-\dfrac{2}{\sqrt{x}}\right)^6$

$\qquad\quad = (1)(x^3) + 6(x^{\frac{5}{2}})(-2x^{-\frac{1}{2}}) + 15(x^2)(-2)^2(x^{-\frac{2}{2}})$

$\qquad\qquad + 20(x^{\frac{3}{2}})(-2)^3(x^{-\frac{3}{2}}) + 15x(-2)^4(x^{-\frac{4}{2}})$

$\qquad\qquad + 6(x^{\frac{1}{2}})(-2)^5(x^{-\frac{5}{2}}) + (-2)^6(x^{-\frac{6}{2}})$

$\qquad\quad = x^3 - 12x^2 + 60x - 160 + \dfrac{240}{x} - \dfrac{192}{x^2} + \dfrac{64}{x^3}$

(d) $\left(2x^3 + \dfrac{\sqrt{y}}{3}\right)^4 = \binom{4}{0}(2x^3)^4 + \binom{4}{1}(2x^3)^3\left(\dfrac{\sqrt{y}}{3}\right) + \binom{4}{2}(2x^3)^2\left(\dfrac{\sqrt{y}}{3}\right)^2$

$\qquad\qquad + \binom{4}{3}(2x^3)\left(\dfrac{\sqrt{y}}{3}\right)^3 + \binom{4}{4}\left(\dfrac{\sqrt{y}}{3}\right)^4$

$\qquad = 16x^{12} + \dfrac{32x^9\sqrt{y}}{3} + \dfrac{8x^6 y}{3} + \dfrac{8x^3 y^{\frac{3}{2}}}{27} + \dfrac{y^2}{81}$

(e) $\left(x^2 + \dfrac{3y}{x}\right)^4 = \binom{4}{0}(x^2)^4 + \binom{4}{1}(x^2)^3\left(\dfrac{3y}{x}\right) + \binom{4}{2}(x^2)^2\left(\dfrac{3y}{x}\right)^2$

$\qquad\qquad + \binom{4}{3}(x^2)\left(\dfrac{3y}{x}\right)^3 + \binom{4}{4}\left(\dfrac{3y}{x}\right)^4$

$\qquad = x^8 + 4(x^6)\left(\dfrac{3y}{x}\right) + 6(x^4)\left(\dfrac{3^2 y^2}{x^2}\right) + 4x^2\left(\dfrac{3^3 y^3}{x^3}\right) + \dfrac{3^4 y^4}{x^4}$

$\qquad = x^8 + 12x^5 y + 54x^2 y^2 + \dfrac{108y^3}{x} + \dfrac{81y^4}{x^4}$

(f) $\left(3x^2 + \dfrac{1}{x}\right)^3 = \binom{3}{0}(3x^2)^3 + \binom{3}{1}(3x^2)^2\left(\dfrac{1}{x}\right) + \binom{3}{2}(3x^2)\left(\dfrac{1}{x}\right)^2 + \binom{3}{3}\left(\dfrac{1}{x}\right)^3$

$\qquad = (1)(3^3)x^6 + (3)(3^2)x^4\left(\dfrac{1}{x}\right) + (3)(3x^2)\left(\dfrac{1}{x^2}\right) + (1)\left(\dfrac{1}{x^3}\right)$

$\qquad = 27x^6 + 27x^3 + 9 + \dfrac{1}{x^3}$

7.2.6. $(x + y + z)^3 = \binom{3}{0}x^3 + \binom{3}{1}x^2(y + z) + \binom{3}{2}x(y + z)^2 + \binom{3}{3}(y + z)^3.$

Using the Binomial Theorem to expand $(y + z)^2$ and $(y + z)^3$,

$(y + z)^2 = \binom{2}{0}y^2 + \binom{2}{1}yz + \binom{2}{2}z^2 = y^2 + 2yz + z^2$

$(y + z)^3 = \binom{3}{0}y^3 + \binom{3}{1}y^2 z + \binom{3}{2}yz^2 + \binom{3}{3}z^3 = y^3 + 3y^2 z + 3yz^2 + z^3.$

Now, substituting back into the original expansion of $(x + y + z)^3$, we have

$(x + y + z)^3 = x^3 + 3x^2(y + z) + 3x(y^2 + 2yz + z^2) + (y^3 + 3y^2 z + 3yz^2 + z^3)$

$\qquad\qquad = x^3 + 3x^2 y + 3x^2 z + 3xy^2 + 6xyz + 3xz^2 + y^3 + 3y^2 z + 3yz^2 + z^3.$

7.2.7. Expanding $(W + B)^8$ using the Binomial Theorem, like in Example 7.2.4, we have

$(W + B)^8 = W^8 + 8W^7 B + 28W^6 B^2 + 56W^5 B^3 + 70W^4 B^4 + 56W^3 B^5 + 28W^2 B^6$

$\qquad\qquad + 8WB^7 + B^8.$

The result is $W^5 B^3$ since this represents 5 white balls and 3 black balls. There are 56 such possible results since the coefficient of $W^5 B^3$ is 56.

7.2.8. In the expansion of $(H + T)^8$, we need the coefficient of the term $H^3 T^5$, which represents the number of ways 3 heads and 5 tails can appear in 8 tosses. Then, $r = 5$, so $\binom{8}{5}H^3 T^5$ and $\binom{8}{5} = 56$. Therefore, there are 56 ways that 3 heads and 5 tails can appear.

7.2.9. The coefficient is $\binom{9}{3} = 84$.

7.2.10. In this case, $n = 5$ and $r = 3$, where the required term is $\binom{n}{r}x^{n-r}y^r = \binom{5}{3}(2x)^2(3y)^3 = \binom{5}{3}(2^2)(3^3)x^2y^3$. Then, the required coefficient is $\binom{5}{3}(2^2)(3^3) = 1080$.

7.2.11. Let $x = 1$, then

$$0 = 0^{10} = (1 - (1))^{10} = \sum_{r=0}^{10} \binom{10}{r} \underbrace{(1)^{10-r}}_{1}(-1)^r$$

$$= \sum_{r=0}^{10} \binom{10}{r}(-1)^r$$

$$= 1 - \binom{10}{1} + \binom{10}{2} + \cdots - \binom{10}{9} + \binom{10}{10}.$$

7.2.12. Set $x = 1$, then $2^8 = \binom{8}{0} + \binom{8}{1} + \cdots + \binom{8}{8}$, and

$$0 = \binom{8}{0} - \binom{8}{1} + \binom{8}{2} - \binom{8}{3} + \binom{8}{4} - \binom{8}{5} + \binom{8}{6} - \binom{8}{7} + \binom{8}{8}$$

$$2^8 = 2\binom{8}{0} + 2\binom{8}{2} + 2\binom{8}{4} + 2\binom{8}{6} + 2\binom{8}{8}$$

$$2^8 = 2\left[\binom{8}{0} + \binom{8}{2} + \binom{8}{4} + \binom{8}{6} + \binom{8}{8}\right]$$

$$2^7 = \binom{8}{0} + \binom{8}{2} + \binom{8}{4} + \binom{8}{6} + \binom{8}{8}$$

$$2^7 - \binom{8}{0} = \binom{8}{2} + \binom{8}{4} + \binom{8}{6} + \binom{8}{8}$$

$$2^7 - 1 = \binom{8}{2} + \binom{8}{4} + \binom{8}{6} + \binom{8}{8}.$$

7.2.13. Using Corollary 7.2.3, we know that

$$\binom{10}{0} + \binom{10}{1} + \binom{10}{2} + \cdots + \binom{10}{10} = 2^{10}.$$

We only require half of these combinations, so $\dfrac{2^{10}}{2} = 2^9 = 512$ is the number of ways to choose an even number of objects from 10 objects.

7.2.14. $\sum_{r=0}^{n} \binom{n}{r} x^{n-r} y^r = \binom{n}{0} x^n + \binom{n}{1} x^{n-1} y + \binom{n}{2} x^{n-2} y^2 + \cdots + \binom{n}{n} y^n$

$$= \binom{n}{n} x^n + \binom{n}{n-1} x^{n-1} y + \cdots + \binom{n}{0} y^n \qquad \text{By Theorem 5.7.8}$$

$$= \binom{n}{0} y^n + \binom{n}{1} x y^{n-1} + \binom{n}{2} x^2 y n - 2 + \cdots + \binom{n}{n} x^n \qquad \text{Rearrange the terms}$$

$$= \sum_{r=0}^{n} \binom{n}{r} x^r y^{n-r}$$

7.2.15. Using the Binomial Theorem,

$$\sum_{i=0}^{20} \binom{20}{i} 4^i = (1+4)^{20} = 5^{20}$$

(since $n = 20$, $x = 1$, $y = 4$). Again, using the Binomial Theorem,

$$\sum_{i=0}^{20} \binom{20}{i} 3^i \cdot 2^{20-i} = (2+3)^{20} = 5^{20}$$

(since $n = 20$, $x = 2$, and $y = 3$).

Therefore,

$$\sum_{i=0}^{20} \binom{20}{i} 4^i - \sum_{i=0}^{20} \binom{20}{i} 3^i \cdot 2^{20-i} = 5^{20} - 5^{20} = 0.$$

7.2.16. This can be easily shown with the use of the Binomial Theorem. Set $x = 1$ and $y = 2$. Then,

$$3^n = (1+2)^n = \sum_{r=0}^{n} \binom{n}{r} (1)^{n-r} (2)^r$$

$$= 1^n 2^0 \binom{n}{0} + 1^{n-1} 2^1 \binom{n}{1} + 1^{n-2} 2^2 \binom{n}{2} + \cdots + 1^{n-n} 2^n \binom{n}{n}$$

$$= \binom{n}{0} + 2 \binom{n}{1} + 2^2 \binom{n}{2} + \cdots + 2^n \binom{n}{n}.$$

7.2.17. Suppose that you are making a sandwich, and there are n sauces. You can specify 3 levels of sauce: none, light, and heavy.

(a) For every sauce, there are 3 choices or fixed amounts you can put on the sandwich. Thus, for n sauces there are

$$\underbrace{3}_{\text{Sauce 1}} \times \underbrace{3}_{\text{Sauce 2}} \times \cdots \times \underbrace{3}_{\text{Sauce } n} = 3^n.$$

(b) Suppose you want to put i sauces on the sandwich ($i \le n$); the number of ways you can select i sauces out of n sauces is $\binom{n}{i}$. Since we have selected the sauces we wish to use, you cannot say the amount of a selected sauce is none. Thus, for each selected sauce, there are only two remaining choices: light and heavy. By the Multiplication Principle, for i selected sauces, there are $2^i \binom{n}{i}$ ways. Finally, the number of selected sauces i can be 0, *or* 1, *or* 2, ..., *or* n, so by the Sum Principle,

$$\sum_{i=0}^{n} 2^i \binom{n}{i} = \binom{n}{0} + 2 \binom{n}{1} + 2^2 \binom{n}{2} + \cdots + 2^n \binom{n}{n}.$$

Since parts (a) and (b) are different (yet equivalent) ways of thinking about the situation, we have shown that

$$\binom{n}{0} + 2\binom{n}{1} + 2^2\binom{n}{2} + \cdots + 2^n\binom{n}{n} = 3^n$$

as required.

7.2.18. This can be easily shown with the use of the Binomial Theorem. Set $x = 1$ and $y = m$. Then,

$$(m+1)^n = (1+m)^n = \sum_{r=0}^{n} \binom{n}{r}(1)^{n-r}m^r$$

$$= 1^n m^0 \binom{n}{0} + 1^{n-1}m^1\binom{n}{1} + 1^{n-2}m^2\binom{n}{2} + \cdots + 1^{n-n}m^n\binom{n}{n}$$

$$= \binom{n}{0} + m\binom{n}{1} + m^2\binom{n}{2} + \cdots + m^n\binom{n}{n}.$$

7.2.19. Suppose that you are making a sandwich, and there are n sauces. You can specify $(m+1)$ levels of sauce: $0, 1, 2, \ldots, m$.

(a) For every sauce, there are $(m+1)$ levels or amounts of sauce you can put on the sandwich. Thus, for n sauces there are

$$\underbrace{m+1}_{\text{Sauce 1}} \times \underbrace{m+1}_{\text{Sauce 2}} \times \cdots \times \underbrace{m+1}_{\text{Sauce } n} = (m+1)^n.$$

(b) Suppose you want to put i sauces on the sandwich ($i \leq n$); the number of ways you can select i sauces out of n sauces is $\binom{n}{i}$. Since we have selected the sauces we wish to use, you cannot say the amount of a selected sauce is none. Thus, for each selected sauce, there are only m remaining choices: $1, 2, \ldots, m$. By the Multiplication Principle, for i selected sauces, there are $m^i \binom{n}{i}$ ways. Finally, the number of selected sauces i can be 0, *or* 1, *or* 2, $\ldots$, *or* n, so by the Sum Principle,

$$\sum_{i=0}^{n} m^i \binom{n}{i} = \binom{n}{0} + m\binom{n}{1} + m^2\binom{n}{2} + \cdots + m^n\binom{n}{n}.$$

Since parts (a) and (b) are different (yet equivalent) ways of thinking about the situation, we have shown that

$$\binom{n}{0} + m\binom{n}{1} + m^2\binom{n}{2} + \cdots + m^n\binom{n}{n} = (m+1)^n$$

as required.

7.2.20. Using the geometric series formula,

$$1 - x + x^2 - x^3 + \cdots + x^8 - x^9 = \frac{1 - x^{10}}{1 + x} = \frac{1 - x^{10}}{y}$$

Since $x = y - 1$, then

$$\frac{1 - x^{10}}{y} = \frac{1 - (y-1)^{10}}{y}$$

We want a_2, which is the coefficient of y^3 term in $-(y-1)^{10}$. By the Binomial Theorem,

$$(-1)(-1)^{10-3}\binom{10}{3} = (-1)(-1)^7\binom{10}{3} = 120.$$

Similarly for a_4, we want the coefficient of y^5 term in $-(y-1)^{10}$. By the Binomial Theorem,

$$(-1)(-1)^{10-5}\binom{10}{5} = (-1)(-1)^5\binom{10}{5} = 252.$$

(This was adapted from the 1986 American Invitational Mathematics Examination problem 11.)

7.2.21. We will prove this using the Principle of Mathematical Induction.

(a) *Base step.* For $n = 1$, we have on the left side,

$$(x+y)^1 = \sum_{r=0}^{1}\binom{1}{r}x^{1-r}y^r$$

$$= \binom{1}{0}x^{1-0}y^0 + \binom{1}{1}x^{1-1}y^1$$

$$= 1x^1y^0 + 1x^0y^1$$

$$= x + y$$

which is the same as the right side. Thus, it is true for $n = 1$.

(b) *Inductive step.*

(i) We *assume* that the proposition is true for $n = k$, $k \in \mathbb{N}$. That is, we assume that

$$(x+y)^k = \sum_{r=0}^{k}\binom{k}{r}x^{k-r}y^r.$$

This is our inductive hypothesis.

(ii) We need to *prove* that the proposition is true for $n = k + 1$. That is, we need to prove that

$$(x+y)^{k+1} \sum_{r=0}^{k+1}\binom{k+1}{r}x^{k+1-r}y^r.$$

Now,

$$(x+y)^{k+1} = (x+y)(x+y)^k$$

$$= x(x+y)^k + y(x+y)^k$$

$$= x\left[\sum_{r=0}^{k}\binom{k}{r}x^{k-r}y^r\right] + y\left[\sum_{r=0}^{k}\binom{k}{r}x^{k-r}y^r\right]$$

$$= \sum_{r=0}^{k}\binom{k}{r}x^{k+1-r}y^r + \sum_{r=0}^{k}\binom{k}{r}x^{k-r}y^{r+1}$$

$$= \binom{k}{0}x^{k+1} + \binom{k}{1}x^k y + \cdots + \binom{k}{k}xy^k$$

$$+ \binom{k}{0}x^k y + \binom{k}{1}x^{k-1}y^2 + \cdots + \binom{k}{k-1}xy^k + \binom{k}{k}y^{k+1}$$

$$= \binom{k}{0}x^{k+1} + \left[\binom{k}{1} + \binom{k}{0}\right]x^k y + \cdots + \left[\binom{k}{k} + \binom{k}{k-1}\right]xy^k$$

$$+ \binom{k}{k}y^{k+1}$$

$$= \binom{k}{0}x^{k+1} + \sum_{r=1}^{k}\left[\binom{k}{r} + \binom{k}{r-1}\right]x^{k+1-r}y^r + \binom{k}{k}y^{k+1}$$

$$\underbrace{\qquad\qquad\qquad\qquad}_{\binom{k+1}{r} \text{ by Pascal's Theorem}}$$

$$= \binom{k}{0}x^{k+1} + \sum_{r=1}^{k}\binom{k+1}{r}x^{k+1-r}y^r + \binom{k}{k}y^{k+1}$$

$$\underbrace{\quad}_{\binom{k+1}{0}} \qquad\qquad\qquad\qquad \underbrace{\quad}_{\binom{k+1}{k+1}}$$

$$= \binom{k+1}{0}x^{k+1} + \sum_{r=1}^{k}\binom{k+1}{r}x^{k+1-r}y^r + \binom{k+1}{k+1}y^{k+1}$$

$$= \sum_{r=0}^{k+1}\binom{k+1}{r}x^{k+1-r}y^r$$

For proofs of the two identities that we used, $\binom{k}{0} = \binom{k+1}{0}$ and $\binom{k}{k} = \binom{k+1}{k+1}$, see Exercise 7.1.9. This concludes the inductive step.

Therefore, $\sum_{r=0}^{k}\binom{k}{r}x^{k-r}y^r$ is true for every natural number n by the Principle of Mathematical Induction.

7.3.1. (a) By Pascal's Theorem, $\binom{20}{4} + \binom{20}{3} = \binom{21}{4} = 5985$.

(b) By the sum of alternating sign property, $\binom{12}{0} - \binom{12}{1} + \binom{12}{2} - \binom{12}{3} + \cdots + \binom{12}{12} = 0$.

(c) By the sum of rows property, $\sum_{r=0}^{16}\binom{16}{r} = 2^{16} = 65\,536$.

(d) By the sum of squares property, $\binom{7}{0}^2 + \binom{7}{1}^2 + \binom{7}{2}^2 + \cdots + \binom{7}{7}^2 = \binom{14}{7} = 3432$.

(e) By the sum of rows property, $\binom{9}{0} + \binom{9}{1} + \binom{9}{2} + \cdots + \binom{9}{9} = 2^9 = 512$.

7.3.2. We need to solve the following equation:

$$\binom{n}{0} + \binom{n}{1} + \binom{n}{2} + \cdots + \binom{n}{n} = 2^n = 512$$

$$\log 2^n = \log 512$$

$$n \log 2 = \log 512$$

$$n = \frac{\log 512}{\log 2}$$

$$n = 9.$$

7.3.3. (a) By Pascal's Theorem, $\binom{90}{50} + \binom{90}{49} = \binom{91}{50}$. Therefore, $n = 91$, and $r = 50$.

(b) By Pascal's Theorem, $\binom{x-1}{y} + \binom{x-1}{y-1} = \binom{x}{y}$. Therefore, $n = x$, and $r = y$.

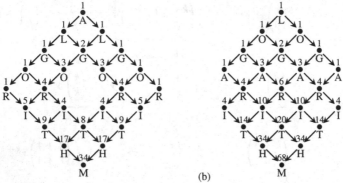

7.3.4. (a)　　　　　　　　　　　　　(b)

Remark: Logarithm and algorithm both rhyme with each other, both have the same number of letters, and they are anagrams.

7.3.5.
$$\binom{n+3}{r+3} = \binom{n+2+1}{r+3}$$
$$= \binom{n+2}{r+2} + \binom{n+2}{r+3}$$
$$= \binom{n+1+1}{r+2} + \binom{n+1+1}{r+3}$$
$$= \binom{n+1}{r+1} + \binom{n+1}{r+2} + \binom{n+1}{r+2} + \binom{n+1}{r+3}$$
$$= \binom{n+1}{r+1} + 2\binom{n+1}{r+2} + \binom{n+1}{r+3}$$
$$= \binom{n}{r} + \binom{n}{r+1} + 2\left[\binom{n}{r+1} + \binom{n}{r+2}\right] + \binom{n}{r+2} + \binom{n}{r+3}$$
$$= \binom{n}{r} + 3\binom{n}{r+1} + 3\binom{n}{r+2} + \binom{n}{r+3}$$

7.3.6. The coefficient is $\binom{7}{5} = 21$.

7.3.7. $(x+y)^6$

7.3.8. (a) 11　　　　　　　　(b) 13　　　　　　　　(c) 13

7.3.9. (a) $(1+x)^7 = x^7 + 7x^6 + 21x^5 + 35x^4 + 35x^3 + 21x^2 + 7x + 1$
(b) $(3-x)^6 = x^6 - 18x^5 + 135x^4 - 540x^3 + 1215x^2 - 1458x + 729$
(c) $(4-3x)^5 = -243x^5 + 1620x^4 - 4320x^3 + 5760x^2 - 3840x + 1024$
(d) $(2y-3)^8 = 256y^8 - 3072y^7 + 16\,128y^6 - 48\,384y^5 + 90\,720y^4 - 108\,864y^3 + 81\,648y^2 - 34\,992y + 6561$

7.3.10. (a) $\dfrac{x^3}{y^3} + \dfrac{3x}{y} + \dfrac{3y}{x} + \dfrac{y^3}{x^3}$

(b) $x^3 + 6x^{\frac{5}{2}}\sqrt{y} + 15x^2 y - 20x^{\frac{3}{2}}y^{\frac{3}{2}} + 15xy^2 - 6\sqrt{x}y^{\frac{5}{2}} + y^3$

(c) $(3x + y)^5 = 243x^5 + 405x^4 y + 270x^3 y^2 + 90x^2 y^3 + 15xy^4 + y^5$

7.3.11. (a) $\dfrac{5005}{19\,683}$ (b) $\dfrac{455}{19\,683}$

8.2.2. No, these are not mutually exclusive. For two events E and F to be mutually exclusive, $E \cap F = \varnothing$. But, we know that in a deck of cards, there is a king of hearts; this is in the intersection of E and F. Hence $E \cap F \neq \varnothing$.

8.2.3. The events E and F are mutually exclusive. To see this, the event $E = \{2, 4, 6\}$, and the event $F = \{1, 3, 5\}$. The intersection $E \cap F = \varnothing$, which means that E and F are mutually exclusive.

8.2.4. The events are not mutually exclusive. The overlapping outcomes between the two events are $\{6, 12\}$.

8.2.5. All the possible events of the sample space are $\varnothing, \{H\}, \{T\}, \{H, T\}$.

8.2.6. The set of all possible events of S is the powerset of S. Thus, there are $2^6 = 64$ possible events.

8.2.7. We define the following events:
- $R =$ a red ball is drawn
- $B =$ a blue ball is drawn
- $G =$ a green ball is drawn
- $Y =$ a yellow ball is drawn

(a) The tree diagram is given in Fig. C.22(a).
The sample space for this experiment is $S = \{RR, RB, RG, RY, BR, BB, BG, BY, GR, GB, GG, GY, YR, YB, YG, YY\}$.

(b) The tree diagram is in Fig. C.22(b). The sample space for this experiment is $S = \{RB, RG, RY, BR, BG, BY, GR, GB, GY, YR, YB, YG\}$.

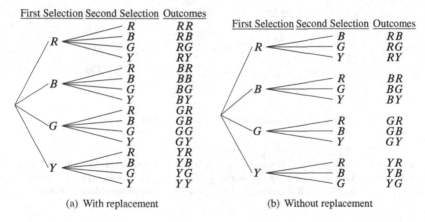

(a) With replacement   (b) Without replacement

Fig. C.22   Tree diagrams for Exercise 8.2.7

8.2.8. The number of elements in the sample space is $13\,983\,816$. See the solution to the Lottery Ticket Problem on p. 74.

8.2.9. (a) $E = \{(1, 2), (2, 2), (3, 2), (4, 2), (5, 2), (6, 2), (1, 4), (2, 4), (3, 4), (4, 4), (5, 4), (6, 4), (1, 6), (2, 6), (3, 6), (4, 6), (5, 6), (6, 6)\}$.

(b) $F = \{(1,1),(2,1),(3,1),(4,1),(5,1),(6,1),(1,3),(2,3),(3,3),(4,3),(5,3),(6,3),$
$(1,5),(2,5),(3,5),(4,5),(5,5),(6,5)\}$.

(c) $G = \{(3,5),(5,3)\}$.

8.3.1. (a) Yes. All of the outcomes have the same probability.

Outcome	Probability
1	$\frac{1}{6}$
2	$\frac{1}{6}$
3	$\frac{1}{6}$
4	$\frac{1}{6}$
5	$\frac{1}{6}$
6	$\frac{1}{6}$

(b) No. The outcomes do not have the same probability. Since the die is non-uniformed weighted, the probability of a particular outcome(s) will be higher than other outcomes.

(c) No. The probability of picking a red ball is $\frac{5}{8}$, while the probability of picking a black ball is $\frac{3}{8}$.

8.3.2. A die is fair if each of the outcomes $\{1,2,3,4,5,6\}$ is equally likely to occur.

8.3.3. (a) $\frac{1}{6}$ (c) $\frac{3}{6} = \frac{1}{2}$ (e) $\frac{2}{6} = \frac{1}{3}$

      (b) $\frac{2}{6} = \frac{1}{3}$ (d) $\frac{3}{6} = \frac{1}{2}$ (f) $\frac{4}{6} = \frac{2}{3}$

8.3.4. (a) $\mathcal{S} = \{HH, TH, HT, TT\}$. The tree diagram is given in Fig. C.23.

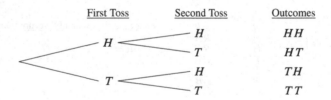

Fig. C.23 Tree diagram for the outcomes of two coin tosses

(b) $E = \{HH\}$. $P(E) = \frac{n(E)}{n(S)} = \frac{1}{4}$

Remark: This same question was asked of British Members of Parliament (MPs) in 2012. Only 40 out of 97 MPs (or 41%) gave the correct answer of $\frac{1}{4}$. The survey was conducted by a research group in the Royal Statistical Society that campaigns to bring awareness and improve the way Britain handles probability and statistics (Easton, 2012).

8.3.5. (a) $\mathcal{S} = \{HHH, HHT, HTH, HTT, THH, THT, TTH, TTT\}$. You can draw a tree diagram to help you.

(b) $E = \{HTT, THT, TTH, TTT\}$. $P(E) = \frac{n(E)}{n(S)} = \frac{4}{8} = \frac{1}{2}$.

8.3.6. (a) $\dfrac{4}{52} = \dfrac{1}{13}$ (b) $\dfrac{12}{52} = \dfrac{3}{13}$ (c) $\dfrac{26}{52} = \dfrac{1}{2}$ (d) $\dfrac{26}{52} = \dfrac{1}{2}$

8.3.7. Let E be the event that three people will receive the correct coat. Upon searching the sample space $\mathcal{S}$, we find that there are no outcomes where three people receive the correct coat. Thus, $E = \{\ \} = \varnothing$;

$n(E) = 0$. Therefore, the probability that exactly three people will receive the correct coat is

$$P(E) = \frac{n(E)}{n(S)} = \frac{0}{24} = 0.$$

Remark: The reason why there are no outcomes with exactly three people who receive the correct coat is because of the following: if three people receive the correct coats, then the remaining coat must belong to the last person.

8.3.8. There are 30 days in the month of June.

(a) $\dfrac{1}{30}$

(b) $\dfrac{15}{30} = \dfrac{1}{2}$

8.3.9. The experiment is drawing a single card from a well-shuffled deck of cards; $n(S) = 52$. Let E be the event of drawing either a 7 or a heart face card. Then, $E = \{7\diamond, 7\heartsuit, 7\clubsuit, 7\spadesuit, J\heartsuit, Q\heartsuit, K\heartsuit\}$; $n(E) = 7$. Therefore, the probability of selecting a 7 or a heart face card is $P(E) = \frac{n(E)}{n(S)} = \frac{7}{52} \approx 0.13461$.

8.3.10. It is *not* reasonable to believe that no lottery ticket will win. We stated it clearly in the assumption that there is a winning ticket. This is better known as the **Lottery Paradox**. You will see this type of argument when people are reasoning about the occurrence of an event with a very small probability; they believe that because of the small probability, it cannot happen at all. In Sec. 8.9.2, we will see how a jury makes the same fallacy.

8.3.11. The probability that the randomly selected letter is B is $\frac{2}{11}$. The probability that the randomly selected letter is one of the vowels, A, E, I, O, or U, is $\frac{4}{11}$.

8.3.12. The tree diagram given in Fig. C.24 lists all of the possible outcomes.

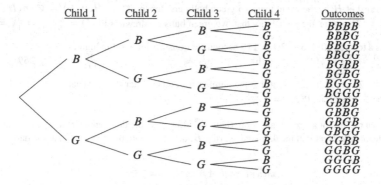

Fig. C.24   Tree diagram for Exercise 8.3.12

Let E be the event that there are more girls than boys. Then

$$E = \{BGGG, GBGG, GGBG, GGGB, GGGG\}.$$

The total number of outcomes is 16; $n(S) = 16$. The number of outcomes where there are more girls than boys is 5; $n(E) = 5$. Therefore, the required probability is $P(E) = \frac{n(E)}{n(S)} = \frac{5}{16}$.

8.3.13. Let E be the event of picking a king of hearts. Then $n(E) = 1$, and $n(S) = 52$. Therefore, $P(E) = 1/52$.

8.3.14. (a) $\dfrac{1}{2}$ (b) $\dfrac{1}{8}$

8.3.15. We are told that $n(S) = 20$.

    (a) Let E be an odd digit is selected the first time. This means on the second time, it can be either an odd or an even digit.

$$n(E) = \underbrace{3}_{\text{odd digits}} \times \underbrace{4}_{\substack{\text{pick amongst} \\ \text{the remaining digits}}} = 12$$

    Therefore, $P(E) = \dfrac{12}{20}$.

    (b) $\dfrac{12}{20}$ (c) $\dfrac{6}{20}$

8.3.16. (a)   (i) The tree diagram is given in Fig. C.22(a) on p. 619. The sample space for this experiment is $S = \{RR, RB, RG, RY, \ BR, BB, BG, BY, \ GR, GB, GG, GY, YR, YB, YG, YY\}$; $n(S) = 16$.

        (ii) Let E be the event that at least one green ball is drawn. Then, $E = \{RG, BG, GR, GB, GG, GY, YG\}$; $n(E) = 7$. Therefore, the probability of at least one green ball is drawn is $P(E) = \frac{n(E)}{n(S)} = \frac{7}{16} = 0.4375$.

        (iii) Let F be the event that two green balls are drawn. Then, $F = \{GG\}$; $n(F) = 1$. Therefore, the probability of two green balls are drawn is $P(F) = \frac{n(F)}{n(S)} = \frac{1}{16} = 0.0625$.

    (b)   (i) The tree diagram is in Fig. C.22(b) on p. 619. The sample space for this experiment is $S = \{RB, RG, RY, BR, BG, BY, GR, GB, GY, YR, YB, YG\}$; $n(S) = 12$.

        (ii) Let G be the event that at least one green ball is drawn. Then, $G = \{RG, BG, GR, GB, GY, YG\}$; $n(G) = 6$. Therefore, the probability that at least one green ball is drawn is $P(G) = \frac{n(G)}{n(S)} = \frac{6}{12} = \frac{1}{2} = 0.5$.

        (iii) Let H be the event that two green balls are drawn. Then, $H = \{\ \} = \varnothing$; $n(H) = 0$. Therefore, the probability that two green balls are drawn is $P(H) = \frac{n(H)}{n(S)} = \frac{0}{12} = 0$.

8.3.17. There is a total of $80 + 5 + 30 = 115$ seconds to consider for a traffic light cycle.

    (a) $\frac{80}{115} = \frac{16}{23} \approx 70\%$ (b) $\frac{5}{115} = \frac{1}{23} \approx 4\%$ (c) $\frac{30}{115} = \frac{6}{23} \approx 26\%$

8.3.18. $n(S) = 12$. Let E be the event that the number rolled is greater than 9. Then $E = \{10, 11, 12\}$; $n(E) = 3$. Therefore, $P(E) = \frac{n(E)}{n(S)} = \frac{3}{12} = \frac{1}{4}$.

8.3.19. Recall that the area of a circle is area $= \pi(\text{radius})^2$. Let A denote the event that the dart lands in circle A, B denote the event that the dart lands in ring B, and C denote that the event that the dart lands in ring C.

$$\text{area of circle } A = \pi(5)^2 = 25\pi$$
$$\text{area of ring } B = \pi(9)^2 = 81\pi$$
$$\text{area of ring } C = \pi(12)^2 = 144\pi$$

Then,

$$P(A) = \frac{\text{the area of circle } A}{\text{the total area of target}} = \frac{25\pi}{144\pi} = \frac{25}{144}$$

$$P(B) = \frac{\text{the area of ring } B - \text{the area of circle } A}{\text{the total area of target}} = \frac{81\pi - 25\pi}{144\pi} = \frac{56\pi}{144\pi} = \frac{56}{144}$$

$$P(C) = \frac{\text{the area of ring } C - \text{the area of ring } B}{\text{the total area of target}} = \frac{144\pi - 81\pi}{144\pi} = \frac{63\pi}{144\pi} = \frac{7}{16}$$

8.4.1. (a) All the possible events of the sample space are $\varnothing, \{H\}, \{T\}, \{H, T\}$.

(b) Since we have assumed that the coin is fair, then we assign equal probability to the two possible outcomes; that is, $P(\{H\}) = \frac{1}{2}$, and $P(\{T\}) = \frac{1}{2}$. Since $\{H\}$ and $\{T\}$ are mutually exclusive events, then by Axiom 3,

$$P(\{H\}) + P(\{T\}) = P(\{H\} \cup \{T\}) = P(\{H, T\}) = 1.$$

Lastly,

$$P(\varnothing) = 1 - P(\{H, T\}) = 1 - \underbrace{P(S)}_{1} = 1 - 1 = 0.$$
by Axiom 1

All of the event probabilities are greater than or equal to 0, so they all satisfy Axiom 2. Therefore, the probabilities of each event

$$P(\varnothing) = 0, \quad P(\{H\}) = \frac{1}{2}, \quad P(\{T\}) = \frac{1}{2}, \quad P(\{H, T\}) = 1$$

satisfies the three axioms of probability.

8.5.1. The events Grumio winning and Grumio not winning are mutually exclusive, but this does not necessarily mean that the events are equally probable.

8.5.2. (a) $P(E \cup F) = P(E) + P(F) - P(E \cap F) = 0.3 + 0.75 - 0.25 = 0.8$
(b) $P(E') = 1 - P(E) = 1 - 0.3 = 0.7$
(c) $P(F') = 1 - P(F) = 1 - 0.75 = 0.25$
(d) $P(E' \cup F') = P\big((E \cap F)'\big) = 1 - P(E \cap F) = 1 - 0.25 = 0.75$
(e) $P(E' \cap F') = P\big((E \cup F)'\big) = 1 - P(E \cup F) = 1 - 0.8 = 0.2$

8.5.3. (a) $P(E \cup F) = 0.85 + 0.35 - 0.25 = 0.95$
(b) $P(E \cup F) = \frac{3}{6} + \frac{3}{6} - \frac{2}{6} = \frac{4}{6} = \frac{2}{3}$.

8.5.4. (a) $P(E \cap F) = P(E) + P(F) - P(E \cup F) = 0.2 + 0.5 - 0.7 = 0.$
(b) $P(E') = 1 - P(E) = 1 - 0.2 = 0.8$
(c) $P\big((E \cup F)'\big) = 1 - P(E \cup F) = 1 - 0.7 = 0.3.$

8.5.5. Since $E \cup F = S$, this means that $P(E \cup F) = P(S) = 1$. By the addition rule (Theorem 8.5.2),

$$P(E \cup F) = 1 = P(E) + P(F) - P(E \cap F)$$
$$1 = P(E) + P(F) - P(E \cap F)$$
$$1 - P(E) - P(F) = -P(E \cap F)$$
$$P(E \cap F) = P(E) + P(F) - 1$$
$$P(E \cap F) = 0.75 + 0.55 - 1$$
$$P(E \cap F) = 0.3$$

8.5.6. (a) $P(E') = 1 - P(E) = 1 - 0.25 = 0.75$
(b) $P(F') = 1 - P(F) = 1 - 0.6 = 0.4$
(c) $P(E \cup F) = P(E) + P(F) - P(E \cap F) = 0.6 + 0.25 - 0 = 0.85$
(d) $P(E \cap F) = 0$
(e) $P(E' \cup F') = P\big((E \cap F)'\big) = 1 - P(E \cap F) = 1 - 0 = 1$
(f) $P(E' \cap F') = P\big((E \cup F)'\big) = 1 - P(E \cup F) = 1 - 0.85 = 0.15$

8.5.7. (a) $P(E \cap F) = 0$
(b) $P(E \cup F) = P(E) + P(F) = 0.6 + 0.2 = 0.8$
(c) $P(E') = 1 - P(E) = 1 - 0.6 = 0.4$

(d) $P\left(E' \cap F\right) = P(F) = 0.2$

8.5.8. (a) $P(E) = P(\{s_1\}) + P(\{s_2\}) + P(\{s_3\}) = \frac{1}{10} + \frac{3}{10} + \frac{2}{10} = \frac{6}{10}$

(b) $P(E' \cup F) = 1 - [P(A) - P(A \cap B)] = 1 - [\frac{6}{10} - P(\{s_1, s_3\})] = 1 - [\frac{6}{10} - (\frac{1}{10} + \frac{2}{10})] = 1 - [\frac{6}{10} - \frac{3}{10}] = 1 - \frac{3}{10} = \frac{7}{10}$

8.5.9. (a)
$$P(T) + P(H) = 1$$
$$P(T) + \frac{1}{2}P(T) = 1$$
$$\frac{3}{2}P(T) = 1$$
$$P(T) = \frac{2}{3}$$

(b) By the complement rule, $P(H) = 1 - \frac{2}{3} = \frac{1}{3}$.

8.5.10. There are $5^4 = 625$ outcomes with no 3s out of the possible $6^4 = 1296$. Hence, the required probability is $1 - \frac{5^4}{6^4} = 1 - \frac{625}{1296} = \frac{671}{1296} \approx 0.5177$.

8.5.11. Let E denote snow, and F denote freezing rain. Then by the addition rule (Theorem 8.5.2),
$$P(E \cup F) = P(E) + P(F) - P(E \cap F)$$
$$P(E) = P(E \cup F) - P(F) + P(E \cap F)$$
$$P(E) = 0.75 - 0.30 + 0.10$$
$$= 0.55.$$

Therefore, there is a 55% chance that it will snow tomorrow.

8.5.12.
$$P(E \cup F \cup G)$$
$$= \frac{n(E \cup F \cup G)}{n(S)}$$
$$= \frac{n(E) + n(F) + n(G) - n(E \cap F) - n(E \cap G) - n(F \cap G) + n(E \cap F \cap G)}{n(S)} \qquad \text{By (3.8)}$$
$$= \frac{n(E)}{n(S)} + \frac{n(F)}{n(S)} + \frac{n(G)}{n(S)} - \frac{n(E \cap F)}{n(S)} - \frac{n(E \cap G)}{n(S)} - \frac{n(F \cap G)}{n(S)} + \frac{n(E \cap F \cap G)}{n(S)}$$
$$= P(E) + P(F) + P(G) - P(E \cap F) - P(E \cap G) - P(F \cap G) + P(E \cap F \cap G) \qquad \text{By (8.3)}$$

8.5.13. Let T be the set of students who play tennis, S be the set of students who play squash, and B be the set of students who play badminton. Then, the number of students who play at least one sport is
$$n(T \cup S \cup B) = n(T) + n(S) + n(B) - n(T \cap S) - n(T \cap B)$$
$$- n(S \cap B) + n(T \cap S \cap B)$$
$$= 36 + 28 + 18 - 22 - 12 - 9 + 4$$
$$= 43.$$

(a) $\dfrac{43}{100}$ (b) $\dfrac{43}{N}$

8.5.14. We are given that in the urn there are 7 white, 5 black and 8 red balls, which gives us a total of 20 balls in the urn.

(a) $P(\text{white ball}) = \frac{\text{number of white balls}}{\text{total number of balls}} = \frac{7}{20}$

(b) P(not a black ball) $= 1 - $ P(black ball)

$$= 1 - \left(\frac{\text{number of black balls}}{\text{total number of balls}} \right)$$

$$= 1 - \frac{5}{20}$$

$$= 1 - \frac{1}{4}$$

$$= \frac{3}{4}$$

8.5.15. Let the event E be a card selected is a 7, and the event F be a card selected is a heart face card. Then,

$$P(E \cup F) = P(E) + P(F) - P(E \cap F) = \frac{4}{52} + \frac{3}{52} - \frac{0}{52} = \frac{7}{52}.$$

Since $P(E \cap F) = 0$, events E and F are mutually exclusive.
Remark: Compare this to the solution of Exercise 8.3.9.

8.5.16. (a) P(person has blood factor A) $= P(A) + P(AB) = 0.42 + 0.03 = 0.45$.
(b) P(person does not have blood factor B) $= P(B') = 1 - P(B) = 1 - (0.09 + 0.03) = 1 - 0.12 = 0.88$

8.5.17. $P\left(E' \cap F'\right) = P\left((E \cup F)'\right) = 1 - P(E \cup F)$

8.6.2. This is an example of inductive reasoning. You are concluding that the die is fair (a general principle) from the evidence of the rolls you have observed. Refer to the definition of inductive reasoning on p. 59.

8.6.3. (a) $\frac{43+20+82+38+67}{1000} = \frac{250}{1000} = 0.25$
(b) $\frac{43}{1000} = 0.043$
(c) $\frac{93+35+3}{1000} = \frac{131}{1000} = 0.131$
(d) $\frac{36}{1000} = 0.036$
(e) $\frac{20+36+62+14+24+50+68+50+64}{1000} = \frac{388}{1000} = 0.388$
(f) $\frac{67+64+53+32+7}{1000} = \frac{223}{1000} = 0.223$
(g) $1 - \frac{67+64+53+32+7}{1000} = 1 - \frac{223}{1000} = \frac{777}{1000} = 0.777$

8.7.1. Since there are 52! different orderings for the cards, and we assume all are equally likely after shuffling; $n(S) = 52!$. There is only one (initial) ordering that we want, so the required probability is $\frac{1}{52!} \approx 1.240 \times 10^{-68}$.

8.7.2. The sample space S consists of all the possible permutations of letters and spaces (with replacement) with two monkeys. Then, the number of sample points in S is

$$n(S) = \underbrace{27 \times 27 \times 27 \times 27 \times 27}_{\text{Monkey 1}} \times \underbrace{27 \times 27 \times 27 \times 27 \times 27}_{\text{Monkey 2}} = 27^{10}.$$

Let E be the event that both monkeys typed "TO BE." There is only one way to do this; $n(E) = 1$. Therefore, the required probability is $P(E) = \frac{1}{27^{10}} \approx 4.857 \times 10^{-15}$.

8.7.3. Let A be the event that the friends will arrive in order of ascending age. There is only one way in which the friends can arrive at the party in ascending age, so $n(A) = 1$. Then the number of sample points is the order in which the friends arrive, thus $n(S) = 5! = 120$. Thus, the probability that the friends will arrive in order of ascending age is $P(A) = \frac{n(A)}{n(S)} = \frac{1}{120}$.

8.7.4. (a) Let E be the event that two lieutenants, two captains, and two majors are selected.

$$P(E) = \frac{\overbrace{\binom{6}{2}}^{\text{Lieutenants}} \times \overbrace{\binom{5}{2}}^{\text{Captains}} \times \overbrace{\binom{2}{2}}^{\text{Majors}}}{\binom{13}{6}} = \frac{15 \times 10 \times 1}{1716} = \frac{150}{1716} = \frac{25}{286} \approx 0.0874$$

(b) Let E be the event that one major is selected as president, one captain as vice president, and one lieutenant as secretary.

$$P(E) = \frac{2 \times 5 \times 6}{13 \times 12 \times 11} = \frac{60}{1716} = \frac{5}{143} \approx 0.035$$

8.7.5. Let E be the event that the committee has at least four professors. Then

$$n(E) = \underbrace{\binom{5}{4} \times \binom{9}{2}}_{\substack{\text{4 professors} \quad \text{2 students}}} + \underbrace{\binom{5}{5} \times \binom{9}{1}}_{\substack{\text{5 professors} \quad \text{1 student}}} = 180 + 9 = 189,$$

and

$$n(S) = \binom{14}{6} = 3003.$$

Therefore, the probability that the committee will include at least four professors is

$$P(E) = \frac{n(E)}{n(S)} = \frac{189}{3003} = \frac{9}{143} \approx 0.0630.$$

8.7.6. $\dfrac{\binom{23}{4}}{\binom{25}{4}} = \dfrac{8855}{12\,650} = 0.7$

8.7.7. $\dfrac{\binom{7}{5}}{\binom{15}{5}} = \dfrac{21}{3003} \approx 0.007$

8.7.8. Let E be the event that the novice picks the forgery. The novice picks the forgery, and then picks 2 from the remaining 19 paintings; the number of ways to do this is $n(E) = 1 \times \binom{19}{2} = 171$. The number of ways to pick three paintings is $n(S) = \binom{20}{3} = 1140$. Thus, the probability that the novice makes a correct guess in selecting the forgery is

$$P(E) = \frac{n(E)}{n(S} = \frac{\binom{19}{2}}{\binom{20}{3}} = \frac{171}{1140} = 0.15.$$

8.7.9. (a) Using the principle of inclusion-exclusion, the number of students who did do the reading, *or* the assignment, *or* both is $15 + 12 - 7 = 20$. Then by the complement rule, $29 - 20 = 9$ students did neither.

(b) Let E be the event that the three students in the group did neither the reading nor the assignment. From part (a), we know that there are 9 such students. Thus, the number of ways to choose 3 students from the 9 unprepared students is $\binom{9}{3}$; that is, $n(E) = \binom{9}{3}$. The total number of ways to form a group of 3 students from all 29 students is $\binom{29}{3}$; that is, $n(S) = \binom{29}{3}$. Then,

$$P(A) = \frac{n(E)}{n(S)} = \frac{\binom{9}{3}}{\binom{29}{3}} = \frac{84}{3654} = \frac{2}{87} \approx 0.022\,99.$$

Therefore, the probability that all three students in a group of three that did neither the reading nor the assignment (or in other words, they are not prepared to discuss the material) is 2.299%.

8.7.10. Let E be the event that the four players are all wingers.

$$P(E) = \frac{n(E)}{n(S)} = \frac{\binom{8}{4}}{\binom{20}{4}} = \frac{70}{4845} \approx 0.014\,5.$$

Therefore, the probability that four players are all wingers is 1.45%.

8.7.11. (a) of the same rank?

$$\frac{\overbrace{\binom{13}{1}}^{\text{rank}}\overbrace{\binom{4}{3}}^{\text{suit}}}{\binom{52}{3}} = \frac{13 \cdot 4}{22\,100} \approx 0.002\,35$$

(b) of the same suit?

$$\frac{\overbrace{\binom{13}{3}}^{\text{rank}}\overbrace{\binom{4}{1}}^{\text{suit}}}{\binom{52}{3}} = \frac{286 \cdot 4}{22\,100} \approx 0.051\,76$$

(c) that are in sequential order? (*e.g.*, A-2-3, 2-3-4, ...) If you assumed/interpreted that three cards must have the same suit, then the set of combinations we want is {A-2-3, 2-3-4, 3-4-5, 4-5-6, 5-6-7, 6-7-8, 7-8-9, 8-9-10, 9-10-J, 10-J-Q, J-Q-K, Q-K-A}. The number of possible combinations is 12. So,

$$\frac{\overbrace{12}^{\text{rank}} \times \overbrace{4}^{\text{suit}}}{\binom{52}{3}} = \frac{48}{22\,100} \approx 0.002\,17$$

Now if you allow each of the three cards to be not of the same suit, then

$$\frac{\overbrace{12}^{\text{rank}}\overbrace{\binom{4}{1}\binom{4}{1}\binom{4}{1}}^{\text{suit}}}{\binom{52}{3}} = \frac{12 \times 4 \times 4 \times 4}{22\,100} = \frac{768}{22\,100} \approx 0.034\,75$$

8.7.12. Let E denote getting a straight flush. For each of the four suits, there are 10 sequences: A-2-3-4-5, 2-3-4-5-6, 3-4-5-6-7, 4-5-6-7-8, 5-6-7-8-9, 6-7-8-9-10, 7-8-9-10-J, 8-9-10-J-Q, 9-10-J-Q-K, 10-J-Q-K-A; $n(E) = 4 \times 10 = 40$. In the sample space, there are $n(S) = \binom{52}{5} = 2\,598\,960$ possible 5-card hands. Therefore, the probability of getting a straight flush is $P(E) = \frac{n(E)}{n(S)} = \frac{40}{2\,598\,960} \approx 0.000\,015\,3$ or 0.001 53%.

8.7.13. $\text{P(at most one 8)} = \dfrac{\overbrace{\binom{48}{7}}^{\text{no 8s}} + \overbrace{\binom{4}{1}\binom{48}{6}}^{\text{one 8}}}{\binom{52}{7}} \approx 0.9173$

8.7.14. (a) $\dfrac{13 \times 13 \times 13 \times 13}{\binom{52}{4}} = \dfrac{28\,561}{270\,725} \approx 0.1055$

(b) $1 - \dfrac{13 \times 13 \times 13 \times 13}{\binom{52}{4}} = 1 - \dfrac{28\,561}{270\,725} = 1 - 0.1055 \approx 0.8945$

8.7.15. The experiment is a child is randomly selecting/stealing 2 candies from a bowl. For the sample space, the order is not important, and so $n(S) = \binom{8}{2} = 28$ ways to select two pieces of candy.

(a) Let event A be stealing two chocolates. Then $n(A) = \binom{5}{2}\binom{3}{0} = 10$. Therefore, the probability of stealing two chocolates is $\text{P}(A) = \dfrac{n(A)}{n(S)} = \dfrac{10}{28} = \dfrac{5}{14}$.

(b) $\text{P(not selecting 2 suckers)} = 1 - \text{P(selecting 2 suckers)} = 1 - \dfrac{\binom{3}{2}}{\binom{8}{2}} 1 - \dfrac{3}{28} = \dfrac{25}{28}$.

8.7.16. (a) $\dfrac{\binom{5}{5} \times \binom{1}{1}}{\binom{59}{5} \times \binom{35}{1}} = \dfrac{1}{5\,006\,386 \times 35} = \dfrac{1}{175\,223\,510} \approx 5.707 \times 10^{-9}$.

(b) $\dfrac{\binom{5}{5} \times \binom{34}{1}}{\binom{59}{5} \times \binom{35}{1}} = \dfrac{1 \times 34}{5\,006\,386 \times 35} = \dfrac{34}{175\,223\,510} \approx 1.9404 \times 10^{-7}$.

8.7.17. Let E be the event of being dealt two jacks and any three other cards. The Euchre deck has 24 cards.

$$P(E) = \dfrac{\overbrace{\binom{4}{2}}^{\text{Jacks}} \times \overbrace{\binom{20}{3}}^{\text{Other cards}}}{\binom{24}{5}} = \dfrac{6 \times 1140}{42\,504} = \dfrac{6840}{42\,504} = \dfrac{285}{1771} \approx 0.1609$$

8.7.18. (a) Not all the possible outcomes are equally likely since there are more balls of one type than another. Consider the outcome of selecting all three black balls, and selecting all three red balls. The probability of selecting all three black balls is

$$\dfrac{\binom{5}{3}}{\binom{20}{3}} = \dfrac{10}{1140} \approx 0.0089,$$

and the probability of selecting all three red balls is

$$\dfrac{\binom{8}{3}}{\binom{20}{3}} = \dfrac{56}{1140} \approx 0.049.$$

The probabilities for the two different outcomes are not equal.

(b) P(all 3 selected balls are red) $= \dfrac{\text{number of ways to select all 3 balls red}}{\text{total possible ways to select 3 balls}}$

select no white   select no black   select all red

$$= \dfrac{\overbrace{\binom{7}{0}}^{} \times \overbrace{\binom{5}{0}}^{} \times \overbrace{\binom{8}{3}}^{}}{\binom{20}{3}}$$

$$= \dfrac{56}{1140} = \dfrac{14}{285} \approx 0.050$$

(c) P(none of the 3 selected balls are red) $= \dfrac{\text{number of ways to select 3 balls with no red}}{\text{total possible ways to select 3 balls}}$

select white or black   select no red

$$= \dfrac{\overbrace{\binom{12}{3}}^{} \times \overbrace{\binom{8}{0}}^{}}{\binom{20}{3}}$$

$$= \dfrac{220}{1140} = \dfrac{22}{114} = \dfrac{11}{57} \approx 0.193$$

8.7.19. The number of subsets of $\{1, 2, \ldots, N\}$ of size t containing the best candidate is $\binom{N-1}{t-1}$. The total number of subsets of $\{1, 2, \ldots, N\}$ of size t is $\binom{N}{t}$. Thus,

$$\text{P(Best candidate is in the first } t \text{ candidates)} = \dfrac{\binom{N-1}{t-1}}{\binom{N}{t}} = \dfrac{t}{N}.$$

8.8.1. (a) $P(F \mid E) = \dfrac{0.5}{0.8} = 0.625$

(b) Rearranging the conditional probability formula, we get

$$P(F \mid E) = \dfrac{P(E \cap F)}{P(E)}$$

$$P(E) = \dfrac{P(E \cap F)}{P(F \mid E)} = \dfrac{0.2}{0.4} = 0.5$$

and similarly,

$$P(E \mid F) = \dfrac{P(F \cap E)}{P(F)}$$

$$P(F) = \dfrac{P(E \cap F)}{P(E \mid F)} = \dfrac{0.2}{0.3} = \dfrac{2}{3} \approx 0.667.$$

(c) $P(E \cap F) = 0.2 \times 0.4 = 0.08$

(d) $P(E \mid F) = \dfrac{0.2}{0.6} = \dfrac{2}{6} = \dfrac{1}{3} \approx 0.333$ and $P(G \mid F) = \dfrac{0.3}{0.6} = \dfrac{3}{6} = \dfrac{1}{2} = 0.5$

8.8.2. Let E be the event that a king is drawn on the first card, and F is a king on the second card. The probability tree diagram is given in Fig. C.25.

(a) P(first card is a king and second card is not a king) $= P(E \cap F') = \dfrac{16}{221}$

(b) P(first card is not a king and second card is not a king) $= P(E' \cap F') = \dfrac{188}{221}$

8.8.3. (a) $A \cap B = \{4\}; n(A \cap B) = 1.$ $P(A \cap B) = \dfrac{n(A \cap B)}{n(S)} = \dfrac{1}{6}$

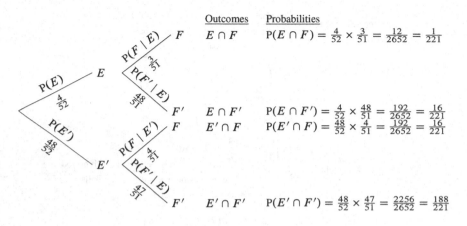

Fig. C.25   The probability tree diagram for Exercise 8.8.2

(b) $A \cap C = \{2, 4\}$; $n(A \cap C) = 2$. $P(A \cap C) = \frac{n(A \cap C)}{n(S)} = \frac{2}{6} = \frac{1}{3}$

(c) $C \cap A = \{2, 4\}$; $n(C \cap A) = 2$. $P(A \mid C) = \frac{P(C \cap A)}{P(C)} = \frac{P(A \cap C)}{P(C)} = \frac{\frac{1}{3}}{\frac{2}{6}} = \frac{2}{5}$

(d) $P(A \mid B) = \frac{P(B \cap A)}{P(B)} = \frac{P(A \cap B)}{P(B)} = \frac{\frac{1}{6}}{\frac{2}{6}} = \frac{1}{2}$

(e) $P(B \mid A) = \frac{P(A \cap B)}{P(A)} = \frac{\frac{1}{6}}{\frac{3}{6}} = \frac{1}{3}$

(f) $P(C \mid A) = \frac{P(A \cap C)}{P(A)} = \frac{\frac{1}{3}}{\frac{3}{6}} = \frac{2}{3}$

(g) $C \cap B = \{3, 4\} = B$; $n(C \cap B) = 2 = n(B)$. $P(B \mid C) = \frac{P(C \cap B)}{P(C)} = \frac{\frac{2}{6}}{\frac{2}{6}} = \frac{2}{5}$

(h) $B \cap C = \{3, 4\}$. $P(C \mid B) = \frac{P(B \cap C)}{P(B)} = \frac{\frac{2}{6}}{\frac{2}{6}} = 1$

8.8.4. See Table 9.1 on p. 394. Let A be the event that the doubles are thrown. Let B be the event that the sum of the dice is even. Then, $n(A \cap B) = 6$, $n(B) = 18$, and $n(S) = 36$, and hence, $P(A \mid B) = \frac{n(A \cap B)}{n(B)} = \frac{6}{18} = \frac{1}{3}$. Therefore, the probability that doubles are thrown given that the sum of the dice is even is $\frac{1}{3}$.

8.8.5. Let E_1 denote the event that the first card is not a king, E_2 denote the event that the second card is not a king, and E_3 denote the event that the third card is not a king.

We need to calculate $P(E_1 \cap E_2 \cap E_3)$, which is the probability that none of the three cards is a king. Using Eq. (8.12), $P(E_1 \cap E_2 \cap E_3) = P(E_1) \times P(E_2 \mid E_1) \times P(E_3 \mid E_1 \cap E_2)$. So, for the first card, the probability that it is not a king is $P(E_1) = \frac{48}{52}$, since there are $52 - 4 = 48$ cards that are not kings in the deck of 52 cards. Given that the first card is not a king, we have 51 cards remaining, $51 - 4 = 47$ of which are not kings, and $P(E_2 \mid E_1) = \frac{47}{51}$. Lastly, given that the first two cards drawn are not kings, there are $50 - 4 = 46$ cards that are not kings in the remaining deck of 50 cards, and $P(E_3 \mid E_1 \cap E_2) = \frac{46}{50}$. Therefore, the probability that all three cards are not kings is $P(E_1 \cap E_2 \cap E_3) = \frac{48}{52} \times \frac{47}{51} \times \frac{46}{50} = \frac{103\,776}{132\,600} = \frac{2171}{2774} \approx 0.782\,62$.

8.8.7. (a) Let E be the event that a 20-year-old Canadian male will live to be 70 years old; that is, $n(E) = 80\,301$. Let S be the set of 20-year-old Canadian males; that is, $n(S) = 99\,047$. Therefore, the required probability is $P(E) = \frac{n(E)}{n(S)} = \frac{80\,301}{99\,047} \approx 0.81$.

(b) Let F be the event that a 20-year-old Canadian male will live to be 100 years old; that is, $n(F) = 2049$. Let S be the set of 20-year-old Canadian males; that is, $n(S) = 99\,047$. Therefore, the required probability is $P(F) = \frac{n(F)}{n(S)} = \frac{2049}{99\,047} \approx 0.02$.

8.8.8. (a) Let E be the event that a 40-year-old Canadian female will live to be 70 years old; that is, $n(E) = 87\,202$. Let S be set of 40-year-old Canadian females; that is, $n(S) = 98\,434$. Therefore, the required probability is $P(E) = \frac{n(E)}{n(S)} = \frac{87\,202}{98\,434} \approx 0.89$.

(b) Let F be the event that a 40-year-old Canadian female will live to be 100 years old; that is, $n(F) = 4708$. Let S be the set of 40-year-old Canadian females; that is, $n(S) = 98\,434$. Therefore, the required probability is $P(F) = \frac{n(F)}{n(S)} = \frac{4708}{98\,434} \approx 0.05$.

8.8.9. (a) $0.043 \times 0.182 = 0.007\,826$

(b) $0.043 \times 0.06 = 0.002\,58$

(c) $0.043(1 - 0.06 - 0.182) = 0.032\,594$

(d) $0.043 \times 0.28 \times 0.504 = 0.006\,068$

8.8.10. Let B be the event that both children are boys, and A be the event that at least one of them is a boy. This means that $B \cap A = \{(b, b)\}$ and $P(B \cap A) = \frac{n(B \cap A)}{n(S)} = \frac{1}{4}$. Also, $A = \{(b, b), (b, g), (g, b)\}$ and $P(A) = \frac{n(A)}{n(S)} = \frac{3}{4}$. Then the desired probability is

$$P(B \mid A) = \frac{P(B \cap A)}{P(A)} = \frac{P(\{(b, b)\})}{P(\{(b, b), (b, g), (g, b)\})} = \frac{\frac{1}{4}}{\frac{3}{4}} = \frac{1}{3}.$$

Remark: This problem has been hotly debated in the mathematical community. For instance, see Laporte *et al.* (1980), Bar-Hillel and Falk (1982), or Carlton and Stansfield (2005).

8.8.11. Let E be the event that the administration assistant had put all of the letters into the correct corresponding envelopes. There is only one correct way to do this; $n(E) = 1$. There are 6! ways to assign letters to the envelopes; $n(S) = 6! = 720$. Therefore, the probability that the letters were put in the correct envelopes is $P(E) = \frac{n(E)}{n(S)} = \frac{1}{720}$. Alternatively, $P(E) = \frac{1}{6} \times \frac{1}{5} \times \frac{1}{4} \times \frac{1}{3} \times \frac{1}{2} \times \frac{1}{1} = \frac{1}{6!} = \frac{1}{720}$.

8.8.12. Let F and E denote, respectively, the events that the first and the second balls drawn are black. Now, given that the first ball selected is black, there are 8 remaining black balls and 6 red balls, and so $P(E \mid F) = \frac{8}{14}$. As $P(F)$ is $\frac{9}{15}$, our desired probability is

$$P(E \cap F) = P(F) \times P(E \mid F) = \frac{9}{15} \times \frac{8}{14} = \frac{72}{210} = \frac{12}{35} \approx 0.3429.$$

8.8.13. Let $E_1, E_2, \ldots, E_{10}$ denote that person 1, person 2, $\ldots$, person 10 received his or her own name, respectively.

(a)
$$1 - P(E_1 \cup E_2 \cup \cdots \cup E_{10}) = 1 - \left(\frac{1}{1!} - \frac{1}{2!} + \frac{1}{3!} - \frac{1}{4!} + \frac{1}{5!} - \frac{1}{6!} + \cdots - \frac{1}{10!} \right)$$
$$= 1 - \frac{28\,319}{44\,800}$$
$$= \frac{16\,481}{44\,800}$$
$$\approx 0.367\,88$$

(b) $P(E_1 \cup E_2 \cup \cdots \cup E_{10}) = \dfrac{1}{1!} - \dfrac{1}{2!} + \dfrac{1}{3!} - \dfrac{1}{4!} + \dfrac{1}{5!} - \dfrac{1}{6!} + \cdots - \dfrac{1}{10!}$

$$= \dfrac{28\,319}{44\,800}$$

$$\approx 0.632\,12$$

(c) $\dfrac{1}{10}$

8.8.14. (a) $(0.5)^4 = 0.0625$

(b) $0.5 \times 0.6 \times 0.7 \times 0.8 = 0.168$

(c) The probability that the mouse did not make it on the first trial is $1 - 0.168 = 0.832$. Then the probability that the mouse fails the first trial but completely it successfully on the second trial is $0.832 \times 0.168 \approx 0.1398$

8.8.15. (a) $\frac{1}{3}$

(b) Some may follow this incorrect line of reasoning: Observing the gold coin eliminates Box 2; Boxes 1 and 3 must be equally likely; thus, the probability of box 3 must be $\frac{1}{2}$.

The question is really asking for the conditional probability of selecting Box 3 given that one of the drawers contains a gold medal.

$$P(\text{Box 3} \mid \text{Gold}) = \frac{P(\text{Box 3} \cap \text{Gold})}{P(\text{Gold})}$$

$$= \frac{P(\text{Box 3}) \times P(\text{Gold} \mid \text{Box 3})}{P(\text{Gold})}$$

$$= \frac{\frac{1}{3} \times \frac{1}{2}}{\frac{1}{2}}$$

$$= \frac{1}{3}$$

This seems counter-intuitive since the chosen coin was equally likely to be any of the three cold coins available, only one of which is in Box 3 (Rosenthal, 2009, pp. 35–36).

8.9.1. To show that two events E and F are mutually exclusive, you need to check that $E \cap F = \varnothing$. To show that two events are independent, you need to check that $P(E) \cdot P(F) = P(E \cap F)$.

8.9.2. (a) Yes. The events are independent since $P(E) \times P(F) = 0.6 \times 0.3 = 0.18 = P(E \cap F)$.

(b) No. Since, $P(E) \times P(F) = 0.2 \times 0.3 = 0.06 \neq 0.05 = P(E \cap F)$.

(c) We need to find $P(E \cap F)$ first.

$$P(E \cup F) = P(E) + P(F) - P(E \cap F)$$

$$0.76 = 0.7 + 0.2 - P(E \cap F)$$

$$P(E \cap F) = 0.7 + 0.2 - 0.76$$

$$= 0.14$$

Now we can check if E and F are independent.

$$P(E) \times P(F) = 0.7 \times 0.2 = 0.14 \text{ and } P(E \cap F) = 0.14$$

Therefore, E and F are independent.

8.9.3. (a) Independent

(b) Dependent

(c) Independent

  (d) Dependent
  (e) Dependent
  (f) Independent

8.9.4. $P(A) = \frac{3}{8} - \frac{1}{2}$, $P(B) = \frac{2}{6} - \frac{1}{3}$, and $P(C) = \frac{2}{6}$.

  (a) $P(A) \times P(B) = \frac{1}{2} \times \frac{1}{3} = \frac{1}{6} = P(A \cap B)$. Therefore, A and B are independent.

  (b) $P(A) \times P(C) = \frac{1}{2} \times \frac{5}{6} = \frac{5}{12} \neq \frac{2}{6} = P(A \cap C)$. Therefore A and C are not independent.

  (c) $P(B) \times P(C) = \frac{1}{3} \times \frac{5}{6} = \frac{5}{18} \neq \frac{2}{6} = P(B \cap C)$. Therefore, B and C are not independent.

8.9.5. Beginning with an eikosogram that has two independent events A and B (Fig. C.26), we can create the eikosograms for the subsequent events in question (Fig. C.27), and we can see that they are all independent.

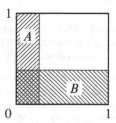

Fig. C.26 Two independent events A and B.

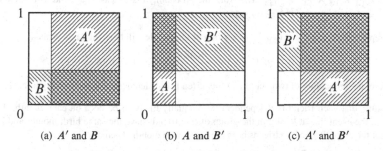

    (a) A' and B (b) A and B' (c) A' and B'

Fig. C.27 The eikosograms for the pairs of events in Exercise 8.9.5.

8.9.6. Recall that if A and B are independent, then $P(A \cap B) = P(A) P(B)$.

  (a) $P(A' \cap B) = P(B) - P(A \cap B) = P(B) - P(A) P(B) = P(B)[1 - P(A)] = P(B) P(A')$.
     Therefore, A' and B are also independent.

  (b) $P(A \cap B') = P(A) - P(A \cap B) = P(A) - P(A) P(B) = P(A)[1 - P(B)] = P(A) P(B')$.
     Therefore, A and B' are also independent.

(c) $P(A' \cap B') = P((A \cup B)')$

$\qquad\qquad = 1 - P(A \cup B)$

$\qquad\qquad = 1 - [P(A) + P(B) - P(A \cap B)]$

$\qquad\qquad = 1 - P(A) - P(B) + P(A \cap B)$

$\qquad\qquad = 1 - P(A) - P(B) + P(A) P(B)$

$\qquad\qquad = [1 - P(A)] - P(B)[1 - P(A)]$

$\qquad\qquad = [1 - P(A)][1 - P(B)]$

$\qquad\qquad = P(A') P(B')$

Therefore, A' and B' are also independent.

8.9.7. Let E be the event of throwing a 6 on the first throw, and F be the event of throwing a 6 on the second throw. Then, $P(E) = \frac{1}{6}$ and $P(F) = \frac{1}{6}$. We assume that E and F are independent. Then, the required probability is the probability of getting a 6 on the first throw *and* a 6 on the second throw: $P(E \cap F) = \frac{1}{6} \times \frac{1}{6} = \frac{1}{36}$.

8.9.8. Let E be the event that the first and second flares does not work, and the event E' that at least one of the flares work. Then, $P(E) = 0.25 \times 0.25 = 0.0625$ and $P(E') = 1 - 0.0625 = 0.9375$ or $\frac{15}{16}$. Therefore, the probability that at least one flare will work is 93.75%.

8.9.9. The problem requires us to find the probability that Brandan will hit his target *at least once*. Notice that the words *at least once* were not used in the problem, but it was implied. We will calculate the required probability using the complement rule (*indirect reasoning*). For Brandan to miss one shot, the probability is $1 - \frac{1}{3} = \frac{2}{3}$. Since each shot is assumed to be independent, then the probability that he misses the target three times is $\frac{2}{3} \times \frac{2}{3} \times \frac{2}{3} = (\frac{2}{3})^3 = \frac{8}{27}$. Therefore, the probability that he will hit his target at least once is $1 - \frac{8}{27} = \frac{19}{27} \approx 0.7037$.

8.9.10. Use *indirect reasoning*. Let E be the event that at least one round will hit the target. Then, E' is the event that no rounds hit the target. The probability that a shot misses is $1 - \frac{13}{14} = \frac{1}{14}$. Then, $P(E') = \frac{1}{14} \times \frac{1}{14} \times \frac{1}{14} = \frac{1}{2744}$. Therefore, the probability that at least one round will hit the target is $P(E) = 1 - \frac{1}{2744} = \frac{2743}{2744} \approx 99.96\%$.

8.9.11. Let A represent the event that two out of a group of ten people will have the same birthday.

$$P(A) = 1 - P(A') = 1 - \frac{P(365, 10)}{365^{10}} \approx 1 - 0.88 = 0.12$$

Therefore, the probability of two out of a group of ten people having the same birthday is 12%.

8.9.12. We assume that each friend is independent, and equally likely to be born on each month. Let B represent the event that at least 2 in the group of five friends have the same birth month, and let B' represent the event that no two friends have the same birth month. Then,

$$n(B') = P(12, 5) = 12 \times 11 \times 10 \times 9 \times 8 = 95\,040$$

and

$$n(S) = 12^5 = 248\,832.$$

Hence,

$$P(B') = \frac{n(B')}{n(S)} = \frac{95\,040}{248\,832} = \frac{55}{144} \approx 0.3819.$$

Therefore, the probability that at least two friends have the same birth month is

$$P(B) = 1 - P(B') = 1 - \frac{55}{144} = \frac{89}{144} \approx 0.6181.$$

8.9.13. This is similar to the Birthday paradox. Let A represent the event that at least 2 friends order the same entrée, and let A' represent the event that no two friends order the same entrée. Then,

$$n(A') = P(14,5) = 14 \times 13 \times 12 \times 11 \times 10 = 240\,240$$

and

$$n(S) = 14^5 = 537\,824.$$

Hence,

$$\mathrm{P}(A') = \frac{n(A')}{n(S)} = \frac{240\,240}{537\,824} = \frac{2145}{4802} \approx 0.4467.$$

Therefore, the chance that at least two friends will order the same entrée is

$$\mathrm{P}(A) = 1 - \mathrm{P}(A') = 1 - \frac{2145}{4802} = \frac{2657}{4802} \approx 0.5533.$$

8.9.14. By using *indirect reasoning*, we can calculate the complement of the probability of the employee not being chosen in any of the four quarters is $0.75 \times 0.75 \times 0.75 \times 0.75 = (0.75)^4$. Therefore, the probability that an employee will selected at least once in a year is $1 - (0.75)^4 \approx 0.684$.

8.9.15. We assume that you and I would select any of the numbers with equal probability. The probability that both numbers are the same is $\frac{1}{10}$. Thus, by the Complement Rule, the probability that both numbers are not the same is $1 - \frac{1}{10} = \frac{9}{10}$. If the two numbers selected are different, then the probability is $\frac{1}{2}$ that my number is higher than yours; the two outcomes are either my number is higher than yours, or it is not. Therefore, by independence the probability that the two numbers are different *and* my number is greater than yours is $\frac{1}{2} \times \frac{9}{10} = \frac{9}{20}$.

8.9.16. Let the event C be that you receive a call, and the event D that you are eating dinner. We are required to find the probability that you receive a call and you are eating dinner; we need to find $\mathrm{P}(C \cap D)$. The probability of randomly selecting a time when you are eating dinner from 5 to 9 pm is $\frac{1\,\text{hour}}{4\,\text{hours}}$; $\mathrm{P}(D) = \frac{1}{4}$. If we assume that these two events are independent, then

$$\mathrm{P}(C \cap D) = \mathrm{P}(C) \times \mathrm{P}(D) = 0.2 \times \frac{1}{4} = 0.2 \times 0.25 = 0.05.$$

8.9.17. We assume that each ant acts independently; that is, each ant will decide for themselves which edge to take. The four ants can only avoid a collision if they all picked to travel in the same direction (clockwise or counter-clockwise). If they do not all pick the same direction, then there is a collision. Thus, each ant has two choices: to move clockwise or counter-clockwise. Then,

$$\mathrm{P}(\text{clockwise}) = \mathrm{P}(\text{counter-clockwise}) = \frac{1}{2} \times \frac{1}{2} \times \frac{1}{2} \times \frac{1}{2} = \left(\frac{1}{2}\right)^4$$

and

$$\mathrm{P}(\text{no collision}) = \mathrm{P}(\text{clockwise}) + \mathrm{P}(\text{counter-clockwise})$$

$$= \left(\frac{1}{2}\right)^4 + \left(\frac{1}{2}\right)^4$$

$$= 2\left(\frac{1}{2}\right)^4$$

$$= \frac{1}{8}.$$

Here is another possible solution:

$$\mathrm{P}(\text{no collision}) = \frac{n(\text{no collision})}{n(\text{number of paths})} = \frac{2}{2^4} = \frac{2}{16} = \frac{1}{8}.$$

8.9.18. (a) P(winning on single number) $= \frac{1}{38}$

(b) P(winning on two numbers) $= \frac{2}{38}$

(c) Let E_i be the event that a black number appears on spin i, where $i = 1, 2, 3, \ldots, 10$. Then, $P(E_i) = \frac{18}{38}$. Since the spins are independent,

$$P(E_1 \cap E_2 \cap E_3 \cap \cdots \cap E_{10}) = P(E_1) \times P(E_2) \times P(E_3) \times \cdots \times P(E_{10})$$

$$= \underbrace{\frac{18}{38}}_{\text{Spin 1}} \times \underbrace{\frac{18}{38}}_{\text{Spin 2}} \times \underbrace{\frac{18}{38}}_{\text{Spin 3}} \times \cdots \times \underbrace{\frac{18}{38}}_{\text{Spin 10}}$$

$$= \left(\frac{18}{38}\right)^{10}$$

$$\approx 0.000\,57.$$

(d) Let F_i be the event that one of the eight numbers that you selected appears on spin i, where $i = 1, 2, 3, \ldots, 12$. Then $P(F_i) = \frac{8}{38} \approx 0.210\,53$, and the complement of F_i that none of the eight numbers appear is

$$P\left(F_i'\right) = 1 - \frac{8}{38} = \frac{30}{38} \approx 0.789\,47$$

Then, assuming that each spin is independent, the required probability is

$$P\left(F_1' \cap F_2' \cap \cdots \cap F_{12}'\right) = P\left(F_1'\right) \times P\left(F_2'\right) \times \cdots \times P\left(F_{12}'\right)$$

$$= \underbrace{\frac{30}{38}}_{\text{Spin 1}} \times \underbrace{\frac{30}{38}}_{\text{Spin 2}} \times \cdots \times \underbrace{\frac{30}{38}}_{\text{Spin 12}}$$

$$= \left(\frac{30}{38}\right)^{12}$$

$$\approx 0.058\,62.$$

(e) If we assume that each wheel spin outcome is independent, then the history of past outcomes will not affect the future outcomes. Therefore, we only need to focus on the probability of the next three consecutive spins coming up 26 black. Let G_i be the event that 26 black appears on spin i, where $i = 1, 2, 3$. Then, the required probability is

$$P(G_1 \cap G_2 \cap G_3) = P(G_1) \times P(G_2) \times P(G_3)$$

$$= \underbrace{\frac{1}{38}}_{\text{Spin 1}} \times \underbrace{\frac{1}{38}}_{\text{Spin 2}} \times \underbrace{\frac{1}{38}}_{\text{Spin 3}}$$

$$= \left(\frac{1}{38}\right)^{3}$$

$$\approx 0.000\,02.$$

(f) Let R_i be the event that red appears on spin i, where $i = 1, 2, 3$. Then $P(R_i) = \frac{18}{38}$. We assume that each spin is independent, so the probability that three reds appears is

$$P(R_1 \cap R_2 \cap R_3) = P(R_1) \times P(R_2) \times P(R_3)$$

$$= \underbrace{\frac{18}{38}}_{\text{Spin 1}} \times \underbrace{\frac{18}{38}}_{\text{Spin 2}} \times \underbrace{\frac{18}{38}}_{\text{Spin 3}}$$

$$= \left(\frac{18}{38}\right)^3$$

$$\approx 0.106\,28.$$

(g) Yes. Each number has a probability of $\frac{1}{38}$.

(h) No. The colors red and black have the same probability of $\frac{18}{38}$, but green only has a probability of $\frac{2}{38}$.

8.9.19. From the Venn diagram in Fig. 8.30, we have the following:

- $P(E) = 0 + 0.10 + 0.04 + 0.06 = 0.20$
- $P(F) = 0.16 + 0.10 + 0.04 + 0.20 = 0.50$
- $P(G) = 0.10 + 0.20 + 0.04 + 0.06 = 0.40$
- $P(E \cap F) = 0.10 + 0.04 = 0.14$
- $P(E \cap G) = 0.06 + 0.04 = 0.10$
- $P(F \cap G) = 0.04 + 0.20 = 0.24$.

Then,

(a) $P(E \cap F \cap G) = 0.04 = 0.20 \times 0.50 \times 0.40 = P(E) \times P(F) \times P(G)$

(b) $P(E \cap F) = 0.14$, but $P(E) \times P(F) = 0.20 \times 0.50 = 0.10$.

(c) $P(E \cap G) = 0.10$, but $P(E) \times P(G) = 0.20 \times 0.40 = 0.08$.

(d) $P(F \cap G) = 0.24$, but $P(F) \times P(G) = 0.50 \times 0.40 = 0.20$.

Remark: This exercise was adapted from George (2004).

8.10.1. $6^3 = 216$

8.10.2. *Take cases.* We can break this problem into four cases. *Case 1*: Getting a six on the first throw: $\frac{1}{6}$. *Case 2*: Getting a six on the second throw: $\frac{5}{6} \times \frac{1}{6} = \frac{5}{36}$. *Case 3*: Getting a six on the third throw: $\frac{5}{6} \times \frac{5}{6} \times \frac{1}{6} = \frac{25}{216}$. *Case 4*: Getting a six on the fourth throw: $\frac{5}{6} \times \frac{5}{6} \times \frac{5}{6} \times \frac{1}{6} = \frac{125}{1296}$. Thus, adding up the cases, we have the required probability: $\frac{1}{6} + \frac{5}{36} + \frac{25}{216} + \frac{125}{1296} = \frac{671}{1296}$.

8.10.3.

$$1 - \left(\frac{35}{36}\right)^n = \frac{1}{2}$$

$$-\left(\frac{35}{36}\right)^n = \frac{1}{2} - 1$$

$$-\left(\frac{35}{36}\right)^n = -\frac{1}{2}$$

$$\left(\frac{35}{36}\right)^n = \frac{1}{2}$$

$$\log\left(\frac{35}{36}\right)^n = \log\left(\frac{1}{2}\right)$$

$$n\log\left(\frac{35}{36}\right) = \log\left(\frac{1}{2}\right)$$

$$n = \frac{\log\left(\frac{1}{2}\right)}{\log\left(\frac{35}{36}\right)} \approx 24.6$$

8.11.1. (a) The completed probability tree diagram is given in Fig. C.28.

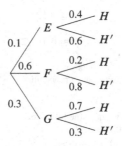

Fig. C.28   Completed probability tree for Exercise 8.11.1

(b)   (i) $P(H \mid E) = 0.4$

(ii) $P(H \mid F) = 0.2$

(iii) $P(H' \mid G) = 0.3$

(iv) $P(E \mid H) = \dfrac{P(E)\,P(H \mid E)}{P(E)\,P(H \mid E) + P(F)\,P(H \mid F) + P(G)\,P(H \mid G)}$

$= \dfrac{(0.1)(0.4)}{(0.1)(0.4) + (0.6)(0.2) + (0.3)(0.7)} \approx 0.108\,11$

(v) $P(F \mid H) = \dfrac{P(F)\,P(H \mid F)}{P(E)\,P(H \mid E) + P(F)\,P(H \mid F) + P(G)\,P(H \mid G)}$

$= \dfrac{(0.6)(0.2)}{(0.1)(0.4) + (0.6)(0.2) + (0.3)(0.7)} \approx 0.324\,32$

(vi) $P(G \mid H) = \dfrac{P(G)\,P(H \mid G)}{P(E)\,P(H \mid E) + P(F)\,P(H \mid F) + P(G)\,P(H \mid G)}$

$= \dfrac{(0.3)(0.7)}{(0.1)(0.4) + (0.6)(0.2) + (0.3)(0.7)} \approx 0.567\,57$

(vii) $P(E \mid H') = \dfrac{P(E)\,P(H' \mid E)}{P(E)\,P(H' \mid E) + P(F)\,P(H' \mid F) + P(G)\,P(H' \mid G)}$

$= \dfrac{(0.1)(0.6)}{(0.1)(0.6) + (0.6)(0.8) + (0.3)(0.3)} \approx 0.095\,24$

(viii) $P(F \mid H') = \dfrac{P(F)\,P(H' \mid F)}{P(E)\,P(H' \mid E) + P(F)\,P(H' \mid F) + P(G)\,P(H' \mid G)}$

$= \dfrac{(0.6)(0.8)}{(0.1)(0.6) + (0.6)(0.8) + (0.3)(0.3)} \approx 0.761\,90$

(ix) $P(G \mid H') = \dfrac{P(G)\,P(H' \mid G)}{P(E)\,P(H' \mid E) + P(F)\,P(H' \mid F) + P(G)\,P(H' \mid G)}$

$= \dfrac{(0.3)(0.3)}{(0.1)(0.6) + (0.6)(0.8) + (0.3)(0.3)} \approx 0.142\,86$

8.11.2. Let W be the event that a white ball is drawn, and let H be the event that the coin comes up heads. The desired probability $P(H \mid W)$ may be calculated as follows:

$$P(H \mid W) = \frac{P(H \cap W)}{P(W)} = \frac{P(W \mid H)P(H)}{P(W)}$$

$$= \frac{P(W \mid H)P(H)}{P(W \mid H)P(H) + P(W \mid H')P(H')}$$

$$= \frac{\frac{2}{9} \cdot \frac{1}{2}}{\frac{2}{9} \cdot \frac{1}{2} + \frac{5}{11} \cdot \frac{1}{2}} = \frac{22}{67} \approx 0.328\,36.$$

8.11.3. Let C_1 be the event that the fair coin was selected. Let C_2 be the event that the coin with two heads was selected. Let C_3 be the event that the coin whose probability of heads appearing is $\frac{1}{3}$. Since the coin is chosen at random, $P(C_1) = P(C_2) = P(C_3) = \frac{1}{3}$.

(a) $P(H) = P(C_1)P(H \mid C_1) + P(C_2)P(H \mid C_2) + P(C_3)P(H \mid C_3)$

$$= \frac{1}{3}\left(\frac{1}{2}\right) + \frac{1}{3}(1) + \frac{1}{3}\left(\frac{1}{3}\right) = \frac{11}{18}$$

(b) $P(C_2 \mid H) = \dfrac{P(C_2 \cap H)}{P(H)}$

$$= \frac{P(C_2) \cap H)}{P(C_1 \cap H) + P(C_2 \cap H) + P(C_3 \cap H)}$$

$$= \frac{P(C_2)P(H \mid C_2)}{P(C_1)P(H \mid C_1) + P(C_2)P(H \mid C_2) + P(C_3)P(H \mid C_3)}$$

$$= \frac{\frac{1}{3}(1)}{\frac{1}{3}\left(\frac{1}{2}\right) + \frac{1}{3}(1) + \frac{1}{3}\left(\frac{1}{3}\right)} = \frac{\frac{1}{3}}{\frac{11}{18}} = \frac{6}{11} \approx 0.545$$

8.11.4. Let G represent that the selected student is a graduate student. Let S represent that the student supports the idea of a café. Then, the required probability is

$$P(G \mid S) = \frac{P(G \cap S)}{P(S)}$$

$$= \frac{P(G)P(S \mid D)}{P(G)P(S \mid D) + P(G')P(S \mid G')}$$

$$= \frac{(0.3)(0.59)}{(0.3)(0.59) + (0.7)(0.77)}$$

$$= \frac{0.177}{0.177 + 0.539} \approx 0.247\,21.$$

Therefore, the probability that a randomly selected participant is a graduate student given that he or she supports the café idea is approximately 25%.

8.11.5. Let D denote that the sports car is defective, and D' denote that the sports car is not defective. The probability that a car came from plant A is $P(A) = \frac{350}{1000} = 0.35$, plant B is $P(B) = \frac{250}{1000} = 0.25$, and plant C is $P(C) = \frac{400}{1000} = 0.40$. From the records, the probability that the sports car will be defective given that it came from plant A is $P(D \mid A) = 0.05$, plant B is $P(D \mid B) = 0.03$, plant C is $P(D \mid C) = 0.07$. Now, we can address the questions we were asked.

(a) We would like to calculate the probability that the sports car came from plant A given that it is defective, $P(A \mid D)$. Using Bayes' Theorem, we have

$$
\begin{aligned}
P(A \mid D) &= \frac{P(A \cap D)}{P(D)} \\
&= \frac{P(A \cap D)}{P(A \cap D) + P(A' \cap D)} \\
&= \frac{P(A \cap D)}{P(A \cap D) + \underbrace{P(B \cap D) + P(C \cap D)}_{P(A' \cap D)}} \\
&= \frac{P(A) P(D \mid A)}{P(A) P(D \mid A) + P(B) P(D \mid B) + P(C) P(D \mid C)} \\
&= \frac{0.35(0.05)}{0.35(0.05) + 0.25(0.03) + 0.40(0.07)} \\
&\approx 0.330\,19.
\end{aligned}
$$

Therefore, there is a 33% chance that the defective sports car came from plant A.

(b) We would like to calculate the probability that the sports car came from plant B given that it is defective, $P(B \mid D)$. Using Bayes' Theorem, we have

$$
\begin{aligned}
P(B \mid D) &= \frac{P(B \cap D)}{P(D)} \\
&= \frac{P(B) P(D \mid B)}{P(A) P(D \mid A) + P(B) P(D \mid B) + P(C) P(D \mid C)} \\
&= \frac{0.25(0.03)}{0.35(0.05) + 0.25(0.03) + 0.40(0.07)} \\
&\approx 0.141\,51.
\end{aligned}
$$

Therefore, there is a 14% chance that the defective sports car came from plant B.

(c) We would like to calculate the probability that the sports car came from plant C given that it is defective, $P(C \mid D)$. Using Bayes' Theorem, we have

$$
\begin{aligned}
P(C \mid D) &= \frac{P(C \cap D)}{P(D)} \\
&= \frac{P(C) P(D \mid C)}{P(A) P(D \mid A) + P(B) P(D \mid B) + P(C) P(D \mid C)} \\
&= \frac{0.40(0.07)}{0.35(0.05) + 0.25(0.03) + 0.40(0.07)} \\
&\approx 0.528\,30.
\end{aligned}
$$

Therefore, there is a 53% chance that the defective sports car came from plant C.

8.11.6. Let E be the event that the message is an email, and E' be the event that the message is spam. Let F be the event that the message is identified as email by the program, and F' be the event that the message is identified as spam by the program.

(a) P(identified as spam | message is spam) $= \mathrm{P}(F' \mid E')$

$$= \frac{\mathrm{P}(E' \cap F)}{\mathrm{P}(E')}$$

$$= \frac{0.334}{0.334 + 0.053}$$

$$\approx 0.863\,05$$

(b) P(identified as email | message is email) $= \mathrm{P}(F \mid E)$

$$= \frac{\mathrm{P}(E \cap F)}{\mathrm{P}(E)}$$

$$= \frac{0.573}{0.573 + 0.040}$$

$$\approx 0.934\,75$$

8.11.7. (a) The probability that the player's game console survives on the first round is $\frac{1}{6}$.

(b) The probability tree is given in Fig. C.29.

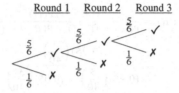

Round 1   Round 2   Round 3

Fig. C.29   Probability tree for Game Console Russian Roulette

(c) The probability that the game console survives on the third round is

$$\frac{5}{6} \times \frac{5}{6} \times \frac{5}{6} = \left(\frac{5}{6}\right)^3 = \frac{125}{216}.$$

(d) The probability that the game console does not survive by the third round is

$$\underbrace{\frac{1}{6}}_{\text{Round 1}} + \underbrace{\frac{5}{6} \cdot \frac{1}{6}}_{\text{Round 2}} = \frac{11}{36}.$$

8.11.8. If the barrel is spun, then it will randomly land on any of the chambers. Thus, the probability of surviving is $\frac{4}{6} = \frac{2}{3} \approx 67\%$.

If the barrel is not spun, then we can use the information that the previous chamber did not contain a bullet. We will number each of the chambers like Fig. C.30, and assume the barrel rotates clockwise. If there was no gunfire in the previous trigger pull, then it must have been chamber 1, 2, 3, or 4. If it was chamber 4, then the next trigger pull will be on chamber 5 and results in death. Thus, the chambers that ensure survival is 1, 2, and 3. Thus, the probability of survival is $\frac{3}{4} = 75\%$. Therefore, you would prefer to just pull the trigger again since it has a higher probability of survival.

Remark: This problem was adapted from Poundstone (2003, pp. 8–9), and his book focuses on the types of problems that employers (*e.g.*, Microsoft) ask candidates during a job interview. I find it curious that employers give their candidates problems to solve during an interview in the first place. If the candidates have a degree, wouldn't employers trust the university's opinion that the candidate can solve problems? Nevertheless, it seems to be an increasingly popular thing to do for employers. And

Fig. C.30   The gun's cylinder that is assumed to rotate clockwise.

when my students ask me if they ever will solve probability problems like the ones in class or in this textbook in the real-world, I can smile and say, "Yes! You might see this not only on your final exam, but in your job interview too!"

8.11.9. Let I denote innocent people; $n(I) = 37$. Let I' denote the robbers; $n(I') = 3$. There are $37 + 3 = 40$ people in total. Thus, $P(I) = \frac{37}{40}$ and $P(I') = \frac{3}{40}$.

Let R denote that the polygraph indicates that the person tested is a robber, while R' denotes that the person is identified not as a robber.

By Bayes' Theorem,

$$
\begin{aligned}
P(I' \mid R) &= \frac{P(R \mid I')\,P(I')}{P(R \mid I')\,P(I') + P(R \mid I)\,P(I)} \\
&= \frac{(0.95)\left(\dfrac{3}{40}\right)}{(0.95)\left(\dfrac{3}{40}\right) + (0.05)\left(\dfrac{37}{40}\right)} \\
&= \frac{0.071\,25}{0.071\,25 + 0.042\,65} \\
&\approx 0.606\,38.
\end{aligned}
$$

8.11.10. Suppose you pick Door 1, and the host forgot which door the prize is behind, and opens Door 3. The probabilities that Door 3 happens to not contain the prize, if the prize is behind Door 1, Door 2, and Door 3 are 1, 1, and 0 respectively. Thus, the probabilities the prize is behind Door 1, Door 2, or Door 3 are $\frac{1}{2}$, $\frac{1}{2}$, and 0 respectively. Therefore, the probability is the same whether you switch or not.

8.11.11. Let A, B, and C be the event that the corresponding felon will be set free, and b be the event that the Warden says he will be executed.

(a) The probability A will live is the following conditional probability:

$$
\begin{aligned}
P(A \mid b) &= \frac{P(b \mid A)\,P(A)}{P(b \mid A)\,P(A) + P(b \mid B) + P(b \mid C)\,P(C)} \\
&= \frac{\frac{1}{2} \times \frac{1}{3}}{\left(\frac{1}{2} \times \frac{1}{3}\right) + \left(0 \times \frac{1}{3}\right) + \left(1 \times \frac{1}{3}\right)} \\
&= \frac{\frac{1}{6}}{\frac{1}{6} + \frac{1}{3}} = \frac{\frac{1}{6}}{\frac{3}{6}} = \frac{1}{6} \times \frac{6}{3} = \frac{1}{3}.
\end{aligned}
$$

So A's probability of survival is not affected by the additional information from the Warden. This highlights that the probability does not change by the information obtained, but by the way in which it was obtained (Bar-Hillel and Falk, 1982, pp. 118–119).

(b) Poor B is absolutely going to die.

$$P(B \mid b) = \frac{P(b \mid B)\,P(B)}{P(b \mid A)\,P(A) + P(b \mid B) + P(b \mid C)\,P(C)}$$

$$= \frac{0 \times \frac{1}{3}}{\left(\frac{1}{2} \times \frac{1}{3}\right) + \left(0 \times \frac{1}{3}\right) + \left(1 \times \frac{1}{3}\right)} = 0$$

(c) The probability C will live is the following conditional probability:

$$P(C \mid b) = \frac{P(b \mid C)\,P(C)}{P(b \mid A)\,P(A) + P(b \mid B) + P(b \mid C)\,P(C)}$$

$$= \frac{1 \times \frac{1}{3}}{\left(\frac{1}{2} \times \frac{1}{3}\right) + \left(0 \times \frac{1}{3}\right) + \left(1 \times \frac{1}{3}\right)} = \frac{\frac{1}{3}}{\frac{3}{6}} = \frac{1}{3} \times \frac{6}{3} = \frac{2}{3}.$$

3.11.12. Accuracy $= \dfrac{TP + TN}{TP + FP + FN + TN}$

$$= \frac{P(\text{HIV}^+ \cap T^+) + P(\text{HIV}^- \cap T^-)}{P(\text{HIV}^+ \cap T^+) + P(\text{HIV}^- \cap T^+) + P(\text{HIV}^+ \cap T^-) + P(\text{HIV}^- \cap T^-)}$$

$$= \frac{0.002\,02 + 0.848\,19}{0.002\,02 + 0.149\,68 + 0.000\,11 + 0.848\,19}$$

$$= 0.850\,21$$

Therefore, the accuracy of the at-home HIV test is 85%.

3.11.13. $\dfrac{\frac{1}{2}}{\frac{1}{2}(1) + \frac{1}{2}\left(\frac{1}{4}\right)} = \dfrac{\frac{1}{2}}{\frac{1}{2} + \frac{1}{8}} = \dfrac{\frac{1}{2}}{\frac{5}{8}} = \frac{1}{2} \times \frac{8}{5} = \frac{4}{5}$

Remark: This problem was adapted from Rosenthal (2008).

3.11.14. Using the hint, we will create a confusion matrix with 1000 women. Of the 1000 women, $0.55 \times 1000 = 550$ women will have a UTI, and $1000 - 550 = 450$ women will not. Of the 550 women with a UTI, there are $0.90 \times 550 = 495$ true positives, and $550 - 495 = 55$ false negatives. There are $0.6 \times 450 = 270$ true negatives, and $450 - 270 = 180$ false positives.

We can create a confusion matrix like the one in Table C.6. From Table C.6, the probability the women has a UTI given that at least one of the tests is positive is $\frac{495}{675} = 0.73 = 73\%$.

The accuracy of the test is $\frac{495+270}{1000} = \frac{765}{1000} = 0.765 = 76.5\%$.

Table C.6 Results of dipstick testing in 1000 women presenting to a general practitioner with painful urination. Reproduced from Doust (2009, Table 1), with permission from BMJ Publishing Group Ltd.

	UTI present	UTI absent	Total
Either nitrites or leukocyte esterase positive	495	180	675
Both tests negative	55	270	325
Total	550	450	1000

8.11.15. Let C be the event that the note is counterfeit, and C' be the event that the note is legitimate. Let N be the event that the test result is negative, and N' be the event that the test result is positive.

We can create a probability tree diagram such as the one in Fig. C.31.

Fig. C.31   probability tree diagram for Exercise 8.11.15.

Using Bayes' Theorem, the probability that a bank note is counterfeit given that a test result is positive is

$$
\begin{aligned}
\mathrm{P}(C \mid N') &= \frac{\mathrm{P}(C)\,\mathrm{P}(N' \mid C)}{\mathrm{P}(C)\,\mathrm{P}(N' \mid C) + \mathrm{P}(C')\,\mathrm{P}(N' \mid C')} \\
&= \frac{\mathrm{P}(C \cap N')}{\mathrm{P}(C \cap N') + \mathrm{P}(C' \cap N')} \\
&= \frac{0.0045}{0.0045 + 0.149\,25} \\
&\approx 0.029\,27.
\end{aligned}
$$

At this point, there are two ways we can calculate the probability with the second positive test:

(a) We can update our probability estimate just like we did in the World Trade Center attacks Example 8.11.2. Thus, our *posterior* probability of a counterfeit note after the first positive test, 2.927%, becomes our *prior* probability before the second positive test; that is, $\mathrm{P}(C) = 0.029\,27$ and $\mathrm{P}(C') = 1 - 0.029\,27 = 0.970\,73$. Then,

$$
\begin{aligned}
\mathrm{P}(C \mid N') &= \frac{\mathrm{P}(C)\,\mathrm{P}(N' \mid C)}{\mathrm{P}(C)\,\mathrm{P}(N' \mid C) + \mathrm{P}(C')\,\mathrm{P}(N' \mid C')} \\
&= \frac{(0.029\,27)(0.9)}{(0.029\,27)(0.9) + (0.970\,73)(0.15)} \\
&\approx 0.153\,19.
\end{aligned}
$$

Therefore, the probability of the note being counterfeit has increased from 2.99% to 15.32%.

(b) Or,

$$
\mathrm{P}\big(C \mid (N' \cap N')\big) = \frac{\mathrm{P}(C)\,\mathrm{P}\big((N' \cap N') \mid C\big)}{\mathrm{P}(C)\,\mathrm{P}\big((N' \cap N') \mid C\big) + \mathrm{P}(C')\,\mathrm{P}\big((N' \cap N') \mid C'\big)}
$$

Since the tests are independent, $P((N' \cap N') \mid C) = P(N' \mid C)P(N' \mid C)$ and $P((N' \cap N') \mid C') = P(N' \mid C')P(N' \mid C')$, then

$$P(C \mid N' \cap N') = \frac{P(C)P(N' \mid C)P(N' \mid C)}{P(C)P(N' \mid C)P(N' \mid C) + P(C')P(N' \mid C')P(N' \mid C')}$$

$$= \frac{(0.005)(0.9)(0.9)}{(0.005)(0.9)(0.9) + (0.995)(0.15)(0.15)}$$

$$= \frac{(0.005)(0.9)^2}{(0.005)(0.9)^2 + (0.995)(0.15)^2}$$

$$\approx 0.153\,19.$$

Therefore, the probability of the note being counterfeit has increased from 2.99% to 15.32%.

8.11.16. (a) The probability tree is given in Fig. C.32.

(b) $P(\text{at least one red}) = \left(\frac{6}{36}\right)\left(\frac{5}{10}\right) + \left(\frac{6}{36}\right)\left(\frac{5}{10}\right)\left(\frac{5}{9}\right) + \left(\frac{6}{36}\right)\left(\frac{5}{10}\right)\left(\frac{5}{9}\right)$

$$+ \left(\frac{6}{36}\right)\left(\frac{5}{10}\right)\left(\frac{4}{9}\right)$$

$$= \left(\frac{6}{36}\right)\left(\frac{5}{10}\right)\left(1 + \frac{5}{9} + \frac{5}{9} + \frac{4}{9}\right)$$

$$= \left(\frac{6}{36}\right)\left(\frac{5}{10}\right)\left(\frac{23}{9}\right)$$

$$= \left(\frac{1}{6}\right)\left(\frac{1}{2}\right)\left(\frac{23}{9}\right)$$

$$= \frac{23}{108}$$

(c) $P(\text{sum} \geq 10 \mid \text{at least one red}) = \dfrac{P(\text{sum} \geq 10 \cap \text{at least one red})}{P(\text{at least one red})}$

$$= \frac{(\frac{6}{36})(\frac{5}{10})(\frac{5}{9}) + (\frac{6}{36})(\frac{5}{10})(\frac{5}{9}) + (\frac{6}{36})(\frac{5}{10})(\frac{4}{9})}{\frac{23}{108}}$$

$$= \frac{(\frac{6}{36})(\frac{5}{10})(\frac{5}{9} + \frac{5}{9} + \frac{4}{9})}{\frac{23}{108}}$$

$$= \frac{(\frac{6}{36})(\frac{5}{10})(\frac{14}{9})}{\frac{23}{108}}$$

$$= \frac{(\frac{1}{6})(\frac{1}{2})(\frac{14}{9})}{\frac{23}{108}}$$

$$= \frac{\frac{14}{108}}{\frac{23}{108}}$$

$$= \frac{14}{23}$$

8.12.1. There are 38 numbers on a roulette wheel; $n(S) = 38$.

(a) The odd numbers are $O = \{1, 3, 5, \ldots, 35\}$; $n(O) = 18$. Therefore, the probability of an odd number is $P(O) = \frac{n(O)}{n(S)} = \frac{18}{38} = \frac{9}{19}$.

(b) The even numbers are $E = \{2, 4, 6, \ldots, 36\}$; $n(E) = 18$. Therefore, the probability of an even number is $P(E) = \frac{n(E)}{n(S)} = \frac{18}{38} = \frac{9}{19}$.

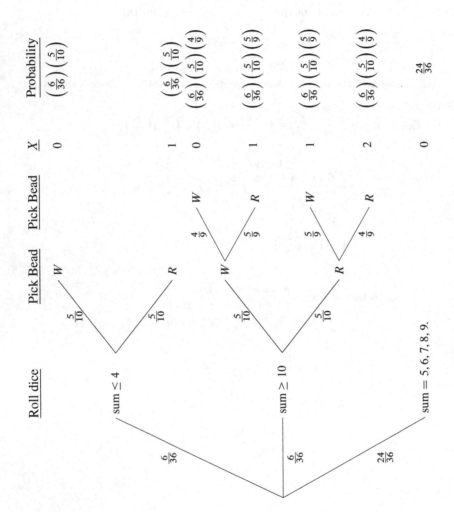

Fig. C.32   Probability Tree for Urn Experiment

(c) The prime numbers on the roulette wheel are $N = \{2, 3, 5, 7, 11, 13, 17, 19, 23, 29, 31\}$. Therefore, the probability of a prime number is $P(N) = \frac{n(N)}{n(S)} = \frac{11}{38}$.

8.12.2. (a) $P(A') = 1 - P(A) = 1 - 0.4 = 0.6$

(b) $P(A \cup B) = P(A) + P(B) - P(A \cap B) = 0.4 + 0.75 - 0.25 = 0.9$

(c) $P(A' \cap B) = P(B) - P(A \cap B) = 0.75 - 0.25 = 0.5$

(d) $P\big((A \cup B)'\big) = 1 - P(A \cup B) = 1 - 0.9 = 0.1$

(e) $P(A \mid B) = \frac{P(A \cap B)}{P(B)} = \frac{0.25}{0.75} = \frac{1}{3}$

(f) $P(B \mid A') = \frac{P(B \cap A')}{P(A')} = \frac{0.5}{0.6} = \frac{5}{6} \approx 0.833$

8.12.3. (a) $P(E \mid F) = \dfrac{P(E \cap F)}{P(F)}$

$P(E \cap F) = P(E \mid F) P(F) = (0.1)(0.6) = 0.06$

(b) $P(E \cup F) = P(E) + P(F) - P(E \cap F) = 0.2 + 0.6 - 0.06 = 0.74$

(c) $P(F \mid E) = \frac{P(E \cap F)}{P(E)} = \frac{0.06}{0.2} = 0.3$

(d) $P(E' \cup F) = P(E') + P(F) - P(E' \cap F)$

$\qquad = P(E') + P(F) - [P(F) - P(E \cap F)]$

$\qquad = 0.8 + 0.6 - (0.6 - 0.06)$

$\qquad = 0.86$

An alternative solution is $P(E' \cup F) = 1 - P(E) + P(E \cap F) = 1 - 0.2 + 0.06 = 0.86$.

8.12.4. (a) $P(A') = 1 - P(A) = 1 - 0.38 = 0.62$

(b) $P(B') = 1 - P(B) = 1 - 0.25 = 0.75$

(c) $P(A \cup B) = P(A) + P(B) - P(A \cap B) = 0.38 + 0.25 - 0.16 = 0.47$

(d) $P(A \mid B) = \dfrac{P(A \cap B)}{P(B)} = \dfrac{0.16}{0.25} = 0.64$

(e) $P(A' \cup B') = P\big((A \cap B)'\big) = 1 - P(A \cap B) = 1 - 0.16 = 0.84$

(f) $P(A' \cap B') = P\big((A \cup B)'\big) = 1 - P(A \cup B) = 1 - 0.47 = 0.53$

8.12.5. There are three cards that we know are not in the deck: $K\heartsuit$, $J\spadesuit$, $Q\heartsuit$. So there are $52 - 3 = 49$ remaining cards. For the dealer to have blackjack, he needs to have one of the four aces. Thus, the probability that the dealer has blackjack is $\frac{4}{49} \approx 0.082 = 8.2\%$.

8.12.6. (a) $\frac{6}{6^5} = \frac{1}{6^4} = \frac{1}{1296} \approx 0.000\,77$

(b) $\frac{\binom{6}{1}\binom{5}{1}}{6^5} = \frac{15 \times 5}{7776} = \frac{75}{7776} = \frac{25}{2592} \approx 0.009\,65$

(c) $\frac{\binom{6}{2}\binom{5}{2}}{6^5} = \frac{15 \times 10}{7776} = \frac{150}{7776} = \frac{25}{1296} \approx 0.019\,29$

8.12.7. *By indirect reasoning.* The probability of the six people not sharing a birth month is $\frac{12 \times 11 \times 10 \times 9 \times 8 \times 7}{12^6} = \frac{P(12,6)}{12^6} = \frac{665\,280}{2\,985\,984} \approx 0.223$. Therefore, the required probability is $1 - 0.223 = 0.777$.

8.12.8. We can use the general formula that was derived in Example 8.8.6 for $n = 7$. Thus, the probability that at least one of the pairs is a couple is

$$1 - \frac{1}{2!} + \frac{1}{3!} - \frac{1}{4!} + \frac{1}{5!} - \frac{1}{6!} + \frac{1}{7!} = 1 - \frac{1}{2} + \frac{1}{6} - \frac{1}{24} + \frac{1}{120} - \frac{1}{720} + \frac{1}{5040} = \frac{177}{280}.$$

8.12.9. The probability that a caller misses the question is $1 - 0.3 = 0.7$. We assume that each caller's probability of succeeding (or failing) is independent. If we want the probability of the fifth caller succeeding, then this means that caller 1, 2, 3, and 4 failed and caller 5 succeeded. Therefore, $0.7 \times 0.7 \times 0.7 \times 0.7 \times 0.3 = (0.7)^4 \times 0.3 = 0.072\,03 = 7.2\%$ chance the fifth caller will win.

8.12.10. Let A be the event the first card is a king, and B be the event the second card is a king. Then, $P(A \cap B) = \frac{4}{52} \times \frac{3}{51} = \frac{12}{2652} = \frac{1}{221} \approx 0.0045.$

8.12.11. Since these events are independent, $P(E \cap F) = P(E) \times P(F) = 0.4 \times 0.3 = 0.12$. Therefore, the required probability is $P(E \cup F) = P(E) + P(F) - P(E \cap F) = 0.4 + 0.3 - 0.12 = 0.58.$

8.12.12. (a)   (i) $P(A' \cap T') = P\big((A \cup T)'\big) = 1 - P(A \cup T)$. Therefore, $P(A \cup T) = 1 - P(A' \cap T') = 1 - 0.5 = 0.5.$

     (ii) $P(A \cup T) = P(A) + P(T) - P(A \cap T)$

$$P(T) = P(A \cup T) - P(A) + P(A \cap T)$$
$$= 0.5 - 0.2 + 0.1$$
$$= 0.3 + 0.1$$
$$= 0.4$$

(b) $P(A \mid T) = \frac{P(A \cap T)}{P(T)} = \frac{0.1}{0.4} = \frac{1}{4} = 0.25$

(c) $P(A) \times P(T) = 0.2 \times 0.4 = 0.08 \neq 0.1 = P(A \cap T)$. Thus, events A and T are not independent.

8.12.13. P(all women) $= \frac{\binom{4}{4}}{\binom{9}{4}} = \frac{1}{126} \approx 0.007\,94$. There was a 0.794% chance that the supervisor selected all female. The probability is small enough to cast doubt on the supervisor's claim. The supervisor picking all the women is unusual under random selection so the supervisor's selection makes us question if it was truly random.

8.12.14. (a) Let J be the event that James is on the committee. Then the number of ways that James is on the committee is $n(J) = \binom{9}{3}$. The sample space would be all the possible committees; $n(S) = \binom{10}{4}$. Therefore, the required probability is

$$P(J) = \frac{\binom{9}{3}}{\binom{10}{4}} = \frac{\frac{9!}{(9-3)!3!}}{\frac{10!}{(10-4)!4!}} = \frac{\frac{9!}{6!3!}}{\frac{10!}{6!4!}} = \frac{9! \; 6!4!}{10! \; 6!3!} = \frac{9! \times 4 \times 3!}{10 \times 9! \times 3!} = \frac{4}{10} = \frac{2}{5}.$$

(b) Let K be the event that James is selected with this additional restriction. The number of committees with James with one more male and two females is

$$n(K) = \underbrace{1}_{\text{James}} \times \underbrace{\binom{6}{1}}_{\text{Male}} \times \underbrace{\binom{3}{2}}_{\text{Females}} = 1 \times 6 \times 3 = 18.$$

The total possible committees with two males and two females is

$$n(S) = \underbrace{\binom{7}{2}}_{\text{Males}} \times \underbrace{\binom{3}{2}}_{\text{Females}} = 21 \times 3 = 63.$$

Therefore, the required probability is

$$P(K) = \frac{n(K)}{n(S)} = \frac{\binom{6}{1} \times \binom{3}{2}}{\binom{7}{2} \times \binom{3}{2}} = \frac{\binom{6}{1}}{\binom{7}{2}} = \frac{2}{7}.$$

8.12.15. (a) $\dfrac{\overbrace{\dbinom{4}{1}}^{\text{Suit}}\overbrace{\dbinom{13}{2}}^{\text{Rank}}}{\dbinom{52}{2}} = \dfrac{312}{1326} = \dfrac{4}{17} \approx 0.235\,29$

(b) Let E denote that the first card is a heart, and F denote that the second card is a heart.

$$P(E \mid F) = \frac{P(E \cap F)}{P(E)\,P(F \mid E) + P(E')\,P(F \mid E')}$$

$$= \frac{\frac{13}{52} \times \frac{12}{51}}{\frac{13}{52} \times \frac{12}{51} + \frac{39}{52} \times \frac{13}{51}} = \frac{4}{17} \approx 0.235\,29$$

9.2.1. (a) Discrete   (c) Discrete   (e) Continuous
       (b) Continuous   (d) Discrete   (f) Discrete

9.2.2. (a) Discrete   (c) Discrete   (e) Discrete
       (b) Continuous   (d) Discrete   (f) Continuous

9.2.3. (a) The probability distribution is given below, and its graph is in Fig. C.33.

x	1	2	3	4	5	6
$P(X = x)$	$\frac{1}{6}$	$\frac{1}{6}$	$\frac{1}{6}$	$\frac{1}{6}$	$\frac{1}{6}$	$\frac{1}{6}$

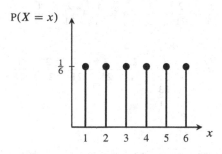

Fig. C.33 Graph of the probability distribution of a die

(b)

$$\sum_{\text{all } x} P(X = x) = \sum_{x=1}^{6} P(X = x)$$

$$= P(X = 1) + P(X = 2) + P(X = 3) + \cdots + P(X = 6)$$

$$= \frac{1}{6} + \frac{1}{6} + \frac{1}{6} + \frac{1}{6} + \frac{1}{6} + \frac{1}{6}$$

$$= 1$$

9.2.4. The probability distribution is given in Table C.7, and its graph is in Fig. C.34.

Table C.7   Probability distribution of people receiving their correct coat

x	0	1	2	3	4
$P(X = x)$	$\frac{9}{24}$	$\frac{8}{24}$	$\frac{6}{24}$	$\frac{0}{24}$	$\frac{1}{24}$

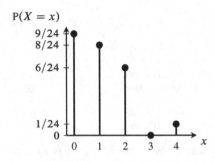

Fig. C.34   Graph of the probability distribution of people receiving their correct coat

9.2.5. The probabilities in the distribution must sum to 1. Then,

$$\frac{1}{21} + \frac{2}{21} + \frac{3}{21} + a + \frac{4}{21} + \frac{2}{21} + \frac{1}{21} = 1.$$

Therefore,

$$a = 1 - \frac{1}{21} - \frac{2}{21} - \frac{3}{21} - \frac{4}{21} - \frac{2}{21} - \frac{1}{21}$$
$$= 1 - \frac{13}{21}$$
$$= \frac{8}{21}$$

9.2.6. If we think of representing an outcome as one permutation of symbols, like H and T, then how ways can we permute $HHHH$ (0 tails), or $THHH$ (1 tail), *etc*. This is similar to Example 5.8.4 on p. 176 with permutations of a word that had repeated letters.

X	"Word"	Number of permutations	$P(X = x)$
0	$HHHH$	$\frac{4!}{0!4!} = \binom{4}{0} = 1$	$\frac{1}{16}$
1	$THHH$	$\frac{4!}{1!3!} = \binom{4}{1} = 4$	$\frac{4}{16}$
2	$TTHH$	$\frac{4!}{2!2!} = \binom{4}{2} = 6$	$\frac{6}{16}$
3	$TTTH$	$\frac{4!}{3!1!} = \binom{4}{3} = 4$	$\frac{4}{16}$
4	$TTTT$	$\frac{4!}{4!0!} = \binom{4}{4} = 1$	$\frac{1}{16}$
$\sum$		16	1

9.2.7. If we think of representing an outcome as one permutation of symbols, like 6 and N (not a 6), then how many ways can we permute $NNNN$ (no 6's), or $6NNN$ (one 6), *etc.* This is similar to Example 5.8.4 on p. 176 with permutations of a word that had repeated letters.

X	"Word"	Number of permutations	$P(X = x)$
0	$NNNN$	$\frac{4!}{0!4!} = \binom{4}{0} = 1$	$\frac{1}{16}$
1	$6NNN$	$\frac{4!}{1!3!} = \binom{4}{1} = 4$	$\frac{4}{16}$
2	$66NN$	$\frac{4!}{2!2!} = \binom{4}{2} = 6$	$\frac{6}{16}$
3	$666N$	$\frac{4!}{3!1!} = \binom{4}{3} = 4$	$\frac{4}{16}$
4	6666	$\frac{4!}{4!0!} = \binom{4}{4} = 1$	$\frac{1}{16}$
Σ		16	1

(a) $P(X = 2) = \dfrac{6}{16} = \dfrac{3}{8}$

(b) $P(X \le 2) = P(X = 0) + P(X = 1) + P(X = 2)$

$$= \frac{1}{16} + \frac{4}{16} + \frac{6}{16}$$

$$= \frac{11}{16}$$

(c) By *indirect reasoning*: $P(X > 2) = 1 - \frac{11}{16} = \frac{5}{16}$. Alternatively, by *direct reasoning*:
$P(X > 2) = P(X = 3) + P(X = 4) = \frac{4}{16} + \frac{1}{16} = \frac{5}{16}$.

(d) $P(1 < X < 4) = P(X = 2) + P(X = 3) = \frac{6}{16} + \frac{4}{16} = \frac{10}{16} = \frac{5}{8}$

9.2.8. The probability distribution is given below.

x	3	4	5	6
$P(X = x)$	$\frac{5}{26}$	$\frac{6}{26}$	$\frac{7}{26}$	$\frac{8}{26}$

$$\frac{5}{26} + \frac{6}{26} + \frac{7}{26} + \frac{8}{26} = \frac{26}{26} = 1$$

Since the probabilities sum to one, $P(X = x) = \frac{x+2}{26}$ for $x = 3, 4, 5, 6$ is a probability distribution.

9.2.9. $P(N > 6 \mid N \le 60) = \dfrac{P(N > 6 \cap N \le 60)}{P(N \le 60)}$

$$= \frac{P(N = 12) + P(N = 60)}{P(N = 2) + P(N = 6) + P(N = 12) + P(N = 60)}$$

$$= \frac{0.1 + 0.3}{0.15 + 0.05 + 0.1 + 0.3}$$

$$= \frac{0.4}{0.6}$$

$$= \frac{2}{3}$$

9.2.10. (a) $P(X = 2) = 0.25$

(b) $P(X \ge 3) = P(X = 3) + P(X = 4) + P(X = 5) = 0.25 + 0.15 + 0.1 = 0.5$

(c) $P(X \text{ is odd}) = P(X = 1) + P(X = 3) + P(X = 5) = 0.15 + 0.25 + 0.1 = 0.5$

(d) $P(1 \leq X \leq 4) = P(X = 1) + P(X = 2) + P(X = 3) + P(X = 4) = 0.15 + 0.25 + 0.25 + 0.15 = 0.8$

9.2.11. We are required to find $P(N \geq 1 \mid N \leq 4)$.

$$
\begin{aligned}
P(N \geq 1 \mid N \leq 4) &= \frac{P(N \geq 1 \cap N \leq 4)}{P(N \leq 4)} \\
&= \frac{P(N = 1) + P(N = 2) + P(N = 3) + P(N = 4)}{P(N = 0) + P(N = 1) + P(N = 2) + P(N = 3) + P(N = 4)} \\
&= \frac{\frac{1}{6} + \frac{1}{12} + \frac{1}{20} + \frac{1}{30}}{\frac{1}{2} + \frac{1}{6} + \frac{1}{12} + \frac{1}{20} + \frac{1}{30}} \\
&= \frac{\frac{1}{3}}{\frac{5}{6}} \\
&= \frac{2}{5}
\end{aligned}
$$

9.2.12. (a) The probability distribution of X is given below.

x	< 130	130 − 150	> 150
$P(X = x)$	$\frac{97}{2300}$	$\frac{2082}{2300}$	$\frac{121}{2300}$

(b) The probability that a randomly selected vehicle going faster than 150 kph is $\frac{121}{2300} \approx 5.26\%$.

9.2.13. Let X be the number of months the student has kept the phone.

(a) $P(X \geq 6) = \frac{132+121+113}{500} = \frac{366}{500} = \frac{183}{250} \approx 0.732$

(b) $P(X < 4) = \frac{4+30}{500} = \frac{34}{500} = \frac{17}{250} \approx 0.068$

(c) $P(4 \leq X < 10) = \frac{100+132+121}{500} = \frac{353}{500} \approx 0.706$

9.2.14. You could repeat a similar experiment found in Example 8.6.4 on p. 333. You could spin repeatedly the Roulette wheel, and keep a record of how many times each number appears in a number of spins. As the number of spins increase, the empirical probabilities of the ball landing on each number should approach $\frac{1}{38}$.

9.3.1. Let X be the amount of the casino's winnings. The probability of a single particular number being selected is $\frac{1}{38}$, and the probability that the player loses his bet is $\frac{37}{38}$. Therefore, the expected casino's winnings is

$$
E(X) = -35[P(X = -35)] + (1)[P(X = 1)] = -35\left(\frac{1}{38}\right) + 1\left(\frac{37}{38}\right) = 0.052\,63.
$$

This can be interpreted to mean that in the long run, the casino expects to gain about 5 cents for every dollar the player bets.

9.3.2. There is a total of 38 numbers on a Roulette wheel. The red numbers in Roulette are 1, 3, 5, 7, 9, 12, 14, 16, 18, 19, 21, 23, 25, 27, 30, 32, 34, and 36. Thus, the probability of the ball landing on red is $\frac{18}{38}$ The black numbers in Roulette is 2, 4, 6, 8, 10, 11, 13, 15, 17, 20, 22, 24, 26, 28, 29, 31, 33, and 35. Thus, the probability of the ball landing on black is $\frac{18}{38}$. There are only two green numbers, 0 and 00; the probability of landing on green is $\frac{2}{38}$.

$$\overbrace{\text{Lands on red}}\qquad\qquad\overbrace{\text{Lands on black}}$$

$$\text{E}(X) = \underbrace{1[\text{P}(X = \text{red})]}_{\text{Win red bet}} + \underbrace{(-1)[\text{P}(X = \text{black})]}_{\text{Lose black bet}} + \underbrace{(-1)[\text{P}(X = \text{red})]}_{\text{Lose red bet}} + \underbrace{1[\text{P}(X = \text{black})]}_{\text{Win black bet}}$$

$$\overbrace{\text{Lands on green}}$$

$$+ \underbrace{(-2)[\text{P}(X = \text{green})]}_{\text{Lose both bets}}$$

$$= \frac{18}{38} - \frac{18}{38} - \frac{18}{38} + \frac{18}{38} - 2\left(\frac{2}{38}\right)$$

$$= -\frac{2}{19}$$

$$= -0.10526$$

Therefore, you would expect to lose 10 cents every time you made this bet.

9.3.3. Let X be the amount of money the gambler wins. Then,

$$\text{E}(X) = (1)\left(\frac{1}{6}\right) + (2)\left(\frac{1}{6}\right) + (3)\left(\frac{1}{6}\right) + (4)\left(\frac{1}{6}\right) + (5)\left(\frac{1}{6}\right) - (6)\left(\frac{1}{6}\right) = \frac{3}{2}.$$

Therefore, a fair price for a gambler to play this game is $1.50.

9.3.4. Zero. If the mathematician is selecting the names/phone numbers from the phone book, then all of the them would be listed. (If you answered this incorrectly, then take this as a reminder to read the question carefully, and answer what the exercise is asking.)

9.3.5. From Table C.7 (on p. 650), the expected value is

$$\text{E}(X) = 0\left(\frac{9}{24}\right) + 1\left(\frac{8}{24}\right) + 2\left(\frac{6}{24}\right) + 3\left(\frac{0}{24}\right) + 4\left(\frac{1}{24}\right) = \frac{8}{24} + \frac{12}{24} + \frac{4}{24} = \frac{24}{24} = 1.$$

So, we expect one person on average to receive the correct coat.

9.3.6. Let X be the gain for the insurance company for a policy.
    (a) $\text{E}(X) = 800 \times (1 - 0.000\,29) + (-300\,000 + 800) \times 0.000\,29 = \$713$.
    (b) $\text{E}(X) = 800 \times (1 - 0.000\,24) + (-300\,000 + 800) \times 0.000\,24 = \$728$.

9.3.7. While it is uncertain what the outcome of a casino game will be, or when a person will die, as long as the expected value is in the favor of the casino or the insurance company, then on average, they will expect to make a profit. For example, the casino expects to make 5 cents per dollar wagered on their roulette table (see Example 9.3.3 on p. 404).

9.3.8. $\text{E}(X) = 1000\left(\frac{1}{5000}\right) + 500\left(\frac{1}{5000}\right) + 250\left(\frac{1}{5000}\right) - 2\left(\frac{4997}{5000}\right) = -1.6488.$ Thus, an individual's expected return on a single ticket is -1.64. If an individual buys five tickets, then the expected return is $-1.64 \times 5 = \$-8.24$.

9.3.9. The expected value of the random variable X is given by $\text{E}(X) = 0(0.1) + 1(0.15) + 2(0.25) + 3(0.25) + 4(0.15) + 5(0.1) = 2.5$. Also, $\text{E}(X^2) = 0^2(0.1) + 1^2(0.15) + 2^2(0.25) + 3^2(0.25) + 4^2(0.15) + 5^2(0.1) = 8.3$.

9.3.10. (a) $6a + 5a + 4a + 2a + a = 1$

$$18a = 1$$

$$a = \frac{1}{18}$$

(b) $\mathrm{E}(X) = 1\left(\frac{6}{18}\right) + 2\left(\frac{5}{18}\right) + 3\left(\frac{4}{18}\right) + 4\left(\frac{2}{18}\right) + 5\left(\frac{1}{18}\right)$

$$= \frac{41}{18}$$

$$\approx 2.2778$$

9.3.11. $\mathrm{E}(X) = 2\left(\frac{1}{6}\right) + 3\left(\frac{1}{3}\right) + 5\left(\frac{1}{4}\right) + 7y + 11z = 4\frac{2}{3}$

$$\frac{31}{12} + 7y + 11z = 4\frac{2}{3}$$

$$7y + 11z = 4\frac{2}{3} - \frac{31}{12}$$

$$7y + 11z = \frac{11}{4}$$

We also know that

$$\frac{1}{6} + \frac{1}{3} + \frac{1}{4} + y + z = 1$$

$$\frac{3}{4} + y + z = 1$$

$$y + z = 1 - \frac{3}{4}$$

$$y + z = \frac{1}{4}$$

We have two equations and two unknowns.

$$7y + 11z = \frac{11}{4}$$

$$y + z = \frac{1}{4}$$

Solving for y and z, we have $z = \frac{1}{4}$ and $y = 0$.

9.3.12. $\mathrm{E}(X) = 0(0.10) + 1(0.40) + 2(0.30) + 3(0.10) + 4(0.05) + 5(0.05) = 1.75$

9.3.13. $100\,000 \times (1 - 0.999\,24) = 76$. Therefore, he should expect to pay a minimum of \$76 for the policy for the year.

9.3.14. Using the data in Table 9.4, we can calculate the expected length of a word. For clarity, we have used a table as shown below to display the calculations.

x	$P(X = x)$	$x\,P(X = x)$	x	$P(X = x)$	$x\,P(X = x)$
1	0.029 981 09	0.029 981 09	13	0.005 176 60	0.067 295 80
2	0.176 507 50	0.353 015 00	14	0.002 222 35	0.031 112 90
3	0.205 108 33	0.615 324 99	15	0.000 759 89	0.011 398 35
4	0.147 865 00	0.591 460 00	16	0.000 203 30	0.003 252 80
5	0.106 997 48	0.534 987 40	17	0.000 097 88	0.001 663 96
6	0.083 876 86	0.503 261 16	18	0.000 038 48	0.000 692 64
7	0.079 388 62	0.555 720 34	19	0.000 011 44	0.000 217 36
8	0.059 430 95	0.475 447 60	20	0.000 008 54	0.000 170 80
9	0.044 373 52	0.399 361 68	21	0.000 000 17	0.000 003 57
10	0.030 764 35	0.307 643 50	22	0.000 001 09	0.000 023 98
11	0.017 608 64	0.193 695 04	23	0.000 000 43	0.000 009 89
12	0.009 577 49	0.114 929 88	$\sum$	1.000 000 00	**4.790 669 73**

Therefore, the expected length of a word in English is approximately 4.79 letters long.

9.3.15. $E(X) = \displaystyle\sum_{\text{all } x} x\,P(X = x)$

$= 0(0.9) + 1500(0.1 \times 0.6) + 5000(0.1 \times 0.3) + 15\,000(0.1 \times 0.1)$

$= 390$

On average, the vehicle owner will spend \$390 in a year on car damage due to accidents.

9.3.16. (a) First, we need to convert Table 9.10 to an empirical probability distribution.

Table C.8 Probability distribution of soda refills
for Exercise 9.3.16

x	0	1	2	3	4
$P(X = x)$	$\frac{3}{15}$	$\frac{6}{15}$	$\frac{4}{15}$	$\frac{2}{15}$	$\frac{1}{15}$

Then, the expected number of refills per customer is

$$\sum_{x=0}^{4} x\,P(X = x) = 0\left(\frac{3}{15}\right) + 1\left(\frac{6}{15}\right) + 2\left(\frac{4}{15}\right) + 3\left(\frac{2}{15}\right) + 4\left(\frac{1}{15}\right) = 1.6.$$

(b) The expected number of refills is 1.6, so the cost of the refills is $1.6 \times 0.50 = 0.80$. We need to also add the cost of the initial fill and cup which is \$1.50. Therefore, a fair price for the manager to charge is \$1.50 + \$0.80 = \$2.30.

(c) It is unlikely that the manager will set his price as \$2.30. The manager needs to make a profit, so the price would likely be higher than \$2.30.

9.4.1. (a) The variance of X, or the variance of the sum of two dice is

$$\text{var}(X) = (2 - 7)^2\left(\frac{1}{36}\right) + (3 - 7)^2\left(\frac{2}{36}\right) + (4 - 7)^2\left(\frac{3}{36}\right)$$
$$+ (5 - 7)^2\left(\frac{4}{36}\right) + (6 - 7)^2\left(\frac{5}{36}\right) + (7 - 7)^2\left(\frac{6}{36}\right)$$

$$+ (8-7)^2 \left(\frac{5}{36}\right) + (9-7)^2 \left(\frac{4}{36}\right) + (10-7)^2 \left(\frac{3}{36}\right)$$
$$+ (11-7)^2 \left(\frac{2}{36}\right) + (12-7)^2 \left(\frac{1}{36}\right)$$
$$= \frac{210}{36}$$
$$= \frac{35}{6} \approx 5.833$$

(b) The standard deviation of the sum of two dice is $\sigma = \sqrt{\mathrm{var}(X)} = \sqrt{\dfrac{35}{6}} \approx 2.4152$.

9.4.2. The expected value of the random variable X is given by $E(X) = 0(0.1) + 1(0.15) + 2(0.25) + 3(0.25) + 4(0.15) + 5(0.1) = 2.5$. Also, $E(X^2) = 0^2(0.1) + 1^2(0.15) + 2^2(0.25) + 3^2(0.25) + 4^2(0.15) + 5^2(0.1) = 8.3$.
   (a) The variance of X is given by $\sigma^2 = E(X^2) - [E(X)]^2 = 8.3 - (2.5)^2 = 2.05$.
   (b) The standard deviation of X is given by $\sigma = \sqrt{2.05} \approx 1.431\,78$.

9.4.3. From Exercise 9.3.15, $E(X) = 390$.

$$\sigma^2 = \sum_{\text{all } x} (x - E(X))^2\, P(X = x)$$
$$\sigma^2 = (0-390)^2 0.9 + (1500-390)^2(0.1 \times 0.6) + (5000-390)^2(0.1 \times 0.3)$$
$$+ (15\,000-390)^2(0.1 \times 0.1)$$
$$\sigma^2 = 2\,982\,900$$
$$\sigma = 1727.107$$

9.4.4. (a) $a = 1 - 0.06 - 0.08 - 0.12 - 0.04 - 0.14 - 0.12 - 0.16 - 0.14 - 0.04$
$$= 0.1$$
   (b) $E(X) = 1(0.06) + 2(0.08) + 3(0.12) + 4(0.04) + 5(0.14) + 6(0.12)$
$$+ 7(0.16) + 8(0.14) + 9(0.1) + 10(0.04)$$
$$= 5.7$$
   (c) $E(X^2) = 1^2(0.06) + 2^2(0.08) + 3^2(0.12) + 4^2(0.04) + 5^2(0.14) + 6^2(0.12)$
$$+ 7^2(0.16) + 8^2(0.14) + 9^2(0.1) + 10^2(0.04)$$
$$= 38.82$$
   Then,

$$\sigma^2 = E(X^2) - [E(X)]^2$$
$$= 38.82 - (5.7)^2$$
$$= 6.33$$
$$\sigma = \sqrt{6.33} \approx 2.5159.$$

9.4.5. (a)   (i) $E(Y) = E(aX + b)$

$$= \sum_{\text{all } x} (ax + b) P(X = x)$$

$$= \sum_{\text{all } x} (ax \, P(X = x) + b \, P(X = x))$$

$$= \sum_{\text{all } x} ax \, P(X = x) + \sum_{\text{all } x} b \, P(X = x)$$

$$= a \underbrace{\sum_{\text{all } x} x \, P(X = x)}_{E(X)} + b \underbrace{\sum_{\text{all } x} P(X = x)}_{1}$$

$$= a \, E(X) + b$$

(ii) $\text{var}(Y) = E\left((Y - E(Y))^2\right)$

$$= E\left((aX + b - \underbrace{E(aX + b)}_{a\,E(X) + b})^2\right)$$

$$= E\left((aX + b - (a\,E(X) + b))^2\right)$$

$$= E\left((aX + b - a\,E(X) - b)^2\right)$$

$$= E\left((aX - a\,E(X))^2\right)$$

$$= E\left((a(X - E(X)))^2\right)$$

$$= E\left(a^2(X - E(X))^2\right)$$

$$= \sum_{\text{all } x} a^2(X - E(X))^2 \, P(X = x)$$

$$= a^2 \underbrace{\sum_{\text{all } x} (X - E(X))^2 \, P(X = x)}_{\text{var}(X)}$$

$$= a^2 \, \text{var}(X)$$

(b) $E(Y) = E(2X + 3)$

$$= 2\,E(X) + 3$$

$$= 2(3) + 3$$

$$9$$

$$\text{var}(Y) = \text{var}(2X + 3)$$

$$= 2^2 \, \text{var}(X)$$

$$= 4(4)$$

$$= 16$$

9.5.1. $P(X \geq 5) = 1 - P(X < 5) = 1 - 0.7564 = 0.2436$

9.5.2. $E(X) = 0(0.25) + 1(0.40) + 2(0.35) = 1.1$

$E(X^2) = 0^2(0.25) + 1^2(0.40) + 2^2(0.35) = 1.8$

$\text{var}(X) = E(X^2) - [E(X)]^2$

$\qquad = 1.8 - (1.1)^2$

$\qquad = 0.59$

9.5.3. Let X be the sum of the two dice. The probabilities for each value of X can be found in Table 9.2 on p. 395.

(a) $P(X = 7) + P(X = 11) = \frac{6}{36} + \frac{2}{36} = \frac{8}{36} = \frac{2}{9}$

(b) $P(X = 2) + P(X = 3) + P(X = 12) = \frac{1}{36} + \frac{2}{36} + \frac{1}{36} = \frac{4}{36} = \frac{1}{9}$

(c) $P(X = 4) + P(X = 5) + P(X = 6) + P(X = 8) + P(X = 9) + P(X = 10)$

$= \frac{3}{36} + \frac{4}{36} + \frac{5}{36} + \frac{5}{36} + \frac{4}{36} + \frac{3}{36}$

$= \frac{24}{36}$

$= \frac{2}{3}$

9.5.4. (a) We know that the probability distribution must add to 1. Thus,

$$\sum_{x=1}^{4} P(X = x) = 1$$

$$0.1 + 0.2 + 0.3 + a = 1$$

$$0.6 + a = 1$$

$$a = 1 - 0.6$$

$$a = 0.4$$

(b) $E(X) = \displaystyle\sum_{x=1}^{4} x\,P(X = x)$

$\qquad = 1(0.1) + 2(0.2) + 3(0.3) + 4(0.4)$

$\qquad = 0.1 + 0.4 + 0.9 + 1.6$

$\qquad = 3.0$

(c) $\text{var}(X) = \displaystyle\sum_{x=1}^{4} (x - E(X))^2\, P(X = x)$

$\qquad = (1 - 3)^2(0.1) + (2 - 3)^2(0.2)$

$\qquad\quad + (3 - 3)^2(0.3) + (4 - 3)^2(0.4)$

$\qquad = 0.4 + 0.2 + 0 + 0.4$

$\qquad = 1$

(d) The standard deviation of X is $\sigma = \sqrt{\text{var}(X)} = \sqrt{1} = 1$.

9.5.5. (a) $S = \{\text{HHH, HHT, HTH, HTT, THH, THT, TTH, TTT}\}$.

(b) No. This is not a uniform sample space, since the outcomes do not all have the same probability of occurring. For example, $P(\text{HHH}) = (\frac{8}{10})(\frac{1}{2})(\frac{1}{2}) = \frac{2}{10}$, while $P(\text{THH}) = (\frac{2}{10})(\frac{1}{2})(\frac{1}{2}) = \frac{1}{20}$.

(c) $P(Y = 0) = \dfrac{2}{10} = \dfrac{4}{20}$

$P(Y = 1) = P(HHT) + P(HTH) + P(THH)$

$$= \left(\frac{8}{10}\right)\left(\frac{1}{2}\right)\left(\frac{1}{2}\right) + \left(\frac{8}{10}\right)\left(\frac{1}{2}\right)\left(\frac{1}{2}\right) + \left(\frac{2}{10}\right)\left(\frac{1}{2}\right)\left(\frac{1}{2}\right)$$

$$= \frac{9}{20}$$

$P(Y = 2) = P(HTT) + P(THT) + P(TTH)$

$$= \left(\frac{8}{10}\right)\left(\frac{1}{2}\right)\left(\frac{1}{2}\right) + \left(\frac{2}{10}\right)\left(\frac{1}{2}\right)\left(\frac{1}{2}\right) + \left(\frac{2}{10}\right)\left(\frac{1}{2}\right)\left(\frac{1}{2}\right)$$

$$= \frac{12}{40}$$

$$= \frac{6}{20}$$

$P(Y = 3) = P(TTT)$

$$= \left(\frac{2}{10}\right)\left(\frac{1}{2}\right)\left(\frac{1}{2}\right)$$

$$= \frac{1}{20}$$

Therefore, the probability distribution of Y is given in the table below.

y	$P(Y = y)$
0	$\frac{4}{20}$
1	$\frac{9}{20}$
2	$\frac{6}{20}$
3	$\frac{1}{20}$

(d) $E(Y) = 10\,P(Y = 3) + 2\,P(Y = 2) - 5\,P(Y = 1) - 5\,P(Y = 0)$ No, you should not play

$$= 10\left(\frac{1}{20}\right) + 2\left(\frac{6}{20}\right) - 5\left(\frac{9}{20}\right) - 5\left(\frac{4}{20}\right)$$

$$= \frac{10}{20} + \frac{12}{20} - \frac{45}{20} - \frac{20}{20}$$

$$= -\frac{43}{20}$$

$$= -2.15$$

this game. It is not fair, and you would expect to lose $2.15 each time you play.

10.1.1. (a) uniform   (c) non-uniform
    (b) uniform   (d) uniform

10.1.2. $P(X = x) = \frac{1}{6}$ where $x \in \{10, 11, 12, 13, 14, 15\}$. The graph is in Fig. C.35.

10.1.3. The probability distribution is $P(X = x) = \frac{1}{j-i+1}$ where $x \in \{i, i + 1, \ldots, j - 1, j\}$, and graph of the distribution is in Fig. C.36.

10.1.4. By looking at Table 9.2 or Fig. 9.1 on p. 395, the probability for each outcome is not the same. Thus, it is not a uniform probability distribution.

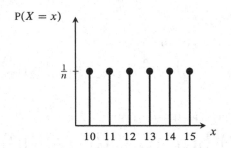

Fig. C.35   The probability distribution for Exercise 10.1.2

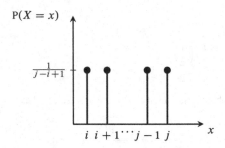

Fig. C.36   The probability distribution for Exercise 10.1.3

10.1.5. Let X be the amount of the prize won.

(a) $E(X) = 500\,000\left(\dfrac{2}{2\,000\,000}\right) + 50\,000\left(\dfrac{2}{2\,000\,000}\right) + 5000\left(\dfrac{10}{2\,000\,000}\right)$

$\quad + 500\left(\dfrac{20}{2\,000\,000}\right) + 50\left(\dfrac{2000}{2\,000\,000}\right) + 5\left(\dfrac{20\,000}{2\,000\,000}\right)$

$\quad = \$0.68$

(b) No, it is not a fair game. For every ticket, you expect to lose $\$1.00 - \$0.68 = \$0.32$.

(c) There are two ways that you can think about this problem.

  (i) For every ticket, $\$1.00 - \$0.68 = \$0.32$ goes to the lottery. Then the amount the charity should receive is $\$0.32 \times 2\,000\,000 = \$640\,000$.

  (ii) The total amount of money is $\$1.00 \times 2\,000\,000 = \$2\,000\,000$. The amount of prizes to be given out is $(2 \times \$500\,000) + (2 \times \$50\,000) + (10 \times \$5000) + (20 \times \$500) + (2000 \times \$50) + (20\,000 \times \$5) = \$1\,360\,000$. Thus, the amount of money that the charity should receive is $\$2\,000\,000 - \$1\,360\,000 = \$640\,000$.

10.1.6. Let $x \in \{1, 2, 3, \ldots, n\}$. Then,

$$E(X) = \sum_{\text{all } x} x\, P(X = x)$$

$$= \sum_{x=1}^{n} x\left(\frac{1}{n}\right)$$

$$= \frac{1}{n} \sum_{x=1}^{n} x$$

$$= \frac{1}{n} \left(\frac{n(n+1)}{2} \right) \qquad \text{By the Gaussian summation formula}$$

$$= \frac{n+1}{2}$$

Next,

$$E(X^2) = \sum_{\text{all } x} x^2 \, P(X = x)$$

$$= \sum_{\text{all } x} x^2 \left(\frac{1}{n} \right)$$

$$= \frac{1}{n} \sum_{x=1}^{n} x^2$$

$$= \frac{1}{n} \left[\frac{n(n+1)(2n+1)}{6} \right] \qquad \text{By Theorem 6.10.3}$$

$$= \frac{(n+1)(2n+1)}{6}$$

Then, the variance is

$$\text{var}(X) = E(X^2) - [E(X)]^2$$

$$= \frac{(n+1)(2n+1)}{6} - \left[\frac{n+1}{2} \right]^2$$

$$= \frac{(n+1)(2n+1)}{6} - \frac{(n+1)^2}{4}$$

$$= \frac{2(n+1)(2n+1)}{12} - \frac{3(n+1)^2}{12}$$

$$= \frac{2(n+1)(2n+1) - 3(n+1)^2}{12}$$

$$= \frac{(n+1)[2(2n+1) - 3(n+1)]}{12}$$

$$= \frac{(n+1)[4n+2 - 3n - 3]}{12}$$

$$= \frac{(n+1)(n-1)}{12} \qquad \text{Difference of Squares}$$

$$= \frac{n^2 - 1}{12}$$

10.2.1. $n = 10$, and $p = 0.7$.

(a) $P(X = 0) = \binom{10}{0} (0.7)^0 (0.3)^{10} = 0.000\,01$

(b) $P(X = 1) = \binom{10}{1} (0.7)^1 (0.3)^9 = 0.000\,14$

(c) $P(X = 2) = \binom{10}{2} (0.7)^2 (0.3)^8 = 0.001\,45$

(d) $P(X > 2) = 1 - P(X \le 2)$

$= 1 - [P(X = 0) + P(X = 1) + P(X = 2)]$

$= 1 - [0.000\,01 + 0.000\,14 + 0.001\,45]$

$= 0.998\,40$

(e) $P(X \ge 2) = 1 - P(X < 2)$

$= 1 - [P(X = 0) + P(X = 1)]$

$= 1 - [0.000\,01 + 0.000\,14]$

$= 0.999\,85$

(f) $P(0 \le X \le 2) = P(X = 0) + P(X = 1) + P(X = 2)$

$= 0.000\,01 + 0.000\,14 + 0.001\,45$

$= 0.001\,60$

(g) $P(0 < X < 2) = P(X = 1) = 0.000\,14$

(h) $P(0 < X \le 2) = P(X = 1) + P(X = 2) = 0.000\,14 + 0.001\,45 = 0.001\,59$

(i) $P(X < 2) = P(X = 0) + P(X = 1) = 0.000\,01 + 0.000\,14 = 0.000\,15$

(j) $P(X \le 2) = P(0 \le X \le 2) = 0.001\,60$

(k) $E(X) = 10(0.7) = 7$

(l) $\text{var}(X) = 10(0.7)(1 - 0.7) = 10(0.7)(0.3) = 2.1$

10.2.2. $n = 15$ and $p = 0.3$.

(a) $P(Y = 2) = \binom{15}{2}(0.3)^2(0.7)^{13} = 0.091\,56$

(b) $P(Y = 1) = \binom{15}{1}(0.3)^1(0.7)^{14} = 0.030\,52$

(c) $P(Y > 1) = 1 - P(Y = 0) = 1 - \binom{15}{0}(0.7)^{15} = 1 - 0.004\,75 = 0.995\,25$

(d) $P(Y < 1) = P(Y = 0) = \binom{15}{0}(0.7)^{15} = 0.004\,75$

(e) $P(Y \le 1) = P(Y = 0) + P(Y = 1) = 0.004\,75 + 0.030\,52 = 0.035\,27$

(f) $P(1 \le Y \le 2) = P(Y = 1) + P(Y = 2) = 0.030\,52 + 0.091\,56 = 0.122\,08$

(g) $E(Y) = 15(0.3) = 4.5$

(h) $\text{var}(Y) = 15(0.3)(1 - 0.3) = 15(0.3)(0.7) = 3.15$

10.2.3. (a) $\displaystyle\sum_{x=0}^{n} x\binom{n}{x}p^x(1-p)^{n-x}$

$$= 0\binom{n}{0}p^0(1-p)^{n-0} + 1\binom{n}{1}p^1(1-p)^{n-1} + \cdots + n\binom{n}{n}p^n(1-p)^{n-n}$$

$$\underbrace{\phantom{0\binom{n}{0}p^0(1-p)^{n-0}}}_{0}$$

$$= 0 + \binom{n}{1}p^1(1-p)^{n-1} + \cdots + n\binom{n}{n}p^n(1-p)^{n-n}$$

$$= \sum_{x=1}^{n} x\binom{n}{x}p^x(1-p)^{n-x}$$

(b) $\displaystyle\sum_{x=0}^{n} x^2 \binom{n}{x} p^x (1-p)^{n-x}$

$= \underbrace{n^2 \binom{n}{0} p^0 (1-p)^{n-0}}_{0} + 1^2 \binom{n}{1} p^1 (1-p)^{n-1} + \cdots + n^2 \binom{n}{n} p^n (1-p)^{n-n}$

$= 0 + \binom{n}{1} p^1 (1-p)^{n-1} + \cdots + n \binom{n}{n} p^n (1-p)^{n-n}$

$= \displaystyle\sum_{x=1}^{n} x \binom{n}{x} p^x (1-p)^{n-x}$

10.2.4. Let X be the random variable that represents the number of people canceling their reservation. Then $X \sim \mathcal{B}(5, 0.2)$.

(a) The probability distribution is given Table C.9 and the graph is in Fig. C.37.

Table C.9   Probability distribution for Exercise 10.2.4

x	0	1	2	3	4	5
$P(X = x)$	0.327 68	0.409 60	0.204 80	0.051 20	0.006 40	0.000 32

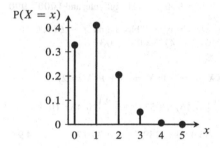

Fig. C.37   The probability distribution for Exercise 10.2.4

(b) $P(X \geq 4) = P(X = 4) + P(X = 5) = 0.006\,40 + 0.000\,32 = 0.006\,72$

10.2.5. (a) Let X be the number of misses if he fires 6 times. Then, $X \sim \mathcal{B}(6, 0.1)$.

$P(X \geq 2) = 1 - P(X < 2)$

$= 1 - [P(X = 0) + P(X = 1)]$

$= 1 - \left[\binom{6}{0}(0.9)^6 + \binom{6}{1}(0.1)(0.9)^5 \right]$

$\approx 1 - [0.531\,44 + 0.354\,29]$

$= 0.885\,73$

(b) Let Y be the number of misses if he fires 20 times; $Y \sim \mathcal{B}(20, 0.1)$. Then, $\mathrm{E}(Y) = 20(0.1) = 2$.

10.2.6. Let X represent the number of defective computer chips out of a total of 30 chips; $X \sim \mathcal{B}(30, 0.07)$.

    (a)   (i) $P(X = 3) = \dbinom{30}{3}(0.07)^3(0.93)^{27} = 0.196\,27$

            (ii) $P(X \geq 3) = 1 - P(X < 3)$

$$= 1 - [P(X = 0) + P(X = 1) + P(X = 2)]$$

$$= 1 - \left[\binom{30}{0}(0.93)^{30} + \binom{30}{1}(0.07)(0.93)^{29} \right.$$

$$\left. + \binom{30}{2}(0.07)^2(0.93)^{28} \right]$$

$$= 1 - [0.113\,37 + 0.255\,99 + 0.279\,39]$$

$$= 0.351\,25$$

    (b) $\mathrm{E}(X) = 30(0.07) = 2.1$

10.2.7. Out of $35\,000$ subscribers, there are $15\,300$ females, which means that there are $19\,700$ male subscribers. There are $0.4 \times 19\,700 = 7880$ males and $0.33 \times 15\,300 = 5049$ females who read the sports section. Altogether, there are $7880 + 5049 = 12\,929$ people who read the sports section. Thus, the probability of picking a person who reads the sports section is $\frac{12\,929}{35\,000} = 0.3694$. Therefore, the expected number of 100 randomly selected subscribers who reads the sports section is $100 \times 0.3694 = 36.94$.

10.2.8. We would expect $1000 \times 0.99 = 990$ pens to be reliable, and $1000 - 990 = 10$ pens to be defective.

10.2.9. Let X be the number of people who type O blood, and $X \sim \mathcal{B}(4, 0.46)$.
    (a) $P(X = 0) = \binom{4}{0}(0.46)^0(0.54)^4 \approx 0.085\,03$
    (b) We need to find $P(X \leq 2)$.

$$P(X \leq 2) = P(X = 0) + P(X = 1) + P(X = 2)$$

$$= \binom{4}{0}(0.46)^0(0.54)^4 + \binom{4}{1}(0.46)^1(0.54)^3 + \binom{4}{2}(0.46)^2(0.54)^2$$

$$\approx 0.085\,03 + 0.289\,73 + 0.370\,22 = 0.744\,98$$

    (c) The expected value is $\mathrm{E}(X) = 4 \times 0.46 = 1.84$.

10.2.10. Let X the number of defective light bulbs, and $X \sim \mathcal{B}(5, 0.005)$. Then, the probability that two out of a set of five bulbs are found to be defective is $P(X = 2) = \binom{5}{2}(0.005)^2(0.995)^3 = 0.000\,25$.

10.2.11. Let X be the number of target hits, and $X \sim \mathcal{B}(10, \frac{1}{7})$.

$$P(X \geq 3) = 1 - P(X < 3)$$

$$= 1 - [P(X = 0) + P(X = 1) + P(X = 2)]$$

$$= 1 - \left[\binom{10}{0}\left(\frac{6}{7}\right)^{10} + \binom{10}{1}\left(\frac{1}{7}\right)\left(\frac{6}{7}\right)^9 + \binom{10}{2}\left(\frac{1}{7}\right)^2\left(\frac{6}{7}\right)^8 \right]$$

$$\approx 1 - [0.214\,06 + 0.356\,76 + 0.267\,57]$$

$$= 0.161\,60$$

10.2.12. Let X be the number of the student's correct answers, and $X \sim \mathcal{B}(20, \frac{1}{4})$. Therefore, the expected value is $E(X) = np = 20(\frac{1}{4}) = 5$ and the variance is $\sigma^2 = np(1-p) = 20(\frac{1}{4})(\frac{3}{4}) = 3.75$.

10.2.13. A majority means that the number of jurors is more than half of the total. This means that we need at least seven or more jurors to have a majority who have correctly determined the guilt or innocence of a defendant. Let X be the number of jurors who make the correct choice. This is a binomial random variable since each juror acts independently, and the probability of making a correct choice is $p = 0.7$. Thus, $X \sim \mathcal{B}(12, 0.7)$. So, the required probability is

$$P(X > 6) = P(X = 7) + P(X = 8) + \cdots + P(X = 12)$$

$$= \binom{12}{7}(0.7)^7(0.3)^5 + \binom{12}{8}(0.7)^8(0.3)^4 + \binom{12}{9}(0.7)^9(0.3)^3$$

$$+ \binom{12}{10}(0.7)^{10}(0.3)^2 + \binom{12}{11}(0.7)^{11}(0.3)^1 + \binom{12}{12}(0.7)^{12}$$

$$\approx 0.158\,50 + 0.231\,14 + 0.239\,70 + 0.167\,79 + 0.071\,18 + 0.013\,84$$

$$= 0.882\,15.$$

10.2.14. (a) Let $X \sim \mathcal{B}(40, 0.15)$.

$$P(X \le 10) = P(X = 0) + \cdots + P(X = 10)$$

$$= \binom{40}{0}(0.85)^{40} + \cdots + \binom{40}{10}(0.15)^{10}(0.85)^{30}$$

$$= 0.001\,50 + 0.010\,60 + \cdots + 0.037\,30$$

$$= 0.970\,08$$

(b) Let $Y \sim \mathcal{B}(40, 0.175)$. $P(Y = 2) = \binom{40}{2}(0.175)^2(0.825)^{38} = 0.015\,97$.

(c) The probability that a candidate achieves at least the minimum score but not the average score estimate is $0.85 - 0.50 = 0.35$. Let $W \sim \mathcal{B}(40, 0.35)$.

$$P(10 < W < 15) = P(X = 11) + P(X = 12) + P(X = 13) + P(X = 14)$$

$$= \binom{40}{11}(0.35)^{11}(0.65)^{29} + \binom{40}{12}(0.35)^{12}(0.65)^{28}$$

$$+ \binom{40}{13}(0.35)^{13}(0.65)^{27} + \binom{40}{14}(0.65)^{14}(0.65)^{26}$$

$$= 0.075\,85$$

10.2.15. (a) Let X be the number of words that requires a correction. Then $X \sim \mathcal{B}(500, 0.02)$. Hence, $E(X) = np = 500 \times 0.02 = 10$ words. Therefore, on average the secretary will spend 10 corrections $\times$ 5 seconds $= 50$ seconds making corrections.

(b) The secretary can type 120 words per minute, or 2 words per second. So,

$$\frac{500 \text{ words}}{2 \frac{\text{words}}{\text{second}}} = 250 \text{ seconds.}$$

Thus, $250 + 50 = 300$ seconds to type a 500-word memo with the corrections.

10.2.16. Let X be the number of users who click on the link.

(a) $X \sim \mathcal{B}(50\,000, 0.1)$. $E(X) = 50\,000 \times 0.1 = 5\,000$ users are expected to click on the link.

(b) The business should expect to pay $5\,000 \times \$0.10 = \$500$ for the advertisement link.

10.2.17. (a) Consider the 10 tables as identical, independent trials with success meaning that the party shows up for their reservation ($p = 0.8$). Let X be the number of parties show up for their reservation; $X \sim \mathcal{B}(11, 0.8)$. The probability to compute is then that of getting 11 successes in 11 such trials, which is given by $\mathrm{P}(X = 11) = \binom{11}{11}(0.8)^{11}(0.2)^0 = 0.8^{11} = 0.08590$.

(b) From the previous answer, we know that we are already close to the 10% risk that is acceptable by the manager. So, we can try accepting 12 reservations. We need to calculate the probability then of 11 reservations showing up, *or* 12 reservations showing up out of 12 reservations. In this case, we let $X \sim \mathcal{B}(0.8, 12)$, and we need to calculate

$$\mathrm{P}(X = 11) + \mathrm{P}(X = 12) = \binom{12}{11}(0.8)^{11}(0.2)^1 + \binom{12}{12}(0.8)^{12}(0.2)^0$$

$$= 12(0.8)^{11}(0.2) + (0.8)^{12}$$

$$\approx 0.274\,88.$$

Therefore, the chance that 11 or 12 parties show up for their reservation if the restaurant takes 12 reservations is approximately 27%. So, the manager should be advised to only take on at most 11 reservations.

10.2.18. The probability that Amélie can scoop those three caramel squares is

$$\frac{\binom{5}{3}}{\binom{10+5}{3}} = \frac{\binom{5}{3}}{\binom{15}{3}} = \frac{10}{455}.$$

We will assume that Amélie must do three independent and identical trials; thus, it will follow a binomial distribution. The probability of success is the same for each trial: $p = \frac{10}{455}$. Let X be the number of times Amélie successfully picks 3 caramel squares from the candy jar; $X \sim \mathcal{B}(3, \frac{10}{455})$. Then, the required probability is

$$\mathrm{P}(X \geq 2) = \mathrm{P}(X = 2) + \mathrm{P}(X = 3)$$

$$= \binom{3}{2}\left(\frac{10}{455}\right)^2\left(\frac{445}{455}\right) + \binom{3}{3}\left(\frac{10}{455}\right)^3$$

$$\approx 0.001\,42 + 0.000\,01$$

$$= 0.001\,43$$

which is quite a low probability for poor Amélie. (The probability that she will not have any candy this week is $1 - 0.001\,43 = 0.998\,57$.)

10.2.19. (a) $(0.984)^5 \approx 0.923$

(b) The probability of incorrectly reading a ZIP code is $p = 1 - 0.923 = 0.077$. So the expected number of misread ZIP codes in 100 pieces of mail is $100 \times 0.077 = 7.7$ or about 8.

10.3.1. Using Eq. (10.9), we have the following:
- $n = 1: 1 + \frac{1}{1!} = 2$
- $n = 2: 1 + \frac{1}{1!} + \frac{1}{2!} = 2\frac{1}{2} = 2.5$
- $n = 3: 1 + \frac{1}{1!} + \frac{1}{2!} + \frac{1}{3!} = 2\frac{2}{3} \approx 2.666\,67$
- $n = 4: 1 + \frac{1}{1!} + \frac{1}{2!} + \frac{1}{3!} + \frac{1}{4!} = 2\frac{17}{24} \approx 2.708\,33$
- $n = 5: 1 + \frac{1}{1!} + \frac{1}{2!} + \frac{1}{3!} + \frac{1}{4!} + \frac{1}{5!} = 2\frac{43}{60} \approx 2.716\,67$
- $n = 6: 1 + \frac{1}{1!} + \frac{1}{2!} + \frac{1}{3!} + \frac{1}{4!} + \frac{1}{5!} + \frac{1}{6!} = 2\frac{517}{720} \approx 2.718\,06.$

Table C.10 The probability distribution of the number of bombs landing in a square.

x	$P(X = x)$	
0	$\dfrac{(0.932)^0 e^{-0.932}}{0!}$	$\approx 0.393\,77$
1	$\dfrac{(0.932)^1 e^{-0.932}}{1!}$	$\approx 0.366\,99$
2	$\dfrac{(0.932)^2 e^{-0.932}}{2!}$	$\approx 0.171\,02$
3	$\dfrac{(0.932)^3 e^{-0.932}}{3!}$	$\approx 0.053\,13$
4	$\dfrac{(0.932)^4 e^{-0.932}}{4!}$	$\approx 0.012\,38$
$\vdots$	$\vdots$	

10.3.2. The probability distribution of X is given in Table C.10.

10.3.3. (a) $P(X = 0) = \dfrac{4^0 e^{-4}}{0!} = e^{-4} \approx 0.018\,32$

(b) $P(X = 1) = \dfrac{4^1 e^{-4}}{1!} = 4e^{-4} \approx 0.073\,26$

(c) $P(X = 2) = \dfrac{4^2 e^{-4}}{2!} = 8e^{-4} \approx 0.146\,53$

(d) $P(X \geq 2) = 1 - P(X < 2)$

$\qquad\qquad = 1 - [P(X = 0) + P(X = 1)]$

$\qquad\qquad = 1 - [e^{-4} + 4e^{-4}]$

$\qquad\qquad = 1 - 5e^{-4}$

$\qquad\qquad \approx 0.908\,42$

(e) $P(X < 2) = P(X = 0) + P(X = 1) = e^{-4} + 4e^{-4} = 5e^{-4} \approx 0.091\,58$

(f) $P(X < 3) = P(X = 0) + P(X = 1) + P(X = 2)$

$\qquad\qquad = e^{-4} + 4e^{-4} + 8e^{-4}$

$\qquad\qquad = 13e^{-4}$

$\qquad\qquad \approx 0.238\,10$

10.3.4. We are required to find λ.

$$\frac{\lambda^4 e^{-\lambda}}{4!} = 3\frac{\lambda^3 e^{-\lambda}}{3!}$$

$$\frac{\lambda^4 e^{-\lambda}}{\lambda^3 e^{-\lambda}} = \frac{3 \times 4!}{3!}$$

$$\lambda e^0 = 3 \times \frac{4 \times \cancel{3!}}{\cancel{3!}}$$

$$\lambda = 12$$

To check our answer, we can substitute the value of $\lambda = 12$ back into the original equation. On the left side, we have $\frac{12^4 e^{-12}}{4!} \approx 0.005\,31$. On the right side, we have $3 \times \frac{12^3 e^{-12}}{3!} \approx 0.005\,31$. Now, we can calculate $P(X = 5) = \frac{12^5 e^{-12}}{5!} \approx 0.012\,74$.

10.3.5. Let the random variable X represent the weekly demand for a laptop.

(a) $P(X \geq 2) = 1 - P(X < 2) = 1 - [P(X = 0) + P(X = 1)]$

$$= 1 - \left[\frac{3^0 e^{-3}}{0!} + \frac{3^1 e^{-3}}{1!} \right]$$

$$= 1 - [e^{-3} + 3e^{-3}]$$

$$= 1 - 4e^{-3}$$

$$\approx 0.800\,85$$

(b) $P(X \leq 5) = P(X = 0) + P(X = 1) + P(X = 2) + P(X = 3) + P(X = 4)$
$$+ P(X = 5)$$

$$= \frac{3^0 e^{-3}}{0!} + \frac{3^1 e^{-3}}{1!} + \frac{3^2 e^{-3}}{2!} + \frac{3^3 e^{-3}}{3!} + \frac{3^4 e^{-3}}{4!} + \frac{3^5 e^{-3}}{5!}$$

$$= e^{-3} + 3e^{-3} + 4.5e^{-3} + 4.5e^{-3} + 3.375e^{-3} + 2.025e^{-3}$$

$$= 18.4e^{-3}$$

$$\approx 0.916\,08$$

(c) $P(X < 5) = P(X = 0) + P(X = 1) + P(X = 2) + P(X = 3) + P(X = 4)$

$$= \frac{3^0 e^{-3}}{0!} + \frac{3^1 e^{-3}}{1!} + \frac{3^2 e^{-3}}{2!} + \frac{3^3 e^{-3}}{3!} + \frac{3^4 e^{-3}}{4!}$$

$$= e^{-3} + 3e^{-3} + 4.5e^{-3} + 4.5e^{-3} + 3.375e^{-3}$$

$$= 16.375e^{-3}$$

$$\approx 0.815\,26$$

10.3.6. Let the random variable X be the number of incoming requests. Then X follows a Poisson distribution with $\lambda = 10$. The probability of receiving 4 or more requests in a particular hour is

$$P(X \geq 4) = 1 - P(X < 3)$$

$$= 1 - [P(X = 0) + P(X = 1) + P(X = 2)]$$

$$= 1 - \left[\frac{10^0 e^{-10}}{0!} + \frac{10^1 e^{-10}}{1!} + \frac{10^2 e^{-10}}{2!} \right]$$

$$= 1 - [e^{-10} + 10e^{-10} + 50e^{-10}]$$

$$= 1 - 61e^{-10}$$

$$\approx 0.997\,23.$$

10.3.7. Let X be the number of typographical errors on this page.

(a) For only one error, the probability is $P(X = 1) = \frac{(0.2)^1 e^{-0.2}}{1!} \approx 0.163\,75$.

(b) For at least two errors, the probability is

$$P(X \geq 2) = 1 - P(X < 2)$$
$$= 1 - [P(X = 0) + P(X = 1)]$$
$$= 1 - P(X = 0) - P(X = 1)$$
$$= 1 - \frac{(0.2)^0 e^{-0.2}}{0!} - \frac{(0.2)^1 e^{-0.2}}{1!}$$
$$\approx 1 - 0.818\,73 - 0.163\,75$$
$$= 0.017\,52.$$

Therefore, the probability that there is more than two error on this page is approximately 1.75%.

10.3.8. Let the random variable X be the number of grammatical errors in a particular page that follows a Poisson distribution with $\lambda = 2$.

(a) $P(X = 0) = \dfrac{2^0 e^{-2}}{0!} = e^{-2} \approx 0.135\,34$

(b) $P(X = 1) = \dfrac{2^1 e^{-2}}{1!} = 2e^{-2} \approx 0.270\,67$

(c) $P(X = 2) = \dfrac{2^2 e^{-2}}{2!} = 2e^{-2} \approx 0.270\,67$

(d) $P(X > 2) = 1 - P(X \leq 2)$
$$= 1 - [P(X = 0) + P(X = 1) + P(X = 2)]$$
$$= 1 - [e^{-2} + 2e^{-2} + 2e^{-2}]$$
$$= 1 - 5e^{-2} \approx 0.323\,32$$

10.3.9. If we assume that X is a Poisson-distributed random variable, then $\lambda = \frac{2049}{100\,000} = 0.020\,49$. Thus, the probability of one male living to be 100 years old in Canada is

$$P(X > 1) = 1 - P(X \leq 1)$$
$$= 1 - [P(X = 0) + P(X = 1)]$$
$$= 1 - \left[\frac{(0.020\,49)^0 e^{-0.020\,49}}{0!} + \frac{(0.020\,49)^1 e^{-0.020\,49}}{1!} \right]$$
$$= 1 - \left[e^{-0.020\,49} + 0.020\,49 e^{-0.020\,49} \right]$$
$$= 1 - 1.020\,49 e^{-0.020\,49}$$
$$\approx 0.000\,21.$$

10.3.10. (a) The expected daily income from the rentals is $\$50 \times 3.1 = \$155$.

(b) Let X be the number of customers who rent a rug cleaner, and X follow a Poisson distribution with $\lambda = 3.1$.

(i) $P(X = 0) = \dfrac{(3.1)^0 e^{-3.1}}{0!} \approx 0.045\,05$

(ii) $P(X = 1) = \dfrac{(3.1)^1 e^{-3.1}}{1!} \approx 0.139\,65$

(iii) $P(X = 2) = \dfrac{(3.1)^2 e^{-3.1}}{2!} \approx 0.216\,46$

(iv) $P(X \geq 3) = 1 - P(X < 3)$

$$= 1 - [P(X = 0) + P(X = 1) + P(X = 2)]$$

$$= 1 - [0.045\,05 + 0.139\,65 + 0.216\,46]$$

$$= 0.598\,84$$

(c) There are only two rug cleaners available.

$$\$50[0(0.045\,05) + 1(0.139\,65) + 2(0.216\,46)] = \$50(0.572\,57)$$

$$= \$28.63$$

10.3.11. (a) $0.000\,23 \times 445\,830 = 102.5409$. Therefore, there are approximately 102 or 103 female prostitutes in Colorado Springs.

(b) $\frac{102.5409 \text{ prostitutes}}{194.7 \text{ square mile}} \approx 0.5267$ prostitutes per square mile.

(c) Let the random variable X represent the number of prostitutes in 4 square miles. Given that X follows a Poisson distribution, then we let $E(X) = \lambda = 0.5267 \times 4 = 2.1066$. So this means that on average, there are 2.1066 prostitutes per 4 square miles. The required probability is

$$P(X \geq 3) = 1 - P(X < 3)$$

$$= 1 - [P(X = 0) + P(X = 1) + P(X = 2)]$$

$$= 1 - \left[\frac{(2.1066)^0 e^{-2.1066}}{0!} + \frac{(2.1066)^1 e^{-2.1066}}{1!} \right.$$

$$\left. + \frac{(2.1066)^2 e^{-2.1066}}{2!} \right]$$

$$\approx 1 - [0.121\,65 + 0.256\,27 + 0.269\,93]$$

$$= 1 - 0.647\,85$$

$$= 0.352\,15$$

Therefore, the probability that the sociologist will find three or more prostitutes in the 4 square miles is approximately 35.215%.

10.3.12. (a) The average number of bacterial colonies in a square is

$$\frac{3+4+2+4+6+3+1+3+2+0+2+1+3+2+0+4}{16} = \frac{40}{16} = 2.5.$$

(b)   (i) $P(X = 0) = \dfrac{(2.5)^0 e^{-2.5}}{0!} = e^{-2.5} \approx 0.082\,08$

(ii) $P(X = 1) = \dfrac{(2.5)^1 e^{-2.5}}{1!} = 2.5 e^{-2.5} \approx 0.205\,21$

(iii) $P(X = 2) = \dfrac{(2.5)^2 e^{-2.5}}{2!} = 3.125 e^{-2.5} \approx 0.256\,52$

(iv) $P(X = 3) = \dfrac{(2.5)^3 e^{-2.5}}{3!} \approx 0.213\,76$

(v) $P(X = 4) = \dfrac{(2.5)^4 e^{-2.5}}{4!} \approx 0.133\,60$

(vi) $P(X = 5) = \dfrac{(2.5)^5 e^{-2.5}}{5!} \approx 0.066\,80$

(vii) $P(X \geq 6) = 1 - P(X < 6)$

$$= 1 - [P(X = 0) + P(X = 1) + \cdots + P(X = 5)]$$

$$= 1 - [0.082\,08 + 0.205\,21 + 0.256\,52 + 0.213\,76$$

$$+ 0.133\,60 + 0.066\,80]$$

$$= 1 - 0.957\,97$$

$$= 0.042\,03$$

(c) See Table C.11.

Table C.11   Solution for Exercise 10.3.12 of the number of bacterial colonies in a square.

Number of bacterial colonies in a square, x	Actual number of squares with x bacterial colonies	Expected number of 16 squares with x bacterial colonies, $16\,P(X = x)$
0	2	$16(0.082\,08) = 1.313\,28$
1	2	$16(0.205\,21) = 3.283\,36$
2	4	$16(0.256\,52) = 4.104\,32$
3	4	$16(0.213\,76) = 3.420\,16$
4	3	$16(0.133\,60) = 2.137\,60$
5	0	$16(0.066\,80) = 1.068\,80$
6 or more	1	$16(0.042\,03) = 0.672\,48$

10.3.13. (a) Let the random variable X be the number of calls in an hour. Then, X follows a Poisson distribution with parameter $\lambda = 7.2$. Therefore, the probability that there is exactly 5 calls in an hour is $P(X = 5) = \frac{(7.2)^5 e^{-7.2}}{5!} \approx 0.120\,38$.

(b) Let the random variable X be the number of calls in an hour. Then, X follows a Poisson distribution with parameter $\lambda = 7.2$. The probability that there is 4 or less calls in an hour is

$$P(X \leq 4) = P(X = 0) + P(X = 1) + P(X = 2) + P(X = 3) + P(X = 4)$$

$$= \frac{(7.2)^0 e^{-7.2}}{0!} + \frac{(7.2)^1 e^{-7.2}}{1!} + \frac{(7.2)^2 e^{-7.2}}{2!} + \frac{(7.2)^3 e^{-7.2}}{3!}$$

$$+ \frac{(7.2)^4 e^{-7.2}}{4!}$$

$$\approx 0.000\,75 + 0.005\,38 + 0.019\,35 + 0.046\,44 + 0.083\,60$$

$$= 0.155\,52.$$

(c) Let the random variable Y be the number of calls in two hours. Then, Y follows a Poisson distribution with parameter $\lambda = 7.2 \times 2 = 14.4$. The probability that there is exactly 5 calls in *two* hours is $P(Y = 5) = \frac{(14.4)^5 e^{-14.4}}{5!} \approx 0.002\,88$.

(d) Let the random variable Z be the number of calls in an hour. Then, Z follows a Poisson distribution with parameter $\lambda = 7.2 \times 0.5 = 3.6$. The probability that there is exactly 2 calls in *half* an hour is $P(Z = 2) = \frac{(3.6)^2 e^{-3.6}}{2!} \approx 0.177\,06$.

10.3.14. First, note that $\sum\limits_{n=1}^{\infty} \frac{n}{n!} = \sum\limits_{n=0}^{\infty} \frac{1}{n!} = e$. Then, using the hint $n^2 = n + n(n-1)$, we have

$$\sum_{n=1}^{\infty} \frac{n^2}{n!} = \sum_{n=1}^{\infty} \frac{n + n(n-1)}{n!}$$

$$= \underbrace{\sum_{n=1}^{\infty} \frac{n}{n!}}_{e} + \underbrace{\sum_{n=1}^{\infty} \frac{n(n-1)}{n!}}_{0 + \sum_{n=2}^{\infty} \frac{n(n-1)}{n!}}$$

$$= e + \sum_{n=2}^{\infty} \frac{n(n-1)}{n(n-1)(n-2)!}$$

$$= e + \sum_{n=2}^{\infty} \frac{1}{(n-2)!}$$

$$= e + \sum_{n=0}^{\infty} \frac{1}{n!}$$

$$= e + e$$

$$= 2e$$

as required (Euler and Sadek, 2009, p. 504).

10.4.1. First, we need to determine the value of p from the expected value $E(X)$. To do this,

$$E(X) = 5 \Leftrightarrow \frac{1}{p} = 5 \Leftrightarrow p = \frac{1}{5}$$

(a) $P(X = 1) = \left(\frac{4}{5}\right)^{1-1} \left(\frac{1}{5}\right) = \frac{1}{5}$

(b) $P(X = 2) = \left(\frac{4}{5}\right)^{2-1} \left(\frac{1}{5}\right) = \frac{4}{5} \cdot \frac{1}{5} = \frac{4}{25}$

(c) $P(X = 3) = \left(\frac{4}{5}\right)^{3-1} \left(\frac{1}{5}\right) = \left(\frac{4}{5}\right)^2 \left(\frac{1}{5}\right) = \frac{16}{125}$

(d) $P(X < 3) = P(X = 1) + P(X = 2) = \frac{1}{5} + \frac{4}{25} = \frac{9}{25}$

(e) $P(X \geq 3) = 1 - P(X < 3) = 1 - [P(X = 1) + P(X = 2)] = 1 - \left[\frac{1}{5} + \frac{4}{25}\right] = 1 - \frac{9}{25} = \frac{16}{25}$

(f) $\text{var}(X) = \frac{1 - \frac{1}{5}}{\left(\frac{1}{5}\right)^2} = \frac{\frac{4}{5}}{\frac{1}{25}} = 20$

10.4.2. (a) The table is below, and the graph is in Fig. C.38(a).

x	1	2	3	4
$P(X = x)$	0.2	0.16	0.128	0.1024

(b) The table is below, and the graph is in Fig. C.38(b).

x	1	2	3	4
$P(X = x)$	0.5	0.25	0.125	0.0625

(c) The table is below, and the graph is in Fig. C.38(c).

x	1	2	3	4
$P(X = x)$	0.7	0.21	0.063	0.0189

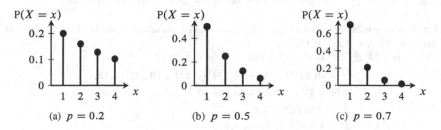

(a) $p = 0.2$ (b) $p = 0.5$ (c) $p = 0.7$

Fig. C.38   Graphs of the probability distributions for Exercise 10.4.2

10.4.3. $\left(\dfrac{5}{6}\right)^3 \left(\dfrac{1}{6}\right) = \dfrac{125}{1296} \approx 0.096\,45$

10.4.4. $\left(\dfrac{4}{6}\right)^4 \left(\dfrac{2}{6}\right) = \dfrac{16}{243} \approx 0.065\,84$

10.4.5. The spins are independent. The probability of winning is $p = \frac{1}{38}$. We want to calculate the expected number of spins before you win, so this will follow a geometric distribution. Thus, $E(X) = \frac{1}{\frac{1}{38}} = 38$.

10.4.6. Let X be the success of the professor opening the door on the kth try. The required probability is

$$P(X = 1) + P(X = 2) + P(X = 3) = \left(\frac{9}{10}\right)^0 \left(\frac{1}{10}\right) + \left(\frac{9}{10}\right)^1 \left(\frac{1}{10}\right) + \left(\frac{9}{10}\right)^2 \left(\frac{1}{10}\right)$$
$$= \frac{1}{10} + \frac{9}{100} + \frac{81}{1000}$$
$$= \frac{271}{1000}$$
$$= 0.271.$$

10.4.7. (a) Let X represent the number of cups of coffee purchased when you win for the first time.
    (i) $P(X = 1) = \left(\frac{11}{12}\right)^0 \left(\frac{1}{12}\right) = \frac{1}{12}$
    (ii) $P(X = 2) = \left(\frac{11}{12}\right) \left(\frac{1}{12}\right) = \frac{11}{144}$
    (iii) $P(X = 1) + P(X = 2) + P(X = 3) + P(X = 4) = \left(\frac{11}{12}\right)^0 \left(\frac{1}{12}\right) + \left(\frac{11}{12}\right) \left(\frac{1}{12}\right) +$
    $\left(\frac{11}{12}\right)^2 \left(\frac{1}{12}\right) + \left(\frac{11}{12}\right)^3 \left(\frac{1}{12}\right) = \frac{1}{12} + \frac{11}{144} + \frac{121}{1728} + \frac{1331}{20\,736} = \frac{6095}{20\,736} \approx 0.293\,93$
(b) $E(X) = \frac{1}{\frac{1}{12}} = 12$
(c) No. She should expect to win every 12 cups, but this is just on average.

10.4.9. Let X be a random variable that follows a geometric distribution with probability of success $p = \frac{1}{4}$. Then, using Eq. (10.18),

$$E(X) - 1 = \frac{1}{\frac{1}{4}} - 1 = 4 - 1 = 3.$$

Thus, the student will expect to answer 3 questions incorrectly before getting one correct.

10.4.10. We assume that people are born with equal probability on each of the 365 days in a year (excluding leap years). Then, we can describe the situation of finding someone with the same birthday as you with a random variable X that follows a geometric distribution with $p = \frac{1}{365}$. Therefore, you would expect to ask $E(X) = \frac{1}{\frac{1}{365}} = 365$ people to find someone who has the same birthday as you.

10.4.11. Let X be the number of calls that Casey will make when he first talks to someone. The probability of talking to someone is $p = 0.45$.

(a)   (i) $P(X \le 3) = P(X = 1) + P(X = 2) + P(X = 3)$

$$= (0.55)^0(0.45) + (0.55)^1(0.45) + (0.55)^2(0.45)$$

$$= 0.45 + 0.2475 + 0.136\,13$$

$$= 0.833\,63$$

(ii) $P(X > 2) = 1 - P(X \le 2)$

$$= 1 - [P(X = 1) + P(X = 2)]$$

$$= 1 - [0.45 + 0.2475]$$

$$= 1 - 0.6975$$

$$= 0.3025$$

(b) $\sigma^2 = \text{var}(X) = \frac{1-0.45}{(0.45)^2} = \frac{0.55}{0.2025} = 2.716\,05.$

(c) $\sigma = \sqrt{\sigma^2} = \sqrt{2.716\,05} = 1.648\,04$

10.4.12. Let X be a random variable that represents the number of goals before Céline misses. In this instance, the probability of a success is Céline *failing* to score a goal; that is, $p = 1 - 0.78 = 0.22$. Then, using Eq. (10.18),

$$E(X) - 1 = \frac{1}{0.22} - 1 \approx 4.545\,45 - 1 = 3.545\,45.$$

On average, Céline will score 3.5 goals before she will miss.

10.4.14. Let the random variable X represent the number of chips checked when the first defective chip is found. So, $P(X = x) = (0.9)^{x-1}(0.1)$. Now, we need to solve for x in the following:

$$P(X \ge x) = 0.05$$

$$P(X = x) + P(X = x + 1) + \cdots = 0.05$$

$$\underbrace{(0.9)^{x-1}(0.1) + (0.9)^x(0.1) + \cdots}_{\text{Geometric series}} = 0.05$$

$$\frac{0.1(0.9)^{x-1}}{1 - 0.9} = 0.05$$

$$0.9^{x-1} = 0.05$$

$$(x - 1)\log 0.9 = \log 0.05$$

$$x = \frac{\log 0.05}{\log 0.9} + 1$$

$$x = 29.433\,16.$$

At least 29.4 chips means that 30 chips must be checked. Therefore, $x = 30$.

Table C.12 Probability distribution of matching the winning lottery numbers

n	$P(Y = r)$
0	$\dfrac{\binom{6}{0}\binom{43}{6}}{\binom{49}{6}} \approx 0.4360$
1	$\dfrac{\binom{6}{1}\binom{43}{5}}{\binom{49}{6}} \approx 0.4130$
2	$\dfrac{\binom{6}{2}\binom{43}{4}}{\binom{49}{6}} \approx 0.1324$
3	$\dfrac{\binom{6}{3}\binom{43}{3}}{\binom{49}{6}} \approx 0.0177$
4	$\dfrac{\binom{6}{4}\binom{43}{2}}{\binom{49}{6}} \approx 9.686 \times 10^{-4}$
5	$\dfrac{\binom{6}{5}\binom{43}{1}}{\binom{49}{6}} \approx 1.845 \times 10^{-5}$
6	$\dfrac{\binom{6}{6}\binom{43}{0}}{\binom{49}{6}} \approx 7.151 \times 10^{-8}$

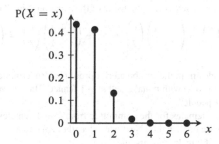

Fig. C.39 The probability distribution of winning the lottery

10.5.1. We give the probability distribution of X in Table C.12.
The graph of the probability distribution is in Fig. C.39.

10.5.2. (a) There are four kinds of items of interest (the suits of the cards) instead of two kinds of items required for the hypergeometric distribution. We can modify the situation so that the cards are recorded as either a particular suit (*e.g.,* hearts) or not.

(b) There is a chance that a person could be polled more than once. This violates the restriction that selection is made without replacement for a hypergeometric distribution. Thus, to enable the use of the hypergeometric distribution, the pollsters need to ensure they only phone a person once.

(c) This could be described best as a geometric distribution, since we are waiting for the first selection of a female. An alternative reason for this failing to be a hypergeometric distribution, the total number of available males and females from which to form the committee is not given. To make this situation follow a hypergeometric distribution, we need to specify the number of selections we will perform; that is, the parameter n needs to be specified. We also need to specify the other parameters r, and N.

10.5.3. $\dbinom{2n}{n} = \dbinom{n+n}{n}$

$$= \sum_{x=0}^{n} \dbinom{n}{x}\dbinom{n}{n-x} \qquad \text{By Vandermonde's identity}$$

$$= \sum_{x=0}^{n} \dbinom{n}{x}\dbinom{n}{x} \qquad\qquad \text{By Theorem 5.7.8}$$

$$= \sum_{x=0}^{n} \dbinom{n}{x}^{2}$$

This is a property that we investigated in Pascal's triangle in Exercise 7.1.15.

10.5.4. (a) Set $r = k-1, s = r$, and $n = r-1$ in Vandermonde's Identity. Then,

$$\underbrace{\dbinom{k-1}{0}\dbinom{r}{r-1}}_{\binom{r}{1}} + \underbrace{\dbinom{k-1}{1}\dbinom{r}{(r-1)-1}}_{\binom{r}{2}} + \cdots + \underbrace{\dbinom{k-1}{r-1}\dbinom{r}{0}}_{\binom{r}{r}} = \dbinom{k-1+r}{r-1}.$$

But $\binom{r}{r-1} = \binom{r}{r-(r-1)} = \binom{r}{r-r+1} = \binom{r}{1}$ by Theorem 5.7.8. Similarly, we can convert the second combination in each term on the left side of the equation using Theorem 5.7.8. Therefore,

$$\dbinom{k-1}{0}\dbinom{r}{1} + \dbinom{k-1}{1}\dbinom{r}{2} + \cdots + \dbinom{k-1}{r-1}\dbinom{r}{r} = \dbinom{k+r-1}{r-1}$$

as required.

(b) We can use the same idea from the combinatorial proof for Vandermonde's Identity, but instead we consider a group of people with r males and $k-1$ females. In how many ways can we form a committee with $r-1$ people?

(i) We first select x females for the committee from $k-1$ females in the group; there are $\binom{k-1}{x}$ ways to accomplish this. Next, we select the remaining $r-x$ committee members from the group of r male students; there are

$$\dbinom{r}{r-1-x} = \underbrace{\dbinom{r}{r-(r-1-x)}}_{\text{By Theorem 5.7.8}} = \dbinom{r}{r-r+1+x} = \dbinom{r}{x+1}$$

ways to do this. By the Multiplication Principle, $\binom{k-1}{x}\binom{r}{x+1}$. Since x could be $0, 1, 2, \ldots, n$, these cases can be combined by the Sum Principle. Therefore, there are altogether

$$\sum_{x=0}^{n} \dbinom{k-1}{x}\dbinom{r}{x+1} = \dbinom{k-1}{0}\dbinom{r}{1} + \dbinom{k-1}{1}\dbinom{r}{2} + \cdots + \dbinom{k-1}{r-1}\dbinom{r}{r}$$

ways to form such committees.

(ii) There are $(k-1)+r = k+r-1$ people, and we need to select $r-1$ of them; there are $\binom{k+r-1}{r-1}$ ways to do this.

Since (i) and (ii) are equivalent ways of answering the question,

$$\dbinom{k-1}{0}\dbinom{r}{1} + \dbinom{k-1}{1}\dbinom{r}{2} + \cdots + \dbinom{k-1}{r-1}\dbinom{r}{r} = \dbinom{k+r-1}{r-1}$$

as required.

10.5.5. (a) $\displaystyle\sum_{x=0}^{n} x \frac{\binom{r}{x}\binom{N-r}{n-x}}{\binom{N}{n}} = \underbrace{0 \frac{\binom{r}{0}\binom{N-r}{n-0}}{\binom{N}{n}}}_{0} + \sum_{x=1}^{n} x \frac{\binom{r}{x}\binom{N-r}{n-x}}{\binom{N}{n}} = \sum_{x=1}^{n} x \frac{\binom{r}{x}\binom{N-r}{n-x}}{\binom{N}{n}}$

(b) $\displaystyle\sum_{x=0}^{n} x^2 \frac{\binom{r}{x}\binom{N-r}{n-x}}{\binom{N}{n}} = \underbrace{0^2 \frac{\binom{r}{0}\binom{N-r}{n-0}}{\binom{N}{n}}}_{0} + \sum_{x=1}^{n} x^2 \frac{\binom{r}{x}\binom{N-r}{n-x}}{\binom{N}{n}} = \sum_{x=1}^{n} x^2 \frac{\binom{r}{x}\binom{N-r}{n-x}}{\binom{N}{n}}$

10.5.6. (a) The probability distribution of X is given in Table C.13.

Table C.13 The probability distribution of the number of kings in a poker hand.

x	$P(X = x)$	
0	$\dfrac{\binom{4}{0}\binom{48}{5}}{\binom{52}{5}}$	$\approx 0.658\,84$
1	$\dfrac{\binom{4}{1}\binom{48}{4}}{\binom{52}{5}}$	$\approx 0.299\,47$
2	$\dfrac{\binom{4}{2}\binom{48}{3}}{\binom{52}{5}}$	$\approx 0.039\,93$
3	$\dfrac{\binom{4}{3}\binom{48}{2}}{\binom{52}{5}}$	$\approx 0.001\,74$
4	$\dfrac{\binom{4}{4}\binom{48}{1}}{\binom{52}{5}}$	$\approx 0.000\,02$

(b) The graph of the probability distribution of X is in Fig. C.40.

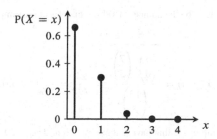

Fig. C.40 Graph of the probability distribution of the number of kings in a poker hand.

(c) $P(X \geq 2) = 1 - P(X < 2) = 1 - P(X = 0) - P(X = 1) = 1 - 0.658\,84 - 0.299\,47 = 0.041\,69$

(d) $P(X \neq 1) = 1 - P(X = 1) = 1 - 0.299\,47 = 0.700\,53$

(e) $E(X) = \frac{5 \times 4}{52} = \frac{20}{52} = \frac{5}{13} \approx 0.384\,62$

(f) $\text{var}(X) = \frac{4(5)}{52}\left[\frac{3(4)}{51} + 1 - \frac{4(5)}{52}\right] = \frac{940}{2873} \approx 0.327\,18$

10.5.7. Let random variable X be the number of defective tablets in the sample. Then, X follows a hypergeometric distribution with parameters $N = 20$, $r = 3$, and $n = 5$ with the probability distribution

$$P(X = x) = \frac{\binom{3}{x}\binom{20-3}{5-x}}{\binom{20}{5}}$$

(a) $P(X = 1) = \dfrac{\binom{3}{1}\binom{20-3}{5-1}}{\binom{20}{5}} = \dfrac{\binom{3}{1}\binom{17}{4}}{\binom{20}{5}} = \dfrac{3 \times 2380}{15\,504} \approx 0.460\,53$

(b) Using *indirect reasoning*, we have

$$1 - P(X \geq 3) = 1 - P(X = 3) = 1 - \frac{\binom{3}{3}\binom{17}{2}}{\binom{20}{5}} = 1 - \frac{136}{15\,504} = \frac{15\,368}{15\,504} \approx 0.991\,23.$$

Using *direct reasoning*, we have

$$P(X < 3) = P(X = 0) + P(X = 1) + P(X = 2)$$

$$= \frac{\binom{3}{0}\binom{17}{5}}{\binom{20}{5}} + \frac{\binom{3}{1}\binom{17}{4}}{\binom{20}{5}} + \frac{\binom{3}{2}\binom{17}{3}}{\binom{20}{5}}$$

$$= \frac{6188}{15\,504} + \frac{3(2380)}{15\,504} + \frac{3(680)}{15\,504}$$

$$= \frac{15\,368}{15\,504} \approx 0.991\,23.$$

10.5.8. (a) Let the random variable X be the number of blue marbles in the sample. Then,

$$P(X = 4) = \frac{\overbrace{\binom{4}{1}}^{\text{blue}}\overbrace{\binom{8}{1}}^{\text{other}}}{\binom{12}{5}} = \frac{4 \times 8}{792} = \frac{4}{99} \approx 0.0404.$$

(b) Let the random variable Y be the number of yellow marbles in the sample. Then,

$$P(Y = 0) = \frac{\overbrace{\binom{5}{0}}^{\text{yellow}}\overbrace{\binom{7}{5}}^{\text{other}}}{\binom{12}{5}} = \frac{21}{792} = \frac{7}{264} \approx 0.0265.$$

10.5.10. Let X be the number of employees that have a staph infection. Then, X follows a hypergeometric distribution with parameters $N = 2000$, $r = 667$, and $n = 10$. Therefore, we expect

$$E(X) = \frac{10(667)}{2000} = 3.335$$

employees to be infected.

10.5.11. (a) Let random variable X be the number of males selected. Then,

$$P(X = 5) = \frac{\binom{20}{5}}{\binom{35+20}{5}} = \frac{\binom{20}{5}}{\binom{55}{5}} = \frac{15\,504}{3\,478\,761} = \frac{304}{68\,211} \approx 0.0045.$$

(b) Let random variable Y be the number of females selected. Then,

$$P(Y=5) = \frac{\binom{35}{5}}{\binom{35+20}{5}} = \frac{\binom{35}{5}}{\binom{55}{5}} = \frac{324\,632}{3\,478\,761} = \frac{1736}{18\,603} \approx 0.0933$$

(c) $P(Y=3) + P(Y=4) + P(Y=5) =$

$$\overbrace{\frac{\binom{35}{3}\binom{20}{2}}{\binom{55}{5}}}^{\text{3 female 2 males}} + \overbrace{\frac{\binom{35}{4}\binom{20}{1}}{\binom{55}{5}}}^{\text{4 female 1 male}} + \overbrace{\frac{\binom{35}{5}\binom{20}{0}}{\binom{55}{5}}}^{\text{5 female 0 males}}$$

$$= \frac{1\,243\,550}{3\,478\,761} + \frac{1\,047\,200}{3\,478\,761} + \frac{324\,632}{3\,478\,761}$$

$$= \frac{2\,615\,382}{3\,478\,761}$$

$$= \frac{518}{689} \approx 0.7518$$

10.5.12. Let E be the event that none of the SAMs strike the friendly jet fighters.

(a) Then, $P(E) = \frac{\binom{8}{4}\binom{4}{0}}{\binom{12}{4}} = \frac{(70)(1)}{495} = \frac{14}{99} \approx 0.141\,41$.

(b) The required probability is $P(E') = 1 - P(E) = 1 - \frac{14}{99} = \frac{85}{99} \approx 0.858\,59$.

10.6.1. (a) 60   (b) 10   (c) $10^2 = 100$

10.6.2. (a) True.

(b) False. The normal distribution is symmetric about its expected value.

(c) False. Approximately 68% of the area is under the curve within one standard deviation of the expected value.

10.6.3. (a) $P(-1 \le Z \le 1) = P(Z \le 1) - P(Z \le -1) = 0.8413 - 0.1587 = 0.6826$

(b) $P(-2 \le Z \le 2) = P(Z \le 2) - P(Z \le -2) = 0.9772 - 0.0228 = 0.9544$

(c) $P(-3 \le Z \le 3) = P(Z \le 3) - P(Z \le -3) = 0.9987 - 0.0013 = 0.9974$

10.6.4. (a) $P(Z < 0.87) = 0.8078$

(b) $P(Z > 1.49) = 1 - P(Z < 1.49) = 1 - 0.9319 = 0.0681$

10.6.5. (a) $P(X < 65) = P\left(Z < \frac{65-70}{5}\right) = P(Z < -1) = 0.1587$

(b) $P(X > 75) = 1 - P(X < 75)$

$$= 1 - P\left(Z < \frac{75-70}{5}\right)$$

$$= 1 - P(Z < 1)$$

$$= 1 - 0.8413$$

$$= 0.1587$$

10.6.6. $P(277 \le X \le 412) = P\left(\dfrac{277 - 322}{45} \le Z \le \dfrac{412 - 322}{45}\right)$

$$= P(-1 \le Z \le 2)$$
$$= P(Z \le 2) - P(Z \le -1)$$
$$= 0.9772 - 0.1587$$
$$= 0.8185$$

10.6.7. Since it is symmetric, $P(0 < Z < a) = \dfrac{0.5160}{2} = 0.2580$. This means that $P(Z < a) = 0.5 + 0.2580 = 0.7580$. According to the table, $P(Z < 0.7) = 0.7580$. Thus, $a = 0.7$. To check,

$$P(-0.7 \le Z \le 0.7) = P(Z \le 0.7) - P(Z \le -0.7)$$
$$= 0.7580 - 0.2420$$
$$= 0.5160.$$

10.6.8. (a) $P(X > 4) = P\left(Z > \dfrac{4 - 1.5}{\sqrt{4.41}}\right)$

$$= P\left(Z > \dfrac{4 - 1.5}{2.1}\right)$$
$$= P(Z > 1.19)$$
$$= 1 - P(Z \le 1.19)$$
$$= 1 - 0.8830$$
$$= 0.1170$$

(b) $P(-3 < X < 1.7) = P\left(\dfrac{-3 - 1.5}{2.1} < Z < \dfrac{1.7 - 1.5}{2.1}\right)$

$$= P(-2.14 < Z < 0.10)$$
$$= P(Z < 0.10) - P(Z < -2.14)$$
$$= 0.5398 - 0.0162$$
$$= 0.5236$$

(c) From the table, $P(Z \le 0.68) = 0.7517$. Then,

$$Z = \frac{X - \mu}{\sigma}$$
$$X = \mu + Z\sigma = 1.5 + (0.68)(2.1) = 2.93.$$

10.6.9. (a) From the Z table, the corresponding value of Z for $P(X < 20) = 0.8413$ is $Z = 1$. Then,

$$Z = \frac{X - \mu}{\sigma}$$
$$\mu = X - Z\sigma$$
$$= 20 - (1)(5)$$
$$= 15.$$

(b) $P(15 < X < 20) = P\left(\dfrac{15 - 15}{5} < Z < \dfrac{20 - 15}{5}\right)$

$= P(0 < Z < 1)$

$= P(Z < 1) - P(Z < 0)$

$= 0.8413 - 0.5$

$= 0.3413$

10.6.10. From the Z table, the corresponding value of Z for $P(X < 105) = 0.6915$ is $Z = 0.5$. Then,

$$Z = \frac{X - \mu}{\sigma}$$

$$\sigma = \frac{X - \mu}{Z} = \frac{106 - 100}{0.5} = \frac{6}{0.5} = 12.$$

10.6.11. Let X denote a student's mark on the discrete mathematics final examination.

(a) $P(X > 75) = P(Z > \frac{75 - 70}{5}) = P(Z > 1) = 1 - P(Z < 1) = 1 - 0.8413 = 0.1587.$
Therefore, the probability of a student's mark greater than 75 is 15.87%.

(b) $P(Z < 60) = P(Z < \frac{60 - 70}{5}) = P(Z < -2) = 0.0228.$ Therefore, the probability of a student's mark is below 60 is 2.28%.

(c) $P(65 < X < 70) = P\left(\dfrac{65 - 70}{5} < Z < \dfrac{70 - 70}{5}\right)$

$= P(-1 < Z < 0)$

$= P(Z < 0) - P(Z < -1)$

$= 0.5000 - 0.1587$

$= 0.3413$

Therefore, the probability of a student's mark is between 65 and 70 is 34.13%.

10.6.12. Let X be a person's IQ score.

(a) $P(85 < X < 115) = P\left(\dfrac{85 - 100}{15} < Z < \dfrac{115 - 100}{15}\right)$

$= P(-1 < Z < 1)$

$= P(Z < 1) - P(Z < -1)$

$= 0.8413 - 0.1587$

$= 0.6826$

Therefore, the probability that a person's IQ is between 85 and 115 is 68.26%.

(b) $P(X > 125) = P\left(Z > \dfrac{125 - 100}{15}\right)$

$= P(Z > 1.67)$

$= 1 - P(Z < 1.67)$

$= 1 - 0.9525$

$= 0.0475$

Therefore, the probability that a person's IQ is above 125 is 4.75%.

(c) $P(X < 50) = P\left(Z < \dfrac{50 - 100}{15}\right)$

$= P(Z < -3.33)$

$= 0.0004$

Therefore, the probability that a person's IQ is below 50 is 0.04%.

10.6.13. Let X denote the amount of protein in a randomly selected shake.

(a)   (i) $P(X < 515) = P\left(Z < \dfrac{515 - 507.5}{4.0}\right) = P(Z < 1.88) = 0.9699$

      (ii) $P(X \geq 515) = 1 - P(X < 515)$

$$= 1 - P\left(Z < \frac{515 - 507.5}{4.0}\right)$$

$$= 1 - P(Z < 1.88)$$

$$= 1 - 0.9699$$

$$= 0.0301$$

      (iii) $P(500 < X < 512) = P\left(\dfrac{500 - 507.5}{4} < Z < \dfrac{512 - 507.5}{4}\right)$

$$= P(-1.88 < Z < 1.13)$$

$$= P(Z < 1.13) - P(Z < -1.88)$$

$$= 0.8708 - 0.0301$$

$$= 0.8407$$

      (iv) $P(X \neq 507.5) = 1$

(b) $P(Z < 1.75) = 0.96$. Then,

$$P(X < x) = P\left(\frac{x - 507.5}{4} < 1.75\right) = 0.96,$$

and

$$\frac{x - 507.5}{4} = 1.75$$

$$x = 1.75(4) + 507.5$$

$$x = 514.5.$$

10.6.14. Let X be the volume of soda in a randomly selected can.

(a) $P(X < 360) = P(Z < \dfrac{360 - 355}{2.5}) = P(Z < 2) = 0.9772$

(b) $P(X > 348) = 1 - P(X < 348)$

$$= 1 - P\left(Z < \frac{348 - 355}{2.5}\right)$$

$$= 1 - P(Z < -2.8)$$

$$= 1 - 0.0026$$

$$= 0.9974$$

(c) $P(350 < X < 360) = P\left(\dfrac{350 - 355}{2.5} < Z < \dfrac{360 - 355}{2.5}\right)$

$$= P(-2 < Z < 2)$$

$$= P(Z < 2) - P(Z < -2)$$

$$= 0.9772 - 0.0228$$

$$= 0.9544$$

(d) $P(X \neq 355) = 1$

10.6.15. Let X be a randomly selected American woman's height. Then the 95th percentile is $P(Z < z) = 0.95$, where $z = 1.65$.

$$z = \frac{X - \mu}{\sigma}$$
$$X = \mu + \sigma Z$$
$$= 63.8 + (0.16)(1.65)$$
$$= 64.064$$

Therefore, Layla's height is 64.1 inches.

10.6.16. (a)   (i) $P(X = 500) = 0$

    (ii) $P(X < 510) = P\left(Z < \frac{510 - 500}{10}\right) = P(Z < 1) = 0.8413$

    (iii) $P(495 < X < 505) = P\left(\frac{495 - 510}{10} < Z < \frac{505 - 510}{10}\right)$

$$= P(-1.5 < Z < -0.5)$$
$$= P(Z < -0.5) - P(Z < -1.5)$$
$$= 0.3085 - 0.0668$$
$$= 0.2417$$

(b) $P(Z < 2.06) = 0.98$. Then, $P\left(\frac{x - 500}{10} < 2.06\right) = 0.98$, and

$$\frac{x - 500}{10} = 2.06$$
$$x = 2.06(10) + 500$$
$$x = 520.6.$$

10.6.17. Let $X \sim \mathcal{N}(10\,000, 7400)$.

$$P(X > 15\,000) = P\left(Z > \frac{15\,000 - 10\,000}{7400}\right)$$
$$= P(Z > 0.68)$$
$$= 1 - P(Z < 0.68)$$
$$= 1 - 0.7517$$
$$= 0.2483$$

Therefore, the probability that a health insurance claim exceeds $15\,000$ is 24.83%.

10.6.18. Let $X \sim \mathcal{N}(50, 100)$ be the demand for blood for the day. To ensure that 95% of the time that the hospital has enough blood, $P(Z < z) = 0.95$, $z = 1.65$. Thus,

$$\frac{x - 50}{10} = 1.65$$
$$x = 1.65(10) + 50$$
$$x = 66.5.$$

The hospital needs to stock 66.5 L of blood to ensure there is only a 5% chance of running by the end of the day.

10.6.19. Discrete mathematics: $\frac{79 - 70}{10} = 0.9$. Calculus: $\frac{85 - 80}{8} = 0.625$. Comparing the standardized scores, the student did better relative to his or her peers in discrete mathematics than calculus.

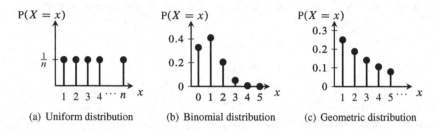

(a) Uniform distribution   (b) Binomial distribution   (c) Geometric distribution

Fig. C.41   Graphs of the probability distributions for Exercise 10.7.1

10.7.1. The graphs of the probability distributions are in Fig. C.41.

10.7.2. (a) Hypergeometric
(b) Binomial
(c) Hypergeometric
(d) Uniform
(e) Hypergeometric
(f) Geometric
(g) Geometric
(h) Binomial

10.7.3. We know that $n = 20$ and $p = 0.25$. Therefore the expected value $E(X) = 20(0.25) = 5$, and the variance is $\text{var}(X) = 20(0.25)(0.75) = 3.75$.

10.7.4. (a) Each of the trials is independent and the probabilities do not change from trial to trial. Each trial outcome can be labeled as a success or a failure.
(b) $P(X = 2) = \binom{5}{2}\left(\frac{2}{10}\right)^2\left(\frac{8}{10}\right)^3 = \frac{128}{625} = 0.2048$
(c) $P(X \geq 1) = 1 - P(X = 0) = 1 - \binom{5}{0}\left(\frac{2}{10}\right)^0\left(\frac{8}{10}\right)^5 = 1 - \left(\frac{8}{10}\right)^5 = \frac{2101}{3125} = 0.672\,32$
(d) $E(X) = np = 5(0.2) = 1.0$
(e) $\sigma = \sqrt{np(1-p)} = \sqrt{5(0.2)(1-0.2)} = \sqrt{.8} \approx 0.894\,43$

10.7.5. Geometric distribution.

10.7.6. Let the random variable X be the number of tornadoes in a 15-day period with $\lambda = 2 \times 3 = 6$. Then,

$$P(X = 0) + P(X = 1) + P(X = 2) = \frac{e^{-6}6^0}{0!} + \frac{e^{-6}6^1}{1!} + \frac{e^{-6}6^2}{2!} \approx 0.0619\,7.$$

10.7.7. Let the random variable X be the amount of waste water released into the lake for the week.

$$P(X < 2.75) = P\left(Z < \frac{2.75 - 2.50}{0.5}\right) = P(Z < 0.5) = 0.6915$$

Therefore, the probability that less than 2.75 tons of waste water is released in a week is 69.15%.

10.7.8. (a) $P(X < 50) = P\left(Z < \dfrac{50 - 68}{12}\right)$

$$= P\left(Z < -\dfrac{18}{12}\right)$$

$$= P\left(Z < -\dfrac{3}{2}\right)$$

$$= P(Z < -1.50)$$

$$= 0.0668$$

Therefore, 6.68% of the students failed the course.

(b) $P(X > 92) = 1 - P(X < 92)$

$$= 1 - P\left(Z < \dfrac{92 - 68}{12}\right)$$

$$= 1 - P\left(Z < \dfrac{24}{12}\right)$$

$$= 1 - P(Z < 2)$$

$$= 1 - 0.9772$$

$$= 0.0228$$

Therefore, 2.28% of the students got an A+.

(c) $P(62 < X < 74) = P(X < 74) - P(X < 62)$

$$= P\left(Z < \dfrac{74 - 68}{12}\right) - P\left(Z < \dfrac{62 - 68}{12}\right)$$

$$= P\left(Z < \dfrac{6}{12}\right) - P\left(Z < -\dfrac{6}{12}\right)$$

$$= P(Z < 0.5) - P(Z < -0.5)$$

$$= 0.6915 - 0.3085$$

$$= 0.3830$$

Therefore, 38.29% of the students obtained a mark between 62% and 74%.

B.1.1. The number

(a) $\frac{3}{4} \in \mathbb{Q}$ (b) $\frac{3}{4} \notin \mathbb{Q}'$ (c) $\frac{3}{4} \notin \mathbb{Z}$ (d) $\frac{3}{4} \in \mathbb{R}$ (e) $\frac{3}{4} \notin \mathbb{N}$

B.3.1. (a) Commutativity   (d) Multiplicative inverse
        (b) Associativity   (e) Additive inverse
        (c) Distributivity   (f) Additive identity

B.3.2. No. For example, $x = 3$ and $y = \frac{1}{3}$. Hence, $3(\frac{1}{3}) = 1$.

B.3.3. No, subtraction does not have the associativity property. For example, $(3 - 4) - 5 \neq 3 - (4 - 5)$.

B.3.4. No, division does not have the commutative property. For example, $\frac{1}{3} \neq \frac{3}{1}$.

B.3.5. We can draw an arbitrary rectangle. We can measure the rectangle in two ways: in Fig. C.42(a) we can represent the area of the rectangle by $w \cdot (x + y + z)$, or we can represent the area of the rectangle as $w \cdot x + w \cdot y + w \cdot z$ like in Fig. C.42(b). Since Fig. C.42(a) and Fig. C.42(b) represent different expressions of area for the same rectangle, therefore, $w \cdot (x + y + z) = w \cdot x + w \cdot y + w \cdot z$.

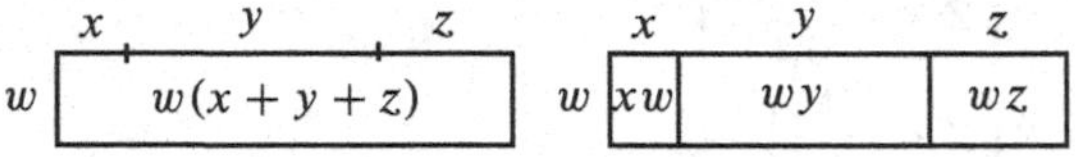

(a) $w \cdot (x + y + z)$ (b) $w \cdot x + w \cdot y + w \cdot z$

Fig. C.42   The extended distributive property

B.3.6. (a) No   (b) No   (c) No   (d) Yes

B.3.7. (a) No   (b) No   (c) Yes   (d) Yes

B.3.8. (a) Yes   (b) No   (c) Yes   (d) Yes

B.3.9. (a) Yes   (b) Yes   (c) Yes   (d) Yes

B.4.1. (a) $-8, 6$ (c) $-2, 6$ (e) $\frac{1}{6}, 1$
        (b) $-3, 7$ (d) $-\frac{10}{3}, 3$ (f) $-\frac{5}{3}, \frac{3}{4}$

B.4.2. $(2ax + b)^2 = (2ax + b)(2ax + b)$

$$= (2ax)(2ax) + (2ax)b + b(2ax) + b^2$$
$$= 4a^2x^2 + 2axb + 2axb + b^2$$
$$= 4a^2x^2 + 4axb + b^2$$

B.5.1. (a) 4   (b) 64   (c) $\frac{1}{16}$ (d) $\frac{1}{6}$ (e) $\frac{9}{2}$

B.5.2. (a) $\frac{1}{y}$ (b) $\frac{64}{9}$ (c) x (d) $\frac{1}{a^{4m}}$

B.5.3. (a) c^6 (h) x^{10} (n) $\frac{x^8 y^{12}}{z^{16}}$
        (b) y^9 (i) a^7 (o) $x^4 y^{10}$
        (c) x^{14} (j) x^6 (p) $9x^8$
        (d) x^3 (k) $x^6 y^{12}$ (q) $10\,000 x^{24}$
        (e) x^4 (l) $\frac{a^{15}}{b^{20}}$ (r) $-1728 m^{15} n^{33}$
        (f) w (m) $a^5 b^6$ (s) $6x^4 y^5$
        (g) x^{15}

B.5.4. (a) $x^{\frac{5}{2}}$ (b) $y^{\frac{10}{3}}$ (c) $(z^2 + 2)^{\frac{1}{2}}$ (d) $(x + 5)^3$

B.5.5. (a) $x = 2$ (c) $x = \frac{5}{4}$ (e) $x = -3$
        (b) $x = -3$ (d) $x = 4, -3$ (f) $x = 0$

B.6.4. $2\log(100) - \log(1000) + \log(10) = \log(100^2) - \log(1000) + \log(10)$

$$= \log\left(\frac{10\,000}{1000}\right) + \log(10)$$
$$= \log\left(\frac{10\,000}{1000} \times 10\right)$$
$$= \log(100)$$
$$= 2$$

B.6.5. (a) $2\log(a) - \log(b) + \log(c) = \log(a^2) - \log(b) + \log(c)$

$$= \log\left(\frac{a^2}{b}\right) + \log(c)$$

$$= \log\left(\frac{a^2 c}{b}\right)$$

(b) $\log(a) + 5\log(b) - \log(c) = \log(a) + \log(b^5) - \log(c)$

$$= \log(ab^5) - \log(c)$$

$$= \log\left(\frac{ab^5}{c}\right)$$

B.6.6. (a) $\log(3 \times 4) = \log(3) + \log(4) \approx 0.4771 + 0.6021 = 1.0792$

(b) $\log(\frac{3}{4}) = \log(3) - \log(4) \approx 0.4771 - 0.6021 = -0.125$

(c) $\log(\sqrt{3}) = \log(3^{1/2}) = \frac{1}{2}\log(3) \approx \frac{0.4771}{2} \approx 0.2386$

B.6.8. (a) $\log_2 64 = 6$ (b) $\log_{10} 0.001 = -3$ (c) $\log_{32} 8 = \frac{3}{5}$

B.6.10. (a) $x = \frac{\log(5)}{\log(2)} \approx 2.3219$ (d) $x = 2$

(b) $x = \frac{\log(7)}{\log(5)} \approx 1.2091$ (e) $x \approx 1.1610$

(f) $x \approx 24.603$

(c) $x = \frac{\log(30)}{\log(6)} \approx 1.8982$

B.6.11. (a) $x = 1.5$ (c) $x = \frac{4+\ln(2)}{3} \approx 1.564\,38$

(b) $x = 5$ (d) $x = \frac{\log(3)}{\log(2)} \approx 1.584\,96$

B.6.12. $\log_b(x) = \frac{\log(x)}{\log(b)} = \frac{1}{\frac{\log(b)}{\log(x)}} = \frac{1}{\log_x(b)}$

B.6.13. $\log_b\left(\sqrt[y]{x}\right) = \log_b\left(x^{\frac{1}{y}}\right)$ Law of Square Root

$$= \frac{1}{y}\log_b(x) \qquad \text{Law of Exponent}$$

B.6.14. Let A be the original amount of the car. Then,

$$A(1 - 0.19)^x = 0.5A$$
$$\cancel{A}(1 - 0.19)^x = 0.5\cancel{A}$$
$$(0.81)^x = 0.5$$
$$x\log(0.81) = \log(0.5)$$
$$x = \frac{\log(0.5)}{\log(0.81)}$$
$$x = 3.289 \text{ years.}$$

B.6.15. $A2^3 = 4500$

$$A = \frac{4500}{8} = 562.5$$

Therefore, there were 562 or 563 locusts.

B.6.16. $15\,000(0.925)^{3.5} = 11\,417.9 \approx 11\,418$

B.6.17. $\frac{3}{\log_2(a)} = \frac{3}{\frac{\log(a)}{\log(2)}} = \frac{3\log(2)}{\log(a)} = \frac{\log(2^3)}{\log(a)} = \frac{\log(8)}{\log(a)} = \frac{1}{\frac{\log(a)}{\log(8)}} = \frac{1}{\log_8(a)}$

B.7.1. (a) 2×10^0
     (b) 3×10^2
     (c) $4.321\,768 \times 10^3$
     (d) -5.3×10^4
     (e) 6.72×10^9
     (f) 7.51×10^{-9}

Bibliography

Adams, D. (1979). *The Hitch Hiker's Guide to the Galaxy* (Thomson).

Aigner, M. and Ziegler, G. M. (2014). *Proofs from THE BOOK*, 5th edn. (Springer), doi:10.1007/978-3-662-44205-0.

Aitken, C. G. G. and Taroni, F. (2004). *Statistics and the Evaluation of Evidence for Forensic Scientists* (Wiley and Sons, New York), doi:10.1002/0470011238.

Anderson, J. (2011a). A few million monkeys randomly recreate every work of Shakespeare, http://www.jesse-anderson.com/2011/10/a-few-million-monkeys-randomly-recreate-every-work-of-shakespeare/.

Anderson, J. (2011b). A million monkeys and Shakespeare, *Significance* **8**, pp. 190–192, doi:10.1111/j.1740-9713.2011.00533.x.

Anellis, I. H. (2012). Peirce's truth-functional analysis and the origin of the truth table, *History and Philosophy of Logic* **33**, pp. 87–97, doi:10.1080/01445340.2011.621702.

Anselin, L. (1988). *Spatial Econometrics: Methods and Models* (Kluwer Academic).

Arias, E. (2006). United States life tables, 2003, *National Vital Statistics Reports* **54**, 14, pp. 1–40.

Artin, E. (1953). Review of N. Bourbaki's Éléments de mathématique, *Bulletin of the American Mathematical Society* **59**, 5, pp. 474–479, doi:10.1090/S0002-9904-1953-09725-7.

Asher, R. (1951). Munchausen's syndrome, *The Lancet* **257**, 6650, pp. 339–341, doi:10.1016/S0140-6736(51)92313-6.

Augustine, S. and Green, R. (2008). *On Christian Teaching*, Oxford World's Classics (Oxford University Press).

Bamford, J. (1982). *The Puzzle Palace* (Houghton-Mifflin).

Bar-Hillel, M. and Falk, R. (1982). Some teasers concerning conditional probabilities, *Cognition* **11**, 2, pp. 109–122, doi:10.1016/0010-0277(82)90021-X.

Baron, M. E. (1969). A note on the historical development of logic diagrams: Leibniz, Euler, and Venn, *The Mathematical Gazette* **53**, 383, pp. 113–125, doi:10.2307/3614533.

Bell, E. T. (1937). *Men of Mathematics: The Lives and Achievements of the Great Mathematicians from Zeno to Poincaré* (Simon and Schuster, New York).

Bellhouse, D. (2005). Decoding Cardano's Liber de Ludo Aleæ, *Historia Mathematica* **32**, 2, pp. 180–202, doi:10.1016/j.hm.2004.04.001.

Bellhouse, D. R. (2000). De vetula: A medieval manuscript containing probability calculations, *International Statistical Review* **68**, 2, pp. 123–136, doi:10.2307/1403664.

Benjamin, A. T. and Quinn, J. T. (2003). *Proofs that Really Count: The Art of Combinatorial Proof* (The Mathematical Association of America).

Bertrand, J. (1889). *Calcul des probabilités* (Gauthier-Villars, Paris).

Bostwick, C. W. (1958). Elementary problems and solutions—e 1321, *American Mathematical Monthly* **65**, 6, p. 446, doi:10.2307/2310728.

Bostwick, C. W., Rainwater, J., and Baum, J. D. (1959). Solution for e 1321, *American Mathematical Monthly* **66**, 2, pp. 141–142, doi:10.2307/2310027.

Buchanan, M. (2007). Statistics: Conviction by numbers, *Nature* **445**, pp. 254–255, doi:10.1038/445254a.

Bundy, A., Jamnik, M., and Fugard, A. (2005). What is a proof? *Philosophical Transactions of the Royal Society A* **363**, 1835, pp. 2377–2391, doi:10.1098/rsta.2005.1651.

Burnet, J. (1908). *Early Greek Philosophy* (Adam and Charles Black, London).

Burton, D. M. (2011). *The History of Mathematics: An Introduction*, 7th edn. (McGraw-Hill, New York).

Bussey, W. H. and Hartwell, G. W. (1914). Solution to algebra problem 399, *The American Mathematical Monthly* **21**, 5, p. 158, doi:10.2307/2972187.

Cain, S. (2012). *Quiet: The Power of Introverts in a World That Can't Stop Talking* (Crown Publishers, New York).

Cajori, F. (1918). Origin of the name "mathematical induction", *The American Mathematical Monthly* **25**, 5, pp. 197–201, doi:10.2307/2972638.

Cajori, F. (1920). The purpose of Zeno's arguments on motion, *Isis* **3**, 1, pp. 7–20.

Calkin, N. and Wilf, H. S. (2000). Recounting the rationals, *The American Mathematical Monthly* **107**, 4, pp. 360–363, doi:10.2307/2589182.

Carlton, M. A. and Stansfield, W. D. (2005). Making babies by the flip of a coin? *The American Statistician* **59**, 2, pp. 180–182, doi:10.1198/000313005X42813.

Carroll, L. (1866). *Alice in Wonderland* (Appleton, New York).

Carroll, L. (1886). *The Game of Logic* (Macmillan, London).

Carroll, L. (1895). What the Tortoise said to Achilles, *Mind* **4**, 14, pp. 278–280, doi:10.1093/mind/IV.14.278.

Carroll, L. (1896). *Symbolic Logic—Part I. Elementary* (Macmillian, London).

Carruccio, E. (1964). *Mathematics and Logic in History and in Contemporary Thought* (Faber and Faber, London).

Carslaw, H. S. (1914). *Plane Trigonometry*, 2nd edn. (Macmillian and Company, Limited, London).

Chang, K.-T. (2010). *Introduction to Geographic Information Systems* (McGraw-Hill).

Chen, C. (2013). The paradox of the proof, Online, `http://projectwordsworth.com/the-paradox-of-the-proof/`.

Cherry, W. H. and Oldford, R. W. (2002). The poverty of Venn diagrams for teaching probability: their history and replacement by eikosograms, `http://sas.uwaterloo.ca/~rwoldfor/papers/venn/eikosograms/paper-ss.pdf`.

Clarke, R. D. (1946). An application of the poisson distribution, *Journal of the Institute of Actuaries* **72**, p. 481.

Cohen, M., Ensley, D., Frantz, M., Hauss, P., Kennedy, J., Mitchell, K., and Oakley, P. (2001). Proof without words: geometric series, *The Mathematics Magazine* **74**, 4, p.

320, doi:10.2307/2691106.

Coleman, A. J., Del Grande, J. J., Duff, G. F. D., Egsgard, J. C., and Kirby, B. J. (1973). *Algebra* (Gage Educational Publishing).

Copi, I. M. (1954). *Symbolic Logic* (Macmillan, New York).

Courant, R. and Robbins, H. (1941). *What is Mathematics? An elementary approach to ideas and methods* (Oxford University Press).

Curry, A. E., Hafetz, J., Kallan, M. J., Winston, F. K., and Durbin, D. R. (2011). Prevalence of teen driver errors leading to serious motor vehicle crashes, *Accident Analysis and Prevention* **43**, 4, pp. 1285–1290, doi:10.1016/j.aap.2010.10.019.

Dauben, J. W. (1977). Georg Cantor and Pope Leo XIII: mathematics, theology, and the infinite, *Journal of the History of Ideas* **38**, 1, pp. 85–108, doi:10.2307/2708842.

David, F. N. (1962). *Games, Gods, and Gambling* (Hafner Publishing Comany, New York).

Dawid, P. (2005). Statistics on trial, *Significance* **2**, 1, pp. 6–8, doi:10.1111/j.1740-9713.2005.00075.x.

Deadman, K. A. (1970). Convergence of geometric series, *The Mathematical Gazette* **54**, 388, pp. 140–141, doi:10.2307/3612095.

Dedekind, R. (1888). *Was sind und was sollen die Zahlen?* (Braunschweig).

Doust, J. (2009). Using probabilistic reasoning, *British Medical Journal* **339**, 7729, pp. 1080–1082, doi:10.1136/bmj.b3823.

Drake, S. (1957). *Discoveries and Opinions of Galileo* (Doubleday).

Dyer, C. (2003). Sally Clark freed after appeal court quashes her convictions, *British Medical Journal* **326**, p. 304, doi:10.1136/bmj.326.7384.304.

Dyer, C. (2005a). GMC hears case against paediatrician in Sally Clark trial, *British Medical Journal* **330**, p. 1463, doi:10.1136/bmj.330.7506.1463.

Dyer, C. (2005b). Meadow defends his role in conviction of Sally Clark, *British Medical Journal* **331**, p. 66, doi:10.1136/bmj.331.7508.66-a.

Dyer, C. (2005c). Professor Roy Meadow struck off, *British Medical Journal* **331**, p. 177, doi:10.1136/bmj.331.7510.177.

Dyer, C. (2005d). Sally Clark pathologist removed from Home Office list, *British Medical Journal* **331**, p. 1355, doi:10.1136/bmj.331.7529.1355-a.

Dyer, C. (2007). Pathologist in Sally Clark case wins removal appeal, *British Medical Journal* **335**, p. 466, doi:10.1136/bmj.39328.472627.4E.

Dyer, C. (2010). Expert witnesses above the parapet, *British Medical Journal* **341**, pp. 178–179, doi:10.1136/bmj.c3672.

Easton, M. (2012). What happened when MPs took a maths exam, Online, http://www.bbc.com/news/uk-19801666.

Ecker, M. W. (1998). A novel approach to geomteric series, *The College Mathematics Journal* **29**, 5, pp. 419–420, doi:10.2307/2687261.

Eddington, A. (1935). *New Pathways in Science* (Cambridge University Press, Cambridge).

Eldridge-Smith, P. and Eldridge-Smith, V. (2010). The pinocchio paradox, *Analysis* **70**, 2, pp. 212–215, doi:10.1093/analys/anp173.

Elkind, D. (1967). Egocentrism in adolescence, *Child Development* **38**, 4, pp. 1025–1034, doi:10.2307/1127100.

Ericsson, K. A., Krampe, R. T., and Tesch-Römer, C. (1993). The role of deliberate practice in the acquisition of expert performance, *Psychological Review* **100**, 3, pp. 363–406, doi:10.1037/0033-295X.100.3.363.

Ernest, P. (1984). Mathematical induction: a pedagogical discussion, *Educational Studies in Mathematics* **15**, 2, pp. 173–189, doi:10.1007/BF00305895.

Euler, R. and Sadek, J. (2009). A 'Sterling' summation method, *The Mathematical Gazette* **93**, pp. 504–508.

Fawcett, T. (2006). An introduction to ROC analysis, *Pattern Recognition Letters* **27**, 8, pp. 861–874, doi:10.1016/j.patrec.2005.10.010.

Feller, W. (1968). *An Introduction to Probability Theory and Its Applications: Volume I* (Wiley and Sons, New York).

Fendel, D. and Resek, D. (1990). *Foundations of Higher Mathematics* (Addison-Wesley).

Fermat, P. (1894). *Œuvres de Fermat* (Gauthier-Villars, Paris).

Feynman, R. and Leighton, R. (1985). *"Surely You're Joking, Mr. Feynman": Adventures of a Curious Character* (W. W. Norton).

Finkelman, P. (2007). Scott v. Sandford: The court's most dreadful case and how it changed history, *Chicago-Kent Law Review* **82**, pp. 3–48.

Finucan, H. M. (1970). Methods of attack on problems about permuatations, *The Mathematical Gazette* **54**, 390, pp. 381–382.

Galilei, G. (1638). *Mathematical Discourses concerning Two New Sciences relating to Mechanicks and Local Motion in Four Dialogues* (Lodewijk Elzevir), translation into English by John Weston in 1730.

Galton, F. (1889). *Natural Inheritance* (Macmillan, London).

Gamow, G. (1961). *One Two Three... Infinity: Facts and Speculations of Science* (Viking Press).

Gardner, M. (1959). Mathematical games: Problems involving questions of probability and ambiguity, *Scientific American* **201**, 4, pp. 174–182, doi:10.1038/scientificamerican1059-174.

George, G. (2004). Testing for the independence of three events, *The Mathematical Gazette* **88**, p. 568.

Gibbins, N. M. (1944). Infinite series for fifith-formers, *The Mathematical Gazette* **28**, 282, pp. 170–172, doi:10.2307/3609544.

Gonseth, F. (1936). *Les mathématiques et la réalité : essai sur la méthode axiomatique* (Félix Alcan).

Grant, T. (2011). Statistics Canada to stop tracking marriage and divorce rates, *The Globe and Mail* http://www.theglobeandmail.com/news/national/statistics-canada-to-stop-tracking-marriage-and-divorce-rates/article4192704/.

Grattan-Guinness, I. (1971). Towards a biography of Georg Cantor, *Annals of Science* **27**, 4, pp. 345–391, doi:10.1080/00033797100203837.

Gulbekian, E. (1987). The origin and value of the stadion unit used by Eratosthenes in the third century B.C. *Archive for History of Exact Sciences* **37**, 4, pp. 359–363, doi:10.1007/BF00417008.

Gunderson, D. S. (2011). *Handbook of Mathematical Induction: Theory and Applications* (CRC Press).

Hald, A. (1990). *A History of Probability and Statistics and their Applications before 1750* (Wiley and Sons, New York).

Hall, G. H. (1967). The clinical application of Bayes' theorem, *The Lancet* **290**, pp. 555–557, doi:10.1016/S0140-6736(67)90514-4.

Halmos, P. R. (1955). Book review of Pólya's *Mathematics and Plausible Reasoning, Bulletin of the American Mathematical Society* **61**, 3, pp. 243–245, doi:10.1090/S0002-9904-1955-09904-X.

Hastie, T., Tibshirani, R., and Friedman, J. (2009). *The Elements of Statistical Learning: Data Mining, Inference, and Prediction* (Springer), doi:10.1007/978-0-387-84858-7.

Hayes, B. (2006). Gauss' day of reckoning, *American Scientist* **94**, 3, pp. 200–205, doi:10.1511/2006.3.200.

Hedges, S. A. (1978). Dice music in the eighteenth century, *Music and Letters* **59**, 2, pp. 180–187.

Heyman, S. (2015). Google Books: A complex and controversial experiment, *The New York Times* http://www.nytimes.com/2015/10/29/arts/international/google-books-a-complex-and-controversial-experiment.html?_r=1.

Hilbert, D. (1931). Die grundlegung der elementaren zahlenlehre, *Mathematische Annalen* **104**, 1, pp. 485–494, doi:10.1007/BF01457953.

Hilbert, D. (2013). *David Hilbert's Lectures on the Foundations of Arithmetic and Logic 1917–1933* (Springer Berlin Heidelberg), doi:10.1007/978-3-540-69444-1, edited by William Ewald and Wilfried Sieg.

Hilbert, D. and Ackermann, W. F. (1959). *Grundzüge der theoretischen Logik* (Springer, Berlin).

Hill, R. (2005). Reflections on the cot death cases, *Signficance* **2**, 1, pp. 13–16, doi:10.1111/j.1740-9713.2005.00077.x.

Hoehn, L. (1975). A more elegant method of deriving the quadratic formula, *The Mathematics Teacher* **68**, 5, pp. 442–443.

Hurley, W. J. (2007). A note on probability trees, *Journal of Modern Applied Statistical Methods* **6**, pp. 645–648.

Ingalls, A. G. (1955). The amateur scientist: About little computers that solve puzzles and an experiment on the "gyroscopic eye", *Scientific American* **192**, pp. 116–122, doi:10.1038/scientificamerican0355-116.

International Telecommunications Union (2009). M.1677: International morse code, www.itu.int/rec/R-REC-M.1677-1-200910-I/.

James, G., Burley, D., Clements, D., Dyke, P., Searl, J., and Wright, J. (2001). *Modern Engineering Mathematics* (Prentice Hall, Upper Saddle River, NJ).

Joyce, G. H. (1908). *Principles of Logic* (Longmans, Green and Company).

Kahn, D. (1996). *The Code Breakers: The Comprehensive History of Secret Communication from Ancient Times to the Internet* (Scribner).

Kasner, E. and Newman, J. (1949). *Mathematics and the Imagination* (G. Bell and Sons, London).

Kemeny, J. G., Snell, J. L., and Thompson, G. L. (1956). *Introduction to Finite Mathematics* (Prentice Hall, Englewood Cliffs).

Kendall, M. G. (1956). Studies in the history of probability and statistics: II. the beginnings of a probability calculus, *Biometrika* **43**, 1/2, pp. 1–14, doi:10.2307/2333573.

Kendall, M. G. (1963). Studies in the history of probability and statistics: XIII. Isaac Todhunter's History of the Mathematical Theory of Probability, *Biometrika* **50**, 1/2, pp. 204–205, doi:10.2307/2333762.

Krantz, S. G. (1997). *A Primer of Mathematical Writing* (American Mathematical Society).

Kurzweil, R. (2013). *How to Create a Mind: The Secret of Human Thought Revealed* (Penguin Press, New York).

Lang, S. (1993). *Real and Functional Analysis* (Springer, New York), doi:10.1007/978-1-4612-0897-6.

Laporte, G., Ouellet, R., and Lefebvre, F. (1980). A paradox in elementary probability theory, *The Mathematical Gazette* **64**, pp. 53–54.

LeCun, Y. (1989). Generalization and network design strategies, Tech. rep., University of Toronto.

Lehrer, J. (2011). Cracking the scratch lottery code, Online, http://www.wired.com/2011/01/ff_lottery/.

Levene, S. and Bacon, C. J. (2004). Sudden unexpected death and covert homicide in infancy, *Archives of Disease in Childhood* **89**, pp. 443–447, doi:10.1136/adc.2003.036202.

Lindberg, D. C. and Numbers, R. L. (1986). Beyond war and peace: A reappraisal of the encounter between Christianity and Science, *Church History* **55**, 3, pp. 338–354, doi:10.2307/3166822.

Linder, F. E. and Grove, R. D. (1947). *Vital Statistics Rates in the United States 1900–1940* (United States Government Printing Office, Washington).

Link, G. (2004). *One Hundred Years of Russell's Paradox: Mathematics, Logic, Philosophy* (Walter De Gruyter Inc).

Mark, G., Gonzalez, V. M., and Harris, J. (2005). No task left bedind? Examining the nature of fragmented work, in *Proceedings of the SIGCHI conference on Human factors in computing systems*, pp. 321–330.

Markowsky, G. (1992). Misconceptions about the golden ratio, *The College Mathematics Journal* **23**, pp. 2–19, doi:10.2307/2686193.

Meadow, R. (1977). Munchausen syndrome by proxy the hinterland of child abuse, *The Lancet* **310**, pp. 343–345, doi:10.1016/S0140-6736(77)91497-0.

Meadow, R. (2002). Personal paper: A case of murder and the BMJ, *British Medical Journal* **324**, pp. 41–43.

Milankovitch, M. (1909). Eine graphische darstellung der geometrischen progressionen, *Zeitschrift für mathematischen und naturwissenschaftlichen* **40**, p. 329.

Miller, G. A. (1956). The magical number seven, plus or minus two: some limits on our capacity for processing information. *Psychological Review* **63**, pp. 81–97.

Minino, A. M., Arias, E., Kochanek, K. D., Murphy, S. L., and Smith, B. L. (2002). Deaths: Final data for 2000, *National Vital Statistics Reports* **50**, 15, pp. 1–120.

Montmort, P. R. d. (1708). *Essay d'Analyse sur les Jeux de Hazard*, 1st edn. (Jacque Quillau).

Moore, G. H. and Garciadiego, A. (1981). Burali-Forti's paradox: A reappraisal of its origins, *Historia Mathematica* **8**, 3, pp. 319–350, doi:10.1016/0315-0860(81)90070-7.

Moore, P. G. (1997). The development of the UK national lottery: 1992–96, *Journal of the Royal Statistical Society. Series A (Statistics in Society)* **160**, 2, pp. 169–185, doi:10.1111/1467-985X.00055.

Murphy, S. L., Xu, J. Q., and Kochanek, K. D. (2013). Deaths: final data for 2010, *National vital statistics reports* **61**, 4, pp. 1–118.

Nash Jr, J. F. (2007). *The Essential John Nash* (Princeton University Press).

National Council of the Churches of Christ in the United States of America (1993). *New Revised Standard Version Bible: Catholic Edition* (Catholic Bible Press), used with permission.

Natural Resources Canada (2001). Atlas of Canada—Capital City Locations and Names of Canada, Used with permission.

Natural Resources Canada (2002), Atlas of Canada—Ontario with Names Reference Map, Used with permission. http://www.nrcan.gc.ca/earth-sciences/geography/atlas-canada/reference-maps/16846.

Office for National Statistics (2013). Deaths Registered in England and Wales (Series DR), 2012, http://www.ons.gov.uk/ons/dcp171778_331565.pdf.

Office for National Statistics (2014). What percentage of marriages end in divorce? http://www.ons.gov.uk/ons/dcp171778_351693.pdf.

Ore, O. (1960). Pascal and the invention of probability theory, *The American Mathematical Monthly* **67**, 5, pp. 409–419, doi:10.2307/2309286.

Oxenford, J. (1875). *Conversations of Goethe with Eckermann and Soret* (George Bell and Sons), translated from the German work by Eckermann and Soret into English by Oxenford.

Poincaré, H. (1912). *Calcul des Probabilités* (Gauthier-Villars, Paris).

Pólya, G. (1945). *How to Solve It: A New Aspect of Mathematical Method* (Princeton University Press).

Pólya, G. (1954). *Induction and Analogy in Mathematics: Volume 1 of Mathematics and Plausible Reasoning* (Princeton University Press).

Pólya, G., Tarjan, R. E., and Woods, D. R. (1983). *Notes on Introductory Combinatorics* (Birkhäuser), doi:10.1007/978-0-8176-4953-1.

Potterat, J. J., Woodhouse, D. E., Muth, J. B., and Muth, S. Q. (1990). Estimating the prevalence and career longevity of prostitute women, *Journal of Sex Research* **27**, 2, pp. 233–243, doi:10.1080/00224499009551554.

Pound, J. (2015). Fifteen of note: 15 musical world records, *BBC Music* **23**, pp. 48–50.

Poundstone, W. (2003). *How Would You Move Mount Fuji?: Microsoft's Cult of the Puzzle— How the World's Smartest Companies Select the Most Creative Thinkers* (Little, Brown and Company).

Prantl, C. (1855). *Geschichte der Logik im Abendlande* (S. Hirzel).

Ramsey, F. P. (1930). On a problem of formal logic, *Proceedings of the London Mathematical Society* **s2-30**, 1, pp. 264–286, doi:10.1112/plms/s2-30.1.264.

Raymond, M. (2009). How 'big' is the Library of Congress? http://blogs.loc.gov/loc/2009/02/how-big-is-the-library-of-congress/.

Reid, C. (1996). *Julia: A Life in Mathematics* (The Mathematical Association of America).

Richards, I. (1984). Proof without words: Sum of integers, *The Mathematics Magazine* **57**, 2, p. 104, doi:10.2307/2689592.

Rosenthal, J. S. (2008). Monty Hall, Monty Fall, Monty Crawl, *Math Horizons* **16**, 1, pp. 5–7.

Rosenthal, J. S. (2009). A mathematical analysis of the Sleeping Beauty problem, *The Mathematical Intelligencer* **31**, 3, pp. 32–37, doi:10.1007/s00283-009-9060-z.

Rosenthal, J. S. (2014). Statistics and the Ontario Lottery Retailer Scandal, *Chance* **27**, 1, pp. 4–9, doi:10.1080/09332480.2014.890864.

Ross, S. M. (2010). *Introduction to Probability Models* (Academic Press).

Rudin, W. (1976). *Principles of Mathematical Analysis* (McGraw-Hill).

Schneps, L. and Colmez, C. (2013). *Math on Trial: How Numbers Get Used and Abused in the Courtroom* (Basic Books).

Selvin, S. (1975). On the Monty Hall problem (letter to the editor), *The American Statistician* **29**, 3, p. 134, doi:10.1080/00031305.1975.10477398.

Sesardic, N. (2007). Sudden infant death or murder? A royal confusion about probabilities, *The British Journal for the Philosophy of Science* **58**, 2, pp. 299–329, doi:10.1093/bjps/axm015.

Shannon, C. E. (1938). A symbolic analysis of relay and switching circuits, *Transactions of the American Institute of Electrical Engineers* **57**, pp. 713–723, doi:10.1109/T-AIEE.1938.5057767.

Silver, N. (2012). *The Signal and the Noise: Why so many predictions fail—but some don't* (Penguin Press, New York).

Sinkov, A. and Feil, T. (2009). *Elementary Cryptanalysis: A Mathematical Approach*, 2nd edn. (Mathematical Association of America).

Smullyan, R. (2015). *The Magic Garden of George B and Other Logic Puzzles* (World Scientific).

Statistics Canada (2008). 30 and 50 year total divorce rates per 1,000 marriages, Canada, provinces and territories, annual (rate per 1,000 marriages), CANSIM (database), Table 101-6511. http://www5.statcan.gc.ca/cansim/a05?lang=eng&id=1016511.

Statistics Canada (2013a). Aboriginal peoples in Canada: First Nations People, Métis and Inuit, Catalogue no. 99-011-X2011001. http://www12.statcan.gc.ca/nhs-enm/2011/as-sa/99-011-x/99-011-x2011001-eng.pdf.

Statistics Canada (2013b). Life tables, Canada, provinces and territories, 2009 to 2011, Catalogue no. 84-537-X. No.005. http://www.statcan.gc.ca/pub/84-537-x/84-537-x2013005-eng.pdf.

Statistics Canada (2014). Leading causes of death, total population, by age group and sex, Canada, Table 102-0561. http://www5.statcan.gc.ca/cansim/a26?lang=eng&retrLang=eng&id=1020561&paSer=&pattern=&stByVal=1&p1=1&p2=37&tabMode=dataTable&csid=.

Stewart, J., Davison, T. M. K., Hamilton, O. M. G., Laxton, J., and Lenz, M. P. (1988). *Finite Mathematics* (McGraw-Hill).

Stigler, S. M. (1983). Who discovered Bayes's theorem? *The American Statistician* **37**, 4a, pp. 290–296, doi:10.1080/00031305.1983.10483122.

Todhunter, I. (1865). *A History of the Mathematical Theory of Probability: From the Time of Pascal to that of Laplace* (Macmillan, London).

Turing, A. M. (1950). Computing machinery and intelligence, *Mind* **59**, 236, pp. 433–460, doi:10.1093/mind/LIX.236.433.

von Bortkewitsch, L. (1898). *Das Gesetz der kleinen Zahlen* (B. G. Teubner, Leipzig).

von Waltershausen, W. S. (1856). *Gauss: zum gedächtniss* (Verlag von S. Hirzel), hdl.handle.net/2027/hvd.hnxxgv.

Walker, J. (1847). *Murray's Compendium of Logic with a Corrected Latin Text, an Accurate Translation, and a Familiar Commentary* (Longman, Brown, Green, and Longmans).

Walker, R. (1977). Three into two won't go, *The Mathematical Gazette* **61**, pp. 25–31, doi:10.2307/3617439.

Watkins, S. J. (2000). Conviction by mathematical error? *British Medical Journal* **320**, pp. 2–3, doi:10.1136/bmj.320.7226.2.

Weber, H. (1893). Leopold Kronecker, *Mathematische Annalen* **43**, 1, pp. 1–25, doi:10.1007/BF01446613.

Whitehead, A. N. and Russell, B. (1910). *Principia Mathematica: Volume 1*, 1st edn. (Cambridge University Press, Cambridge).

Whitehead, A. N. and Russell, B. (1912). *Principia Mathematica: Volume 2*, 1st edn. (Cambridge University Press, Cambridge).

Whitehead, A. N. and Russell, B. (1913). *Principia Mathematica: Volume 3*, 1st edn. (Cambridge University Press, Cambridge).

Whitworth, W. A. (1870). *Choice and Chance*, 2nd edn. (Deighton, Bell and Co.).

Wilson, R. (1979). Analyzing the daily risks of life, *Technology Review* **81**, 4, pp. 41–46.

Wyatt, J. P., Squires, T., Norfolk, G., and Payne-James, J. J. (2011). *Oxford Handbook of Forensic Medicine* (Oxford University Press).

Zaloga, S. J. (2005). *V-1 Flying Bomb 1942–52: Hilter's infamous "doodlebug"* (Osprey Publishing).

Zerjal, T., Xue, Y., Bertorelle, G., Wells, R. S., Bao, W., Zhu, S., Qamar, R., Ayub, Q., Mohyuddin, A., Fu, S., Li, P., Yuldasheva, N., Ruzibakiev, R., Xu, J., Shu, Q., Du, R., Yang, H., Hurles, M. E., Robinson, E., Gerelsaikhan, T., Dashnyam, B., Mehdi, S. Q., and Tyler-Smith, C. (2003). The genetic legacy of the Mongols, *The American Journal of Human Genetics* **72**, 3, pp. 717–721, doi:10.1086/367774.

Index

Basic Discrete Mathematics

Printed in the United States
By Bookmasters